The Effects of LED Light Spectra and Intensities on Plant Growth

The Effects of LED Light Spectra and Intensities on Plant Growth

Editors

Valeria Cavallaro
Rosario Muleo

MDPI • Basel • Beijing • Wuhan • Barcelona • Belgrade • Manchester • Tokyo • Cluj • Tianjin

Editors

Valeria Cavallaro
CNR
National Research Council
Institute of BioEconomy
Catania
Italy

Rosario Muleo
Department of Agricultural
and Forestry Science
University of Tuscia
Viterbo
Italy

Editorial Office
MDPI
St. Alban-Anlage 66
4052 Basel, Switzerland

This is a reprint of articles from the Special Issue published online in the open access journal *Plants* (ISSN 2223-7747) (available at: www.mdpi.com/journal/plants/special_issues/light).

For citation purposes, cite each article independently as indicated on the article page online and as indicated below:

LastName, A.A.; LastName, B.B.; LastName, C.C. Article Title. *Journal Name* **Year**, *Volume Number*, Page Range.

ISBN 978-3-0365-7129-4 (Hbk)
ISBN 978-3-0365-7128-7 (PDF)

Cover image courtesy of Valeria Cavallaro

Contents

About the Editors

Valeria Cavallaro

Valeria Cavallaro has been a full-time researcher of the Italian National Research Council (CNR) since 1st April, 1986. Her areas of expertise are the influence of abiotic stress (salinity, drought) and seed pre-treatments (scarification, osmopriming, etc.) on the seed germination of different Mediterranean species (tomato, Cynara spp, forage grasses, carob, etc.); biotechnology in the rapid propagation and regeneration of improved varieties (micropropagation of globe artichoke, biomass species, carob, etc.); and the use of arbuscolar mychorrhizal inoculum to improve the acclimatisation process of micropropagated plants and the tolerance to nutritional and abiotic stresses.

She taught "Plant Genetics" and "Field crops" for the Agricultural Sciences Faculty of Catania University. She has participated in national and international projects and organized diverse training courses on micropropagation and sustainable agriculture for students and high school teachers. She is the author of one book, several book chapters, and 50 articles published in national and international peer-reviewed journals.

Rosario Muleo

Rosario Muleo is a Full Professor at the University of Tuscia (Viterbo). His research interests are plant interactions with the environment. Particularly, he is studying the role of light and the photoperiod on juvenility, branching, apical dominance, propagation, and flowering induction, as well as the optogenetic regulation of PR genes. He developed genetic and epigenetic molecular markers (EpyHRMAssay). Other research topics include the biodiversity and identification of neutral and functional markers in olive trees; genes and the physiological regulation of the synthesis and accumulation of secondary metabolites in fruits; genetic and epigenetic woody fruit crop adaptation in extreme waterlogging and cold conditions; and the identification of functional olive miRNA capable of cross-kingdom interaction with human oncogenes in an active role in anti-tumoral response.

In 2014, Rosario Muleo became the director of the Master's program in Agricultural and Environmental Sciences. He participated in national and international projects. He is the author of several book chapters and an editor of scientific books. He is the author of a total of 301 publications.

plants

MDPI

Editorial

The Effects of LED Light Spectra and Intensities on Plant Growth

Valeria Cavallaro [1,*] and Rosario Muleo [2,*]

1 Institute of BioEconomy (IBE), National Research Council of Italy, 95126 Catania, Italy
2 Tree Physiology and Fruit Crop Biotechnology Laboratory, Department of Agriculture and Forest Sciences (DAFNE), University of Tuscia, 01100 Viterbo, Italy
* Correspondence: valeria.cavallaro@cnr.it (V.C.); muleo@unitus.it (R.M.)

Light is an electromagnetic radiation that occurs in a narrow range of over an extremely wide range of wavelengths, from gamma rays with wavelengths to radio waves measured in meters. Light detected by human eyes and plant photoreceptors is a group of wavelengths from about 700 nanometers (nm) for red light down to about 400 nm for violet light: the visible light. However, even the wavelengths form the spectral regions adjacent to the visible ones are referred to as light also, infrared at the red-light and ultraviolet at the violet light end.

We know that light is the source of energy for the primary sustenance process of life on our earth planet: the photosynthesis which is life's adopted strategy for capturing and incorporating energy, and under this context in which light is primarily experienced, explored and exploited. However, anyone knows that light is necessary to the aesthetic appreciation of the visual world, and through the sense of sight, light is a primary tool for perceiving the world and communicating within it. Plants also perceive information from the ambient and communicate with other organism using the light and have developed a plethora of photoreceptors that permit the communication with the surrounding ambient.

The physical properties of light as spectral quality, irradiance, intensity, and photoperiod play a deep role on the morphogenesis, growth, and metabolism of many biochemical pathways in plants.

Nowadays, Light-Emitting Diodes (LEDs) have been demonstrated to offer interesting prospects for use in plant lighting designs in controlled environment agriculture (greenhouses) and growth chambers for in vitro cultures. As compared to the previously used light sources, LEDs possess advantages such as wavelength specificity, less heat radiation, longest durability, much lower power consumption, and the possibility to manipulate the spectral qualities of the emitted light.

In high-technology greenhouses (for instance vertical agriculture) artificial light may assume both assimilative and control function. In the first case, light provides the opportunity to optimize photosynthetic efficiency under low (high latitudes) or short day (winter months) solar radiation, enhancing or accelerating the efficiency of plant production. The control activity of light, on the other hand, has the function of guiding growth and development e.g., by promoting changes in the morphology of the plant (e.g., elongation of the stem, branching), or the internal rhythms as the transition from one stage of development to the next (e.g., from vegetative to reproductive), or the synthesis and accumulation of plant metabolites to adapt the plant to adverse environmental conditions, increasing plant fitness, and the nutraceutical properties of the products.

In vitro culture is regulated by different factors, and among them light is the most important. LED illumination system for in vitro cultures should provide light in the spectral region that is involved in photosynthesis and photomorphogenic responses without wasting energy on non-productive wavelengths. The combined effects of light and growth regulators or other components of the culture media is another important issue. Even on

Citation: Cavallaro, V.; Muleo, R. The Effects of LED Light Spectra and Intensities on Plant Growth. *Plants* **2022**, *11*, 1911. https://doi.org/10.3390/plants11151911

Received: 27 June 2022
Accepted: 18 July 2022
Published: 23 July 2022

in vitro cultures, LED light may regulate gene expression and physiological behaviour that in turn influences metabolite production.

In this special issue, many of the light concerns had been addressed. Very briefly, the effects on biomass production and photosynthetic efficiency of different light spectra (induced by LEDs) have been reported on *Lactuca sativa* L. [1], *Cucumis sativus* seedlings [2], *Ocimum basilicum* L. [3], two microgreens (*Amaranthus tricolor* L. and *Brassica rapa* L. subsp. oleifera (DC.) Metzg) [4], *Glycine max* (L.) Merr.) [5] and *Medicago Sativa* L. seedlings [6]. The influence of light intensity on *Lactuca sativa* and *Brassica rapa* var. *nipposinica* in vertical farms [7] and of daytime or edge-of-daytime intra-canopy illumination on the fruit set of *Capsicum annuum* [8] were also studied.

Two papers covered the topic of new technologies (LED-Sourced CoeLux®System) [9] and of a new (IoT-Enabled) systems to control light [10].

Interestingly, one article concerned the interactions between light quality and plant pathogens [11]. Two articles regarded some important physiological traits regulation by light, and in particular the effects of supplemental lighting spectra given to widen photoperiod on *Solanum lycopersicum* L. [12] and/or night interrupting on *Chrysanthemum* [13].

The effects of lighting on the production of secondary metabolites in confined environment and in vitro is attracting a growing attention by the researchers. In this special issue, seven articles deal with the influence of LED lighting on important aromatic components in *Mentha canadensis* L. [14], in *Thymus vulgaris* L. [15], and on some nutraceutical and pharmacological components of *Scutellaria baicalensis* [16], *Triticum aestivum* L., *Hordeum vulgare* sprouts [17], *Fritillaria cirrhosa* D. Don, [18], different in vitro cultures [19] and two microgreens [4].

One paper concerned the effects of adding Silicon-Containing Fertilizer [20] on light stress. Finally, a complete review concerning Light and Plant Growth Regulators on in vitro proliferation have been presented [21].

Even if lighting conditions represent an important tool to enhance plant productivity in confined environment, and many issues need still to be explored in this view, this special issue represents an overview of some of the most internationally studied topics.

Conflicts of Interest: The authors declare no conflict of interest.

References

1. Tarakanov, I.G.; Tovstyko, D.A.; Lomakin, M.P.; Shmakov, A.S.; Sleptsov, N.N.; Shmarev, A.N.; Litvinskiy, V.A.; Ivlev, A.A. Effects of Light Spectral Quality on Photosynthetic Activity, Biomass Production, and Carbon Isotope Fractionation in Lettuce, *Lactuca sativa* L. Plants. *Plants* **2022**, *11*, 441. [CrossRef]
2. Claypool, N.B.; Lieth, J.H. Green Light Improves Photosystem Stoichiometry in Cucumber Seedlings (Cucumis sativus) Compared to Monochromatic Red Light. *Plants* **2021**, *10*, 824. [CrossRef] [PubMed]
3. Rahman, M.M.; Vasiliev, M.; Alameh, K. LED Illumination Spectrum Manipulation for Increasing the Yield of Sweet Basil (*Ocimum basilicum* L.). *Plants* **2021**, *10*, 344. [CrossRef] [PubMed]
4. Toscano, S.; Cavallaro, V.; Ferrante, A.; Romano, D.; Patané, C. Effects of different light spectra on final biomass production and nutritional quality of two microgreens. *Plants* **2021**, *10*, 1584. [CrossRef]
5. Hitz, T.; Graeff-Hönninger, S.; Munz, S. Modelling of Soybean (*Glycine max* L. Merr.) Response to Blue Light Intensity in Controlled Environments. *Plants* **2020**, *9*, 1757. [CrossRef]
6. Tang, W.; Guo, H.; Baskin, C.C.; Xiong, W.; Yang, C.; Li, Z.; Song, H.; Wang, T.; Yin, J.; Wu, X.; et al. Effect of Light Intensity on Morphology, Photosynthesis and Carbon Metabolism of Alfalfa (Medicago Sativa) Seedlings. *Plants* **2022**, *11*, 1688. [CrossRef]
7. Jayalath, T.C.; van Iersel, M.W. Canopy Size and Light Use Efficiency Explain Growth Differences between Lettuce and Mizuna in Vertical Farms. *Plants* **2021**, *10*, 704. [CrossRef] [PubMed]
8. Tiwari, V.; Kamara, I.; Ratner, K.; Many, Y.; Lukyanov, V.; Ziv, C.; Gilad, Z.; Esquira, I.; Charuvi, D. Daytime or Edge-of-Daytime Intra-Canopy Illumination Improves the Fruit Set of Bell Pepper at Passive Conditions in the Winter. *Plants* **2022**, *11*, 424. [CrossRef]
9. Beatrice, P.; Terzaghi, M.; Chiatante, D.; Scippa, G.S.; Montagnoli, A. Morpho-Physiological Responses of *Arabidopsis thaliana* L. to the LED-Sourced CoeLux®System. *Plants* **2021**, *10*, 1310. [CrossRef]
10. Afzali, S.; Mosharafian, S.; van Iersel, M.W.; Mohammadpour Velni, J. Development and Implementation of an IoT-Enabled Optimal and Predictive Lighting Control Strategy in Greenhouses. *Plants* **2021**, *10*, 2652. [CrossRef]

11. Sgamma, T.; Forgione, I.; Luziatelli, F.; Iacona, C.; Mancinelli, R.; Thomas, B.; Ruzzi, M.; Muleo, R. Monochromic radiations provided by light emitted diode (LED) modulate infection and defense response to fire blight in pear trees. *Plants* **2021**, *10*, 1886. [CrossRef]
12. Lanoue, J.; Thibodeau, A.; Little, C.; Zheng, J.; Grodzinski, B.; Hao, X. Light Spectra and Root Stocks Affect Response of Greenhouse Tomatoes to Long Photoperiod of Supplemental Lighting. *Plants* **2021**, *10*, 1674. [CrossRef]
13. Park, Y.G.; Jeong, B.R. How Supplementary or Night-Interrupting Low-Intensity Blue Light Affects the Flower Induction in Chrysanthemum, A Qualitative Short-Day Plant. *Plants* **2020**, *9*, 1694. [CrossRef] [PubMed]
14. Ueda, T.; Murata, M.; Yokawa, K. Single Wavelengths of LED Light Supplement Promote the Biosynthesis of Major Cyclic Monoterpenes in Japanese Mint. *Plants* **2021**, *10*, 1420. [CrossRef] [PubMed]
15. Tabbert, J.M.; Schulz, H.; Krähmer, A. Increased Plant Quality, Greenhouse Productivity and Energy Efficiency with Broad-Spectrum LED Systems: A Case Study for Thyme (*Thymus vulgaris* L.). *Plants* **2021**, *10*, 960. [CrossRef] [PubMed]
16. Yeo, H.-J.; Park, C.-H.; Park, S.-Y.; Chung, S.-O.; Kim, J.-K.; Park, S.-U. Metabolic Analysis of Root, Stem, and Leaf of Scutellaria baicalensis Plantlets Treated with Different LED Lights. *Plants* **2021**, *10*, 940. [CrossRef]
17. Muthusamy, M.; Kim, J.H.; Kim, S.H.; Kim, J.Y.; Heo, J.W.; Lee, H.; Lee, K.-S.; Seo, W.D.; Park, S.; Kim, J.A.; et al. Changes in Beneficial C-glycosylflavones and Policosanol Content in Wheat and Barley Sprouts Subjected to Differential LED Light Conditions. *Plants* **2020**, *9*, 1502. [CrossRef] [PubMed]
18. Chen, C.-C.; Lee, M.-R.; Wu, C.-R.; Ke, H.-J.; Xie, H.-M.; Tsay, H.-S.; Agrawal, D.C.; Chang, H.-C. LED Lights Affecting Morphogenesis and Isosteroidal Alkaloid Contents in Fritillaria cirrhosa D. Don—An Important Chinese Medicinal Herb. *Plants* **2020**, *9*, 1351. [CrossRef]
19. Hashim, M.; Ahmad, B.; Drouet, S.; Hano, C.; Abbasi, B.H.; Anjum, S. Comparative Effects of Different Light Sources on the Production of Key Secondary Metabolites in Plants In Vitro Cultures. *Plants* **2021**, *10*, 1521. [CrossRef]
20. Semenova, N.A.; Smirnov, A.A.; Grishin, A.A.; Pishchalnikov, R.Y.; Chesalin, D.D.; Gudkov, S.V.; Chilingaryan, N.O.; Skorokhodova, A.N.; Dorokhov, A.S.; Izmailov, A.Y. The Effect of Plant Growth Compensation by Adding Silicon-Containing Fertilizer under Light Stress Conditions. *Plants* **2021**, *10*, 1287. [CrossRef]
21. Cavallaro, V.; Pellegrino, A.; Muleo, R.; Forgione, I. Light and Plant Growth Regulators on In Vitro Proliferation. *Plants* **2022**, *11*, 844. [CrossRef] [PubMed]

plants

Article

Effect of Light Intensity on Morphology, Photosynthesis and Carbon Metabolism of Alfalfa (*Medicago sativa*) Seedlings

Wei Tang [1], Haipeng Guo [1], Carol C. Baskin [2,3], Wangdan Xiong [1], Chao Yang [1], Zhenyi Li [1], Hui Song [1], Tingru Wang [1], Jianing Yin [1], Xueli Wu [1], Fuhong Miao [1], Shangzhi Zhong [1], Qibo Tao [1], Yiran Zhao [1] and Juan Sun [1,*]

1 College of Grassland Science, Qingdao Agricultural University, Qingdao 266109, China; a20052123@126.com (W.T.); 20202203019@stu.qau.edu.cn (H.G.); xiongwd@qau.edu.cn (W.X.); yangchao@qau.edu.cn (C.Y.); lizhenyily@163.com (Z.L.); biosonghui@outlook.com (H.S.); xiaoxiang0203@126.com (T.W.); 20202103045@stu.qau.edu.cn (J.Y.); xueli0510@163.com (X.W.); miaofh@qau.edu.cn (F.M.); zhongsz@qau.edu.cn (S.Z.); taoqibo1992@163.com (Q.T.); zhaoyiran@qau.edu.cn (Y.Z.)

2 Department of Biology, University of Kentucky, Lexington, KY 40506-0225, USA; carol.baskin@uky.edu

3 Department of Plant and Soil Science, University of Kentucky, Lexington, KY 40546-0312, USA

* Correspondence: sunjuan@qau.edu.cn

Abstract: To understand how light intensity influences plant morphology and photosynthesis in the forage crop alfalfa (*Medicago sativa* L. cv. Zhongmu 1), we investigated changes in leaf angle orientation, chlorophyll fluorescence, parameters of photosynthesis and expression of genes related to enzymes involved in photosynthesis, the Calvin cycle and carbon metabolism in alfalfa seedlings exposed to five light intensities (100, 200, 300, 400 and 500 μmol m^{-2} s^{-1}) under hydroponic conditions. Seedlings grown under low light intensities had significantly increased plant height, leaf hyponasty, specific leaf area, photosynthetic pigments, leaf nitrogen content and maximal PSII quantum yield, but the increased light-capturing capacity generated a carbon resource cost (e.g., decreased carbohydrates and biomass accumulation). Increased light intensity significantly improved leaf orientation toward the sun and upregulated the genes for Calvin cycle enzymes, thereby increasing photosynthetic capacity. Furthermore, high light (400 and 500 μmol m^{-2} s^{-1}) significantly enhanced carbohydrate accumulation, accompanied by gene upregulation and increased activity of sucrose and starch-synthesis-related enzymes and those involved in carbon metabolism. Together, these results advance our understanding of morphological and physiological regulation in shade avoidance in alfalfa, which would guide the identification of suitable spatial planting patterns in the agricultural system.

Keywords: alfalfa; light intensity; photosynthesis; growth; adaption

Citation: Tang, W.; Guo, H.; Baskin, C.C.; Xiong, W.; Yang, C.; Li, Z.; Song, H.; Wang, T.; Yin, J.; Wu, X.; et al. Effect of Light Intensity on Morphology, Photosynthesis and Carbon Metabolism of Alfalfa (*Medicago sativa*) Seedlings. *Plants* **2022**, *11*, 1688. https://doi.org/10.3390/plants11131688

Academic Editors: Valeria Cavallaro and Rosario Muleo

Received: 31 May 2022
Accepted: 23 June 2022
Published: 25 June 2022

1. Introduction

Light is one of the most important environmental factors influencing plant growth and development. Changes in light intensity, light quality and the photoperiod have impacts on plant morphology and metabolism [1]. Subsequently, plants can exhibit numerous adaptative strategies in response to the light environment [2]. When grown in the shade, many shade-intolerant plants (e.g., *Arabidopsis thaliana*) exhibit a well-known shade avoidance syndrome (SAS) that increases their adaptive and competitive ability [3]. The SAS is triggered by a reduction in light intensity perceived by photoreceptor cryptochromes, which in turn control adaptive responses [4]. These SAS responses range from development changes, such as increased leaf hyponasty, specific leaf area and ratio of palisade/spongy tissues; hypocotyl, petiole and stem elongation; reduced tillering (monocots)/branching (dicots); and increased internode length [5]. Physiological changes, such as decreased leaf carbon assimilation and enzyme activity, also occur [6]. The morphological changes in response to shading allow the plant to elongate and thereby gain access to unfiltered sunlight [7].

However, plant elongation due to shading comes at a cost. Plant carbon resources must be redirected to stems or petioles to promote their elongation at the expense of production of new leaves. Additionally, excessive stem elongation leads to plant lodging or mechanical injury, which decreases plant fitness [8]. In crop production, shading occurs for the low-tier plants, which decreases light intensity and changes the light quality to a low ratio of red light, especially in intercropping system, [9]. Subsequently, these plants respond to shade by inducing a series of adaptive morphological and physiological changes at the cost of assimilated resources, which eventually negatively affects yield [10]. Thus, gaining a better understanding of how crops adapt and respond to shade stress could help guide the design of crop cultivation in agriculture systems.

A range of light levels is a common approach for exploring how shading stress affects pigment accumulation and the photosynthetic capacity of leaves [1]. Light intensity can directly affect light harvesting by plants and lead to changes in the abundance of chlorophyll pigments and differences in the health status of PSII. Rascher et al. (2010) [11] found that low light led to higher levels of Chl a, b, an improved maximal PSII quantum yield (F_v/F_m) and an early onset of nonphotochemical quenching (NPQ), which increased light-capturing capacity. Similar results were obtained for seedlings of Chinese cabbage (*Brassica campestris*) [12] and sweet pepper (*Capsicum annuum*) [13]. These significant differences in photochemical efficiency can be viewed as adaptations to low light; therefore, their regulatory mechanisms have long been important areas of research.

Photosynthesis allows plants to convert light energy into chemical energy. The Calvin cycle is a series of biochemical redox reactions that take place in the stroma of chloroplasts, and they play a vital role in photosynthetic carbon fixation [14]. In shade-intolerant species, low photosynthesis due to low light reduces expression of genes and activity of the Calvin cycle enzymes involved in CO_2 fixation and regeneration of rubisco-1, 5-bisphosphate (RuBP), thereby decreasing the potential for carbon assimilation in plants [15]. RuBP carboxylase or oxygenase (Rubisco) is the rate-limiting step of photosynthesis, and it catalyzes CO_2 fixation in C_3 plants [16]. Previously, it was reported that shade-associated with downregulation of the net photosynthetic rate was due to reduction in the amount or activity of Rubisco [17]. Photosynthesis is also catalyzed by other key enzymes, e.g., Rubisco activase (RCA) and fructose-1, 6-bisphosphatase (FBPase) [18]. Recent studies on soybean (*Glycine max*) and tomato (*Lycopersicon esculentum*) have shown that gene expression of the key enzymes involved in the Calvin cycle was downregulated in low but not high light [18,19]. However, the specific effects of light intensity on the photosynthesis processes in plants remain largely unknown. Therefore, levels of gene expression of the key enzymes of the Calvin cycle of plants grown at different light intensities need to be studied to elucidate the molecular mechanism of plant response to shading stress.

Alfalfa (*Medicago sativa* L.) is a high-quality forage for dairy cows and other livestock because of its high dry matter accumulation and high protein and soluble sugar content [20,21]. With increasing demand for food and the decreasing availability of arable land, grass/legume forage intercropping is gaining in popularity as a sustainable practice for low-input or resource-limited agricultural systems, such as maize–alfalfa and oat–alfalfa [22]. However, intercropped alfalfa plants often suffer from shade stress due to the reduced amount of intercepted sunlight. Subsequently, shading increases plants height and internodal distance, and reduces stems strength, which makes alfalfa plants susceptible to lodging, thereby reducing forage yield [23]. The SAS effects on alfalfa could be of high practical importance for intercropping systems, but the minimum amount of light required for alfalfa growth and development has received little research attention. To date, only one study has indicated that shade-intolerant alfalfa plants will delay flowering when grown in the shade (i.e., low ratio of red to far-red light) [24]. Research-based information is lacking on the effect of shading on the growth and physiological metabolism of *M. sativa* seedlings. Thus, it is important to investigate the adaptability of alfalfa responses to low light intensity, which would be useful information for determining proper plant spacing and strip configuration in intercropping systems.

The objective of our research was to determine how light intensity affects alfalfa seedling morphology and photosynthetic characteristics, as well as the key enzymes involved in the Calvin cycle and carbon metabolism coupled with expression of these genes. Here, alfalfa seedlings were exposed to five levels of light intensity for 14 days in a climate room, and their morphological and physiological responses were investigated. We hypothesized that a brief exposure to low light would increase leaf hyponasty and stem elongation but downregulate expression of genes for the key enzymes involved in the Calvin cycle and carbon metabolism, resulting in a synergistic decrease in photosynthetic rates and accumulation of dry matter.

2. Results

2.1. Morphological Characteristics

Light treatment had a significant effect on alfalfa morphological characteristics (i.e., plant height, specific leaf area, abaxial leaf petiole angle and stem diameter) ($p < 0.001$) (Figure 1, Tables 1 and S2). Maximum plant height, specific leaf area and abaxial leaf petiole angle were measured in L100; these are the traits that decreased with increased light intensity. However, the highest and lowest stem diameters were measured for plants at L500 and L100, respectively. In addition, shoot dry matter (SDM), root dry matter (RDM) and the root-to-shoot ratio (RSR) were significantly affected by light treatments ($p < 0.001$) (Table S3). The SDM, RDM and RSR of alfalfa plants in L500 were significantly higher than those in L100. For the most part, the RSR did not differ significantly between L300, L400 and L500 (Table 1).

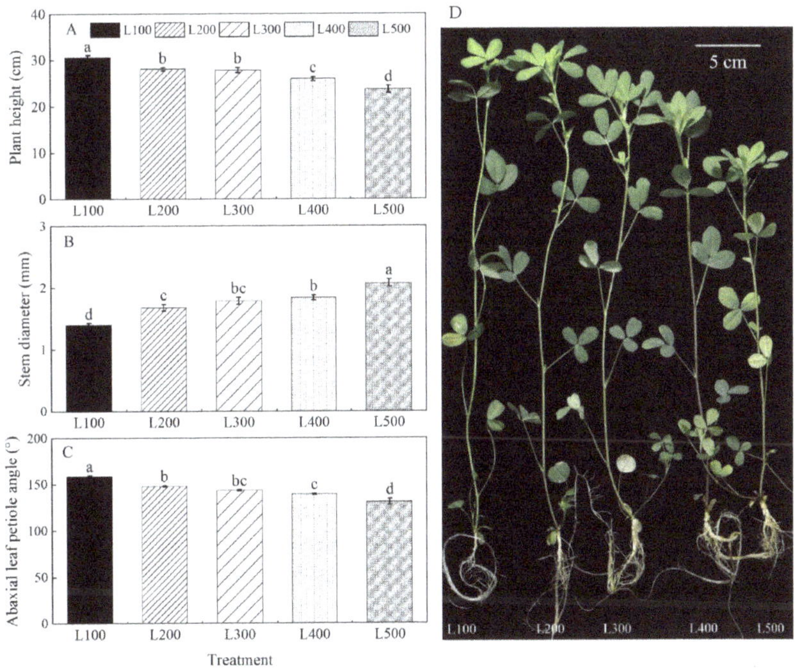

Figure 1. Changes in phenotype and plant traits of alfalfa as affected by light treatments. The plant height (**A**), stem diameter (**B**), abaxial leaf petiole angle (**C**) and plant phenotype (**D**) of alfalfa plants under different light intensity treatments. L100, L200, L300, L400 and L500 refer 100, 200, 300, 400 and 500 µmol m^{-2} s^{-1}, respectively. Vertical bars indicate 1 s.e. of the mean ($n = 4$). Different lowercase letters on the different bar mean significant differences ($p < 0.05$).

Table 1. Effect of different light intensity treatments on specific leaf area (SLA, cm^2 mg^{-1}), shoot dry matter (SDM, mg $plant^{-1}$), root dry matter (RDM, mg $plant^{-1}$) and root-to-shoot ratio (RSR) of alfalfa plants.

Treatment [a]	SLA	SDM	RDM	RSR
L100	0.474a (0.023)	205.4d (11.8)	22.0e (1.3)	0.108c (0.007)
L200	0.371b (0.021)	227.6d (17.7)	30.6d (2.2)	0.137b (0.013)
L300	0.269c (0.015)	320.4c (21.5)	54.2c (2.6)	0.170a (0.004)
L400	0.125d (0.004)	379.4b (10.6)	70.0b (3.2)	0.185a (0.007)
L500	0.117d (0.002)	461.4a (15.8)	82.2a (1.8)	0.179a (0.007)

[a] L100, L200, L300, L400 and L500 refer 100, 200, 300, 400 and 500 μmol m^{-2} s^{-1}, respectively. Within a column, values followed by different letters are significantly different (p <0.05). Values within parentheses are the standard errors of the means (n = 4).

Root morphology parameters, including root length (RL), surface area (RSA), volume (RV) and diameter (RD), varied among light treatments (Table S3). These parameters increased with increasing light up to L500 compared to the L100 treatment (Table 2). Increased light significantly increased RL by 22.8 to 182.5%, RSA by 26.0 to 353.2%, RV by 42.2 to 925.9% and RD by 4.3 to 84.5%. RD did not differ significantly from L300 to L500.

Table 2. Effect of light intensity treatments on root length (RL, cm), root surface area (RSA, cm^2), root volume (RV, cm^3) and root diameter (RD, mm) of alfalfa plants.

Treatment [a]	RL	RSA	RV	RD
L100	206.4e (7.2)	18.6e (0.7)	0.229e (0.011)	0.243c (0.013)
L200	397.1d (6.2)	44.4d (1.0)	0.787d (0.009)	0.325b (0.006)
L300	426.8c (8.7)	58.0c (2.2)	1.298c (0.081)	0.418a (0.012)
L400	474.6b (11.2)	67.0b (2.8)	1.650b (0.083)	0.430a (0.009)
L500	583.1a (5.7)	84.4a (2.7)	2.349a (0.179)	0.449a (0.010)

[a] L100, L200, L300, L400 and L500 refer 100, 200, 300, 400 and 500 μmol m^{-2} s^{-1}, respectively. Within a column, values followed by different letters are significantly different (p < 0.05). Values within parentheses are the standard errors of the means (n = 4).

2.2. *Leaf Pigment and Nitrogen Content*

Chlorophyll a (Chl a), Chlorophyll b (Chl b), carotenoids (Car), Chl a + b, Chl a/b and leaf nitrogen content (LN) were significantly affected by light treatment (Table S4). Increased light intensity from L100 to L500 decreased Chl a, Chl b, Chl a + b and Car contents, while Chl a/b increased (Table 3). Chl a, Chl b, Chl a + b and Car contents for plants in the L500 treatment were decreased by 27.8%, 49.5%, 32.9% and 25.9% ($p < 0.01$), respectively, compared to L100, but those for plants grown at L400 and L500 did not differ significantly. Chl a/b was 17.5% ($p < 0.01$) higher in L500 than in L100. In addition, increased light intensity decreased LN content, and at L500, LN content decreased by 50.6% ($p < 0.001$) compared to L100.

Table 3. Effect of light treatments on Chlorophyll a (Chl a, μg cm^{-2}), Chlorophyll b (Chl b, μg cm^{-2}), carotenoids (Car, μg cm^{-2}), Chl a + b (μg cm^{-2}), Chl a/b and leaf nitrogen content (LNC, mg/g) of alfalfa plants.

Treatment [a]	Chl a	Chl b	Car	Chl a + b	Chl a/b	LNC
L100	36.5a (1.7)	13.1a (0.3)	6.91a (0.24)	49.6a (1.8)	2.80b (0.14)	38.0a (1.1)
L200	34.1ab (2.0)	11.5ab (0.9)	6.19b (0.12)	45.6ab (2.9)	2.98ab (0.07)	28.0b (0.5)
L300	31.0bc (1.7)	10.3bc (0.8)	6.15b (0.21)	41.9bc (2.4)	3.10ab (0.12)	18.7c (0.2)
L400	29.7bc (1.1)	9.4c (0.4)	6.11b (0.13)	39.1c (1.4)	3.18a (0.12)	17.4c (0.6)
L500	28.6c (1.2)	8.7c (0.5)	5.49c (0.29)	37.3c (1.6)	3.29a (0.12)	18.8c (0.2)

[a] L100, L200, L300, L400 and L500 refer 100, 200, 300, 400 and 500 μmol m^{-2} s^{-1}, respectively. Within a column, values followed by different letters are significantly different (p < 0.05). Values within parentheses are the standard errors of the means (n = 4).

2.3. Photosynthetic and Chlorophyll Fluorescence Characteristics

The photosynthetic characteristics of alfalfa plants varied among light treatments (Table S5, Figure 2). The maximum net photosynthetic rate (P_n), transpiration rate (T_r) and stomatal conductance (g_s) values of alfalfa plants were at L400 and L500, whereas intercellular CO_2 concentration (C_i) was highest at L100 to L300. On average, the net photosynthetic rates, T_r and g_s of alfalfa plants, were significantly increased by 230, 62 and 52%, respectively, ($p < 0.01$) at L400 and L500 compared to L100. However, intercellular CO_2 concentration at L400 and L500 decreased by 8.9 and 10.1% ($p < 0.001$), respectively, compared to L100. The photosynthetic characteristics of alfalfa leaves did not differ significantly at L400 and L500. The increased P_n at L400 and L500 suggests that high light intensity was positively related to increased g_s and T_r, but negatively related to decreased C_i in alfalfa plants.

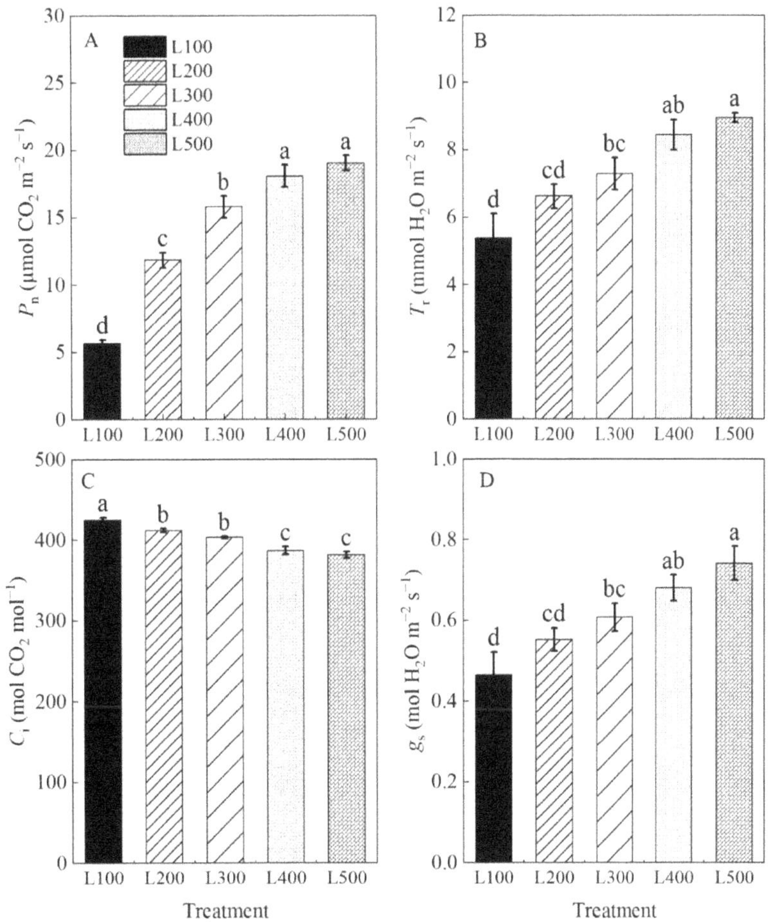

Figure 2. Photosynthetic characteristics of alfalfa leaves under different light treatments. L100, L200, L300, L400 and L500 refer 100, 200, 300, 400 and 500 μmol m^{-2} s^{-1}, respectively. Net photosynthetic rate (P_n) (**A**), transpiration rate (T_r) (**B**), intercellular CO_2 concentration (C_i) (**C**), stomatal conductance and (g_s) (**D**). Vertical bars indicate 1 s.e. of the mean (n = 4). Different lowercase letters on the different bar mean significant differences ($p < 0.05$).

Chlorophyll fluorescence characteristics, including maximal PSII quantum yield (F_v/F_m), effective PSII quantum yield (Φ_{PSII}), non-photochemical quenching (NPQ) and the electron transport rate (ETR), were significantly affected by treatment (Table S6, Figure 3). Figure 3 shows the difference in absorbed radiation energy of alfalfa leaves in response to light treatments. The F_v/F_m, Φ_{PSII} and NPQ of alfalfa plants grown in the low-light treatments were significantly higher than those in the high-light treatments. Furthermore, L100 increased the F_v/F_m, Φ_{PSII}, and NPQ by 12.0, 24.9 and 60.8%, respectively, but it decreased the ETR by 71.2% ($p < 0.001$) compared to L500. These results indicate that the original activity of the PSII reaction center was increased, and the transformation efficiency of primary light energy was improved in the low-light-intensity adaption of alfalfa.

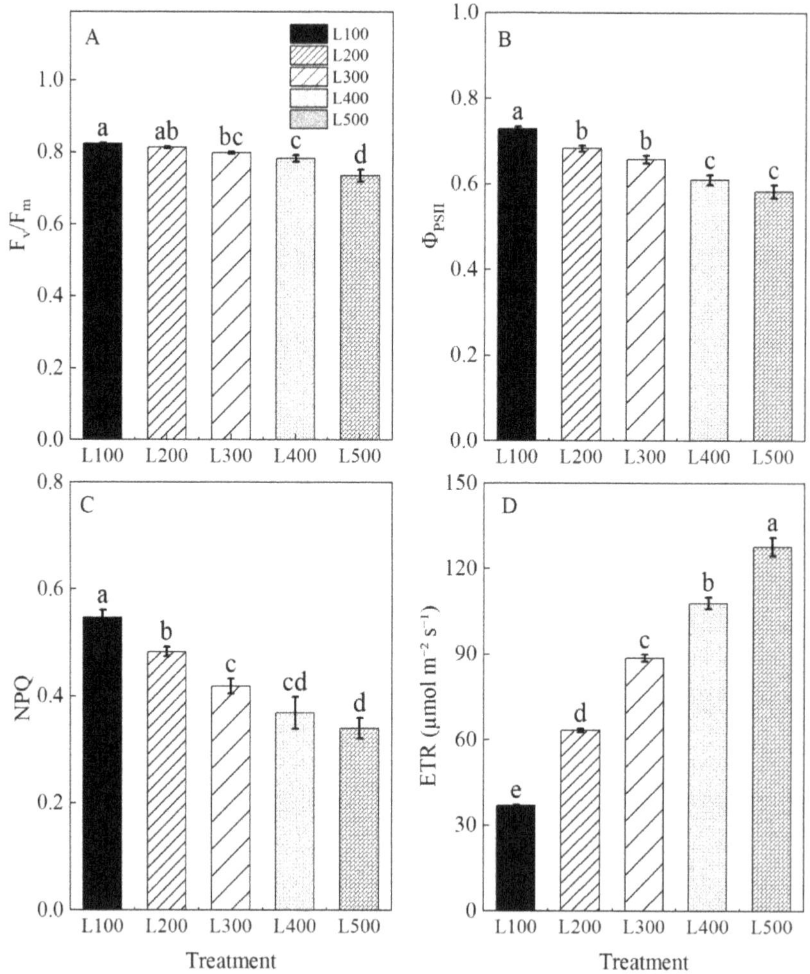

Figure 3. Chlorophyll fluorescence characteristics of alfalfa leaves under different light treatments. L100, L200, L300, L400 and L500 refer 100, 200, 300, 400 and 500 μmol m^{-2} s^{-1}, respectively. Maximal PSII quantum yield (Fv/Fm) (**A**), effective PSII quantum yield (Φ_{PSII}) (**B**), non-photochemical quenching (NPQ) (**C**) and electron transport rate (ETR) (**D**). Vertical bars indicate 1 s.e. of the mean (n = 4). Different lowercase letters on the different bar mean significant differences ($p < 0.05$).

2.4. Leaf Non-Structural Carbohydrate Contents

Soluble sugar (SS), sucrose and starch (St) were significantly affected by light treatments ($p < 0.001$) (Table S7). As expected, the content of SS, sucrose and St in leaves increased significantly with increased light intensity (Figure 4). The highest SS, sucrose and St contents in leaves were measured in the high-light treatments (i.e., L400 and L500) compared to low-light treatments (i.e., L100 and L200).

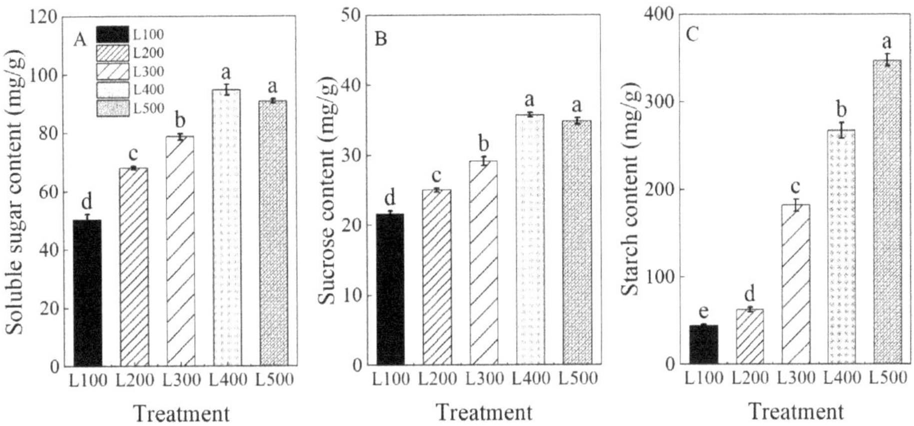

Figure 4. Changes in carbon balance of alfalfa plants under different light treatments. L100, L200, L300, L400 and L500 refer 100, 200, 300, 400 and 500 μmol m^{-2} s^{-1}, respectively. Soluble sugar content (**A**), sucrose content (**B**) and starch content (**C**). Vertical bars indicate 1 s.e. of the mean ($n = 4$). Different lowercase letters on the different bar mean significant differences ($p < 0.05$).

2.5. Gene Expression and Enzymatic Activity

The expression levels of genes encoding sucrose synthase (*SS*), sucrose phosphate synthase (*SPS*), starch synthase (*AGPase, SSS, SBE* and *SP*) and those involved in the Calvin cycle (such as *RCA, RbcL, RbcS, FBPase, TK* and *PGK*) were quantitatively analyzed, and they were significantly affected by the light treatments (Table S8). The relative expression levels of these genes were upregulated with increasing light intensity up to L500 compared to the L100 treatment (Figure 5A–L). In addition, the relative expression of *RbcS* in the L400 treatment was 2.6 ($p < 0.001$) times higher than that in the L100 treatment.

The activity of ribulose-1,5-bisphosphate carboxylase/oxygenase activase (RCA), Rubisco, fructose-1, 6-bisphosphatase (FBPase), thioredoxin reductase (TRXs), sucrose synthase (SS), sucrose phosphate synthase (SPS), adenosine diphosphate glucose pyrophosphorylase (AGPase), soluble starch synthase (SSS), starch-branching enzyme (SBE) and starch phosphorylase (SP) of alfalfa plants varied with the light treatment (Table S9). Rubisco, RCA, FBPase, TRXs, SS, SPS, AGPase, SSS, SBE and SP activities of alfalfa plants increased gradually with increasing light intensity from L100 to L500, and the highest values were at L400 and L500 (Figure 6A–J). On average, the activities of RCA, Rubisco, FBPase, TRXs, SS, AGPase, SSS, SBE and SP were higher ($p < 0.001$) at L500 than at L100. In addition, SPS activity of alfalfa plants was the highest at L400, which was 40.8% higher than at L100 ($p < 0.001$).

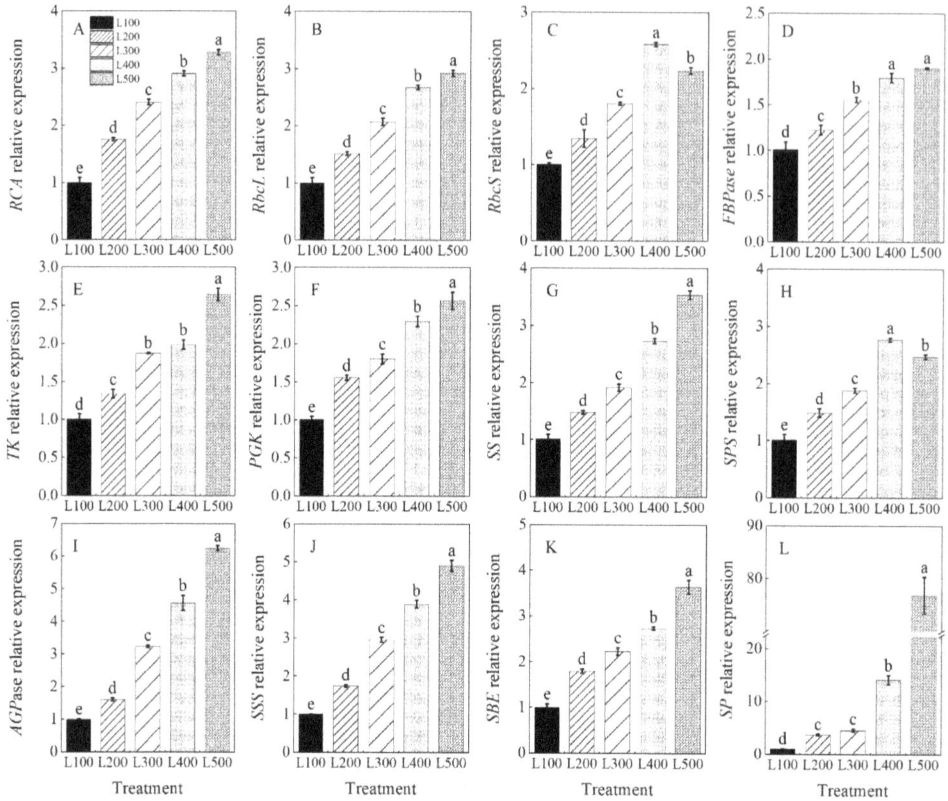

Figure 5. Changes in level of gene expression of alfalfa plants growing in different light treatments. L100, L200, L300, L400 and L500 refer 100, 200, 300, 400 and 500 μmol m^{-2} s^{-1}, respectively. Rubisco activase (*RCA*, (**A**)), Rubisco large subunit (*RbcL*, (**B**)), Rubisco small subunit (*RbcS*, (**C**)), Fructose-1,6-bisphosphatase (*FBPase*, (**D**)), Transketolase (*TK*, (**E**)), Phosphoglycerate kinase (*PGK*, (**F**)), sucrose synthase (*SS*, (**G**)), sucrose phosphate synthase (*SPS*, (**H**)), ADP-glucose pyrophosphorylase (*AGPase*, (**I**)), soluble starch synthase (*SSS*, (**J**)), starch-branching enzyme (*SBE*, (**K**)) and starch phosphorylase (*SP*, (**L**)). Vertical bars indicate 1 s.e. of the mean ($n = 3$). Different lowercase letters on the different bar mean significant differences ($p < 0.05$).

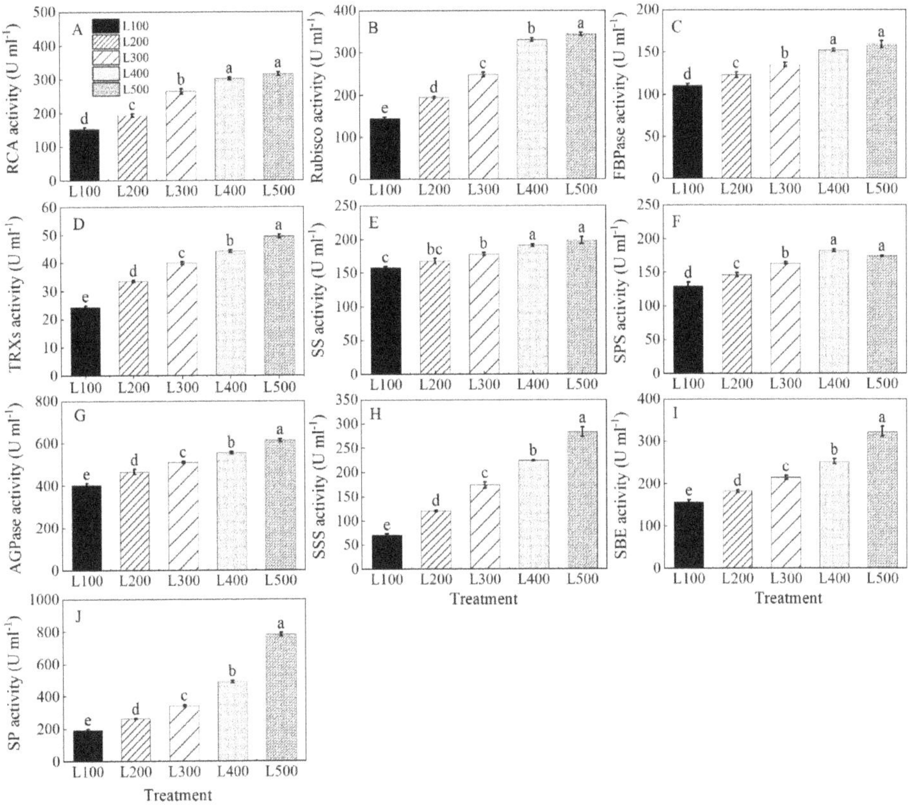

Figure 6. Changes in enzymatic activity of alfalfa plants growing in different light treatments. L100, L200, L300, L400 and L500 refer 100, 200, 300, 400 and 500 μmol m^{-2} s^{-1}, respectively. Ribulose-1,5-bisphosphate carboxylase/oxygenase activase (RCA, (**A**)), ribulose-1,5-bisphosphate carboxylase/oxygenase (Rubisco, (**B**)), fructose-1, 6-bisphosphatase (FBPase, (**C**)), thioredoxin reductase (TRXs, (**D**)), sucrose synthase (SS, (**E**)), sucrose phosphate synthase (SPS, (**F**)), ADP-glucose pyrophosphorylase (AGPase, (**G**)), soluble starch synthase (SSS, (**H**)), starch-branching enzyme (SBE, (**I**)) and starch phosphorylase (SP, (**J**)), Vertical bars indicate 1 s.e. of the mean (n = 4). Different lowercase letters on the different bar mean significant differences ($p < 0.05$).

3. Discussion

The shade avoidance syndrome is an adaptive response that increases fitness in a shaded environment by reshaping the plant morphology and modifying physiological processes [25]. Our study examined the morphology, photosynthesis, carbohydrate metabolism and the expression of genes related to photosynthesis and carbon metabolism in leaves of alfalfa seedlings grown under low-to-high light intensities. Alfalfa seedlings displayed a degree of morphological adaptation, photosynthetic tolerance and carbon balance to the light intensity attenuation. Nevertheless, excessively low light significantly accelerated stem elongation and inhibited the photosynthetic process (e.g., net photosynthetic rate and Rubisco activity) of the seedlings. Low light intensity also negatively impacted production of photoassimilates (e.g., soluble sugar and starch), which in turn restricted growth and dry matter accumulation. Therefore, our results reveal the effects of simulated shade on phenotypic, physiological and expressional regulation in alfalfa, and thus provide insight into shade regulation in intercropping systems. The implications of these results are considered below.

3.1. Light Intensity Affects Morphological Characteristics

When the shading level is increased, plants adjust through a series of growth responses, such as increasing plant height, leaf hyponasty, leaf area and specific leaf area [5,26]. In our study, plant height in the low-light treatments was significantly higher than that in the high-light treatments, whereas the reverse occurred with stem diameter. When plants are shaded, more carbohydrates are used to increase stem length than to increase stem diameter. Increased plant height may result in an increase in the amount of light received by the leaves [27]. In agriculture production, shading increases plant height and reduces stem diameter and eventually increases lodging, which hinders the transportation of nutrients, water and photosynthetic products and causes huge yield losses [28]. Additionally, light intensity affects leaf position and expansion, which play important roles in the process of irradiation interception and photosynthesis [19]. There was a greater increase in leaf hyponasty and specific leaf area of alfalfa leaves in the low-light (100–300 μmol m^{-2} s^{-1}) than in high-light (400–500 μmol m^{-2} s^{-1}) treatments, which increased light interception by the leaves. This finding is in agreement with Song et al. (2015) [29], who found that increased leaf area and leaf angle could optimize the absorbed light for carbon fixation, which in turn increased photosynthetic capacity, thereby counteracting the stress of growing in low light. Additionally, the abaxial leaf petiole angle and specific leaf area (SLA) can increase under low photosynthetic photon flux density (PPFD) compared to high-PPFD interception conditions [5]. Thus, plants grown under high light have a decreased SLA, which in turn mitigates or prevents leaf internal structure damage caused by excessive light intensity. Therefore, morphological changes in resource-harvesting organs can contribute to increased photosynthetic efficiency, which helps override light limitation stress.

Similarly, we found that increased light intensity significantly changed the morphology of alfalfa seedlings by increasing dry matter accumulation in the shoot and root, resulting in more robust seedlings. We also confirmed results from a previous study by Pan et al. (2020) [18], which showed that increased light intensity significantly increases dry matter accumulation of each organ, indicating that in turn additional photosynthates are partitioned among all organs. It is possible that increased leaf growth (source) drives root growth (sink) and thus increases the ability of plants to acquire more water and nutrients, which could be an optimal way to maintain a source–sink balance under high-light conditions [30]. Our results also are in agreement with previous research on peanut (*Arachis pintoi*) [31], suggesting that plants allocate more resources to the part that is acquiring the resource that is currently the most limiting [32]. In addition, the morphological differences in alfalfa seedlings undergoing different light treatments may be due to alterations in the molecular regulation networks or endogenous plant hormones [33,34], which deserve further investigation.

3.2. Effect of Light Intensity on Chlorophyll Content and Chlorophyll Fluorescence Characteristics

The chlorophyll content of leaves is an important part of the light-harvesting system, and it is affected by shading [35]. According to our study, significant changes were observed in Chla, Chl b, Chl a + b and Car contents, which increased in low light. The chlorophyll content of leaves of maize (*Zea mays*) plants grown in low light intensity was significantly higher than that of leaves grown in high light [36], which agrees with our results. Increased Chl b content could be a typical response to low-light conditions that allows shade-intolerant plants to capture more photosynthetically efficient blue light, thereby stimulating adaptive photomorphogenesis and alleviating the negative impacts of shade stress on photosynthetic activities [37]. Yi et al. (2020) [38] also found that adequate CO_2 assimilation and fixation promoted sugar accumulation and decreased pigment–protein complexes in leaves in high light intensity, resulting in senescence and chlorophyll degradation. Furthermore, decreased chlorophyll content could prevent excess light from damaging the photosynthetic metabolic process, which would enhance plant fitness under high-light conditions [39].

Chlorophyll fluorescence parameters can reflect the photosynthetic regulation ability of plants, and the efficiency of photochemistry can be used to evaluate the physiological responses of plants to environment stress [11]. Fv/Fm is quantum photochemical yield, and it is the ratio of number of quanta transferred to the QA acceptor to number of quanta absorbed by PSII. The high Fv/Fm value observed in L100-treated alfalfa seedlings indicates that resistance to photoinhibition was improved. Sun et al. (2014) [17] reported that Fv/Fm also increased significantly with light intensity attenuation in cucumber (*Cucumis sativus*) leaves, which displayed a decreased degree of photoinhibition and an increase in the openness and electron transport efficiency of PSII. In addition, the efficiency of PSII photochemistry (ΦPSII) can be used to reveal the physiological state of plants, and non-photochemical quenching (NPQ) is linearly related to excited energy dissipation of plants [40]. In our study, increased ΦPSII was accompanied by a corresponding increase in NPQ in the leaves grown under low light. It is possible that heat dissipation increases enough to protect the PSII photosystem from photoinhibition in the leaves grown in a low-light environment [38]. Our results suggest that the original activity of the PSII reaction center was increased, and the transformation efficiency of primary light energy was improved in the low-light adaption of alfalfa. However, the ETR was significantly higher in L400 and L500, further indicating that increased light intensity could enhance the electron transport from PSII to PSI. Similar results were also found in soybean under optimum light conditions (400 and 500 μmol m^{-2} s^{-1}) in a growth chamber [19].

3.3. *Effect of Light Intensity on Photosynthetic Characteristics and Carbohydrate Accumulation*

The net photosynthetic rate (P_n), transpiration rate (T_r) and stomatal conductance (g_s) gradually increased with an increase in light intensity, whereas the reverse occurred in the intercellular CO_2 concentration (C_i) of alfalfa plants. The main factors influencing P_n were g_s and C_i, both of which are indispensable for determining the primary cause of change in P_n [36,41]. These results suggest that increased P_n under high light intensity could be due to increased stomatal opening, which would increase net CO_2 assimilation and water vapor exchange, thus promoting photosynthesis [42]. Transpiration acts as a driving force behind the absorption and transportation of water and inorganic irons to the above-ground part of the plant [43]. Additionally, loss of water through the stomata is an important heat dissipation mechanism [37]. We found that under optimum light conditions (L400 and L500), the P_n, T_r and g_s of alfalfa can be increased and the C_i reduced, which in turn enhanced photosynthesis in alfalfa plants.

As in previous studies [44], we found that sucrose, starch and total soluble sugar contents in alfalfa leaves were significantly improved with increased light intensity. Our results indicate that increased light (L400 and L500) increased specific leaf weight and the net leaf-level photosynthetic rate, which improved the number of photosynthates stored in the leaves. However, low light intensity (L100) could cause carbohydrate loss due to inhibition of photosynthesis and inhibit plant growth. Low light intensity decreased electron transfer and net photosynthetic rates, thereby exerting a negative impact on accumulation of photosynthetic products by the seedlings [19]. Furthermore, carbohydrates also serve as carbon reserves (e.g., sucrose and starch) and are stored in plant organs. Sucrose is one of the main sources of carbon and energy in plants. In our study, sucrose content was significantly higher under high light than low light, suggesting that plants grown in high light possessed stronger photosynthesizing leaves (source tissues) that in turn increased the sucrose produced by photosynthesis for supplying the demand of growing tissues [1]. In addition, starch reserves provide an immediate available energy source that may act as a buffer when environmental conditions are not optimal for photosynthesis (e.g., shade and cloudy days) [45,46]. Less carbon was partitioned to starch synthesis at low light intensity than at high light intensity [47,48]. Our results agree with those of Dayer et al. [45] and Jian et al. (2019) [49], who found that the assimilate demand of plants exceeds the photosynthetic rate under shaded conditions, suggesting that degradation of starch reserves into soluble sugars could be used to support metabolism during a period of moderate shading stress [50].

3.4. Effect of Light Intensity on Enzymatic Activity

In C3 plants, photosynthesis is mostly regulated by the activity of Calvin cycle enzymes, including RCA, Rubisco, FBPase and TRXs, which are recognized as very early and fast responses of plants to shading stress [51,52]. Our results show that Rubisco activity in high-light-treated alfalfa plants was significantly higher than that in plants grown in low-light conditions. Similar results were reported by Feng et al. [19], suggesting that the activity of Rubisco increases with increasing light intensity, which could increase carbon assimilation and RuBP regeneration in the Calvin cycle [17]. We also found that decreased P_n in alfalfa grown under low light intensity was accompanied by reductions in RAC and Rubisco activity and the transcriptional levels of most genes (*RCA, RbcL, RbcS, FBPase, TK* and *PGK*) involved in the Calvin cycle. Our results are in accordance with those of Zhang et al. (2020) [53], suggesting that restriction of CO_2 carboxylation in the Calvin cycle is a result of impaired activity of RCA. The RCA could remove inhibitors bound to Rubisco, and thus a decline in the activity of RCA indirectly causes the decreased CO_2 assimilation rate in low-light-grown seedlings [54]. Further, the activation state of RCA, which is controlled by the redox state of the cell, is sensitive to light intensity, and the proper regulation of RCA activity is also vital for acclimation to light fluctuation in *Arabidopsis* [55,56]. Therefore, depression of photosynthetic capacity induced by low light could be attributed to deceleration of the Calvin cycle [53].

Light intensity also plays a vital role in regulating the enzymes related to sucrose and starch biosynthesis [57,58]. The relative expression levels of *SPS, SS, AGPase, SSS, SBE* and *SP* were enhanced, and their encoding enzymes showed higher activities in the high-light treatments than low-light treatments, resulting in improved production of sucrose and starch [19]. Similar results also are reported for *Arabidopsis* [59] and soybean [19], suggesting that changes in light intensity equally promote the activities of SS, SPS and SSS and increase the sucrose and starch content, which improve plant growth and development. Therefore, the enzymatic activities of sucrose synthesis and starch synthesis enzymes play a vital role in regulating carbohydrate production, which is important in controlling storage of carbon reserves and growth of cells and tissues in plants under low light [60,61]. Therefore, the enzymatic activities for increasing sucrose and starch contents in alfalfa plants were the most effective in the L400 and L500 treatments.

4. Conclusions

In the present study, we studied the impacts of low and high levels of light on the morphology, photosynthesis characteristics and carbon metabolism of alfalfa seedlings and found that they are sensitive to shade. Increased light intensity (400 to 500 $\mu mol\ m^{-2}\ s^{-1}$) enhanced the growth and dry matter accumulation, photosynthesis, carbon assimilates (sucrose and starch) and leaf enzymatic activities of enzymes related to the Calvin cycle by upregulating the important corresponding synthase genes, which positively improved carbon balance. In addition, alfalfa seedlings displayed a shade avoidance syndrome that increased their adaptive ability to compensate for low-light limitation (L100) but at the expense of dry matter and carbohydrate accumulation. The results allow us to understand the morphology, physiology and molecular behavior of plants exposed to different light intensities. Thus, gaining a more complete mechanistic picture of how alfalfa plants adapt and respond to light levels would provide useful support for guiding spatial arrangement of the alfalfa canopy in an intercropping system, thereby improving food production and ensuring higher yields.

5. Materials and Methods

5.1. Plant Material and Growth Conditions

The experiment was conducted in LED climate rooms located in the basement of the College of Grassland Science, Qingdao Agricultural University, Qingdao, China. Light intensity and spectral irradiance (λ = 350–800 nm) were measured by HR550 (Hipoint Inc., Gaoxiong, Taiwan), and the spectral distributions are shown in Figure S1. The photoperiod was 12 h with a 25 °C day temperature, 20 °C night temperature and a relative humidity of 60%.

M. sativa L. cv. Zhongmu 1 was chosen for the studies on phenotypic responses to growth conditions. Before the experiment, alfalfa seeds were surface-sterilized by 75% ethanol for 1 min and rinsed with deionized water for 5 min and germinated on wet sterile Whatman No. 1 filter paper in a daily 8 h light from white fluorescent tubes (Sanpai Corporation, Shanghai, China) with a mean photon flux density of 60 μmol m^{-2} s^{-1} (400–700 nm) at 25 °C. After seed coat rupture and cotyledon expansion at 5 days, 10 uniform-sized sprouting alfalfa seedlings were transplanted into a separate plugged hole in a foam sheet floating in a 3.3 L plastic container filled with half-strength Hoagland's solution. These containers were placed in the LED climate room with a light intensity of 500 μmol m^{-2} s^{-1}. When the first trifoliate leaf was well-developed, the pre-cultured seedlings were transferred to five light treatments. Photosynthetic photon flux density (PPFD) was 100 μmol m^{-2} s^{-1} (L100), 200 μmol m^{-2} s^{-1} (L200), 300 μmol m^{-2} s^{-1} (L300), 400 μmol m^{-2} s^{-1} (L400) and 500 μmol m^{-2} s^{-1} (L500), and light quality was the same in all treatments. The highest light intensity was chosen as it was used to grow alfalfa seedlings under laboratory conditions, and the lowest light intensity can be considered comparable to the natural shade under a closed oat forage canopy under clear-sky conditions (82–116 μmol m^{-2} s^{-1}, unpublished analysis of solar radiation penetrating closed canopy in the alfalfa–oat intercropping system by W. Tang). Four containers were placed in each light treatment, and each container had 10 seedlings. The plastic containers were moved daily to avoid boundary effects, and Hoagland's solution was renewed every 3 days and kept aerated by air-spraying. Every treatment was performed with four replicates.

5.2. Plant Morphology Parameters

After 14 d of treatment, four plants from each treatment were randomly selected, and the main growth parameters were measured: plant height, stem diameter, abaxial leaf petiole angle and leaf area. Leaf area was determined using an Li-3000 leaf area meter (Li-Cor Inc.). Specific leaf area (SLA) was calculated as: SLW = leaf area / leaf dry weight. Root morphology parameters were digitized with a LA-S scanner and analyzed using WinRhizo software (LA-S, Wanseng, China). After this, shoot and root samples were heated at 105 °C for 30 min and then dried to a constant weight in a fan oven at 75 °C. Dry matter was expressed as mg $plant^{-1}$. Root-to-shoot ratio was also calculated.

5.3. Photosynthetic Pigment Content and Leaf Nitrogen Content

After 14 d of treatment, the third trifoliate leaf on alfalfa seedlings was collected for photosynthetic pigment and leaf nitrogen content measurement. Chlorophyll a (Chl a), Chlorophyll b (Chl b) and carotenoids (Car) were extracted from all leaf samples. Six leaf discs (0.6 cm in diameter) were cut from the middle part of each middle leaflet with a puncher, and they were placed in 25 mL of 95% acetone in the dark for 24 h, at which time the sample was colorless. Concentration of Chl a, Chl b and Car was measured at wavelengths of A_{663}, A_{645} and A_{470} nm, respectively, using a UV spectrophotometer (UV-2700, Shimadzu, Kyoto, Japan), and calculated according to Pan et al. (2020) [18].

The dried leaf tissues were ground to obtain homogenous samples. A subsample of about 0.1 g was then digested with 5 mL of concentrated H_2SO_4 for 2 h in 420 °C, and K_2SO_4 and $CuSO_4 \cdot 5H_2O$ (K_2SO_4:$CuSO_4 \cdot 5H_2O$ = 10:1) as the catalyzer. Finally, the leaf nitrogen concentration was measured by an automatic flow injection analyzer (AA3, SEAL).

5.4. Photosynthesis Parameters

After 14 d of treatment, Li-6800 portable photosynthesis system (LI-COR Inc., Lincoln, NE, USA) was used for photosynthetic parameter measurement on the third fully expanded leaf of alfalfa seedlings. All parameters, including net photosynthetic rate (P_n), transpiration rate (T_r), intercellular CO_2 concentration (C_i) and stomatal conductance (g_s), were measured under steady photosynthetic photon flux density in the leaf chamber, which was set to the same level as the relevant treatments. An Li-COR standard red–blue chamber set at 25 °C and a CO_2 concentration of 460 μmol mol^{-1} were used.

5.5. Chlorophyll Fluorescence Measurements

After measuring the rate of photosynthesis, a Chlorophyll a fluorescence measurement was performed with the third fully expanded leaf on the alfalfa seedlings. Before measurement, each seedling was held in a dark chamber for 30 min prior to being submitted to the chlorophyll fluorescence procedure using an Li-6800 portable photosynthesis system. Fluorescence parameters characterizing the state of the photosynthetic apparatus were calculated on the basis of induction fluorescence curves obtained using data from the JIP test, which is usually used to evaluate the state of PSII. On the basis of induction fluorescence curves (OJIP curves), the following parameters, which characterize the maximal PSII quantum yield (F_v/F_m), effective PSII quantum yield (Φ_{PSII}, $(F_m{}' - F_t)/F_m$), non-photochemical quenching (NPQ, $F_m/F_m{}' - 1$) and electron transport rate (ETR), were determined. Here, F_v is the value of variable fluorescence, equal to the difference between Fm and F_0; F_0 is the minimum amplitude of fluorescence (F); and F_m is the maximum amplitude of fluorescence. F_m and $F_m{}'$ are the maximum Chl fluorescence levels under dark- and light-adapted conditions, respectively. F_v is the photoinduced change in fluorescence, and F_t is the level of fluorescence before a saturation impulse is applied. F_0 is the initial Chl fluorescence level. All parameters were calculated according to the methods reported by Pashkovskiy et al. (2021) [62].

5.6. Soluble Sugar, Sucrose and Starch Content

After 14 d of treatment, the fourth trifoliate leaf of alfalfa seedlings was collected for measurement of soluble sugar, sucrose and starch contents. Dried leaf tissues from all plants were ground to obtain homogenous samples, and subsamples were used to determine content of soluble sugar, sucrose and starch.

Soluble sugar and sucrose were extracted from the powdered sample (50 mg) three times, using 80% (*v*/*v*) ethanol at 80 °C. The supernatants were pooled and then diluted with 80% ethanol to 25 mL for the measurement of soluble sugar and sucrose content. Soluble sugar content was determined using the anthrone–sulfuric acid reagent method and calculated based on absorbance at a wavelength of 620 nm using the UV spectrophotometer [63]. Sucrose content was measured using the resorcinol method and estimated on the basis of the absorbance at a wavelength of 480 nm using the UV spectrophotometer [64]. The residue obtained after extraction was analyzed for starch, using the perchloric acid digestion method. Following extraction, starch content was determined photometrically in the presence of anthrone–sulfuric acid reagent and estimated on the basis of absorbance at a wavelength of 620 nm using the UV spectrophotometer [63].

5.7. Enzyme Activity

The second fully expanded leaf of alfalfa seedlings was harvested and used for enzymatic assays. The activity of enzymes, including ribulose-1,5-bisphosphate carboxylase/oxygenase (Rubisco, EC 4.1.1.39), ribulose-1,5-bisphosphate carboxylase/oxygenase activase (RCA, EC was not found), fructose-1, 6-bisphosphatase (FBPase, EC 3.1.3.11), thioredoxin reductase (TRXs, EC 1.8.1.9), sucrose synthase (SS, EC 2.4.1.13), sucrose phosphate synthase (SPS, EC 2.4.1.14), ADP-glucose pyrophosphorylase (AGPase, EC 2.7.7.27), soluble starch enzyme (SSE, EC 2.4.1.21), starch-branching enzyme (SBE, EC 2.4.1.18) and starch phosphorylase (SP, EC 2.4.1.1), was determined using plant-enzyme-linked immunosorbent assay (ELISA) kits. A frozen leaf sample (0.1 g) was homogenized in 1 mL of phosphate buffer (0.01 mol L^{-1}, pH = 7.4) using a cold mortar and pestle and centrifuged at 5000× *g* and 4 °C for 10 min. The clear supernatant was then stored at 4 °C for 24 h pending analyses for the activity of enzymes. Firstly, 50 μL of standard or sample was added to the appropriate well of a microplate (except the blank wells). Secondly, 100 μL of HRP conjugate reagent was added, and the wells were covered with an adhesive plate membrane and incubated for 60 min at 37 °C. Thirdly, the liquid was discarded, and the wells were washed with 350 μL of wash buffer, and this procedure was repeated five times. Fourthly, a mixture of 50 μL of substrate A and 50 μL of substrate B was added to each well, mixed gently and incubated at 37 °C for 15 min in the dark. Finally, 50 μL of stop solution was added to each well, and the optical density was measured within 15 min at 450 nm using a microtiter plate reader (Infinite MPlex, Tecan, Austria). All the activity of enzymes was calculated using the methods reported by Pan et al. (2020) [18].

The protein concentration of each enzyme extraction solution was measured according to Li et al. (2020) [65]. The results are expressed as U/mL of protein.

5.8. Real-Time Quantitative PCR Verification

The second fully expanded leaf was harvested from five seedlings growing in each light treatment and used to determine RNA abundance. All the leaves were labeled and frozen in liquid nitrogen immediately. RNA was extracted using the TRIzol™ Plus RNA Purification Kit (TaKaRa Biotechnology, Dalian, China). Reverse transcription and amplification of cDNA were performed using Super Script III First-Strand Synthesis Super Mix for qRT-PCR (Vazyme, Nanjing, China). Real-time quantitative PCR was conducted in Real-Time PCR System (CFX96, Bio-rad, USA), and $2^{-\Delta\Delta CT}$ method was used for data analysis [66]. The actin was selected as the reference gene. All target genes and target genes primers are listed in Supporting Information Table S1.

5.9. Statistical Analysis

All data analyses were conducted using one-way analysis of variance with the ANOVA packages of SPSS® version 17.0 (SPSS Inc., Chicago, IL, USA). The homoscedasticity of the variables was determined using the Levene test. When the *F*-test indicated statistical significance was $p < 0.05$, Duncan's new multiple range test for least significant difference (l.s.d.) was used to determine least significant range between means.

Supplementary Materials: The following supporting information can be downloaded at: https://www.mdpi.com/article/10.3390/plants11131688/s1. Figure S1. The spectral distribution of LED lights under different light intensity treatments. Table S1. List of primers for characterizing alfalfa genes (5′–3′) measured under different light treatments. *RCA*, *RbcL*, *RbcS*, *FBPase*, *TK* and *PGK* involve in Rubisco activase, Rubisco large subunit, Rubisco small subunit, Fructose-1,6-bisphosphatase, Transketolase, and Phosphoglycerate kinase, respectively. SS and SPS genes involve in sucrose synthase and sucrose phosphate synthase, respectively. *AGPase*, *SSS*, *SBE*, and *SP* involve in ADP-glucose pyrophosphorylase, soluble starch synthase, starch branching enzyme and starch phosphorylase, respectively. Table S2. Results of one-way ANOVA of effect of light intensity treatment on plant height, stem diameter, abaxial leaf petiole angle, specific leaf weight, shoot dry matter, root dry matter and root to shoot ratio of alfalfa plants. Table S3. Results of one-way ANOVA of effect of light intensity treatment on root length (RL), root surface (RS), root volume (RV), and root diameter (RD) of alfalfa plants. Table S4. Results of one-way ANOVA of effect of light intensity treatment on Chlorophyll a (Chl a), Chlorophyll b (Chl b), carotenoids (Car), Chl a + b, Chl a/b and leaf nitrogen content (LNC) of alfalfa plants. Table S5. Results of one-way ANOVA of effect of light intensity treatment on net photosynthetic rate (P_n), transpiration rate (T_r), intercellular CO_2 concentration (C_i), and stomatal conductance (g_s) of alfalfa plants. Table S6. Results of one-way ANOVA of effect of light intensity treatment on maximal PSII quantum yield (F_v/F_m), effective PSII quantum yield (ΦPSII), non-photochemical quenching (NPQ) and electron transport rate (ETR) of alfalfa plants. Table S7. Results of one-way ANOVA of effect of light intensity treatment on soluble sugar (SS), sucrose, and starch (St) content of alfalfa plants.

Author Contributions: Conceptualization, J.S. and W.T.; methodology, W.T. and H.G.; software and validation, C.Y. and Z.L., investigation, T.W. and S.Z.; formal analysis, Y.Z.; resources, F.M. and Q.T.; data curation, H.S. and Z.L.; writing—original draft preparation, W.T.; writing—review and editing, C.C.B. and W.X.; visualization, J.Y. and X.W.; supervision, J.S. and W.T. All authors have read and agreed to the published version of the manuscript.

Funding: This project has received funding from the China Forage and Grass Research System (CARS-34), First Class Grassland Science Discipline Program of Shandong Province, China (1619002), Natural Science Foundation of Shandong Province (ZR202103040824) and Doctoral Scientific Research Startup of Qingdao Agricultural University (No. 6631120005 and No. 6631120007).

Institutional Review Board Statement: Not applicable.

Informed Consent Statement: Not applicable.

Data Availability Statement: The data presented in this study are available in the article and supplementary material.

Conflicts of Interest: The authors declare that they have no conflicts of interest.

References

1. Yang, F.; Feng, L.; Liu, Q.; Wu, X.; Fan, Y.; Raza, M.A.; Cheng, Y.; Chen, J.; Wang, X.; Yong, T.; et al. Effect of interactions between light intensity and red-to- far-red ratio on the photosynthesis of soybean leaves under shade condition. *Environ. Exp. Bot.* **2018**, *150*, 79–87. [CrossRef]
2. Wu, Y.S.; Gong, W.Z.; Yang, W.Y. Shade Inhibits leaf size by controlling cell proliferation and enlargement in soybean. *Sci. Rep.* **2017**, *7*, 9259. [CrossRef]
3. Ciolfi, A.; Sessa, G.; Sassi, M.; Possenti, M.; Salvucci, S.; Carabelli, M.; Morelli, G.; Ruberti, I. Dynamics of the shade-avoidance response in *Arabidopsis*. *Plant Physiol.* **2013**, *163*, 331–353. [CrossRef]
4. Casal, J.J. Photoreceptor signaling networks in plant responses to shade. *Annu. Rev. Plant Biol.* **2013**, *64*, 403–427. [CrossRef]
5. Hitz, T.; Hartung, J.; Graeff-Honninger, S.; Munz, S. Morphological response of soybean (*Glycine max* (L.) Merr.) cultivars to light intensity and red to far-red ratio. *Agronomy* **2019**, *9*, 428. [CrossRef]
6. Mauser, H.; King, W.A.; Gready, J.E.; Andrews, T.J. CO_2 fixation by Rubisco: Computational dissection of the key steps of carboxylation, hydration, and C-C bond cleavage. *J. Am. Chem. Soc.* **2001**, *123*, 10821–10829. [CrossRef] [PubMed]
7. Hussain, S.; Pang, T.; Iqbal, N.; Shafiq, I.; Skalicky, M.; Brestic, M.; Safdar, M.E.; Mumtaz, M.; Ahmad, A.; Asghar, M.A.; et al. Acclimation strategy and plasticity of different soybean genotypes in intercropping. *Funct. Plant Biol.* **2020**, *47*, 592–610. [CrossRef] [PubMed]
8. Raza, M.A.; Feng, L.Y.; Iqbal, N.; Ahmed, M.; Chen, Y.K.; Bin Khalid, M.H.; Din, A.M.U.; Khan, A.; Ijaz, W.; Hussain, A.; et al. Growth and development of soybean under changing light environments in relay intercropping system. *PeerJ* **2019**, *7*, 7262. [CrossRef]
9. Chen, T.T.; Zhang, H.J.; Zeng, R.E.; Wang, X.Y.; Huang, L.P.; Wang, L.D.; Wang, X.W.; Zhang, L. Shade effects on peanut yield associate with physiological and expressional regulation on photosynthesis and sucrose metabolism. *Int. J. Mol. Sci.* **2020**, *21*, 5284. [CrossRef]
10. Patel, S.; Bartel, C.A.; Lenssen, A.W.; Moore, K.J.; Berti, M.T. Stem density, productivity, and weed community dynamics in corn-alfalfa intercropping. *Agronomy* **2021**, *11*, 1696. [CrossRef]
11. Rascher, U.; Liebig, M.; Lüttge, U. Evaluation of instant light-response curves of chlorophyll fluorescence parameters obtained with a portable chlorophyll fluorometer on site in the field. *Plant Cell Environ.* **2010**, *23*, 1397–1405. [CrossRef]
12. Hu, L.L.; Liao, W.B.; Dawuda, M.M.; Yu, J.H.; Lv, J. Appropriate NH_4^+: NO_3^- ratio improves low light tolerance of mini Chinese cabbage seedlings. *BMC Plant Biol.* **2017**, *17*, 22. [CrossRef] [PubMed]
13. Sui, X.L.; Mao, S.L.; Wang, L.H.; Zhang, B.X.; Zhang, Z.X. Effect of low light on the characteristics of photosynthesis and chlorophyll a fluorescence during leaf development of sweet pepper. *J. Integr. Agric.* **2012**, *11*, 1633–1643. [CrossRef]
14. Sheth, B.P.; Thaker, V.S. In silico analyses of Rubisco enzymes from different classes of algae. *Int. Res. J. Biol. Sci.* **2014**, *3*, 11–17.
15. Sharkey, T.D.; Bernacchi, C.J.; Farquhar, G.D.; Singsaas, E.L. Fitting photosynthetic carbon dioxide response curves for C3 leaves. *Plant Cell Environ.* **2007**, *30*, 1035–1040. [CrossRef]
16. Redondo-Gomez, S.; Mateos-Naranjo, E.; Davy, A.J.; Fernandez-Munoz, F.; Castellanos, E.M.; Luque, T.; Figueroa, M.E. Growth and photosynthetic responses to salinity of the salt-marsh shrub *Atriplex portulacoides*. *Ann. Bot.* **2007**, *100*, 555–563. [CrossRef] [PubMed]
17. Sun, J.L.; Sui, X.L.; Huang, H.Y.; Wang, S.X.; Wei, Y.X.; Zhang, Z.X. Low light stress down-regulated Rubisco gene expression and photosynthetic capacity during cucumber (*Cucumis sativus* L.) leaf development. *J. Integr. Agric.* **2014**, *13*, 997–1007. [CrossRef]
18. Pan, T.; Wang, Y.; Wang, L.; Ding, J.; Cao, Y.; Qin, G.; Yan, L.; Xi, L.; Zhang, J.; Zou, Z. Increased CO_2 and light intensity regulate growth and leaf gas exchange in tomato. *Physiol. Plant.* **2020**, *168*, 694–708. [CrossRef]

19. Feng, L.Y.; Raza, M.A.; Li, Z.C.; Chen, Y.K.; Khalid, M.H.B.; Du, J.B.; Liu, W.G.; Wu, X.L.; Song, C.; Yu, L.; et al. The influence of light Intensity and leaf movement on photosynthesis characteristics and carbon balance of soybean. *Front. Plant Sci.* **2019**, *9*, 1952. [CrossRef]
20. Liu, M.G.; Mu, L.; Lu, Y.X.; Yang, H.M. Forage accumulation and radiation use of alfalfa under deficit irrigation. *Crop Sci.* **2021**, *61*, 2190–2202. [CrossRef]
21. Vasileva, V.; Kostov, O. Effect of mineral and organic fertilization on alfalfa forage and soil fertility. *Emir. J. Food Agric.* **2015**, *27*, 678–686. [CrossRef]
22. Xu, R.X.; Zhao, H.M.; Liu, G.B.; Li, Y.; Li, S.J.; Zhang, Y.J.; Liu, N.; Ma, L. Alfalfa and silage maize intercropping provides comparable productivity and profitability with lower environmental impacts than wheat-maize system in the North China plain. *Agric. Syst.* **2022**, *195*, 103305. [CrossRef]
23. Xu, R.X.; Zhao, H.M.; Liu, G.B.; You, Y.L.; Ma, L.; Liu, N.; Zhang, Y.J. Effects of nitrogen and maize plant density on forage yield and nitrogen uptake in an alfalfa-silage maize relay intercropping system in the North China Plain. *Field Crops Res.* **2021**, *263*, 108068. [CrossRef]
24. Lorenzo, C.D.; Iserte, J.A.; Lamas, M.S.; Antonietti, M.S.; Gagliardi, P.G.; Hernando, C.E.; Dezar, C.A.A.; Vazquez, M.; Casal, J.J.; Yanovsky, M.J.; et al. Shade delays flowering in *Medicago sativa*. *Plant J.* **2019**, *99*, 7–22. [CrossRef]
25. Pan, T.H.; Ding, J.J.; Qin, G.G.; Wang, Y.L.; Xi, L.J.; Yang, J.W.; Li, J.M.; Zhang, J.; Zou, Z.R. Interaction of supplementary light and CO_2 enrichment improves growth, photosynthesis, yield, and quality of tomato in autumn through spring greenhouse production. *HortScience* **2019**, *54*, 246–252. [CrossRef]
26. Larbi, A.; Vazquez, S.; El-Jendoubi, H.; Msallem, M.; Abadia, J.; Abadia, A.; Morales, F. Canopy light heterogeneity drives leaf anatomical, eco-physiological, and photosynthetic changes in olive trees grown in a high-density plantation. *Photosynth. Res.* **2015**, *123*, 141–155. [CrossRef]
27. Gao, Z.; Khalid, M.; Jan, F.; Rahman, S.U.; Jiang, X.; Yu, X. Effects of light-regulation and intensity on the growth, physiological and biochemical properties of *Aralia elata* (Miq.) seedlings. *S. Afr. J. Bot.* **2019**, *121*, 456–462. [CrossRef]
28. Wu, Y.S.; Gong, W.Z.; Yang, F.; Wang, X.C.; Yong, T.W.; Yang, W.Y. Responses to shade and subsequent recovery of soybean in maize-soybean relay strip intercropping. *Plant Prod. Sci.* **2016**, *19*, 206–214. [CrossRef]
29. Song, K.S.; Jeon, K.S.; Choi, K.S.; Chang, H.K.; Kim, J.J. Characteristics of photosynthesis and leaf growth of *Peucedanum japonicum* by leaf mold and shading level in forest farming. *Korean J. Med. Crop Sci.* **2015**, *23*, 43–48. [CrossRef]
30. Dong, J.; Li, X.; Duan, Z.Q. Biomass allocation and organs growth of cucumber (*Cucumis sativus* L.) under elevated CO_2 and different N supply. *Korean J. Med. Crop Sci.* **2016**, *62*, 277–288.
31. da Cruz, P.J.R.; Santos, M.V.; da Silva, L.D.; Ferreira, E.A.; Magalhaes, M.A.; Martuscello, J.A.; de Fonseca, D.M. Morphogenetic, physiological, and productive of forage peanut responses to shading. *Pesquisa Agropecuária Brasileira* **2020**, *55*, e01746. [CrossRef]
32. Poorter, H.; Niklas, K.J.; Reich, P.B.; Oleksyn, J.; Poot, P.; Mommer, L. Biomass allocation to leaves, stems and roots: Meta-analyses of interspecific variation and environmental control. *New Phytol.* **2012**, *193*, 30–50. [CrossRef] [PubMed]
33. Bawa, G.; Feng, L.Y.; Chen, G.P.; Chen, H.; Hu, Y.; Pu, T.; Cheng, Y.J.; Shi, J.Y.; Xiao, T.; Zhou, W.G.; et al. Gibberellins and auxin regulate soybean hypocotyl elongation under low light and high-temperature interaction. *Physiol. Plant.* **2020**, *170*, 345–356. [CrossRef] [PubMed]
34. Li, Y.M.; Gao, M.F.; He, R.; Zhang, Y.T.; Song, S.W.; Su, W.; Liu, H.C. Far-red light suppresses glucosinolate profiles of Chinese kale through inhibiting genes related to glucosinolate biosynthesis. *Environ. Exp. Bot.* **2021**, *188*, 104507. [CrossRef]
35. Wang, Y.B.; Huang, R.D.; Zhou, Y.F. Effects of shading stress during the reproductive stages on photosynthetic physiology and yield characteristics of peanut (*Arachis hypogaea* Linn.). *J. Integr. Agric.* **2021**, *20*, 1250–1265. [CrossRef]
36. Ren, B.Z.; Cui, H.Y.; Camberato, J.J.; Dong, S.T.; Liu, P.; Zhao, B.; Zhang, J.W. Effects of shading on the photosynthetic characteristics and mesophyll cell ultrastructure of summer maize. *Sci. Nat.* **2016**, *103*, 67. [CrossRef]
37. Wang, X.; Chen, G.D.; Du, S.J.; Wu, H.X.; Fu, R.; Yu, X.B. Light Intensity Influence on Growth and Photosynthetic Characteristics of *Horsfieldia hainanensis*. *Front. Ecol. Evol.* **2021**, *9*, 636804. [CrossRef]
38. Yi, Z.H.; Cui, J.J.; Fu, Y.M.; Liu, H. Effect of different light intensity on physiology, antioxidant capacity and photosynthetic characteristics on wheat seedlings under high CO_2 concentration in a closed artificial ecosystem. *Photosynth. Res.* **2020**, *144*, 23–34. [CrossRef]
39. Neda, L.; Michael, K.; Jörg, F. Interaction effect between elevated CO_2 and fertilization on biomass, gas exchange and C/N ratio of European beech (*Fagus sylvatica* L.). *Plants* **2016**, *5*, 38.
40. Zhong, X.M.; Shi, Z.S.; Li, F.H.; Huang, H.J. Photosynthesis and chlorophyll fluorescence of infertile and fertile stalks of paired near-isogenic lines in maize (*Zea mays* L.) under shade conditions. *Photosynthetica* **2014**, *52*, 597–603. [CrossRef]
41. Gao, J.; Liu, Z.; Zhao, B.; Liu, P.; Zhang, J.W. Physiological and comparative proteomic analysis provides new insights into the effects of shade stress in maize (*Zea mays* L.). *BMC Plant Biol.* **2020**, *20*, 60. [CrossRef] [PubMed]
42. Loconsole, D.; Cocetta, G.; Santoro, P.; Ferrante, A. Optimization of LED lighting and quality evaluation of romaine lettuce grown in an innovative indoor cultivation system. *Sustainability* **2019**, *11*, 841. [CrossRef]
43. Jafarikouhini, N.; Pradhan, D.; Sinclair, T.R. Basis of limited-transpiration rate under elevated vapor pressure deficit and high temperature among sweet corn cultivars. *Environ. Exp. Bot.* **2020**, *179*, 104205. [CrossRef]
44. Michalska, J.; Zauber, H.; Buchanan, B.B.; Cejudo, F.J.; Geigenberger, P. NTRC links built-in thioredoxin to light and sucrose in regulating starch synthesis in chloroplasts and amyloplasts. *Proc. Natl. Acad. Sci. USA* **2009**, *106*, 9908–9913. [CrossRef]

45. Dayer, S.; Murcia, G.; Prieto, J.A.; Duran, M.; Martinez, L.; Piccoli, P.; Pena, J.P. Non-structural carbohydrates and sugar export in grapevine leaves exposed to different light regimes. *Physiol. Plant.* **2021**, *171*, 728–738. [CrossRef]
46. Fernandez, O.; Ishihara, H.; George, G.M.; Mengin, V.; Flis, A.; Sumner, D.; Arrivault, S.; Feil, R.; Lunn, J.E.; Zeeman, S.C. Foliar starch turnover occurs in long days and in falling light at the end of the day. *Plant Physiol.* **2017**, *174*, 00601. [CrossRef]
47. Breuer, G.; Martens, D.E.; Draaisma, R.B.; Wijffels, R.H.; Lamers, P.P. Photosynthetic efficiency and carbon partitioning in nitrogen-starved *Scenedesmus obliquus*. *Algal Res.* **2015**, *9*, 254–262. [CrossRef]
48. Schmitz, J.; Heinrichs, L.; Scossa, F.; Fernie, A.R.; Oelze, M.L.; Dietz, K.J.; Rothbart, M.; Grimm, B.; Flugge, U.I.; Hausler, R.E. The essential role of sugar metabolism in the acclimation response of *Arabidopsis thaliana* to high light intensities. *J. Exp. Bot.* **2014**, *65*, 1619–1636. [CrossRef]
49. Jian, Y.; Wu, G.L.; Zhou, D.H.; Hu, Z.Q.; Quan, Z.X.; Zhou, B.Y. Effects of shading on carbohydrates of *Syzygium samarangense*. *Not. Bot. Horti Agrobot.* **2019**, *47*, 1252–1257. [CrossRef]
50. Mengin, V.; Pyl, E.T.; Moraes, T.A.; Sulpice, R.; Krohn, N.; Encke, B.; Stitt, M. Photosynthate partitioning to starch in *Arabidopsis thaliana* is insensitive to light intensity but sensitive to photoperiod due to a restriction on growth in the light in short photoperiods. *Plant Cell Environ.* **2017**, *40*, 2608–2627. [CrossRef]
51. Kono, T.; Mehrotra, S.; Endo, C.; Kizu, N.; Matusda, M.; Kimura, H.; Mizohata, E.; Inoue, T.; Hasunuma, T.; Yokota, A.; et al. A RuBisCO-mediated carbon metabolic pathway in methanogenic archaea. *Nat. Commun.* **2017**, *8*, 14007. [CrossRef] [PubMed]
52. Tyagi, A.K.; Gaur, T. Light regulation of nuclear photosynthetic genes in higher plants. *Crit. Rev. Plant Sci.* **2003**, *22*, 417–452. [CrossRef]
53. Zhang, J.Y.; Cun, Z.; Chen, J.W. Photosynthetic performance and photosynthesis-related gene expression coordinated in a shade-tolerant species *Panax notoginseng* under nitrogen regimes. *BMC Plant Biol.* **2020**, *20*, 273. [CrossRef] [PubMed]
54. Carmo-Silva, A.E.; Salvucci, M.E. The regulatory properties of rubisco activase differ among species and affect photosynthetic induction during light transitions. *Plant Physiol.* **2013**, *161*, 1645–1655. [CrossRef]
55. Zhang, N.; Kallis, R.P.; Ewy, R.G.; Portis, A.R. Light modulation of Rubisco in *Arabidopsis* requires a capacity for redox regulation of the larger Rubisco activase isoform. *Proc. Natl. Acad. Sci. USA* **2002**, *99*, 3330–3334. [CrossRef]
56. Kaiser, E.; Galvis, V.C.; Armbruster, U. Efficient photosynthesis in dynamic light environments: A chloroplast's perspective. *Biochem. J.* **2019**, *476*, 2725–2741. [CrossRef]
57. Eliyahu, E.; Rog, I.; Inbal, D.; Danon, A. ACHT4-driven oxidation of APS1 attenuates starch synthesis under low light intensity in *Arabidopsis* plants. *Proc. Natl. Acad. Sci. USA* **2015**, *112*, 12876–12881. [CrossRef]
58. Ding, Z.H.; Zhang, Y.; Xiao, Y.; Liu, F.F.; Wang, M.H.; Zhu, X.G.; Liu, P.; Sun, Q.; Wang, W.Q.; Peng, M.; et al. Transcriptome response of cassava leaves under natural shade. *Sci. Rep.* **2016**, *6*, 31673. [CrossRef]
59. Bahaji, A.; Sanchez-Lopez, A.M.; De Diego, N.; Munoz, F.J.; Baroja-Fernandez, E.; Li, J.; Ricarte-Bermejo, A.; Baslam, M.; Aranjuelo, I.; Almagro, G.; et al. Plastidic phosphoglucose isomerase is an important determinant of starch accumulation in mesophyll cells, growth, photosynthetic capacity, and biosynthesis of plastidic cytokinins in *Arabidopsis*. *PLoS ONE* **2015**, *10*, e0119641. [CrossRef]
60. Yang, L.Y.; Chen, J.J.; Sun, X.M.; Li, J.X.; Chen, N.L. Inhibition of sucrose and galactosyl-sucrose oligosaccharide metabolism in leaves and fruits of melon (*Cucumis melo* L.) under low light stress. *Sci. Hortic.* **2019**, *244*, 343–351. [CrossRef]
61. Wang, J.; Shi, K.; Lu, W.P.; Lu, D.L. Effects of post-silking shading stress on enzymatic activities and phytohormone contents during grain development in spring maize. *J. Plant Growth Regul.* **2021**, *40*, 1060–1073. [CrossRef]
62. Pashkovskiy, P.; Kreslavski, V.D.; Ivanov, Y.; Ivanova, A.; Kartashov, A.; Shmarev, A.; Strokina, V.; Kuznetsov, V.V.; Allakhverdiev, S.I. Influence of light of different spectral compositions on the growth, photosynthesis, and expression of light-dependent genes of scots pine seedlings. *Cells* **2021**, *10*, 3284. [CrossRef] [PubMed]
63. Tang, W.; Baskin, C.C.; Baskin, J.M.; Nan, Z.B. Plastic film mulching improves seed germination, seedling development and potential for perenniality of *Vicia unijuga* under subalpine climate conditions. *Crop Pasture Sci.* **2020**, *71*, 592–609. [CrossRef]
64. Shi, H.R.; Wang, B.; Yang, P.J.; Li, Y.B.; Miao, F. Differences in sugar accumulation and mobilization between sequential and non-sequential senescence wheat cultivars under natural and drought conditions. *PLoS ONE* **2016**, *11*, e0166155. [CrossRef] [PubMed]
65. Li, Y.; Xin, G.F.; Liu, C.; Shi, Q.H.; Yang, F.J.; Wei, M. Effects of red and blue light on leaf anatomy, CO_2 assimilation and the photosynthetic electron transport capacity of sweet pepper (*Capsicum annuum* L.) seedlings. *BMC Plant Biol.* **2020**, *20*, 318. [CrossRef]
66. Kenneth, J.L.; Thomas, D.S. Analysis of relative gene expression data using real-time quantitative PCR and the $2^{-\Delta\Delta CT}$ method. *Methods* **2002**, *25*, 402–408.

plants MDPI

Review

Light and Plant Growth Regulators on In Vitro Proliferation

Valeria Cavallaro [1,*], Alessandra Pellegrino [1], Rosario Muleo [2,*] and Ivano Forgione [2]

[1] Institute of BioEconomy (IBE), National Research Council of Italy, 95126 Catania, Italy; alessandra.pellegrino@cnr.it
[2] Tree Physiology and Fruit Crop Biotechnology Laboratory, Department of Agriculture and Forest Sciences (DAFNE), University of Tuscia, 01100 Viterbo, Italy; ivano.forgione@unitus.it
* Correspondence: valeria.cavallaro@cnr.it (V.C.); muleo@unitus.it (R.M.)

Abstract: Plant tissue cultures depend entirely upon artificial light sources for illumination. The illumination should provide light in the appropriate regions of the electromagnetic spectrum for photomorphogenic responses and photosynthetic metabolism. Controlling light quality, irradiances and photoperiod enables the production of plants with desired characteristics. Moreover, significant money savings may be achieved using both more appropriate and less consuming energy lamps. In this review, the attention will be focused on the effects of light characteristics and plant growth regulators on shoot proliferation, the main process in in vitro propagation. The effects of the light spectrum on the balance of endogenous growth regulators will also be presented. For each light spectrum, the effects on proliferation but also on plantlet quality, i.e., shoot length, fresh and dry weight and photosynthesis, have been also analyzed. Even if a huge amount of literature is available on the effects of light on in vitro proliferation, the results are often conflicting. In fact, a lot of exogenous and endogenous factors, but also the lack of a common protocol, make it difficult to choose the most effective light spectrum for each of the large number of species. However, some general issues derived from the analysis of the literature are discussed.

Keywords: light spectra; light fluence rate; photoperiod; growth regulators; in vitro culture

Citation: Cavallaro, V.; Pellegrino, A.; Muleo, R.; Forgione, I. Light and Plant Growth Regulators on In Vitro Proliferation. *Plants* **2022**, *11*, 844. https://doi.org/10.3390/plants11070844

Academic Editors: Paul Devlin and Eva Darko

Received: 14 February 2022
Accepted: 17 March 2022
Published: 22 March 2022

Publisher's Note: MDPI stays neutral with regard to jurisdictional claims in published maps and institutional affiliations.

1. Introduction

Plants, like any other living organisms on planet Earth, are strongly influenced by environmental cues. Unlike animals, plants are sessile and at the mercy of their surrounding environment. Consequently, they have evolved mechanisms that perceive and respond to environmental changes and adapt their development and growth accordingly. Light plays a pivotal role in a plant's life, not only for photosynthetic energy production but also for its regulative role of molecular, biochemical and morphological processes that underlie plant growth and development [1–3]. Fluence rate, regions of wavelength electromagnetic spectrum, duration and direction are the key attributes of light that drive photosynthesis and photomorphogenesis through mechanisms including the selective activation of various light receptors [4–9]. Plant light photoreceptors have evolved in articulated biochemistry structure that capture photons and detect many of the light physical properties. Subsequently, through interactive pathways the photoreceptors interpret information from incoming light and traduce them in biochemical and biological responses able to regulate plant growth and development. A discrete number of photosensor families have evolved in plants. The phytochrome (PHY) family receptors monitor the red (R, 600–700 nm) and far red (FR, 700–750 nm) light regions [10–12]. PHY can be present in two states and the active state (P_{fr}) is formed due to absorption of red light by the inactive state (P_r) [13]. The wavelength region of light from UV-A to blue (B, 320–500 nm) is perceived by three small families of photoreceptors [14] that mediate plant responses. All three photoreceptor families contain flavin adenine dinucleotide (FAD) as a chromophore: three cryptochromes (CRY) with CRY1 and CRY2 acting in the nucleus, whereas CRY3 is probably acting in the mitochondrion and chloroplast [15,16], two phototropins (PHOT) [9,11,17] and the members of the *Zeitlupe* family (ztl, fkf1 and lkp2) [18]. In addition, PHY has also been found to mediate various blue responses [19]. The UV Resistance Locus 8, monitoring ultraviolet B wavelengths (UV-B, 280–315 nm), regulates both developmental and UV-protective outcomes [20–22].

PHYs act in detecting mutual plant shading through the change in the R:FR ratio and appropriately redirect growth and development through the modulation of apical dominance and of axillary meristems formation according to survival [23–25]. CRY1 is thought to be the CRY responsible for the B high-irradiance response, inhibiting stem plant growth and reducing internode elongation, whereas CRY2 is likely responsible for the inhibition because of the B low-irradiance response [19]; collectively, in plants they perform important traits such as flowering and plant stem elongation [26]. PHOT1 and PHOT2 are involved in auxin polar transport, modulation of auxin sensing and phototropism [27–29].

Micropropagation is considered an effective large-scale in vitro plant multiplication of selected insect/disease/virus-free plants in a short time, all year round, and is a reliable method for in vitro preservation of threatened plant species. The micropropagation technology differs strongly from all other agamic propagation methods since the plants, cultured frequently as microcuttings, can remain under constant environmental conditions for a long time. The habitat of an in vitro culture is strongly restricted, and plants switch from an ontogenetic processing that starts from similar juvenility traits to a much deeper juvenility state [30]. Photoperiod, light intensity, light quality, temperature and relative humidity are factors that in the in vitro habitat are subjected to scarce fluctuations that alter the periodic and oscillator systems upon which plants depend; therefore, plants remain under largely invariable conditions. Although, currently, we cannot establish whether the mutations that are detected in the genomes of in vitro growing plants appear during in vitro culture, however, we could hypothesize that under pressure of these unnatural conditions, plants develop adaptive mechanisms to survive in limited spaces. These adaptive mechanisms involve epigenetic modifications that are programmed to confer plasticity to in vitro plants [31].

Tissue culture is also used in genetic improvement procedures with the aim of selecting subjects under the conditions of selected stress pressure, although in most cases the conditions do not reproduce the real ones. Evolution, in fact, diversifies and adapts species to better achieve suitability to the environmental conditions prevailing at a given time and habitat; a chain of genetic adjustments is selected at the same time as the periodic physiological events that generally occur during plant's life [32].

In vitro propagation proved to be particularly valuable for vegetatively propagated plants such as *Solanum tuberosum* L., *Allium sativum* L., *Musa acuminata*, *Saccharum officinarum* L., different ornamentals, orchids and fruit trees and energy crops [33,34]. Currently, micropropagation has also attracted growing attention from researchers as an efficient alternative way for rapid and controlled production of bioactive phyto-chemicals or food ingredients from medicinal and aromatic plants.

However, the effectiveness of a micropropagation protocol depends on the proliferation rate and stability, i.e., the number of explants, such as microshoots and single nodes, obtained from a single donor plant [35]. In addition, adventitious roots induction and the subsequent extra vitro acclimation of plantlets determine the success of a commercial propagation protocol [2]. The multiplication of shoots is based on the concomitance of two iterative processes: the induction and formation of phytomer, which includes lateral meristems formation (axillary buds) from the apical meristem (apex) and the subsequent outgrowth of the axillary buds into new shoots [36]. In this contest, artificial light plays a crucial role in successful in vitro plant production, together with other factors such as medium composition, gas exchange in the culture vessel, temperature and specific physiological outcomes of plant explant, i.e., the species-specific physiologic adaptation to the in vitro conditions previously described. Illumination should provide light in the appropriate spectral regions for promoting photosynthetic metabolism and photomorphogenic responses [37,38]. Controlling light quality (wavelength ranges), irradiances (photon flux) and light regime (photoperiod) enables the production of plants with desired characteristics [35,39].

From the outset, the lighting systems used in in vitro plant growth had been fluorescent tubes (Fls), high pressure sodium (HPS), metal halide (MH) and incandescent lamps (IL) with varying wavelengths from 400 to 700 nm. Among these, Fls have been the most popular in tissue culture rooms and consume approximately 65% of total electricity in tissue culture labs [40]. The Fls have high amounts of photons in the infrared and red ranges, gradually dropping toward blue. Due to the presence of phosphor coating, white FLs also have a continuous visible spectrum with peaks near 400–450 nm (violet-blue), 540–560 nm (green-yellow) and 620–630 nm (orange-red). The main inconveniences tied to the use of these lamps are: (i) a significant portion of the spectral output emitted (from 350 to 750 nm) [41] is not utilized by the plant cultures since they are abundant in green (G) and yellow (Y) light, which are less efficient for plants and usually lack FR light [35,41], (ii) light irradiation may cause photo-inhibition of growth and differentiation [42] and photooxidative damage in plants [43] and (iii) the dissipation of a large amount of energy as heat [44].

In recent years, light-emitting diodes (LEDs) have attracted increasing attention as potential light sources for various applications of plant tissue culture [40]. The advantages of LED lights over conventional lighting systems mainly consist in the higher photosynthetic photon efficacy (PPE) as compared to the previously used HPS or Fls. The maximum PAR efficiency of LED lamps ranges between 80 and 100%, while Fls provide only 20–30% [45,46]. The precision in converting electrical energy to photons of specific wavelengths at the desired photosynthetic photon flux density (PPFD) with negligible heat loss makes LEDs more energy-efficient than all other available artificial lighting sources. Based on the manufacturers' specifications, the LED lamps require about 32% less energy than the Fls per μmol m^2 s^{-1} of photons delivered to the plants [34] and 10–25% total energy saving can be realized when considering climate modification by the transition from HPS to LED [47]. Moreover, LED lamps possess a longer operating lifetime (>50,000 h), negligible heat emissions and, consequently, an indirect reduction in refrigeration costs, a more robust and easy-to-handle plastic body, no emissions of greenhouse gases (CO_2) for their production and they produce no mercury pollution [46,48].

The narrow waveband emission and dynamic control of light intensity in LED-based illumination systems allow the choice of spectral quality to match the absorption range of a specific photoreceptor and thus to regulate the photosynthetically and photomorphogenic responses required for the cultivation of each species in vitro [41]. For these reasons, the use of LED lamps in the in vitro culture systems is a useful tool for photobiological studies since they allow the control of irradiance and the emission of specific spectral patterns [41]. With the rapid advancement of the technology, the reduction of LED prices and the diverse studies that show more vigorous in vitro plants cultivated under these lighting conditions, the replacement of Fls with LED lamps has attracted considerable attention around the globe [9].

Numerous studies reported the applications of LEDs, as compared to white Fls, in promoting in vitro organogenesis, growth and morphogenesis from various plant species such as *Gossypium hirsutum*, *Anthurium andreanum*, *Brassica napus*, *Musa acuminata* and so on [49–52]. The impact of LED lighting on somatic embryogenesis has also been explored for a few plant species [53–58].

Although there are a discrete number of studies, many tissue culture laboratories hesitate to replace conventional lighting systems with LEDs out of apprehension of an unpredictable and aberrant in vitro, which may damage consolidated production protocols [59].

Moreover, light quality influences the biological effectiveness of the growth regulators added to the culture substrate, as well as the endogenous hormonal balance in the tissues [60], which must be readdressed after the substitution of the old ones with LED lamps.

Keeping this in mind, in this review, the attention will focus on the literature on the effects of light on shoot proliferation, a main process of in vitro propagation. The effects of the light spectrum on the balance of endogenous growth regulators will be also presented.

2. Effects of Spectral Quality of Light on In Vitro Proliferation

The spectral quality of light significantly influences the shoot biological response. Since plant photoreceptors responsible for plant development and photosynthesis are known to be primarily and most significantly stimulated by red (RL) and blue (BL) regions of the light spectrum, most of the studies evaluated the impact of monochromatic RL (660 nm), BL (460 nm) and combined BL (440–480 nm) with RL (630–665 nm) lights. Scarce is the information available on the effects of the far-red (FRL), green (GL) and yellow (YL) regions of the spectrum [44]. For each light spectrum, the evaluated effects concern the proliferation rate and characters related to development, morphology and plantlet quality, i.e., shoot length, fresh and dry weight and photosynthetic pigment accumulation. In fact, the light treatments yielding higher chlorophyll and carotenoid contents (relevant components of the light-harvesting antenna of photosystems) are generally linked with improved fresh and dry matter accumulation and shoot growth [50,61–66]. The main results obtained on flowering plant species are presented in Tables 1 and 2.

Table 1. Summary of the use of LED lighting on in vitro propagation of herbaceous and shrub species.

Studied Species/Explant Type	Light Intensity and Photoperiod	Light Spectra	Growth Regulators in Medium	Results on In Vitro Proliferation	Morphogenetic Response	Authors and Year
Nicotiana tabacum L. var. Wisconsin 38)/Callus	mW cm^{-2}: 0, 0.0028; 0.024; 0.13; 0.37; 0.60; 0.80 photoperiod 16 h	8 narrow band lights: 371, 419.5, 467, 504, 550, 590, 660, 750 nm, 4 commercial broad band-Fl lamps	For shoot differentiation: 2 mg L^{-1} K, 2 mg L^{-1} IAA, 80 mg L^{-1} adenine sulfate dihydrate	Near UV at low intensity (0.024 mw/cm^2) and BL at higher intensities, callus growth and shoot initiation.	Higher carotenoids, porphyrins, associated with the high irradiance response.	[67] No LEDs
Vitis vinifera L. hybrid 'Remaily Seedless'/Node shoots (axillary bud proliferation)	μW cm^{-2}: 1500 for RL 1600 for BL light	RL BL No LED	BAP at 5 μM	BL = more shoots in the medium containing the lower concentration of manganese sulphate.	BL = larger shoots and more vigorous plantlets.	[68] No LEDs
Saintpaulia ionantha Wendl cv. Sona/leaves and *Lycopersicon esculentum* Mill./Cotyledons cv. UC 105		Continuous light and daily light pulses		RL ad WL = highest bud regeneration in *L. esculentum*, BL in *S. ionantha*		[69] No LEDs
Vitis vinifera L. hybrid 'Remaily Seedless'/Leaf axillary buds	10-h and 16-h photoperiods	WL of various spectral irradiances, BL and RL light.	Apex removal from the explant was evaluated.	BL = best for shoot production. Under W, shoot production was greater with ratios of BL:RL of 0.6 to 0.9.		[70] No LEDs

Table 1. *Cont.*

Studied Species/Explant Type	Light Intensity and Photoperiod	Light Spectra	Growth Regulators in Medium	Results on In Vitro Proliferation	Morphogenetic Response	Authors and Year
Solanum tuberosum L., cv. Miranda/Three- to four-node shoots (15 mm)	160 µmol m^{-2} s^{-1} 18 h (LD) or 10 h (SD). photoperiod	RL, BL	With or without 1 mg L^{-1} IAA or 1 mg L^{-1} K.	BL and K = better tuber production. RL and IAA application = high root/shoot ratio. Darkening strongly promoted tuber formation	Under BL, K increased total fresh weight and root (>stolons)/shoot ratio).	[71] No LEDs
Lavandin (*Lavandula officinalis* Chaix ×*Lavandula latifolia* Villars cv. Grosso)/Node explants	µmol m^{-2} s^{-1}: Fl high fluence (HF) = 66 Fl low fluence (LF) = 7 RL (HF) = 7 RL (LF) = 1 FrL (HF) = 8 FrL (LF) = 2 BL (HF) = 13 BL (LF) = 1.5 UVL (HF) = 62 UVL (LF) = 5	D control WL RL Fr L FrD (25 min Frh + 30 d D) FrRD (25 min Frh + 10 min R high + 30 d D) BL UV (UV A and B)	BA (l µM), putrescine (Put, 1 and 10 µM)	Low fluence RL = higher shoot number in presence or absence of BA. At low fluence rates also WL and BL enhanced shoot number on BA-free medium. 10 µM putrescine + Ba improved proliferation.	Rl and D positively affected shoot length.	[72] No LEDs
Begonia gracilis Kunth/Direct somatic embryogenesis from petiole explants.	45 µMol m^{-2} s^{-1}	RL and D	0.5 mg L^{-1} kinetin	Somatic embryo production was higher under RL that in the dark.		[73] two cycles
Azorina vidalii (Wats.) Feer (Dwarf shrub)	50 µmol m^{-2} s^{-1}: 16 h photoperiod	High and low ratios of BL + RL (2.3; 0.9) or RL + FRL (1.1; 0.6). Control: Fl	in vitro shoots no growth regulators	High ratio of RL/FRL light or BL/RL = the highest number of axillary shoots as compared to control.	Low ratio RL/FRL = maximum plant length and leaf area	[74] three months
Rhododendron spp./Axillary buds *Disanthus cercidifolius* Maxim./Shoot. *Crataegus oxyacantha* L./Axillary bud	µmol m^{-2} s^{-1}: 11, 25, 55, 106 and 161 for *Disanthus* and *Crataegus;* 16, 26, 60 and 120 for Rhododendron	RL, GL and BL	*Rhododendron* 2.5 µM 2iP. *Disanthus cercidifolius* 3 µM BAP *Crataegus oxyacantha* 2.5 µM BAP and 0.5 µM IBA.	RL promoted axillary branching. All cultures grew well at low levels of irradiance	RL promoted shoot extension.	[75] No LEDS
Solanum lycopersicum cv. UC 105 an aurea (au) mutant and its isogenic wild type/Organogenesis from hypocotyls	µmol m^{-2} s^{-1}: Fl = 50 0, 2.5 and 5 the other light treatments. 16 h photoperiod	D and Fl for aseptic seed germination RL, FRL, BL for regeneration.	Hormone free medium	All genotypes germinated under Fl. The wild type even under dark. Under RL, FRL and BL, hypocotyls showed a position-dependent regeneration.		[76] two cycles No LEDs
Petunia x *atkinsiana* 'Surfinia White' cv.'Revolution'/ Leaf explants	19–21 µmol m^{-2} s^{-1}	WL, RL, BL, GL	0.1 mg L^{-1} NAA, 1 mg L^{-1} BAP	Organogenesis was carried out in darkness. WL, GL and RL = the highest number of adventitious shoots.	Blue = the longest shoots and the biggest leaf area.	[77]

Table 1. *Cont.*

Studied Species/Explant Type	Light Intensity and Photoperiod	Light Spectra	Growth Regulators in Medium	Results on In Vitro Proliferation	Morphogenetic Response	Authors and Year
Lilium oriental hybrid 'Pesaro'/In vitro-raised bulbs	70 mmol m^{-2} s^{-1} 12 h photoperiod	D, Fl, RL, BL, RL + BL (1:1).	1.0 mg L^{-1} BA + 0.3 mg L^{-1} NAA	Fl, BL, and BL + Rl enhanced, plant regeneration as compared to D.	Bulblets under R + B were bigger in size, in fresh and dry weight.	[78]
Begonia erythrophylla J. Neuman/Petiole explants.	µmol m^{-2} s^{-1}: WL, RL, and BL, and RL + BL = 35 Fr = 5 Continuous light	D, WL, R, B, RL + BL (1:1), FR	0.54 mM NAA, 4.44 mM BA	RL or WL, as pre-treatments, promoted competence. RL or WL during culture, enhanced shoot number.	White light produced best developed and expanded shoots.	[79] No LEDS
Cymbidium Twilight Moon cv.'Day Light'/ PLB segments.	45 µmol m^{-2} s^{-1} 16-h photoperiod	RL, RL + BL (3:1), RL + BL (50:50), RL + BL (1:3), BL. Control = Fl (PGF)	For callus induction from PLBs: 0.1 mg L^{-1} NAA and 0.01 mg L^{-1} TDZ For callus proliferation: 0.1 mg L^{-1} NAA and 0.01 mg L^{-1} TDZ. For PLBs production from callus: no growth regulators.	RL determined more callus induction; RL + BL (3:1) and PGF more callus proliferation RL + BL (1:3) more PLBs formation		[80]
Lactuca sativa L./Cotyledon explants	35 µmol m^{-2} s^{-1}	D, WL, RL, BL, BL + RL	0.44mM BA, 0.54mM NAA	Light improved organogenesis as compared to D. RL and WL light promoted shoot production.		[81] No LEDS
Fragaria × *ananassa* Duch. cv. Toyonoka/Leaf discs	2000 lux	GL, RL, BL and YL Fl as control	1.5 mg L^{-1} TDZ and 0.4 mg L^{-1} IBA.	Red and Green films determined the highest percentage of shoot regeneration and the max number of shoots per explant	RL and GL = a lower chlorophyll a/b ratio and higher antioxidant enzymes activity.	[82] No LEDs
Euphorbia milii Des Moul./Inflorescences *Spathiphyllum cannifolium (Dryand. ex Sims) Schott*/In vitro shoots	µmol m^{-2} s^{-1}: 50 for *Euphorbia*: 35 for *Spatifillum* 16 h photoperiod	LEDS: RL, BL, RL + BL (1:1); BL + FrL (1:1); RL + FrL (1:1) Fl = Control	For *E. miliii* 1 mg L^{-1} BA, and 0.3 mg L^{-1} IBA. For *S. cannifolium* 3 mg L^{-1} BA, and 1 mg L^{-1} IBA.	*S. cannifolium* = best shoot proliferation under RL, RL + FRL.	For *E.milii.* BL = higher fresh and dry weight, and leaf number. For *Spatifillum.* BL= the highest chlorophyll and carotenoid contents. In both species, RL= higher plantlet length and higher fresh and dry weights.	[83]
Two species of Petunia: *Petunia × atkinsiana (Sweet) D. Don* and *P. axillaris* (Lam.)/Leaf tissue	50 µmol m^{-2} s^{-1} 16-h photoperiod	Fl, D	5.7 µM IAA and 2.25 µM Zeatin.	*Petunia × atkinsiana* did not regenerate in darkness. Both species regenerate under light.		[84]

Table 1. *Cont.*

Studied Species/Explant Type	Light Intensity and Photoperiod	Light Spectra	Growth Regulators in Medium	Results on In Vitro Proliferation	Morphogenetic Response	Authors and Year
Vitis vinifera L. cvs: Hybrid Franc, Ryuukyuu-ganebu (a wild grape native to Japan) and Kadainou R-1/Nodal segments	50 µmol m^{-2} s^{-1} 16-h photoperiod	RL and BL PGF light was used as control	PGR-free medium	No differences or slight differences on proliferation due to light treatments	RL = longest shoots. BL = higher chlorophyll content, leaf and stomata number per explant.	[85]
Phalaenopsis hybrid cv. Cassandra Rose/PLBs from in vitro germinated seeds and flower-stalk nodes.		RL, RL + BL (9:1, 8:2), RL + WL (1:1) Fl		RL + BL (8:2) = the highest PLBs development. RL + BL (9:1) = the highest shoots number. Shoot tips had higher PLBs induction under RL and BL.	RL and BL =the highest PLBs fresh weight. LED lights = more fresh weight, Height and leaf length.	[86]
Oncidium Sweets Sugar/Shoot apex		Fl (control), RL, BL		RL promoted PLB induction from shoot apex with the highest proliferation rate; BL the highest differentiation.	RL determined the highest content of carbohydrates. BL the highest protein content and enzyme activity.	[87]
Cymbidium finlaysonianum Lindl./PLBs	16 h photoperiod	RL, Fl.		RL increased PLBs proliferation and number		[88] No LEDs
OncidiumGower Ram-sey/Embryogenic calli	50 µmol m^{-2} s^{-1}	D, Fl, BL, RL or RL + BL + Fr (RBFr)	0.1 mg L^{-1} NAA and 0.4 mg L^{-1} BA	PLB formation and plantlet conversion was higher under (RBFr) LEDs and Fl.	RBFr enhanced leaf number and expansion, root, chlor. contents, fresh and dry weight.	[89]
Oncidium Gower Ramsey/Shoot tips	11 µmol m^{-2} s^{-1}	Fl(control)RL, BL, YL and GL.	For PLBs induction, 1.0 mg L^{-1} BA, For PLB proliferation: 1.0 mg L^{-1} BA, 0.5 mg L^{-1} NAA.	RL enhanced PLB induction and multiplication, but low differentiation BL promoted PLbs differentiation into shoots	RL = the highest PLBs fresh weight and starch content. BL = higher chlorophyll, carotenoids and soluble protein content.	[90]
Cymbidium finlaysonianum Lindl., *Cymbidium Waltz* cv.'Idol', and *Phalaenopsis* cv:'1327'/protocorm-like bodies (PLBs)		RL, BL and YL fluorescent films		RL and YL increased the number of PLBs of *C. Waltz.*, RL, BL and YL increased the formation of shoots. RL and BL increased PLBs number in *Phalaenopsis*.	RL, BL and YL increased the fresh weight of PLBs in *C.finlaysonianum.*	[91] No LEDS

Table 1. *Cont.*

Studied Species/Explant Type	Light Intensity and Photoperiod	Light Spectra	Growth Regulators in Medium	Results on In Vitro Proliferation	Morphogenetic Response	Authors and Year
Dendrobium officinale Kimura & Migo/PLBs	70 μmol m^{-2} s^{-1} 16 h photoperiod	D, Fl, RL, BL; RL + BL (1:1); RL + BL (2:1); and RL + BL (1:2).	0.5 g L^{-1} NAA, 0.2 g L^{-1}, 6-BA	BL, RL + BL (1:1) and RL + BL (1:2) = higher percentage of PLBs producing shoots and the number of shoots produced per PLB	BL and different RL + BL ratios enhanced chlorophyll and carotenoids. BL, Fl, and RL + BL (1:2) produced higher dry matter.	[92] three cycles
Cymbidium insigne Rolfe/PLBs		WL, RL, BL and GL	Chondroitin sulfate The medium was added with Chitosan H or hyaluronic acid (HA9)	GL and 0.1 (mg L^{-1}) and Chitosan H determined the highest PLBs and shoot formation.	Fresh weight of PLBs was higher at HA9 (1 mg L^{-1}) treatment with GL.	[93]
Ficus benjamina L. cv Exotica		BL, RL and FR. Fl as control	0.5 mg L^{-1} IAA and 2 mg L^{-1} BA.	BL increased shoot number, and callus growth.	RL determined an increase in shoot length.	[94]
Cymbidium Waltz cv 'Idol'/5 mm protocorm-like bodies (PLBs)	50 μmol m^{-2} s^{-1} 16 h photoperiod	Fl, RL, BL, GL, Fl + GL, RL + GL, BL + GL. The last three treatment were subjected to 1d green exposure every 7d.	No growth regulators	RL + GL and BL promoted the highest PLB formation. Fl + GL and increased shoot formation from PLBs.	Fl gave the highest fresh weight. B + G the highest SOD activity.	[95]
Brassica napus L. cv Wes-tar/Cotyledons from germinated seeds.	60 μmol m^{-2} s^{-1} 12 h photoperiod	Fl, BL, BL + RL (B:R = 3:1, 1:1, 1:3) RL.	For induction: 2,4-D in the dark; for shoots differentiation: 0.8 mg L^{-1} BA, 0.5 mg L^{-1} NAA; for shoots proliferation 1.0 mg L^{-1} BA.	The proliferation rate was greater under BL and BL:RL = 3:1 than under Fl	BL:RL (3:1) = higher fresh dry mass, chlorophyll a, soluble sugar, stem diameter, leaf stomata surface, than under Fl. Starch was higher in plantlets cultured under R light as compared to Fl.	[51]
Linum usitatissimum L., cv. 'Szafir/Hypocotyls	50 μmol m^{-2} s^{-1}	Light (Fl) or D conditions	0.05 mg L^{-1} 2,4-D and 1 mg L^{-1} BA	Shoot multiplication was about twice higher in light-grown cultures than those in darkness.	Fresh and dry mass and cyanogenic potential of light-grown cultures was about twice higher than those in the dark	[96] two cycles
Solanum tuberosum L. cvs Agrie Dzeltenie, Maret, Bintje, Désirée and Anti/Shoot tips from in vitro plantets	40 μmol m^{-2} s^{-1}	Fl, warm WL light BL, RL, RL + BL (9:1 RB) and RL + BL + FR (70:10:20 RBF)	0.5 mg L^{-1} zeatin riboside, 0.2 mg L^{-1}, GA_3 and 0.5 mg L^{-1} IAA.	RL + BL (9:1) doubled the regeneration percentage of all cultivars after cryoconservation		[97]

Table 1. *Cont.*

Studied Species/Explant Type	Light Intensity and Photoperiod	Light Spectra	Growth Regulators in Medium	Results on In Vitro Proliferation	Morphogenetic Response	Authors and Year
Abeliophyllum distichum Nakai,/Apical and axillary buds	40 µmol m^{-2} s^{-1}	BL, RL + BL (1:1 RB), RL, Fl	BA 1.0 mg L^{-1}, IBA 0.5 mg L^{-1}	BL and RL + BL promoted shoot proliferation.	RL increased shoot length.	[98]
Dendrobium kingianum Bidwill ex Lindl./PLBs	50 µmol m^{-2} s^{-1} 16 h photoperiod	RL, BL, RL + BL (1:1), GL and WL, Fl = control	MS medium supplemented with 412.5 mg/L ammonium nitrate, 950 mg/L potassium nitrate	BL and RL determined the highest PLBs number. RL and WL increased the percentage of shoot formation.	BL increased chlorophyll percentage, RL determined the highest fresh weight.	[99]
Cymbidium Waltz cv 'Idol'	16 h photoperiod	GL, RL, BL	N- acetylglu-cosamine (NAG) 0, 0.01, 0.1, 1, and 10 mg L^{-1}	GL and RL + NAG determined the highest PLB formation rate RL or GL + NAG determined high shoot formation (80%)	Fresh weight of PLBs was highest at 0.01 mg L^{-1} NAG under green LED	[100]
Saccharum officinarum L., variety RB92579/in vitro grown plantlets	µmol m^{-2} s^{-1}: (1) 72 (2) 60 (3) 57 (4) 53 (5) 77 16 h photoperiod	(1) BL + RL (70:30) (2) BL + RL (50:50) (3) BL + RL (40:60) (4) BL + RL (30:70) (5) WL	1.3 µM BAP.	BL + RL (70:30) gave the highest multiplication followed by 50:50. WL the lowest one.	BL + RL (70:30) and (50:50) = the highest total fresh weight. WL = the highest total chlorophyll content	[101]
Scrophularia takesimensis Nakai/Leaf, petiole, and stem explants	45 µmol m^{-2} s^{-1} 16 h photoperiod	Fl, RL, BL	2.0 mg L^{-1} BA and 1.0 mg L^{-1} IAA	Fl = the highest number of shoots per leaf, petiole and stem explants	RL gave better shoot growth followed by Fl and BL.	[102]
Curculigo orchioides Gaertn./Leaf explants	60 µmol m^{-2} s^{-1}	BL, RL, RL + BL (1:1). Fl as control.	4 mg L^{-1} BA	BL determined the highest percentage of shoot organogenesis and shoot buds per explant.		[103]
Fragaria x ananassa Duch. cv. *'Camarosa'*/ Encapsulated shoot tips	50 µmol m^{-2} s^{-1} 16 h photoperiod	Fl (control) RL + BL (9:1 R9B1); RL + BL (7:3 R7B3); RL + BL (1:1 R5B5); RL + BL (3:7 R3B7);	Hormone free medium for plantlets development, and 4.9 µM IBA or 6.7 µM BA plus 2.3 µM K for shoots proliferation	RL + BL (1:9) were most effective *for* in vitro sprouting of encapsulated strawberry shoot tips.	R7B3 promoted shoot length, chlorophyll content, fresh and dry biomass accumulation.	[104]
Panax vietnamensis Ha et Grushv/Callus	20–25 µmol m^{-2} s^{-1} 16 h photoperiod	D, Fl, BL, GL, YL, RL, WL, and RL + BL: 90:10, 80:20, 70:30, 60:40, 50:50, 40:60, 30:70, 20:80, 10:90.	For embryogenic callus differentiation: 1 mg L^{-1} BA, 0.5 mg L^{-1} NAA. For plantlets differentiation: 0.5 mg L^{-1} BA, 0.5 mg L^{-1} NAA	YL most effective for callus production. RL + BL (6:4) was the most effective for differentiating the highest number of plants per explant from embryogenic callus.	YL gave the highest values of callus fresh and dry weight, followed by RL + BL (60:40). This last light gave the highest values of plantlet height, fresh and dry weight.	[105]

Table 1. *Cont.*

Studied Species/Explant Type	Light Intensity and Photoperiod	Light Spectra	Growth Regulators in Medium	Results on In Vitro Proliferation	Morphogenetic Response	Authors and Year
Vanilla planifolia Andrews./Axillary buds axillary bud cuttings	25 µmol m^{-2} s^{-1} 16 h photoperiod	BL, RL, RL + BL (1:1), WL, Fl	9.55 µM BA	Fl, WL and RL + BL gave best results on shoot proliferation	Fl, WL and BL + RL determined higher shoot growth, plant height, leaves number, fresh weight, dry weight and chlorophyll content	[106]
Gerbera jamesonii Bolus ex Hooker f. cv Rosalin/In vitro propagated shoots	140 ± 10 µmol m^{-2} s^{-1}	RL, BL, and their various mixtures. Fl was used as control	1 mg L^{-1} BAP and 0.1 mg L^{-1} NAA	Fl lamps, BL, WL and RL + BL (70:30) = the highest number of shoots/explant and 70% R + 30%.	The same treatments also yielded the highest values in terms of shoot length, plant fresh and dry weight.	[107]
Cymbidium dayanum Rchb.f. and *Cymbidium finlaysonianum Lindl.*/PLBs	50 µmol m^{-2} s^{-1} 16 h photoperiod	RL, BL, GL Fl.	(0, 0.1, 1 and 10 mg L^{-1}), chondroitin sulfate	GL and BL + different concentrations of chondroitin sulfate promoted PLBs and shoots formation in the two species		[108]
Bacopa monnieri L. (Water hyssop)/Full, upper and lower, leaf cuttings.		WL, RL + BL (4:1, 3:1, 2:1,1:1)	0.25, 0.50 and 1.0 mg L^{-1} BA	WL was most effective in enhancing shoot regeneration.	Shoot length was increased by RL:BL (1:1) + 0.25 BA	[109]
Vaccinium ashei Reade cv Titan	50 µmol m^{-2} s^{-1} 16 h photoperiod	Fl, RL, RL + BL (80:20) (R8B2), RL + BL (50:50 (R5B5), BL.	1 mg L^{-1} zeatin riboside. Ventilated and non-ventilated vessels	No differences in shoot number between the different light treatments.	R8B2 and ventilated vessels were the most suitable for plant growth.	[110]
Anthurium andreanum Lind./Nodal segments	25 µmol m^{-2} s^{-1} 16 h light photoperiod	Fl, WL, RL, BL, BL + RL.	No growth regulators during the light treatments	BL + RL gave the highest number of adventitious shoots.	WL LEDs and BL LEDs, showed the greatest plantlet length and number of leaves. BL gave the greatest growth and chlorophyll content.	[111]
Saccharum officinarum L. variety RB867515)	(1) 72; (2) 60; (3) 53; (4) 77; (5) 46. 16 h photoperiod	BL:RL= (1) 70:30, (2) 50:50, (3) 30:70, (4) WL, (5) Fl	1.3 µM BAP.	BL:RL = 50:50 promoted proliferation	BL:RL = 50:50 promoted the highest stem length, fresh mass production, leaf number.	[112]
Staphylea pinnata L./in vitro regenerated shoots	35 µmol m^{-2} s^{-1} 16 h photoperiod	Fl, RL + BL (50:50:1), RL + BL + FR (49:49:2) RL + BL + WL (40:40:20)	5 µM BA, 0.5 µM NAA	Treatment with RB and RBFR resulted in increased multiplication rate as compared to Fl.	RB and RBFR increased leaf chlorophyll content and carotenoids. RBW light increased the number of newly developed leaves.	[113]

Table 1. *Cont.*

Studied Species/Explant Type	Light Intensity and Photoperiod	Light Spectra	Growth Regulators in Medium	Results on In Vitro Proliferation	Morphogenetic Response	Authors and Year
Stevia rebaudiana Bertoni/Nodal segments measuring 0.5–1 cm in length	40–50 µmol m^{-2} s^{-1}. 16 h light photoperiod	Fl (Control), BL, RL, RL + BL (1:1), WL	1 mg L^{-1} BA.	RL = higher proliferation rate	Under BL + RL, maximum shoot elongation and leaf number	[114]
Vanilla planifolia Andrews/Nodal segments measuring 0.5–1 cm in length	40 µmol m^{-2} s^{-1} 16 h light photoperiod	Fl (control) BL, RL, RL + BL (1:1), WL	2.1 mg L^{-1} BA	No differences in shoot multiplication.	BL enhanced leaf number and area. RL + BL enhanced shoot lengtht and chlorophyll content Fl determined higher fresh and dry weight and carotenoids.	[115]
Dendrobium sonia,/Mature PLBs	µmol m^{-2} s^{-1}: W 17.7 B 22.5 Y 24.6 R 15.6 16 h photoperiod	WL (control), BL, YL, and RL.	11.1 µM BAP and 11.42 µM IAA	YL induced early PLB formation, shoot differentiation and initiation, higher number of shoots per explant.	Under YL, higher leaf area and fresh weight, longer shoots under the other lights.	[116]
Nicotiana tabacum L. and *Artemisia annua*/In vitro-grown plantlets	35 µMoles cm^{-1} s^{-1}	WL, RL + BL (1:1), RL + BL (3:1) RL + BL (1:3)	no growth regulators	In *Nicotiana* more shoots under 1:1 RL + BL In *Artemisia* under RL + BL (3:1)	In both species, RL + BL (3:1) determined taller shoots, and higher fresh weight.	[34]
Saccharum officinarum var. RB98710 (Sugar-cane)/shoot segments	50 µmol m^{-2} s^{-1} for FL, 80 µmol m^{-2} s^{-1} for LED 16-h photoperiod	Fl, WL, RL + BL (82:18).	For callus induction in the dark two substrates: C1 = 9 µM 2,4-D and 1.1 µM BA; C2 = 13.6 µM 2,4-D + 2.2 µM BAP. For shoot regeneration: hormone free medium.	LED were ineffective on somatic embryo regeneration but successful on shoot multiplication from somatic embryo.	Root length, number of leaves, shoot fresh and dry biomass did not differ between treatments.	[117] six subcultures
Gerbera jamesonii Bolus ex. Hook f. cv. Dura/in vitro propagated shoots	40 µmol m^{-2} s^{-1} 16-h photoperiod	BL, RL + BL1 (50:50), RL + BL2 (70:30), RL + BL + WL (40:40:20), RL + BL + FR (49: 49:2), RL, Fl (Control)	5 µM BA (1,1 mg L^{-1}) and 0.5 µM NAA (0.1 mg L^{-1})	RB1 and RB2 determined a higher shoot multiplication rate as compared to the control	RL = the greatest shoot elongation; BL = the highest leaf dry weight; RB2 = higher concentrations of total chlorophyll and carotenoids; RB1 = high leaf number.	[118]
Lippia gracilis Schauer./Apical and nodal segments	42 µmol m^{-2} s^{-1} 16 h photoperiod	WL, RL, BL, RL + BL (2.5:1 and 1:2.5)	no growth regulators	No influence of the light intensity nor of quality on shoot number both on nodal and apical segments.	RL and WL = best results on leaf and dry weights. B = higher photosynthetic pigment production in plantlets from apical explants, WL of those from nodal explants.	[119]

Table 1. *Cont.*

Studied Species/Explant Type	Light Intensity and Photoperiod	Light Spectra	Growth Regulators in Medium	Results on In Vitro Proliferation	Morphogenetic Response	Authors and Year
Myrtus communis L./Axillary shoots	35 μmol m^{-2} s^{-1} 16 h photoperiod	BL; RL:BL (70:30); RL; Fl = control.	0.5 μM L^{-1} NAA and different concentrations of BA: 1, 2.5 and 5 μM.	RL and 5 μM BA resulted in the highest multiplication rate.	At 5 μM BA, RL determined the higher dry weight; BL = a greater leaves number, BL and RL:BL increased the FW compared to Fl.	[120]
Chrysanthemum × morifolium Ramat., Ficus benjamina L., *Gerbera jamesonii Bolus f., Heuchera hybrida, and Lamprocapnos spectabilis (L.) Fukuhara.*	62–65 μM m^{-2} s^{-1} 16 h photoperiod	Fl (control), *NS1* lamps (BL + GL + RL + FRL-21:38:35: 6) *G2 lamps* (BL + GL + RL+ FRL-8:2:65:25), *AP673L* (BL + GL + RL + FRL- 12:19:61:8), *AP67* (BL + GL + RL + FRL-14:16: 53: 17)	No PGRs for *C. grandiflorum;* 4.0 mg L^{-1} BA and 30 mg L^{-1} adenine sulfate for *F. benjamina;* 3.0 mg L^{-1} K. for *G. jamesonii;* 0.1 mg L^{-1} BA and 0.1 mg L^{-1} IAA for *H. hybrida;* 0.25 mg L^{-1} BA and 0.25 mg L^{-1} IAA for *L. spectabilis*	Except for *F. benjamina*, RL and G2 lamp gave highest or similar propagation ratios as compared to Fl. NS1 lamps was also efficient for *G. jamesonii, H. hybrida* and *L. spectabilis*	The highest chlorophyll content was recorded under Fl and AP673L in all species, in NS1 in two species.	[35]
Oryza sativa L. cultivar Nipponbare.	50 μmol m^{-2} s^{-1}. 12 h photoperiod	Fl, BL BL:RL = 3:1 BL:RL = 1:1; B:R = 1:3; RL;	For callus induction:2.0 mg L^{-1} 2,4-D. For callus differentiation: 1.0 mg L^{-1} 2,4-D. For shoot differentiation 0.5 mg L^{-1} K, 2 mg L^{-1} BA, 0.25 mg L^{-1} NAA	BL = decreased time for callus proliferation, differentiation and regeneration, and highest frequency of plantlet differentiation, and regeneration.	BL:RL = 1:1 highest seedling growth, chlorophyll, and carotenoid contents and photosynthetic rates.	[121]

Abbreviations: white (WL), blue (BL), red (RL), far-red (FRL), dark (D), fluorescent light (Fl), NAA (1-Naphthaleneacetic acid), BA (6-Benzylaminopurine), IAA (Indole 3- Acetic Acid), 2,4-D (2,4-dichlorophenoxyacetic acid), PLB-Protocorm-Like Bodies.

Table 2. Summary of the use of LED lighting in in vitro propagation of woody species.

Studied Species/Explant Type	Light Intensity and Photoperiod	Light Spectra	Growth Regulators in Medium	Results on In Vitro Proliferation	Morphogenetic Response	Authors and Year
Pseudotsuga menziesii Mirb. Douglas fir embryo	0.01–0.71 W/cm^2 16 h photoperiod	8 different narrow bandwidth Fl having maxima each at one of the following wavelengths 371, 420, 467, 504, 550, 590, 660, and 740 nm.	Embryo from seeds; For callus induction: 800 pg L^{-1} IAA, 1 mg L^{-1} IBA, 1 mg L^{-1} BA, 1 mg L^{-1} AS-isopentyladenine After four weeks, 0.5 mM BA and 0.25 mM zeatin were added. No growth regulators for growing buds.	Callus and adventitious bud formation on the embryo-derived callus was maximum at (0.42 mW/cm^{-2}) under RL (660 nm).		[122]

Table 2. *Cont.*

Studied Species/Explant Type	Light Intensity and Photoperiod	Light Spectra	Growth Regulators in Medium	Results on In Vitro Proliferation	Morphogenetic Response	Authors and Year
Woody ornamental plants. Organogenesis (axillary bud proliferation)		Fl (control), high pressure sodium lamps (HPS), BL and RL	Light pipe modified growth chambers	HPS increased shoot number as compared to FL. RL increased shoot number over control.		[123]
Spirea nipponica Maxim/Shoot explants from 8 to 10 week-old stock cultures	WL: low fluence 15.0–23.0; high fluence 47.0–62.0 μmol m^{-2} s^{-1}; RL+FR: low fluence) 8.7–15.9 μmol m^{-2} s^{-1} 16 h photoperiod	WL, RL + Fr	BA 0.25, 0.4, or 0.5 mg L^{-1}.	RL + FR = improved proliferation especially by 0.5 Ba addition. RL + FR followed by high fluence WL improved proliferation at lower BA levels.	RL + Fr favourably influenced shoot length and growth	[124] No LEDS
'Mr.S 2/5' clone of *Prunus domestica Ehrh.*/Cuttings;	WL = 38.0 BL = 9.1 RL = 19.6 FR = 7.2 μMol m^{-2} s^{-1}	WL BL RL FR	Ba 0.6 mg L^{-1}	In intact cuttings, WL gave the highest shoot proliferation In decapitated seedlings, all lights gave 100% bud outgrowth.	BL and WL = a higher number of nodes; RL = longer internodes. Shoots produced in RL were longer in decapitated seedlings.	[125] all experiments were repeated twice
Cydonia oblonga Mill/Leaves from the second to the fourth node of the apical portion of in vitro shoots	BL, WL and RL = 20 ± 1; FR = 1.2 R + B 10 + 10 B + Fr= 20 + 1.2 Fr + RL = 0.5 + 1.6 (μmol m^{-2} s^{-1})	D, BL, WL, FRL, RL, RL+Bl, BL+FRL RL+FRL After All light treatments, further 20 days of WL light exposure.	4.7 μM K and 0.5 μM NAA	Somatic embryogenesis was highest under RL treatment.		[126] No LEDS
Prunus avium L. cv 'Hedelfinger'and one of its somatoclones/Leaves	~9 μMol 16 h photoperiod	WL, RL, BL, FR, D	2 mg dm^3 TDZ+ 2,4-D or IAA	WL and BL = the highest node number. BL and FR = the highest shoot outgrowth from buds.	RL = highest shoot length under. WL and BL and WL= high chlorophyll.	[127] no LEDS
Malus domestica [Suckow] Borkh. genotype MM106/Shoot tips from in vitro cultures	~40 μmol m^{-2} s^{-1} 16 h photoperiod	WL, RL, BL, GL, YL, UV-AL, D	8.86 (2 mg L^{-1}) μM BA, 0.53 (0.06 mg L^{-1}) μM Ga_3, 0.3 μM (0.1 mg L^{-1}) IBA	GL and WL gave the higher total number of shoots at the end of the fourth culturing cycle.	Leader stem height was greater under D, RL and YL.	[128] No LEDs Four cycles
Populus alba × P. berolinensis/Stems from in vitro shoots	40 μmol m^{-2} s^{-1} 16 h photoperiod	GL, RL, BL and YL. Fl (control)	0.02 mg· L^{-1} NAA, and 0.1 mg· L^{-1} TDZ.	Fl and YL exhibited better effects on shoot regeneration		[129] no LEDs
Musa spp. cv.'Grande naine' AAA)/Meristematic shoot tips	40 μmol m^{-2} s^{-1} 16 h photoperiod	WL, Fl	16.8 μM BAP, 3.8 μM IAA, 1 mg L^{-1} on a temporary immersion system (TIS)	WL under TIS enhanced shoot proliferation.		[130]

Table 2. *Cont.*

Studied Species/Explant Type	Light Intensity and Photoperiod	Light Spectra	Growth Regulators in Medium	Results on In Vitro Proliferation	Morphogenetic Response	Authors and Year
Populus x euramericana selected clones, 'I-476' and 'Dorskamp'/Petioles (5-mm long) from in vitro plants	60 μM m^{-2} s^{-1} 16 h photoperiod	Fl, BL, RL, RL +BL (1:1 and 7:3), and RL + BL + GL (7:2:1)	0.44 μM BA	Highest shoot regeneration on RL + BL (1:1) for 'I-476', on BL +RL (7:3) for *'Dorskamp'* as compared to Fl.	High RL (100% or 7:3) = higher shoot length and leaf area BL or RL +BL (7:3) = higher stem diameter	[131]
Malus domestica [Suckow] Borkh rootstock cvs. Budagovsky 9 (B.9), Geneva 30 (G.30), and Geneva 41 (G.41). I exp = single-node segments	BL = 5.7 RL = 6.6 WL = 25 μMol·m^{-2}·s^{-1}	WL, RL, BL for both experiments	1.0 mg· L^{-1} BA, 0.1 mg· L^{-1} IBA, and 0.5 mg· L^{-1} GA_3. II exp: cv. G.30 with and without gibberellic acid (GA3).	RL increased the number of shoots in B.9 and G.30 as compared to WL.	RL increased the length, and the number of elongated shoots of B.9 and G.30. GA_3 promoted shoot growth of G.30 under RL and BL.	[132] No LEDS
Phoenix dactylifera L. cv. 'Alshakr' (Date palm)/shoot buds	20–25 μmol m^{-2} s^{-1} 14 h photoperiod	FL (control), RL +BL (18:2) (CRB-LED)	1 mg L^{-1} (NAA), 0.5 mg L^{-1} (BA) and 0.5 mg L^{-1} kinetin (K)	CRB enhanced the percentage of buds producing shoots and average shoots formation compared to FL	CRB-LED enhanced total soluble carbohydrates, starch, free amino acids, and peroxidase activity	[133]
Camellia oleifera C. Abel/Axillary buds	50 m^{-2} s^{-1} 16 h photoperiod	RL, BL, RL + BL, (4:1) RL + BL (1:4), WL was used as control	3.0 mg L^{-1} BA + 0.02 mg L^{-1} IBA	RL + BL (4:1) = the highest proliferation coefficient.	RL + BL (4:1) = good chlorophyll content, the thickest leaves, high stomatal density.	[134]

Abbreviations: white (WL), blue (BL), red (RL), far-red (FRL), dark (D), fluorescent light (Fl), NAA (1-Naphthaleneacetic acid), BA (6-Benzylaminopurine), IAA (Indole 3- Acetic Acid), 2,4-D (2,4-dichlorophenoxyacetic acid).

2.1. Red Light Effects

2.1.1. Red Light Effects on Shoot Proliferation

Some authors agree on the positive role of RL [123], and high-ratio RL:FRL [35] on shoot proliferation [135]. RL significantly enhanced the adventitious bud formation and development in *Gerbera jamesonii* [136], in *Lactuca sativa.* [137], in *Spathiphyllum cannifolium* [83], in *Stevia rebaudiana* [114] and in *Mirtus communis* [120]. RL was effective for bud formation and outgrowth in *Pseudotsuga menziesii* embryo cultures [122]. In contrast, as compared to the cultivation under WL or combined RL with BL, under monochromatic RL or BL, Bello-Bello et al. [106] observed a decrease in the proliferation ratio in *Vanilla planifolia* Andrews and Estrada et coll. [111] and Lotfi et al. [59] found the same decrease in *Anthurium andreanum* and in *Pyrus communis* L., respectively. Somatic embryo germination and conversion of three southern pine species [53] and *Cydonia oblonga* [126] were positively affected by application of RL.

Positive effects of RL illumination have been ascertained in many orchids. In *Cymbidium Waltz* 'cv Idol', the highest protocorm-like bodies (PLBs) formation rate (100%) was found in the culture media containing 0.01 and 0.1 mg L^{-1} N- acetylglucosamine (NAG) under RL, although a promotive role was observed under GL, but at 1 mg L^{-1} NAG [100]. In a study of Mengxi et al. [90], the highest PLBs induction rate, propagation coefficient and fresh weight of *Oncidium Gower Ramsey* were observed under RL treatment, which agrees with observations on the Cattleya hybrid [138]. However, in this last species, monochromatic RL resulted in an impaired leaf growth and chlorophyll content. Moreover, in *Oncidium Gower Ramsey*, even if R-LEDs promoted PLB induction, it was observed that

BL emitted by LEDs promoted a differentiation of PLBs [90]. Hamada et al. [88] found that R fluorescent lamps increased the PLB proliferation of *Cymbidium finlaysonianum*, even if used only during the early stage of the culture. The R spectrum was effective for *Cymbidium* callus proliferation [80] but not for the successive propagation. The combination of RL and FRL wavelengths determined the highest number of somatic embryos in *Doritaenopsis 'Happy Valentine'* [54].

The action mechanisms promoted by RL has been investigated by different authors. In *Vitis vinifera*, the axillary shoot development could be due to the release of apical dominance caused by BL, as suggested by Chée [68] and Chée and Pool [70]. Similarly, Burritt and Leung [79] observed that the inhibitory influence of FRL on shoot proliferation is reversible, whereas exposure to BL permanently reduces explant's competence for new shoot formation. They suggested that PHY and an independent BL photoreceptor, probably CRY, regulate shoot production from *Begonia* × *erythrophylla* petiole explants. RL has been shown to exert effects on plants proliferation through the PHY, which, in the active form, would alter the endogenous hormonal balance increasing in the quantity of cytokinin (CK) in tissue, counteracting the action of auxin and thus determining an increase in the development of lateral shoots [139,140].

Moreover, research on the effects of PHY on in vitro multiplication of shoots of the *Prunus domestica* rootstock GF655-2 [141] demonstrated that the actions of WL, BL and FRL on shoot proliferation were fluence-rate dependent, while RL was effective both at 37 μmol m^{-2} s^{-1} and at 9 μmol m^{-2} s^{-1}. The increase in light intensity had, instead, a positive effect on the production of axillary shoots in a *Prunus domestica* Mr.S.2/5 shoot exposed to RL and BL. However, if the number of shoots produced was expressed as a percentage of the total number of axillary buds, the rate of bud outgrowth for each shoot under RL was significantly higher than that detected under BL [142].

The effects of RL on proliferation are also largely dependent on the growth regulators, mainly cytokinins (CKs) applied to the culture medium, and they were found to be indispensable in the outgrowth of lateral buds in *Prunus domestica* rootstock shoots [142]. The same was true for *Spiraea nipponica* where the interaction between CKs and RL resulted in an enhancement of the shoot proliferation rate [123]. Plantlets of this species exposed to RL and FRL resulted in more marked growth than under WL [123]. Interesting interactions resulted from the growth of this species under low RL:FRL photon fluence followed by high-fluence WL and the benzyl aminopurine (BA) levels [123]. More detailed information on the interactions between light and growth regulators will be provided in paragraph 5.

2.1.2. Red Light Effects on Shoot Morphology

Stem elongation, leaf growth and chlorophyll reduction are frequently observed under RL and are all supposed to be associated with shade-avoidance syndrome (SAS) [8].

Shoot and internode elongation: It is mostly reported that RL enhances the elongation of primary and axillary shoots when there is an actively growing apex [74,75], and it determines changes in the plant anatomies [143] of multiple species [36]. The RL effect on stem elongation is species dependent. RL increases shoots and internode lengths in *Pelargonium* × *hortorum* [144], *Vitis vinifera* [85,145], *Rehmannia glutinosa* [65,146], *Gerbera jamesonii* [118], *Abeliophyllum distichum* [98], *Vaccinium ashei reade* [110,147], *Ficus benjamina* [94], *Cymbidium spp.* [148], *Plectranthus amboinicus* [48] and *Fragaria* × *ananassa* plantlets [149]. The promotive effect of RL was also found on the elongation of secondary and tertiary shoots of *Malus domestica* rootstock MM106 [128], and on in vitro zygotic embryo germination and seedling growth in chestnut, whereas BL suppresses them [150]. In *Populus americana*, cultivar 'I-476', shoot length and leaf area of in vitro plants were greatest when exposed to RL, whereas on the other poplar cultivar, 'Dorskamp', BL plus RL were more effective [131]. An increase in the shoot elongation caused by internode elongation under red LEDs may result in stem fragility because of excessive elongation of the internode, as occurred in the third internode from the apex of *Dendranthema grandiflorum* Kitam cv.Cheonsu [42] and in *Rehmannia glutinosa* [146]. Following these results, it is required to adjust the ratio of RL when mixed with BL or Fl. In *Fragaria × ananassa* under R-LEDs, leaf petioles were elongated but the leaves turned yellowish green, revealing an irregular in vitro growth [149].

RL also caused thin elongated shoots and the formation of small leaves in *Solanum tuberosum* cv. Miranda, while BL produced short shoots with regular leaf development and many micro-tubers. The micro-tuber development was reversed when the IAA was added to the medium [71]. According to Kim et al. [42], synergistic interactions among CRYs and PHYs may promote or inhibit stem elongation in various ways in different species.

Differences in the response of the different species in the response to the RL:FRL ratios may be explained by the different habitats in which the species evolved. It has been proposed from studies on the elongation of shoots of *Vitis vinifera* [70], *Disanthus cercidifolius* and *Crataegus oxyacantha* axillary shoots [75] that this enhancement is PHY-mediated through the control of enzyme-affected auxin degradation, such that the extremely photolabile auxin would be conserved in cultures illuminated with RL and degraded in cultures under BL. In addition, other plant hormones may be modulated by light and by PHY directly (see paragraph 5).

Fresh and dry weight: The greatest mean fresh and dry weight of each cluster of the *Malus domestica* rootstock M9 was observed under RL and it was 83% greater than that observed under WL [135]. Gains in fresh weight were observed in *Vaccinium ashei* [110] and *cattleya* [138]. Dry weight was positively affected by RL in *Myrtus communis* L. [120], in *Euphorbia milii* and *Spathiphyllum cannifolium* [83] and in *Plectranthus amboinicus* [48]. Furthermore, increased growth of in vitro cultured plants provided by RL was also shown in *Scrophularia takesimensis* [102], *Lippia gracilis* [119] and *Vitis vinifera* [145]. Likewise, dry weight increased under RL, probably by the promotion of starch accumulation [50].

Chlorophyll content: R-LED increases chlorophyll content in *Musa acuminata* [52], *Passiflora edulis* [151] and *Rehmannia glutinosa*, although less than B-LED [65]. Most authors agree that RL, as compared to other light spectra, promoted leaf growth [74,131,152] but decreased the chlorophyll and carotenoids content of in vitro plantlets [83,90,148,153,154]. On the contrary, Cybularz-Urban et al. [138] found that in *Cattleya* plantlets grown in vitro RL caused the collapse of some of the mesophyll cells and a reduction of leaf blades, meaning that, in the absence of BL and/or WL/GL, the regular development of cells and leaf tissues is blocked. Similar results were found in cultures of birch [154] where the total content of chlorophyll under BL was twice that detected under RL. Smaller amounts of chlorophyll a and carotenoids were also detected in cultures of *Azorina vidalii* [74] under RL, FRL and RL:FRL. Other authors wrote that prolonged RL illumination may result in the 'RL syndrome', which is characterized by low photosynthetic capacity, low maximum quantum yield of chlorophyll fluorescence (Fv/Fm), carbohydrate accumulation and impaired growth. It was observed, also, that thylakoid disarrangement in the chloroplast is proportional to the increasing incidence of RL [155]. This damage may be reduced by adding BL to the light spectrum [156]. Regulation of carbohydrate metabolism by light quality has been well documented [41,157]. RL emitted by LED seemed to promote the accumulation of soluble sugar, starch and carbohydrate in upland *Gossypium hirsutum* L. and *Brassica napus* [50,51,158] and in *Oncidium* [16,87]. RL probably may inhibit the translocation of photosynthetic products, thereby increasing the accumulation of starch [50,154]. Moreover, Li et al. [50] suggested that plantlets with lower chlorophyll content utilize the chlorophyll more efficiently than plantlets with higher chlorophyll content under R-LEDs.

2.2. Blue Light Effects

2.2.1. Blue Light Effects on Shoot Proliferation

The effects of BL are often reported to be antagonistic of RL ones, although the studies reported in literature concerning the role played by BL on new meristem formation are not always consistent. The positive effects of BL on the stimulation of shoot production and growth of *Nicotiana tabacum* during in vitro culture were reported, but at a higher light intensity [67], and the authors hypothesized photoinactivation of IAA. Five weeks of exposure to BL induced the highest shoot production from *Nicotiana tabacum* callus [159].

Monochromatic BL increased shoot number in *Ficus benjamina* [94], the number of shoots and nodes in *Vitis vinifera* L. *hybrid* [68,70], the number of adventitious buds in *Hyacinthus orientalis* L. [160] and the percentage of organogenesis and the mean number of buds per explant in *Curculigo orchioides* [103]. Higher percentages of BL in the light spectrum were also effective on in vitro shoot induction and proliferation of *Anthurium andreanum* [49], *Gerbera jamesonii* 'Rosalin' [107], *Remnania glutinosa* [65] and *Saintpaulia ionantha* [69]. In various species, positive results on proliferation from adding different ratios of B to the R spectrum have been described and will be widely discussed in sub-paragraph 2.3.1. The proliferation rate was greater in *Brassica napus* plantlets when cultured under monocromatic BL and BL plus RL [51]. In lavandin, on a BA-free medium, shoot number was enhanced under BL, WL and RL at low photon fluence rates [72]. In *Oryza sativa* [121] under B-LED illumination, the time required for callus proliferation, differentiation and regeneration was the shortest and the frequency of plantlet initiation, differentiation and regeneration was the highest. Concerning orchids, in *Dendrobium officinale*, the monochromatic BL and RL:BL (1:2) emitted by LEDs determined a higher percentage of protocorm-like bodies (PLBs) producing a higher number (1.5 fold) of shoots [92], in *Cattleya intermedia* × *C. aurantiaca* the number of shoots regenerated from PLBs was enhanced by BL [161]. In *Oncidium*, RL promoted PLB induction from shoot apex and the higher content of carbohydrate but the lowest differentiation rate, while the highest differentiation rate and protein content were observed under B-LED [87]. BL increased node and total shoot number as compared to RL, FRL and dark in *Prunus avium* cv 'Hedelfinger' and one of its somaclones [127]. In contrast, on *Begonia erythrophylla* petiole explants, RL played a role in meristem initiation and BL and FRL were antagonistic to meristem formation, but BL was important for primordia development [79]. In *Gerbera jamesonii* [118], inhibition of shoot multiplication and a reduced plant height was observed under BL compared to what resulted from all other light treatments, and a decrease of lateral shoots number was observed on *Malus domestica* [135] as compared to RL. The same study demonstrated that BL inhibited the rate of proliferation, increasing the apical dominance. Inhibition of meristematic tissue proliferation by BL has also been observed for the embryogenic tissue of Norway spruce [162]. The conflicting reports found in the literature might not only be attributed to species effects, but also to the different types of explants and to the stage of the organogenic process. Hunter and Burritt [81], working on different *Lactuca sativa* L. genotypes, observed a significant decrease under monochromatic BL in shoot proliferation as compared to RL or WL. They argued that RL is required for the formation of shoot primordia, whereas BL is inhibitory to primordia initiation. The effects of RL and BL on this species depended on the stage of the organogenic process in which *Lactuca sativa* plantlets were exposed to the different lights. Exposure to BL during the critical first few days of culture, when meristems are being initiated, results in a significant reduction in the number of shoots produced as compared to exposure to RL and WL. Furthermore, this suppression of meristem initiation is permanent and not reversible afterward by culturing plants under RL. Observations with a scanning electron microscope (SEM) clarified that the lowest shoot development under BL was attributable to the production of much more callus as compared to those cultured under WL or RL, demonstrating that rapid cell division occurred, although the organized center of cell division required for primordia formation was reduced. Moreover, the same authors observed that explants exposed to continuous RL developed numerous small shoot primordia, which occurred more slowly than those detected on tissue exposed to WL. Based on the literature, they stated that the stimulatory effects of RL as compared to WL is genotype dependent, but the inhibitory effect of BL is more widely diffused. Callus formation as affected by continuous BL illumination was observed also in *Pyrus communis*, where callus weight doubled as compared to BL plus RL and BL plus FRL [59]. In *Ficus benjamina*, BL induced a huge formation of callus at the basal section of shoots [94]. Other studies have shown that the timing of exposure to different light regimes is also critical for shoot development *in vitro*. For example, at least 2 wks under RL were required to improve shoot numbers from *Pseudotsuga menziesii* callus, and the length of time in which

RL promoted shoot production lasted only 2–3 wks [122]. It was suggested that PHY plays an inductive role in organogenesis of *Lactuca sativa* L., as suggested by Kadkade and Seibert [137], in contrast to antagonistic role of BL, probably via CRYs.

In a series of research projects carried out with different rootstocks of *Malus domestica*, *Prunus domestica* and *Prunus persica*, M9, MM106, Mr.S.2/5, and GF677, respectively [125,128,142,163], it was demonstrated that BL induced, in the starting explant and in the developed shoots, a greater number of nodes with shorter internodes than those observed in RL and in dark. It should be noted that the percentage of nodes that formed lateral shoots was higher in the presence of RL as compared to the BL one. In the *Malus domestica* M9 rootstock, the percentage of sprouted buds under RL was double that under BL [135].

Based on these results, shoot multiplication can be defined as the result of two events: the induction and formation of new buds from the apical meristem and their sprouting through the reduction or the suppression of apical dominance [2,36]. BL would increase the number of axillary buds but, in contrast, it exerts an inhibitory action on buds sprouting (increase in apical dominance). RL, on the other hand, would reduce the apical dominance even though it reduces the formation of new axillary buds. The lower outgrowth of buds in the presence of BL compared to RL would indicate a role in a specific photoreceptor(s) of BL, which would act as an antagonist of the PHY. Photomorphogenetic events detected in the presence of RL and BL would agree with an antagonistic model of stem branching, modulated by light through the PHYs and the photoreceptors of BL, which would interact with each other according to a dynamic model. Moreover, Muleo et al. [142] also showed that the internode extension inhibition under BL exposure and the concomitant positive effect of BL in enhancing axillary bud formation (neoformed nodes) was dependent on the photon fluence rate, but not on PHY photoequilibrium or on concomitant exposure to RL. A quantitative BL threshold was found near 30 μmol m^{-2} s^{-1} (400–500 nm); up to this value, internode extension decreased [142].

Plants, thus, possess a complex and dynamic light response and memory system that involves reactive oxygen species and hormonal signaling, which are used to optimize light acclimation and immune defenses [164]. Thus, regulating the spectral quality, particularly by the B-LED, improves the antioxidant defense line and is directly correlated with the enhancement of phytochemicals in *Rehmannia glutinosa* [65]. Mengxi et al. [90] found higher values of superoxide dismutase (SOD), peroxidase (POD) and catalase (CAT) activities in leaves under B-spectrum irradiance and concluded that B-LED may be more satisfactory for activating different defensive systems to reduce excessive amounts of reactive oxygen species. However, in two important *Dianthus caryophyllus* cultivars, 'Green Beauty' and 'Purple Beauty', RL treatment also increased the activities of antioxidant enzymes and nutrient contents [165]. The B-LED illumination also significantly increased the antioxidant enzyme activities in leaves and roots in *Amaranthus tricolor* and *Brassica rapa* L. subsp. *oleifera* [166]. In the in vitro cultured *Pyrus communis* plantlets, it was detected that the gene encoding the pathogenesis-related protein PR10 is regulated daily by the body clock of a plant, while *PR1* was expressed without clear evidence of circadian regulation [167]. In the same studies, a specific function was played by PHYB and CRY1 photoreceptors, considering that in transgenic plants the first photoreceptor enhanced the gene expression of *PR1* 5- to 15-fold, and CRY1 enhanced plant resistance to the *Erwinia amylovora* bacterial infection [167]. *Prunus avium* rootstock plantlets, overexpressing the PHYA gene and grown in vitro, displayed a strong resistance to bacterial canker (*Pseudomonas syringae* pv. morsprunorum), highlighting a role of light quality and quantity in the regulation of plant resistance to bacterial disease [168]. Therefore, light quality through the regulative network of photoreceptors plays a relevant role in the endogenous rhythms of gene expression and pathogen attacks.

2.2.2. Blue Light Effects on Plantlet Morphology

BL is mostly considered to be able to increase leaf growth, photosynthetic pigment synthesis, chloroplast development and stomatal opening, soluble proteins and carbohydrates and dry matter content and to inhibit stem and root elongation, while RL enhances stem growth and carbohydrate accumulation [41,50,87,158]. In *Scrophularia kakudensis*, BL imposed a stressful environment that resulted in the activation of several proteins related to stress tolerance, photosynthesis, gene regulation, post-translational modification and secondary metabolism [169]. The improvement in the leaf characteristics induced by the addition of BL to RL seem to indicate a better quality of micropropagated plantlets, which in turn may also improve acclimation [2,170].

Plant height: A few papers report positive effects of BL on shoot length, while most studies agree on its inhibition of plantlet elongation. The blue spectrum was recognized to inhibit stem growth in *Oncidium* [90], in *Pelargonium* × *hortorum* [144], in *Dendranthema grandiflorum* [42] and in *Zantedeschia jucunda* [171], especially as compared to RL or RL:FRL. In different tree species, *Prunus domestica* Mr.S.2/5 and *Malus domestica* MM106 and M9, inhibition of internode elongation was also detected [128,135,142]. In contrast, BL (470 nm) and RL (660 nm) illumination were found effective for increasing shoot length in *Achillea millefolium* [172] and *Dendrobium Sonia*, where, however, BL significantly reduced multiplication as compared to YL [116].

In some cases, BL is necessary to contrast the excessive effects of RL on shoot length assuring good plantlet development. Nhut et al. [149] observed that *Fragaria x ananassa* plantlet growth was inhibited under BL, whereas an irregular plantlet growth and development was observed in the absence of BL. In the experiment of Jao et al. [171], a shorter stem of plant and a higher chlorophyll content was found in the RL plus BL treatment, highlighting that BL may be involved in the regulation of both plant height and chlorophyll development.

BL induces the production of short shoots with good leaf development and many micro-tubers in *Solanum tuberosum*. Under BL, kinetin not only strongly stimulated tuber formation, but also increased the total fresh weight and root(+stolons)/shoot ratio [71].

Fresh and dry weight: In *Dendrobium officinale*, compared to other light treatments (dark, Fl and R-LEDs), B-LEDs, alone or with R-LEDs (1:2), induced higher dry matter accumulations of PLBs and shoots [92]. Increased biomass production in cultures of *A. millefolium* [172] was noted under monochromatic B-LED or R + B-LEDs. Monochromatic BL determined higher fresh and dry weight and leaf number per plantlets in *Euphorbia milii*, *Spathiphyllum cannifolium* [83] and *Rehmannia glutinosa* [146].

It is noteworthy that monochromatic BL had a negative effect on the dry matter production of *Lippia gracilis* [119], *Plectranthus amboinicus* [48], *Gossypium hirsutum* [50] and *Vanilla planifolia* [106], as well as in the sensitive cv Dopey of *Rhododendron* where it also reduced leaf chlorophyll content [75]. In most cases, however, RL was the most effective in all these species.

Many authors, however, agree on the most positive effects obtained on fresh and/or dry weight of plantlets by adding different ratios of BL to RL as compared to only monochromatic BL (see the next chapter) [62,65,90,173,174]. Moreover, Kurilčik et al. [174] demonstrated that the influence on shoot length and weight of the BL component of a mixed light is tied to the photon flux density (PFD) of the FRL component. Once more, these results indicate the species-specific effects of BL on in vitro plantlet growth [51]. Cioć et al. [120] evidenced the relationship of BL and growth regulators. B-LED illumination and a high BA content in the substrates stimulated the growth of a greater number of *Mirtus communis* L. leaves (BL and RL plus BL) and increased the fresh weight as compared to Fls, but did not affect the dry weight, whereas RL with low amount of BA enhanced both proliferation and shoot growth. Moreover, in *Oncidium*, the amounts of soluble protein in the PLBs and leaves were the highest in the BL treatment, which suggests that the B spectrum was advantageous for protein synthesis [87,90].

Leaf morphology and functionality: BL is considered an important regulator of leaf expansion; however, differences have been ascertained among the different species. BL induced the largest number of leaves per plant, and the largest leaf thickness and area in *Altenanthera brasiliana* [175] and *Platycodon grandiflorum* [158] and a similar response on leaf area was demonstrated in *Gossypium hirsutum* [50] and *Brassica napus* [51]. BL enhanced leaf chloroplast area and the translocation of carbohydrates from chloroplasts in *Betula pendula* [154]. In contrast, less leaf area was observed in *Pyrus communis* under monochromatic BL, as compared to RL, RL plus FRL and RL plus BL [59] and in *Azorina vidalii* [74], as compared to RL plus FRL. Furthermore, CRYs are known to regulate chloroplast development in response to BL [176].

Photosynthetic pigments accumulation: Several studies have reported that B irradiation resulted in higher chlorophyll contents and carotenoids in the in vitro plantlets as compared to RL and FL. Cultures of *Euphorbia milii* [61], *Doritaenopsis* [63], *Oncidium* [16,87], *Stevia rebaudiana* [114], *Dendrobium officinale* [92], *Prunus avium* cv 'Hedelfinger' and in its somatoclone [127], *Zantedeschia jucunda* [171], *Tripterospermum japonicum* [62], *Chrysanthemum* [174], *Anthurium andreanum* [111], *Phalaenopsisis* [177], *Brassica napus* [51] and *Vaccinium ashei reade* [147] exhibited higher total chlorophyll content under monochromatic B-LEDs or combinations of R- plus B-LEDs as compared to cultures exposed to R-LED or Fls treatments. The chlorophyll content, leaf and stomata number per explant were also highest on plants cultured under BL in *Vitis vinifera* [85] and in *Gossypium hirsutum* [50].

BL and UV irradiation enhanced chlorophyll content in *Hyacinthus orientalis* L. [160] and chlorophyll a+b content, but not the carotenoid content, in leaves of *Pyrus communis* [59]. Photosynthetic capacity was highest in *Betula pendula* Roth [154] and in chrysanthemum (*Dendranthema grandiflorum*) [42] when the plantlets were exposed to BL as compared to RL. In *Dendrobium kingianum*, the average number of PLBs and the chlorophyll content were highest under B-LEDs, in contrast to the explants cultured under R-LEDs where the highest shoot formation and fresh weight were observed [99]. Likewise, a study of *Oncidium* PLBs by Mengxi et al. [90] showed that chlorophyll a and b and carotenoid levels and the greatest growth were detected under B-LEDs. On the contrary, a reduction in chlorophyll levels in plants grown under BL was observed in *Vanilla planifolia* [106]. Thus, according to Li et al. [51], the chlorophyll content of in vitro plantlets grown under different light qualities varies within plant species or cultivars. Moreover, even if BL, as compared to RL or different RL:BL ratios, reduced leaf expansion and hence leaf area in *Azorina vitalii*, the chlorophyll and carotenoid content per unit leaf area was higher than RL:FRL [74].

Changes in chlorophyll biosynthesis induced by changes in spectral quality may provide advantages regarding plant growth [178]. The species-specific responses to the B spectrum, in terms of photosynthetic pigments, are probably tied to the different environments in which the different species developed and to the type of explant used for in vitro initiation. In *Lippia gracilis*, plantlets that originated from apical explants had higher pigment production under the BL spectrum, whereas those from nodal explants showed higher production under WL, followed by the BL conditions [119]. These studies indicate that BL provides important environmental information and mostly promotes higher photosynthetic efficiency.

2.3. Combined Blue and Red Light Effects

2.3.1. Blue and Red Light Effects on Shoot Proliferation

Many studies have been carried out on the effects of combining BL and RL. A mixture of photon quantity of BL plus RL may combine the advantages of monochromic RL and BL and may overcome the individual disadvantages of these lights. However, a large amount of research regarded the assessment of the best proportion of photon quantity of BL and RL, since different behaviors have been ascertained between species and varieties [50]. In some cases, the same ratio between RL and BL is effective (RL:BL = 1:1); in other cases, higher percentages of RL as compared to BL or vice versa are effective.

A large number of studies demonstrated the promoting role of R- plus B-LEDs in various combinations on shoot regeneration and the growth of the regenerated plants: BL:RL = 1:1 in *Lilium oriental* [78], RL:BL = 9:1 in the recovery of *Solanum tuberosum* plantlets after cryoconservation [97], RL:BL = 9:1 [104] and RL:BL = 7:3 in *Fragaria x ananassa* [149], RL:BL = 7:3 in *Saccharum officinarum* [101] and RL:BL = 1:1 in upland *Gossypium hirsutum* L. [50] and *Abeliophyllum distichum* [98]. In *Gerbera jamesonii* [118], the highest shoot multiplication rate (40% higher proliferation as compared to plantlets grown under Fls) was observed under RL:BL = 50:50 and RL:BL = 70:30. In *Anthurium andreanum*, shoot propagation was promoted by exposure to RL:BL illumination and higher growth under BL [111]. In the same species, following Budiarto [49], the number of regenerated shoots was greater when exposed to higher percentages of B than R-LEDs (RL:BL = 25:75). In *Brassica napus* L. as well, proliferation was greater under higher percentages of BL (BL:RL = 3:1 light, [51]. Good results on shoot proliferation have been also reported in *Azorina vidalii* using high RL and BL combinations (2,3; BL:RL, [74] or high RL:FRL ratios (1,1)). For *Panax vietnamensis* [105], the most effective plant formation was obtained when embryogenic calli were cultured under the combination of 60% RL and 40% BL and was reported to be two times higher than under Fl [105]. Concerning woody species, better results on proliferation were obtained on *Phoenix dactylifera* with an RL:BL ratio equal to 18:2 [133], on *Pyrus communis* with an RL:BL ratio equal to 1:1 [59] and on *Populus x euramericana* with an RL:BL combination of both 70:30 and 50:50 [131] as compared to monochromic lights and Fl.

Concerning orchids, it seems that higher RL percentages as compared to BL ones are effective. A combination of R:B = 9:1 gave the highest shoot proliferation in *Phalaenopsis* protocorms [86]. In *Cymbidium*, 100% R-LED was the most effective for callus induction, but callus proliferation was best under 75% R-LED plus 25% B-LED treatment. PLB formation from callus was obtained in 25% R-LED plus 75% B-LED [80].

The composite light of R- and FR-abundant G2 LEDs (8% BL, 2% GL, 65% RL and 25% FRL-Valoya Oy, Helsinki, Finland) resulted effective in *C. grandiflorum*, *G. jamesonii*, *H. hybrida* and *Lamprocapnos spectabilis* giving similar or higher propagation of the Fls. However, in this case, the influence of FRL and GL must be considered and will be discussed in the following chapters [35].

2.3.2. Blue and Red Light Effects on Plantlet Morphology

Many studies confirmed the effectiveness of R- and B-LEDs in enhancing growth and photosynthesis in many plant species. B- and R-LEDs were developed to grow in vitro plants because chlorophyll a and b show a maximum absorption at their respective wavelengths (460 and 660 nm). The same light ratios were effective on proliferation and in promoting the quality of plantlet characteristics.

Plantlet elongation: Various combinations of R- and B-LEDs proved to determine the best results for stem length and leaf growth for *Saccharum officinarum* [112], *Stevia rebaudiana* [114], *Populus x euramericana* cv 'Dorskamp' [131], *Pyrus communis* [59], *Fragaria x ananassa* [104] and *Dendrobium officinale* [92]. Sivakumar et al. [179] showed that continuous RL plus BL or intermittent BL significantly stimulated shoot elongation of sweet Solanum tuberosum plantlets in vitro. Hahn et al. [146], on *Rehmannia glutinosa*, found that shoot lengths under either B- or R-LEDs were greater than under mixed LED or Fls, but the plantlets overgrew and appeared fragile, whereas plantlets under mixed LED or Fls were healthy, with normal shoot lengths. Thus, normal plant growth was clearly related to the presence of monochromatic BL or RL. According to some authors, the synergistic interactions between CRY and PHY could either promote or inhibit the shoot elongation in different plant species.

Plantlet growth: The composite spectra of R- and B-LEDs positively regulated fresh and, in most cases, also dry matter accumulation. As compared to the cultures raised under Fls or monochromatic lights, in most cases LEDs supplying higher RL ratios (from 70–90%) as compared to the BL ones were effective in enhancing the in vitro growth of different species such as banana [180], grape [145], *Fragaria x ananassa* a [149], *Vaccinium corymbosum* [147], *Tripterospermum japonicum* [62], *Eucalyptus citriodora* [181], *Phoenix dactylifera* [133] and *Lippia alba* [66]. Highest growth was observed under Fl and under a mixture of BL and RL in *Withania somnifera* plantlets [182]. Highest fresh and dry weights were obtained when plantlets were cultured under an equal BL and RL combination (50:50) in different species such as *Chrysanthemum* [42], *Lilium* [78], *Doritaenopsis* [63], *Pyrus communis* [59], *Saccharum officinarum* ([112], upland *Gossypium hirsutum* L. [50], *Vanilla planifolia* [106] and *Solanum tuberosum* [183]. As for proliferation, higher BL rates as compared to the other species are necessary to obtain the best growth in *Brassica napus* [51]. Similarly, to proliferation, higher RL ratios enhanced plant growth and the development of different orchids: *Cymbidium* [148] and *Phalaenopsis* [86]. RL plus BL and FRL or RL plus FRL light significantly enhanced the fresh and dry weights of *Oncidium* plantlets [89].

Differently from other cultures in which the same lights resulted in optimal proliferation and plantlet growth, according to Mengxi et al. [90], in *Oncidium*, the highest induction rate, propagation and fresh weight appeared in the RL treatment, whereas the largest dry weight per plantlet were obtained under B:R = 20%:80% and B:R = 30%:70%, respectively. Differently from other orchids, the in vitro growth of plantlets of the *Calanthe* hybrid was efficiently enhanced under a mixture of BL plus RL (0.7:1) and inhibited by RL plus FRL [184].

Leaf number and area: In *Gerbera jamesonii* [118], monochromatic RL and BL treatments resulted in a reduced leaf area, whereas leaf number was enhanced by exposure to RL:BL = 1:1.

R and B mixed LED treatments in various combinations improved leaf number and sometimes length of in vitro cultures of *Fragaria x ananassa* [149] and *Doritaenopsis* [63], leaf area of *Populus x euramericana* [131] and leaf growth of *Stevia rebaudiana* [114].

Photosynthetic pigment levels: Many studies showed that optimizing the RL:BL ratio may improve photosynthesis. The positive effect of the appropriate B-:R-LEDs combination on the synthesis of photosynthetic pigments was reported in several studies [51,92]. An appropriate mixture of B- and R-LEDs, compared with solely monochromatic BL or RL, is more effective to increase the chlorophyll a/b ratio and/or carotenoids content of the in vitro grown plants of *Tripterospermum. japonicum* [62], *Lippia alba* [66] and *Staphylea pinnata* [113]. On *Fragaria x ananassa* mixotrophic cultures, the chlorophyll content was the greatest under RL:BL = 70:30 and the least under 100% RL [149].

Plant growth and development caused by increasing the net photosynthetic rate was also observed in *Chrysanthemum* (*Dendranthema grandiflorum*) under mixed R-:B-LED treatments and has been attributed to the adjustment of the spectral energy distribution of RL:BL to chlorophyll absorption [42]. RL or BL plus RL treatments were found more effective in grape for net photosynthetic rates [145] as compared to BL alone. Differences in chlorophyll content in *Artemisia* and *Nicotiana tabacum* plants were ascertained. In plants grown under WL, significantly less chlorophyll content than plants growing in RL:BL (3:1) or RL:BL (1:1) was determined [34]. In *Gossypium hirsutum* L., chlorophyll content, leaf thickness and leaf and stomata area were higher in plantlets cultured under BL; however, the best growth was provided by BL:RL = 1:1 [50]. In addition, in the Colt rootstock of *Prunus avium* exposed to BL and BL plus RL dichromatic light, the leaves had a greater accumulation of chlorophyll [170].

A ratio of BL:RL = 1:1 emitted by LED light facilitated the growth and produced the highest chlorophyll, carotenoid contents and photosynthetic rates in *Oryza sativa* seedlings, but not callus proliferation, differentiation and regeneration, which were enhanced by BL [121].

Different from the other species, higher BL rates as compared to RL (3:1) are necessary in *Brassica napus* L. (cv Westar) to increase chlorophyll concentrations compared to the other LED treatments and Fl. Therefore, the response of chlorophyll content of in vitro plantlets to different light qualities may vary among plant species or cultivars [51].

In different orchid species, BL plus RL was reported as the most efficient treatment on the synthesis of photosynthetic pigments. Shin et al. [63], in *Doritaenopsis*, showed that mixtures of RL plus BL stimulated photosynthesis and chlorophyll accumulation. In *Dendrobium officinale*, chlorophyll a and b and carotenoid contents were the highest in protocorm-like bodies incubated under RL:BL LEDs = 66.6:33.3 [92]. Moreover, in *Oncidium* plantlets, it was demonstrated that the RL and BL combined with FRL or RL plus FRL radiation significantly enhanced chlorophyll content [89].

2.4. White Light Effects

2.4.1. White Light Effects on Shoot Proliferation

The use of monochromatic or combined R- or B- LEDs may determine a mismatch with the photosynthetic spectrum. The application of the broad band WL may overcome this problem [44].

Shoot number: The best proliferation in *Vanilla planifolia* Andrews [106] was obtained under WL and RL plus BL. Fls and WL increased the *Gerbera jamesonii* 'Rosalin' propagation ratio [107]. Similarly, W-LEDs (NS1 lamps of Valoya Oy, Helsinki, Finland) determined by the combination of 20% BL, 39% GL, 35% RL, 5% FRL and G2 LED lamps, enriched in RL and FRL, were as effective as Fls on shoot propagation of *Gerbera jamesonii*, *Heuchera × hybrida*, and *Lamprocapnos spectabilis*. In the same study, the propagation ratio for *Ficus benjamina* was significantly higher under Fls as compared to all tested LEDs. These positive results were attributed to the absence of UV or cool light in the LEDs [35]. Similarly, the most positive effects of Fls on propagation were observed in *Saccharum officinarum* [112] and in *Spathiphyllum cannifolium*, where, however, high citokinins (3 mg L^{-1} BA) were applied [83]. White LED exposure improved the shoot proliferation as compared to Fls but also to RL or RL plus BL lamps in *Musa* spp. [130], *Bacopa monnieri* [109] and *Malus domestica* genotype MM106 [128]. An exposure to low-level WL after 10 days in the dark (to induce organogenesis) determined the regeneration of well-proportioned shoots within 3–4 weeks in transgenic *Petunia x atkinsiana* [77]. In *Prunus domestica* subsp. *insititia*, however, the effect of the light differed in relation to the concentration of CK applied. At the optimal BA concentration (2.7 mM), WL (66 μmol m^{-2} s^{-1}) provided better responses on proliferation than RL, BL and FRL, if the CK concentration was below the optimal level, the production of axillary shoots was greater in the RL. The highest BA concentration (13.3 mM) decreased proliferation in monochromatic lights, as BL, RL and FRL, but not in WL [141].

The regeneration of buds from cotyledons of *Lycopersicon esculentum* was high under continuous RL and WL [69]. In *Anthurium* [111], proliferation obtained in WL was similar to Fl. Muleo and Thomas [125] working on *Prunus cerasifera*, obtained better effects on shoot proliferation in intact microcuttings (with apical bud) under WL. Although apical dominance was weakest in the RL and FRL treatments, the highest proliferation of new shoots was detected under WL because of the shorter internodes and high number of new nodes in that treatment as compared to RL, FRL and dark [125].

In contrast, WL, which establishes a similar P_{fr}/P_{tot} ratio to RL, did not reduce apical dominance compared with dark. WL would also excite blue-absorbing photoreceptors and the effects of BL on apical dominance were similar to those of WL. It seems, therefore, that the cytokine ratio may be enhanced in woody species under WL to obtain higher proliferation; however, in some species, after a long cultivation time under WL the rate of newly formed sprouts was reduced regardless of the cytokinin concentration but increased when plantlets were exposed to RL [2]. Moreover, under a low BA addition to the substrate (0.5 mg L^{-1}), after one month permanence under an R-enriched light (12% BL, 19% GL, 61% RL and 8% FRL), significant enhancement in shoot proliferation in *Ananas comosus* was observed after it was transferred under WL (Cavallaro et al. unpublished data). More than one cycle permanence under the enriched RL, however, determined callus formation on the basis of the shoots, the loss of leaves and impaired growth in *Euphorbia milii* and in *Ceratonia siliqua* L. [185].

2.4.2. White Light Effects on Plantlet Morphology

In *Phalaenopsisis* and *Anthurium andreanum*, treatments with Fls, W-LEDs (460 and 560 nm) and the combination of B- and R-LEDs showed the greatest plantlet length and number of leaves [177]. Shoot fresh and dry weight, plant height, number of leaves, number and length of roots were greater under Fls and W-LEDs in *Vanilla planifolia* [106].

Enhanced chlorophyll biosynthesis was also noted in *Vanilla planifolia* [106] and in different *Saccharum officinarum* varieties [101,112] under W-LED illumination. Exposure to WL was also beneficial for the accumulation of carotenoid pigments in *Saccharum officinarum* [112]. For the apical and nodal segments of *Hyptis suaveolens*, the best growth parameters were provided by W-LED light and RL:BL combinations [186].

2.5. Green Light Effects on Shoot Proliferation and Plantlet Morphology

GL has received less attention from the scientific community because it is a misconception that GL mainly plays a role in stomatal regulation, driving photosynthesis through chloroplast gene expression and so contributing to carbon gain. GL's role in plant growth and development was controversial because it was supposed that, in conveying information, physiological responses were scarce. Since photons of the RL and BL spectrum are depleted by the absorption of plant tissues, the light reflected from and transmitted through the tissues is enriched in photons of the GL wavelength region that efficiently penetrate farther into the body of a plant [187]. Under this condition, GL carries signals for acclimation to irradiance on a whole plant, providing information for fine-tuning developmental acclimation to shade and acting as a secondary antagonistic regulator to the well-known RL:FRL and BL responses [188]. Unlike for RL and BL, a green-light-specific photoreceptor has yet to be discovered [189]. The most accredited GL sensor is the CRY-DASH, which reverts the physiological effect of CRY [190] because many physiological responses regulated by CRY are reversible by GL [191]. Tanada [192] hypnotized the existence of the heliochrome, an FRL:GL reversible receptor acting in complement to PHY. Therefore, GL effects share several attributes that are specific to the receptor antagonists of the physiological actions of RL or BL photoreceptors [128,135,193]. Consequently, GL penetration of the plant canopy potentially increases plant growth by increasing photosynthesis of the leaves in the lower canopy more efficiently than either BL or RL [194].

GL positively influenced shoot branching on the first- and second-order branches of Mr.S.2/5 *Prunus domestica* rootstock and determined a higher internode number and shoot elongation in GF677 *Prunus persica* rootstock [142]. Based on these results, Morini and Muleo [2] hypothesized that GL had a negative effect on apical dominance, similar to RL and YL.

Kim et al. [195] reported that adding 24% of GL to R- plus B-LEDs illumination increased *Lactuca sativa* L. biomass by 47%, even if the total PPFD was the same in both lighting treatments. They attributed the growth-stimulation effect of GL on its ability to penetrate deeper into leaves and canopies. In *Achillea millefolium*, the concentrations of chlorophyll a, chlorophyll b, b/a ratio and carotenoids were higher in plantlets under GL. The highest levels of pigments observed in the GL may indicate plant stress, which can be a way to compensate for the lack of photosynthetically active light [172].

In a study on the *Cymbidium insigne* orchid, the highest PLB formation, shoot formation rate (90%) and root formation rate (50%) were found among explants cultured in a medium supplemented with 0.1 mg L^{-1} chitosan H under GL. After 11 weeks of culture, the fresh weight of PLBs was higher in the treatment with hyaluronic acid (0.1 mg/L) under GL [93]. GL and BL also enhanced in vitro PLB production in *Cymbidium dayanum* and *Cymbidium finlaysonianum* with the addition of chondroitin sulfate [108]. In *Gerbera jamesonii*, GL and RL illumination resulted in a highest number of axillary shoots and leaves number in the medium with 5 mg L^{-1} kinetin. However, in the same medium, a high fresh weight was obtained in WL [136].

On *Cymbidium* Waltz 'cv Idol', the highest shoot formation (80%) was observed in the medium containing 0.1 mg L^{-1} N- acetylglucosamine (NAG), under RL and 1 mg L^{-1}

under GL; the fresh weight of PLBs was highest at 0.01 mg L^{-1} NAG under GL [100]. In the same orchid, six times of breaking the weekly light by 1 day of G-lighting during R-LED illumination showed optimal numbers and formation rates of PLBs. Optimal shoot formation was obtained by treatments of Fl+interval lighting of G-LED and B-LED+G interval lighting [95].

In combination with RL and BL, GL also positively affects plant growth, including leaf growth and early stem elongation [196,197], and is involved in the orientation of chloroplasts and in regulation of the stomatal opening [198].

In *Solanum tuberosum* plantlets in vitro, the addition of GL to the combined RL and BL increased stem diameter and leaf area, and the amounts of chlorophyll, soluble sugar, soluble protein and starch. The addition of GL to the combined RL and BL contributed to the growth and development of *Solanum tuberosum* plantlets more than the combined of RL and BL without GL [64].

Further research is necessary to understand the role of radiation oscillating around 550 nm, since the studies in this field are very limited and are mainly conducted in combination with other spectral wavelength radiations under in vivo conditions.

2.6. Yellow Light Effects on Shoot Proliferation and Plantlet Morphology

The reduction of apical dominance seems to be the main effect determined by YL and by the GL [128,135]. YL applied to cultures of *Prunus domestica* rootstocks Mr.S.2/5 and GF677 reduced apical dominance [199]; in *Malus domestica* rootstock M9, this light induced a production of axillary shoots greater than that detected under BL and FRL but still lower than that detected under RL [135]. Similar to the RL, the YL and GL induced a greater elongation of the internodes and outgrowth axillary shoots than the BL; in particular, the YL stimulated longer internodes in *Prunus domestica* rootstocks Mr.S.2/5 [142]. YL illumination induced higher proliferation in *Populus alba* × *P. berolinensis* [129].

YL irradiation followed by the RL one induced higher shoot proliferation (98%), a higher number of shoots per explants and early PLB formation, differentiation and shoot initiation in *Dendrobium sonia* [116]. YL elicited response of callus multiplication in *Vitis vinifera* [200]. YL also determined a higher leaf area and fresh weight and a lower shoot length in *Dendrobium sonia* [116]. YL showed a smaller increase in mean fresh weight as compared to BL but less than RL [135].

The YL positively affected growth in *Lactuca sativa* [201]. Based on current knowledge, the behavior of in vitro cultures subjected to YL would not be attributable to the actions of PHYs and BL photoreceptors.

2.7. Far Red Light Effects on Shoot Proliferation

Sunlight emits almost as much FR radiation as R radiation. Leaves absorb most RL but reflect or transmit most FRL [202]. As stated before, plants under a canopy or the lower leaves of plants spaced close together receive a greater proportion of FRL than RL radiation, i.e., a reduced RL:FRL ratio. Plants perceive this filtering of light and, in response, redirect growth and development according to the survival strategies of shade avoidance, increasing apical dominance and typically elongating in an attempt to capture available light [25]. In contrast, once sunlight has been reached, PHY and UVR8 inhibit shade avoidance. Several studies suggest that multiple plant photoreceptors converge on a shared signaling network to regulate responses to shade [203]. PHYs are the receptors of RL and FRL and are mainly involved in this perception, but plants shaded within a canopy also perceive reduced BL and possibly enriched green light through CRYs [190]. The detection of canopy gaps may be further facilitated by BL sensing phototropins and the UV-B photoreceptor, UVR8. Moreover, Zhen and van Iersel [204] reported that adding FRL consistently increased net photosynthesis of *Lactuca sativa* L. as compared to RL and BL. They attributed this effect to the increased quantum yield of photosystem II (ΦPSII).

The commonly applied Fl but also the R:B LEDs usually lack FRL, which is important for plant development, stem elongation and PHY activity, whereas they are abundant in GL and YL, which are less efficient for plants [35].

PHY in its active form, as may occur under high RL or RL:FR ratio, seems to alter the endogenous hormonal balance, reducing the apical dominance and increasing the shoot proliferation rate through enhancing lateral shoot development. On the contrary, low RL:FRL ratio or FRL alone reduces in vitro proliferation [2,205].

FRL appeared to increase node formation and decrease internode extension (but to a less degree than BL) as compared to the effects of RL. With dichromatic BL plus FRL, the effects on these two variables induced by BL were found to be slightly modified, indicating that the active form of PHY was only partially able to influence CRY-regulated physiological functions. While the effects of RL and BL and the RL:FRL effects during in vitro phases have been extensively examined, the effects of FRL alone have been less studied [59]. A high RL:FRL ratio or a low BL:RL ratio stimulated the sprouting of axillary buds in *Azorina vidalii* [74] and *Vaccinium corymbosum*, where, however, the presence of UV in the lighting device influenced shoot length differently in different cultivars [206]. Even in *Spirea nipponica*, shoot proliferation was greater when explants were exposed to combinations of high-ratio RL and FRL [124]. In a study on *Oncidium* [89], the best results on PLB formation were obtained under R+B+FR LEDs. This study also indicated that this combined radiation or RL:FRL radiation significantly enhanced leaf expansion, number of leaves and roots, chlorophyll contents and fresh and dry weight. The highest propagation ratios for *Chrysanthemum* × *morifolium*, *Heuchera* × *hybrida*, *Gerbera jamesonii* and *Lamprocapnos spectabilis* were reported under light emitted by RL- and FRL-abundant G2 LEDs [35]. The G2 spectrum was favorable in most of the species tested, probably because of the high GL:BL and RL:FRL ratios, which provide a higher portion of active PHYs [207].

Under a constant fraction of RL and BL, root number, length of roots and stems and fresh weight of the plantlets was related to the FRL component of the total PPFD in the *Chrysanthemum morifolium*. At the higher intensity of FRL tested (9 μmol m^{-2} s^{-1} of the total 43 μmol m^{-2} s^{-1} of PPFD), a reduction of the previous morphogenic characters was observed [174].

On the *Prunus domestica* rootstock GF655-2 cultured in vitro in the presence of BA, at a photon fluence rate of 20 μMol m^{-1} s^{-1}, FRL irradiation significantly promoted shoot proliferation as compared to the dark [141]. At a lower photon fluence rate of 9 μMol m^{-1} s^{-1} the response was lower than the other lights and similar to that detected in the dark. Based on the data obtained in their experiments, the authors concluded that the proliferation rate induced under BL, FRL and WL strongly depended on the photon fluence rate, while no statistically significant differences could be found in the effects of RL irradiation at different photon fluence rates. In *Pyrus communis*, FRL was advantageous for shoot number, but shoot quality was inferior because of low shoot weight, hyperhydricity and chlorosis as indicated by the low total chlorophyll and carotenoid content [59]. Werbrouck et al. [94] reported the negative effect of FRL on in vitro biomass production of *F. benjamina* showing a reduction in the total number of shoots and in both shoot cluster and callus weight.

A reduced RL:FRL ratio (1:1.1) had an inhibitory effect on the growth of two *Calanthe* hybrids [184].

In microcuttings of a *Prunus cerasifera* rootstock, BL and WL produced a higher number of nodes, with shorter internodes compared to RL or FRL or dark. Differently, the proportion of nodes producing outgrowing of lateral shoots was higher in RL followed by FRL than in WL, BL or dark because of the weakening of apical dominance induced by the former two lights [125]. However, the highest proliferation of new shoots was seen in WL because of the high number of new nodes. Even here, as evidenced also by Baraldi et al. [141], the effectiveness of FRL required prolonged exposures and was dependent on photon fluence rates [125]. On M9 rootstock of *Malus domestica*, the development of phytomers appeared to be primarily caused by the active form of PHY, with a marginal effect from BL. Shoot growth, which combines internode elongation, development of the phytomer and branching, was highest under RL and the lowest under BL and FRL, showing the largely positive role of PHY photoequilibrium. FRL was the most inhibiting light type, reducing the proliferation rate compared with BL. Under FRL, reduced stem elongation was due to the very small number of phytomers formed [135].

3. Effects of Light Intensity

The selection of the optimal light intensity to support in vitro proliferation and growth is also important for an optimization of the processes. Among others, light intensity regulates the dimension of leaves and stems, as well as their morphogenic pathway, and is involved in pigment formation and hyperhydricity [208].

In vitro cultures are subjected to a much lower light intensity as compared to those grown under open field conditions. The permanent low light conditions in vitro have been considered a limiting factor for photosynthesis and for supporting plant morphogenesis in vitro, so it is necessary, in most cases, to supply sucrose to the medium [209]. In vitro plants are also very susceptible to high light conditions [210] and prone to photoinhibition [211]. Too high irradiation can severely damage the photosynthetic apparatus and photosynthetic pigment synthesis [48,212], leading to the formation of harmful free oxygen radicals and damage to cells [213].

In Table 3, the research that mainly addressed the effects of different light intensities is shown, but only in a few of the studies shoot proliferation is examined.

Table 3. Effects of different light intensities on shoot proliferation in increasing light-intensity order.

Species	Tested Intensities	Best Yielding Intensity (μmol m^{-2} s^{-1})	Main Parameters Affected and Notes	Authors
Disanthus cercidifolius, *Rhododendron* spp., and *Crataegus oxyacantha*	11, 25, 55, 106 and 161 μmol m^{-2} s^{-1}	11–27	Better growth and leaf chlorophyll content	[75]
Acer saccharum Marshall	4, 16 and 40 μmol m^{-2} s^{-1}	4 and 16	Low intensity overcomes recalcitrance.	[214]
Achillea millefolium L.	13; 27; 35; 47 and 69 μmol m^{-2} s^{-1}	27 μmol m^{-2} s^{-1}	Higher dry mass of shoots and roots, shoot length	[172]
Withania somnifera (L.)	15, 30, 60, and 90 μmol m^{-2} s^{-1}	30 μmol m^{-2} s^{-1}	Greater growth and development.	[182]
Chrysanthemum morifolium Ramat. 'Ellen'	25, 40, 55, 70, 55 μmol m^{-2} s^{-1}	40 μmol m^{-2} s^{-1}	Better plantlet growth	[174]
Vaccinium corymbosum)	55 to 240 μmol m^{-2} s^{-1} for 7 to 60 days		Higher irradiances (≥55 = 210 μmol m^{-2} s^{-1}) improved proliferation only with short time applications (7 days).	[215]
Spathiphyllum cannifolium Culture Pack", on rockwool system, with CO_2 enrichment	45, 60, 75 μmol m^{-2} s^{-1} 80% RL + 20% BL LED	60 μmol m^{-2} s^{-1}	Best growth	[216]
Fragaria × *ananassa* Duchesne	45, 60, 75 μmol m^{-2} s^{-1}	60 μmol m^{-2} s^{-1}	Better shoot growth	[149]
Plectranthus amboinicus (Lour.) Sprengof	26, 51, 69, 94 and 130 μmol m^{-2} s^{-1}	69 μmol m^{-2} s^{-1} and to a lesser extend 94	Higher shoot number, leaf area, total dry weight and carvacrol content	[48]

Table 3. *Cont.*

Species	Tested Intensities	Best Yielding Intensity (μmol m^{-2} s^{-1})	Main Parameters Affected and Notes	Authors
Phaius tankervilliae (Banks ex L'Herit) and *Vanda coerulea* Giff	28, 37, 56, 74 and 93 μmol m^{-2} s^{-1}	74 μmol m^{-2} s^{-1}	Better plantlet growth	[217]
Pyrus spp. *rootstock* BP10030	from 10 to 80 μmol m^{-2} s^{-1} 16 and 24 h photoperiod	from 10 to 80 μmol m^{-2} s^{-1} 16 h photoperiod = greatest shoot number	10 μmol m^{-2} s^{-1} better for initial explant growth. Increasing irradiance to max higher growth 24 h = the highest shoot fresh and dry weight.	[218]
Lippia gracilis Schauer	26, 51, 69, 94, or 130 μmol m^{-2} s^{-1}	94 μmol m^{-2} s^{-1}	higher number of segments, leaf, shoot, root, and total weight plantlet^{-1}	[119]
Momordica grosvenorii Swingle	25, 50, 100, or 200 μMol·m^{-2}·s^{-1}, and an increased CO_2 concentration	increasing intensities up to 100 μmol m^{-2} s^{-1}	Better plantlet growth	[219]
Actinidia deliciosa (A. Chev.) C.F. Liang & A.R.	30 to 250 μmol m^{-2} s^{-1} and an increased CO_2 concentration	120 μmol m^{-2} s^{-1}	better plantlet growth and proliferation	[220]
Rosa hybrida	0, 4, 17, 66, and 148 μE m^{-2} s^{-1}	17 μE m^{-2} s^{-1} 148 μE m^{-2} s^{-1}	At the highest intensity best proliferation. At 17 μE m^{-2} s^{-1} lower propagation but better leaves	[221]

The optimal value of the PFD for plantlets changes from species to species and the predominant in vivo light conditions may give an indication of the requirements for optimal culture growth in vitro [75]. In *Alocasia amazonica* [222] and *Momordica grosvenori* [219], shoot length increased with the reduction in light intensity, an adaptation mechanism indicating that these species can survive in low light-intensity environments. In *Lippia gracilis*, the weight increase of plantlets grown under high light intensities indicates that this species originates in a semiarid environment where high irradiance (HI) incoming light occurs [119]. Evidence has been previously presented [178] that plants adapted to an environment with incoming HI present better photosynthetic rates and high growth rates under intense light. In an extensive study on the photosynthetic pigments, Lazzarini et al. [119] concluded that the increase in chlorophyll b content under low irradiance (LI) is indicated as an important marker of plant adaptation to shaded environments because this pigment is more efficient for capturing the photons of the higher wavelengths of the spectrum that are mainly present. Furthermore, it is worth noting that the type of explant also influences the amount of photosynthetic pigment: leaves of plantlets generated from apical explants had higher amounts of chlorophyll a, total chlorophyll and carotenoids regardless of light conditions, whereas the amount of chlorophyll b resulted in more plantlets generated from the lateral buds of nodal segments. Moreover, an increase in the synthesis of carotenoids was observed in plants grown under high light intensities and was associated with the photoprotection exerted by these pigments within the photosystems. In *Lippia gracilis*, this increase led to better efficiency of the photosynthetic activity and, hence, the higher production of dry weight observed under these conditions [119]. In three different species, *Disanthus cercidifolius*, *Rhododendron* cultivars and *Crataegus oxyacantha*, low levels of irradiance (11 μmol m^{-2} s^{-1}) were optimal for in vitro growth, while higher irradiance determined a decrease in shoot development and leaf chlorophyll content in *Disanthus* and

Rhododendron cultivars, which are shade-tolerant species in their natural habitat. Plantlets of *Crataegus* generated from in vivo plants adapted to higher levels of irradiance resulted in tolerance to a wide range of irradiances in vitro. Only shoot extension was inhibited at the highest levels tested, whereas leaf chlorophyll content was unaffected. These differences were attributed to a differential adaptation to light determined by the natural habitats of these plants and of the possible direct effect of irradiance upon plant growth regulators in the culture system [75]. Different effects of rising light intensity were observed in *Plectranthus amboinicus* grown in vitro. In this species, intensities below or above the optimum (69 μmol m^{-2} s^{-1}) led to the lowest growth. In fact, photosynthesis was inefficient under low light intensity (26 μmol m^{-2} s^{-1}) but increased light intensities led to reduced concentrations of a, b and total chlorophyll, and carotenoids and thus of growth [48]. In *Withania somnifera* and *Achillea millefolium*, the treatments with the highest light intensity (60 and 69 μmol m^{-2} s^{-1}, respectively) showed the highest levels of photosynthetic pigments but not the highest growth. Alvarenga et al. [172] concluded that the significant increase observed in chlorophyll and carotenoids under high light conditions would indicate that these pigments have the photoprotective function, as assumed by Biswal et al. [223], since they may be inefficient in absorbing light and increasing photosynthetic efficiency. They also attributed the damage of excess light to the photosynthetic apparatus to the production of free radicals, which may degrade these pigments [45,213]. Kurilčik et al. [174] on *Chrysanthemum (Chrysanthemum morifolium)*, noticed that the maximal PFD (85 μmol m^{-2} s^{-1}) used in their experiment induces light abnormalities on the leaf surface. In ginger [224], the growth was restrained when the light reached 180 μmol m^{-2} s^{-1} and the chlorophyll content decreased as the light intensity increased.

However, a different sensibility to light intensity seems to affect proliferation rate and the plantlet growth, and in most cases lower plant intensities are required for proliferation.

Based on the observation of the examined papers for this review, in Figure 1, the light intensities were grouped in ranges and the frequency of their use is shown. From this study, it emerged that whatever the light spectrum, the most used light intensities range from 20 to 80 μMoles m^{-2} s^{-1} and the most used intensity for proliferation is 50 (μmoles m^{-2} s^{-1}).

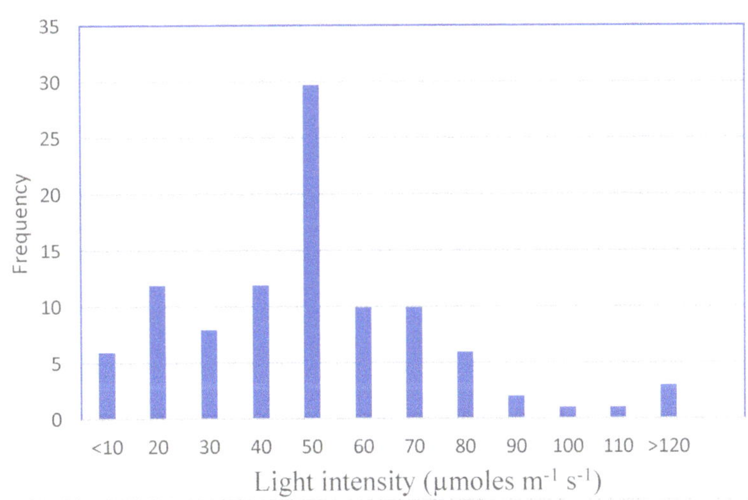

Figure 1. Frequency of light intensities used in literature for proliferation.

In *Rubus* spp, rising WL fluence rates from 0 to 81 μmol m^{-2} s^{-1} did not improve the organogenesis from cotyledons [225]. In *Vaccinium corymbosum*, exposure at rising intensities from 55 up to 210 μmol m^{-2} s^{-1} improved proliferation and rooting ratios only with short time applications (7 days). Longer exposure of the leaves (14 and 28 days) determined inhibition of growth and the red color of leaves and sprouts, and less vigorous plants after in vivo transferring [215].

However, a better multiplication under increasing irradiance, from 10 to 80 μmol m^{-2} s^{-1}, resulted in *Pyrus communis* [218], in *L. gracilis* at 94 μmol m^{-2} s^{-1} [119] and in *Rosa hybrida* from 4 to 148 μmol m^{-2} s^{-1} [221]. In this last species, higher irradiance (66 and 148 μmol m^{-2} s^{-1}) showed better effects on shoot proliferation, but leaf chlorosis was observed and better results on shoot growth were obtained at 17 μmol m^{-2} s^{-1} [221]. The chlorosis occurring at the higher levels of irradiance may be due to photochemical oxidation, photoinhibition or chloroplast damage [226].

In *Castanea sativa*, Sáez et al. [227] highlighted a correlation between light intensity and the addition of sugar to the growth medium. They demonstrated that HI (150 μmol m^{-2} s^{-1}) and high sugar amounts (30 g L^{-1}) produced an increase in photosynthetic activity and chlorophyll content and determined a higher proliferation rate and biomass production. However, a high proliferation rate was obtained even under LI with a higher sugar content in the medium. Thus, HI but also LI may be beneficial during the in vitro culture, but this is only possible in the presence of sucrose added to the culture medium.

Kozai [228], in *Cymbidium*, doubled in vitro growth by adding CO_2 to the culture vessels at high PFD (230 μmol m^{-2} s^{-1}), demonstrating that CO_2 limitation may have a relevant role in enhancing the growth when high PFDs are adopted. The same was also true for *Actinidia deliciosa* where the proliferation rate and dry and fresh weight increased up to 120 μmol m^{-2} s^{-1} but decreased at higher rates. The biomass produced was also affected by light intensity, since both dry and fresh weight increased at the PPFD up to 120 μmol m^{-2} s^{-1}, while only dry weight increases thereafter up to the highest value of 250 μmol m^{-2} s^{-1}.

The photosynthetic rate was nearly four times higher when raising CO_2 up to 1450 and 4500 μL L^{-1} compared to the lowest CO_2 concentration tested (330 μL L^{-1}) [220].

In fact, it has been shown that, just a few hours after the light was turned on, CO_2 underwent a drastic reduction in concentration and sub-optimal CO_2 availability has been correlated with reduced photosynthetic ability [229]. Thus, exogenous enrichments of this gas in the culture vessels improves photosynthesis at high PFDs [230,231].

Finally, most studies on the effects of light intensities have been carried out under Fl or W-LED. However, some studies revealed a relationship between the light spectrum and the intensity that affects plant growth and development. In the presence of BA, WL, BL and FRL, action on proliferation was dependent on the fluence rate [141].

Phytochrome has been shown to induce a high-irradiance response and low-irradiance response in *Prunus domestica* rootstock Mr.S. 2/5 [142]. Similar results were also obtained with the rootstock GF 677 in which the newly formed shoots were fewer but longer under the two intensities of RL (15 and 40 μmol m^{-2} s^{-1}) than those treated with WL. In addition, the low intensity RL (15 μmol m^{-2} s^{-1}) induced higher shoot multiplication as compared to the higher irradiance (40 μmol m^{-2} s^{-1}). The formation of new shoots in the two species was affected differently by the increase in the RL irradiance, and shoot formation was found to increase in the cultures of Mr.S. 2/5 and decrease in those of GF 677. This result could be related to a species-specific response on which would depend different PHY regulation strategies [2].

4. Effects of Photoperiod

An organism's life has evolved adaptation mechanisms that are related to environmental variations. Some of these variations exhibit regular cyclicality such as light:dark cycles, others fluctuate, such as temperature; however, all of them induce significant changes in the physiology and metabolism of most organisms, occurring in their life trajectory as characterized by the night and day cycle [232,233]. Plants possess the circadian clock, an endogenous time-keeping device that triggers and regulates physiological events in accordance with predicted daily changes in the environment. The input of light into the circadian clock is led by a set of photoreceptors such as the ZTL-type and UVR8 receptors [234].

Photosynthesis and stomatal movements are controlled by the circadian clock [235,236]. Among several physiological processes that include chromatin-regulation, diurnal rhythmic gene expression generates networks of genes that act specifically throughout the day or the night [237–240]. The circadian clock is an endogenous oscillator with a duration of approximately 24 h, and it is coordinated by external factors such as temperature and light. These external factors are relatively constant during the micropropagation procedure since there is no change in photoperiodism and thermoperiodism. During the shoot multiplication phase of in vitro cultures, photoperiod regimes of 16 h of light and 8h of dark are usually adopted. Plantlets in vitro are mixotrophic organisms, therefore nutrients such as carbohydrates are absorbed from the medium. In plantlets exposed to a 16:8 h photoperiod, the photosynthetic activity is intense at the onset of the light cycle and decreases rapidly thereafter. The block of CO_2 assimilation depends on the rapid and progressive lower concentration of CO_2 in the culture vessels. The CO_2 availably in the culture vessels is largely generated during the respiration of sucrose supplied with the growth medium, since the gas exchange between the inside and the outside the vessel is almost absent (Abbot and [220,230,241,242]. The modification of the photoperiodic regime from a 16 h photoperiod cycle to a 4 h photoperiod cycle promoted the increment of fresh and dry weights of shoot clusters, and the number of neo-formed shoots from initial shoot explants in two *Prunus persica* rootstocks [243]. An analogous response was found in the *Prunus persica* cultivars Suncrest, Belle of Georgia and Evergreen when cultured in the presence of 10μM of BA in the medium [244]. However, Morini et al. [245] have found that the photosynthetic activity was only extended until 4 h after the beginning of the illumination, although the concentration of CO_2, (under the 16/8 h regime) was not a limiting factor since at the end of the light period its availability was still much higher than that outside the vessel. From the same authors, the reduction of photosynthetic capacity was attributed to a reduced efficiency of the chloroplasts coupled with the lengthening of the light period. The promotive role of the 4 h photoperiod cycle on the shoot proliferation rate was hypothesized to be dependent on the diverse regime of photo-equilibrium of photoreceptors that promoted the reduction in apical dominance and development of axillary buds.

However, in studies carried out on other species, subjected to a 16 h photoperiod, low concentration of CO_2 into the vessels was observed: *Pfaffia glomerata* [246], Solanum tuberosum [247,248], *Carica papaya* [249], *Castanea sativa* [227], *Vitis vinifera* [250,251], *Fragaria x ananassa* [252], *Hyptis marrubioides* and *Hancornia speciosa* [253].

5. Light and Plant Growth Regulators

Some in vitro studies highlighted the effects of light spectra on the effectiveness of endogenous- and exogenous-applied growth regulators.

5.1. Light Effects on Endogenous Growth Regulators

Endogenous auxins and CKs are the most involved growth regulators in regulating apical dominance [254]. Apical dominance and its correlative inhibition are determined by the synthesis of auxins by the apex [255]. In the classical model, it is hypothesized that these hormones are synthesized by the apex and transported downward into axillary buds, with subsequent direct downregulation of outgrowth, or indirect regulation via other mechanisms such as nutrient diversion, expression of genes that control the growth of axillary buds, adjustment of the auxin/cytokinin ratio, including activation of strigolactones capable of modifying the hormonal balance, and the apical dominance [256,257]. On the other side, an increase of CK quantity in tissues leads to a marked growth of axillary buds, counteracting the action of auxins. Studies on transgenic plants have shown that regulation of apical dominance by plant hormones is not determined by the absolute concentration of hormones but by the ratio between them [258]. In vitro shoot proliferation is strongly dependent on the ability of CKs to counteract apical dominance, i.e., the physiological control exerted by the apex over the induction and development of the new lateral meristems in axillary buds along the axis of the growing explant. Light acts mainly as a morphogenic signal in the triggering of bud outgrowth and initial steps in the light signaling pathway induce changes in the levels of cytokinin-like substances [259–261]. The effect of light in modulating endogenous CKs levels is well-known and has been demonstrated in several species such as *Rosa hybrida* and *Chlorella minutissima* (*Chlorophyta: Trenouxiophyceae*) [262,263]. In *Rosa hybrida*, in dark, inhibition of bud outgrowth is suppressed solely by the application of CKs. In contrast, application of sugars has a limited effect. Exposure of plants to WL induces a rapid (after 3–6 h) up-regulation of *RhIPT3* and *RhIPT5* genes involved in CK synthesis, of the *RhLOG8* gene involved in CK activation and of the *RhPUP5* gene involved in CK putative transporter and induces the repression of the *RhCKX1* gene involved in CK degradation in the node. This leads to the accumulation of CKs in the node and to the triggering of bud outgrowth [263]. In *C. minutissima* [262], a rise in endogenous auxin and CK and a decrease over time in gibberellin concentrations was observed in the actively growing cultures under light:dark conditions (L:D) and continuous dark+glucose (CD+G) but no increase was determined under continuous dark (CD). The L:D cultures had the largest CK increase.

It has been known for several years [264] that the bands of the light spectrum that have been shown to promote morphogenetic processes through the activation of the various photoreceptors are mainly represented by RL, FRL and BL. As stated in paragraph 2.1, RL increases the quantity of cytokinin in tissue, counteracting the action of auxins and thus determining an increase in the development of lateral shoots [139,140]. RL also regulates the synthesis of carotenoids and strigolactones [265]. Previous studies reported that RL decreased the IAA concentrations in maize epidermal cells [266].

The interaction between CK and PHY would induce, in the latter, an extension of the active form (P_{fr}) even in conditions of dark and FRL [2]. In addition, other plant hormones may be modulated by light and by phytochrome directly. Among these are gibberellins [65] and brassinosteroids [267], another important category of growth regulators affecting cell elongation and cell division. Thus, RL may promote stem growth by regulating the biosynthesis of gibberellin or induce the expression of an auxin inhibitor gene to promote stem and root lengthening in grape [8]. In contrast, BL seems to affect more the auxin content (indoleacetic acid-IAA in particular). In fact, it was demonstrated that BL induced higher IAA content than RL in the leaves of the balloon flower [158] and thus it is more effective in promoting leaf growth. Significantly higher IAA contents occurred in the leaves under the BL:RL = 3:1 and BL:RL 1:3 and induced larger leaf areas compared to RL. Thus, BL appeared more beneficial for increasing IAA concentrations and for promoting better leaf growth than RL. However, in tobacco, a species in which BL stimulated shoot proliferation, contrasting effects of BL have been reported, since it was hypothesized that at higher intensities it determines the photoinactivation of IAA [67]. These mechanisms, both related to apical dominance and bud dormancy, are masked by WL, a condition under which cryptochrome and phytochrome are activated.

5.2. *Effects of Light on Exogenous Applied Growth Regulators*

In *Prunus domestica subsp. insititia*, clone GF655-2, BA, a promotive effect on proliferation was repressed under dark, whereas no proliferation was observed under light conditions without BA. It is noteworthy that at the highest BA supplied, the proliferation rate increased under the broadband WL, whereas it decreased under the monochromatic sources RL, BL and FRL [141]. Light and BA also proved to be indispensable factors in adventitious shoot formation from *Pinus radiata* cotyledons [268]. In *Spirea nipponica*, the interaction between CKs (0.25 mg L^{-1}) and RL resulted in an enhancement of the shoot proliferation rate [123]. The same indications on the interaction between light quality and CK content were obtained on multiplication and growth during in vitro culture of *Myrtus communis* L. [120] and *Spirea nipponica* [124]. The highest number of shoots was obtained under RL or R:FR-LEDs with the higher CK concentrations tested in the media (5 μg L^{-1} i.e 1.1 and 0.5 mg L^{-1}, respectively).

At lower BA levels (0.4 mg L^{-1}), 4 weeks of RL:FRL at low fluence followed by 1 week of WL at higher fluence rate produced almost the same proliferation levels and optimal growth [124]. If the CK concentration was below the optimal level, the production of axillary shoots was greater in the RL; at higher CK concentration, the multiplication rate decreased [2]. The effect of light spectrum differs, however, in relation to the concentration of CK applied: at the optimal concentration, WL provided responses better than those obtained with RL and BL. Thus, the quantity of applied CKs may decrease under RL. Analogously, CK incorporation into the culture medium annulled the promoting effect of RL in axillary bud proliferation from azalea apices and adventitious bud regeneration from *Petunia* spp. leaf segments [269,270]. Probably, light quality and hormone application may affect the morphogenesis of in vitro plants, in part because of changes in sink strength and, as a consequence, to redistribution of active growth [71].

Panizza et al. [72] analyzed the effect of spectral composition on axillary proliferation of lavandin *(Lavandula officinalis* Chaix • *L. latifolia* ViUars cv. Grosso) in relation to the application of exogenous BA, putrescine (Put) and endogenous ethylene production. The effect of BA was predominant over the light quality, whereas in BA-free medium, shoot number was enhanced under BL, WL and RL at low photon fluence rates. BA, however, could reduce the inhibiting effect of BL and UVL at high photon fluence rates. Exogenous Put stimulated axillary bud proliferation under some light treatments in the presence of BA, although the short fluence RL treatment was critical to allow the positive effect of Put on shoot formation. A positive correlation between biotic ethylene production and shoot formation was evidenced under FRL at a high photon fluence rate in the presence of BA. In the BA-free medium, further evidence of the correlation between biotic ethylene and the proliferation process was given since the biotic emanation increased under those radiation treatments (RL, BL and WL), which also improved shoot number. The authors conclude that in the evaluation of the responsiveness of a tissue to radiation in vitro, great care should also be devoted to radiation-induced changes in the abiotic environment (e.g., ethylene release) [72].

6. Discussion and Conclusions

Several papers on different species concern the effects of light on in vitro proliferation to assess the light properties capable of enhancing the efficiency of the micropropagation process, also ensuring consistent energy savings, as compared to traditionally used Fls lamps, or the broad range of WL. However, the results are often conflicting. Many authors ascribe these results to the different responses to light of plant species, cultivars or even explant types [119], plant stage development [122], medium composition [143] and micro-environmental characteristics such as PPFD [174] and vessel ventilation [146]. However, a large cause of variability may be tied also to the difficulty in applying uniform intensities along the shelves, and/or the use of the right spectral composition for each light quality.

Moreover, the lack of sufficient in vitro experimental protocols like those available for in vivo study, which would make the effects of light clearer, limits the comparability of the experiments [34]. The issues of major concern, among others, in this regard are (i) the short timescale in which these experiments are carried out (mostly a propagation cycle), (ii) the quality and quantity of exogenous applied growth regulators, (iii) the narrow range of light intensity values within which the efficiency of axillary multiplication of explants occurs and (iv) the mixotrophic state of plantlets. Concerning the first issue, the short-time experiments strongly limit the comprehension of the effects of light spectra on the stability of proliferation and plantlet growth during subsequent multiplication cycles (see particularly the RL effects). Concerning the second one, due to the interaction of light with endogenous growth regulators (particularly CKs), attention must be paid to the doses of the exogenous growth regulators applied. It seems from the examined literature, in fact, that RL effects are visible under low CK concentrations in the medium, whereas WL effects are even visible under high CKs doses [83]. Too high CKs quantities mask the effects of RL or may determine growth alteration. Concerning light intensities, excessive LIs or HIs may determine low growth rates, photoinhibition and may mask light spectra effects. Moreover, information on how the mixotrophic metabolism of a plantlet grown in vitro affects the development and morphology of the microcutting is scarce.

In this review, several research are presented regarding the different response of species and cultivars to different light spectra, intensities and photoperiod and it seems that some general indications arise from the different studies. Concerning the optimal irradiance intensity, it has been hypothesized that the prevailing light conditions under the natural habitats of some species can be used to indicate their requirements for optimal in vitro growth [75]. Evidence have been presented that plants adapted to an environment characterized by high light intensities present better photosynthetic rates and high growth rates under in vitro intense light, whereas shade-tolerant plants are damaged by high intensities. A survey of the tested literature revealed that in most species, whatever the light spectrum, the most used light intensities range from 20 to 80 μMoles m^{-2} s^{-1} and the most used intensity for proliferation is 50 μMoles m^{-2} s^{-1}. Better growth, however, have been registered especially in plants adapted to high intensities (see *Saccharum officinarum*, *Actinidia deliciosa*, *Lippia gracilis*, etc.) at intensities up to or exceeding 80 μMoles m^{-2} s^{-1}. Significant improvements on in vitro fresh and dry weights of shoot clusters, and the number of neo-formed shoots from initial shoot explants were obtained, also modifying the photoperiodic regime from a 16 h photoperiod to a 4 h photoperiod cycle, thus permitting the plantlets to replace the CO_2 [243,244]. In fact, in plantlets exposed to the 16:8 h photoperiod, the photosynthetic activity is intense at the onset of the light cycle and decrease rapidly thereafter because of the rapid and progressive lower concentration of CO_2 in the culture vessels. Moreover, the promotive role of the 4 h photoperiod cycle on the shoot proliferation rate was hypothesized to be dependent on the diverse regime of photo-equilibrium of photoreceptors that promoted the reduction in apical dominance and development of axillary buds [243]. In this view, also adding CO_2 [220] or aerating the vessels [146] proved to be effective in enhancing in vitro growth.

Concerning light spectra, RL alone or high RL:FRL ratios seem to enhance shoot proliferation, as well as PLB and callus formation, in many species. The main effects of RL are tied to the promotive role of phytochrome in the synthesis of CK in tissue, which counteracts the actions of auxins, increasing the development of lateral shoots. RL also regulates the synthesis of carotenoids and, in particular, strigolactones that seem to regulate apical dominance by modification of auxin fluxes [271]. The stimulatory effects of RL seem to be exerted during the beginning of the multiplication phases. However, different reports indicated that RL alone is not able to activate the pathway of chlorophyll synthesis and may determine excessive stem elongation and leaf disorders, the so-called Red Light Syndrome [36]. In fact, when plants are grown under 100% monochromatic RL a strong decrease in photosynthetic capacity, rates of electron transport, dark-adapted Fv/Fm and leaf thickness, as well as unresponsive stomata and reduced leaf pigmentation occurs [272]. BL is effective in increasing callus formation and the number of axillary buds but exerts an inhibitory action on buds sprouting (increase in apical dominance). It has been demonstrated that this light mostly controls some morphological characteristics such as shoot length and enhances chlorophyll synthesis and chloroplast development. RL, on the other hand, would remove the apical dominance but seem to reduce the formation of new axillary buds. Hence, a minimum threshold of BL is necessary for normal plant growth [146]. Moreover, regulating the spectral quality particularly by the BL improves the antioxidant defense line and is directly correlated with the enhancement of phytochemicals [65,90,166] or with the regulation of gene expression [167]. All these reasons would explain why the RL:BL illumination resulted effectively in a wide range of species. Moreover, more recently, an abundance of evidence has indicated the role of GL in carrying information about the environment to the plants, because it is involved in the shade avoidance response, but also in regulating different biological, morphological and biological processes in vitro and in vivo [189]. The addition of GL to the combined RL and BL contributed to the proliferation, the growth and development of some in vitro cultures. In a few cases, even the addition of

YL seems to improve plant proliferation and growth. In addition, the absence of ultraviolet light may determine foliar intumescence and could become a serious limitation for some crops lighted solely by narrow-band LEDs [273]. Thus, the use of monochromatic or combined R- or B-LEDS may determine a mismatch with the photosynthetic spectrum. The application of the broad band WL may overcome this problem [44]. In some species, better results have been obtained under W-LEDs [109,112,130]. Even if WL is not as effective as RL in overcoming apical dominance, high proliferation rates are obtained when CKs are added to the medium. In most cases, the best propagation was obtained at higher CK ratio [141]. It seems that the CK ratio may be enhanced in woody species under WL to obtain high and stable proliferation. However, in some species, after long-time cultivation under WL the rate of newly formed sprouts was reduced regardless of the CK concentration but increased when RL was applied to the crops [2]. Thus, in some cases, an early phase of RL irradiation of at least 2 weeks [122], followed by growth under a WL, may be advisable. The use of an initial stimulatory effect of RL or RL enriched followed by the WL may also improve proliferation and somatogenesis [126] in species that are particularly difficult to regenerate in vitro and/or with an high sensibility to higher concentration of CKs in the medium, such as *Euphorbia milii* and *Ceratonia siliqua* L. (Cavallaro et al., unpublished data). Moreover, the exposition to a period of RL:FRL followed by the WL may enable a reduction in exogenous growth regulator concentrations, mainly CKs added to the medium [124], which may be unnaturally high in vitro. This reduction may be favorable for enhancing the following phases of the in vitro process (rooting and acclimation). Finally, currently, lamps with a more optimal spectral composition of WL enriched in the most useful wavelengths (BL, RL and GL) are already available on the market [185,274] for vertical farming systems and could be interesting for in vitro production after appropriate investigation.

Author Contributions: Conceptualization, V.C. and R.M. contributed to the conception and the design of the review; V.C. and A.P., contributed to the drafting of tables; writing—original draft Preparation, V.C. and R.M. contributed to define the original draft; writing—review & editing, V.C., I.F., A.P. and R.M. wrote and edited the final version of the manuscript. All authors have read and agreed to the published version of the manuscript.

Funding: This research received no external funding.

Conflicts of Interest: The authors declare no conflict of interest.

References

1. Smith, H. Sensing the light environment: The functions of the phytochrome family. In *Photomorphogenesis in Plants*; Kendrick, R.E., Kronenberg, G.H.M., Eds.; Springer: Dordrecht, The Netherlands, 1994; pp. 377–416.
2. Morini, S.; Muleo, R. Il ruolo della qualità della luce nei processi di sviluppo e differenziazione delle colture in vitro. *Italus Hortus Rev.* **2012**, *19*, 37–49.
3. Kozai, T. Why LED lighting for urban agriculture? In *LED Lighting for Urban Agriculture*; Kozai, T., Fujiwara, K., Runkle, E., Eds.; Springer: Singapore, 2016; pp. 3–18. [CrossRef]
4. Whitelam, G.C.; Halliday, K.J. Photomorphogenesis: Phytochrome takes a partner! *Curr. Biol.* **1999**, *9*, 225–227. [CrossRef]
5. Muneer, S.; Kim, E.J.; Park, J.S. Influence of green, red and blue light emitting diodes on multiprotein complex proteins and photosynthetic activity under different light intensities in lettuce leaves (*Lactuca sativa* L.). *Int. J. Mol. Sci.* **2014**, *15*, 4657–4670. [CrossRef] [PubMed]
6. Higuchi, Y.; Hisamatsu, T. Light acts as a signal for regulation of growth and development. In *LED Lighting for Urban Agriculture*; Kozai, T., Fujiwara, K., Runkle, E., Eds.; Springer: Singapore, 2016; pp. 57–73.
7. Pedmale, U.V.; Huang, S.C.; Zander, M.; Cole, B.J.; Hetzel, J.; Ljung, K.; Chory, J. Cryptochromes interact directly with PIFs to control plant growth in limiting blue light. *Cell* **2016**, *164*, 233–245. [CrossRef]
8. Li, C.-X.; Xu, Z.-G.; Dong, R.-Q.; Chang, S.-X.; Wang, L.-Z.; Khalil-Ur-Rehman, M.; Tao, J.-M. An RNA-Seq analysis of grape plantlets grown in vitro reveals different responses to blue, green, red led light, and white fluorescent light. *Front. Plant Sci.* **2017**, *8*, 78. [CrossRef]
9. Batista, D.S.; Felipe, S.H.S.; Silva, T.D.; de Castro, K.M.; Mamedes-Rodrigues, T.C.; Miranda, N.A.; Ríos-Ríos, A.M.; Faria, D.V.; Fortini, E.A.; Chagas, K.; et al. Light quality in plant tissue culture: Does it matter? *Vitr. Cell. Dev. Biol. Plant* **2018**, *54*, 195–215. [CrossRef]
10. Chen, M.; Chory, J.; Fankhauser, C. Light signal transduction in higher plants. *Annu. Rev. Genet.* **2004**, *38*, 87–117. [CrossRef]
11. Banerjee, R.; Batschauer, A. Plant blue-light receptors. *Planta* **2005**, *220*, 498–502. [CrossRef]

12. Chen, M.; Chory, J. Phytochrome signaling mechanisms and the control of plant development. *Trends Cell Biol.* **2011**, *21*, 664–671. [CrossRef]
13. Vierstra, R.D.; Zhang, J. Phytochrome signaling: Solving the Gordian knot with microbial relatives. *Trends Plant Sci.* **2011**, *16*, 417–426. [CrossRef]
14. Galvão, V.C.; Fankhauser, C. Sensing the light environment in plants: Photoreceptors and early signaling steps. *Curr. Opin. Neurobiol.* **2015**, *34*, 46–53. [CrossRef] [PubMed]
15. Chaves, I.; Pokorny, R.; Byrdin, M.; Hoang, N.; Ritz, T.; Brettel, K.; Essen, L.-O.; van der Horst, G.T.J.; Batschauer, A.; Ahmad, M. The cryptochromes: Blue light photoreceptors in plants and animals. *Annu. Rev. Plant Biol.* **2011**, *62*, 335–364. [CrossRef] [PubMed]
16. Liu, H.; Liu, B.; Zhao, C.; Pepper, M.; Lin, C. The action mechanisms of plant cryptochromes. *Trends Plant Sci.* **2011**, *16*, 684–691. [CrossRef]
17. Christie, J.M. Phototropin blue-light receptors. *Annu. Rev. Plant Biol.* **2007**, *58*, 21–45. [CrossRef]
18. Suetsugu, N.; Wada, M. Evolution of three LOV blue light receptor families in green plants and photosynthetic stramenopiles: Phototropin, ZTL/FKF1/LKP2 and aureochrome. *Plant Cell Physiol.* **2013**, *54*, 8–23. [CrossRef] [PubMed]
19. Lin, C. Plant blue-light receptors. *Trends Plant Sci.* **2000**, *5*, 337–342. [CrossRef]
20. Rizzini, L.; Favory, J.-J.; Cloix, C.; Faggionato, D.; O'Hara, A.; Kaiserli, E.; Baumeister, R.; Schäfer, E.; Nagy, F.; Jenkins, G.I.; et al. Perception of UV-B by the *Arabidopsis* UVR8 protein. *Science* **2011**, *332*, 103–106. [CrossRef]
21. Jenkins, G.I. The UV-B Photoreceptor UVR8: From structure to physiology. *Plant Cell* **2014**, *26*, 21–37. [CrossRef]
22. Tilbrook, K.; Arongaus, A.B.; Binkert, M.; Heijde, M.; Yin, R.; Ulm, R. The UVR8 UV-B Photoreceptor: Perception, signaling and response. *Arab. Book* **2013**, *11*, e0164. [CrossRef]
23. Aphalo, P.J.; Ballare, C.L. On the importance of information-acquiring systems in plant-plant interactions. *Funct. Ecol.* **1995**, *9*, 5–14. [CrossRef]
24. Gilbert, I.R.; Jarvis, P.G.; Smith, H. Proximity signal and shade avoidance differences between early and late successional trees. *Nature* **2001**, *411*, 792–795. [CrossRef] [PubMed]
25. Ballaré, C.L. Keeping up with the neighbours: Phytochrome sensing and other signalling mechanisms. *Trends Plant Sci.* **1999**, *4*, 97–102. [CrossRef]
26. Cao, S.; Luo, X.; Xu, D.; Tian, X.; Song, J.; Xia, X.; Chu, C.; He, Z. Genetic architecture underlying light and temperature mediated flowering in Arabidopsis, rice, and temperate cereals. *New Phytol.* **2021**, *230*, 1731–1745. [CrossRef]
27. Zhao, X.; Wang, Y.-L.; Qiao, X.-R.; Wang, J.; Wang, L.-D.; Xu, C.-S.; Zhang, X. Phototropins function in high-intensity blue light-induced hypocotyl phototropism in Arabidopsis by altering cytosolic calcium. *Plant Physiol.* **2013**, *162*, 1539–1551. [CrossRef] [PubMed]
28. Kami, C.; Allenbach, L.; Zourelidou, M.; Ljung, K.; Schütz, F.; Isono, E.; Watahiki, M.K.; Yamamoto, K.T.; Schwechheimer, C.; Fankhauser, C. Reduced phototropism in *Pks* mutants may be due to altered auxin-regulated gene expression or reduced lateral auxin transport. *Plant J.* **2014**, *77*, 393–403. [CrossRef]
29. Christie, J.M.; Yang, H.; Richter, G.L.; Sullivan, S.; Thomson, C.E.; Lin, J.; Titapiwatanakun, B.; Ennis, M.; Kaiserli, E.; Lee, O.R.; et al. Phot1 inhibition of ABCB19 primes lateral auxin fluxes in the shoot apex required for phototropism. *PLoS Biol.* **2011**, *9*, e1001076. [CrossRef]
30. Sgamma, T.; Pape, J.; Massiah, A.; Jackson, S. Selection of reference genes for diurnal and developmental time-course real-time PCR expression analyses in lettuce. *Plant Methods* **2016**, *12*, 21. [CrossRef]
31. Cirilli, M.; Delfino, I.; Caboni, E.; Muleo, R. EpiHRMAssay, in tube and in silico combined approach for the scanning and epityping of heterogeneous DNA methylation. *Biol. Methods Protoc.* **2017**, *2*, bpw008. [CrossRef]
32. Hoffmann, A.A.; Sgrò, C.M. Climate change and evolutionary adaptation. *Nature* **2011**, *470*, 479–485. [CrossRef]
33. Cavallaro, V.; Tringali, S.; Patanè, C. Large-scale in vitro propagation of giant reed (*Arundo donax* L.), a promising biomass species. *J. Hortic. Sci. Biotechnol.* **2011**, *86*, 452–456. [CrossRef]
34. Shukla, M.R.; Singh, A.S.; Piunno, K.; Saxena, P.K.; Jones, A.M.P. Application of 3D printing to prototype and develop novel plant tissue culture systems. *Plant Methods* **2017**, *13*, 6. [CrossRef] [PubMed]
35. Miler, N.; Kulus, D.; Woźny, A.; Rymarz, D.; Hajzer, M.; Wierzbowski, K.; Nelke, R.; Szeffs, L. Application of wide-spectrum light-emitting diodes in micropropagation of popular ornamental plant species: A study on plant quality and cost reduction. *Vitr. Cell. Dev. Biol. Plant* **2019**, *55*, 99–108. [CrossRef]
36. Morini, S.; Muleo, R.; Jain, S.; Ishii, K. Micropropagation of woody trees and fruits. In *Effect of Light Quality on Micropropagation of Woody Species*; Jain, S.M., Ishii, K., Eds.; Springer: Dordrecht, The Netherlands, 2003; pp. 3–35. [CrossRef]
37. Bula, R.J.; Morrow, R.C.; Tibbitts, T.W.; Barta, D.J.; Ignatius, R.W.; Martin, T.S. Light-emitting diodes as a radiation source for plants. *HortScience* **1991**, *26*, 203–205. [CrossRef] [PubMed]
38. Seabrook, J.E.A. Light effects on the growth and morphogenesis of potato (*Solanum tuberosum*) in vitro: A review. *Am. J. Potato Res.* **2005**, *82*, 353–367. [CrossRef]
39. Bantis, F.; Ouzounis, T.; Radoglou, K. Artificial LED lighting enhances growth characteristics and total phenolic content of *Ocimum basilicum*, but variably affects transplant success. *Sci. Hortic.* **2016**, *198*, 277–283. [CrossRef]
40. Yeh, N.; Chung, J.-P. High-brightness LEDs energy efficient lighting sources and their potential in indoor plant cultivation. *Renew. Sustain. Energy Rev.* **2009**, *13*, 2175–2180. [CrossRef]

41. Dutta Gupta, S.; Jatothu, B. Fundamentals and applications of light-emitting diodes (LEDs) in vitro plant growth and morphogenesis. *Plant Biotechnol. Rep.* **2013**, *7*, 211–220. [CrossRef]
42. Kim, S.-J.; Hahn, E.-J.; Heo, J.-W.; Paek, K.-Y. Effects of LEDs on net photosynthetic rate, growth and leaf stomata of *chrysanthemum* plantlets in vitro. *Sci. Hortic.* **2004**, *101*, 143–151. [CrossRef]
43. Kasahara, M.; Kagawa, T.; Sato, Y.; Kiyosue, T.; Wada, M. Phototropins mediate blue and red light-induced chloroplast movements in *Physcomitrella patens*. *Plant Physiol.* **2004**, *135*, 1388–1397. [CrossRef]
44. Dutta Gupta, S.; Agarwal, A. Influence of LED Lighting on in vitro plant regeneration and associated cellular redox balance. In *Light Emitting Diodes for Agriculture*; Dutta Gupta, S., Ed.; Springer: Singapore, 2017; pp. 273–303. ISBN 978-981-10-5806-6.
45. Darko, E.; Heydarizadeh, P.; Schoefs, B.; Sabzalian, M.R. Photosynthesis under artificial light: The shift in primary and secondary metabolism. *Philos. Trans. R. Soc. B-Biol. Sci.* **2014**, *369*, 1640. [CrossRef]
46. Gupta, S.D.; Agarwal, A. *Artificial Lighting System for Plant Growth and Development: Chronological Advancement, Working Principles, and Comparative Assessment*; Gupta, S.D., Ed.; Springer: Singapore, 2017; pp. 1–26. ISBN 978-981-10-5806-6.
47. Martineau, V.; Lefsrud, M.; Naznin, M.T.; Kopsell, D.A. Comparison of light-emitting diode and high-pressure sodium light treatments for hydroponics growth of Boston lettuce. *HortScience* **2012**, *47*, 477–482. [CrossRef]
48. Silva, S.T.; Bertolucci, S.K.V.; da Cunha, S.H.B.; Lazzarini, L.E.S.; Tavares, M.C.; Pinto, J.E.B.P. Effect of light and natural ventilation systems on the growth parameters and carvacrol content in the in vitro cultures of *Plectranthus amboinicus* (Lour.) Spreng. *Plant Cell Tissue Organ Cult.* **2017**, *129*, 501–510. [CrossRef]
49. Budiarto, K. Spectral quality affects morphogenesis on *Anthurium* plantlet during in vitro culture. *AJAS* **2010**, *32*, 234–240.
50. Li, H.; Xu, Z.; Tang, C. Effect of light-emitting diodes on growth and morphogenesis of upland cotton (*Gossypium hirsutum* L.) Plantlets in vitro. *Plant Cell Tissue Organ Cult.* **2010**, *103*, 155–163. [CrossRef]
51. Li, H.; Tang, C.; Xu, Z. The effects of different light qualities on rapeseed (*Brassica napus* L.) plantlet growth and morphogenesis in vitro. *Sci. Hortic.* **2013**, *150*, 117–124. [CrossRef]
52. Do Nascimento Vieira, L.; de Freitas Fraga, H.P.; dos Anjos, K.G.; Puttkammer, C.C.; Scherer, R.F.; da Silva, D.A.; Guerra, M.P. Light-emitting diodes (LED) increase the stomata formation and chlorophyll content in *Musa acuminata* (AAA)'Nanicão Corupá'in vitro plantlets. *Theor. Exp. Plant Physiol.* **2015**, *27*, 91–98. [CrossRef]
53. Merkle, S.A.; Montello, P.M.; Xia, X.; Upchurch, B.L.; Smith, D.R. Light quality treatments enhance somatic seedling production in three southern pine species. *Tree Physiol.* **2006**, *26*, 187–194. [CrossRef] [PubMed]
54. Park, S.Y.; Yeung, E.C.; Paek, K.Y. Endoreduplication in Phalaenopsis is affected by light quality from light-emitting diodes during somatic embryogenesis. *Plant Biotechnol. Rep.* **2010**, *4*, 303–309. [CrossRef]
55. Kim, Y.W.; Moon, H.K. Enhancement of somatic embryogenesis and plant regeneration in Japanese red pine (*Pinus densiflora*). *Plant Biotechnol. Rep.* **2014**, *8*, 259–266. [CrossRef]
56. Botero Giraldo, C.; Urrea Trujillo, A.I.; Naranjo Gomez, E.J. Regeneration potential of *Psychotria ipecacuanha* (Rubiaceae) from thin cell layers. *Acta Biol. Colomb.* **2015**, *20*, 181–192. [CrossRef]
57. Chen, C.-C.; Agrawal, D.C.; Lee, M.-R.; Lee, R.-J.; Kuo, C.-L.; Wu, C.-R.; Tsay, H.-S.; Chang, H.-C. Influence of LED light spectra on in vitro somatic embryogenesis and LC–MS analysis of chlorogenic acid and rutin in *Peucedanum japonicum* Thunb: A medicinal herb. *Bot. Stud.* **2016**, *57*, 9. [CrossRef] [PubMed]
58. Mai, N.T.; Binh, P.T.; Khoi, P.H.; Hung, N.K.; Ngoc, P.B.; Ha, C.H.; Binh, H.T.T. Effects of light emitting diodes-LED on regeneration ability of *Coffea canephora* mediated via somatic embryogenesis. *Acad. J. Biol.* **2016**, *38*, 228–235. [CrossRef]
59. Lotfi, M.; Mars, M.; Werbrouck, S. Optimizing pear micropropagation and rooting with light emitting diodes and trans-cinnamic Acid. *Plant Growth Regul.* **2019**, *88*, 173–180. [CrossRef]
60. Ding, Z.; Galván-Ampudia, C.S.; Demarsy, E.; Łangowski, Ł.; Kleine-Vehn, J.; Fan, Y.; Morita, M.T.; Tasaka, M.; Fankhauser, C.; Offringa, R.; et al. Light-mediated polarization of the PIN3 auxin transporter for the phototropic response in Arabidopsis. *Nat. Cell Biol.* **2011**, *13*, 447–452. [CrossRef] [PubMed]
61. Dewir, Y.; Chakrabarty, D.; Hahn, E.; Paek, K. The effects of paclobutrazol, light emitting diodes (LEDs) and sucrose on flowering of Euphorbia millii plantlets in vitro. *Eur. J. Hortic. Sci.* **2006**, *71*, 240.
62. Moon, H.K.; Park, S.-Y.; Kim, Y.W.; Kim, C.S. Growth of Tsuru-rindo (*Tripterospermum japonicum*) cultured in vitro under various sources of light-emitting diode (LED) irradiation. *J. Plant Biol.* **2006**, *49*, 174–179. [CrossRef]
63. Shin, K.S.; Murthy, H.N.; Heo, J.W.; Hahn, E.J.; Paek, K.Y. The effect of light quality on the growth and development of in vitro cultured Doritaenopsis plants. *Acta Physiol. Plant.* **2008**, *30*, 339–343. [CrossRef]
64. Ma, X.; Wang, Y.; Liu, M.; Xu, J.; Xu, Z. Effects of green and red lights on the growth and morphogenesis of potato (*Solanum tuberosum* L.) plantlets in vitro. *Sci. Hortic.* **2015**, *190*, 104–109. [CrossRef]
65. Manivannan, A.; Soundararajan, P.; Halimah, N.; Ko, C.H.; Jeong, B.R. Blue LED light enhances growth, phytochemical contents, and antioxidant enzyme activities of Rehmannia glutinosa cultured in vitro. *Hortic. Environ. Biotechnol.* **2015**, *56*, 105–113. [CrossRef]
66. Batista, D.S.; de Castro, K.M.; da Silva, A.R.; Teixeira, M.L.; Sales, T.A.; Soares, L.I.; das Graças Cardoso, M.; de Oliveira Santos, M.; Viccini, L.F.; Otoni, W.C. Light quality affects in vitro growth and essential oil profile in Lippia alba (Verbenaceae). *Vitr. Cell. Dev. Biol. Plant* **2016**, *52*, 276–282. [CrossRef]
67. Seibert, M.; Wetherbee, P.J.; Job, D.D. The effects of light intensity and spectral quality on growth and shoot initiation in tobacco callus. *Plant Physiol.* **1975**, *56*, 130–139. [CrossRef] [PubMed]

68. Chée, R. In vitro culture of Vitis: The effects of light spectrum, manganese sulfate and potassium iodide on morphogenesis. *Plant Cell Tissue Organ Cult.* **1986**, *7*, 121–134. [CrossRef]
69. Lercari, B.; Tognoni, F.; Anselmo, G.; Chapel, D. Photocontrol of in vitro bud differentiation in *Saintpaulia ionantha* leaves and lycopersicon esculentum cotyledons. *Physiol. Plant.* **1986**, *67*, 340–344. [CrossRef]
70. Chee, R.; Pool, R. Morphogenic responses to propagule trimming, spectral irradiance, and photoperiod of grapevine shoots recultured in vitro. *J. Am. Soc. Hortic. Sci.* **1989**, *114*, 350–354.
71. Aksenova, N.P.; Konstantinova, T.N.; Sergeeva, L.I.; Macháčková, I.; Golyanovskaya, S.A. Morphogenesis of potato plants in vitro. I. Effect of light quality and hormones. *Plant Growth Regul.* **1994**, *13*, 143–146. [CrossRef]
72. Panizza, M.; Tognoni, F.; Lercari, B. Axillary bud proliferation and ethylene production as controlled by radiation of different spectral composition and exogenous phytohormones. *Biol. Plant.* **1994**, *36*, 553–563. [CrossRef]
73. Castillo, B.; Smith, M.A.L. Direct somatic embryogenesis from *Begonia gracilis* explants. *Plant Cell Rep.* **1997**, *16*, 385–388. [CrossRef]
74. Da Silva, M.; Debergh, P. The effect of light quality on the morphogenesis of in vitro cultures of *Azorina vidalii* (Wats.) Feer. *Plant Cell Tissue Organ Cult.* **1997**, *51*, 187–193. [CrossRef]
75. Marks, T.; Simpson, S. Effect of irradiance on shoot development in vitro. *Plant Growth Regul.* **1999**, *28*, 133–142. [CrossRef]
76. Lercari, B.; Moscatelli, S.; Ghirardi, E.; Niceforo, R.; Bertram, L. Photomorphogenic control of shoot regeneration from etiolated and light-grown hypocotyls of tomato. *Plant Sci.* **1999**, *140*, 53–62. [CrossRef]
77. Michalczuk, B.; Michalczuk, L. The effect of light quality on regeneration rate and plantlet development in transgenic petunia 'Revolution'(Surfinia type). *Acta Hortic.* **2000**, *530*, 397–402. [CrossRef]
78. Lian, M.-L.; Murthy, H.; Paek, K.-Y. Effects of light emitting diodes (LEDs) on the in vitro induction and growth of bulblets of *Lilium oriental* hybrid 'Pesaro'. *Sci. Hortic.* **2002**, *94*, 365–370. [CrossRef]
79. Burritt, D.J.; Leung, D.W. Adventitious shoot regeneration from Begonia x erythrophylla petiole sections is developmentally sensitive to light quality. *Physiol. Plant.* **2003**, *118*, 289–296. [CrossRef]
80. Huan, L.V.T.; Tanaka, M. Effects of red and blue light-emitting diodes on callus induction, callus proliferation, and protocorm-like body formation from callus in *Cymbidium orchid. Environ. Control Biol.* **2004**, *42*, 57–64. [CrossRef]
81. Hunter, D.C.; Burritt, D.J. Light quality influences adventitious shoot production from cotyledon explants of Lettuce (*Lactuca sativa* L.). *In Vitro Cell. Dev. Biol. Plant* **2004**, *40*, 215–220. [CrossRef]
82. Yonghua, Q.; Shanglong, Z.; Asghar, S.; Lingxiao, Z.; Qiaoping, Q.; Kunsong, C.; Changjie, X. Regeneration mechanism of Toyonoka strawberry under different color plastic Films. *Plant Sci.* **2005**, *168*, 1425–1431. [CrossRef]
83. Dewir, Y.H.; Chakrabarty, D.; Kim, S.J.; Hahn, E.J.; Paek, K.Y. Effect of light-emitting diode on growth and shoot proliferation of *Euphorbia millii* and *Spathiphyllum cannifolium. Hortic. Environ. Biotechnol.* **2005**, *46*, 375–379.
84. Moshe, R.; Dalia, E. On the effect of light on shoot regeneration in petunia. *Plant Cell Tissue Organ Cult.* **2007**, *89*, 49–54.
85. Poudel, P.R.; Kataoka, I.; Mochioka, R. Effect of red and blue-light-emitting diodes on growth and morphogenesis of grapes. *Plant Cell Tissue Organ Cult.* **2008**, *92*, 147–153. [CrossRef]
86. Wongnok, A.; Piluek, C.; Techasilpitak, T.; Tantivivat, S. Effects of light emitting diodes on micropropagation of *Phalaenopsis orchids. Acta Hortic.* **2007**, *788*, 149–156. [CrossRef]
87. Xu, Z.; Cui, J. Effects of different spectral energy distribution on tissue culture of *Oncidium* in vitro. *J. Beijing For. Univ.* **2009**, *31*, 45–50.
88. Hamada, K.; Shimasaki, K.; Nishimura, Y.; Egawa, H.; Yoshida, K. Effect of red fluorescent films on the proliferation of *Cymbidium finlaysonianum* Lindl. PLB cultured in vitro. *Hortic. Environ. Biotechnol.* **2009**, *50*, 319–323.
89. Chung, J.-P.; Huang, C.-Y.; Dai, T.-E. Spectral effects on embryogenesis and plantlet growth of Oncidium 'Gower Ramsey'. *Sci. Hortic.* **2010**, *124*, 511–516. [CrossRef]
90. Mengxi, L.; Zhigang, X.; Yang, Y.; Yijie, F. Effects of different spectral lights on Oncidium PLBs induction, proliferation, and plant regeneration. *Plant Cell Tissue Organ Cult.* **2011**, *106*, 1–10. [CrossRef]
91. Hamada, K.; Shimasaki, K.; Nishimura, Y.; Sasaoka, H.; Nishimura, K. Effects of red, blue and yellow fluorescent films on proliferation and organogenesis in *Cymbidium* and *Phalaenopsis* PLB in vitro. *Acta Hortic.* **2011**, *907*, 385–388. [CrossRef]
92. Lin, Y.; Li, J.; Li, B.; He, T.; Chun, Z. Effects of light quality on growth and development of protocorm-like bodies of dendrobium officinale in vitro. *Plant Cell Tissue Organ Cult.* **2011**, *105*, 329–335. [CrossRef]
93. Nahar, S.; Shimasaki, K.; Haque, S. Effect of different light and two polysaccharides on the proliferation of protocorm-like bodies of Cymbidium cultured in vitro. *Acta Hortic.* **2012**, *956*, 307–313. [CrossRef]
94. Werbrouck, S.; Buyle, H.; Geelen, D.; Van Labeke, M.C. Effect of red-, far-red- and blue-light-emitting diodes on in vitro growth of *Ficus benjamina. Acta Hortic.* **2012**, *961*, 533–538. [CrossRef]
95. Kaewjampa, N.; Shimasaki, K. Effects of green LED lighting on organogenesis and superoxide dismutase (SOD) activities in protocorm-like bodies (PLBs) of *Cymbidium cultured* in vitro. *Environ. Control Biol.* **2012**, *50*, 247–254. [CrossRef]
96. Siegień, I.; Adamczuk, A.; Wróblewska, K. Light affects in vitro organogenesis of *Linum usitatissimum* L. and its cyanogenic potential. *Acta Physiol. Plant.* **2013**, *35*, 781–789. [CrossRef]
97. Edesi, J.; Kotkas, K.; Pirttilä, A.M.; Häggman, H. Does light spectral quality affect survival and regeneration of potato (*Solanum tuberosum* L.) shoot tips after cryopreservation? *Plant Cell Tissue Organ Cult.* **2014**, *119*, 599–607. [CrossRef]

98. Lee, N.N.; Choi, Y.E.; Moon, H.K. Effect of LEDs on shoot multiplication and rooting of rare plant *Abeliophyllum distichum* Nakai. *J. Plant Biochem. Biotechnol.* **2014**, *41*, 94–99. [CrossRef]
99. Habiba, S.U.; Shimasaki, K.; Ahasan, M.M.; Alam, M.M. Effects of different light quality on growth and development of protocorm-like bodies (PLBs) in Dendrobium kingianum cultured in vitro. *Bangladesh Res. Public. J.* **2014**, *10*, 223–227.
100. Kamal, M.M.; Shimasaki, K.; Akter, N. Effect of light emutting diode (LED) lamps and N-Acetylglucosamine (NAG) on organogenesis in protocorm-like bodies (PLBs) of a *Cymbidium hybrid* cultured in vitro. *Plant Tissue Cult. Biotechnol.* **2014**, *24*, 273–277. [CrossRef]
101. Silva, M.M.A.; de Oliveira, A.L.B.; Oliveira-Filho, R.A.; Gouveia-Neto, A.S.; Camara, T.J.R.; Willadino, L.G. Effect of blue/red LED light combination on growth and morphogenesis of *Saccharum officinarum* plantlets in vitro. In Proceedings of the SPIE—The International Society for Optical Engineering, San Francisco, CA, USA, 4 March 2014; Volume 8947. [CrossRef]
102. Jeong, B.R.; Sivanesan, I. Adventitious shoot regeneration, in vitro flowering, fruiting, secondary metabolite content and antioxidant activity of *Scrophularia takesimensis* Nakai. *Plant Cell Tissue Organ Cult.* **2015**, *123*, 607–618. [CrossRef]
103. Dutta Gupta, S.; Sahoo, T. Light emitting diode (LED)-induced alteration of oxidative events during in vitro shoot organogenesis of *Curculigo orchioides* Gaertn. *Acta Physiol. Plant.* **2015**, *37*, 233. [CrossRef]
104. Hung, C.D.; Hong, C.-H.; Jung, H.-B.; Kim, S.-K.; Van Ket, N.; Nam, M.-W.; Choi, D.-H.; Lee, H.-I. Growth and morphogenesis of encapsulated strawberry shoot tips under mixed LEDs. *Sci. Hortic.* **2015**, *194*, 194–200. [CrossRef]
105. Nhut, D.T.; Huy, N.P.; Tai, N.T.; Nam, N.B.; Luan, V.Q.; Hien, V.T.; Tung, H.T.; Vinh, B.T.; Luan, T.C. Light-emitting diodes and their potential in callus growth, plantlet development and saponin accumulation during somatic embryogenesis of *Panax vietnamensis* Ha *et* Grushv. *Biotechnol. Biotechnol. Equip.* **2015**, *29*, 299–308. [CrossRef]
106. Bello-Bello, J.J.; Martínez-Estrada, E.; Caamal-Velázquez, J.H.; Morales-Ramos, V. Effect of LED light quality on in vitro shoot proliferation and growth of Vanilla (*Vanilla planifolia* Andrews). *Afr. J. Biotechnol.* **2016**, *15*, 272–277.
107. Gök, K.M.; Şan, B.; Bayhan, A.K. Micropropagation of Gerbera (*Gerbera jamesonii bolus*) under different color of light-emitting diodes. *Süleyman Demirel Üniversitesi Fen Bilimleri Enstitüsü Dergis* **2016**, *20*, 468. [CrossRef]
108. Nahar, S.J.; Haque, S.M.; Kazuhiko, S. Application of chondroitin sulfate on organogenesis of two *Cymbidium* spp. under different sources of lights. *Not. Sci. Biol.* **2016**, *8*, 156–160. [CrossRef]
109. Karataş, M.; Aasim, M.; Dazkirli, M. Influence of light-emitting diodes and benzylaminopurin on adventitious shoot regeneration of water hyssop (*Bacopa monnieri* (L.) Pennell) in vitro. *Arch. Biol. Sci.* **2016**, *68*, 501–508. [CrossRef]
110. Hung, C.D.; Hong, C.-H.; Kim, S.-K.; Lee, K.-H.; Park, J.-Y.; Dung, C.D.; Nam, M.-W.; Choi, D.-H.; Lee, H.-I. In vitro proliferation and ex vitro rooting of microshoots of commercially important rabbiteye blueberry (*Vaccinium ashei reade*) Using spectral lights. *Sci. Hortic.* **2016**, *211*, 248–254. [CrossRef]
111. Martínez-Estrada, E.; Caamal-Velázquez, J.H.; Morales-Ramos, V.; Bello-Bello, J. Light emitting diodes improve in vitro shoot multiplication and growth of *Anthurium andreanum* Lind. *Propag. Ornam. Plants* **2016**, *16*, 3–8.
112. Silva, M.M.; de Oliveira, A.L.B.; Oliveira-Filho, R.A.; Camara, T.; Willadino, L.; Gouveia-Neto, A. The effect of spectral light quality on in vitro culture of sugarcane. *Acta Sci. Biol. Sci.* **2016**, *38*, 157–161. [CrossRef]
113. Szewczyk-Taranek, B.; Pawłowska, B.; Prokopiuk, B.; Żupnik, M. Effectiveness of LED and fluorescent light on in vitro shoot proliferation of *Staphylea pinnata*. *Acta Hortic.* **2015**, *1155*, 375–380.
114. Ramírez-Mosqueda, M.A.; Iglesias-Andreu, L.G.; Bautista-Aguilar, J.R. The effect of light quality on growth and development of in vitro plantlet of *Stevia rebaudiana* Bertoni. *Sugar Tech.* **2017**, *19*, 331–336. [CrossRef]
115. Ramírez-Mosqueda, M.; Iglesias-Andreu, L.; Luna-Sánchez, I. Light quality affects growth and development of in vitro plantlet of *Vanilla planifolia* Jacks. *S. Afr. J. Bot.* **2017**, *109*, 288–293. [CrossRef]
116. Billore, V.; Jain, M.; Suprasanna, P. Monochromic radiation through light-emitting diode (LED) positively augments in vitro shoot regeneration in Orchid (*Dendrobium sonia*). *Can. J. Biotech.* **2017**, *1*, 50–58. [CrossRef]
117. Ferreira, L.T.; de Araújo Silva, M.M.; Ulisses, C.; Camara, T.R.; Willadino, L. Using LED lighting in somatic embryogenesis and micropropagation of an elite sugarcane variety and its effect on redox metabolism during acclimatization. *Plant Cell Tissue Organ Cult.* **2017**, *128*, 211–221. [CrossRef]
118. Pawłowska, B.; Żupnik, M.; Szewczyk-Taranek, B.; Cioć, M. Impact of LED light sources on morphogenesis and levels of photosynthetic pigments in *Gerbera jamesonii* grown in vitro. *Hortic. Environ. Biotechnol.* **2018**, *59*, 115–123. [CrossRef]
119. Lazzarini, L.E.S.; Bertolucci, S.K.V.; Pacheco, F.V.; dos Santos, J.; Silva, S.T.; de Carvalho, A.A.; Pinto, J.E.B.P. Quality and intensity of light affect *Lippia gracilis* Schauer plant growth and volatile compounds in vitro. *Plant Cell Tissue Organ Cult.* **2018**, *135*, 367–379. [CrossRef]
120. Cioć, M.; Szewczyk, A.; Żupnik, M.; Kalisz, A.; Pawłowska, B. LED lighting affects plant growth, morphogenesis and phytochemical contents of *Myrtus communis* L. in vitro. *Plant Cell Tissue Organ Cult.* **2018**, *132*, 433–447. [CrossRef]
121. Yu, L.; Song, C.; Sun, L.; Li, L.; Xu, Z.; Tang, C. Effects of light-emitting diodes on tissue culture plantlets and seedlings of Rice (*Oryza sativa* L.). *J. Integr. Agric.* **2020**, *19*, 1743–1754. [CrossRef]
122. Kadkade, P.; Jopson, H. Influence of light quality on organogenesis from the embryo-derived callus of Douglas fir (*Pseudotsuga menziesii*). *Plant Sci. Lett.* **1978**, *13*, 67–73. [CrossRef]
123. Norton, C.R.; Herrington, T.; Phillips, D.; Norton, M. Light quality and light pipe in the micropropagation of woody ornamental plants grown in vitro. *Acta Hort.* **1988**, *227*, 413–416. [CrossRef]

124. Herrington, E.; McPherson, J.C. Light quality growth promotion of *Spiraea nipponica*: The influence of a low photon fluence rate and transfer time to a higher fluence rate. *Plant Cell Tissue Organ Cult.* **1993**, *32*, 161–167. [CrossRef]
125. Muleo, R.; Thomas, B. Effects of light quality on shoot proliferation of *Prunus cerasifera* in vitro are the result of differential effects on bud induction and apical dominance. *J. Hortic. Sci.* **1997**, *72*, 483–499. [CrossRef]
126. D'Onofrio, C.; Morini, S.; Bellocchi, G. Effect of light quality on somatic embryogenesis of quince leaves. *Plant Cell Tissue Organ Cult.* **1998**, *53*, 91–98. [CrossRef]
127. Piagnani, C.; Iacona, C.; Intrieri, M.C.; Muleo, R. A New somaclone of *Prunus avium* shows diverse growth pattern under different spectral quality of radiation. *Biologia Plant.* **2002**, *45*, 11–17. [CrossRef]
128. Muleo, R.; Morini, S. Light quality regulates shoot cluster growth and development of MM106 apple genotype in vitro culture. *Sci. Hortic.* **2006**, *108*, 364–370. [CrossRef]
129. Wang, H.; Liu, H.; Wang, W.-J.; Zu, Y. Effects of thidiazuron, basal medium and light quality on adventitious shoot regeneration from in vitro cultured stem of *Populus alba* × *P. berolinensis*. *J. For. Res.* **2008**, *19*, 257–259. [CrossRef]
130. Wilken, D.; Jiménez Gonzalez, E.; Gerth, A.; Gómez-Kosky, R.; Schumann, A.; Claus, D. Effect of immersion systems, lighting, and TIS designs on biomass increase in micropropagating banana (Musa Spp. Cv.'Grande Naine'AAA). *Vitr. Cell. Dev. Biol. Plant.* **2014**, *50*, 582–589. [CrossRef]
131. Kwon, A.; Cui, H.-Y.; Lee, H.; Shin, H.; Kang, K.-S.; Park, S.-Y. Light quality affects shoot regeneration, cell division, and wood formation in elite clones of *Populus euramericana*. *Acta Physiol. Plant.* **2015**, *37*, 65. [CrossRef]
132. Geng, F.; Moran, R.; Day, M.; Halteman, W.; Zhang, D. In vitro shoot proliferation of apple rootstocks 'B. 9', 'G. 30', and 'G. 41'grown under red and blue light. *HortScience* **2015**, *50*, 430–433. [CrossRef]
133. Al-Mayahi, A.M.W. Effect of red and blue light emitting diodes "CRB-LED" on in vitro organogenesis of date palm (*Phoenix dactylifera* L.) cv. Alshakr. *World J. Microbiol. Biotechnol.* **2016**, *32*, 160. [CrossRef] [PubMed]
134. He, C.; Zeng, Y.; Fu, Y.; Wu, J.; Liang, Q. Light quality affects the proliferation of in vitro cultured plantlets of *Camellia oleifera* Huajin. *PeerJ* **2020**, *8*, e10016. [CrossRef] [PubMed]
135. Muleo, R.; Morini, S. Physiological dissection of blue and red light regulation of apical dominance and branching in M9 apple rootstock growing in vitro. *J. Plant Physiol.* **2008**, *165*, 1838–1846. [CrossRef] [PubMed]
136. Gabryszewska, E.; Rudnicki, R. The influence of light quality on the shoot proliferation and rooting of *Gerbera jamesonii* in vitro. *Acta Agrobot.* **1995**, *48*, 105–111. [CrossRef]
137. Kadkade, P.; Seibert, M. Phytochrome-regulated organogenesis in lettuce tissue culture. *Nature* **1977**, *270*, 49–50. [CrossRef]
138. Cybularz-Urban, T.; Hanus-Fajerska, E.; Swiderski, A. Effect of light wavelength on in vitro organogenesis of a *Cattleya hybrid*. *Acta Biol. Crac. Ser. Bot.* **2007**, *49*, 113–118.
139. Morelli, G.; Ruberti, I. Shade avoidance responses. Driving auxin along lateral routes. *Plant Physiol.* **2000**, *122*, 621–626. [CrossRef]
140. Finlayson, S.A.; Krishnareddy, S.R.; Kebrom, T.H.; Casal, J.J. Phytochrome regulation of branching in Arabidopsis. *Plant Physiol.* **2010**, *152*, 1914–1927. [CrossRef]
141. Baraldi, R.; Rossi, F.; Lercari, B. In vitro shoot development of Prunus GF 655–2: Interaction between light and benzyladenine. *Physiol. Plant.* **1988**, *74*, 440–443. [CrossRef]
142. Muleo, R.; Morini, S.; Casano, S. Photoregulation of growth and branching of plum shoots: Physiological action of two photosystems. *Vitr. Cell. Dev. Biol. Plant* **2001**, *37*, 609–617. [CrossRef]
143. Schuerger, A. Anatomical features of pepper plants (*Capsicum annuum* L.) grown under red light-emitting diodes supplemented with blue or far-red light. *Ann. Bot.* **1997**, *79*, 273–282. [CrossRef] [PubMed]
144. Appelgren, M. Effects of light quality on stem elongation of Pelargonium in vitro. *Sci. Hortic.* **1991**, *45*, 345–351. [CrossRef]
145. Heo, J.W.; Shin, K.S.; Kim, S.K.; Paek, K.Y. Light quality affects in vitro growth of grape Teleki 5BB. *J. Plant Biol.* **2006**, *49*, 276–280. [CrossRef]
146. Hahn, E.-J.; Kozai, T.; Paek, K.-Y. Blue and red light-emitting diodes with or without sucrose and ventilation affect in vitro growth of *Rehmannia glutinosa* plantlets. *J. Plant Biol.* **2000**, *43*, 247–250. [CrossRef]
147. Hung, C.D.; Hong, C.-H.; Kim, S.-K.; Lee, K.-H.; Park, J.-Y.; Nam, M.-W.; Choi, D.-H.; Lee, H.-I. LED Light for in vitro and ex vitro efficient growth of economically important highbush blueberry (*Vaccinium corymbosum* L.). *Acta Physiol. Plant.* **2016**, *38*, 152. [CrossRef]
148. Tanaka, M.; Takamura, T.; Watanabe, H.; Endo, M.; Yanagi, T.; Okamoto, K. In vitro growth of cymbidium plantlets cultured under superbright red and blue light-emitting diodes (LEDs). *J. Hortic. Sci. Biotechnol.* **1998**, *73*, 39–44. [CrossRef]
149. Nhut, D.T.; Takamura, T.; Watanabe, H.; Okamoto, K.; Tanaka, M. Responses of strawberry plantlets cultured in vitro under superbright red and blue light-emitting diodes (LEDs). *Plant Cell Tissue Organ Cult.* **2003**, *73*, 43–52. [CrossRef]
150. Park, S.Y.; Kim, M.J. Development of zygotic embryos and seedlings is affected by radiationspectral compositions from light emitting diode (LED) system in Chestnut (*Castanea crenata* S. et Z.). *J. Korean For. Soc.* **2010**, *99*, 750–754.
151. Chen, Y.-C.; Chang, C.; Lin, H.-L. Topolins and red light improve the micropropagation efficiency of passion fruit (*Passiflora edulis* Sims) 'Tainung No. 1'. *Hortscience* **2020**, *55*, 1337–1344. [CrossRef]
152. Wu, H.-C.; Lin, C.-C. Red light-emitting diode light irradiation improves root and leaf formation in difficult to propagate *Protea cynaroides* L. plantlets in vitro. *HortScience* **2012**, *47*, 1490–1494. [CrossRef]
153. Tibbitts, T.; Morgan, D.; Warrington, I. Growth of lettuce, spinach, mustard, and wheat plants under four combinations of high-pressure sodium, metal halide, and tungsten halogen lamps at equal PPFD. *J. Am. Soc. Hort. Sci.* **1983**, *108*, 622–630.

154. Sæbø, A.; Krekling, T.; Appelgren, M. Light quality affects photosynthesis and leaf anatomy of birch plantlets in vitro. *Plant Cell Tissue Organ Cult.* **1995**, *41*, 177–185. [CrossRef]
155. Maluta, F.A.; Bordignon, S.R.; Rossi, M.L.; Ambrosano, G.M.B.; Rodrigues, P.H.V. In vitro culture of sugarcane exposed to different light sources. *Pesqui. Agropecu. Bras.* **2013**, *48*, 1303–1307. [CrossRef]
156. Miao, Y.; Chen, Q.; Qu, M.; Gao, L.; Hou, L. Blue light alleviates 'red light syndrome' by regulating chloroplast ultrastructure, photosynthetic traits and nutrient accumulation in cucumber plants. *Sci. Hortic.* **2019**, *257*, 108680. [CrossRef]
157. Kowallik, W. Blue light effects on respiration. *Annu. Rev. Plant Physiol.* **1982**, *33*, 51–72. [CrossRef]
158. Liu, M.; Xu, Z.; Guo, S.; Tang, C.; Liu, X.; Jao, X. Evaluation of leaf morphology, structure and biochemical substance of balloon flower (*Platycodon grandiflorum* (Jacq.) A. DC.) plantlets in vitro under different light spectra. *Sci. Hortic.* **2014**, *174*, 112–118. [CrossRef]
159. Weis, J.; Jaffe, M. Photoenhancement by blue light of organogenesis in tobacco Pith Cultures. *Physiol. Plant.* **1969**, *22*, 171–176. [CrossRef]
160. Bach, A.; Świderski, A. The effect of light quality on organogenesis of *Hyacinthus orientalis* L. in vitro. *Acta Biol. Crac. Ser. Bot.* **2000**, *42*, 115–120.
161. Cybularz-Urban, T.; Hanus-Fajerska, E.; Bach, A. Callus induction and organogenesis in vitro of Cattleya from protocorm-like bodies (PLBs) under different light conditions. *Acta Sci. Pol.-Hortorum Cultus* **2015**, *14*, 19–28.
162. Latkowska, M.; Kvaalen, H.; Appelgren, M. Genotype dependent blue and red light inhibition of the proliferation of the embryogenic tissue of Norway spruce. *In Vitro Cell. Dev. Biol. Plant* **2000**, *36*, 57–60. [CrossRef]
163. Muleo, R.; Diodati, A.; Sgamma, T.M.; Morini, S. Ruolo del fotoperiodo e della qualità della luce nella micropropagazione delle specie arboree da frutto. *Italus Hortus* **2009**, *16*, 221–225.
164. Karpinski, S.; Szechynska-Habda, M. Evidence for light wavelength-specific systemic photoelectrical signalling and cellular light memory in *Arabidopsis thaliana*. *Plant Cell* **2010**, *22*, 1–18.
165. Manivannan, A.; Soundararajan, P.; Park, Y.G.; Wei, H.; Kim, S.H.; Jeong, B.R. Blue and red light-emitting diodes improve the growth and physiology of in vitro grown carnations 'green beauty' and 'purple beauty'. *Hortic. Environ. Biotechnol.* **2017**, *58*, 12–20. [CrossRef]
166. Toscano, S.; Cavallaro, V.; Ferrante, A.; Romano, D.; Patané, C. Effects of different light spectra on final biomass production and nutritional quality of two microgreens. *Plants* **2021**, *10*, 1584. [CrossRef]
167. Sgamma, T.; Forgione, I.; Luziatelli, F.; Iacona, C.; Mancinelli, R.; Thomas, B.; Ruzzi, M.; Muleo, R. Monochromic radiations provided by light emitted diode (LED) modulate infection and defense response to fire blight in pear trees. *Plants* **2021**, *10*, 1886. [CrossRef]
168. Cirvilleri, G.; Spina, S.; Iacona, C.; Catara, A.; Muleo, R. Study of rhizosphere and phyllosphere bacterial community and resistance to bacterial canker in genetically engineered phytochrome a cherry plants. *J. Plant Physiol.* **2008**, *165*, 1107–1119. [CrossRef] [PubMed]
169. Manivannan, A.; Soundararajan, P.; Park, Y.G.; Jeong, B.R. Physiological and proteomic insights into red and blue light-mediated enhancement of in vitro growth in *Scrophularia kakudensis* a potential medicinal plant. *Front. Plant Sci.* **2021**, *11*, 607007. [CrossRef] [PubMed]
170. Iacona, C.; Muleo, R. Light quality affects in vitro adventitious rooting and ex vitro performance of cherry rootstock colt. *Sci. Hortic.* **2010**, *125*, 630–636. [CrossRef]
171. Jao, R.-C.; Lai, C.-C.; Fang, W.; Chang, S.-F. Effects of red light on the growth of Zantedeschia plantlets in vitro and tuber formation using light-emitting diodes. *HortScience* **2005**, *40*, 436–438. [CrossRef]
172. Alvarenga, I.C.A.; Pacheco, F.V.; Silva, S.T.; Bertolucci, S.K.V.; Pinto, J.E.B.P. In vitro culture of *Achillea millefolium* L.: Quality and intensity of light on growth and production of volatiles. *Plant Cell Tissue Organ Cult.* **2015**, *122*, 299–308. [CrossRef]
173. Nhut, D.; Takamura, T.; Watanabe, H.; Tanaka, M. Efficiency of a novel culture system by using light-emitting diode (led) on in vitro and subsequent growth of micropropagated banana plantlets. *Acta Hortic.* **2003**, *616*, 121–127. [CrossRef]
174. Kurilčik, A.; Miklušytė-Čanova, R.; Dapkūnienė, S.; Žilinskaitė, S.; Kurilčik, G.; Tamulaitis, G.; Duchovskis, P.; Žukauskas, A. In vitro culture of *Chrysanthemum* plantlets using light-emitting diodes. *Open Life Sci.* **2008**, *3*, 161–167. [CrossRef]
175. Macedo, A.F.; Leal-Costa, M.V.; Tavares, E.S.; Lage, C.L.S.; Esquibel, M.A. The effect of light quality on leaf production and development of in vitro cultured plants of *Alternanthera brasiliana* Kuntze. *Environ. Exp. Bot.* **2011**, *70*, 43–50. [CrossRef]
176. Richter, G.; Wessel, K. Red light inhibits blue light-induced chloroplast development in cultured plant cells at the mRNA level. *Plant Mol. Biol.* **1985**, *5*, 175–182. [CrossRef]
177. Bello-Bello, J.J.; Pérez-Sato, J.A.; Cruz-Cruz, C.A.; Martínez-Estrada, E. Light-Emitting Diodes: Progress in Plant Micropropagation. In *Chlorophyll*; Jacob-Lopes, E., Zepka, L.Q., Queiroz, M.I., Eds.; InTech: London, UK, 2017. [CrossRef]
178. Larcher, W. *Plant Ecophysiology (Ecofisiologia Vegetal)*; RIMA Artes e Textos (Por): São Carlos, Brazil, 2006; pp. 525–550.
179. Sivakumar, G.; Heo, J.; Kozai, T.; Paek, K. Effect of continuous or intermittent radiation on sweet potato plantlets in vitro. *J. Hortic. Sci. Biotechnol.* **2006**, *81*, 546–548. [CrossRef]
180. Nhut, D.T.; Hong, L.; Watanabe, H.; Goi, M.; Tanaka, M. Growth of banana plantlets cultured in vitro under red and blue light-emitting diode (LED) irradiation source. *Acta Hortic.* **2002**, *575*, 117–124. [CrossRef]
181. Nhut, D.; Takamura, T.; Watanabe, H.; Murakami, A.; Murakami, K.; Tanaka, M. Sugar-free micropropagation of *Eucalyptus citriodora* using light-emitting diodes (LEDs) and film-rockwool culture system. *Environ. Control Biol.* **2002**, *40*, 147–155.

182. Lee, S.-H.; Tewari, R.K.; Hahn, E.-J.; Paek, K.-Y. Photon flux density and light quality induce changes in growth, stomatal development, photosynthesis and transpiration of *Withania somnifera* (L.) Dunal. Plantlets. *Plant Cell Tissue Organ Cult.* **2007**, *90*, 141–151. [CrossRef]
183. Jao, R.-C.; Fang, W. Effects of frequency and duty ratio on the growth of potato plantlets in vitro using light-emitting diodes. *HortScience* **2004**, *39*, 375–379. [CrossRef]
184. Baque, M.A.; Shin, Y.-K.; Elshmari, T.; Lee, E.-J.; Paek, K.-Y. Effect of light quality, sucrose and coconut water concentration on the microporpagation of *Calanthe hybrids* ('Bukduseong' × 'Hyesung' and 'Chunkwang' × 'Hyesung'). *Aust. J. Crop Sci.* **2011**, *5*, 1247–1254.
185. Cavallaro, V.; Avola, G.; Fascella, G.; Pellegrino, A.; La Rosa, S.; Di Silvestro, I.; Ierna, A. Effetti della qualità della luce indotta da lampade a LED sull'accrescimento e la moltiplicazione in vitro di specie ornamentali diverse per l'ambiente Mediterraneo. *Acta Italus Hortus* **2018**, *21*, 83–99.
186. Andrade, H.; Braga, A.; Bertolucci, S.; Hsie, B.; Silva, S.; Pinto, J. Effect of plant growth regulators, light intensity and LED on growth and volatile compound of *Hyptis suaveolens* (L.) Poit in vitro plantlets. *Acta Hortic.* **2015**, *1155*, 277–284. [CrossRef]
187. Brodersen, C.R.; Vogelmann, T.C. Do changes in light direction affect absorption profiles in leaves? *Funct. Plant Biol.* **2010**, *37*, 403–412. [CrossRef]
188. Smith, H.L.; McAusland, L.; Murchie, E.H. Don't ignore the green light: Exploring diverse roles in plant processes. *J. Exp. Bot.* **2017**, *68*, 2099–2110. [CrossRef]
189. Golovatskaya, I.; Karnachuk, R.A. Role of green light in physiological activity of plants. *Russ. J. Plant Physiol.* **2015**, *62*, 727–740. [CrossRef]
190. Folta, K.M.; Maruhnich, S.A. Green light: A signal to slow down or stop. *J. Exp. Bot.* **2007**, *58*, 3099–3111. [CrossRef]
191. Liu, B.; Yang, Z.; Gomez, A.; Liu, B.; Lin, C.; Oka, Y. Signaling mechanisms of plant cryptochromes in *Arabidopsis thaliana*. *J. Plant Res.* **2016**, *129*, 137–148. [CrossRef] [PubMed]
192. Tanada, T. The photoreceptors in the high irradiance response of plants. *Physiol. Plant.* **1997**, *101*, 451–454. [CrossRef]
193. Wang, Y.; Folta, K.M. Contributions of green light to plant growth and development. *Am. J. Bot.* **2013**, *100*, 70–78. [CrossRef] [PubMed]
194. Sun, J.; Nishio, J.N.; Vogelmann, T.C. Green light drives CO_2 fixation deep within leaves. *Plant Cell Physiol.* **1998**, *39*, 1020–1026. [CrossRef]
195. Kim, H.-H.; Goins, G.D.; Wheeler, R.M.; Sager, J.C. Green-light supplementation for enhanced lettuce growth under red-and blue-light-emitting diodes. *HortScience* **2004**, *39*, 1617–1622. [CrossRef]
196. Folta, K.M. Green light stimulates early stem elongation, antagonizing light-mediated growth inhibition. *Plant Physiol.* **2004**, *135*, 1407–1416. [CrossRef]
197. Johkan, M.; Shoji, K.; Goto, F.; Hahida, S.-N.; Yoshihara, T. Effect of green light wavelength and intensity on photomorphogenesis and photosynthesis in *Lactuca sativa*. *Environ. Exp. Bot.* **2012**, *75*, 128–133. [CrossRef]
198. Frechilla, S.; Talbott, L.D.; Bogomolni, R.A.; Zeiger, E. Reversal of blue light-stimulated stomatal opening by green light. *Plant Cell Physiol.* **2000**, *41*, 171–176. [CrossRef]
199. Loreti, F.; Muleo, R.; Morini, S. Effect of light quality on growth of in vitro cultured organs and tissues. *Proc. Inter. Plant Prop. Soc.* **1990**, *40*, 615–623.
200. Zhang, Z.; Li, S.; Li, W. Effects of different light qualities on the multiplication of the callus from *Vitis vinifera* L. and resveratrol content. *Plant Physiol. Commun.* **2008**, *44*, 106.
201. Dougher, T.A.; Bugbee, B. Evidence for yellow light suppression of lettuce growth. *Photochem. Photobiol.* **2001**, *73*, 208–212. [CrossRef]
202. Casal, J.J. Photoreceptor signaling networks in plant responses to shade. *Annu. Rev. Plant Biol.* **2013**, *64*, 403–427. [CrossRef] [PubMed]
203. Fraser, D.P.; Hayes, S.; Franklin, K.A. Photoreceptor crosstalk in shade avoidance. *Curr. Opin. Plant Biol.* **2016**, *33*, 1–7. [CrossRef]
204. Zhen, S.; van Iersel, M.W. Far-red light is needed for efficient photochemistry and photosynthesis. *J. Plant Physiol.* **2017**, *209*, 115–122. [CrossRef] [PubMed]
205. Tucker, D. Effects of far-red light on the hormonal control of side shoot growth in the tomato. *Ann. Bot.* **1976**, *40*, 1033–1042. [CrossRef]
206. Noe, N.; Eccher, T.; Del Signore, E.; Montoldi, A. Growth and proliferation in vitro of *Vaccinium corymbosum* under different irradiance and radiation spectral composition. *Biol. Plant.* **1998**, *41*, 161–167. [CrossRef]
207. Pham, V.N.; Kathare, P.K.; Huq, E. Phytochromes and phytochrome interacting factors. *Plant Physiol.* **2018**, *176*, 1025–1038. [CrossRef]
208. Debergh, P.; Aitken-Christie, J.; Cohen, D.; Grout, B.; Von Arnold, S.; Zimmerman, R.; Ziv, M. Reconsideration of the term 'vitrification' as used in micropropagation. *Plant Cell Tissue Organ Cult.* **1992**, *30*, 135–140. [CrossRef]
209. Gago, J.; Martínez-Núñez, L.; Landín, M.; Flexas, J.; Gallego, P.P. Modeling the effects of light and sucrose on in vitro propagated plants: A multiscale system analysis using artificial intelligence technology. *PLoS ONE* **2014**, *9*, e85989. [CrossRef]
210. Semorádová, Š.; Synková, H.; Pospíšilová, J. Responses of tobacco plantlets to change of irradiance during transfer from in vitro to ex vitro conditions. *Photosynthetica* **2002**, *40*, 605–614. [CrossRef]

211. Kadleček, P.; Rank, B.; Tichá, I. Photosynthesis and photoprotection in *Nicotiana tabacum* L. in vitro grown plantlets. *J. Plant Physiol.* **2003**, *160*, 1017–1024. [CrossRef] [PubMed]
212. Singh, P.; Patel, R. Factors influencing in vitro growth and shoot multiplication of pomegranate. *Bioscan* **2014**, *9*, 1031–1035.
213. Taiz, L.; Zeiger, E. *Fisiologia Vegetal*; Universitat Jaume I: Castellón, Spain, 2006; Volume 10.
214. Singh, A.S.; Jones, A.M.P.; Shukla, M.R.; Saxena, P.K. High Light intensity stress as the limiting factor in micropropagation of sugar maple (*Acer saccharum* Marsh.). *Plant Cell Tissue Organ Cult.* **2017**, *129*, 209–221. [CrossRef]
215. Noè, N.; Eccher, T. Influence of irradiance on in vitro growth and proliferation of *Vaccinium corymbosum* (highbush blueberry) and subsequent rooting in vivo. *Physiol. Plant.* **1994**, *91*, 273–275. [CrossRef]
216. Nhut, D.; Takamura, T.; Watanabe, H.; Tanaka, M. Artificial light source using light-emitting diodes (LEDs) in the efficient micropropagation of Spathiphyllum plantlets. *Acta Hortic.* **2005**, *692*, 137–142. [CrossRef]
217. Soontornchainaksaeng, P.; Chaicharoen, S.; Sirijuntarut, M.; Kruatrachue, M. In vitro studies on the effect of light intensity on plant growth of *Phaius tankervilliae* (Banks Ex L'Herit.) Bl. and *Vanda coerulea* Griff. *Sci. Asia* **2001**, *27*, 233–237. [CrossRef]
218. Wang, Q. The effect of light, darkness and temperature on micropropagation of the pear rootstock BP10030. *J. Hort. Sci.* **1992**, *67*, 869–876. [CrossRef]
219. Zhang, M.; Zhao, D.; Ma, Z.; Li, X.; Xiao, Y. Growth and photosynthetic capability of *Momordica grosvenori* plantlets grown photoautotrophically in response to light intensity. *HortScience* **2009**, *44*, 757–763. [CrossRef]
220. Infante, R.; Magnanini, E.; Righetti, B. The role of light and CO_2 in optimising the conditions for shoot proliferation of *Actinidia deliciosa* in vitro. *Physiol. Plant.* **1989**, *77*, 191–195. [CrossRef]
221. Bressan, P.; Kim, Y.; Hyndman, S.; Hasegawa, P.; Bressan, R. Factors affecting in vitro propagation of rose. *J. Am. Soc. Hort. Sci.* **1982**, *107*, 979–990.
222. Jo, U.; Murthy, H.; Hahn, E.; Paek, K. Micropropagation of *Alocasia amazonica* using semisolid and liquid cultures. *In Vitro Cell. Dev. Biol. Plant* **2008**, *44*, 26–32. [CrossRef]
223. Biswal, A.K.; Pattanayak, G.K.; Pandey, S.S.; Leelavathi, S.; Reddy, V.S.; Govindjee; Tripathy, B.C. Light intensity-dependent modulation of chlorophyll *b* biosynthesis and photosynthesis by overexpression of chlorophyllide *a* oxygenase in Tobacco. *Plant Physiol.* **2012**, *159*, 433–449. [CrossRef] [PubMed]
224. Zhou, M.; Guan, Q.; Wei, Y.; Zhang, Z. Effects of Sucrose Concentration and light intensity on growth and photosynthesis of ginger plantlets in vitro. *Chin. J. Appl. Environ. Biol.* **2008**, *14*, 356–361.
225. Fiola, J.A.; Hassan, M.A.; Swartz, H.J.; Bors, R.H.; McNicols, R. Effect of thidiazuron, light fluence rates and kanamycin on in vitro shoot organogenesis from excised Rubus cotyledons and leaves. *Plant Cell Tissue Organ Cult.* **1990**, *20*, 223–228.
226. Krause, G.H. Photoinhibition of photosynthesis. An evaluation of damaging and protective mechanisms. *Physiol. Plant.* **1988**, *74*, 566–574. [CrossRef]
227. Sáez, P.L.; Bravo, L.A.; Sánchez-Olate, M.; Bravo, P.B.; Ríos, D.G. Effect of Photon Flux Density and exogenous sucrose on the photosynthetic performance during in vitro culture of *Castanea sativa*. *Am. J. Plant Sci.* **2016**, *7*, 2087. [CrossRef]
228. Kozai, T. Effects of CO_2 enrichment and sucrose concentration under high photosynthesis photon fluxes on growth of tissue-cultured *Cymbidium* plantlets during the preparation stage. In Proceedings of the Congress 'Plant Micropropagation in Horticultural Industries', Arlon, Belgium, 10–14 August 1987; pp. 135–141.
229. Kozai, T.; Oki, H.; Fujiwara, K. Photosynthetic characteristics of *Cymbidium plantlet* in vitro. *Plant Cell Tissue Organ Cult.* **1990**, *22*, 205–211. [CrossRef]
230. Desjardins, Y.; Laforge, F.; Lussier, C.; Gosselin, A. Effect of CO_2 enrichment and high photosynthetic photon flux on the development of autotrophy and growth of tissue-cultured strawberry, raspberry and asparagus plants. *Acta Hortic.* **1988**, *230*, 45–53. [CrossRef]
231. Kozai, T.; Koyama, Y.; Watanabe, I. Multiplication of potato plantlets in vitro with sugar free medium under high photosynthetic photon flux. *Acta Hortic.* **1988**, *230*, 121–128. [CrossRef]
232. McClung, C.R. Plant circadian rhythms. *Plant Cell* **2006**, *18*, 792–803. [CrossRef]
233. Bell-Pedersen, D.; Cassone, V.M.; Earnest, D.J.; Golden, S.S.; Hardin, P.E.; Thomas, T.L.; Zoran, M.J. Circadian rhythms from multiple oscillators: Lessons from diverse organisms. *Nat. Rev. Genet.* **2005**, *6*, 544–556. [CrossRef] [PubMed]
234. Devlin, P.F.; Kay, S.A. Cryptochromes are required for phytochrome signaling to the circadian clock but not for rhythmicity. *Plant Cell* **2000**, *12*, 2499–2509. [CrossRef] [PubMed]
235. Dodd, A.N.; Salathia, N.; Hall, A.; Kévei, E.; Tóth, R.; Nagy, F.; Hibberd, J.M.; Millar, A.J.; Webb, A.A. Plant circadian clocks increase photosynthesis, growth, survival, and competitive advantage. *Science* **2005**, *309*, 630–633. [CrossRef] [PubMed]
236. Loudon, A.S. Circadian biology: A 2.5 Billion year old clock. *Curr. Biol.* **2012**, *22*, R570–R571. [CrossRef] [PubMed]
237. Harmer, S.L.; Hogenesch, J.B.; Straume, M.; Chang, H.-S.; Han, B.; Zhu, T.; Wang, X.; Kreps, J.A.; Kay, S.A. Orchestrated transcription of key pathways in Arabidopsis by the circadian clock. *Science* **2000**, *290*, 2110–2113. [CrossRef] [PubMed]
238. Barneche, F.; Malapeira, J.; Mas, P. The impact of chromatin dynamics on plant light responses and circadian clock function. *J. Exp. Bot.* **2014**, *65*, 2895–2913. [CrossRef]
239. Hsu, P.Y.; Harmer, S.L. Wheels within wheels: The plant circadian system. *Trends Plant Sci.* **2014**, *19*, 240–249. [CrossRef]
240. Mermet, J.; Yeung, J.; Hurni, C.; Mauvoisin, D.; Gustafson, K.; Jouffe, C.; Nicolas, D.; Emmenegger, Y.; Gobet, C.; Franken, P. Clock-dependent chromatin topology modulates circadian transcription and behavior. *Genes Dev.* **2018**, *32*, 347–358. [CrossRef]

241. Abbott, A.; Belcher, A. Analysis of gases in culture flasks. In *Report Long Ashton Research Station, Dawson & Goodall Ltd.*; The Mendip Press: Bristol, UK, 1980; p. 79.
242. Fujiwara, K. Measurements of carbon dioxide gas concentration in closed vessels containing tissue cultured plantlets and estimates of net photosynthetic rates of the plantlets. *J. Agric. Methods* **1987**, *43*, 21–30. [CrossRef]
243. Morini, S.; Muleo, R.; Sciutti, R.; Fortuna, P. Effect of different light-dark cycles on growth of fruit tree shoots cultured in vitro. *Adv. Hort. Sci.* **1990**, *3*, 1000–1004.
244. Zimmerman, T.W.; Scorza, R. Benzyladenine and shortened light/dark cycles improve in vitro shoot proliferation of peach. *HortScience* **1994**, *29*, 698. [CrossRef]
245. Morini, S.; Muleo, R.; Sciutti, R.; Fortuna, P. Relationship between evolution of CO_2 and growth of plum shoot tips cultured in vitro under different light/dark regimes. *Physiol. Plant.* **1993**, *87*, 286–290. [CrossRef]
246. Iarema, L.; da Cruz, A.C.F.; Saldanha, C.W.; Dias, L.L.C.; Vieira, R.F.; de Oliveira, E.J.; Otoni, W.C. Photoautotrophic propagation of *Brazilian ginseng* [*Pfaffia glomerata* (Spreng.) Pedersen]. *Plant Cell Tissue Organ Cult.* **2012**, *110*, 227–238. [CrossRef]
247. Badr, A.; Angers, P.; Desjardins, Y. Metabolic profiling of photoautotrophic and photomixotrophic potato plantlets (*Solanum tuberosum*) provides new insights into acclimatization. *Plant Cell Tissue Organ Cult.* **2011**, *107*, 13–24. [CrossRef]
248. Kubota, C.; Kozai, T. Growth and net photosynthetic rate of *Solanun tuberosum* in vitro under forced and natural ventilation. *HortScience* **1992**, *27*, 1312–1314. [CrossRef]
249. Schmildt, O.; Netto, A.T.; Schmildt, E.R.; Carvalho, V.S.; Otoni, W.C.; Campostrini, E. Photosynthetic capacity, growth and water relations in 'golden'papaya cultivated in vitro with modifications in light quality, sucrose concentration and ventilation. *Theor. Exp. Plant Physiol.* **2015**, *27*, 7–18. [CrossRef]
250. During, H. CO_2 assimilation and photorespiration of grapevine leaves: Responses to light and drought. *Vitis* **1988**, *27*, 199–208.
251. Düring, H.; Harst, M. Stomatal behaviour, photosynthesis and photorespiration of in vitro—Grown grapevines: Effects of light and CO_2. *Vitis* **1996**, *35*, 163–168.
252. Fujiwara, K.; Kozai, T.; Watanabe, I. Development of a photoautotrophic tissue culture system for shoots and/or plantlets at rooting and acclimatization stages. *Acta Hortic.* **1988**, *230*, 153–158. [CrossRef]
253. Costa, A.C.; Rosa, M.; Megguer, C.A.; Silva, F.G.; Pereira, F.D.; Otoni, W.C. A Reliable methodology for assessing the in vitro photosynthetic competence of two *Brazilian savanna* species: Hyptis marrubioides and Hancornia speciosa. *Plant Cell, Tissue Organ Cult.* **2014**, *117*, 443–454. [CrossRef]
254. Tamas, I.A. Hormonal regulation of apical dominance. In *Plant Hormones*; Springer: Berlin/Heidelberg, Germany, 1995; pp. 572–597.
255. Chatfield, S.P.; Stirnberg, P.; Forde, B.G.; Leyser, O. The hormonal regulation of axillary bud growth in Arabidopsis. *Plant J.* **2000**, *24*, 159–169. [CrossRef] [PubMed]
256. Crawford, S.; Shinohara, N.; Sieberer, T.; Williamson, L.; George, G.; Hepworth, J.; Müller, D.; Domagalska, M.A.; Leyser, O. Strigolactones enhance competition between shoot branches by Dampening auxin transport. *Development* **2010**, *137*, 2905–2913. [CrossRef]
257. Luisi, A.; Lorenzi, R.; Sorce, C. Strigolactone may interact with gibberellin to control apical dominance in pea (*Pisum sativum*). *Plant Growth Regul.* **2011**, *65*, 415–419. [CrossRef]
258. Klee, H.J.; Lanahan, M.B. Transgenic plants in hormone biology. In *Plant Hormones*; Springer: Berlin/Heidelberg, Germany, 1995; pp. 340–353.
259. Wareing, P.; Thompson, A.G. Rapid effect of red light on hormone levels. In *Light and Plant Development*; Smith, H., Ed.; Butterworths Publ. Co.: London, UK, 1976; pp. 284–295.
260. Köhler, K.; Dörfler, M.; Göring, H. The influence of light on the cytokinin content of *Amaranthus* seedlings. *Biol. Plant.* **1980**, *22*, 128–134. [CrossRef]
261. Lercari, B.; Micheli, P. Photoperiodic regulation of cytokin levels in leaf blades of *Allium cepa* L. *Plant Cell Physiol.* **1981**, *22*, 501–505.
262. Stirk, W.; Bálint, P.; Tarkowská, D.; Novák, O.; Maróti, G.; Ljung, K.; Turečková, V.; Strnad, M.; Ördög, V.; Van Staden, J. Effect of light on growth and endogenous hormones in *Chlorella minutissima* (Trebouxiophyceae). *Plant Physiol. Biochem.* **2014**, *79*, 66–76. [CrossRef]
263. Roman, H.; Girault, T.; Barbier, F.; Péron, T.; Brouard, N.; Pěnčík, A.; Novák, O.; Vian, A.; Sakr, S.; Lothier, J. Cytokinins are initial targets of light in the control of bud outgrowth. *Plant Physiol.* **2016**, *172*, 489–509. [CrossRef]
264. Thimann, K. *Hormone Action in the Whole Life of Plants*; University of Massachusetts Press: Amherst, MA, USA, 1977.
265. Koltai, H.; Kapulnik, Y. Strigolactones as mediators of plant growth responses to environmental conditions. *Plant Signal. Behav.* **2011**, *6*, 37–41. [CrossRef]
266. Jones, A.M.; Cochran, D.S.; Lamerson, P.M.; Evans, M.L.; Cohen, J.D. Red light-regulated growth: I. changes in the abundance of indoleacetic acid and a 22-Kilodalton Auxin-binding protein in the maize mesocotyl. *Plant Physiol.* **1991**, *97*, 352–358. [CrossRef]
267. Mandava, N.B. Plant growth-promoting brassinosteroids. *Annu. Rev. Plant Physiol.* **1988**, *39*, 23–52. [CrossRef]
268. Villalobos, V.M.; Leung, D.W.; Thorpe, T.A. Light-cytokinin interaction in shoot formation in cultured cotyledon explants of radiata pine. *Physiol. Plant* **1984**, *61*, 497–504. [CrossRef]
269. Economou, A.; Read, P. Effect of red and far-red light on *Azalea* microcutting production in vitro and rooting in vivo. In Proceedings of the 6th International Congress Plant Tissue and Cell Culture, Minneapolis, MN, USA, 3–8 August 1986; Volume 431.

270. Economou, A.S.; Read, P.E. Light treatments to improve efficiency of in vitro propagation systems. *Hort. Sci.* **1987**, *22*, 751–754.
271. Teichmann, T.; Muhr, M. Shaping plant architecture. *Front. Plant Sci.* **2015**, *6*, 233. [CrossRef]
272. Trouwborst, G.; Hogewoning, S.W.; van Kooten, O.; Harbinson, J.; van Ieperen, W. Plasticity of photosynthesis after the 'red light syndrome'in cucumber. *Environ. Exp. Bot.* **2016**, *121*, 75–82. [CrossRef]
273. Craver, J.K.; Miller, C.T.; Williams, K.A.; Bello, N.M. Ultraviolet radiation affects intumescence development in ornamental sweetpotato (*Ipomoea batatas*). *HortScience* **2014**, *49*, 1277–1283. [CrossRef]
274. Nguyen, T.K.L.; Cho, K.M.; Lee, H.Y.; Cho, D.Y.; Lee, G.O.; Jang, S.N.; Lee, Y.; Kim, D.; Son, K.-H. Effects of white LED lighting with specific shorter blue and/or green wavelength on the growth and quality of two lettuce cultivars in a vertical farming system. *Agronomy* **2021**, *11*, 2111. [CrossRef]

plants

MDPI

Article

Effects of Light Spectral Quality on Photosynthetic Activity, Biomass Production, and Carbon Isotope Fractionation in Lettuce, *Lactuca sativa* L., Plants

Ivan G. Tarakanov [1,*], Daria A. Tovstyko [1], Maxim P. Lomakin [1], Alexander S. Shmakov [1], Nikolay N. Sleptsov [1], Alexander N. Shmarev [2], Vladimir A. Litvinskiy [3] and Alexander A. Ivlev [1]

[1] Department of Plant Physiology, Russian State Agrarian University—Moscow Timiryazev Agricultural Academy, Timiryazevskaya Str., 49, 127550 Moscow, Russia; tov.dasha@mail.ru (D.A.T.); max124c41@gmail.com (M.P.L.); 5456685@gmail.com (A.S.S.); inkss@mail.ru (N.N.S.); aa.ivlev@list.ru (A.A.I.)

[2] Institute of Basic Biological Problems, Russian Academy of Sciences, 142290 Pushchino, Moscow Region, Russia; shurik_bx_04@mail.ru

[3] Borissiak Paleontological Institute, Russian Academy of Sciences, 123, Profsoyuznaya Str., 117647 Moscow, Russia; vl.litvinskiy@gmail.com

* Correspondence: ivatar@yandex.ru; Tel.: +7-499-976-2054

Abstract: The optimization of plant-specific LED lighting protocols for indoor plant growing systems needs both basic and applied research. Experiments with lettuce, *Lactuca sativa* L., plants using artificial lighting based on narrow-band LEDs were carried out in a controlled environment. We investigated plant responses to the exclusion of certain spectral ranges of light in the region of photosynthetically active radiation (PAR); in comparison, the responses to quasimonochromatic radiation in the red and blue regions were studied separately. The data on plant phenotyping, photosynthetic activity determination, and PAM fluorometry, indicating plant functional activity and stress responses to anomalous light environments, are presented. The study on carbon isotopic composition of photoassimilates in the diel cycle made it possible to characterize the balance of carboxylation and photorespiration processes in the leaves, using a previously developed oscillatory model of photosynthesis. Thus, the share of plant photorespiration (related to plant biomass enrichment with ^{13}C) increased in response to red-light action, while blue light accelerated carboxylation (related to ^{12}C enrichment). Blue light also reduced water use efficiency. These data are supported by the observations from the light environments missing distinct PAR spectrum regions. The fact that light of different wavelengths affects the isotopic composition of total carbon allowed us to elucidate the nature of its action on the organization of plant metabolism.

Keywords: *Lactuca sativa*; LEDs; plant factory; photosynthesis; chlorophyll fluorescence; carbon isotope discrimination

Citation: Tarakanov, I.G.; Tovstyko, D.A.; Lomakin, M.P.; Shmakov, A.S.; Sleptsov, N.N.; Shmarev, A.N.; Litvinskiy, V.A.; Ivlev, A.A. Effects of Light Spectral Quality on Photosynthetic Activity, Biomass Production, and Carbon Isotope Fractionation in Lettuce, *Lactuca sativa* L., Plants. *Plants* **2022**, *11*, 441. https://doi.org/10.3390/plants11030441

Academic Editors: Valeria Cavallaro and Rosario Muleo

Received: 23 December 2021

Accepted: 3 February 2022

Published: 5 February 2022

Publisher's Note: MDPI stays neutral with regard to jurisdictional claims in published maps and institutional affiliations.

1. Introduction

The application of light-emitting diodes (LED) in horticultural lighting systems provides new possibilities for light intensity and light spectrum fine regulation along with a significant reduction in energy consumption [1–3]. A breathtaking possibility to modulate the LED lighting spectrum can also help in promoting the accumulation of important plant metabolites, which are often associated with nutraceutical properties, as has been shown in various crops, including lettuce [4]. The set-up of plant-specific light protocols for their cultivation is a critical phase in improving the sustainability of indoor growing systems [2].

Besides photosynthesis, plants are capable of perceiving and processing information with light signals from their biotic and abiotic surroundings for optimal growth and development [5]. Reviews of studies on light quality effects on plant growth and development can be found elsewhere [6–8]. Red and blue are generally recognized as the most important light regions necessary for plant development and growth [3]. However, other wavelengths (such as those corresponding to yellow or green colors) could also have a role in affecting the quality of crops [6]. Blue light is involved in a wide range of plant processes such as phototropism, photomorphogenesis, stomatal opening, and leaf photosynthetic functioning [9]. Most studies assessing the effects of blue light (blue LEDs) on the leaf or whole plants have either compared their response to a broadband light source with the response to blue-deficient light [10] or plants grown under red light alone [11,12]. On the other hand, red LEDs emit a narrow spectrum of light (660 nm) that is close to the maximum absorbance for both chlorophyll and phytochromes. The absorption of blue and red light (LEDs) by plants has been measured as 90% [13], which indicates that plant development and physiology is strongly influenced by blue and red light [6]. The effects of green light tend to reverse the processes established by red and/or blue light. In this way, green light may be functioning in a manner similar to far-red light, informing the plant of photosynthetically unfavorable conditions and triggering adaptative responses [14]. Many studies have been reported on several crops grown under deficiency/efficiency or using a combination of red and blue light at different wavelengths [15,16] to investigate their effects on plant growth and development. While red light promotes biomass accumulation, growth, and photosynthesis in lettuce, blue LED light is effective in stimulating photomorphogenesis and adaptive phenomena such as the stomata-opening/closing-regulation mechanism, as well as biomass accumulation and chlorophyll and anthocyanin biosynthesis [3,17]. A positive growth response to the combination of blue and red light was confirmed in Batavia lettuce plants [18]. Green LED light regulates leaf expansion, stem stretching, and stomatal conductance. Moreover, it has been shown that green LED light addition leads to greater dry mass accumulation and growth stimulation [19].

The plant perceives light environment signals by means of photosynthetic apparatus (PSA) and specific photoreceptors sensitive to different light spectral regions. Blue and red light are not equal in their effects on photoreceptors: red light is perceived in addition to PSA by phytochromes only, and blue light is absorbed by both phytochromes and blue-light receptors (cryptochromes, phototropins) [20]. Blue light influences a greater number of photoreceptors and is functionally more versatile. It is most effective in stimulating the transcription of photosynthesis-related genes (via cryptochromes and phytochromes) [21]. Interestingly, barley plants grown with monochromatic red light demonstrated specific organization of chloroplast membranes (shaggy-formed grana) and light-harvesting complexes (increased energy transfer to PSI, possibly due to spillover promoted by this particular granum structure) [20]. These specific responses can be related to contradictory information from the photoreceptors; the signals from the phytochromes and photosynthetic apparatus indicate the incidence of light, while the lack of a signal from the blue-light receptors can be misinterpreted as darkness [20]. Most of the negative monochromatic red-light effects can be avoided by the addition of blue light [22–24]. Furthermore, a combination of red and blue light in certain cases can result in synergetic effects in biomass accumulation [25,26] Plant photosynthesis and growth, directly or indirectly, can also be mediated by the photoreceptor response. Additionally, chloroplasts play an important role in photoreceptor-mediated control of photomorphogenic responses [27]. The main obstacle in the transition to LED lighting in crop production is that it involves a complex system change beyond lighting (e.g., plant light recipes, which are species- and often cultivar-dependent), resulting in serious associated costs [28]. Lighting systems using specific wavelengths are capable of target compound biosynthesis fortification; however, special attention has to be paid to the stress the artificial light may cause in the photosynthesis and biomass accumulation [29]. To explore the action mode of different light spectrum regions, various experimental approaches are used. Thus, in the studies on the blue-light effects, plant responses to a broadband light source with a response to blue-region-deficient light were compared [10] with plants grown under red light alone [11]. So, the experimental set up can include studies on the effects of monochromatic irradiation. Additionally, plant responses to photosynthetically active radiation (PAR) missing distinct spectrum regions can be investigated [30].

In our studies with lettuce plants, we have used both screens mentioned above, emphasizing research on light spectral quality effects on the carbon isotope composition of plant biomass (Section 2.3). It is known that plant cells are able to fractionate carbon isotopes in the light and in the dark [31,32]. The carbon isotope composition of plant leaf biomass is mainly related to the light processes, CO_2 assimilation, and photorespiration [32,33]. The ^{12}C enrichment of plant biomass during CO_2 assimilation occurs at Calvin cycle entry during RuBP carboxylation. During photorespiration, carbon isotope fractionation occurs with the opposite sign, thus reducing the effect of CO_2 assimilation and enriching biomass with ^{13}C. The isotope effect of photosynthetic assimilation and photorespiration are coupled by a key photosynthetic enzyme, Rubisco, that oscillates from CO_2 assimilation to photorespiration and back [34,35]. The effects of monochromatic light and other unique artificial light treatments on the carbon isotope fractioning have not been investigated until now.

2. Materials and Methods

2.1. Plant Material

Lettuce, *Lactuca sativa* L., plants of the Aficion RZ cultivar were used in our studies. This is Batavia-type lettuce, leaves with strongly wavy edge, light green. Batavia lettuce is highly appreciated in the market due to the variability in shape, color, texture, and taste. As for the nutritional value, it is a source of vitamin A, niacin, riboflavin, thiamine, Ca, Fe, K, Mn, Se, and β-carotene [36]. Aficion cultivar is widely grown in greenhouses and vertical farms with artificial lighting.

2.2. Cultivation Conditions

Plants were grown in growth chambers (Urbangrower 150, China; Figure 1) with various light treatments according to experimental layout described in Section 2.3. Each chamber had dimensions of 1.50 × 0.90 × 2.00, 2.7 m^3, with gloss white walls. Chambers were supplied with fans; day/night temperature was 20/18 °C, with less than 1 °C variation over time and 1 °C variation among chambers.

Figure 1. Plant-growing chambers with various light environments.

Plants were grown in 2 L vegetational vessels (3 plants in each container). Seeds were sown directly into the commercial neutralized peat-based substrate "Agrobalt-C" (Pindstrup, Pskov region, Russia) with pH 6.0–6.5 and complete macro- and micronutrient supply including 150 mg L^{-1} [NH_4^+ andNO_3^-], 270 mg L^{-1} P_2O_5, and 300 mg L^{-1}

K_2O. Substrate humidity was maintained at 70% of full water capacity, watering up to calculated weight.

2.3. Light Treatments

Plant chambers were illuminated with lamps consisting of various light-emitting diode (LED) bars specifically designed to provide a custom spectrum in each chamber. Fixtures consisted of light modules with tunable light-emitting diodes varying in wavelength and spectral composition of the emitted light over wide ranges (Figure 2).

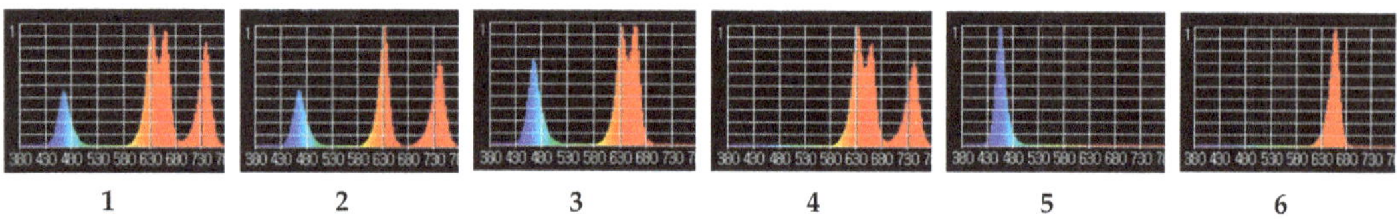

Figure 2. Light treatments: (**1**) "460 + 640 + 660 + 730"—4-peak reference treatment; (**2**) "460 + 640 + 730" —3-peak treatment missing red-light R_{660} region; (**3**) "460 + 640 + 660"—3-peak treatment missing far-red-light FR_{730} region; (**4**) "640 + 660 + 730"—3-peak treatment missing blue-light B_{460} region; (**5**) "450"—monochromatic blue-light B_{450} region; (**6**) "659" monochromatic red-light R_{659} region.

Four types of high-performance narrow-band 3-Watt LEDs (Estar Technology, Changchun, China) were used: short-wave red ($\Delta\lambda_{0.5}$ = 623 ÷ 641 nm, λ_{max} = 632 nm), long-wave red ($\Delta\lambda_{0.5}$ = 646 ÷ 674 nm, λ_{max} = 660 nm), far-red ($\Delta\lambda_{0.5}$ = 727 ÷ 751 nm, λ_{max} = 739 nm), and blue ($\Delta\lambda_{0.5}$ = 452 ÷ 477 nm, λ_{max} = 465 nm). The control light treatment included all 4 types of LEDs, and in each of the other regimes one of them was excluded (except short-wave red) in order to elucidate the wavelength that affected distinct crop physiological processes. Short-wave red was used as an additional background spectral region to provide chlorophyll *a* excitation in the absence of long-wave red light. The same daily light integral (DLI) of 9.72 mol m^{-2} d^{-1} was maintained in all the treatments with photosynthetic photon flux density (PPFD) 150 µmol m^{-2} s^{-1}, photoperiod 18 h. Spectra of the resulting lamp systems were measured with a spectrometer UPRtek PG100N (Taiwan). To measure the PPFD in the PAR region, an LI-191R quantum sensor with an LI-250A data logger (LI-COR Biosciences, NE, USA) was used. It was measured at the top of the plant canopy (the distance from the light source was $\geq$50 cm), and each chamber was adjusted to maintain PPFD at ±5%. To provide uniform PPFD, plant pots were moved and rotated within the marked uniform light platform every second day.

In the experiment on the red and blue monochromatic light effects, two types of tunable LEDs (Cree, USA) were used: red ($\Delta\lambda_{0.5}$ = 647 ÷ 671 nm, λ_{max} = 659 nm) and blue ($\Delta\lambda_{0.5}$ = 438 ÷ 462 nm, λ_{max} = 450 nm).

To provide easy reading of the figure legends, wavelengths representing figures for combined-spectra regions (blue, short-wave red, long-wave red, and far-red) are "rounded" to 460, 640, 660, and 730, respectively.

2.4. Plant Growth Parameters Analyses

Four plants per each treatment were destructively harvested 30 days after emergence. The number of leaves (>1 cm) per plant was counted, and total leaf area was measured using a leaf area meter LI-3000A (LI-COR Biosciences, NE, USA). Shoot fresh weight was measured using an electronic balance. Subsequently, shoots were oven-dried to a constant weight at 70 °C for dry weight determination. Specific leaf weight (SLW) was calculated by dividing leaf weight by leaf area (dry weight per unit leaf area).

2.5. Photosynthesis and Transpiration

Plant leaf photosynthetic rate and transpiration analyses were carried out using an LI-6400XT Portable Photosynthesis System (LI-COR Biosciences, NE, USA) with a standard leaf chamber of 2 × 3 cm. During the measurements, CO_2 concentration was maintained at 400 ± 12.0 μmol mol^{-1}, air temperature 21–23 °C, and air humidity 60 ± 4.0%.

Photosynthesis and transpiration were measured at the same light intensity as the growth light. Photosynthetic water use efficiency was calculated as the ratio of the rate of carbon assimilation (photosynthesis) to the rate of transpiration.

The light response curve, i.e., the photosynthetic rate as a function of incident light intensity, was measured for four different leaves using an infrared gas analyzer (IRGA) with an LI-6400XT standard lighting chamber; the internal LED lamp in the IRGA machine was used as a light source.

2.6. *Chlorophyll a Fluorescence Determination*

Chlorophyll *a* fluorescence in PSII was measured using Junior-PAM fluorimeter (Heinz Walz, Germany). Minimum (Fo) and maximum (Fm) fluorescence rates were determined after 15 min leaf exposition in darkness. Maximum quantum efficiency of PSII Fv − Fm = (Fm − Fo)/Fm was calculated after [37]. Relative PSII operating efficiency (ΦPSII) of the light-adapted leaves was calculated as ΦPSII = (F′m − Ft)/F′m. Chlorophyll a non-photosynthetic quenching (NPQ) was calculated as NPQ = (Fm − F′m)/F′m. Photochemical electron transport rate (ETR) was calculated as ETR = (Φ_{PSII})·PPFD·0.5. Fluorescence parameters were determined in 4–6 biological replicates.

2.7. *Carbon Isotope Discrimination (Δ) Measurements*

Plant sampling was conducted with 6 h intervals during 24 h cycle: at the start of 18 h photoperiod and at 6, 12, and 18 h. Plant leaves were sampled in 4 replicates and dried in an oven at 65 °C.

Dry plant material was milled using Vibromill vibrator. After milling in powder, samples were weighed in tin containers and introduced by means of an autosampler into the elemental analyzer (FlashEA, Thermo Electron, Waltham, MA, USA), where in presence of O_2 and catalysts, they were quantitatively burnt to CO_2. The formed gas was separated from other combustion gases on a chromatographic column and transferred via ConFlo interface to the isotope ratio mass-spectrometer (Delta V, Thermo) for analysis. Each sample was analyzed three times.

For calibration of IRMS instrument and control of results' accuracy we used IAEA standards of L-glutamic acid (USGS40, USGS41).

The values of the isotope ratio are expressed in δ‰ according to the formula

$$\delta‰ = \left(R_{sample} - R_{standard}/R_{standard}\right) \times 1000$$

where R_{sample} and $R_{standard}$ are the $^{13}C/^{12}C$ ratios of the sample and the standard, respectively. By international convention, the standard used for the analysis was the carbon belemnite from the PeeDee formation (VPDB).

2.8. *Statistical Analysis of Experimental Data*

For each light treatment, four replications were tested during plant phenotyping (sampling). Statistical analysis of physiological parameters was performed using one-way analysis of variance (ANOVA) followed by Duncans' multiple range test with MS Excel software and AGROS software (version 2.11, Moscow, Russia). In the graphs, means ± standard errors (SE) are presented; means followed by the same letter were not different at $p \leq 0.05$.

3. Results

3.1. *Growth Responses*

Lettuce plant growth (leaf biomass accumulation) was significantly retarded in the monochromatic blue light (Figures 3 and 4). It was also reduced in the combined-spectrum environment missing blue light. Monochromatic red light was especially favorable for biomass accumulation. Red light's absence was unfavorable for dry biomass accumulation, though it did not affect fresh biomass yield.

Figure 3. Representative photo of plants from each light treatment, 20 days after emergence: (1) "460 + 640 + 660 + 730"—4-peak reference treatment; (2) "460 + 640 + 730"—3-peak treatment missing red-light R_{660} region; (3) "460 + 640 + 660"—3-peak treatment missing far-red-light FR_{730} region; (4) "640 + 660 + 730"—3-peak treatment missing blue-light B_{460} region; (6) "450"—monochromatic blue-light B_{450} region; (7) "659" monochromatic red-light R_{659} region.

In the treatment missing blue light, plants demonstrated a tendency towards bolting. The highest bolting resistance was observed in response to single-blue-light treatment or to the combined spectrum missing far-red light. Additionally, leaf size was reduced in the single-blue-light treatment and in response to the combined spectrum missing far-red light. In the blue-light treatment, leaf size reduction resulted in dramatically reduced total leaf area. However, in the combined spectrum missing far-red light, there was no reduction in leaf area due to the increased total leaf number; also, the highest specific leaf weight was observed under these conditions. Blue-light knockout in the combined spectrum resulted in a considerable leaf area increase.

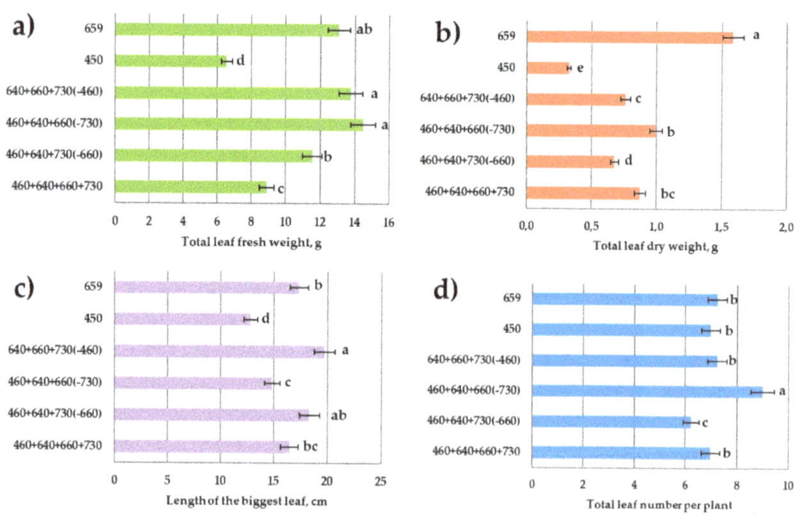

Figure 4. *Cont.*

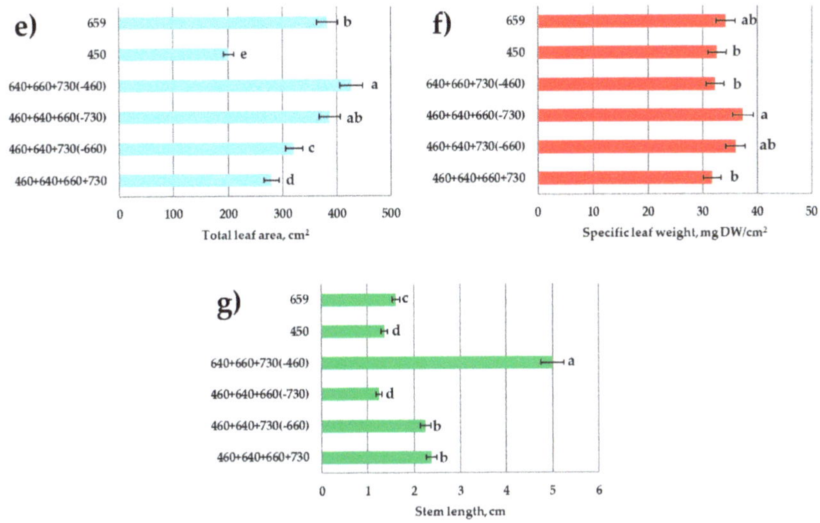

Figure 4. Growth parameters of lettuce plants in response to various light treatments. Sampling 30 days after emergence. Means ± standard error (SE); means followed by the same letter were not different at $p \leq 0.05$. (**a**) Total leaf fresh weight; (**b**) total leaf dry weight; (**c**) length of the biggest leaf; (**d**) total leaf number per plant; (**e**) total leaf area; (**f**) specific leaf weight; (**g**) stem length. Light treatments from the bottom of y-axis: "460 + 640 + 660 + 730"—4-peak reference treatment; "460 + 640 + 730"—3-peak treatment missing red-light R_{660} region; "460 + 640 + 660"—3-peak treatment missing far-red-light FR_{730} region; "640 + 660 + 730"—3-peak treatment missing blue-light B_{460} region; "450"—monochromatic blue-light B_{450} region; "659" monochromatic red-light R_{659} region.

3.2. Photosynthesis and Transpiration

The highest net photosynthesis was observed in plants grown in the combined-spectrum light environment missing red or blue light (Figure 5). Interestingly, the net photosynthesis in the reference treatment was lower than in all the other treatments.

It has been shown in previous studies that blue and red light induce stomatal opening via different pathways [38]. In our experiment, the highest stomatal conductance and transpiration were observed under monochromatic blue light. Red- or far-red-light absence in the combined spectrum decreased these parameters as compared to blue-light treatment, though more significant response was observed in the absence of blue light. Water use efficiency (WUE, photosynthesis/transpiration ratio) was extremely low under blue light (mostly due to the highest transpiration rate) and increased by three times in the treatments with red or blue light.

As for the light response curve determination, the lowest photosynthesis intensity at saturating PPFD was observed in response to red light (Figure 6). Here, low light intensity at light response curve saturation was found, as well. This kind of response is typical for plants originating from the shaded habitats. The highest photosynthesis at saturating light intensity was observed in response to blue light and the combined spectrum without red light R_{660}; in part, the absence of the long-wave red light was compensated for here by short-wave red light R_{640}.

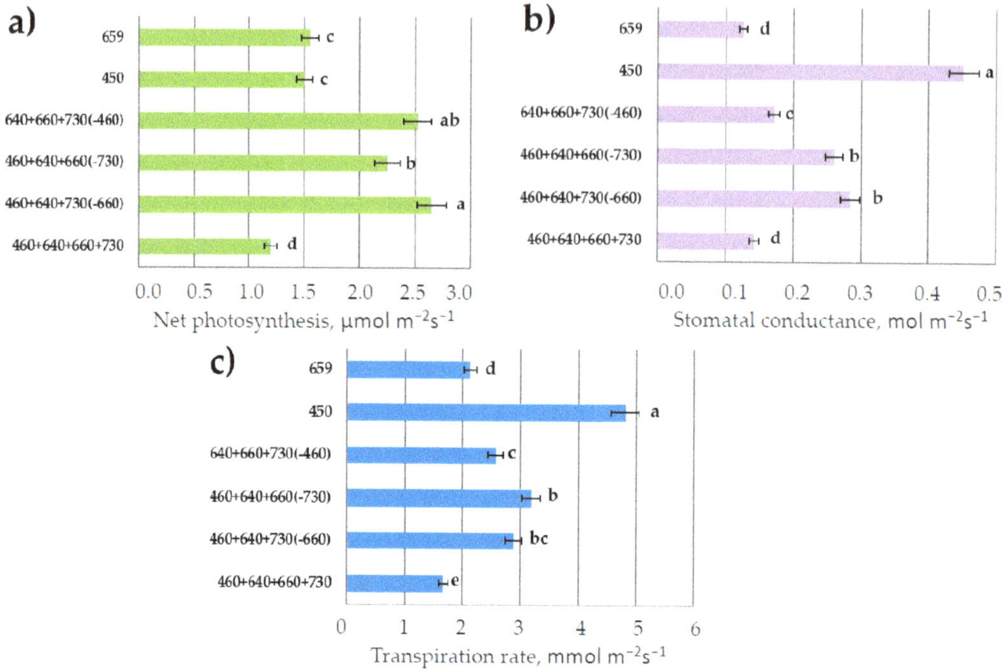

Figure 5. CO_2—H_2O leaf exchange in lettuce plants in response to various light treatments. (**a**) Net photosynthesis; (**b**) stomatal conductance; (**c**) transpiration rate. Means ± standard error (SE); means followed by the same letter were not different at $p \leq 0.05$. For light treatments legend see Figure 4.

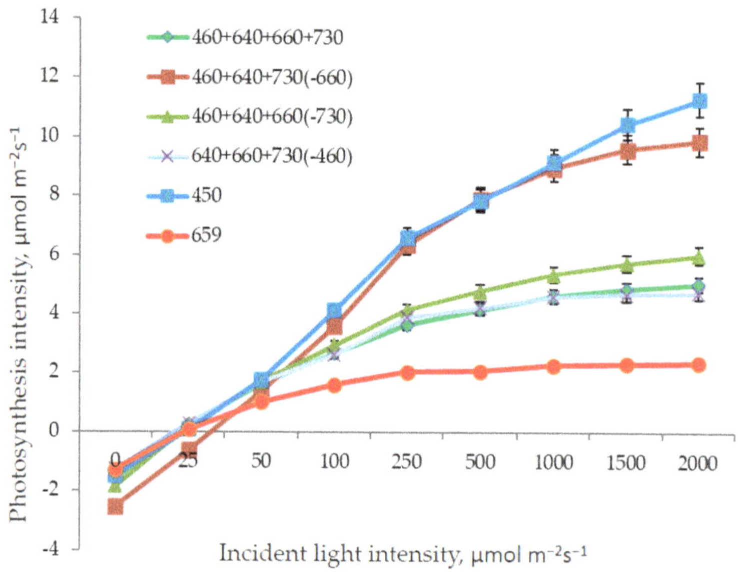

Figure 6. Light response curves in lettuce plants in response to various light treatments Means ± standard error (SE). For light treatments legend see Figure 4.

3.3. Chlorophyll *a* Fluorescence

The maximum quantum efficiency of PSII photochemistry (Fv/Fm) was comparable in all the red + blue spectral treatments and single red (Figure 7). Monochromatic blue light favored the increase in Fv/Fm. There were variations in the level of relative operating efficiency of PSII, but the differences among the treatments were not significant. Higher effectiveness of the photochemical processes was observed in response to monochromatic blue light and in treatments without red or far-red light (changes of the photochemical electron transport, ETR). Chlorophyll *a* non-photosynthetic quenching (NPQ) was relatively higher in the monochromatic-blue-light treatment.

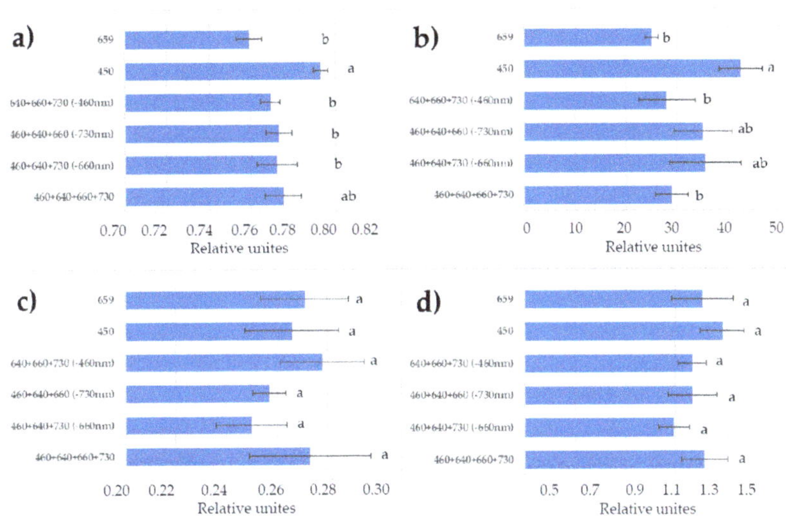

Figure 7. (**a**) Maximum quantum efficiency (Fv/Fm) of photosystem II (PSII); (**b**) photochemical electron transport rate (ETR); (**c**) relative PSII operating efficiency (ΦPSII); (**d**) chlorophyll *a* non-photosynthetic quenching (NPQ) in the leaves of lettuce plants in response to various light treatments. Means ± standard error (SE); means followed by the same letter were not different at $p \leq 0.05$. For light treatments legend see Figure 4.

3.4. Carbon Isotopes Discrimination

Sampling of plant material was carried out with 6 h intervals. Different light treatments showed multidirectional effects on the carbon isotope composition of leaf biomass, resulting in isotopic shifts in opposite directions (Figure 8). The strongest effects were observed in plants in response to monochromatic red and blue light as compared to the combined reference spectrum. Isotopic changes occurred in opposite directions. Thus, blue-light treatment resulted in ^{12}C enrichment of the leaf biomass; after 6 h of illumination it was 2.56‰ "lighter" in relation to the biomass in control treatment. Red light, on the contrary, induced ^{13}C enrichment of the leaf biomass; after 6 h of illumination it was 2.34‰ "heavier" in relation to the biomass in control experiment. In all the other treatments with combined spectrum, blue lightpresence in the spectrum resulted in a stable isotopic shift towards the enrichment of biomass with the ^{12}C isotope. Additionally, in a combined-spectrum environment missing blue light, the presence of red light resulted in biomass enrichment with the ^{13}C isotope.

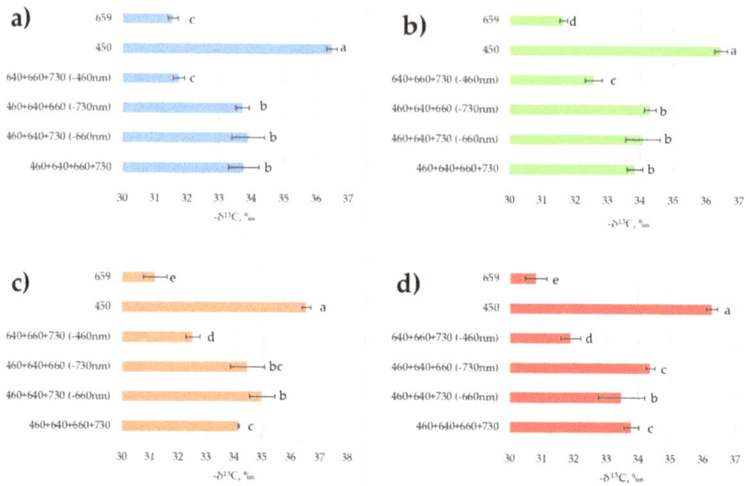

Figure 8. Carbon isotope composition of the leaves in lettuce plants grown in various light environments during 24 h cycle. Carbon isotope composition is given in PDBV $\delta^{13}C$ units. Means ± standard error (SE); means followed by the same letter were not different at $p \leq 0.05$. (**a**) After 6 h of illumination; (**b**) after 12 h of illumination; (**c**) after 18 h of illumination; (**d**) at the end of night after 6 h of darkness. For light treatments legend see Figure 4.

The results of the carbon isotopic differences in leaf biomass during the transition from light to dark are of particular interest. One could expect a strong rearrangement of the metabolic fluxes between day and night periods. Indeed, as we can see from Figure 8, during the light period, the leaf biomass became enriched with the ^{12}C isotope as compared to the biomass carbon composition detected at the end of the dark period. These isotopic differences occurred in all the lighting modes. Isotopic differences were quite distinct though not very great.

The data obtained are consistent with the results of Gessler et al. [39], who studied daily variations in the carbon isotope composition of the *Ricinus communis* plants and had found similar daily variations not only in the carbon of the plant leaf biomass but also in the carbon of its water-soluble and water-insoluble fractions and phloem sap.

4. Discussion

Studies on the light action in plants using various applications of LED techniques provide new insights into plant photobiology. The plant photosynthesis action spectrum matches with blue and red regions of photosynthetically active radiation in a natural environment [40–42]. In an artificial-light environment, the joint application of red and blue light usually results in increased plant photosynthesis and productivity [11,12,22]. Additionally, blue light is thought to participate in the acclimation of leaf photosynthesis to irradiance during growth [10,43]. These two spectral regions were the basic variables in our photobiological studies.

In our experimental set-up, we applied combined-spectrum treatments within two ranges of red light (R_{640}, R_{660}) trying to separate direct light effects on the PSA and light-induced photomorphogenetic responses controlled by the phytochromes. Indeed, far-red-light absence in the combined spectrum resulted in axial organ growth inhibition as compared to the treatments with far-red light (Figure 4g). Leaf blade elongation was also retarded (Figure 4c) due to the blocking of phytochrome-mediated shade-avoidance syndrome. Interestingly, the total leaf number increased significantly in this treatment, providing the growth of the light-harvesting leaf area of the plant. It is still unclear whether this response was observed due to the decreased plastochrone in the FR-deficient treatment or if other more sophisticated compensation mechanisms were involved. Similar results with stem and leaf growth inhibition were observed in response to monochromatic blue light (Figure 4c,g). Actually, the most serious inhibition of leaf blade growth in comparison with the other treatments was found in response to blue light. A reduction in leaf growth in response to blue light decreased plant biomass accumulation significantly.

Data on the decreased total leaf fresh weight yield in the treatment combining all four spectral regions in comparison with monochromatic red were unexpected. However, there are other data suggesting that lettuce biomass under monochromatic red was greater than under mixed red and blue light [44]. Comparable responses in other species were observed by Wollaeger and Runkle [45]. So, the synergetic or antagonistic effects of red and blue light on lettuce are still confused, and more studies need to be conducted [46].

As far as plant growth was inhibited in monochromatic blue light, net photosynthesis was also at a low rate in comparison with the combined-spectra treatments missing distinct spectral regions. The photosynthesis rate in the treatment missing long-wave red light R_{660} was one of the highest due to the compensation by short-wave red R_{640}; photosynthesis at saturating light intensity (light response curve) was also very high (Figure 5).

Net photosynthesis in the reference treatment was lower than in all the other treatments. This is most likely because in more stressful environments lacking distinct spectral regions, compensation mechanisms were activated. Additionally, the red-light PPFD share in the combined-spectrum treatment was much lower than PPFD in the monochromatic-red-light treatment. On the other hand, we observed plant acclimation to the abnormal light environments as a long-term process (sampling 30 days after emergence). This is most likely because an increased assimilate demand and increased sink capacity were the drivers of photosynthesis in monochromatic red light. We shall try to investigate this phenomenon in the future studies. Blue-light treatment significantly increased stomatal conductance and transpiration rate in plants and decreased their WUE; comparable results were obtained in tomato plants [47].

The main points of carbon isotope fractionation during photosynthesis are located at the crossings of the central metabolic pathways; therefore, the isotopic effects are reflected in the carbon isotope composition of biomass, fractions, and of the overwhelming number of metabolites [35]. The first carbon isotope fractionation point is located at the entry of the pentose phosphate reduction cycle (Calvin cycle) and is associated with the reaction of enzymatic carboxylation of ribulosebisphosphate (RuBP). As a result, the assimilated carbon atoms are enriched in ^{12}C in relation to the environmental CO_2. The enzyme that controls carboxylation, Rubisco, has the properties of oxygenase and is able to simultaneously redirect a part of the carbon flux assimilated in Calvin cycle to glycolate cycle, where it is partly oxidized to CO_2 and released back into the environment, creating so-called photorespiration flux. A probable mechanism of switching the functions of the enzyme is maintained by the changing ratio of CO_2/O_2 concentrations in the cell [48]. Due to such organization of photosynthesis, the activities of the Calvin cycle and glycolate cycle are separated in time, and the fluxes of carbon substrates resulting from assimilation and photorespiration become independent and discrete, that is, represented as separate portions [49]. In our experiment, we observed increased stomatal conductance in response to blue-light treatment (Figure 5). As a result, an increased CO_2 supply could enhance Rubisco carboxylating activity and it was followed with leaf tissue ^{12}C enrichment (Figure 8). These results are consistent with the data of other authors that have shown that $\delta^{13}C$ correlates negatively with stomatal conductance [50].

The most intensive lettuce leaf tissue enrichment with ^{13}C was observed in the treatments with monochromatic red light followed by in the combined-spectra environment missing blue light; in the last case this response could be attributed to the contradictory information from the blue- and red-light photoreceptors, as it was mentioned in Section 1. On the contrary, the biomass of plants subjected to blue-light treatment was enriched with the ^{12}C isotope. We have to stress here that plants were subjected to the long-term (during the whole growing cycle) light treatment. Therefore, chloroplast genesis could be affected significantly in the absence of blue light, as it was observed earlier [51]. On the contrary, monochromatic blue light was more favorable for chloroplast development and functioning [20,51]. In our studies, monochromatic-blue-light treatment maintained better plant photosynthetic performance, i.e., the highest maximum quantum efficiency (Fv/Fm) and a higher electron transport rate (ETR). Data from the light response curves show that photosynthesis at saturating light intensity in the blue-light-grown plants was four times higher than in the red-light-grown plants. Taking into consideration the facts discussed above, a possible explanation could be based on the variability in plant adaptations to the abnormal light environments during long-term 30-day exposure. That has resulted in the disturbance of PSA but to a lesser extent in the case of blue-light treatment as compared to red-light treatment.

The second point of carbon isotope fractionation is connected with increased photorespiration when observations show that plant biomass becomes enriched with ^{13}C [52]. This means that photorespiration is accompanied by an isotope effect of opposite sign than photoassimilation. Numerous studies on isotope fractionation in plants and artificial mutants have proved that the glycine decarboxylase reaction of the glycolate cycle was another place where the isotope effect is observed [33,53].

The third point of carbon isotope fractionation relates to post-photosynthetic metabolism and is associated with the end of the glycolytic chain where pyruvate dehydrogenase reaction proceeds. The observed proximity of the carbon isotope composition of the total plant biomass to assimilatory carbon pool suggests that the glycolytic chain and the majority of metabolites (lipids, proteins, lignins, and some carbohydrates), whose synthesis occurs via glycolytic chain, are supplied with the substrates of the assimilatory pool [54]. At the same time, the syntheses of soluble carbohydrates, organic acids, some amino acids, and other metabolite sis mainly bound to the "heavy" photorespiratory carbon pool. Because of the strict temporal and spatial organization in a cell, noticeable mixing of carbon fluxes does not occur, and various isotope distinctions exist [55].

The idea of the Rubisco oscillating mode of action has been analyzed extensively [32,34,35,56] and theoretically it was shown that oscillations can exist under real photosynthetic cell conditions. In the present paper, we returned to this idea. We assumed the presence of an isotopically "light" assimilatory pool and isotopically "heavy" pool of metabolites appear during photosynthesis as a result of dual function of Rubisco.

In our experiment, in all cases, leaf biomass at the end of light period was enriched in ^{12}C as compared to the leaf biomass at the end of dark period. Isotopic differences were quite distinct though not very great. Possible explanations of these differences could be given from our earlier paper [57] based on the model of oscillatory photosynthesis discussed above. Plant tissues enrichment with ^{12}C isotope during the light period was due to the fact that at this time lipids, proteins, lignins, and other structural components were synthesized mainly in the leaf. The isotopically "light" assimilatory pool was the substrate source for them. During the dark period, the outflow of assimilates to generative organs and heterotrophic tissues occurred. The outflow of assimilates occurred mainly in the form of sucrose and other water-soluble carbohydrates and metabolites, the isotope-heavy photorespiratory fund being their carbon source. Different sources of substrates for the synthesis of structural units and transport agents induced isotopic differences in daily variations of leaf biomass. Similar isotopic shifts were observed by other researchers while studying the isotopic differences between photosynthetic and heterotrophic organs and tissues [58].

We can conclude that blue light enhanced the assimilation function of the leaf, while red light enhanced the photorespiratory function. The simultaneous presence of blue and red light compensated for their mutual effects, and therefore the effects of light from the other spectral regions on the isotopic shifts became indistinguishable from the control. It was shown that duration of illumination (6, 12, and 18 h) had a weak effect on the isotope composition of biomass.

5. Conclusions

In our studies, variations in incident light spectral quality simulated with LEDs influenced growth and development in lettuce plants in several ways via direct effects on photosynthesis and control of plant photomorphogenetic responses. PSA structure and growth activity were significantly affected in the distinct light treatments, and these changes influenced source–sink relations in plants through the assimilate demand, etc. (indirect light effects on photosynthesis).

Our studies have shown that monochromatic blue light retarded lettuce plant growth, and monochromatic red light accelerated it. In plants exposed to blue light, the assimilating leaf area growth was retarded (both source and sink simultaneously!), and even an increased photosynthesis rate could not compensate for this delay. In the combined spectrum, far-red-light action was also important as far as it had triggered the shade-avoidance response and enhanced plant assimilate demand.

For the first time, it was found that the light of different PAR spectral regions affected the carbon isotope composition of leaf biomass. The strongest and most opposite in direction effects of monochromatic blue and red light were observed. Continuous blue-light treatment resulted in the ^{12}C enrichment of lettuce plant leaf biomass by about 3‰, whereas continuous red-light treatment resulted in ^{13}C enrichment of the same value. The effects of light of the other PAR spectral regions studied were considerably less significant. Daily variations in the leaf tissue carbon isotope composition were not significant.

Further research is needed to assess light-induced isotopic effects in plants and the mechanisms underlying them. These studies also could provide significant starting points for the development of the dynamic (changing in time) lighting regimes combining the advantages of the distinct spectra studied above at certain periods of plant growth. Thus, plant acclimation and photosynthetic improvements in response to added far-red and green-light wavelengths to the main red–blue spectrum have already been studied along with the changing red-to-blue-light ratio [59].

It is known that photorespiration can serve as an energy sink preventing the overreduction in the photosynthetic electron transport chain and photoinhibition, especially under stress conditions that lead to reduced rates of photosynthetic CO_2 assimilation and provides metabolites for other metabolic processes, e.g., glycine for the synthesis of glutathione, which is also involved in stress protection [60–62]. Therefore, another area of interest could be studies on plant stress responses and stress tolerance mechanisms including light-induced stress, e.g., extremely high PPFD or abnormal spectral environment adaptation.

Author Contributions: Conceptualization, I.G.T. and A.A.I.; methodology, I.G.T., V.A.L. and A.N.S.; software, N.N.S.; validation, I.G.T.; formal analysis, I.G.T.; investigation, I.G.T., D.A.T., M.P.L., A.S.S., A.N.S. and V.A.L.; resources, M.P.L.; data curation, I.G.T.; writing—original draft preparation, I.G.T. and A.A.I.; writing—review and editing, I.G.T. and A.A.I.; visualization, D.A.T.; supervision, I.G.T.; project administration, I.G.T.; funding acquisition, I.G.T. All authors have read and agreed to the published version of the manuscript.

Funding: This research was conducted with the support of the Ministry of Science and Higher Education of the Russian Federation in accordance with agreement № 075-15-2020-905, 16 November 2020, and a grant in the form of subsidies from the federal budget of the Russian Federation. The grant was provided for state support for the creation and development of a world-class scientific center: "Agrotechnologies for the Future".

Data Availability Statement: Data sharing is not applicable to this article.

Acknowledgments: We would like to thank Ivan S. Chuksin for his technical assistance in setting up plant sampling during carbon isotope determination.

Conflicts of Interest: The authors declare no conflict of interest. The funders had no role in the design of the study; in the collection, analyses, or interpretation of data; in the writing of the manuscript; or in the decision to publish the results.

References

1. Bian, Z.; Jiang, N.; Grundy, S.; Lu, C. Uncovering LED light effects on plant growth: New angles and perspectives-LED light for improving plant growth, nutrition and energy-use efficiency. *Acta Hortic.* **2017**, *1227*, 491–498. [CrossRef]
2. Loconsole, D.; Cocetta, G.; Santoro, P.; Ferrante, A. Optimization of LED Lighting and Quality Evaluation of Romaine Lettuce Grown in an Indoor Cultivation System. *Sustainability* **2019**, *11*, 841. [CrossRef]
3. Chen, X.L.; Yang, Q.C.; Song, W.P.; Wang, L.C.; Guo, W.Z.; Xue, X.Z. Growth and nutritional properties of lettuce affected by different alternating intervals of red and blue LED irradiation. *Sci. Hortic.* **2017**, *223*, 44–52. [CrossRef]
4. Lee, M.J.; Son, K.H.; Oh, M.M. Increase in biomass and bioactive compounds in lettuce under various ratios of red to far-red LED light supplemented with blue LED light. *Hortic. Environ. Biotechnol.* **2016**, *57*, 139–147. [CrossRef]
5. Fankhauser, C.; Chory, J. Light control of plant development. *Annu. Rev. Cell Dev. Biol.* **1997**, *13*, 203–229. [CrossRef] [PubMed]
6. Olle, M.; Viršile, A. The effects of light-emitting diode lighting on greenhouse plant growth and quality. *Agric. Food Sci.* **2013**, *22*, 223–234. [CrossRef]
7. Ouzounis, T.; Rosenqvist, E.; Ottosen, C.O. Spectral Effects of Artificial Light on Plant Physiology and Secondary Metabolism: A Review. *Hort. Sci.* **2015**, *50*, 1128–1135. [CrossRef]
8. Berkovich, Y.A.; Konovalova, I.O.; Smolyanina, S.O.; Erokhin, A.N.; Avercheva, O.V.; Bassarskaya, E.M.; Kochetova, G.V.; Zhigalova, T.V.; Yakovleva, O.S.; Tarakanov, I.G. LED crop illumination inside space greenhouses. *Reach. Rev. Hum. Space Explor.* **2017**, *6*, 11–24. [CrossRef]
9. Whitelam, G.; Halliday, K. *Light and Plant Development*; Blackwell: Oxford, UK, 2007.
10. Matsuda, R.; Ohashi-Kaneko, K.; Fujiwara, K.; Kurata, K. Effects of blue light deficiency on acclimation of light energy partitioning in PSII and CO_2 assimilation capacity to high irradiance in spinach leaves. *Plant Cell Physiol.* **2008**, *49*, 664–670. [CrossRef]
11. Yorio, N.C.; Goins, G.D.; Kagie, H.R.; Wheeler, R.M.; Sager, J.C. Improving spinach, radish, and lettuce growth under red light-emitting diodes (LEDs) with blue light supplementation. *Hort. Sci.* **2001**, *36*, 380–383. [CrossRef]
12. Matsuda, R.; Ohashi-Kaneko, K.; Fujiwara, K.; Goto, E.; Kurata, K. Photosynthetic characteristics of rice leaves grown under red light with or without supplemental blue light. *Plant Cell Physiol.* **2004**, *45*, 1870–1874. [CrossRef]
13. Terashima, I.; Fujita, T.; Inoue, T.; Chow, W.S.; Oguchi, R. Green light drives leaf photosynthesis more efficiently than red light in strong white light: Revisiting the enigmatic question of why leaves are green. *Plant Cell Physiol.* **2009**, *50*, 684–697. [CrossRef] [PubMed]
14. Folta, K.M.; Maruhnich, S.A. Green light: A signal to slow down or stop. *J. Exp. Bot.* **2007**, *58*, 3099–3111. [CrossRef] [PubMed]
15. Fan, X.X.; Xu, Z.G.; Liu, X.Y.; Tang, C.M.; Wang, L.W.; Han, X.L. Effect of light intensity on the growth and leaf development of young tomato plants grown under a combination of red and blue light. *Sci. Hort.* **2013**, *153*, 50–55. [CrossRef]
16. Lin, K.H.; Huang, M.Y.; Huang, W.D.; Hsu, M.H.; Yang, Z.W.; Yang, C.M. The effects of red blue and white light-emitting diodes on the growth development and edible quality of hydroponically grown lettuce (*Lactuca sativa* L Var Capitata). *Sci. Hort.* **2013**, *150*, 86–91. [CrossRef]
17. Li, Q.; Kubota, C. Effects of supplemental light quality on growth and phytochemicals of baby leaf lettuce. *Environ. Exp. Bot.* **2009**, *67*, 59–64. [CrossRef]
18. Camejo, D.; Frutos, A.; Mestre, T.; Carmen Pinero, M.; Rivero, R.M.; Martinez, V. Artifcial light impacts the physical and nutritional quality of lettuce plants. *Hortic. Environ. Biotechnol.* **2020**, *61*, 69–82. [CrossRef]
19. Kim, H.H.; Wheeler, R.; Sager, J.; Norikane, J. Photosynthesis of lettuce exposed to different short-term light qualities. *Environ. Control Biol.* **2005**, *43*, 113–119. [CrossRef]
20. Kochetova, G.V.; Belyaeva, O.B.; Gorshkova, D.S.; Vlasova, T.A.; Bassarskaya, E.M.; Zhigalova, T.V.; Avercheva, O.V. Long-term acclimation of barley photosynthetic apparatus to narrow-band red and blue light. *Photosythetica* **2018**, *56*, 851–860. [CrossRef]
21. Liere, K.; Weihe, A.; Börner, T. The transcription machineries of plant mitochondria and chloroplasts: Composition, function, and regulation. *J. Plant Physiol.* **2011**, *168*, 1345–1360. [CrossRef]
22. Goins, G.D.; Yorio, N.C.; Sanwo, M.M.; Brown, C.S. Photomorphogenesis, photosynthesis, and seed yield of wheat plants grown under red light-emitting diodes (LEDs) with and without supplemental blue lighting. *J. Exp. Bot.* **1997**, *48*, 1407–1413. [CrossRef] [PubMed]
23. Lee, S.-H.; Tewari, R.K.; Hahn, E.-J.; Paek, K.-Y. Photon flux density and light quality induce changes in growth, stomatal development, photosynthesis and transpiration of Withania somnifera (L.) Dunal. plantlets. *Plant Cell Tissue Organ Cult.* **2007**, *90*, 141–151. [CrossRef]
24. Shin, K.S.; Murthy, H.N.; Heo, J.W.; Hahn, E.J.; Paek, K.Y. The effect of light quality on the growth and development of in vitro cultured Doritaenopsis plants. *Acta Physiol. Plant* **2008**, *30*, 339–343. [CrossRef]
25. Prikupets, L.B.; Boos, G.V.; Terekhov, V.G.; Tarakanov, I.G. Research into influence from different ranges of PAR radiation on efficiency and biochemical composition of green salad foliage biomass. *Light Eng.* **2018**, *26*, 38–47. [CrossRef]
26. Prikupets, L.B.; Boos, G.V.; Terekhov, V.G.; Tarakanov, I.G. Optimization of lighting parameters of irradiation in light culture of lettuce plants using LED emitters. *Light Eng.* **2019**, *27*, 43–54. [CrossRef]
27. Kreslavski, V.D.; Carpentier, R.; Klimov, V.V.; Allakhverdiev, S.I. Transduction mechanisms of photoreceptor signals in plant cells. *J. Photochem. Photobiol. C Photochem. Rev.* **2009**, *10*, 63–80. [CrossRef]
28. Sipos, L.; Boros, L.F.; Csambalik, L.; Szekely, G.; Jung, A.; Balazs, L. Horticultural lighting system optimization: A review. *Sci. Hortic.* **2020**, *273*, 109631. [CrossRef]

29. Litvin, A.G.; Currey, C.J.; Wilson, L.F. Effects of supplemental light source on basil, dill, and parsley growth, morphology, aroma, and flavor. *J. Am. Soc. Hort. Sci.* **2020**, *145*, 18–29. [CrossRef]
30. Shulgina, A.A.; Kalashnikova, E.A.; Tarakanov, I.G.; Kirakosyan, R.N.; Cherednichenko, M.Y.; Polivanova, O.B.; Baranova, E.N.; Khaliluev, M.R. Influence of Light Conditions and Medium Composition on Morphophysiological Characteristics of Stevia rebaudiana Bertoni In Vitro and In Vivo. *Horticulturae* **2021**, *7*, 195. [CrossRef]
31. Voronin, V.; Ivlev, A.A.; Oskolkov, V.; Boettger, T. Intra-seasonal dynamics in metabolic processes of 13C/12C and 18O/16O in components of Scots pine twigs from southern Siberia interpreted with a conceptual framework based on the Carbon Metabolism Oscillatory Model. *BMC Plant Biol.* **2012**, *30*, 76. [CrossRef] [PubMed]
32. Ivlev, A.A.; Igamberdiev, A.Y.; Dubinsky, A.Yu. Isotopic composition of carbon metabolites and metabolic oscillations in the course of photosynthesis. *Biophysics.* **2010**, *49* (Suppl. 1), 3–16.
33. Igamberdiev, A.U.; Ivlev, A.A.; Bykova, N.V.; Threlkeld, Ch.; Lea, P.J.; Gardestrom, P. Decarboxylation of glycine contributes to carbon isotope fractionation in photosythetic organisms. *Photosynth. Res.* **2001**, *67*, 177–184. [CrossRef] [PubMed]
34. Roussel, M.R.; Ivlev, A.A.; Igamberdiev, A.Y. Oscillations of the internal CO_2 concentration in tobacco leaves transferred to low CO_2. *J. Plant Physiol.* **2007**, *164*, 1188–1196. [CrossRef] [PubMed]
35. Ivlev, A.A. Oscillatory nature of metabolism and carbon isotope distribution in photosynthesizing cells. In *Advances in Photosynthesis-Fundamental Aspects*; Najafpour, M.M., Ed.; Intech Publishers: Balkan, Croatia, 2012; pp. 341–366.
36. Mampholo, B.M.; Maboko, M.M.; Soundy, P.; Sivakumar, D. Phytochemicals and overall quality of leafy lettuce (Lactuca sativa L.) varieties grown in closed hydroponic system. *J Food Qual.* **2016**, *39*, 805–815. [CrossRef]
37. Goltsev, V.N.; Kalaji, H.M.; Paunov, M.; Bąba, W.; Horaczek, T.; Mojski, J.; Kociel, H.; Allakhverdiev, S.I. Variable chlorophyll fluorescence and its use for assessing physiological condition of plant photosynthetic apparatus. *Russ. J. Plant Physiol.* **2016**, *63*, 869–893. [CrossRef]
38. Shimazaki, K.; Doi, M.; Assmann, S.M.; Kinoshita, T. Light regulation of stomatal movement. *Annu. Rev. Plant Biol.* **2007**, *58*, 219–247. [CrossRef] [PubMed]
39. Gessler, A.; Tcherkez, G.; Peuke, A.D.; Ghashghaie, J.; Farquhar, G. Experimental evidence for diel variations of the carbon isotope composition in leaf, stem and phloem sap organic matter in Ricinus communis. *Plant Cell Environ.* **2008**, *31*, 941–953. [CrossRef] [PubMed]
40. McCree, K.J. The action spectrum, absorptance and quantum yield of photosynthesis in crop plants. *Agric. Meteorol.* **1972**, *9*, 191–216. [CrossRef]
41. Inada, K. Action spectra for photosynthesis in higher plants. *Plant Cell Physiol.* **1976**, *17*, 355–365.
42. Evans, J.R. The dependence of quantum yield on wavelength and growth irradiance. *Aust. J. Plant Physiol.* **1987**, *14*, 69–79. [CrossRef]
43. Walters, R.G. Towards an understanding of photosynthetic acclimation. *J. Exp. Bot.* **2005**, *411*, 435–447. [CrossRef] [PubMed]
44. Son, K.H.; Oh, M.M. Leaf shape, growth, and antioxidant phenolic compounds of two lettuce cultivars grown under various combinations of blue and red light-emitting diodes. *Hort. Sci.* **2013**, *48*, 988–995. [CrossRef]
45. Wollaeger, H.M.; Runkle, E.S. Growth of impatiens, petunia, salvia, and tomato seedlings under blue, green, and red lightemitting diodes. *Hort. Sci.* **2014**, *49*, 734–740.
46. Chen, X.L.; Li, Y.; Wang, L.C.; Guo, W.Z. Red and blue wavelengths affect the morphology, energy use efficiency and nutritional content of lettuce (Lactuca sativa L.). *Sci. Rep.* **2021**, *11*, 8374. [CrossRef] [PubMed]
47. Lanoue, J.; Leonardos, E.D.; Grodzinski, B. Effects of light quality and intensity on diurnal patterns and rates of photo-assimilate translocation and transpiration in tomato leaves. *Front. Plant Sci.* **2018**, *9*, 756. [CrossRef] [PubMed]
48. Ivlev, A.A. Carbon isotope effect (13C/ 13C) in biological systems. *Sep. Sci. Technol.* **2001**, *36*, 1815–1910. [CrossRef]
49. Ivlev, A.A. On the descretness of CO_2 assimilation by C-plants in the light. *Biofizika* **1989**, *34*, 887–901.
50. Sanchez-Bragado, R.; Serret, M.D.; Marimon, R.M.; Bort, J.; Araus, J.L. The Hydrogen Isotope Composition δ2H Reflects Plant Performance. *Plant Physiol.* **2019**, *180*, 793–812. [CrossRef]
51. Voskresenskaya, N.P. *Regulatory Role of Blue Light in Photosynthesis*; Nauka: Moscow, Russia, 1982; pp. 203–220.
52. Ivlev, A.A. On the flows of light and heavy carbon during photosynthesis and photorespiration coupling. *Russ. J. Plant Physiol.* **1993**, *40*, 752–758.
53. Igamberdiev, A.U.; Mikkelsen, T.N.; Ambus, P.; Bauwe, H.; Lea, P.J.; Gardeström, P. Photorespiration contributes to stomatal regulation and carbon isotope fractionation: A study with barley, potato and Arabidopsis plants deficient in glycine decarboxylase. *Photosynth. Res.* **2004**, *81*, 139–152. [CrossRef]
54. Ivlev, A.A. Carbon Isotope Heterogeneity in Photosynthesizing Biomass and Perspectives of Its Application in Biological Studies. In *Photosynthesis: Functional Genomics, Physiological Processes and Environmental Benefits*; Khan, N., Ed.; Nova Science Publishers: New York, NY, USA, 2015; pp. 73–104.
55. Schmidt, H.-L.; Kexel, H.; Butzenlechner, M.; Schwarz, S.; Gleixner, G.; Thimet, S.; Werner, R.A.; Gensler, M. 2 Non-statistical isotope distribution in natural compounds: Mirror of their biosynthesis and key for their origin assignment. In *Stable Isotopes in the Biosphere*; Wada, E., Yoneyama, T., Minagawa, M., Ando, T., Fry, B.D., Eds.; Kyoto University Press: Kyoto, Japan, 1995; pp. 17–35.
56. Dubinsky, A.Y.; Ivlev, A.A. Computational analysis of the possibility of the oscillatory dynamics in the processes of CO_2 assimilation and photorespiration. *BioSystems* **2011**, *103*, 285–290. [CrossRef] [PubMed]

57. Ivlev, A.A.; Pichouzkin, V.I.; Tarakanov, I.G. Soil salinity effect on carbon isotope composition of plant biomass. *Adv. Stud. Biol.* **2013**, *5*, 223–234. [CrossRef]
58. Cernusak, L.A.; Tcherkez, G.; Keitel, C.; Cornwell, W.K.; Santiago, L.S.; Knohl, A.; Barbour, M.M.; Williams, D.G.; Reich, P.B.; Ellsworth, D.S.; et al. Why are non-photosynthetic tissues generally 13C enriched compared with leaves in C3 plants? Review and synthesis of current hypotheses. *Funct. Plant Biol.* **2009**, *36*, 199–213. [CrossRef]
59. Tarakanov, I.G.; Kosobryukhov, A.A.; Tovstyko, D.A.; Anisimov, A.A.; Shulgina, A.A.; Sleptsov, N.N.; Kalashnikova, E.A.; Vassilev, A.V.; Kirakosyan, R.N. Effects of Light Spectral Quality on the Micropropagated Raspberry Plants during Ex Vitro Adaptation. *Plants* **2021**, *10*, 2071. [CrossRef] [PubMed]
60. Wingler, A.; Lea, P.J.; Quick, W.P.; Leegood, R.C. Photorespiration: Metabolic pathways and their role in stress protection. *Philos. Trans. R. Soc. Lond B Biol. Sci.* **2000**, *355*, 1517–1529. [CrossRef]
61. Fernie, A.R.; Bauwe, H.; Eisenhut, M.; Florian, A.; Hanson, D.T.; Hagemann, M.; Keech, O.; Mielewczik, M.; Nikoloski, Z.; Peterhänsel, C.; et al. Perspectives on plant photorespiratory metabolism. *Plant Biol.* **2013**, *15*, 748–753. [CrossRef] [PubMed]
62. Biel, K.J.; Fomina, I.R.; Yensen, R.G. *Complex Biological Systems: Adaptation and Tolerance to Extreme Environments*; Gorod: Krasnoyarsk, Russia, 1982; p. 334.

plants MDPI

Article

Daytime or Edge-of-Daytime Intra-Canopy Illumination Improves the Fruit Set of Bell Pepper at Passive Conditions in the Winter

Vivekanand Tiwari [1,†], Itzhak Kamara [1,†], Kira Ratner [1], Yair Many [1], Victor Lukyanov [2], Carmit Ziv [3], Ziva Gilad [4], Itzhak Esquira [5] and Dana Charuvi [1,*]

1 Institute of Plant Sciences, Agricultural Research Organization (ARO), Volcani Center, Rishon LeZion 7505101, Israel; vivekcas805@gmail.com (V.T.); itzahk@volcani.agri.gov.il (I.K.); kiratner@volcani.agri.gov.il (K.R.); yairm@volcani.agri.gov.il (Y.M.)
2 Institute of Soil, Water and Environmental Sciences, ARO, Volcani Center, Rishon LeZion 7505101, Israel; viclukyanov@gmail.com
3 Institute Postharvest and Food Sciences, ARO, Volcani Center, Rishon LeZion 7505101, Israel; carmit.ziv@volcani.agri.gov.il
4 Jordan Valley Research and Development Authority, Mobile Post 9190600, Israel; ziva@mop-bika.org.il
5 Faculty of Sciences and Technology, Tel-Hai College, Upper Galilee, Kiryat Shmona 1220800, Israel; esquirai@gmail.com
* Correspondence: charuvi@volcani.agri.gov.il
† These authors contributed equally to the work.

Abstract: Optimal light conditions ensure the availability of sufficient photosynthetic assimilates for supporting the survival and growth of fruit organs in crops. One of the growing uses of light-emitting diodes (LEDs) in horticulture is intra-canopy illumination or LED-interlighting, providing supplemental light for intensively cultivated crops directly within their canopies. Originally developed and applied in environmentally controlled greenhouses in northern latitude countries, this technique is nowadays also being tested and studied in other regions of the world such as the Mediterranean region. In the present work, we applied intra-canopy illumination for bell pepper grown in passive high tunnels in the Jordan Valley using a commercial LED product providing cool-white light. The study included testing of daytime ('LED-D') and edge-of-daytime ('LED-N') illumination, as well as a detailed characterization of fruit set and fruit survival throughout the growth season. We found that both light regimes significantly improved the fruit set and survival during winter, with some benefit of LED-N illumination. Notably, we found that western-facing plants of illuminated sections had a higher contribution toward the increased winter fruit set and spring yield than that of illuminated eastern-facing plants. Greater plant height and fresh weight of western-facing plants of the illuminated sections support the yield results. The differences likely reflect higher photosynthetic assimilation of western-facing plants as compared to eastern-facing ones, due to the higher daily light integral and higher canopy temperature of the former. This study provides important implications for the use of intra-canopy lighting for crops grown at passive winter conditions and exemplifies the significance of geographical positioning, opening additional avenues of investigation for optimization of its use for improving fruit yield under variable conditions.

Keywords: light-emitting diodes (LEDs); intra-canopy illumination; interlighting; bell pepper; photosynthesis; fruit set; daily light integral (DLI)

Citation: Tiwari, V.; Kamara, I.; Ratner, K.; Many, Y.; Lukyanov, V.; Ziv, C.; Gilad, Z.; Esquira, I.; Charuvi, D. Daytime or Edge-of-Daytime Intra-Canopy Illumination Improves the Fruit Set of Bell Pepper at Passive Conditions in the Winter. *Plants* **2022**, *11*, 424. https://doi.org/10.3390/plants11030424

Academic Editors: Valeria Cavallaro and Rosario Muleo

Received: 25 December 2021
Accepted: 2 February 2022
Published: 4 February 2022

1. Introduction

Light is one of the most important factors for crop production. Light supplementation is a common practice in greenhouses, particularly during winter in northern latitude countries. Nonetheless, even in regions that do not 'suffer' from a severe lack of light, the crop canopy can be light-limited due to self-shading, its geographical position, or cultivation in wintertime. Over the past two decades or so, light-emitting diodes (LEDs) have been replacing the conventional (fluorescent; incandescent; high-pressure sodium; metal halide) lighting sources. Nowadays, the use of LEDs in horticulture is quite widespread, including in middle latitudes, and has numerous advantages over other types of lighting [1–5]. One of these is the possibility to illuminate plants directly within their canopies without resulting in heat-induced damage, termed 'intra-canopy illumination' or 'interlighting' [6]. LED interlighting has been mostly applied for high-wire intensive crops, such as tomato, pepper, and cucumber, grown in environmentally controlled greenhouses, in addition to, or as an alternative to, overhead illumination during winter [7,8]. Use of intra-canopy illumination for these crops, in which tall and dense canopies can result in excessive shading, has been shown to improve plant growth and fruit yield (number and/or size) as well as affect fruit quality [9–11]. Over the past few years, studies and testing of interlighting have spread to additional regions of the world [12–14].

Pepper (*Capsicum annuum*) belongs to the Solanaceae (nightshade) family and is one of the world's most consumed fruit—in raw or cooked form, as well as processed into spices, condiments, or coloring agents [15]. Pepper fruit is an important source of vitamin A, -C, and -E, flavonoids, carotenoids, and additional antioxidant metabolites, highlighting its importance for human nutrition and health [16]. Bell pepper is cultivated as an annual crop all over the world, with long growth periods spanning different seasons. In countries with extremely cold winters and limited natural light, e.g., Canada, the Netherlands, or northern regions of the United States of America, bell pepper is grown in environmentally controlled greenhouses, with high-quality fruit harvest in spring and summer. In tropical, semi-arid climates such as Mexico, the crop is grown (for the most part) with minimal or without climate control. In other regions of the world, such as the Mediterranean basin, bell pepper is grown as a protected crop, with the advantage of being able to produce high quality fruit during winter.

Light limitation in winter is a major factor limiting fruit yield. Even in countries with mostly mild winters, the daily light integral (DLI) in winter is low [12] and temperatures fall below those that are optimal for growth and photosynthesis of bell pepper. Furthermore, the crop cultivation technique can also result in considerable shading. Growth in the "Spanish" trellis system has the benefits of higher yields of large fruit size, a lower rate of blossom-end rot, and 75% lower labor costs required for pruning, over the "V" system for bell pepper cultivation [17]. Nonetheless, a lack of pruning of branches or leaves reduces light penetration into the canopy, resulting in disadvantageous non-uniform light distribution and reduced photosynthesis [18].

Bell pepper grown under passive (protected) conditions is characterized by waves of fruit production, with variable time kinetics for fruit growth, development, and ripening along the growth season. The pattern of waves is determined by the environmental conditions, which affect the photosynthetic efficiency, as well as by the on-plant fruit load, together influencing source–sink relations and ultimately organ (flower bud, flower, young fruits) development and/or survival. Sweet pepper is generally characterized by high organ abortion rates, affected by various factors, as reviewed in [19], with light being a predominant one. Experiments in which bell pepper plants were subjected to low light by shading, or where adjacent leaves were removed, showed that these conditions correlated with reduced sugar accumulation in the flower, increasing flower abscission and reducing fruit set [20]. Additional studies also showed that source and sink strengths are major determinants of organ abortion in pepper [21,22], and that the fruit sink strength can affect the photosynthetic characteristics of proximal leaves [23]. Extending the photoperiod, up to a certain extent, was reported to increase pepper fruit yield [24,25]. Improving photosynthesis, by providing optimal light, CO_2, and temperature conditions, would improve the source strength, reduce organ abortion [19], and support the development of more fruit, thus increasing the yield.

Various studies, conducted at different conditions, have shown that light supplementation of the pepper plant canopy can improve the yield. Increasing the photosynthetic photon flux at different heights of the pepper plant canopy (using HPS lamps at the time) resulted in a 23% increase in the total fruit yield [26]. Later studies demonstrated the improvement of pepper fruit yield using LED-interlighting applied in environmentally controlled greenhouses [10,27]. We recently reported that application of daytime intra-canopy LED illumination for bell pepper grown in passive tunnels in the Jordan Valley results in increased yield during the spring months, due to a higher number of fruit [12]. In a follow-up experiment, we found indications for a higher number of fruitlets in western-facing plants of the double-row beds that were illuminated, as compared to non-illuminated plants on the same side. On the other hand, there was no difference in the number of fruitlets in eastern-facing plants with or without illumination. In the current study, we applied intra-canopy illumination for bell pepper using a commercial product and aimed to (1) compare the effects of daytime- ('LED-D') vs. edge-of-daytime ('LED-N') illumination; and (2) characterize fruit set and fruit survival under the two illumination regimes for eastern- and western-facing plants, as compared to non-illuminated ones.

2. Materials and Methods

2.1. Plants and Growth Conditions

The study was carried out at the 'Zvi' R&D Experimental Station in the Jordan Valley (31°59′49.0″ N, 35°27′09.3″ E) from August 2019 to May 2020. Crop management followed the routine practices of the region. Prior to planting, soil sanitation was carried out by solar fumigation for 4 weeks, with streaming of metam sodium (40 mL m^{-2}) into the soil via drip irrigation during the last week. Red bell pepper seedlings (*Capsicum annuum* L. cv. Cannon, Zeraim Gedera/Syngenta, Revadim, Israel) were planted on 20 August 2019, in a high tunnel (10 m wide × 4 m high × 45 m long) in the local soil, a well-drained (EC < 2.0 dS m^{-1}) clay (30%)–limestone (50%) marl soil. Planting was in double-row beds of width 0.8 m, and center-to-center distance between beds of 1.75 m, with an overall plant density of 2.9 m^{-2}. Plant training applied the 'Spanish' trellis system, with lateral horizontal wires supporting the canopy vertically, and without pruning.

Drip irrigation (emitter flow rate 1.6 L h^{-1}) was provided every 20 cm along each plant row, with a total of ~8000 m^3 ha^{-1} per growth season (from planting in Aug. until May), similarly to commercial plots. Crop irrigation varied according to evapotranspiration calculated (Penman-Monteith FAO56, [28]) from the local meteorological data: ~40 m^3 ha^{-1} d^{-1} from planting to mid-Dec., ~10 m^3 ha^{-1} d^{-1} from mid-Dec. to the end of Feb., and 60–70 m^3 ha^{-1} d^{-1} in March to May. Fertigation with N-P-K (6:3:9, ICL, Tel-Aviv, Israel) was provided at a concentration that varied between 1 to 1.5 L m^{-3} until Feb., and 0.5 L m^{-3} afterward. Fe (5 kg ha^{-1} of Sequestrene Fe 6%, Syngenta) and Mn (15 L ha^{-1} of Koratin-Mn 18 g L^{-1}, ICL Israel) were provided 3 times, at the beginning of November, mid-December, and the beginning of March.

At the time of planting, the tunnel was covered by a 50-mesh insect-proof screen with a black shade net (40%) on top of it. On 19 September 2019, the shade net was removed, and on 17 November 2019, the mesh screen was replaced by a polyethylene sheet (Ginegar Plastic Products Ltd., Kibbutz Ginegar, Israel). To prevent fruit heat damage, on 16 March 2020, the plastic sheet was removed and the 50-mesh screen together with the black shade net (40%) were placed on top of the tunnel until the end of the experiment. The fruit yield was followed in the spring, with harvesting according to the commercial standard of picking at >60% red color. Both 'class 1' (export-quality fruit) and 'class 2' (for local market) were included in the spring yield.

2.2. Supplemental Intra-Canopy Illumination

The supplemental LED illumination was assembled from Crops IP67 tubes (Bioled Eco Light Systems Ltd., Tzova, Israel), providing cool-white (CW; 5700K) light at 32 W/m. CW was chosen as it was found to be preferable for bell pepper in our earlier study [12]. For simplicity, we refer to the LED tubes as 'Bioled' in the text. Two LED tubes affixed back-to-back were installed between the two adjacent rows of the beds (Figure 1). Two illumination regimes of 12 h were provided: daytime ('LED-D', 6:00 to 18:00) and edge-of-daytime ('LED-N', 4:00–10:00 and 16:00–22:00). The experimental setup encompassed four replicate sections (5.4-m-long each) for each of the two intra-canopy light treatments and for the non-illuminated control (Figure 2). The illumination period began 70 days after planting (28 October 2019), when the canopy height was ~1.5 m. Fixtures were installed at a height of 70–80 cm aboveground at the start of the illumination period, raised to 90–100 cm in the middle of December 2019, and raised again to 110–120 cm in the middle of March 2020.

Figure 1. Intra-canopy supplemental illumination. (**A**) Installation of the 'Bioled' light fixtures at the center of the beds (September 2019, prior to start of illumination treatments). (**B**) Side-view of intra-canopy back-to-back LED illumination (picture acquired in March 2020).

Spectra and photosynthetic photon flux densities (PPFD) were recorded using a portable spectroradiometer (EPP2000C, StellarNet, Inc., Tampa, FL, USA) with a cosine-corrected head (Apogee Instruments Inc., Logan, UT, USA) and an LI-250A quantum sensor (LI-COR, Lincoln, NE, USA), respectively. Air temperature within the canopy was recorded using HOBO temperature data loggers (Onset Computer Corporation, Bourne, MA, USA) hung in proximity (10–15 cm) to the LED fixtures or at the same height in control sections.

2.3. Chlorophyll Content and Fluorescence

Measurements were conducted non-destructively on attached leaves of the inner canopy. Chlorophyll (Chl) content was assayed using an MC-100 Chl measurement system (Apogee, Chesapeake, VA, USA). Chl-*a* fluorescence emission was measured using a portable pulse-amplitude-modulated fluorometer (PAM-2000, Heinz Walz GmbH, Pfullingen, Germany) at its default setting designed to determine Fv/Fm ('Da-2000' program). In brief, leaves were subjected to dark adaptation for 20 min using dark leaf clips (DLC-8, Walz), and then initial Chl-*a* fluorescence (F0) and maximum Chl-*a* fluorescence in dark (Fm) were recorded after applying a saturating light pulse for 0.8 s. The Fv/Fm = [(Fm − F0)/Fm] values were calculated by the program and recorded.

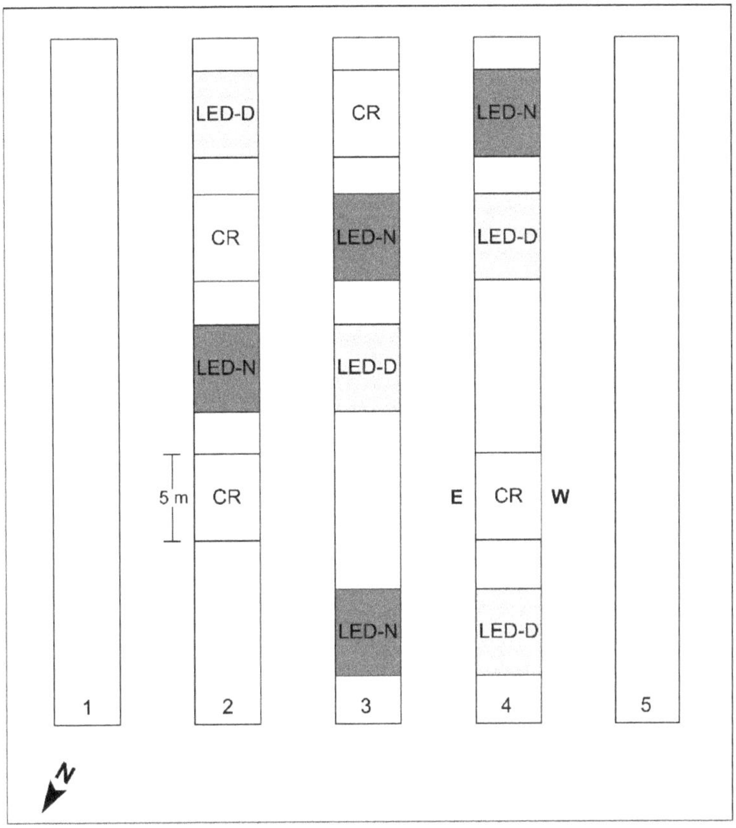

Figure 2. Schematic map of the experimental tunnel. The experiment was carried out in the three central double-row beds (2, 3, and 4) of the tunnel. Intra-canopy lighting was applied at the center of the beds between the two rows (Figure 1A), along 5 m-long sections. The illumination was applied either during daytime ('LED-D') or the two edges of daylight period ('LED-N'), with non-illuminated sections of the same length as controls ('CR'). Each treatment had four replicates. A spacing of at least 2 m was kept between sections. 'E' and 'W' denote the eastern- and western-facing rows, with regard to the results presented in further Figures and Tables. Note that beds and experimental sections are not drawn to scale.

2.4. Gas-Exchange Measurements

Gas-exchange measurements were conducted on attached leaves of the inner or outer canopy using a portable LCi photosynthesis system (ADC BioScientific Ltd., Hoddesdon, UK) with a clear top chamber at ambient conditions. For the inner canopy, leaves were sampled 10–15 cm above the upper LED fixtures, at a height of 110–120 cm above the ground, and at the same height in non-illuminated plots. For outer canopy measurements, leaves from the eastern- and western-facing canopy were probed in the morning or afternoon during peak photosynthetically active radiation (PAR) intensities on each side. Positioning of the LCi chamber was adjusted according to the leaf being measured (to keep the leaf attached) and to allow natural sunlight or light from the LEDs to reach the leaf.

2.5. Fruit Set Quantification

In each of the experimental replicate sections (four sections for each treatment) shown in Figure 2, ten plants were selected for the analysis: five in the eastern-facing row and five in the western-facing row of the same bed. Fruitlets were counted, labelled, and screened for survival on eleven dates throughout the season. Survival of fruit labelled on one date were assayed on the next fruitlet labelling date. On the last day of the experiment (7 May 2020), fruitlets remaining on the plants were counted.

2.6. Daily Light Integral (DLI) Recording

PAR (400–700 nm) was recorded using LI-190SB-L quantum sensors (LI-COR, Lincoln, NE, USA) installed at the eastern- and western-facing canopy (at a height of 1 m above ground), as well as above the canopy (height of 3 m) at 90 degrees. Data were recorded every ten minutes and logged by a Campbell system (CR10X—Campbell Scientific, Logan, UT, USA).

2.7. Statistical Analysis

Statistical analysis was performed using one-way analysis of variance (ANOVA), followed by Tukey's multiple comparisons as a post hoc test. For comparisons of two groups, the Student's *t*-test was used. The level of significance is provided in the figure legends/table.

3. Results

3.1. Intra-Canopy Illumination

The 'Bioled' light fixtures utilized as intra-canopy illumination provided cool-white (CW) light (Figure 3A). The effect of the added illumination on the light intensity between the double-row beds was assessed when the canopy height was ~2 m. In control ('CR') non-illuminated sections, the light intensity of the inner canopy below 1.5 m was generally <50 µmol photons m^{-2} s^{-1}, and for the most part even <20 µmol photons m^{-2} s^{-1} (Figure 3B). In the illuminated sections, a region of almost 1 m in height, from 30 to 120 cm aboveground, exhibited significantly higher light intensities, reaching an average intensity of 225 µmol photons m^{-2} s^{-1} in proximity to the fixtures (Figure 3B). At a canopy height of 1.5 to 1.8 m, the light intensities in the CR and illuminated sections were similar, and the considerable higher intensity at 1.8 m is due to sunlight penetration at this height.

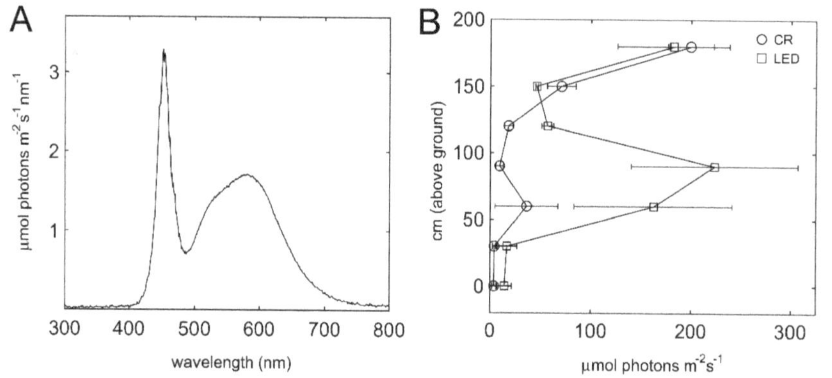

Figure 3. Light spectrum and intensity within the canopy. (**A**) Spectra of the cool-white 'Bioled' light fixtures. (**B**) Light intensity within the canopy in control vs. illuminated sections, recorded in December 2019. Intensities were measured with the light sensor directly below or above the fixtures at the indicated distances above the ground. Values shown are means ± SD of three control (CR) and three illuminated (LED) plots.

The effects of the supplemental illumination, provided by Bioled fixtures, on the photosynthetic parameters and gas-exchange activity of inner canopy leaves were characterized in LED-D plots (Table 1). The chlorophyll content ('Chl'; measured non-destructively) was somewhat higher (~8%) in the illuminated leaves as compared to control ones. The CO_2

assimilation rates ('A') in leaves of LED sections were ~3.3-fold higher than non-illuminated leaves. The stomatal conductance ('Gs') and transpiration rate ('E') were, respectively, 5.2- and 3.5-fold higher in illuminated leaves vs. control ones. As the temperature of the leaves probed for the gas-exchange recordings was the same in both control and LED, the higher Gs and E can be attributed to the supplemental light. The average light intensity, recorded during the measurements ('PAR'), was ~3.7-fold higher in LED than in CR.

Table 1. Photosynthetic and gas-exchange parameters of the inner canopy †.

Parameter	CR	LED
Chl (μmol m^{-2})	571 ± 59 b	616 ± 57 a
Fv/Fm	0.81 ± 0.01 a	0.81 ± 0.01 a
A (μmol CO_2 m^{-2} s^{-1})	2.01 ± 0.76 b	6.56 ± 1.87 a
Gs (mol H_2O m^{-2} s^{-1})	0.030 ± 0.017 b	0.157 ± 0.036 a
E (mmol H_2O m^{-2} s^{-1})	0.530 ± 0.307 b	1.853 ± 0.264 a
Ci (μmol CO_2 mol^{-1})	283 ± 48 b	351 ± 32 a
T (°C)	27.7 ± 1.6 a	27.6 ± 0.6 a
PAR (μmol photons m^{-2} s^{-1})	23 ± 6 b	86 ± 36 a

† Measurements were recorded non-destructively on inner canopy leaves from control (CR) and illuminated sections (LED). Recordings were made on leaves found 10–20 cm above the LED fixtures, and at the same height in control sections. For chlorophyll (Chl) and Fv/Fm measurements, n = 15 and 18 leaves, respectively. For gas-exchange measurements, n = 12 leaves from LED-D or CR sections. A, CO_2 assimilation rate; Gs, stomatal conductance; E, transpiration rate; Ci, intercellular CO_2. Leaf temperature (T) and light intensity (PAR) were recorded during the gas-exchange measurement. Values shown represent means ± SD; distinct letters denote statistical significant differences ($p < 0.05$) between CR and LED.

The effect of the illumination on air temperature within the canopy, in vicinity of the LED fixtures, was recorded along the season. The daily minimal (T-min) and maximal (T-max) air temperatures in the canopy of the three treatments are shown in Figure S1. Three examples for raw air temperature data on representative days depict how air temperature in the canopy is affected by operation of the illumination in LED-D and LED-N (Figure S2). In CR sections, T-min typically occurs between 04:00 and 07:00, and T-max between 12:00 and 15:00. The timing of T-max depends on the time of year and on whether the day is sunny (Figure S2A) or cloudy (Figure S2B). In LED-D sections, where the illumination operated from 06:00 to 18:00, the air temperature was higher by ~4.5 °C than the CR during the operation time. Accordingly, T-max was higher to a similar extent on most days in LED-D (Figure S1). In LED-N sections, higher air temperatures within the canopy were observed at the edges of daytime, in line with the operation times of the illumination for this treatment. As compared to CR, the air temperature within the canopy in LED-N was higher by 3.9 to 5.2 °C from 04:00 to 8:00 and by 4.2 to 5.6 °C from 16:00 to 22:00 (Figure S2). On some days, T-max in LED-N was higher than in CR and occurred around 10:00, just prior to the end of the first illumination period (Figure S2B,C). This is also evident in the whole season graph for T-max (Figure S1). Although LED-N operated from 04:00 to 10:00, the effect of the illumination on daily T-min was minor (Figures S1 and S2). The increased air temperature observed in LED treatments is mostly limited to the regions surrounding the light fixtures. Even though the increase in air temperature may not necessarily result in considerable increases in foliage temperature, it should still be kept in mind as a factor that can affect the plant physiology.

3.2. Supplemental Illumination Results in Increased Fruit Set in the Winter

We previously (2016–2018) found that using intra-canopy illumination in our experimental conditions increases the pepper fruit spring yield by ~30% [12]. In another experiment carried out with the Bioled CW fixtures used for daytime illumination (2018–2019), we quantified the fruit set accumulation during two months in the winter (Figure S3). We found that plants that were illuminated (at their inner canopy) had 46% more fruitlets than the control (whole 'total' bars). However, after assaying the fruitlet survival, illuminated plants remained with ~16% more fruit than control (colored part of 'total' bars). Notably, the increase in fruit number arose from western (W)-facing illuminated plants. These plants had 80% more fruitlets as compared to the W-facing control plants (whole 'western' bars). After assaying the fruit survival, there was 33% more fruit on the W-facing illuminated plants vs. control (colored part of bars). Conversely, the fruit set and fruit survival of the eastern-('E') facing plants was nearly identical in illuminated and control plants. These earlier results provided the rationale for the current analysis of fruit set, as described below and in Section 3.3.

To gain detailed insight into the fruit set behavior of illuminated vs. non-illuminated plants, in the current study, we followed the fruit set and survival in LED-D and LED-N light treatments and in the CR throughout the entire experiment (Figure 4). Panels A–C of this figure depict the average count of fruitlets and the surviving fraction from the four replicate sections of each treatment. In each section, the number of fruitlets were summed for ten plants (five from the E-facing row and five from the W-facing row; see Figure 2). Whole bars (means $\pm$ SD), including white and colored parts, show the average number of fruitlets counted on the indicated date. The colored parts of the bars indicate the surviving fruit, assayed two weeks later.

We were specifically interested in the winter period, during which an improved fruit set would lead to an increased yield in spring months. Examining the fruit set in CR sections, relatively low levels are evident between the end of December to the end of March (Figure 4A). This three-month period is marked by the black brackets in Figure 4A–C. Notably, during this period, plants from both LED-D (Figure 4B) and LED-N (Figure 4C) exhibited higher fruit set and survival, specifically during the coldest part of the winter (Figure S4). Figure 4D depicts the cumulative fruit set and fruit survival during the aforementioned time period. Fruit set (dashed lines) was considerably higher in both LED-D and LED-N, respectively, by 55% and 74%, as compared to the CR. Likewise, the number of surviving fruit (solid lines) in LED-D and LED-N were 51% and 67% higher than the CR.

3.3. *Fruit Set and Survival Are Enhanced in Illuminated Western-Facing Plants*

Following the data we obtained in the earlier experiment for winter fruit set and survival in E- and W-facing plants (Figure S3), we also assessed the data shown in Figure 4D separately for E and W. Figure 5 shows the fruit set and survival in the three treatments of the study (CR, LED-D, LED-N) in E- and W-facing plants. For each treatment, E is shown by the lighter-colored lines and W by the darker-colored lines of the same shade. In illuminated sections of either LED-D or LED-N, the fruit set was higher in W-facing plants as compared to E-facing ones of the same treatment (Figure 5A). Nonetheless, in CR sections, the fruit set was nearly identical in E- and W-facing plants. The highest number of fruitlets was observed in LED-N-W plants (Figure 5A, orange line), 76% higher than CR-W. From the light treatments, LED-D-E had the lowest, but still considerably high, number of fruitlets, 48% higher than CR-E. The number of surviving fruit (Figure 5B) reflects that of the fruit set. Only W-facing plants, of both LED-D and LED-N, exhibited a significantly higher number of surviving fruit as compared to the CR. These were 78% and 62% higher for LED-N-W and LED-D-W, respectively. The trend for a higher number of fruit in LED-N vs. LED-D was observed in both fruit set and fruit survival, although the differences between the two were not statistically significant.

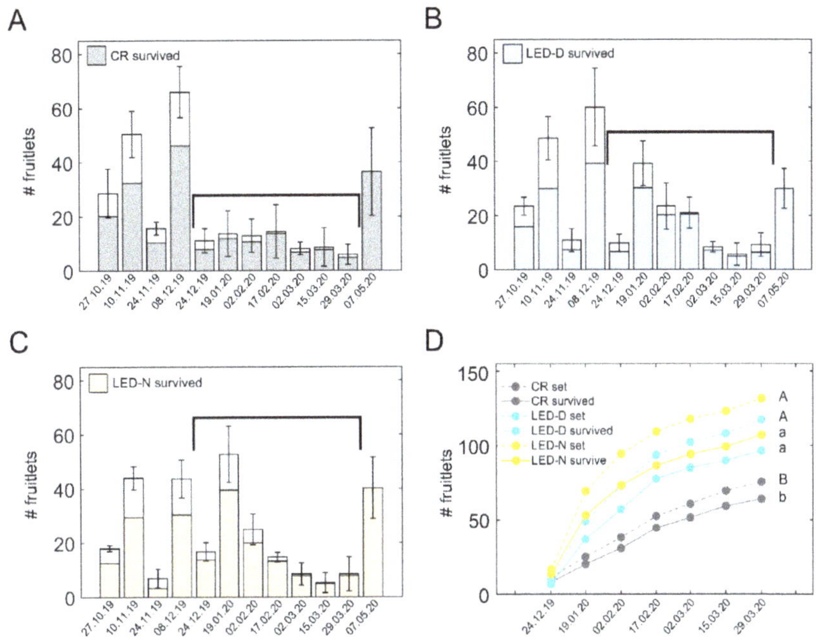

Figure 4. Supplemental intra-canopy illumination increases fruit set and survival in the winter. Fruitlets were labelled along the season in (**A**) control non-illuminated (CR) sections, (**B**) sections illuminated during daytime (LED-D), and (**C**) sections illuminated at the edge of day (LED-N). Whole bars denote the average number ± SD of fruitlets from four sections. In each section, the number of fruitlets was summed for ten plants: five in the eastern-facing row and five in the western-facing row of the same bed. Each section was from a different replicate in the experimental plot, see Figure 2) labelled on the noted dates. Colored portion of the bar shows the fraction of surviving fruits from the total. At the end of the experiment (07.05.20 bar), fruitlets remaining on the plants were counted without follow-up for survival. Black brackets denote the winter time period between the two big fruit set waves of control non-illuminated plants (**A**). (**D**) Cumulative number of fruitlets (dashed lines) and surviving fruit (solid lines) during the winter, corresponding to the period marked by the brackets in (**A**–**C**). Distinct upper- and lower-case letters denote statistically significant differences ($p < 0.05$) for the total number of labelled fruitlets and for surviving fruit, respectively.

3.4. Daily Light Integral and Photosynthetic Activity of the Eastern- and Western-Facing Canopy

To better understand the differential effect of the intra-canopy illumination on E- and W-facing plants, we also probed the natural light conditions and photosynthetic activity at the outer parts of the canopy in these plants. PAR was recorded at the E- and W-facing outer canopy and the daily light integral (DLI) was calculated from the recorded values. Note the positioning of the experimental tunnel, with 'eastern' plants inclined (~25°) toward the north and 'western' plants toward the south (Figure 2). Figure 6 depicts PAR recordings and the derived DLI at the E- and W-facing canopy and above the canopy on two representative sunny days during the winter. On these days, the DLI inside the tunnel, covered by the polyethylene sheet, was 21 and 24 mol photons m^{-2} d^{-1}. PAR sensors at the E- and W-facing canopy were positioned such that they mimic light capture by the canopy. The recordings made at the E- and W-facing canopies show that the DLI at the latter was 2.5-fold (January) and 2-fold (February) higher (Figure 6). Higher DLI values, at both sides of the canopy, were recorded in February as compared to January, as expected when days become longer toward the spring. For E-facing plants, the peak in light intensity was around 9:00–9:30, while for W-facing plants, the peak was between 14:00 and 15:00. Furthermore, the light intensity during peak times was much higher for W-facing plants.

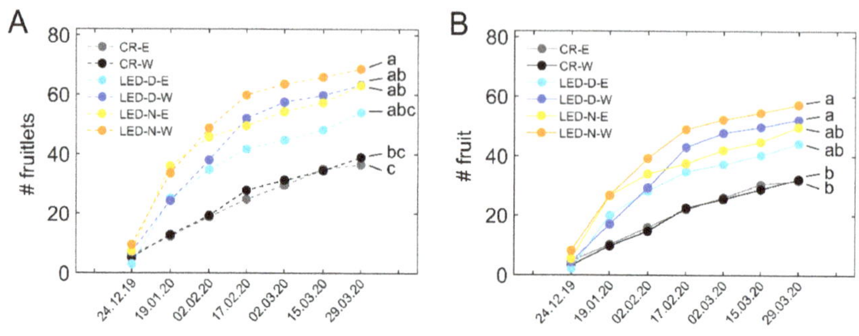

Figure 5. Fruit set and survival in eastern- and western-facing plants. (**A**) Cumulative number of fruitlets labelled during the winter in the eastern (E)- and western (W)-facing plants of control non-illuminated (CR) sections, and in the E- and W-facing sections illuminated during daytime (LED-D) or edge of day (LED-N). Values shown represent the average number ± SD of fruitlets from four E or W sections, each from a different replicate in the experimental plot, see Figure 2). For each E or W section, fruitlets were summed for five plants. (**B**) Cumulative surviving fruit from the E- and W-facing plants of the different treatments, corresponding to the fruitlets that were labelled (**A**). Distinct letters denote statistically significant differences ($p < 0.05$) between the six groups. Data shown in this figure are the same data shown in Figure 4D, separated to E and W.

The photosynthetic activity of plants on the two sides was probed during light peak morning and afternoon hours, using gas exchange measurements of attached outer-canopy leaves (Figure 7). CO_2 assimilation rates were similar (~6 μmol m^{-2} s^{-1}) for the outer canopy side not subjected to direct sunlight: in the morning for W-facing plants and in the afternoon for E-facing plants (Figure 7A). However, the assimilation rates of the W-facing canopy in the afternoon were ~25% higher than those of the E-facing canopy in the morning (Figure 7A). This is due to both the higher light intensity (Figure 7F) and higher leaf temperature (Figure 6E) on the W in the afternoon as compared with E in the morning. Cooling of the canopy via evapotranspiration is prominent for W-facing plants in the afternoon (Figure 7C). The resultant enhanced gas exchange (Figure 7B) contributes toward CO_2 assimilation in these plants. Lower intercellular CO_2 is supportive of the higher assimilation in W-facing plants during the afternoon (Figure 7D).

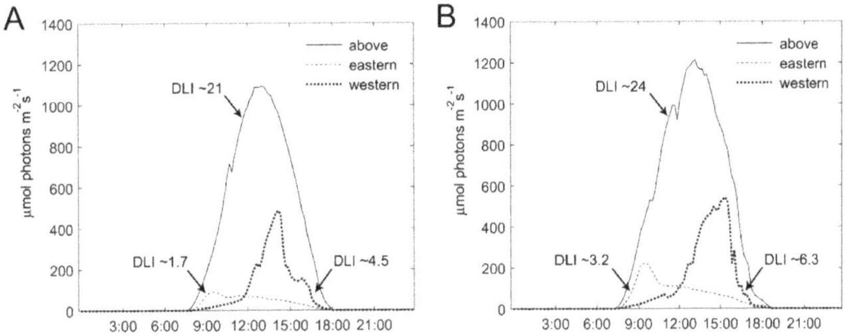

Figure 6. Photosynthetically-active radiation of sunlight at the outer canopy of eastern- and western-facing plants. Photosynthetically-active radiation (PAR; 400–700 nm) shown for two sunny days: (**A**) 6 January 2020 and (**B**) 17 February 2020, was recorded above the canopy (at a height of ~3 m), and at the outer canopy of eastern- and western-facing rows at a height of 1 m. Derived values of the daily light integral (DLI, in mol photons m^{-2} d^{-1}) at the different positions are denoted by arrows.

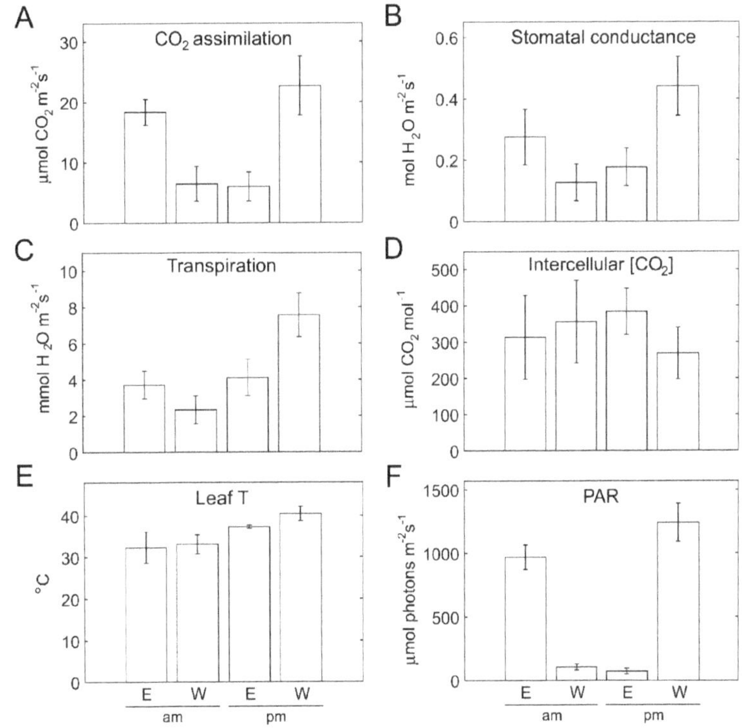

Figure 7. Gas-exchange parameters of the outer canopy of eastern- and western-facing plants in the morning vs. afternoon. (**A**) CO_2 assimilation rate, (**B**) stomatal conductance, (**C**) transpiration rate, and (**D**) intercellular CO_2 of leaves of the outer canopy in eastern (E)- and western (W)-facing plants assayed in the morning (am, 8:50–9:30) and afternoon (pm, 13:50–14:30). (**E**) Leaf temperature and (**F**) light intensity (PAR) were recorded during the gas-exchange measurements. Values shown represent means ± SD of 14 leaves (am) and 10 leaves (pm) measured non-destructively.

3.5. Spring Yield and Plant Biomass

The yield in spring from the experimental sections was summed for six harvests (9 March to 5 May 2020), shown at the top two rows of Table 2. Fruits were picked from the E- and W-facing sides of each section separately and normalized to the number of plants present in each side. For E-facing plants, the yield in LED-D was 30% (kg/plant) and 27% (#/plant) higher as compared to the CR, although these did not pass the significance test. In contrast, the differences of yield in LED-N vs. CR sections for E-facing plants were quite small (~12%). The W-facing spring yield for LED-D was 26% (kg/plant) and 17% (#/plant) higher than CR (not significant). Notably, for LED-N, the W-facing spring yield was 43% higher than the CR both by weight and number of fruits. These results are in agreement with those obtained from the fruit survival quantification (Figure 5B), which show that a significantly higher number of fruits were obtained on the W-facing side.

Table 2. Spring fruit yield and plant biomass.

Parameter	E-Facing			W-Facing		
	CR	LED-D	LED-N	CR	LED-D	LED-N
Yield † (kg/plant)	1.65 ± 0.37 a	2.14 ± 0.46 a	1.89 ± 0.29 a	1.86 ± 0.25 B	2.34 ± 0.34 AB	2.66 ± 0.48 A
Yield † (#/plant)	7.74 ± 1.74 a	9.84 ± 2.03 a	8.73 ± 1.51 a	9.12 ± 0.89 B	10.67 ± 1.40 AB	13.05 ± 1.51 A *
Fresh weight ‡ (kg)	2.00 ± 0.27 a	2.24 ± 0.54 a	2.24 ± 0.59 a	2.04 ± 0.40 B	2.54 ± 0.69 A	2.65 ± 0.63 A *
Height ‡ (m)	2.85 ± 0.32 a	2.89 ± 0.26 a	2.86 ± 0.20 a	2.70 ± 0.30 B	3.03 ± 0.20 A *	3.05 ± 0.23 A *

† Spring yield included six fruit harvests from 9 March 2020 to 5 May 2020 and represent means ± SD from 4 replicate sections. Values were normalized to the number of plants in each side (E, eastern- or W, western-facing) of each section. ‡ Plant biomass (fresh weight and height) were measured at the end of the experiment (7 May 2020) on the same plants assayed for fruit set during the season; means ± SD are shown for n = 20 plants (5 plants from the 4 replicate sections of each treatment in E- or W-facing plants). For each parameter, distinct letters denote statistically significant differences ($p < 0.05$) between the three groups facing the same side (lower- or upper-case letters for E- or W-facing, respectively). Asterisks denote $p < 0.01$.

At the end of the experiment, the plants followed for fruit set and survival were removed and their height and weight were recorded (Table 2). Interestingly, no significant differences were observed for the biomass of E-facing illuminated and CR plants. In contrast, the illumination resulted in considerably heavier and taller plants in the W-facing side. Plants from LED-D were 25% heavier and 12% taller than CR plants, and those from LED-N were 30% heavier and 13% taller than CR. These findings are in line with the fruit set and yield data and support the results of higher assimilation of W-facing plants.

4. Discussion

4.1. Intra-Canopy Illumination at Passive Conditions

High-cost energy inputs are typically not employed in protected crop cultivation in regions with mostly mild winters, such as the Mediterranean area. However, with the ongoing technological improvements, increasing efficiency of LED lighting, decreasing costs, as well as the potential use of photovoltaic systems as energy sources, commercial application of supplemental illumination at passive conditions can also be envisioned [3,29]. Thus, reports of the use of supplemental illumination in regions previously uncommon are becoming available [13,30].

One of the common applications for high-wire intensive crops is intra-canopy illumination (LED-interlighting) [1], feasible due to the relatively low heat-output of LEDs. Some of the available commercial interlighting LED fixtures provide red (R) and blue (B) light. Improvement of growth, yield, and/or quality using R and B LEDs have been demonstrated for tomato, pepper, and cucumber. R/B interlighting accelerated the ripening of tomato fruit and improved the yield by (+16%) due to increased fruit weight and size [14]. In sweet pepper, interlighting improved the yield and/or quality, with the increase in yield

(~16%) arising from a higher number of fruit [10,27]. Recently, R/B interlighting used for mini-cucumber in tropical climate conditions in Brazil resulted in an increased yield (+13% in commercial yield) [13]. In other studies, custom-designed lighting has also been used, with various spectral compositions and ratios of B, R, far-red, and white light applied within the canopy of tomato plants [31,32].

Earlier, we tested several spectral compositions of intra-canopy illumination and found that cool-white (CW) light was preferable for improving the spring yield of winter-grown bell pepper in the Jordan Valley [12]. One evident advantage of using a single type of LED, as opposed to a combination of different wavelength-emitting chips, is a uniform light spectrum applied to the inner canopy. In the current work, we utilized commercial Bioled fixtures providing CW light as intra-canopy illumination for pepper. The photosynthetic and gas-exchange parameters of the inner canopy foliage illuminated with Bioled increased by 3.3- to 5.2-fold, similar to those we observed earlier with other LED fixtures used for bell pepper [12,33]. Using the Bioled product, we extended our previous investigations to testing the application of daytime (LED-D) vs. edge-of-daytime (LED-N) illumination, combined with a detailed characterization of fruit set kinetics during the growth season.

4.2. Illumination during Different Times of the Day

While LEDs generate relatively lower heat than other light sources, their use as intra-canopy illumination still inputs heat into the canopy (Figures S1 and S2). At non-controlled growth conditions, it may be disadvantageous to add heat into the canopy during daytime. This is especially true for our area of the Mediterranean, where day temperatures can be quite high. In contrast, low minimal temperatures in the winter may inhibit fruit set. Therefore, it may in fact be more beneficial to illuminate during nighttime or at the edges of daytime, in order to increase the air temperature when they are lowest at night and dawn, while not affecting (increasing) the temperature within the canopy during the hottest hours of the day, at least on sunny days.

The effects of supplemental illumination may differ when provided at different times along the day. Tewolde et al. utilized LED interlighting for supplementing single-truss tomatoes during daytime (4 a.m. to 4 p.m.) or nighttime (10 p.m. to 10 a.m.) [34]. Interestingly, they showed that daytime illumination increased photosynthetic capacity and yield (+27%) only in winter, while nighttime illumination increased photosynthesis and yield in both winter (+24% yield) and summer (+12% yield). Only the winter nighttime illumination was found to be economical in this study [34]. Aside from the plant physiological considerations, night/edge-of-day illumination can also be more cost-effective when powered by electricity, as energy costs may be lower as compared to daytime in some regions [34,35].

With both light regimes applied here, during daytime (LED-D) or edge-of-daytime (LED-N), the fruit set was improved in the winter, with a slight advantage to LED-N. In our conditions and growth season (over winter), bell pepper is characterized by several waves of fruit set during the season. The changing natural light and temperature conditions along the season result in quite different kinetics of fruit development and ripening for fruit set at different times in the growth period. Thus, following the second big wave of fruit set seen in the non-illuminated sections (Figure 4A, Dec. 8), fruits grow and remain on-plant for 2.5 to 3 months. This results in a heavy fruit load, which consequently inhibits additional significant fruit set until the plants are released by the harvest (Figure 4A—black brackets). This is in addition to the prevailing low light and temperature conditions during the period of winter, which may also limit fruit set. Notably, in plants with supplemental illumination, either LED-D or LED-N, higher fruit set occurred in the same time frame when it was quite low in the non-illuminated sections (Figure 4B,C). These results indicate that the added productivity, attained with the supplemental illumination, accounts for the plants' ability to support additional fruit.

It has been shown earlier that prolonging the photoperiod (using top HPS lamps) in sweet pepper can increase the fruit yield [24,25]. As pepper is a day-neutral plant, the increase in the number of fruits under prolonged days is likely not due to flower induction per se. LED-N exhibited some benefit over LED-D, which may possibly be related to extension of the daily photoperiod in the former. Normally, a fraction of the light-driven photosynthetic assimilates are partitioned toward starch synthesis, utilized by the plant during the dark period. Daytime starch synthesis and its nighttime degradation are highly coordinated to balance the plant metabolic and growth needs, preventing unwanted night starvation responses [36,37]. Modulation of photoperiod length affects starch metabolism, and can therefore affect growth and development [38]; however, the response may differ in different plant species.

When the day length was extended for pepper plants by top lighting, Dorais et al. showed that the daily photosynthate translocation rate increased two- and three-fold for, respectively, photoperiods of 18- and 24 h, compared to 12 h [25]. Under these extended days, fruit yield (kg/plant) increased by 33% (18 h) and 27% (24 h). In our study, the intra-canopy illumination is applied to a fraction of the (inner) canopy, while sunlight affects a different region of the (mostly outer) canopy. Still, the plants illuminated with the LED-N regime are subjected to longer days, of 18 h, as compared to control and LED-D. Prolonging the photoperiod with intra-canopy illumination may have a different effect on the plants than the more conventional overhead illumination, yet both can result in yield increases. The reason that higher yield can be achieved with LED-N than with LED-D may be the availability of sugars from photosynthetic assimilation during the dark hours, and thus alteration of starch metabolism; this direction requires further exploration. The differences between LED-D and LED-N were more pronounced when the two geographical sides of the tunnel were considered, as discussed below. With respect to the idea of the availability of assimilates at night, it would be worthwhile to also test nighttime intra-canopy illumination in our system.

4.3. Eastern- vs. Western-Facing Plants

Differences in photosynthesis, growth, and metabolite profiles pertaining to geographical position, reflecting differences in light and temperature conditions, have been documented in grapevine. Assessment of the photosynthetic activity of grapevine leaves at two microsites showed that only east-facing leaves at the (slightly) cooler site were restricted and exhibited lower carbon gain, leading to differential shoot growth [39]. In another study, the diurnal dynamics of the metabolic profile was shown to differ for grape berries positioned toward north-east vs. south-west, implicating that harvest time during the day should be considered [40].

Our findings demonstrate the differential effect of the intra-canopy illumination on eastern- and western-facing plants. Although the illumination was applied symmetrically within the double-row beds (Figure 1A), the effect on W-facing plants was greater. Thus, the fruit set was consistently higher in W-facing plants as compared to E-facing ones of illuminated sections, as compared to the non-illuminated CR sections (Figure 5). Expectedly, the environmental conditions exhibited by the outer canopy differ along the day for E- and W-facing plants. This is exemplified by gas-exchange measurements of outer canopy leaves of E- and W-facing plants during morning and afternoon hours (Figure 7). In the winter, W-facing plants were subjected to higher light intensities in the afternoon hours. This would likely lead to relatively higher canopy temperatures on this side, which, at least for sunny days, persisted for a longer part of the day as compared to E-facing plants. Elevated temperatures can result in higher transpiration and higher stomatal conductance and thus a higher availability of CO_2, promoting assimilation. We note that vapor pressure deficit (VPD) greatly varies at non-controlled growth conditions, such as the ones in this experiment. Nonetheless, our data suggest that an increased VPD at higher temperatures was not a consistent limiting factor for stomatal conductance and transpiration in W-facing plants. Therefore, it is probable that these plants accumulate more assimilates compared to E-facing ones in winter and early spring. Nonetheless, the fruit set and survival in non-illuminated sections did not differ between the two sides (Figure 5). The differential effect on the two sides was observed only when supplemental illumination is applied. The above indicate that the threshold for supporting additional fruit set and fruit in W-facing plants can be reached earlier, i.e., with less added energy, as compared to E-facing plants. The spring fruit yield and final plant biomass were compared separately for E- and W-facing plants, and significant differences were indeed observed only for W-facing plants. The differences in fruit yield were statistically significant only for LED-N, in line with the higher fruit set observed for this illumination regime.

In conclusion, using cool-white Bioled lighting, we showed that both daytime (LED-D) and edge-of-daytime (LED-N) intra-canopy illumination improved pepper fruit set and fruit survival during the winter at passive conditions. Some additional benefit of LED-N was observed, possibly relating to a longer photoperiod at these conditions. The differential effect of the intra-canopy illumination on eastern- and western-facing plants exemplifies the importance of greenhouse positioning and crop orientation, e.g., the model by [41], and opens additional avenues of investigation for optimizing the use of supplemental illumination under passive growth conditions. These, of course, will likely differ for different crops, as well as for crops grown in different geographical regions of the world.

Supplementary Materials: The following are available online at https://www.mdpi.com/article/10.3390/plants11030424/s1, Figure S1: Daily minimal and maximal air temperature within the canopy, Figure S2: Air temperature within the canopy on three representative days, Figure S3: Supplemental intra-canopy illumination improves fruit set in the winter (experiment during 2018–2019), Figure S4: Daily minimal and maximal air temperature.

Author Contributions: Conceptualization, K.R., Z.G., I.E. and D.C; funding acquisition, C.Z., Z.G. and D.C.; investigation, V.T., I.K., K.R., Y.M., V.L., C.Z., Z.G., I.E. and D.C.; supervision, Z.G. and D.C.; writing—original draft, V.T. and D.C.; writing—review and editing, V.T., I.K., C.Z., Z.G., I.E. and D.C. All authors have read and agreed to the published version of the manuscript.

Funding: This research was funded by the Chief Scientist of the Ministry of Agriculture (grant No. 20-01-0188), the Israeli Plants Production and Marketing Board, and the Jewish National Fund (KKL).

Data Availability Statement: The data presented in this study are available upon request from the corresponding author.

Acknowledgments: We thank Meir Achiam and the team at the Jordan Valley R&D for crop management and maintenance, and Maya Cohen (ARO) and David Silverman (Agricultural Extension Service) for their assistance and fruitful discussions. This work is dedicated to the memory of Kira Ratner, who passed away in September 2021.

Conflicts of Interest: The authors declare no conflict of interest.

References

1. Mitchell, C.A.; Dzakovich, M.P.; Gómez, C.; Lopez, R.; Burr, J.F.; Hernández, R.; Kubota, C.; Currey, C.J.; Meng, Q.; Runkle, E.S.; et al. *Light-Emitting Diodes in Horticulture*; John wiley & Sons, Inc.: Hoboken, NJ, USA, 2015; Volume 43, pp. 1–88.
2. Bantis, F.; Smirnakou, S.; Ouzounis, T.; Koukounaras, A.; Ntagkas, N.; Radoglou, K. Current status and recent achievements in the field of horticulture with the use of light-emitting diodes (LEDs). *Sci. Hortic.* **2018**, *235*, 437–451. [CrossRef]
3. Sipos, L.; Boros, I.F.; Csambalik, L.; Székely, G.; Jung, A.; Balázs, L. Horticultural lighting system optimalization: A review. *Sci. Hortic.* **2020**, *273*, 109631. [CrossRef]
4. Palmitessa, O.D.; Pantaleo, M.A.; Santamaria, P. Applications and development of leds as supplementary lighting for tomato at different latitudes. *Agronomy* **2021**, *11*, 835. [CrossRef]
5. Runkle, E.S.; Meng, Q.; Park, Y. LED applications in greenhouse and indoor production of horticultural crops. *Acta Hortic.* **2019**, *1263*, 17–29. [CrossRef]
6. Massa, G.D.; Kim, H.H.; Wheeler, R.M.; Mitchell, C.A. Plant productivity in response to LED lighting. *HortScience* **2008**, *43*, 1951–1956. [CrossRef]
7. Tewolde, F.T.; Shiina, K.; Maruo, T.; Takagaki, M.; Kozai, T.; Yamori, W. Supplemental LED inter-lighting compensates for a shortage of light for plant growth and yield under the lack of sunshine. *PLoS ONE* **2018**, *13*, e0206592. [CrossRef] [PubMed]
8. Gómez, C.; Mitchell, C.A. Supplemental lighting for greenhouse-grown tomatoes: Intracanopy LED Towers vs. overhead HPS lamps. *Acta Hortic.* **2014**, *1037*, 855–862. [CrossRef]

9. Hao, X.; Zheng, J.; Little, C.; Khosla, S. LED inter-lighting in year-round greenhouse mini-cucumber production. *Acta Hortic.* **2012**, *956*, 335–340. [CrossRef]
10. Guo, X.; Hao, X.; Khosla, S.; Kumar, K.G.S.; Cao, R.; Bennett, N. Effect of LED interlighting combined with overhead HPS light on fruit yield and quality of year-round sweet pepper in commercial greenhouse. *Acta Hortic.* **2016**, *1134*, 71–78. [CrossRef]
11. Pepin, S.; Fortier, E.; Béchard-Dubé, S.A.; Dorais, M.; Ménard, C.; Bacon, R. Beneficial effects of using a 3-D LED interlighting system for organic greenhouse tomato grown in Canada under low natural light conditions. In Proceedings of the Second International Symposium on Organic Greenhouse Horticulture, Avignon, France, 28–31 October 2013; International Society for Horticultural Science (ISHS): Leuven, Belgium, 2014; pp. 239–246.
12. Joshi, N.C.; Ratner, K.; Eidelman, O.; Bednarczyk, D.; Zur, N.; Many, Y.; Shahak, Y.; Aviv-Sharon, E.; Achiam, M.; Gilad, Z.; et al. Effects of daytime intra-canopy LED illumination on photosynthesis and productivity of bell pepper grown in protected cultivation. *Sci. Hortic.* **2019**, *250*, 81–88. [CrossRef]
13. De Freitas, I.S.; Roldán, G.Q.; Macedo, A.C.; Mello, S.D.C. The responses of photosynthesis, fruit yield and quality of mini-cucumber to led-interlighting and grafting. *Hortic. Bras.* **2021**, *39*, 86–93. [CrossRef]
14. Paucek, I.; Appolloni, E.; Pennisi, G.; Quaini, S.; Gianquinto, G.; Orsini, F. LED lighting systems for horticulture: Business growth and global distribution. *Sustainability* **2020**, *12*, 7516. [CrossRef]
15. Lemos, V.C.; Reimer, J.J.; Wormit, A. Color for life: Biosynthesis and distribution of phenolic compounds in pepper (capsicum annuum). *Agriculture* **2019**, *9*, 81. [CrossRef]
16. Wahyuni, Y.; Ballester, A.R.; Sudarmonowati, E.; Bino, R.J.; Bovy, A.G. Secondary metabolites of Capsicum species and their importance in the human diet. *J. Nat. Prod.* **2013**, *76*, 783–793. [CrossRef]
17. Jovicich, E.; Cantliffe, D.J.; Stoffella, P.J. Fruit yield and quality of greenhouse-grown Bell Pepper as influenced by density, container, and trellis System. *Horttechnology* **2004**, *14*, 507–513. [CrossRef]
18. Dueck, T.A.; Grashoff, C.; Broekhuijsen, G.; Marcelis, L.F.M. Efficiency of light energy used by leaves situated in different levels of a sweet pepper canopy. *Acta Hortic.* **2006**, *711*, 201–205. [CrossRef]
19. Wubs, A.M.; Heuvelink, E.; Marcelis, L.F.M. Abortion of reproductive organs in sweet pepper (*Capsicum annuum* L.): A review. *J. Hortic. Sci. Biotechnol.* **2009**, *84*, 467–475. [CrossRef]
20. Aloni, B.; Karni, L.; Zaidman, Z.; Schaffer, A.A. Changes of carbohydrates in pepper (*Capsicum annuum* L.) flowers in relation to their abscission under different shading regimes. *Ann. Bot.* **1996**, *78*, 163–168. [CrossRef]
21. Marcelis, L.F.M.; Heuvelink, E.; Baan Hofman-Eijer, L.R.; Den Bakker, J.; Xue, L.B. Flower and fruit abortion in sweet pepper in relation to source and sink strength. *J. Exp. Bot.* **2004**, *55*, 2261–2268. [CrossRef]
22. Aloni, B.; Karni, L.; Zaidman, Z.; Schaffer, A.A. The relationship between sucrose supply, sucrose-cleaving enzymes and flower abortion in pepper. *Ann. Bot.* **1997**, *79*, 601–605. [CrossRef]
23. González-Real, M.M.; Liu, H.Q.; Baille, A. Influence of fruit sink strength on the distribution of leaf photosynthetic traits in fruit-bearing shoots of pepper plants (*Capsicum annuum* L.). *Environ. Exp. Bot.* **2009**, *66*, 195–202. [CrossRef]
24. Demers, D.A.; Gosselin, A.; Chris Wien, H. Effects of Supplemental Light Duration on Greenhouse Sweet Pepper Plants and Fruit Yields. *J. Am. Soc. Hortic. Sci.* **1998**, *123*, 202–207. [CrossRef]
25. Dorais, M.; Yelle, S.; Gosselin, A. Influence of extended photoperiod on photosynthate partitioning and export in tomato and pepper plants. *N. Zeal. J. Crop Hortic. Sci.* **1996**, *24*, 29–37. [CrossRef]
26. Hovi-Pekkanen, T.; Näkkilä, J.; Tahvonen, R. Increasing productivity of sweet pepper with interlighting. *Acta Hortic.* **2006**, *711*, 165–170. [CrossRef]
27. Jokinen, K.; Särkkä, L.E.; Näkkilä, J. Improving sweet pepper productivity by LED interlighting. *Acta Hortic.* **2012**, *956*, 59–66. [CrossRef]
28. Allen, R.G.; Pereira, L.S.; Raes, D.; Smith, M. *Crop Evapotraspiration Guidelines for Computing Crop Water Requirements-FAO Irrigation & Drainage Paper 56*; FAO: Rome, Italy, 1998; Volume 300, p. D05109.
29. Kusuma, P.; Pattison, P.M.; Bugbee, B. From physics to fixtures to food: Current and potential LED efficacy. *Hortic. Res.* **2020**, *7*, 56. [CrossRef]
30. Palmitessa, O.D.; Paciello, P.; Santamaria, P. Supplemental LED increases tomato yield in mediterranean semi-closed greenhouse. *Agronomy* **2020**, *10*, 1353. [CrossRef]
31. Gómez, C.; Mitchell, C.A. In search of an optimized supplemental lighting spectrum for greenhouse tomato production with intracanopy lighting. *Acta Hortic.* **2016**, *1134*, 57–62. [CrossRef]
32. Song, Y.; Jiang, C.; Gao, L. Polychromatic supplemental lighting from underneath canopy is more effective to enhance tomato plant development by improving leaf photosynthesis and stomatal regulation. *Front. Plant Sci.* **2016**, *7*, 1832. [CrossRef]
33. Ratner, K.; Joshi, N.C.; Yadav, D.; Many, Y.; Kamara, I.; Esquira, I.; Achiam, M.; Gilad, Z.; Charuvi, D. Application of LED-interlighting for improving the yield of passive tunnel-grown bell pepper. *Acta Hortic.* **2020**, *1268*, 19–26. [CrossRef]
34. Tewolde, F.T.; Lu, N.; Shiina, K.; Maruo, T.; Takagaki, M.; Kozai, T.; Yamori, W. Nighttime supplemental LED inter-lighting improves growth and yield of single-truss tomatoes by enhancing photosynthesis in both winter and summer. *Front. Plant Sci.* **2016**, *7*, 448. [CrossRef] [PubMed]
35. Sase, S.; Mito, C.; Okushima, L. Effect of Overnight Supplemental Lighting with Different Spectral LEDs on the Growth of Some Leafy Vegetables. *Acta Hortic.* **2012**, *956*, 327–334. [CrossRef]

36. Stitt, M.; Zeeman, S.C. Starch turnover: Pathways, regulation and role in growth. *Curr. Opin. Plant Biol.* **2012**, *15*, 282–292. [CrossRef] [PubMed]
37. MacNeill, G.J.; Mehrpouyan, S.; Minow, M.A.A.; Patterson, J.A.; Tetlow, I.J.; Emes, M.J. Starch as a source, starch as a sink: The bifunctional role of starch in carbon allocation. *J. Exp. Bot.* **2017**, *68*, 4433–4453. [CrossRef]
38. Gibon, Y.; Pyl, E.; Sulpice, R.; Lunn, J.E.; Höhne, M.; Günther, M.; Stitt, M. Adjustment of growth, starch turnover, protein content and central metabolism to a decrease of the carbon supply when Arabidopsis is grown in very short photoperiods. *Plant Cell Environ.* **2009**, *32*, 859–874. [CrossRef]
39. Hendrickson, L.; Ball, M.C.; Wood, J.T.; Chow, W.S.; Furbank, R.T. Low temperature effects on photosynthesis and growth of grapevine. *Plant Cell Environ.* **2004**, *27*, 795–809. [CrossRef]
40. Reshef, N.; Fait, A.; Agam, N. Grape berry position affects the diurnal dynamics of its metabolic profile. *Plant Cell Environ.* **2019**, *42*, 1897–1912. [CrossRef]
41. Van Der Meer, M.; De Visser, P.H.B.; Heuvelink, E.; Marcelis, L.F.M. Row orientation affects the uniformity of light absorption, but hardly affects crop photosynthesis in hedgerow tomato crops. *Silico Plants* **2021**, *3*, diab025. [CrossRef]

plants

MDPI

Article

Development and Implementation of an IoT-Enabled Optimal and Predictive Lighting Control Strategy in Greenhouses

Shirin Afzali [1], Sahand Mosharafian [1], Marc W. van Iersel [2] and Javad Mohammadpour Velni [1,*]

1 School of Electrical and Computer Engineering, University of Georgia, 111 Boyd Graduate Studies Research Center, 200 D.W. Brooks Drive, Athens, GA 30602, USA; shirin.afzali@uga.edu (S.A.); sahandmosharafian@uga.edu (S.M.)

2 Department of Horticulture, University of Georgia, 1111 Miller Plant Sciences Building, Athens, GA 30602, USA; mvanier@uga.edu

* Correspondence: javadm@uga.edu

Abstract: Global population growth has increased food production challenges and pushed agricultural systems to deploy the Internet of Things (IoT) instead of using conventional approaches. Controlling the environmental parameters, including light, in greenhouses increases the crop yield; nonetheless, the electricity cost of supplemental lighting can be high, and hence, the importance of applying cost-effective lighting methods arises. In this research paper, a new optimal supplemental lighting approach was developed and implemented in a research greenhouse by adopting IoT technology. The proposed approach minimizes electricity cost by leveraging a Markov-based sunlight prediction, plant light needs, and a variable electricity price profile. Two experimental studies were conducted inside a greenhouse with "Green Towers" lettuce *(Lactuca sativa)* during winter and spring in Athens, GA, USA. The experimental results showed that compared to a heuristic method that provides light to reach a predetermined threshold at each time step, our strategy reduced the cost by 4.16% and 33.85% during the winter and spring study, respectively. A paired t-test was performed on the growth parameter measurements; it was determined that the two methods did not have different results in terms of growth. In conclusion, the proposed lighting approach reduced electricity cost while maintaining crop growth.

Keywords: Internet of Things (IoT); optimal control; supplemental lighting in greenhouses; image processing

Citation: Afzali, S.; Mosharafian, S.; van Iersel, M.W.; Mohammadpour Velni, J. Development and Implementation of an IoT-Enabled Optimal and Predictive Lighting Control Strategy in Greenhouses. *Plants* **2021**, *10*, 2652. https://doi.org/10.3390/plants10122652

Academic Editors: Valeria Cavallaro and Rosario Muleo

Received: 1 November 2021
Accepted: 27 November 2021
Published: 2 December 2021

1. Introduction

The global population is predicted to grow to around 9.15 billion by 2050, which will increase the amount of food that needs to be produced [1]. Furthermore, the limitations of natural resources and productive land, as well as climate constraints raise concerns about food security. The rising demand for food has attracted researchers' attention towards the application of Internet of Things (IoT) technology in agriculture. The IoT is a network of physical objects that transfer data to other devices over the Internet [2]. Applying the IoT in controlled environment agriculture (CEA) has reduced human effort, time, and cost and resulted in yield improvements [3]. The IoT integrates several technologies such as wireless sensor networks (WSNs), radio-frequency identification (RFID), cloud and edge computing, and human–computer interaction (HCI) [4].

The application of the IoT in agriculture includes monitoring, control of agriculture machinery, tracking and tracing, precision agriculture, and greenhouse production [3]. Monitoring and acquiring data about some environmental factors in crop farming such as temperature, humidity, solar radiation, pest movement, and rainfall help to understand the patterns and maximize farm production [5]. Other than crop farming, monitoring the water quality, water level, and temperature levels in aquaponics is another application of the IoT [6]. Furthermore, factors to be monitored in forestry [7] and livestock [8] are other applications of the IoT in agriculture. Although some strategies have been developed in the area of monitoring, developing cost-effective methods is still an open area [3]. Both manned and autonomous vehicles (such as unmanned aerial vehicles (UAVs)) can collect useful data for farmers; they can also be remotely controlled through the IoT system [9]. Applying the IoT in tracking and tracing can improve agricultural companies' supply chain. Tracking is related to capturing, collecting, and storing data along the supply chain from upstream to downstream, while tracing enables distinguishing the product from downstream to upstream [10].

Another application of the IoT is in precision agriculture, that is a management strategy that collects real-time data such as crop maturity, weather, air quality, etc., and then analyzing the data to improve crop yields and reduce cost [11]. IoT technology could be employed in greenhouses to maximize profit, reduce cost and labor, and save energy. Several studies have considered applying WSNs in greenhouses for monitoring [12–14]. In this study, we focus on the application of IoT technology in greenhouse production.

Supplemental lighting improves plant growth and contributes to higher yields in CEA, in particular greenhouses. During rainy days or winter months, the overall amount of sunlight that plants receive might not be sufficient for plant growth and development. Therefore, supplemental lighting is often necessary for greenhouse fruit and vegetable production. However, the electricity needed for greenhouse supplemental lighting amounts to about 30% of the operating costs [15]. As a result, enhancing the cost effectiveness of lighting with modern technologies plays an important role in the CEA industry.

Among various lighting types, light-emitting diodes (LEDs) have proven to be more effective, due to their dimming capability, which enables changing the intensity of the output light to any continuous level [16]. On the other hand, the output light of high-intensity discharge (HID) lamps only takes some quantized levels, thereby not considered as the primary choice for optimal lighting strategies. Researchers have proposed rule-based supplemental lighting control methods since 1994; the authors in [17] developed a rule-based approach for HID lamps to reach a predefined light target within a specified photoperiod. Light and shade system implementation (LASSI) is another control method with an on/off control for HID lamps in combination with a sunlight prediction and movable shades to achieve a constant DLI [18]. Another version of LASSI improved the sunlight prediction, resulting in increasing the accuracy of lighting control [19].

In [20], the DynaLight system, an on/off control method for HID lamps, was developed. This system considers a leaf photosynthesis model, variable electricity pricing, and weather forecast to reduce electricity cost while reaching a minimum daily sum of photosynthesis. Nevertheless, the weather forecasts were provided twice daily at an hourly resolution [20], which is not precise enough for proper optimization. Another version of DynaLight for controlling HID lamps is called DynaLight IND, which is a multi-objective optimization platform for optimizing artificial lighting in greenhouses [21]. For DynaLight IND, an evolution strategy was proposed and compared with the original genetic algorithm (GA) in DynaLight. The simulation results showed that the new version improved the GA's evolution efficiency and increased the computation speed compared to the original DynaLight. However, other than simulations, there is a need to conduct practical experiments to validate the properties of DynaLight IND [21].

The adaptive lighting method [22] is another rule-based control approach that reduces cost by adjusting the duty cycle of LEDs to reach a specified threshold based on photosynthetic photon flux density (PPFD) levels. Two general approaches were taken in previous studies on supplemental lighting control with LEDs: either using real-time sunlight intensity measurements to supply fixed PPFD levels [23,24] or assuming prior information about sunlight intensity throughout the day [23]. Both of these perspectives are faulty since sunlight intensity throughout the day is unknown and using the current PPFD alone does not ensure reaching the daily light integral (DLI) or daily photosynthesis.

In our prior studies [25,26], we developed optimal supplemental lighting strategies for both LEDs and HID lamps in greenhouses, which significantly reduced the electricity cost. We formulated the supplemental lighting control problem as a constrained convex optimization problem, and we aimed at minimizing the electricity cost of supplemental lighting, while considering sunlight prediction, plant light needs, and variable electricity pricing in our model. This strategy used a Markov model to predict future sunlight intensities. The simulation studies showed that the proposed method could reduce electricity cost significantly [25]. To control the light intensity of HID lamps, we formulated a discrete constrained optimization problem, which was solved using the method of multipliers and a reinforcement learning (RL) algorithm, considering the Markov-based sunlight prediction [26]. Although we made sure to provide sufficient light for good crop growth as the constraint of the optimization problem, it is necessary to implement the method in the greenhouse and monitor plant growth through experimental studies.

In the present paper, we implemented our proposed lighting strategy for LEDs in a research greenhouse equipped with IoT technology. Lettuce was grown under two treatments with different lighting control strategies over two seasons with low and high natural light (winter and spring). To evaluate the proposed lighting control approach in terms of plant growth, we collected data related to growth during both experiments. The main contributions of this work are as follows: (1) We implemented an optimal supplemental lighting control strategy in a greenhouse equipped with IoT technology. (2) We evaluated the advantages of our proposed optimal lighting method based on not only the electricity cost, but also the plant growth through experimental studies.

2. Materials and Methods

2.1. Preliminaries and Problem Formulation

Photosynthesis is a photon-driven process. Thus, photosynthetic light levels are measured as the photosynthetic photon flux density (PPFD) with units of μmol m^{-2} s^{-1}. The PPFD is the light-photon numbers in the photosynthetically active wavelength range (400–700 nm) per square meter per second. The daily light integral (DLI) in mol m^{-2} d^{-1} is the integral of the PPFDs over 24 h. To ensure sufficient growth for plants in greenhouses, it is recommended that they receive a minimum amount of DLI during the photoperiod, which is the time period each day (up to 24 h) during which plants receive light [23]. To formulate the optimization problem, the relation between two important parameters in plant photosynthesis was considered. These parameters are the PPFD and electron transport rate (ETR), which is the number of electrons transported through photosystem II per square meter of leaf area per second (with units of μmol m^{-2} s^{-1}). The daily photochemical integral (DPI) in mol m^{-2} d^{-1} is the integral of the ETRs over 24 h. Furthermore, the ETR and PPFD generally have an exponential rise to a maximum relation. The authors in [23] derived a relationship between the ETR and PPFD as:

$$ETR = a\left(1 - e^{-k \times PPFD}\right),$$

where a is the asymptote of the ETR and k is the initial slope of the ETR divided by a. For "Green Towers" lettuce, which was used in this study, $a = 121$ μmol m^{-2} s^{-1}, and $k = 0.00277$ [23].

For many greenhouse crops, to guarantee high-quality production and adequate growth, a minimum DLI is suggested. In some cases, a specific photoperiod must also be achieved. However, the DPI and plant growth depend on the combination of the DLI and photoperiod; longer photoperiods with the same DLI result in a higher DPI and more biomass [27,28]. Therefore, the DPI is a better predictor of plant growth and better suited for lighting optimization algorithms than the DLI. "Green Towers" lettuce requires a DPI of 3 mol m^{-2} d^{-1}, corresponding approximately to a DLI of 17 mol m^{-2} d^{-1} under ambient sunlight conditions [23]. The optimization problem was formulated to minimize the total amount of supplemental lighting cost to reach a specified DPI within a specified photoperiod.

The theory behind the experiments considered in this study is based on the constrained nonlinear optimization problem presented in [25], which is as follows:

$$\begin{aligned}
\min_{\overline{x}} f(\overline{x}) &= \sum_{t=1}^{T} \frac{C_t}{k}\left[\ln\left(\frac{a}{a-\overline{x}_t-\overline{s}_t}\right)-s_t\right] \\
\text{subject to:} & \sum_{t=1}^{T} (\overline{x}_t+\overline{s}_t) \geq \frac{\overline{D}}{m} \\
\overline{x}_t &\geq 0\ ;\ t=1,2,\cdots,T \\
\overline{x}_t &\leq \overline{U}_{LED}\ ;\ t=1,2,\cdots,T,
\end{aligned} \tag{1}$$

where $\overline{x}_t$ is the ETR resulting from supplemental light provided by the LEDs at time step t, $\overline{s}_t$ is the ETR resulting from sunlight, s_t is the PPFD received from the Sun, $\overline{U}_{LED}$ is the maximum ETR that can be achieved with LEDs, C_t is the electricity price in cents/kWh, $\overline{D}$ is the minimum DPI needed for the plant during the entire photoperiod, m is the length of each time step in seconds, and T is the number of time steps. The first constraint in (1) guarantees supplying sufficient light to the plants to reach the recommended DPI, and the other constraints define the ETR bounds according to the PPFD of LEDs.

A photoperiod of 16 h is common for greenhouse lettuce production and used to illustrate the performance of the control strategy. The optimization problem was solved at each time step (with the length of m seconds) during the allowed photoperiod for each day, and supplemental light was provided up to the optimal PPFD calculated for that time step. The process was repeated every m seconds time step, for a total number of $T = 16 \times 3600/\mathrm{m}$ when a 16 h photoperiod was used. The Markov-based predictive values are substituted in (1), instead of the actual future sunlight intensities (which are not obtainable in real time). For a detailed description on sunlight prediction using Markov chains, we refer to our previous work [25]. Consequently, (1) can be demonstrated as:

$$\begin{aligned}
\underset{\overline{x}}{\text{minimize}} & \sum_{t=i}^{T} \frac{C_t}{k}\left[\ln\left(\frac{a}{a-\overline{x}_t-\overline{s}_t}\right)-s_t\right] \\
\text{subject to:} & \sum_{t=i}^{T} (\overline{x}_t+\overline{s}_t) \geq \frac{\overline{D}}{m} - \sum_{t=1}^{i-1} (\overline{x}_t+\overline{s}_t), \\
\overline{x}_t &\geq 0; \quad t=i,i+1,\cdots,T, \\
\overline{x}_t &\leq \overline{U}_{LED}; \quad t=i,i+1,\cdots,T.
\end{aligned} \tag{2}$$

The optimization problem (2) was solved once before sunrise and once after sunset. Throughout the day, (2) was solved repeatedly at each time step. The interested reader is referred to [25] for more details on how to calculate the optimal lighting strategy.

2.2. IoT Structure

Four essential components of an IoT ecosystem include: (1) IoT devices; (2) communication technology; (3) the Internet; and (4) data storage and processing [3]. The description of each component is provided as follows:

(1) IoT devices: IoT devices consist of embedded systems, which include microprocessors, field programmable gate arrays (FPGAs), and input/output interfaces that interact with sensors and actuators [3]. In this study, a quantum light sensor was used to measure sunlight irradiance in the greenhouse, and an infrared (IR) night vision camera was deployed to monitor growth and calculate the projected canopy size (PCS) (a morphological indicator of plant growth). A Raspberry Pi was also used as the microprocessor, which was connected to the camera and the light sensor through an analog-to-digital converter (ADC). Data were stored on the Raspberry Pi SD card

and then transferred through WiFi. Reliability, portability, security, and cost were taken into account while choosing the IoT devices;

(2) Communication technology: Communication technology in IoT systems can be categorized based on the standards, spectrum, and application scenarios. In the deployment of wireless connectivity, the range of the communication distance, data rate, security and resilience, and cost of gateway modems are some of the parameters that should be considered [3]. For our study, WiFi was chosen as the wireless technology with a WLAN and WPA2 security;

(3) Internet: The connection between IoT devices and the Internet allows transferring data towards a remote infrastructure for storage, processing, and further analysis. Security, accessibility, and support for real-time data should be taken into account while transferring data [3];

(4) Data storage and processing units: Data collected in agriculture applications can be in different forms including images, text, audio, and video. Cloud IoT platforms have been used to store big data collected from sensors. Some of the commercial platforms that provide data storage and analysis are Onfarm systems, Cropx, KAA, Farmx, Easyfarm, and Farmlogs [3]. In our experiment, data in the form of text and images were stored on the Raspberry Pi SD card and retrieved remotely using WPA2 and RealVNC software (RealVNC is available at https://www.realvnc.com/en/, accessed on 20 November 2020).

2.3. Experimental Setup

To evaluate the performance of the proposed lighting control approach in terms of plant growth, as well as electricity cost, two experiments were conducted inside of a research greenhouse for "Green Towers" lettuce in Athens, GA, USA. The experimental unit consisted of a group of 15 plants in 10 cm^2 pots on each bench. Those pots were filled with a soil-less substrate (80% peat: 20% perlite (v/v) (Fafard 1P; SunGro Horticulture, Agawam, MA, USA)). Plants were grown under two supplemental lighting treatments (the proposed method and a heuristic one) and were manually irrigated every two or three days with a nutrient solution containing 100 mg L^{-1} N made with a water-soluble fertilizer (15N–2.2P–12.45K, Peters Excel 15–5–15 Cal-Mag Special; Everris NA Inc., Dublin, OH, USA).

We compared the lighting electricity cost of our proposed method to two supplemental lighting strategies: (1) baseline, which is also optimal and solves the optimization problem (2) assuming perfect prior knowledge of sunlight throughout the day; (2) heuristic, in which the goal was to supply enough supplemental light to reach a minimum PPFD, unless the PPFD from sunlight alone exceeded the threshold. Thus, by the end of photoperiod, the plants will have received enough supplemental light to reach a DPI greater than or equal to the suggested quantity. In this method, sunlight prediction and real-time electricity price are not taken into account. Since the assumption of perfect prior knowledge of sunlight is not realistic nor practical, only the heuristic method and the proposed method were implemented in the greenhouse to grow lettuce under those treatments and compare plant growth. However, the simulation results related to the electricity cost for the baseline method are provided and represent a theoretical optimal scenario.

Our control algorithms were implemented on the Raspberry Pi 3 Model B (a low-cost, small microprocessor) using the Python programming language. Raspberry Pi operates as a control hardware to decide on how much supplemental lighting the plants need based on the predicted sunlight. CVXPY, a domain-specific language (DSL) that enables solving convex optimization problems with the high-level features of Python [29,30], was used to calculate the optimal lighting (CVXPY is available at http://www.cvxpy.org, accessed on 1 January 2020).

To develop the Markov model for sunlight prediction, sunlight data collected from the NREL database [31] were taken as representative of the sunlight intensity outside our greenhouse. Often, 40–70% of the sunlight PPFD represents the PPFD measured at a leaf surface in a greenhouse because of the building materials of the greenhouse ceiling and the sunlight radiation angle [32]. For the development of the Markov model, we considered 40% as the light transmission rate since the greenhouse had an internal shade cloth, which was used to achieve low sunlight levels and facilitate lighting research. This model was used in both experiments to predict sunlight based on real-time PPFD measurements by a light sensor inside the greenhouse, underneath the shade cloth. These measurements represent the amount of sunlight that reached the plants. The parameters used in our optimization problem were similar to those considered in our prior study [25] and for the type of LEDs used in the experiments (given in Table 1).

Table 1. Model and optimization parameters used in this work.

Variable	Value	Variable	Value
a	121 μmol m^{-2} s^{-1}	m	900 s
D	3 mol m^{-2} d^{-1}	T	64
$\overline{U}_{LED}$	86.21 μmol m^{-2} s^{-1}	k	0.00277

To measure the sunlight PPFD over the plant canopy in the greenhouse, an analog full-spectrum quantum sensor (SQ-500-SS, Apogee instruments, Logan, UT, USA) with an improved spectral range of 389–692 nm ± 5 nm was used. SQ-500-SS is a self-powered PAR light sensor with a 0–40 mV output and calibration factor of 100 μmol m^{-2} s^{-1} mV. Sunlight intensity was measured by the light sensor using a Python script every three minutes, then averaged to be representative of sunlight at each time step (15 min). Since the Raspberry Pi only reads digital inputs, the analog output of the sensor was converted to digital using ADS1115, which is a high-precision 16 bit ADC. ADS1115 uses the I2C communication protocol to read analog values; therefore, I2C should be enabled on the Raspberry Pi before interfacing.

The supplemental lighting source used in our setup was the GE Arize™ element L1000 LED grow light bars with a 3:1 red-to-blue light ratio (HPPB4) and a power of 627 W. Several methods of controlling the output signal of LED drivers are available; the 0–10 V dimming approach was used for our purposes. In this method, there is a near-linear relationship between the dimming voltage of the LED driver and the output light of the LED. To control the output light of the LEDs, two GPIO pins of the Raspberry Pi (GPIO 13 and GPIO 18) were used to generate two pulse-width modulation (PWM) signals. One PWM signal controls the LEDs under the heuristic treatment, and the other one controls the LEDs under the prediction-based treatment. These PWM signals of the Raspberry Pi need to be amplified and filtered to become compatible with the dimming voltage of the LED driver; thus, a signal-conditioning circuit (using the TL081 operational amplifier (op-amp)) was designed to supply the proper signal to the LED drivers. Based on the calculated supplemental lighting strategy, the duty cycle of the PWM signals was determined by the Raspberry Pi and converted to the compatible dimming voltage (control signal) for LED drivers by the signal-conditioning circuit. At each time step (15 min), the control signals for the two treatments were adjusted, thereby changing the output light of the LEDs.

Every morning, the Raspberry Pi automatically runs the Python code, which reads real-time light sensor data, predicts future sunlight, solves the optimization problem, and calculates the supplemental lighting for both treatments at each time step (every 15 min) for a 16 h photoperiod. Once the photoperiod is over and plants receive sufficient light, the LEDs are turned off.

To monitor the plant growth and PCS for the two treatments, Arducam day–night vision cameras were installed above the benches. The camera was mounted facing downward above the crop. These cameras have an IR cut filter that switches automatically. The filter is on for daylight color accuracy and off for IR night vision. Therefore, the camera can take pictures at night in complete darkness using its small IR LEDs. Moreover, this camera is an inexpensive choice (less than USD 30) for monitoring plant growth. Each camera was connected to a Raspberry Pi, which was set to automatically take pictures every night when the LEDs were off, so that there would be no sunlight or LED light interfering with the plant images. Figure 1 shows the imaging setup.

Figure 1. Imaging setup built in the greenhouse.

We accessed the data related to supplemental lighting, sunlight, cost, and plant images through real-time remote monitoring of the Raspberry Pi using WPA2 and RealVNC software. Plant images were processed using PlantCV, which is an image analysis package for plant phenotyping [33] (PlantCV is available at https://plantcv.readthedocs.io/en/stable/, accessed on 20 November 2020). The PCS was measured by using a reference with a known area and counting the number of detected leaf pixels.

Shown in Figure 2 is the experimental setup in the greenhouse. The light sensor was installed above the LED fixtures so that it was not affected by the LED light and was connected to the ADC. The ADC converts the analog measurements to digital, and hence, the Raspberry Pi can read them. After predicting sunlight and solving the optimization problem, the duty cycle of the PWM signals was determined and amplified. These control signals were connected to the dimming wires of the LED driver, thereby changing the output light of the LED. The camera was installed above the crops and connected to the Raspberry Pi through a ribbon cable. The materials used in the proposed setup are presented in Table 2.

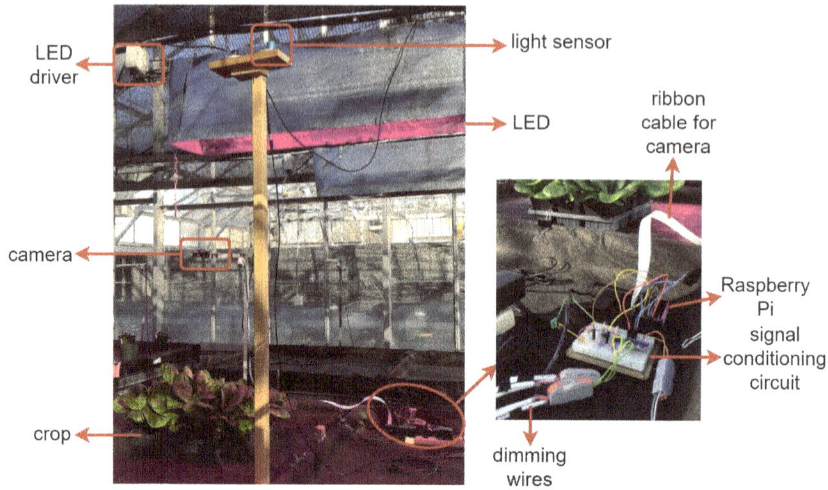

Figure 2. Experimental setup in the greenhouse.

Table 2. Main materials used in our experimental setup.

Item	Details
plant	"Green Towers" lettuce
lighting source	GE Arize^{TM} element L1000 LED grow light bar
light sensor	SQ-500-SS
microprocessor	Raspberry Pi 3 Model B
signal-conditioning circuit	TL081 op-amp, 24V DC power supply adapter, resistors, and capacitors
ADC	ADS1115
camera	Arducam automatic day/night camera module (model number: B003503)

The first experiment ran from 11 December 2020 through 28 January 2021, when sunlight levels were low. The daily minimum temperature, max temperature, minimum vapor pressure deficit (VPD), and maximum VPD inside the greenhouse were 12.4 ± 3.4 °C, 22.4 ± 2.8 °C, 0.46 ± 0.15 kPa, and 1.81 ± 0.63 kPa (mean ± SD), respectively. The CO_2 level was at the ambient level; the average DLI from sunlight under the shade cloth was 2.22 ± 0.6 mol/m^{2}d, and the daily amounts are provided in Figure 3. The setup in Figure 2 was installed, and three lettuce replicates for the heuristic method and three for the proposed optimal method with six LEDs were considered. Figure 4 depicts the greenhouse section in the first experiment. On each bench, 15 plants were grown as 1 experimental unit. This section of the greenhouse was surrounded by windows except for the west wall, where the cooling pad was. Therefore, the benches on the west side may received less sunlight late in the day. It is also possible that the temperatures may have been slightly lower on the west side due to the cooling pad. However, very little cooling was required in winter. Ambient light and temperature can affect plant growth, and the treatments were blocked to account for such effects. Further details are provided in the Results and Discussion Section. Plants were harvested 49 d after seeding, and the proposed method was evaluated in terms of both the electricity cost and plant growth.

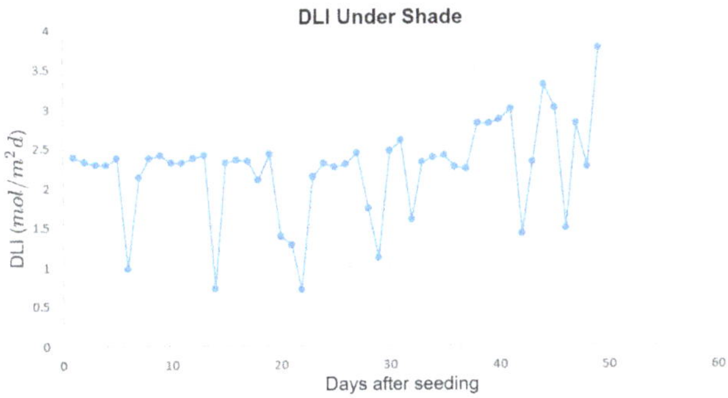

Figure 3. DLI from sunlight under the shade cloth during the first experiment.

For the imaging part, only two cameras were installed above the two benches, one controlled by the heuristic method (Bench Number 8 in Figure 4) and the other one by the prediction-based method (Bench Number 7 in Figure 4). Thus, we used two Raspberry Pis in this experiment, one connected to a camera for taking pictures of one replicate of the heuristic method and another to take pictures of the plant under the prediction-based treatment and to control the output light of all six LED fixtures. During the first experiment, the photoperiod started at 4:30 a.m. and ended at 8:30 p.m. At 8:30 p.m., the LEDs turned off, and the camera automatically took images at 11:30 pm when there was no supplemental light in the greenhouse. We obtained a variable hourly electricity price profile from a website (https://www.ieso.ca/power-data, accessed 1 December 2019) and scaled the numbers to account for the U.S. electricity prices based on government rates. The electricity price profile considered in this experiment is given in Figure 5 and was the same throughout the study.

Figure 4. Experimental units in the greenhouse under two different lighting treatments in the first experiment. Plants on the pink benches were grown using the prediction-based lighting method, and those on the blue benches were grown using the heuristic method.

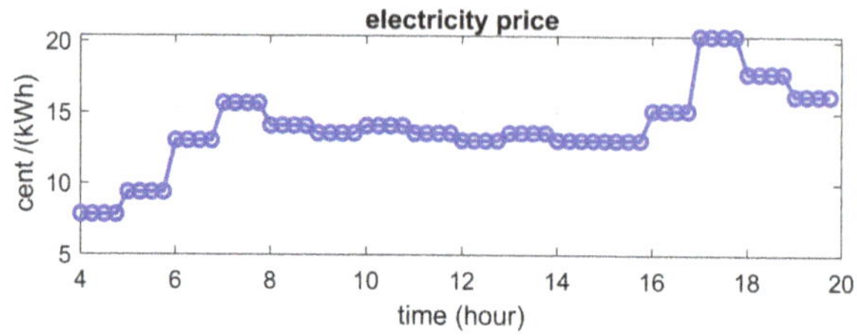

Figure 5. Variable electricity price in cents/kWh used in the first experiment.

Remark 1. *The units of f in (1) are cents $(kWh)^{-1}$ $\mu mol\ m^{-2}\ s^{-1}$; therefore, a conversion factor is needed to convert these units to cents/m^2. Converting the PPFD of GE LEDs to kW m^{-2}, 2.9 $\mu mol\ J^{-1}$ (or equivalently 2.9×10^3 $\mu mol\ (kW)^{-1}\ s^{-1}$) was performed [34]. Assuming the length of each time step (15 min or 0.25 h), the conversion factor q is defined as:*

$$q = \frac{1}{2.9 \times 10^3} \times 0.25. \quad (3)$$

Therefore, $q \times f$ represents the electricity cost of supplemental lighting with units of cents/m^2.

Other than the PCS, the leaf chlorophyll content index (CCI), anthocyanin content index (ACI), specific leaf area (SLA), and shoot dry weight were measured to compare plant growth for the two lighting approaches. The CCI and ACI were measured using chlorophyll and anthocyanin meters (CCM-200 plus and ACM-200 plus; Apogee Instruments, Logan, UT, USA) on the uppermost fully expanded leaves at two different times, once partway through the study and again near the end. Each time, 10 measurements on each experimental unit (group of 15 plants) were collected. The average value of these measurements was considered for each unit. A higher CCI can increase the light absorptance [35], and the anthocyanins in leaves have a protective role against intense light [35].

The SLA was calculated by dividing the leaf area of a plant by the shoot dry weight. For each experimental unit (15 plants), 3 plants were selected, and the total leaf area was measured using a leaf area meter (LI-3100 leaf area meter; LI-COR Biosciences, Lincoln, NE, USA). When drying the plants in the oven, these three plants were placed separately from the other plants. After drying them at 70 °C for 7 d, the shoot dry weight of each group of plants was also measured, thereby achieving the SLA per plant. Furthermore, dividing the total dry weight over the number of pots for each unit (15), we obtained the shoot dry weight per plant.

A paired t-test (two-tailed form) was performed for each growth parameter to determine if the two lighting methods had different results. In this test, each replicate was considered to be a pair, and the mean values for the two treatments were compared. If the p-value was less than the significance level (here, 0.05), the null hypothesis was rejected, which means there was a significant difference between the two methods.

The second experiment was conducted using the same crop species in the same place, from 2 April 2021 through 18 May 2021, when sunlight levels were higher than in the first experiment. The minimum temperature, max temperature, minimum vapor pressure deficit (VPD), and maximum VPD inside the greenhouse were 18.7 ± 1.5 °C, 25.8 ± 1.5 °C, 0.52 ± 0.21 kPa, and 1.86 ± 0.75 kPa (mean $\pm$ SD), respectively. The CO_2 level was at the ambient level; the average DLI from sunlight under the shade cloth was 7.45 ± 3.11 mol/m^2d, and the daily amounts are given in Figure 6. Not only were the sunlight levels much higher than in the first experiment, they were also more variable, thus increasing the potential benefits of an optimized lighting control approach.

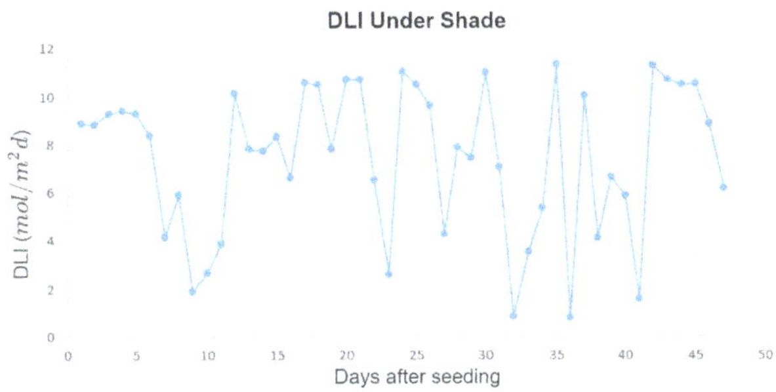

Figure 6. DLI from sunlight under the shade cloth during the second experiment.

This time, we experimented on more replicates, i.e., five replicates for each treatment, which is represented in Figure 7. Plants were harvested 47 d after seeding. The same hardware setup was utilized for this experiment; however, the PCS was monitored for three replicates of each treatment using six Arducam cameras. The cameras were installed above Bench Numbers 1 and 2, 7 and 8, and 14 and 15 (see Figure 7). Moreover, the lighting control started at 5 am every day and ended at 9 pm. This change was made to ensure that the photoperiod fully encompassed the natural photoperiod. Another different factor compared to the first experiment was the use of a fixed electricity price (13.19 cents/kWh), rather than real-time pricing. Considering real-time pricing is advantageous to reduce the electricity cost of lighting. However, for the second experiment, we intended to use the electricity price available to our research greenhouse, which is fixed. Measurements of the growth parameters were repeated for the second experiment, and paired t-tests (two-tailed form) were performed to accept or reject the null hypothesis.

Figure 7. Experimental units in the greenhouse under two different lighting treatments in the second experiment. Pink color is an indicator of the prediction-based lighting method, while blue is for the heuristic method.

3. Results and Discussion

First, we discuss the results of the first experiment. The lighting control performance and sunlight prediction throughout different days with different sunlight levels are shown in Figures 8–10. Figure 8 is representative of a rainy day with very low sunlight irradiance. January 14th's irradiance was close to a day with regular sunlight during that period, and Jan 28th had a relatively high sunlight irradiance (presented in Figures 9 and 10, respectively). The cost increase on each day compared to the baseline was calculated and shown in Table 3. Moreover, the average of actual sunlight, its prediction, and supplemental light for different methods were calculated throughout this experiment and displayed in

Figure 11. Based on the results in Table 3, the cumulative cost increase for the heuristic method was 5.45%, while for the prediction-based method, it was 1.06% compared to the baseline. Therefore, our prediction-based lighting approach showed about a 4.16% electricity cost reduction compared to the heuristic method throughout the first experiment.

Table 3. Cost of lighting control strategies in $/m^2 per day for different days during the first experiment.

Day	Baseline Cost	Prediction-Based Method Cost Increase Compared to Baseline	Heuristic Method Cost Increase Compared to Baseline
December 24	0.39	0.0027 (0.7%)	0.0174 (4.46%)
January 14	0.33	0.0040 (1.21%)	0.0178 (5.39%)
January 28	0.27	0.0038 (1.41%)	0.0286 (10.59%)
cumulative cost	16.14	0.17 (1.05%)	0.88 (5.45%)

In the baseline approach, perfect prior knowledge of sunlight throughout the day is assumed, which is not practical and only represents a theoretical optimal scenario. A predictive model is not able to predict sunlight with 100% accuracy; therefore, the sunlight information in the baseline approach is always more accurate than that in the proposed method. In other words, the baseline is an ideal optimal scenario with the least electricity cost, which is not practical. However, the prediction-based method is optimal and practical, which resulted in a very close solution to the baseline (the global optimal solution of the lighting problem). The heuristic lighting strategy generally provided more light than the minimum DPI and resulted in extra supplemental lighting cost; thus, it is not optimal.

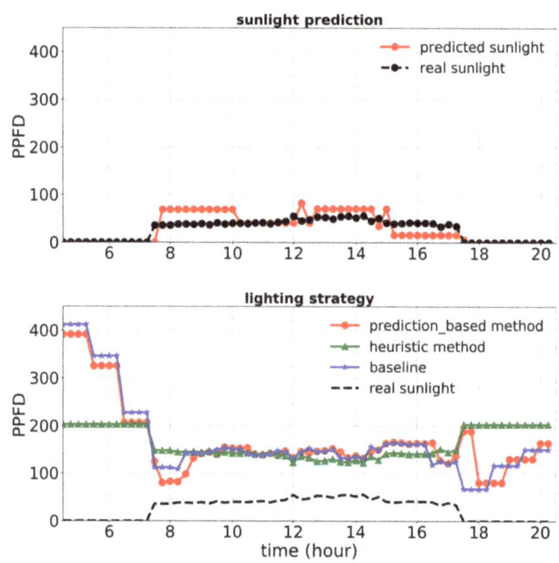

Figure 8. Performance of lighting control strategies and sunlight prediction for December 24th, a day with a low sunlight level during the first experiment.

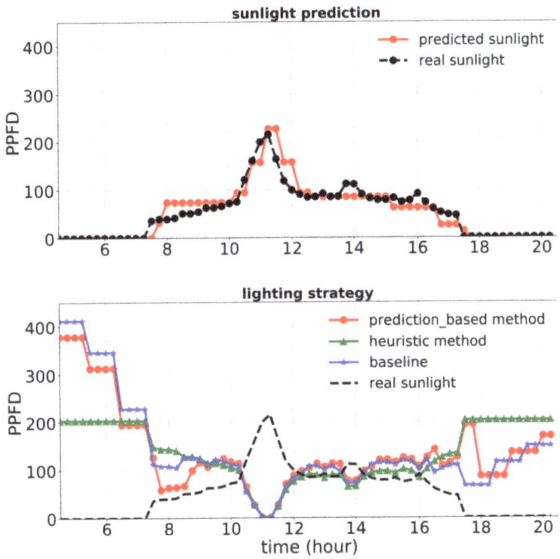

Figure 9. Performance of lighting control strategies and sunlight prediction for January 14th, a day with a moderate sunlight level during the first experiment.

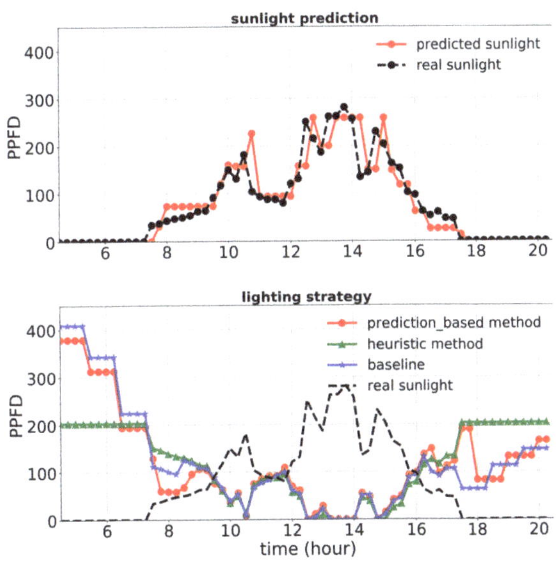

Figure 10. Performance of lighting control strategies and sunlight prediction for January 28th, a day with a high sunlight level during the first experiment.

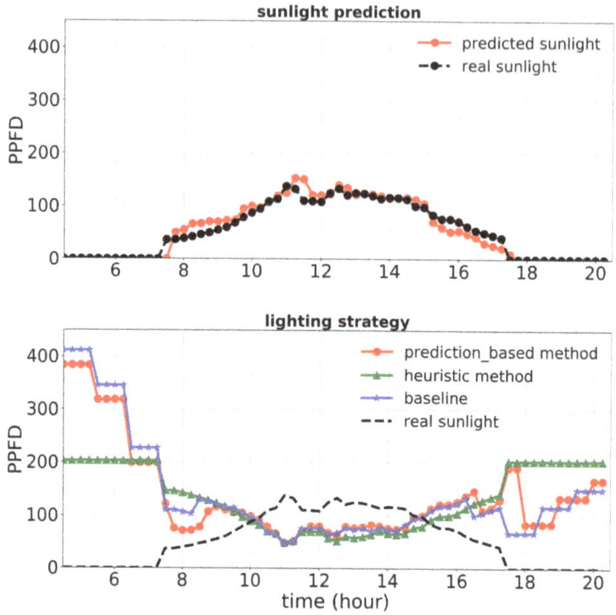

Figure 11. Performance of lighting control strategies and sunlight prediction for the whole period of the first experiment (49 d) as an average.

Some of the images taken by our setup (see Figure 1) are shown in Figure 12. The top images were taken of the plants under prediction-based lighting, and the bottom images were taken of the plants under the heuristic lighting method. The left pairs in Figure 12 were captured 32 d after seeding (January 11th), and the right pairs were captured at the end of the experiment (January 28th). Arducam took the left images of each pair at night in darkness, and then, the images were transferred to a computer through the VNC viewer. Hence, we could access the images remotely and in real time via WiFi. Using the PlantCV package, the images were analyzed, and the PCS was measured by thresholding, thereby resulting in the right images of each pair in Figure 12.

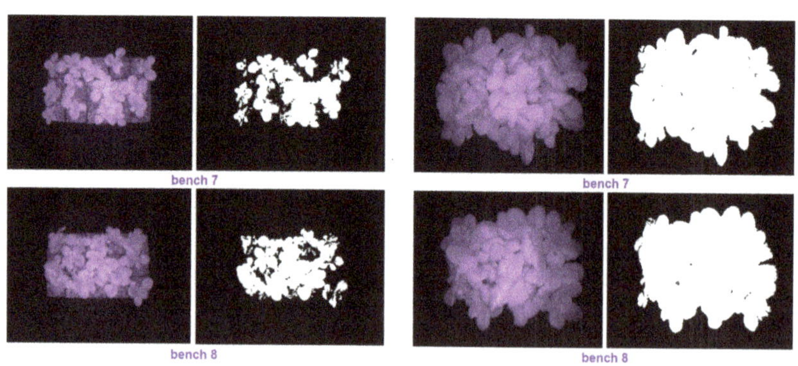

Figure 12. The original and the processed images on Day 32 and the last day of the first experiment.

The results of the statistical analysis on all growth parameters are provided in Table 4. The p-value for all parameters was greater than 0.05; consequently, the means of the two treatments were almost equal for each growth parameter (especially shoot dry weight, which was the most important growth parameter). Therefore, the proposed approach did not affect the plant growth adversely or positively. We calculated another parameter, which was the total dry weight divided by the total electricity cost (in cents/m^2), for each approach. This parameter includes both cost and growth information, which for the prediction-based method was 0.0797 (g/cent) and for the heuristic method was 0.0738 (g/cent), a 7.4% reduction. Hence, the proposed optimal lighting approach resulted in a reduction of the electricity cost and did not affect plant growth adversely.

Table 4. Results of the paired t-test on the growth parameters during the first experiment.

Measurement	Mean (Prediction-Based)	Mean (Heuristic)	Standard Deviation (Prediction-Based)	Standard Deviation (Heuristic)	Standard Error (Prediction-Based)	Standard Error (Heuristic)	p-Value	t-Value
CCI/January 11th	16.36	17.1	2.84	1.68	1.64	0.97	0.40	1.06
CCI/January 28th	24.06	22.93	1.94	1.42	1.12	0.82	0.33	1.26
ACI/January 11th	4.86	4.91	0.45	0.13	0.26	0.08	0.82	0.25
ACI/January 28th	5.56	5.58	0.64	0.17	0.37	0.1	0.96	0.06
SLA per plant (cm^2/g)	114.17	117.67	3.66	11.7	2.11	6.76	0.54	0.73
shoot dry weight per plant (g)	2.89	2.79	0.41	0.36	0.24	0.21	0.8	0.28

Figures 13–15 show the performance of the lighting strategies and sunlight prediction throughout different days with different sunlight levels in the second experiment. Figure 13 is representative of rainy days with low sunlight irradiance compared to the mean irradiance in April. April 7th irradiance was close to a day with regular sunlight during that period, and May 6th had high sunlight irradiance (presented in Figure 14 and Figure 15, respectively). The cost increase in each day compared to the baseline was calculated and shown in Table 5. Moreover, the average of actual sunlight, its prediction, and supplemental light for different methods were calculated throughout this experiment and displayed in Figure 16. Based on the results in Table 5, the cumulative cost increase for the heuristic method was 62.86%, while for the prediction-based method, it was 7.74% compared to the baseline. Our prediction-based lighting approach showed about a 33.85% electricity cost reduction compared to the heuristic method throughout the second experiment.

Some of the images taken by our setup (see Figure 1) during the second experiment are illustrated in Figure 17. The left pairs were captured 33 d after seeding (May 4th), and the right pairs were captured at the end of the experiment (May 18th). The PCS was measured using PlantCV package, and we monitored the plant growth remotely via WiFi in real-time, the same method as the first experiment. The average of the PCS for those replicates that had a camera was measured for each treatment and shown in Figure 18.

Other than the lighting approach, different temperatures and ambient light in the greenhouse had at most minor effects on our results, since our treatments were blocked to account for east to west environmental gradients within the greenhouse.

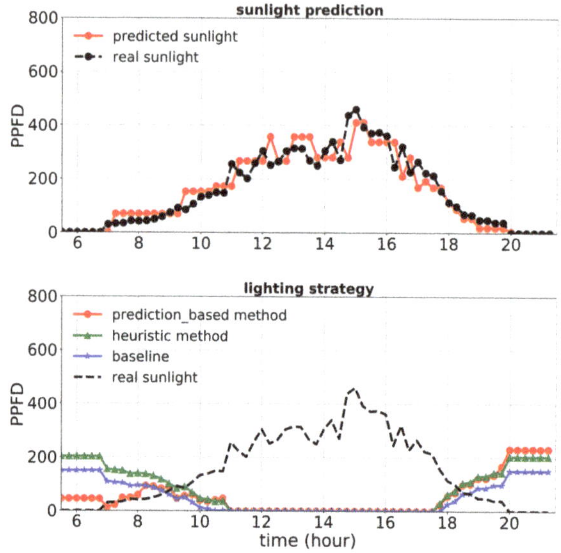

Figure 13. Performance of lighting control strategies and sunlight prediction for April 15th, a day with a low sunlight level during the second experiment.

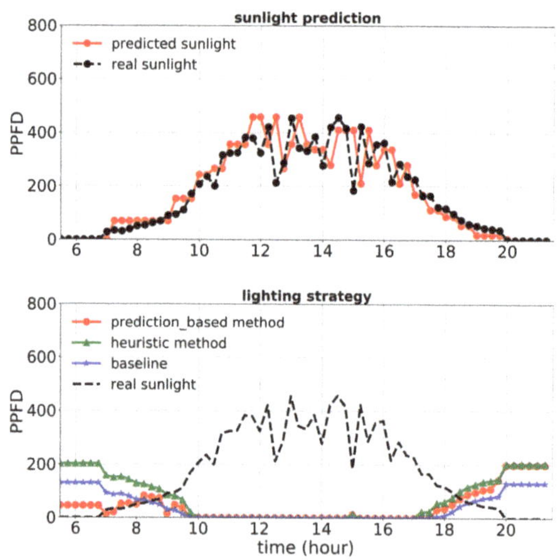

Figure 14. Performance of lighting control strategies and sunlight prediction for April 7th, a day with a moderate sunlight level during the second experiment.

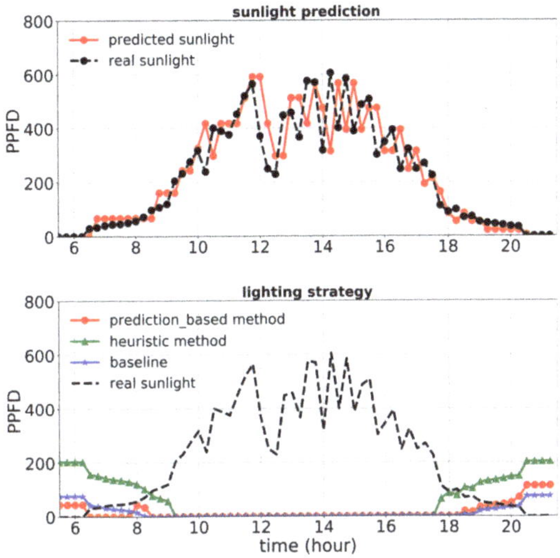

Figure 15. Performance of lighting control strategies and sunlight prediction for May 6th, a day with a high sunlight level during the second experiment.

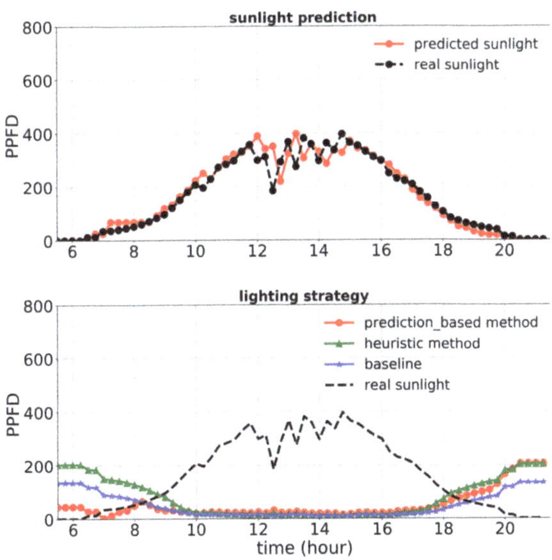

Figure 16. Performance of lighting control strategies and sunlight prediction for the whole period of the second experiment (47 d) as an average.

Table 5. Cost of lighting control strategies in $/m^2 per day for different days during the second experiment.

Day	Baseline Cost	Prediction-Based Method Cost Increase Compared to Baseline	Heuristic Method Cost Increase Compared to Baseline
April 7	0.16	0.0085 (5.31%)	0.14 (87.5%)
April 15	0.21	0.0135 (6.43%)	0.11 (52.38%)
May 6	0.06	0.0015 (2.5%)	0.21 (350%)
cumulative cost	9.49	0.73 (7.69%)	5.96 (62.8%)

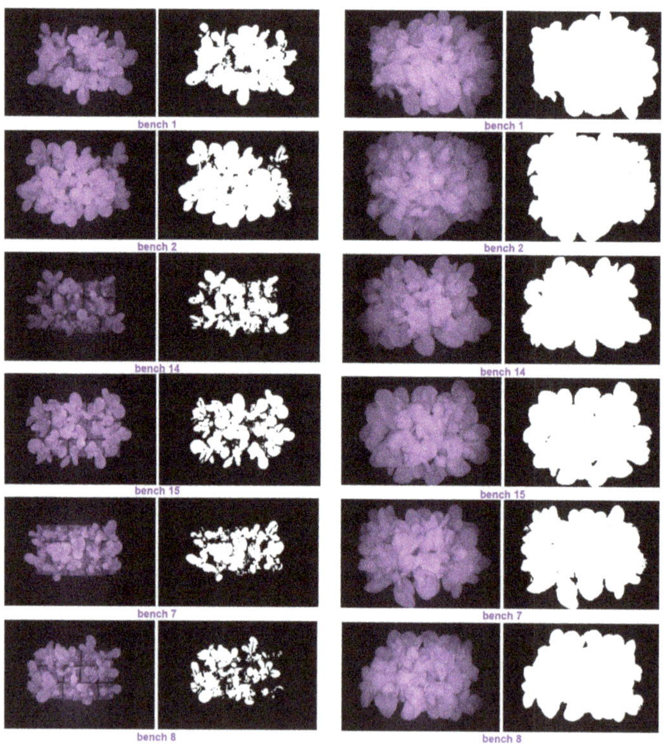

Figure 17. The original and the processed images on Day 33 and the last day of the second experiment.

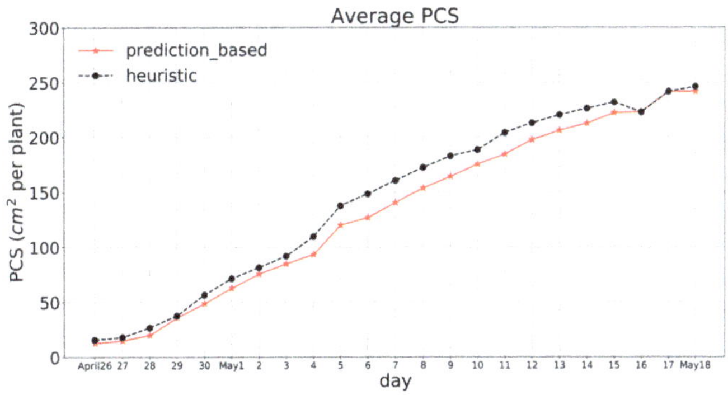

Figure 18. Average PCS for the two treatments during the last 23 d of the second experiment.

The results of the statistical analysis on the growth parameters for the second experiment are provided in Table 6. The p-value for all parameters was greater than 0.05, except for the SLA ($t(4) = 3.37, p = 0.03$). The SLA was greater for the proposed strategy ($\mu = 131.24$, $\sigma = 8.1\ \text{cm}^2/\text{g}$ per plant), compared to the heuristic method ($\mu = 116.69$, $\sigma = 7.34\ \text{cm}^2/\text{g}$ per plant). However, this difference was not necessarily due to a true biological effect. It was more likely due to a Type I error. Thus, the statistical analysis did not show meaningful differences in plant growth between the two treatments. The total shoot dry weight divided by the total cost for the prediction-based method was 0.2166 (g/cent), while for the heuristic method, it was 0.1475 (g/cent), 32% lower. Hence, the proposed strategy resulted in a higher efficiency than the heuristic method in terms of electricity cost together with plant growth.

Table 6. Results of the paired t-test on the growth parameters during the second experiment.

Measurement	Mean (Prediction-Based)	Mean (Heuristic)	Standard Deviation (Prediction-Based)	Standard Deviation (Heuristic)	Standard Error (Prediction-Based)	Standard Error (Heuristic)	p-Value	t-Value
CCI/May 7th	9.81	10.54	0.80	0.73	0.36	0.33	0.19	1.98
CCI/May 18th	12.19	11.80	1.40	0.76	0.63	0.34	0.62	0.54
ACI/May 7th	3.63	3.67	0.22	0.14	0.1	0.06	0.77	0.32
ACI/May 18th	4.1	4.02	0.16	0.25	0.07	0.11	0.49	0.75
SLA per plant (cm^2/g)	131.24	116.69	8.1	7.34	3.62	3.28	0.03	3.37
PCS per plant (cm^2)	241.93	246.06	14.42	32.99	8.33	19.05	0.78	0.33
shoot dry weight per plant (g)	2.95	3.04	0.45	0.42	0.2	0.19	0.06	2.64

Both experiments validated that the prediction-based method can reduce cost while maintaining plant growth. As shown in Figures 3 and 6, the sunlight levels were much lower during the first study compared to the second study; therefore, more supplemental light was provided to reach the minimum DPI during the first study, which resulted in a higher electricity cost (for both approaches). Since the dry weight at the end of the experiments was almost the same, the total dry weight/total cost was lower for the first study. The difference in the DLI from sunlight also affected the electricity cost savings of the proposed method for the two experiments. During the first study, the DLI was very low, and that resulted in providing so much supplemental light to satisfy the light requirements of the plants with both lighting approaches (the heuristic and the proposed method). The heuristic method provided the greatest benefits when the light levels are variable, both in the short term (15 min) and longer term (day to day). Thus, there was not a huge difference in the cost of the lighting methods. However, during the second study, the DLI levels were much higher, and much less supplemental lighting was needed to reach the minimum lighting requirement. Hence, the proposed method reduced the cost by a higher percentage in the second study (33.85% compared to 4.16%).

The cost profile in the second experiment was a fixed price, since we wanted to consider the actual conditions in our research greenhouse. The simulation results in our previous work [25] showed that the proposed method with variable electricity pricing contributed to a cost reduction in different months of the year. If we used a variable cost profile for the second study, as well as the first one, the cost reduction in the second study would have been even higher.

The proposed strategy for controlling supplemental lighting significantly differs from the previous approaches. Most of the previous approaches that were mentioned in the Introduction Section are rule-based methods and do not use sunlight prediction [15,17,22]. Among all the methods proposed for HID lamps, the DynaLight system and DynaLight IND are the most efficient ones since they use weather forecast, a photosynthesis model, and a variable electricity profile. However, these methods are still not optimal because providing the weather forecasts twice daily is not enough to obtain the optimal lighting strategy, and for the improved version of this system (DynaLight IND), practical experiments should be conducted to validate its efficiency [20,21]. On the other hand, our proposed method was evaluated through both simulations and practical experiments.

A study on an adaptive control approach for LEDs [15] showed that a method similar to the heuristic method reduces the cost of supplemental lighting, compared to two other rule-based methods. In the present paper, we concluded that the prediction-based approach reduced the cost significantly compared to the heuristic method. Therefore, our strategy outperformed the other methods by taking advantage of the sunlight predictive model, the minimum DPI requirement, and variable electricity pricing. The shortcoming of our method is that the sunlight predictive model uses the mean value of the historical data at the beginning of the day, and for rainy days with too low sunlight levels, this may result in overestimating sunlight. Therefore, this approach may not perform as expected in this case.

There has been much research on growing lettuce with 24 h photoperiods [36]. Nonetheless, this practice has not (yet) been adopted by the industry, where photoperiods of 16–20 h are more commonly used. To make our research relevant to the greenhouse industry, we decided to use a photoperiod that is commonly used in commercial greenhouses. The energy savings that can be realized may indeed depend on the photoperiod that is used. Our lighting optimization protocol was formulated in such a way that any photoperiod can be used, and it is not specific to a 16 h photoperiod.

Finally, we note that lighting can provide a substantial amount of heat to a greenhouse; but the effects of more efficient lighting methods and greenhouse heating requirements are often ignored. A major challenge is that the interaction between lighting and heating is location specific and requires advanced models to estimate [37]. We thus opted to focus solely on lighting costs in this study.

4. Conclusions

In this paper, a new optimal approach was developed and implemented for supplemental lighting control in greenhouse environments using IoT technology. This method solves an optimization problem to satisfy plant light needs using Markov-based sunlight prediction and variable electricity pricing. Two experimental studies during two different seasons in the same greenhouse were conducted, where the proposed lighting approach was validated in terms of electricity cost reduction (4.16% reduction during the winter study and 33.85% during the spring study) while maintaining plant growth.

Author Contributions: Conceptualization, S.A., S.M., J.M.V. and M.W.v.I.; methodology, S.A., S.M., J.M.V. and M.W.v.I.; software, S.A.; validation, S.A.; formal analysis, S.A.; investigation, S.A., S.M., J.M.V. and M.W.v.I.; data curation, S.A.; writing—original draft preparation, S.A.; writing—review and editing, S.A., J.M.V. and M.W.v.I.; supervision, J.M.V.; funding acquisition, J.M.V. All authors have read and agreed to the published version of the manuscript.

Funding: This work was financially supported by USDA-NIFA-SCRI Award Number #2018-51181-28365, Project Lighting Approaches to Maximize Profits.

Institutional Review Board Statement: Not applicable.

Informed Consent Statement: Not applicable.

Data Availability Statement: Sunlight data that were used to develop the predictive model are available at https://www.nrel.gov/grid/solar-resource/confrrm.html (accessed on 1 December 2019). The code developed for implementing the lighting strategy can be found here: https://github.com/velnilab/optimal-lighting (released 19 November 2021).

Acknowledgments: The authors would like to thank Theekshana C. Jayalath for his technical assistance.

Conflicts of Interest: The authors declare no conflict of interest. The funders had no role in the design, execution, interpretation, nor writing of the study.

Abbreviations

The following abbreviations are used in this manuscript:

ACI	Anthocyanin content index
ADC	Analog-to-digital converter
CCI	Leaf chlorophyll content index
CEA	Controlled environment agriculture
DLI	Daily light integral
DPI	Daily photochemical integral
DSL	Domain-Specific Language
ETR	Electron transport rate
FPGA	Field programmable gate array
GA	Genetic algorithm
HCI	Human–computer interaction
HID	High-intensity discharge
IoT	Internet of Things
IR	Infrared
LASSI	Light and shade system implementation
LED	Light-emitting diode
PCS	Projected canopy size
PPFD	Photosynthetic photon flux density
PWM	Pulse-width modulation

RFID	Radio-frequency identification
RL	Reinforcement learning
SLA	Specific leaf area
UAV	Unmanned aerial vehicle
VPD	Vapor pressure deficit
WSN	Wireless sensor network

References

1. Alexandratos, N.; Bruinsma, J. World Agriculture towards 2030/2050: The 2012 Revision. 2012. Available online: https://doi.org/10.22004/ag.econ.288998 (accessed on 20 November 2020).
2. Khanna, A.; Kaur, S. Evolution of Internet of Things (IoT) and its significant impact in the field of Precision Agriculture. *Comput. Electron. Agric.* **2019**, *157*, 218–231. [CrossRef]
3. Elijah, O.; Rahman, T.A.; Orikumhi, I.; Leow, C.Y.; Hindia, M.N. An overview of Internet of Things (IoT) and data analytics in agriculture: Benefits and challenges. *IEEE Internet Things J.* **2018**, *5*, 3758–3773. [CrossRef]
4. Manrique, J.A.; Rueda-Rueda, J.S.; Portocarrero, J.M. Contrasting internet of things and wireless sensor network from a conceptual overview. In Proceedings of the 2016 IEEE international conference on Internet of Things (iThings) and IEEE Green Computing and Communications (GreenCom) and IEEE Cyber, Physical and Social Computing (CPSCom) and IEEE SMART data (SmartData), Chengdu, China, 15–18 December 2016; pp. 252–257.
5. Zhao, J.C.; Zhang, J.F.; Feng, Y.; Guo, J.X. The study and application of the IoT technology in agriculture. In Proceedings of the 2010 3rd International Conference on Computer Science and Information Technology, Chengdu, China, 9–11 July 2010; Volume 2, pp. 462–465.
6. Odema, M.; Adly, I.; Wahba, A.; Ragai, H. Smart aquaponics system for industrial Internet of Things (IIoT). In Proceedings of the International Conference on Advanced Intelligent Systems and Informatics, Cairo, Egypt, 9–11 September 2017; pp. 844–854.
7. Yu, Z.; Xugang, L.; Xue, G.; Dan, L. IoT forest environmental factors collection platform based on ZIGBEE. *Cybern. Inf. Technol.* **2014**, *14*, 51–62. [CrossRef]
8. Shinde, T.A.; Prasad, J.R. IoT based animal health monitoring with naive Bayes classification. *Int. J. Emerg. Trends Technol.* **2017**, *1*, 252–257.
9. Tripicchio, P.; Satler, M.; Dabisias, G.; Ruffaldi, E.; Avizzano, C.A. Towards smart farming and sustainable agriculture with drones. In Proceedings of the 2015 International Conference on Intelligent Environments, Prague, Czech Republic, 15–17 July 2015; pp. 140–143.
10. Mainetti, L.; Mele, F.; Patrono, L.; Simone, F.; Stefanizzi, M.L.; Vergallo, R. An RFID-based tracing and tracking system for the fresh vegetables supply chain. *Int. J. Antennas Propag.* **2013**, *2013*, 531364. [CrossRef]
11. Lerdsuwan, P.; Phunchongharn, P. An energy-efficient transmission framework for IoT monitoring systems in precision agriculture. In Proceedings of the International Conference on Information Science and Applications, Macau, China, 20–23 March 2017; pp. 714–721.
12. Liu, H.; Meng, Z.; Cui, S. A wireless sensor network prototype for environmental monitoring in greenhouses. In Proceedings of the 2007 International Conference on Wireless Communications, Networking and Mobile Computing, Shanghai, China, 21–25 September 2007; pp. 2344–2347.
13. Zhang, Q.; Yang, X.L.; Zhou, Y.M.; Wang, L.R.; Guo, X.S. A wireless solution for greenhouse monitoring and control system based on ZigBee technology. *J. Zhejiang-Univ.-Sci. A* **2007**, *8*, 1584–1587. [CrossRef]
14. Pekoslawski, B.; Krasinski, P.; Siedlecki, M.; Napieralski, A. Autonomous wireless sensor network for greenhouse environmental conditions monitoring. In Proceedings of the 20th International Conference Mixed Design of Integrated Circuits and Systems-MIXDES 2013, Gdynia, Poland, 20–22 June 2013; pp. 503–507.
15. van Iersel, M.W.; Gianino, D. An adaptive control approach for light-emitting diode lights can reduce the energy costs of supplemental lighting in greenhouses. *HortScience* **2017**, *52*, 72–77. [CrossRef]
16. Narra, P.; Zinger, D.S. An effective LED dimming approach. In Proceedings of the Conference Record of the 2004 IEEE Industry Applications Conference, 2004. 39th IAS Annual Meeting, Seattle, WA, USA, 3–7 October 2004; Volume 3, pp. 1671–1676.
17. Carrier, M.; Gosselin, A.; Gauthier, L. Description of a crop growth model for the management of supplemental lighting in greenhouses. *HortTechnology* **1994**, *4*, 383–389. [CrossRef]
18. Albright, L.; Both, A.J.; Chiu, A. Controlling greenhouse light to a consistent daily integral. *Trans. ASAE* **2000**, *43*, 421. [CrossRef]
19. Seginer, I.; Albright, L.D.; Ioslovich, I. Improved strategy for a constant daily light integral in greenhouses. *Biosyst. Eng.* **2006**, *93*, 69–80. [CrossRef]
20. Mærsk-Møller, H.M.; Jørgensen, B.N. A software product line for energy-efficient control of supplementary lighting in greenhouses. In Proceedings of the The International Conference on Green Computing, Chengdu, China, 4–5 August 2011.
21. Qu, Y.; Clausen, A.; Jørgensen, B.N. A multi-objective optimization platform for artificial lighting system in commercial greenhouses. *Energy Informatics* **2021**, *4*, 1–20. [CrossRef]
22. Pinho, P.; Hytönen, T.; Rantanen, M.; Elomaa, P.; Halonen, L. Dynamic control of supplemental lighting intensity in a greenhouse environment. *Light. Res. Technol.* **2013**, *45*, 295–304. [CrossRef]

23. Weaver, G.M.; van Iersel, M.W.; Velni, J.M. A photochemistry-based method for optimising greenhouse supplemental light intensity. *Biosyst. Eng.* **2019**, *182*, 123–137. [CrossRef]
24. Schwend, T.; Beck, M.; Prucker, D.; Peisl, S.; Mempel, H. Test of a PAR sensor-based, dynamic regulation of LED lighting in greenhouse cultivation of Helianthus annuus. *Eur. J. Hortic. Sci.* **2016**, *81*, 152–156. [CrossRef]
25. Mosharafian, S.; Afzali, S.; Weaver, G.M.; van Iersel, M.; Velni, J.M. Optimal lighting control in greenhouse by incorporating sunlight prediction. *Comput. Electron. Agric.* **2021**, *188*, 106300. [CrossRef]
26. Afzali, S.; Mosharafian, S.; van Iersel, M.W.; Velni, J.M. Optimal Lighting Control in Greenhouses Equipped with High-intensity Discharge Lamps Using Reinforcement Learning. In Proceedings of the 2021 American Control Conference (ACC), New Orleans, LA, USA, 25–28 May 2021; pp. 1414–1419.
27. Elkins, C.; van Iersel, M.W. Longer photoperiods with the same daily light integral increase daily electron transport through photosystem II in lettuce. *Plants* **2020**, *9*, 1172. [CrossRef]
28. Weaver, G.; van Iersel, M.W. Longer photoperiods with adaptive lighting control can improve growth of greenhouse-grown 'Little Gem' lettuce (Lactuca sativa). *HortScience* **2020**, *55*, 573–580. [CrossRef]
29. Diamond, S.; Boyd, S. CVXPY: A Python-embedded modeling language for convex optimization. *J. Mach. Learn. Res.* **2016**, *17*, 2909–2913.
30. Agrawal, A.; Verschueren, R.; Diamond, S.; Boyd, S. A rewriting system for convex optimization problems. *J. Control. Decis.* **2018**, *5*, 42–60. [CrossRef]
31. NREL. National Renewable Energy Laboratory (NREL). Available online: https://www.nrel.gov/grid/solar-resource/confrrm.html (accessed on 1 December 2019).
32. Zabeltitz, C. Light Transmittance of Greenhouses. In *Integrated Greenhouse Systems for Mild Climates*; Springer: Berlin/Heidelberg, Germany, 2011; pp. 137–143.
33. Gehan, M.A.; Fahlgren, N.; Abbasi, A.; Berry, J.C.; Callen, S.T.; Chavez, L.; Doust, A.N.; Feldman, M.J.; Gilbert, K.B.; Hodge, J.G.; et al. PlantCV v2: Image analysis software for high-throughput plant phenotyping. *PeerJ* **2017**, *5*, e4088. [CrossRef]
34. Arize. GE Arize Element L1000 Specification Sheet. Available online: https://images.salsify.com/images/rf2bqhdet0yfnygoabjh/HORT100-Arize-Element-L1000-Spec-Sheet.pdf (accessed on 20 November 2020).
35. Jayalath, T.C.; van Iersel, M.W. Canopy Size and Light Use Efficiency Explain Growth Differences between Lettuce and Mizuna in Vertical Farms. *Plants* **2021**, *10*, 704. [CrossRef] [PubMed]
36. Koontz, H.; Prince, R. Effect of 16 and 24 h daily radiation (light) on lettuce growth. *Hort Sci. Publ. Am. Soc. Hortic. Sci.* **1986**, *21*, 123–124.
37. Katzin, D.; Marcelis, L.F.; van Mourik, S. Energy savings in greenhouses by transition from high-pressure sodium to LED lighting. *Appl. Energy* **2021**, *281*, 116019. [CrossRef]

plants

MDPI

Article

Monochromic Radiations Provided by Light Emitted Diode (LED) Modulate Infection and Defense Response to Fire Blight in Pear Trees

Tiziana Sgamma [1,†], Ivano Forgione [2,†], Francesca Luziatelli [3], Calogero Iacona [4], Roberto Mancinelli [2], Brian Thomas [5], Maurizio Ruzzi [3] and Rosario Muleo [2,*]

1 Biomolecular Technology Group, Leicester School of Allied Health Science, De Montfort University, Leicester LE1 9BH, UK; tiziana.sgamma@dmu.ac.uk
2 Department of Agricultural and Forestry Sciences, Tuscia University, Via S. C. De Lellis, snc., 01100 Viterbo, Italy; ivano.forgione@unitus.it (I.F.); mancinel@unitus.it (R.M.)
3 Department for Innovation in Biological, Agro-Food and Forest Systems (DIBAF), Tuscia University, Via S. C. De Lellis, snc., 01100 Viterbo, Italy; f.luziatelli@unitus.it (F.L.); ruzzi@unitus.it (M.R.)
4 Department of Agriculture, Food and Environment (DAFE), University of Pisa, Via del Borghetto, 80, 56124 Pisa, Italy; calogero.iacona@unipi.it
5 School of Life Sciences, University of Warwick, Coventry CV4 7AL, UK; Brian.Thomas@warwick.ac.uk
* Correspondence: muleo@unitus.it
† These authors have contributed equally to this work as first authors.

Abstract: Pathogenesis-related (PR) proteins are part of the systemic signaling network that perceives pathogens and activates defenses in the plant. Eukaryotic and bacterial species have a 24-h 'body clock' known as the circadian rhythm. This rhythm regulates an organism's life, modulating the activity of the phytochromes (*phys*) and cryptochromes (*crys*) and the accumulation of the corresponding mRNAs, which results in the synchronization of the internal clock and works as zeitgeber molecules. Salicylic acid accumulation is also under light control and upregulates the *PR* genes expression, increasing plants' resistance to pathogens. *Erwinia amylovora* causes fire blight disease in pear trees. In this work, four bacterial transcripts (*erw1-4*), expressed in asymptomatic *E. amylovora*-infected pear plantlets, were isolated. The research aimed to understand how the circadian clock, light quality, and related photoreceptors regulate *PR* and *erw* genes expression using transgenic pear lines overexpressing *PHYB* and *CRY1* as a model system. Plantlets were exposed to different circadian conditions, and continuous monochromic radiations (Blue, Red, and Far-Red) were provided by light-emitting diodes (LED). Results showed a circadian oscillation of *PR10* gene expression, while *PR1* was expressed without clear evidence of circadian regulation. Bacterial growth was regulated by monochromatic light: the growth of bacteria exposed to Far-Red did not differ from that detected in darkness; instead, it was mildly stimulated under Red, while it was significantly inhibited under Blue. In this regulatory framework, the active form of phytochrome enhances the expression of *PR1* five to 15 fold. An ultradian rhythm was observed fitting the zeitgeber role played by CRY1. These results also highlight a regulating role of photoreceptors on the expression of *PRs* genes in non-infected and infected plantlets, which influenced the expression of *erw* genes. Data are discussed concerning the regulatory role of photoreceptors during photoperiod and pathogen attacks.

Keywords: light quality; photosensors; host-pathogen interaction; resistance genes; gene regulation; bacterial growth; *Erwinia amylovora*; circadian rhythms; optogenetics

Citation: Sgamma, T.; Forgione, I.; Luziatelli, F.; Iacona, C.; Mancinelli, R.; Thomas, B.; Ruzzi, M.; Muleo, R. Monochromic Radiations Provided by Light Emitted Diode (LED) Modulate Infection and Defense Response to Fire Blight in Pear Trees. *Plants* **2021**, *10*, 1886. https://doi.org/10.3390/plants10091886

Academic Editor: Magda Pál

Received: 24 August 2021
Accepted: 8 September 2021
Published: 12 September 2021

1. Introduction

Erwinia amylovora causes fire blight, a disease of agronomic and economic importance that affects many Rosaceae species, primarily pear and apple trees. Bacteria penetrate the plant mainly through the flowers and can also enter leaf tissue through wounds [1]. During plant-pathogen interaction, a dialogue occurs between the two organisms: the plant synthesizes molecules for the signaling system and defense; in contrast, the pathogen synthesizes molecules to break down the barriers of the host and mimic plant hormones. In the beginning, under favorable climatic conditions, the pathogen inserts through the intercellular spaces of parenchyma. Afterward, it colonizes the xylem vessels causing extensive damage with the final death of the plant [2]. The necrotic parts of the plant become brown as if they had been burned by fire [3].

During the infection process, pathogenesis-related (PR) proteins are part of an articulated systemic signaling network active in plants to perceive pathogens and activate defenses. The necrotic lesion induces the expression of a set of pathogenesis-related genes. *PR1* gene, one of the 17 *PR* gene families, has frequently been used as a marker for systemic acquired resistance (SAR) in many plant species [4–6]. In the immunization stage, necrosis-causing pathogen when infects a leaf usually provokes the formation of a localized dry, necrotic lesion that limits pathogen spread and provides local resistance. This step is also referred to as hypersensitive response (HR) [7]. Accumulation of salicylic acid (SA) is also associated with this stage. SA is an endogenous plant hormone whose levels increase after pathogen infection. SA can induce the expression of *PR1* [8]. Increases in SA and SA-inducible PR proteins are associated with disease resistance at several levels, not just with the SAR response. A phloem-mobile signal then moves from the immunized leaf to the rest of the plant to establish SAR. The perception of the mobile signal in the uninoculated leaves results in the expression of the same set of *PR* genes as induced around the primary infection site. When the plant is challenged with a second virulent pathogen, the plant responds as if that was an avirulent one because of the rapid accumulation of the *PR1*-transcripts [9]. However, there are indications that several of the *PR* genes are expressed at basal levels in plants without any pathogen attack. Moreover, studies showed that SA is also crucial for sustaining basal levels of genes associated with resistance responses, including *PR1*, and keeping the defense system primed in the absence of pathogen attacks [4,10–13]. Remarkably, when *PR* genes are not expressed this leads to a higher susceptibility to infectious agents [5].

The PR10s defense-related proteins are a ubiquitous class of intracellular in contrast to the extracellular nature of most PR proteins [14]. Most of them are induced upon microbial attack by fungal elicitors, wounding, and stress stimuli, as with most of the other PR-protein families. PR10 proteins are also expressed in a tissue-specific manner during development and some PR10 proteins show constitutive expression patterns [15,16]. PR10s have been attributed a ribonuclease-like function due to sequence homology with ribonucleases (RNase) [17]. However, only some PR10 proteins have been proposed to possess RNase activity [18]. In addition, they have also been shown to respond to plant hormones, including jasmonic acid (JA), and abiotic stresses such as salt and drought [18].

Plant defense responses at the site of the bacterial infection are elevated, accumulation of SA occurs and the transcription of *PR1* is induced in light [19]. The PR1 light dependency and the execution of HR confirm that these responses are closely associated and that light regulation already takes place early in this SA-dependent signaling pathway [7]. Phytochromes are crucial photoreceptors and are involved in the modulation of the *PR1* expression by light. The absence of both PHYA and PHYB strongly reduces the expression of *PR* genes upon treatment with SA, with a more significant influence of PHYB deficiency [5,7,20]. Phytochrome signaling strongly modulates the response of endogenous SA [5]. There is a strict light dependency of gene expression of *PRs* and the HR process. HR lesions are often correlated with the induction of *PR* genes and are also light modulated. HR is strongly reduced by the absence of phytochromes and amplified in an SA-dependent manner in the *psi2* mutant [5].

Moreover, photoreceptor proteins such as cryptochromes (CRYs), which are Blue light (BL) photoreceptors homologous to photolyases, seem to be involved in pathogen response. Proteomics study identified proteins with altered expression related to defense, stress, and detoxification in *cry1* mutant [21]. CRY1 positively regulates SAR, indeed, in *Arabidopsis*, the inactivation of the *CRY1* gene has a mild influence on the SA accumulation and determines a reduction of the *PR1* expression; in contrast, the overexpression of this gene *CRY1* significantly enhances the expression of *PR1* [22]. Furthermore, other studies showed that in mutants of COP1 (constitutive photomorphogenesis 1), COP9, and DET1 (De-etiolated 1), which are part of the CRY1 signaling pathway, *PR* genes were highly up-regulated [23,24].

Prokaryotes have evolved a repertoire of photosensory proteins that determine changes in the external light and regulate cell physiology in a light-dependent manner [25]. Bacterial photoreceptors include proteins with a bilin-type chromophore (bacteriophytochromes) for sensing red light (RL) and far-red light (FRL) [26]. Moreover, they include proteins with photosensory domains for BL such as BLUF (BL sensing using flavin adenine dinucleotide [FAD]), LOV (light, oxygen, or voltage), PYP (photoactive yellow protein), and cryptochrome/photolyase (Cry/PHR) superfamilies, green- or blue-light-absorbing microbial rhodopsin [27,28]. In plant-associated bacteria, the number of candidate photoreceptors varies: *Pseudomonas syringae* pv *syringae* B728a and *Pseudomonas syringae* pv *tomato* DC3000 have two bacteriophytochromes (BphP1 and BphP2) and one LOV domain-containing histidine kinase (LOV-HK) [29]; anoxygenic phototrophs, such as Methylobacteria, can contain between 3 and 16 photosensory proteins [30].

In the Enterobacteriaceae, there is only one report indicating the presence of a *bphP* gene in *Enterobacter cloacae* [31]. At the same time, several studies show that these bacteria are sensitive to irradiation treatments with wavelengths in the range of visible, violet, and blue light [32–34]. So far, no gene encoding photosensory protein has been identified yet in the plant pathogenic Enterobacteriaceae species *E. amylovora*, and there is no evidence that the growth, phenotype, or virulence of this pathogen is affected by the light.

In this work, four *E. amylovora* genes (*erw*) were isolated that could be used as a marker to monitor the initial phase of the infection in asymptomatic plants. To investigate if the circadian internal clock, the light quality, and the related photoreceptors autonomously regulate the abundance of *PR1* and *PR10* transcripts, in vitro-cultured plantlets of Iranian pear cultivar *Dar Gazi-wild type* (*wt*), *Dar Gazi-phyB* (transgenic plant overexpressing *Arabidopsis phyB*, and *Dar Gazi-cry1* (transgenic plant overexpressing tomato *CRY1*) were exposed to different circadian experimental conditions, and continuous BL, RL, and FRL, emitted by light-emitting diodes (LED).

Transcriptional changes in host and pathogen gene expression during early *E. amylovora* infection indicated that both plant *PR*s and bacterial *erw* genes were temporarily expressed and differentially regulated. The results reported in this work indicate that photoreceptor-mediated signals regulate the expression of specific plant and pathogen genes in pear plantlets infected by *E. amylovora*.

2. Results

2.1. Effect on the Monochromatic Light on the Growth of E. amylovora *In Vitro*

To evaluate the effect of the monochromatic light on the cell growth of *E. amylovora* strain Ea273, the microorganism was cultivated in shake flasks under light and temperature-controlled conditions. All cultures were inoculated at the same initial optical density (OD600 = 0.1), with cells from precultures grown in the darkness and in the late exponential phase of growth to minimize the impact of the inoculum on the lag phase of growth. The results reported in Figure 1 indicate that the monochromatic light has had several effects on the growth of this microorganism. Significant differences were observed using OD_{600} as an indirect measure of the cell growth and comparing the OD_{600} values at the end of the growth (on, overnight; 18–24 h after the inoculum). Under the same cultivation conditions (inoculum, temperature, rotation speed, incubation time), the highest cell density (OD_{600} = 4) was achieved for cultivating strain Ea273 under continuous RL (Figure 1). This value was approximately 1.4-fold higher than the one obtained in darkness and 1.2- and 1.3-times higher than the one obtained under FRL and BL. Under BL, the deceleration/decline phase started at least 3 h earlier (8 h after the inoculum), but no significant difference was observed compared to the OD_{600} values at t = 11 and t = on (Figure 1). The latent growth phase was shorter (approximatively 1 h) under FRL compared to darkness (Figure 1). The growth profile under RL showed a prolonged exponential phase (up to t = 8), a prolonged latent phase (up to t = 4), and a higher growth rate in the late phase of growth after the diauxic shift (between 10 and 11 h after the inoculum; Figure 1). These results indicate that monochromatic lighting can modulate the growth pattern of *E. amylovora* independently from the presence of the plant stimuli.

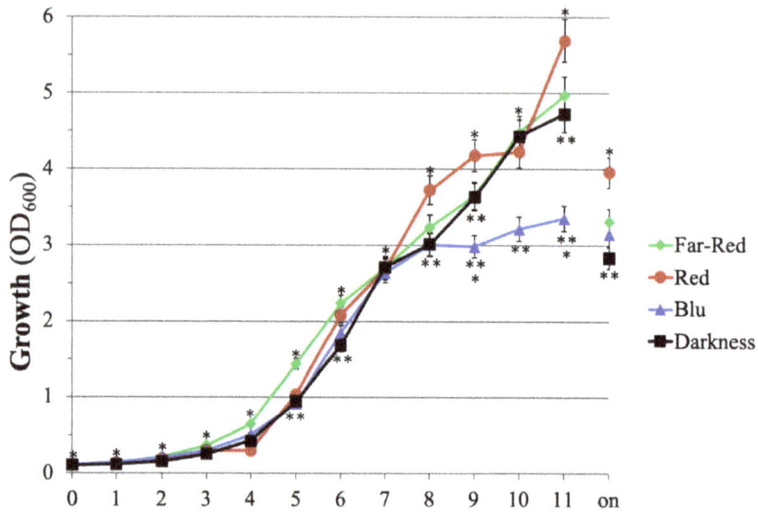

Figure 1. The effect of monochromatic light on the growth of *E. amylovora* Ea273 in shake flask cultures. The data are representative of three independent experiments with three biological and two technical replicates. Error bars represent the SD. The number of asterisks adjacent to the symbols, at the same time point, indicates significant differences between different growth conditions (Student's *t*-test, $p \leq 0.05$).

2.2. *E. amylovora* Causes Tissue Necrosis in In Vitro Pear Dar Gazi Plantlets

The in vitro grown Iranian pear cultivar Dar Gazi wild type (wt) inoculated with *E. amylovora*, showed oxidative stress in the central cylinder and the cortex of the basal portion of plantlets (Figure 2a). The stems resulted in characteristic signs of HR that cause rapid cell death in the vicinity of the infection point (Figure 2a). These observations are consistent with Electrical Conductivity (EC) measurements, where high values reflect the plasma membrane disruption. An ion leakage three-fold higher was detected in *Dar Gazi-wt* infected vs non-infected (Figure 2b).

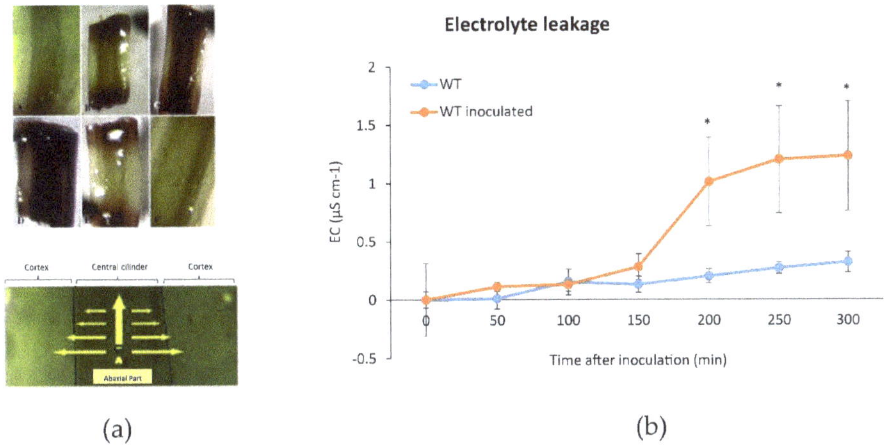

Figure 2. Oxidative activity in inoculated pear plantlets. (**a**) A schematic representation of the oxidative activity in inoculated plantlets with necrotic sections in the stem (upper panel) and necrotic progression from the central cylinder to the cortex, indicated by yellow arrows in the leaf (lower panel); (**b**) Representative data of the electrolyte leakage of inoculated and non-inoculated *Dar Gazi-wt*. Electrolytic conductivity dramatically increased in plants after bacterial inoculation. Error bars represent the SD of three independent experiments, each with three biological replicates. Asterisks indicate significant differences of inoculated vs non-inoculated plants (Student's *t*-test, $p \leq 0.05$).

2.3. Molecular Marker for E. amylovora Infection

A previous study demonstrated that the expression of the chloroplastic gene psbA in the pear cultivar Harrow Sweet is linked to the effects of *E. amylovora* infection [35]. Analysis of psbA expression in inoculated and not-inoculated pear Dar Gazi shoots revealed the presence of unexpected amplicons when we used, as a template, cDNA synthesized with a psbA-specific primer using mRNA extracted from inoculated plants (Figure 3a). The bacterial retrotranscript products were not detected when the cDNA was prepared from non-inoculated shoots or when the cDNA was synthesized using oligo d(T)8-12. Sequencing the PCR products indicated that all of the sequences belonged to *E. amylovora* (sequence identity > 99%). The 1056-bp amplicon, named *erw1*, contains gene sequences encoding: the C-terminal domain of a putative cyclopropane-fatty-acyl-phospholipid synthase (CFAS), an enzyme with synthase and methyltransferase activity involved in the fatty acid biosynthesis; the N-terminal domain of a predicted lipoprotein with an unknown function containing a DUF3833 domain. The 925-bp amplicon (*erw2*) contains the sequence encoding of the predicted Major Facilitator Superfamily (MFS) transporter. These transporters facilitate the transport across cytoplasmic or internal membranes and represent one of the two major classes of transport proteins involved in the protection against endogenous and exogenous toxic compounds in fungi [36]. The 519-bp amplicon (*erw3*) corresponds to the 3′-half of the *erw1* amplicon and contains the sequences encoding the DUF3833 domain-containing protein. The 384-bp amplicon (*erw4*) contains sequences encoding an MFS transporter of the sugar porter (SP) family, the most prominent family of MFS transporter [37]. Gene-specific primers were designed to amplify the same four genes: *erw1*, CFAS; *erw2*, MFS transporter; *erw3*/ *erw1*, DUF3833 protein; erw4, SP MFS transporter. Gene expression analysis was carried out on mRNA extracted from different sections of the asymptomatic pear plantlets (24 h after the infection). The qPCR results indicated that *erw2* was expressed in the basal section up to the middle section (4-9 mm), while *erw1* was expressed in the low- and mid-section (Figure 3b). In contrast, the expression of *erw3* and *erw4* occurred mainly in the mid- and high-section of the plantlet (Figure 3b). This data indicates that, for improving the interaction with the different colonized tissues, *E.*

amylovora modulates the expression of its genes during the internal movement through the vascular system.

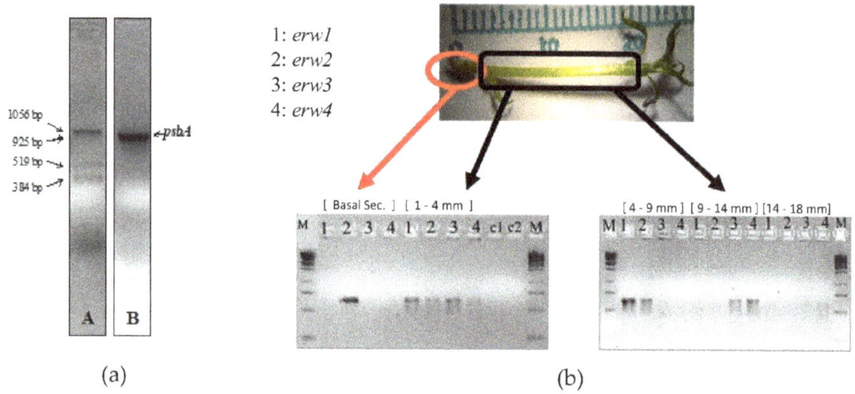

Figure 3. Identification of erw genes: (**a**) Electrophoretic profile of the retrotranscription products (*erw1-4*) obtained from the cDNA synthesized using a psbA gene-specific primer with mRNA from *E. amylovora*-infected plantlets, panel; (**b**) Spatial differential expression in pear tissue of the *E. amylovora* genes revealed using the *erw* genes-specific primers.

2.4. *PR1 and PR10 Expression in Dar Gazi-wt*

To highlight the role of the internal clock in regulating the in vitro expression of *PR1* and *PR10* genes in Iranian pear cultivar *Dar Gazi-wt*, AtPHYB and LeCRY1 overexpressed lines; plantlets were initially exposed to a photoperiod of 16 h/8 h (light/darkness, Figure S1). In *Dar Gazi-wt*, the expression of the *PR1* gene was not oscillatory, keeping an almost constant level of expressed transcripts throughout the day. The level of *PR1* transcripts was, less than *PR10* transcripts during the day, irrespective of the lighting conditions. *PR10* showed an oscillatory state that would seem to be influenced by the circadian rhythm. The results reported in Figure 4 shows a peak expression after 2 h of exposure to darkness, a tendency to decrease after 6 h of darkness, a strong reduction in the first 2 h of exposure to light and faint up-regulation after 10 h of exposure to light, followed by and a subsequent down-regulation of expression (Figure 4).

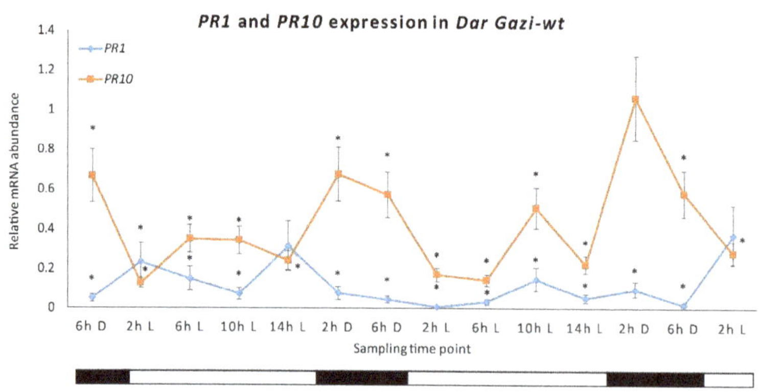

Figure 4. *PR1* and *PR10* expression in *Dar Gazi-wt*. The results are presented after normalization with *ef1A*. The average was generated by two biological replicates run in triplicate. Error bars represent SD. Within the sampling time point, the asterisk indicated a statistically significant difference compared to the highest values of each gene (Student's *t*-test, $p \leq 0.05$). The bars under the horizontal axis show the light and dark periods, respectively.

When plantlets were exposed to continuous light (Figure S2), the expression of the *PR1* was approximately doubled after 24 h (Figure 5a) while the expression of the *PR10*, instead, decreased to around zero. This behavior has prevented the peak of expression to be visible after 2 h of exposure to darkness, although a small peak after 10 h of light was observed. Therefore, the expression profile of *PR10* would seem to be independent of the internal clock since the course no longer follows the oscillations previously seen during alternating darkness and light. In fact, in the absence of environmental time cues, circadian rhythms should persist with a period close to 24 h. Under conditions of continuous darkness (Figure 5b and Figure S2), *PR10* expression was stimulated and showed an oscillatory profile that partially resembles what had been observed under photoperiodic conditions. Under continuous darkness, *PR1 r*emains at lower levels than *PR10*, showing the same expression behavior detected during constant light.

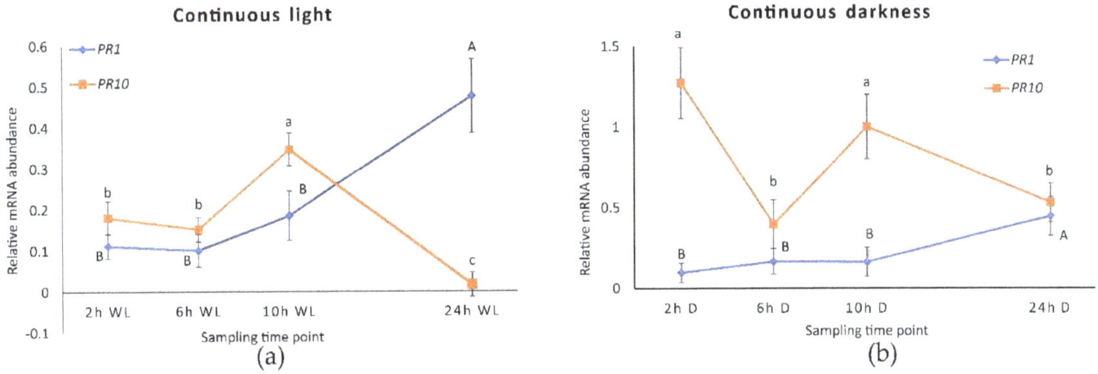

Figure 5. *PR1* and *PR10* expression in *Dar Gazi-wt* plantlets: (**a**) in 24 h continuous light (WL); (**b**) in 24 h continuous darkness. The results are presented after normalization with *ef1A*. Data shown as the average of two biological replicates run in triplicate, with error bars representing SD. For each single gene expression pattern, values with different letters significantly differ according to the analysis of variance (ANOVA) and least significant difference (LSD) tests ($p \leq 0.05$). Uppercase and lowercase letters are referred to as *PR1* and *PR10*, respectively.

2.5. PR1 and PR10 Expression in CRY1 and PHYB Overexpressing Lines in WL

Data for the *PR1* expression in the plantlets of *Dar Gazi-cry1* line indicate that the photoreceptor CRY1 plays a role in the regulatory system of this gene (Figures 6 and S3). Under darkness, in the plantlets of this line, the detected transcripts increased up to 3 times those detected in the plantlets of Dar Gazy-wt. Moreover, a semi-oscillatory rhythm would seem to be evocated by the increased presence of CRY1 in the plantlet tissues, strongly upregulating the expression of *PR1*. From these results, it was evident that BL plays a role as overexpressed CRY1 emphasizes this aspect. The role of RL turns out to be different than that of BL, as can be seen in the plantlets of the PHYB-overexpressing line (Figure 6). The peak expression of *PR1* during the darkness period was approximately 8-fold greater in the transformed lines relative to the wt-line, comparable to that detected in the plantlets of the *Dar Gazi-cry1*. During the light period, the behavior of gene expression in the plantlets of *Dar Gazi-phyB* is similar to that seen in plantlets of the *Dar Gazi-wt*. During darkness, even in the plantlets of *Dar Gazi-phyB* the *PR1* transcripts level was significantly higher than in the plantlets of *Dar Gazi-wt* (Figure 6).

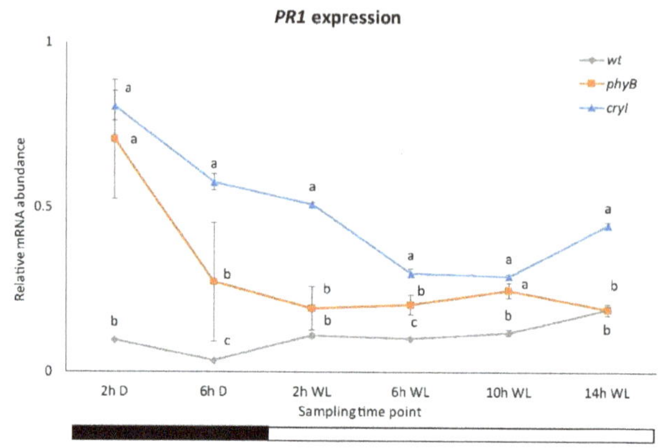

Figure 6. *PR1* expression in *Dar Gazi-wt*, *Dar Gazi-phyB*, and *Dar Gazi-cry1*. Results are presented after normalization with *ef1A*. Data shown as the average of two biological replicates run in triplicate, with error bars representing SD. At each time point, values with different letters significantly differ according to the analysis of variance (ANOVA) and least significant difference (LSD) tests ($p \leq 0.05$). The bars under the horizontal axis show the light and dark periods.

The analysis of *PR10* gene expression indicated that, in the plantlets of the three *Dar Gazi* lines, the overexpression of each photoreceptor gene drastically reduces the amount of transcript detected (Figures 7 and S3). Furthermore, the oscillatory rhythm detect in the plantlets of *Dar Gazi-wt* results was almost repressed. However, in the overexpressing of PHYB plantlets, expression was maintained in the first 2 h of darkness.

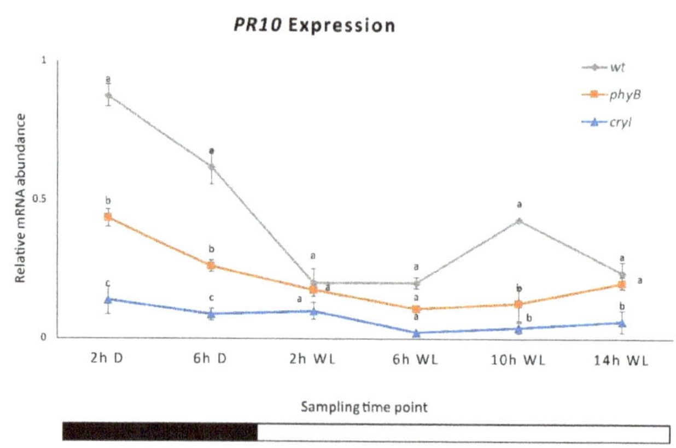

Figure 7. *PR10* expression in *Dar Gazi-wt*, *Dar Gazi-phyB*, and *Dar Gazi-cry1*. Results are presented after normalization with *ef1A*. Data shown as the average of two biological replicates run in triplicate, with error bars representing SD. At each time point, values with different letters significantly differ according to the analysis of variance (ANOVA) and least significant difference (LSD) tests ($p \leq 0.05$). The bars under the horizontal axis show the light and dark periods.

Under continuous light, although the transcription rate of the *PR1* gene in the plantlets of the *Dar Gazi-cry1* was higher than that in plantlets of the *Dar Gazi-wt*, the behavior of transcription was different than under photoperiodic conditions (Figure 8a). A faint increase in the amount of transcript was detected after 24 h of exposure to continuous light. On the other hand, the *PR1* gene expression course in the plantlets of the *Dar Gazi-phyB*, under exposure to constant light, was similar to that detected under photoperiodic conditions (Figure 8a).

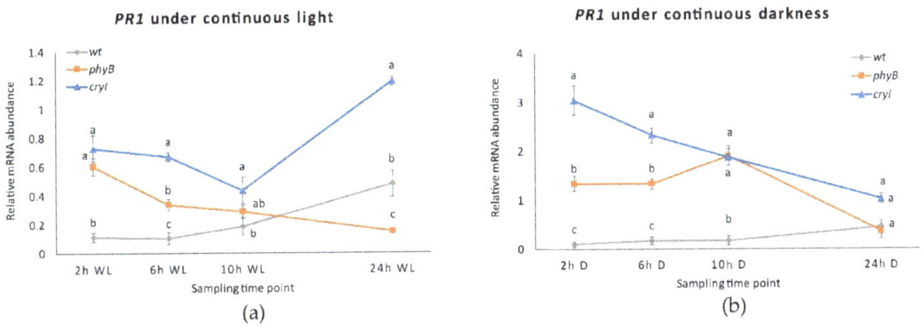

Figure 8. *PR1* expression in *Dar Gazi-wt*, *Dar Gazi-phyB*, and *Dar Gazi-cry1*: (**a**) in 24 h continuous light, panel; (**b**) in 24 h continuous darkness. Results are presented after normalization with *ef1A*. Data shown as the average of two biological replicates run in triplicate, with error bars representing SD. At each time point, values with different letters significantly differ according to the analysis of variance (ANOVA) and least significant difference (LSD) tests ($p \leq 0.05$).

In conditions of continuous darkness, the transcription levels of the *PR1* gene were strongly increased in the tissue of *Dar Gazi-cry1* and *Dar Gazi-phyB*. In contrast, the level of transcript detected in the *Dar Gazi-wt* was very low but did not differ from that seen in continuous light and under photoperiodic conditions (Figure 8b). The highest amount of transcript in the *Dar Gazi-cry1* plantlets was found after 2 h of exposure to the darkness, thereafter, the amount of transcript decreased (Figure 8b). The highest amount of transcript in the plantlets *Dar Gazi-phyB* was found after 10 h of exposure to darkness, but after 24 h, the amount of transcript was the lowest (Figure 8b). Thus, the darkness condition induces always-high *PR1* gene transcription levels in the plantlets of the *Dar Gazi-cry1*. A similar trend was also observed in the plantlet of *Dar Gazi-phyB*, even if at a reduced level.

Under continuous light and darkness, the *PR10* gene expression level was dramatically reduced in plantlets of both transgenic lines (Figure 9a). In the plantlets of the *Dar Gazi-wt*, an oscillatory behavior was detected, more pronounced in continuous darkness than continuous light (Figure 9b). The results suggest that the overexpression of the photoreceptors, irrespective of light conditions, strongly inhibits the expression of the *PR10* gene. The amount of transcript detected in the plantlets of the *Dar Gazi-wt* indicates that the physiological expression of photoreceptors could play a relevant role in permitting the oscillatory expression of the *PR10* gene.

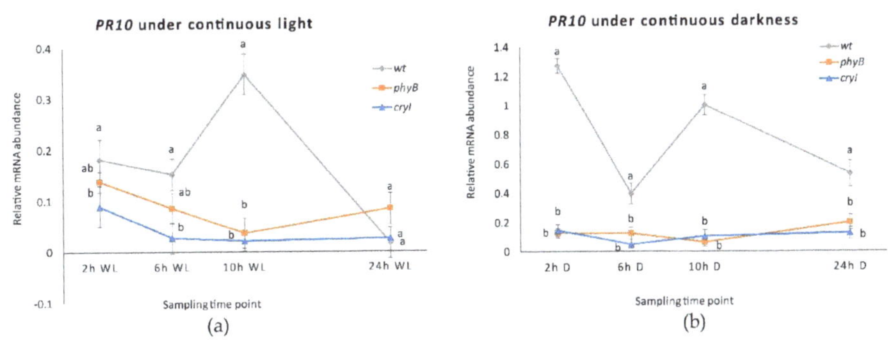

Figure 9. *PR10* expression in *Dar Gazi-wt*, *Dar Gazi-phyB*, and *Dar Gazi-cry1*: (**a**) in 24 h continuous light, panel; (**b**) in 24 h continuous darkness. Results are presented after normalization with *ef1A*. Data shown as the average of two biological replicates run in triplicate, with error bars representing SD. At each time point values with different letters significantly differ according to the analysis of the variance (ANOVA) and least significant difference (LSD) tests ($p \leq 0.05$).

2.6. *PR1 and PR10 Expression in Dar Gazi-cry1 and Dar Gazi-phyB in RL, FRL, and BL*

Studying the role of photoreceptors in the regulation of the expression of the PRs, phytochrome has a pivotal role in regulating the internal clock and the perception of the photoperiod. The expression level of the *PR1* gene in plantlets *Dar Gazi-phyB* exposed to continuous RL increases to the highest rate (Figure 10a and Figure S3). Moreover, the expression of this gene shows an oscillating trend. On the other hand, in the tissue of *Dar Gazi-wt* and *Dar Gazi-cry1* plantlets, the transcript level was constant, at a very low level of expression. Therefore, the photoconversion of phytochrome from the inactive (P_r) to the active form (P_{fr}) should play a permissive role (Figure 10a,b).

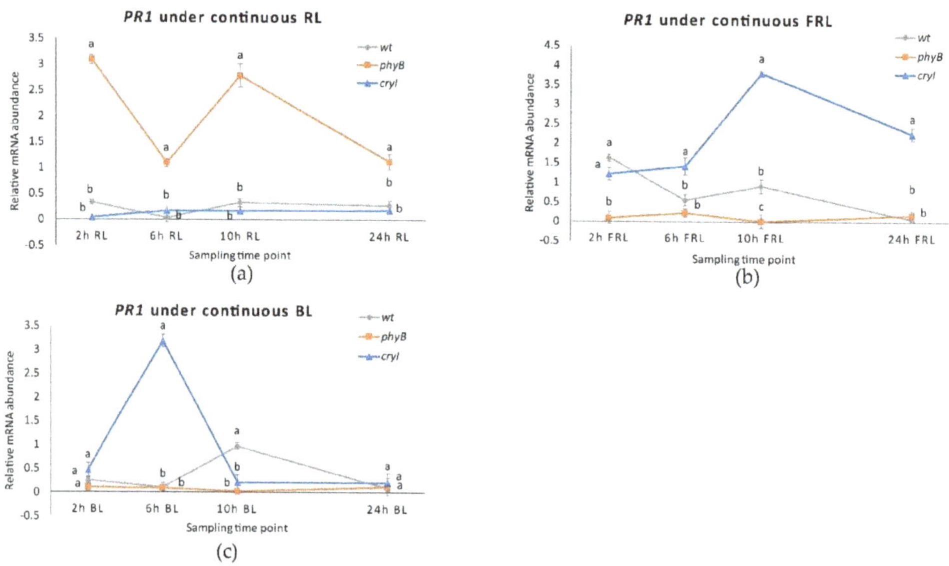

Figure 10. *PR1* expression in plantlets of *Dar Gazi-wt*, *Dar Gazi-phyB*, and *Dar Gazi-cry1*: (**a**) in 24 h continuous RL, panel; (**b**) in 24 h continuous FRL, panel; (**c**) in 24 h continuous BL, panel. Results are presented after normalization with *ef1A*. Data shown are the average of two biological replicates run in triplicate, with error bars representing SD. At each time point values with different letters significantly differ according to the analysis of the variance (ANOVA) and least significant difference (LSD) tests ($p \leq 0.05$).

As determined by exposing plantlets to continuous FRL (Figure 10b), the inactive form of phytochrome inhibits the expression of the *PR1* gene in *Dar Gazi-phyB* plantlets. The inactive form of phytochrome, and probably the amount of PHYB protein, either generates or allows an oscillatory behavior of the expression of the *PR1* gene in the plantlets of wt-line and the plantlets of *Dar Gazi-cry1*. The expression of *PR1* in *Dar Gazi-cry1* increases at a high level after 10 h of continuous FRL. Results, therefore, show that *PR1* expression was promoted by CRY1 activity the and the circadian rhythms are present again.

Under continuous BL conditions (Figure 10c), the highest level of *PR1* expression in *Dar Gazi-cry1* plantlets was reached after 6 h of exposure to light. An oscillatory behavior appeared in plantlets of *Dar Gazi-wt*, while in plantlets of *Dar Gazi-phyB* a very low expression rate without any oscillatory behavior was observed.

The regulation of *PR10* expression under continuous RL was very similar in *Dar Gazi* -wt and *Dar Gazi-cry1* plantlets (Figure 11a). An oscillatory transcriptional behavior was observed in both lines of plantlets, although when this behavior is compared to photoperiodic conditions (Figure 11a). In the *Dar Gazi-phyB* plantlets, the trend of oscillatory behavior is different In particular, after the 6th hour, an autonomous behavior was observed (Figure 11a).

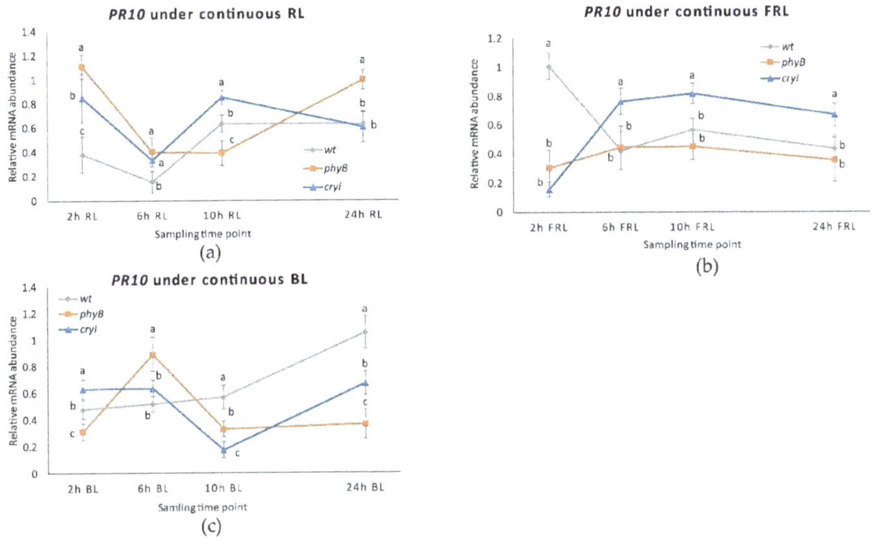

Figure 11. *PR10* expression in *Dar Gazi-wt*, *Dar Gazi-phyB*, and *Dar Gazi-cry1*: (**a**) in 24 h continuous RL, panel; (**b**) in 24 hh continuous FRL, panel; (**c**) in 24 h continuous BL, panel. Results are presented after normalization with *ef1A*. Data shown are the average of two biological replicates run in triplicate, with error bars representing SD. At each time point values with different letters significantly differ according to the analysis of the variance (ANOVA) and least significant difference (LSD) tests ($p \leq 0.05$).

When exposed to continuous FRL, the transcriptional profile of *PR10* in plantlets of *Dar Gazi-wt* resemble an oscillatory behavior analogous to that observed in photoperiodic conditions (Figure 11b). This is not the case for the plantlets of the two transgenic lines, that show a different behavior, but are analogous between themselves (Figure 11b). The rate of gene expression observed in *Dar Gazi-cry1* indicates that the presence of CRY1 is required for the upregulation of this gene.

An analogous oscillatory behavior appears in plantlets of both transgenic lines when exposed to continuous BL (Figure 11c), while in *Dar Gazi-wt* not oscillatory behavior was observed. Comparing the rate of the *PR10* gene expression of the plantlets of *Dar Gazi-cry1* under FRL with BL, it is surprising that the behavior was not the same. The hypothesis could eventually explain this divergent behavior, that the gene expression's promoting role is mainly regulated by phytochrome, and only partly co-regulate by cryptochromes.

2.7. CRY1 Overexpressing Line Is More Resistant to Fire Blight

Plantlets were observed for 96 h after the inoculation of *E. amylovora* to detect necrotic tissues. Necrosis symptoms appeared only in the shoot apex of *Dar Gazi-wt* plantlets after 36 h from the inoculation (Figure 12). After 96 h, the progress of necrosis that affected the entire stem was visible. In the plantlets of *Dar Gazi-phyB*, necrosis was detected in several leaf nodes throughout the stem only after 48 h from the inoculation (Figure 12). Surprisingly, *Dar Gazi-cry1* plantlets better tolerated the pathogen infection showed necrotic tissues after 72/96 h (Figure 12).

Figure 12. Sintomatology of bacterial infection. White arrows indicate the necrotic stem area.

Gene expression analysis showed that transcript levels of PRs were more significant in photoreceptor over-expressing plantlets than in *Dar Gazi-wt* plantlets, indicating an increased capacity to counteract the infection (Figure 13). At the same time, the expression profile of erw genes indicated that only *erw1*, coding for CAFS, was expressed during the first 12 h after the infection. In addition, erws transcript levels were also greater in plantlets over-expressing photoreceptors than in *Dar Gazi-wt* plantlets, suggesting a dynamic interaction occurring during the bacterial invasion of host tissues (Figure 13).

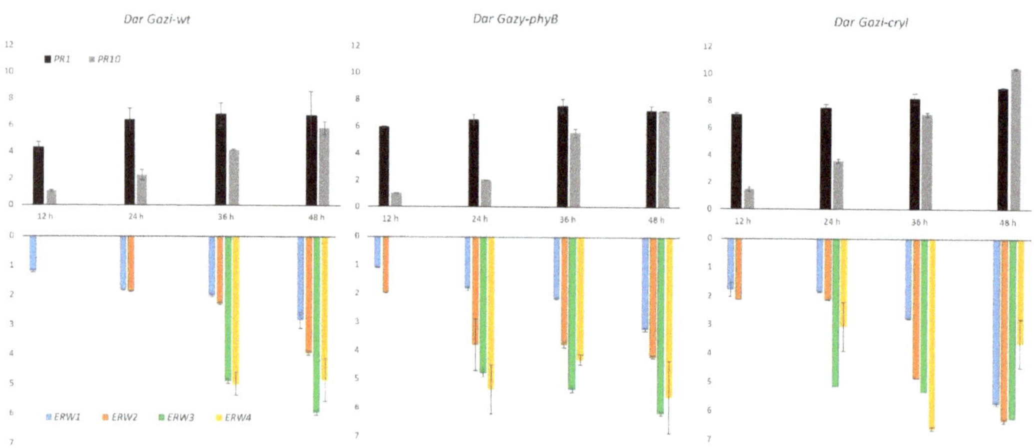

Figure 13. *PR1*, *PR10*, and *erw1-4* expression in *Dar Gazi-wt*, *Dar Gazi-phyB*, and *Dar Gazi-cry1* plants, grown under 16/8 h light/darkness, at 12, 24, 36, and 48 h after the pathogen inoculation. Results are presented after normalization with ef1A. Data shown are the average of two biological replicates run in triplicate, with error bars representing SD.

3. Discussion

Plants have evolved to coordinate their activities with the day-night cycle by Earth's rotation. Direct responses to light and darkness are essential, but, in addition, biological clocks have evolved to time biological processes. Circadian rhythms result from the interaction between the internal oscillatory system and the receptors of environmental cues such as photoreceptors that usually help reset the biological clock to a 24-h day-night cycle. Many environmental (i.e., temperature) and internal cues (i.e., starvation) function as zeitgebers for the rhythms, but photoperiod and light quality are among the most important ones in plants. There is no other environmental factor in any climatic region of comparable importance for the immediate control of annual and daily cycles [38].

3.1. Regulation of PR Genes by Circadian Rhythms and Photoreceptors

This research shows that the expression of the *PR10* gene is partially regulated by the internal biological clock, while photoreceptors mainly control the *PR1* gene. The expression of the *PR10* gene in plantlets exposed to continuous RL and FRL under a 16/8 h photoperiod (Figure 7) maintained an oscillatory pattern, which appeared to be controlled by the circadian rhythm (Figure 11a,b). In contrast, the transcript of the *PR1* gene appeared to be independent of the oscillator and dependent on the photoreceptor's activities. These agree with the observations by Genoud et al. [5] on the effect of single and multiple nil mutants in the light perception (*phyA* and *phyB*) and the light-signal processing (*psi2*, phytochrome signaling) on the interaction between *A. thaliana* and the pathogen *Pseudomonas syringae* pv. *tomato* single and multiple mutants' nil in light perception (*PHYA* and *PHYB*) and light-signal processing (*psi2*, phytochrome signaling).

In these mutants, the growth of an incompatible bacterial strain of this pathogen was enhanced in the double mutant *phyAphyB* and decreased in the *psi2* mutant under darkness and dim light conditions [20]. The last mutant increased the light signal transduction regulated by PHYA and PHYB [39]. Similarly, the results of this work demonstrated that the overexpression of *PHYB* and *CRY1* is associated with an upregulation of the Dar Gazi *PR1* (Figures 6 and 8).

Salicylic acid (SA) induces pathogen-related gene expression and accumulation of related proteins, and its production also depends on the light regime [40]. Phytochromes are required for the expression of the PR1 protein [19]. In *Dar Gazi-phyB*, the expression of *PR1* was dependent on the phytochrome. When the plantlets were exposed to continuous RL an up-regulation of *PR1* was observed (Figure 10a); in contrast, in plantlets exposed to continuous FRL, the expression of this gene was inhibited (Figure 10b). When the plantlets were exposed to FRL and BL, a co-participation of the cryptochrome into the regulation system was also observed (Figure 10b,c).

Although in this study, free SA was not measured, it is known that the SA levels oscillate throughout the day in a circadian rhythm [41], so a fine coordinated regulation of *PR1* gene between this hormone and photoreceptors pathways could be strongly hypothesized, and it will be the challenge for the further investigation. It has been shown that transcription of *PR* genes during plant defense involves a key transcriptional regulator of SA signaling known as Nonexpressor of Pathogenesis-related protein 1 (NPR1). The inactive NPR1 oligomers monomerize in the cytosol after an SA-induced change of the cell's redox state, and a circadian oscillation occurs, peaking at night [42]. The state of monomers allows NPR1 to be translocated to the nucleus where they interact with TIMING OF CAB2 EXPRESSION 1 (TOC1), an evening circadian clock gene, and TGACG-BINDING FACTORs (TGAs), leading to the expression of defense-related genes involved in the set-up of plant immune defense, including *PR* genes [42–45]. The oscillatory rhythms of TOC1 mRNA expression were associated with parallel oscillations in histone acetylation [46]. NPR1 forms an activator complex with histone acetyltransferases (HATs) HAT1 and HAT5. Through NPR1–TGA interaction, the complex is recruited to chromatin finally relaxing genomic DNA and facilitating *PR*s transcription activation [47].

The obtained data suggest that the *PR*s clock-associated regulation is co-regulated by the photoreceptors phytochrome and cryptochrome, maybe functionally as elements of regulator-Zeitlupe systems. In fact, in overexpressing *PHYB* gene plantlets, a circadian oscillation is observed when exposed to a continuous RL (Figures 9 and 10). The results indicate that this behavior is red/far-red reversible. When plantlets are exposed to FRL and BL, a co-participation of cryptochrome into the regulation system was also observed.

In overexpressing *phyA* cherry plants, Cirvilleri et al. [48] concluded that the induction of *PR*s gene is strictly dependent on light quantity and quality, inducing plant resistance to *Pseudomonas syringae* pv. *mors-prunorum*. Therefore, the relationship between biological clock and overexpression of *PHYB* and *CRY1* was tested under different light qualities to unravel their role on *PR*s gene expression pear cv Dar Gazi.

In a previous study, a possible link between *PR1* and light was postulated [40]. However, until now there has been a gap of knowledge around the role of the internal clock in the *PR1* regulome. The effects of many biotic and abiotic stress, including pathogen infection, salt tolerance, UV irradiation, and ozone stress, have been investigated in *PR10* gene expression [49]. These stresses have been shown to activate *PR10* gene expression, suggesting their importance during plant defense responses. Plant hormones and related signaling molecules have been reported to regulate *PR10* gene expression, including jasmonic acid, salicylic acid, abscisic acid [50], kinetin, and auxin [51].

3.2. Light Plays a Role in E. amylovora Growth

Under standard laboratory conditions, a non-photosynthetic micro-organism is grown in the darkness, and the possible effects of light on its growth and physiology are neglected. This practice is strongly consolidated, and, for this reason, microbial culture equipment (static and shaken incubators) is not provided with a light control system in the standard configuration. In contrast, there are several studies on non-photosynthetic bacteria associated with humans, plants, and animals (i.e., *Pseudomonas aeruginosa*, *Pseudomonas syringae*, and *Xanthomonas*) indicating that chemical (quorum sensing), and light (photosensing) signals affect the growth pattern, infectivity, and virulence of these bacteria through common regulatory pathways [30,52] Data presented in this work provide the first evidence that the spectral distribution of the light affects the growth of *E. amylovora* under laboratory conditions. This preliminary result provides novel prospects in studying the impact of spectral quality on the lifestyle of this phytopathogen and its interactions with the plant host.

3.3. Cryptochrome Increase the Defense against the Attack of E. amylovora

In this paper, four *E. amylovora* genes were identified that can be used to monitor the diffusion of the pathogen in pear vascular tissues during asymptomatic and symptomatic periods (Figure 3). Quantitative expression analyses revealed several interesting features: the expression of the *erw* genes was modulated during the internal movement of the pathogen through the plant vascular system (Figure 3); the activation of these genes occurred at specific times during the infection, the temporal expression pattern was dependent upon the pear genotype (Figure 13). In both *Dar Gazi* transgenic lines, the expression of *erw2*, *erw3* and *erw4* was advanced by 12 h compared to *Dar Gazi-wt*, from 24 h to 12 h for *erw2*; 36 h to 24 h for *erw3* and *erw4* (Figure 13). Twelve h after the infection, the transcript levels of *erw1* in *Dar Gazi-wt* and *Dar Gazi-phyB* were similar and increased in the same proportion between 12 h and 24 h (Figure 13). These data suggested that, in infected *Dar Gazi-wt* and *Dar Gazi-phyB* tissues, the growth pattern and the number of *E. amylovora* cells per plant mass unit were comparable. The early activation of *erw2* and the increased *erw2/erw1* ratio at 24 h in *Dar Gazi-phyB* vs. *Dar Gazi-wt* were dependent on the overexpression of *PHYB* in the transgenic line (Figure 13).

In contrast, 12 h after the infection, the mRNA expression level of *erw1* in the transgenic *Dar Gazi-cry1* line was about two-fold higher than in *Dar Gazi-wt*, and this difference remained constant up to 48 h (Figure 13). These data indicated that the overexpression of *CRY1* stimulated *E. amylovora* growth in pear tissues and altered the expression of the other erw genes. Noteworthy, in *Dar Gazi-phyB*, the expression of *erw2* and *erw4* remained constant between 12 h and 48 h (Figure 13). At 48 h, there was no significant difference in the expression levels of the four *erw* genes between *Dar Gazi-wt* and *Dar Gazi-phyB* (Figure 13). In contrast, in the *Dar Gazi-cry1* line, the transcript levels of *erw1*, *erw2*, and *erw3* significantly increased between 36 h and 48 h, reaching the maximum relative abundance. In comparison, the expression of *erw4* had a maximum at 36 h and decreased between 36 h and 48 h (Figure 13). These data indicated that the plant-pathogen interactions occurring during the pathogen invasion were differentially affected by the alterations of the phytochrome- and cryptochrome-modulated signals resulting from the overexpression of *PHYB* and *CRY1*.

In the pear tissue, pathogen invasion generated oxidative stress in the central cylinder and the cortex accompanied by a widespread disruption of the plasma membrane that developed from the basal portion to the apex of the plantlets (Figure 2a). In *Dar Gazi-wt*, the necrosis symptoms appeared 12–36 h earlier than in transgenic lines (Figure 12).

Interestingly, in the plantlets over-expressing the photoreceptors, the transcript levels of *PR1* and *PR10* were higher than in *Dar Gazi-wt* (Figure 13). Independently from the pathogen load estimated by the *erw* genes expression data, it should be noted that there was a correlation between the PR transcript level and the appearance of necrosis symptoms. Delayed symptoms occurred in the *Dar Gazi* lines, such as *Dar Gazi-cry1* (Figure 12), in which higher *PR1* and *PR10* transcription levels were observed (Figure 13). It was demonstrated that a wide range of endogenous and exogenous (a)biotic factors, including pathogen attack, accumulation of salicylic acid, and abiotic stress, can regulate temporally and spatially the expression of *PR* genes [53,54] and the secretion and accumulation of the corresponding proteins in the apoplastic space or the vacuoles [55]. The results of this work demonstrate that the accumulation of the PR proteins can interfere with the dynamic interaction occurring during the *E. amylovora* invasion and delay the infection of the pear host tissues.

It is known that the protein product of the Far-red Insensitive 219/Jasmonate Resistant1 (FIN219/JAR1) functions as a jasmonic acid (JA)-conjugating enzyme responsible for the synthesis of the Jasmonic Acid-isoleucine (JA-Ile), the physiologically active form [56]. Under BL, FIN219 plays a role in the regulation of phenotype development and bacterial resistance [57,58] and how it occurs under FRL, it interacts with CONSTITUTIVE PHOTOMORPHOGENIC 1 (COP1), down-regulating also the levels of *COP1* and up-regulating the levels of *HY5* [59]. COP1 is involved in the negative control of nitrate reductase activity in Arabidopsis *cop1* mutant, reducing the availability of nitrogen [60]. The availability of nitrogen resulted responsible for both *Arabidopsis* resistance to *E. amylovora*, i.e., under nitrogen limitation, the resistance decreased due to the lower apoplastic reactive oxygen species (ROS) accumulation and increased expression of *E. amylovora hrps* genes [61,62]. Moreover, cryptochromes may work together with phytochromes to modulate plant defense responses. In *Arabidopsis*, CRY1 positively regulates the inducible resistance to *P. syringae* pv. *tomato*. The local resistance is down-regulated in the *cry1* mutant; in contrast, in plants overexpressing *CRY1*, the *PR1* gene expression is enhanced, and the resistance is significantly up-regulated [22]. These results agree with the increased expression level observed in *Dar Gazi-cry1* compared to in *Dar Gazi-wt*, where a significant increase of expression was already detected at 12 h from inoculation, for both *PR1* and *PR10* genes.

Although, many key molecular factors involved in the plant-pathogen interaction, from the plant perception of the pathogen (P/MAMPs, PRRs) to the activation of the PAMP-triggered immunity (PTI), and the Effector triggered immunity (ETI), are already known [63,64], in *E. amylovora*-infected plants, the regulation of the photoreceptors by the interaction with the major phytohormones, SA, JA, and ethylene remains to be explored.

3.4. Agronomic Relevance

In a fruit orchard, the canopy dimension dynamically changes, and, consequently, the spectral distribution of the incoming radiation varies widely, as the light penetrates and scatters within the tree canopy due to the structure and optical properties of plant organs [65,66]. In general, the spectral modifications of light inside the tree canopy have a crucial role in growth partitioning among fruit and shoots, affecting the allocation to developing fruits in plant growth and fruit quality [67]. The effects of modification of the CRYs and PHYs abundance and photosensitivity of plants in response to the changing light on cross talks during host-pathogen interaction remain to be studied in fruit trees, and the molecular mechanisms underlying the interaction of monochromatic light with plant and bacteria remain poorly understood because they are influenced by environmental conditions. Results obtained in experiments in vitro, with pure cultures of *E. amylovora* Ea273 strain (Figure 1), and in vivo, with infected transgenic *Dar Gazi* lines (Figure 12), clearly indicated that the quality of the light and the photoreceptor-mediated signals affect the growth of the pathogen and its infectivity and aggressiveness. In this respect, the use of the LED technology can be valuable to develop new procedures for sustainable and non-invasive control of this pathogen.

These findings also have great economic importance because *PR1* is used as a look-out pathogen presence. During the period of fruit conservation in dark conditions, an interruption of these light conditions through BL flesh could repress the insurgence, the development, and bacterial proliferation. Even if there are not many studies on *PR*s and woody fruit crop plants, it has recently been presented that genetically engineered phytochrome A cherry plants showed the highest level of tolerance to *Pseudomonas syringe* pv *mors-prunorum*, when compared to the wild type plants [48].

Finally, one of the four plant food allergens, the Bet v 1 superfamily, contains ten pathogenesis-related proteins [68]. Our findings could be further explored to study the regulation of this allergen-related protein and the relative reduction of its presence and accumulation by modulating the lighting during the post-harvest fruit conservation.

4. Materials and Methods

4.1. Plant Material, Medium Composition, Growth Conditions, and Bacterial Strain

An in vitro-cultured plantlets system of *Pyrus communis* L. cv Dar Gazi was used to evaluate if the internal clock autonomously regulates the abundance of *PR1* and *PR10* transcripts. Plantlets of three different lines: *Dar Gazi-wt*, *Dar Gazi-phyB*, and *Dar Gazi-cry1* were submitted to different circadian experimental conditions and continuous BL, RL- and FRL conditions. Fluorescent WL was used as a control.

The two transgenic lines *Dar Gazi-phyB* and *Dar Gazi-cry1* were obtained starting from leaf explants of in vitro established cv *Dar Gazi-wt* co-cultivated for 20 min on MS liquid basal medium with two different *A. tumefaciens* strains (A, B) prepared as described below.

The disarmed *A. tumefaciens* (A) strain EHA 105, contained the helper plasmid pTiBo542 and the binary vector pROKB (kindly provided by Whitelam, Leicester University, England), harboring the *neomycin phosphotransferase II* (*nptII*) gene under the control of *nos* promoter and the *A. thaliana* cDNA *PHYB* gene under the control of the cauliflower mosaic virus 35S RNA (CaMV 35S) promoter. The disarmed *A. tumefaciens* (B) strain EHA 105, contained the helper plasmid pTiBo542 and the binary vector pBI12 (also provided by Whitelam), harboring the *neomycin phosphotransferase II* (*nptII*) gene under the control of *nos* promoter and *Lycopersicum esculentum* cDNA *CRY1* gene under the control of the CaMV 35S promoter. Vectors were introduced into EHA 105 using freeze-thaw transformation of *Agrobacterium* and *Escherichia coli* as described by [69]. For both transformation experiments *A. tumefaciens*, was cultured overnight at 28 °C on a shaker at 80 rpm in 10 mL liquid Luria-Bertani (LB) [70] medium, prepared with 1% LB containing Bacto-Tryptone, 0.5% Bacto-yeast extract, 1% NaCl, 0.1% glucose with 100 mg L^{-1} of kanamycin added when the temperature arrived at 45° C for selecting bacteria carrying the binary plasmid. After 10 min of centrifugation at 3200× *g*, the pellet was resuspended in MS liquid medium with 3% (*w*/*v*) sucrose, and subsequent dilutions were done to reach a final concentration of around 0.3 (OD600).

After the co-cultivation, the leaf explants were transferred to a regeneration medium containing 100 mg L^{-1} acetosyringone (40-hydroxy-3,5-dimethoxyacetophenone) and incubated in a controlled environment chamber at 23 °C for two days. Cefotaxime (200 mg L^{-1}) was added to all media, to eliminate *Agrobacterium*. The transformed green shoots were picked out from callus tissue (assisted by a stereoscope) and moved to QL0 medium [71] with 10 mg L^{-1} of kanamycin added (Figure S4a). The final selection was carried out onto

QL0 medium added with 100 mg L^{-1} of kanamycin, subculturing every 8–10 days to select putative transgenic lines (Figure S4b).

4.2. Molecular Confirmation of the Transgene Insertion

The selected shoots were subcultured four/five times for rapid and clonal multiplication onto the PQL1 media contained the same composition present in PQL0 and enriched with 0.7 mg L^{-1} BAP.

Based on the assumption that genes encoding for the same proteins in different species show conserved domain with a high degree of identity, divergent regions between pear *PHYB* gene and *AtPHYB*, and between pear *CRY1* and *LeCRY1* were selected to design specific primers. The selected regions were checked against bacteria genes as well and they did not match any homologous eukaryotic sequences of genes present in data banks. *AtPHYB* and *LeCRY1* sequences, expected product sizes and annealing temperatures used to detect each gene are presented in Table S1.

PCR amplification tests were conducted to test to validate the insertion of both genes on plantlets of selected lines (Figure S4c). For DNA extraction procedure, from 100 mg leafy shoot tissues, and PCRs chemicals and amplification profile have been using the procedure reported in previous work [72]. Amplification products were visualized on agarose gels (1.2%, *w*/*v*) and 10 μg mL^{-1} of ethidium bromide (Figures S6 and S7).

4.3. E. amylovora In Vitro Experiments

E. amylovora strain Ea273 was obtained from American Type Culture Collection (ATCC number 49946). The strain was stocked at −80 °C in LB plus glycerol 25% (*v*/*v*) and precultured in LB broth at 30 °C under agitation (150 rpm) in the absence of light. The growth was monitored by turbidimetric measurements (OD_{600}).

Seed cultures in the late exponential phase of growth [OD_{600} of 4.5–4.8] were used to inoculate 100 mL of LB medium (initial OD_{600} of 0.1), to evaluate the effect of the light quality on the growth. The inoculated broth was grown in 500-mL Erlenmeyer flasks in an INFORS HT Multitron incubator equipped with monochromatic LED lights of the appropriate spectral wavelength. The growth was carried out under constant temperature (30 °C) and agitation (180 rpm), in continuous light or darkness, and was monitored over a 24-h period. All experiments were carried out in triplicate and included three biological and two technical replicates.

4.4. Growth Conditions and Sampling

PR1 and *PR10* expression of each line, *Dar Gazi-wt*, *Dar Gazi-phyB*, and *Dar Gazi-cry1*, was evaluated in plants grown in vitro according to Abdollahi et al. (2004) [2] at 22 °C and under a 40 μmol m^{-2} s^{-1} WL. For the light experiments, plants were then placed under different light conditions: WL (100 μmol $m^{-2}s^{-1}$), RL (25 μmol m^{-2} s^{-1}), FRL (25 μmol $m^{-2}s^{-1}$) or BL (25 μmol m^{-2} s^{-1}) obtained using specific LED lamps. Light quality and quantity were measured with an EPP 2000 Fiber Optic Spectrometer (StellarNet Inc., Tampa, Florida, USA). Plants were harvested as reported in supplemented Figures S1–S3.

4.5. RNA Isolation and Quantification

A pool of plants of the three different lines was ground with mortar and pestle in liquid nitrogen. According to the manufacturer's instructions, two independent total RNA extractions were performed from each pool using the kit NucleoSpin RNA plant (Macherey-Nagel), according to the manufacturer's instructions. Total RNA was treated using Invitrogen™ TURBO DNA-free™ Kit (Thermo Fisher Scientific, Milano, Italy) to remove DNA contamination. The nucleic acid purity was analyzed by Thermo Scientific™ NanoDrop™ 2000/2000c Spectrophotometer (Thermo Fisher Scientific) and samples with 260/280 and 260/230 nm absorbance ratios greater than 1.8 nm were used for the following experiments.

4.6. cDNA Synthesis and qRT-PCR

According to the manufacturer's instructions, total RNA was retro-transcripted employing gene-specific primer and random primers of Invitrogen™ SuperScript™ II Reverse Transcriptase (Thermo Fisher Scientific). cDNA was used as a template for qRT-PCR reactions. The PCR reactions were performed in technical triplicates with the LightCycler 480 SYBR Green I Master reagent using the LightCycler® 480 Instrument (Roche, Italy) in 96-well reaction plates. PCR conditions were: one cycle at 95 °C for 5 min, followed by 40 cycles of 95 °C for 15 s and 60 °C for 30 s. At the end of the PCR, to confirm the presence of a unique amplicon, the melting curve was evaluated and a single peak in every reaction was observed. Relative template abundance was quantified using the standard curve method [73] and the *Elongation Factor 1-Alpha* (Accession: AY338249.1) was used as a reference gene for expression normalization. PCR efficiency was estimated using six-point, 10-fold, diluted standard curves. Means from two independent replicates were subjected to SD calculation and Student's t-test. The primers were designed using the Primer3 software web version 4.1.0 (Table 1).

Table 1. The sequences of primers of PRs pear genes and erw genes, used in the qRT-PCR reactions.

Primer Name	Forward	Reverse	Amplicon Size (bp)
ef1A	GTTCGAGAAGGAGGCTGCTGAG	CGAACTTCCACAGGGCAATGTCA	119
PR1	CTCGAGCAGCAGTAGGCGTTG	CATGTTGGTTGGCGTAGTTTTGT	180
PR10	AGGAGACATTGAAATTAAGGAAGAA	AGTTGTATGCGTCGGGGTGGT	167
erw1	GCGATTACCATCAGCGAAGAAC	CCCATCTCAAACTGGTCAACAAC	161
erw2	GCTGGTGCTTGCTGTTGTTTC	GGACGCTTTCAGTTCGTGTGT	103
erw3	CTGTTACTGACGCTTTGCCTGT	CCGCTGTAATCCTGTACCATCC	140
erw4	ACCCTGTTCGTCTGTTTCCTTG	CGATCCACTCTTGTTGATGAGG	130

4.7. Ion Leakage Assay

Dar Gazi-wt infected with *E. amylovora* were collected 24 h after the inoculation and cut at the basal side which was submerged in the media. Plants were washed in de-ionized water twice and placed in 25 mL of de-ionized water. Each time point had triplicate samples for infected and non-infected plants. Solution conductivity was measured using a handheld conductivity meter, Type RS 180-7127 (RS Components), at the indicated times after plant collection.

5. Conclusions

The main findings of this study shed light on the role of light quality and reveal a possible mechanistic control of photoreceptors on the signaling transduction that activates the plant genetic resources to respond to the *E. amylovora* pathogen attack via a large array of transcription factors. CRY1 has an agonistic role in the activation of *PR* genes, during the interaction of host-pathogen, and an antagonistic role in the *E. amylovora* growth. These results suggest that a possible escape signal joined to circadian and ultradian rhythms could be connected to the regulome of PR proteins synthesis under BL and their photosensor, which might play a relevant role in plants grown in an orchard. Moreover, the results provide new knowledge on fire blight control methods targeting the plant light regulation systems. The sensitivity of the *E. amylovora* to monochromatic radiation could use LED technology for sustainable and non-invasive pathogen control. New scenarios in plant pathology control systems through th77e defense gene activation by light and negative regulation of pathogen virulence could be operational in the frame of optogenetic control [74].

To our knowledge, no previous study has addressed the effect of the environmental spectral quality's radiation constraints on the activation of the genes involved in plant-pathogen interaction. It remains to be found whether or not SA, JA, and ethylene genes play a role in the blue-light signaling (COP1, HYH, SPA1) and if they have a regulatory relationship with CRY1. Future studies could focus on the use of light quality, in particular BL, as an elicitor to set up protection methods for fruit crop trees, in the nursery and orchard, against fire blight.

Supplementary Materials: The following are available online at https://www.mdpi.com/article/10.3390/plants10091886/s1, Figure S1: Representative scheme of the sampling time point used to evaluate the PR genes expression in Dar Gazi-wt grown under white light, Figure S2: Representative scheme of the sampling time point used to evaluate the *PR* genes expression in *Dar Gazi-wt* exposed for 24 h under continuous lightness (a) and continuous darkness (b), Figure S3: Representative scheme of the sampling time point used to evaluate the *PR* genes expression in *Dar Gazi-wt, Dar Gazi-phyB* and *Dar Gazi-cry1* plants grown in exposed for 24 h under continuous WL, RL, FRL or BL, Figure S4: Dar Gazi events of regenerations after 25 days in dark condition (a). Shoots were transferred to the medium, containing kanamycin, to select putative transgenic lines, and subcultured four/five times for rapid and clonal multiplication (b). The medium used for the in vitro selection of regenerated buds was enriched with 100 mg/L Kanamycin. (c) Plantlets of transgenic lines during proliferation state, Figure S5: Detection of *ef1A* gene fragments in cDNA of *Dar Gazi-wt, Dar Gazi-phyB,* and *Dar Gazi-cry1* plants (a). M: Ladder; WT: *Dar Gazi-wt;* phyB: *Dar Gazi-phyB*; cryI: *Dar Gazi-cry1;* B: Blank. Detection of *NptII* gene fragments in cDNA of *Dar Gazi-wt, Dar Gazi-phyB,* and *Dar Gazi-cry1* plants (b). M: Ladder; WT: *Dar Gazi-wt;* phyB: *Dar Gazi-phyB;* cryI: *Dar Gazi-cry1;* B: Blank, Figure S6: Detection of *AtPHYB,* using specific primers for *A. thaliana PHYB* that amplify the *AtPHYB* gene fragments only in cDNA of *Dar Gazi-phyB* plants (a). M: Ladder; WT: *Dar Gazi-wt;* phyB: *Dar Gazi-phyB;* cryI: *Dar Gazi-cry1;* B: Blank. Detection of *LeCRYI,* using specific primers for *L. esculentum CRYI* that amplify the cryILE gene fragments in cDNA of *Dar Gazi-cry1* plants (b). M: Ladder; WT: *Dar Gazi-wt;* phyB: *Dar Gazi-phyB;* cryI: *Dar Gazi-cry1;* B: Blank, Table S1: Sequence of primers used to validate the molecular insertion of *AtPHYB* and *LeCRY1* in pear genome.

Author Contributions: T.S., C.I., B.T., M.R. and R.M. (Rosario Muleo) contributed to the conception and design of the study. T.S., I.F., C.I., F.L., M.R. and R.M. (Rosario Muleo) contributed to defining the methodology. T.S., I.F., F.L., R.M. (Roberto Mancinelli), M.R. and R.M. (Rosario Muleo) contributed to data formal analysis. T.S., C.I., F.L., B.T., M.R. and R.M. (Rosario Muleo) contributed to the investigation. T.S., I.F. and R.M. (Rosario Muleo) wrote the first draft of the manuscript. T.S., I.F., C.I., R.M. (Roberto Mancinelli), B.T., M.R. and R.M. (Rosario Muleo) wrote and edited the final version of the manuscript. All authors contributed to the article and approved the submitted version. All authors have read and agreed to the published version of the manuscript.

Funding: This research received no external funding.

Institutional Review Board Statement: Not applicable.

Informed Consent Statement: Not applicable.

Data Availability Statement: The data presented in this study are available on request from the corresponding author.

Conflicts of Interest: The authors declare no conflict of interest.

References

1. Thomson, S.V. The role of the stigma in fire blight infections. *Phytopathology* **1986**, *76*, 476. [CrossRef]
2. Abdollahi, H.; Rugini, E.; Ruzzi, M.; Muleo, R. In vitro system for studying the interaction between *Erwinia amylovora* and genotypes of pear. *Plant Cell Tiss. Org. Cult.* **2004**, *79*, 203–212. [CrossRef]
3. Dellagi, A.; Brisset, M.N.; Paulin, J.P.; Expert, D. Dual role of desferrioxamine in *Erwinia amylovora* pathogenicity. *Mol. Plant-Microbe Interact.* **1998**, *11*, 734–742. [CrossRef] [PubMed]
4. Garbay, B.; Tautu, M.T.; Costaglioli, P. Low level of pathogenesis-related protein 1 mRNA expression in 15-day-old *Arabidopsis cer6-2* and *cer2 eceriferum* mutants. *Plant Sci.* **2007**, *172*, 299–305. [CrossRef]
5. Genoud, T.; Buchala, A.J.; Chua, N.-H.; Métraux, J.-P. Phytochrome signalling modulates the SA-perceptive pathway in *Arabidopsis*, modulation of SA pathway by phytochrome Signal. *Plant J.* **2002**, *31*, 87–95. [CrossRef]
6. Sels, J.; Mathys, J.; De Coninck, B.M.A.; Cammue, B.P.A.; De Bolle, M.F.C. Plant pathogenesis-related (PR) proteins, a focus on PR peptides. *Plant Physiol. Biochem.* **2008**, *46*, 941–950. [CrossRef] [PubMed]

7. Zeier, J.; Pink, B.; Mueller, M.J.; Berger, S. Light conditions influence specific defence responses in incompatible plant-pathogen interactions, uncoupling systemic resistance from salicylic acid and PR-1 accumulation. *Planta* **2004**, *219*, 673–683. [CrossRef]
8. Mitsuhara, I.; Iwai, T.; Seo, S.; Yanagawa, Y.; Kawahigasi, H.; Hirose, S.; Ohkawa, Y.; Ohashi, Y. Characteristic expression of twelve rice *PR1* family genes in response to pathogen infection, wounding, and defense-related signal compounds (121/180). *Mol. Genet. Genom.* **2008**, *279*, 415–427. [CrossRef]
9. Cameron, R.K.; Paiva, N.L.; Lamb, C.J.; Dixon, R.A. Accumulation of salicylic acid and *PR-1* gene transcripts in relation to the systemic acquired resistance (SAR) response induced by *Pseudomonas syringae* pv. *Tomato in Arabidopsis*. *Physiol. Mol. Plant P.* **1999**, *55*, 121–130. [CrossRef]
10. Ali, S.; Ganai, B.A.; Kamili, A.N.; Bhat, A.A.; Mir, Z.A.; Bhat, J.A.; Tyagi, A.; Islam, S.T.; Mushtaq, M.; Yadav, P.; et al. Pathogenesis-related proteins and peptides as promising tools for engineering plants with multiple stress tolerance. *Microbiol. Res.* **2018**, *212–213*, 29–37. [CrossRef]
11. Okushima, Y.; Koizumi, N.; Kusano, T.; Sano, H. Secreted proteins of tobacco cultured BY2 cells, identification of a new member of pathogenesis-related proteins. *Plant Mol. Biol.* **2000**, *42*, 479–488. [CrossRef]
12. Sinha, M.; Singh, R.P.; Kushwaha, G.S.; Iqbal, N.; Singh, A.; Kaushik, S.; Kaur, P.; Sharma, S.; Singh, T.P. Current overview of allergens of plant pathogenesis related protein families. *TSWJ* **2014**, *2014*, 543195. [CrossRef] [PubMed]
13. Guevara-Morato, M.A.; de Lacoba, M.G.; García-Luque, I.; Serra, M.T. Characterization of a pathogenesis-related protein 4 (PR-4) induced in *Capsicum chinense* L3 plants with dual RNase and DNase activities. *J. Exp. Bot.* **2010**, *61*, 3259–3271. [CrossRef]
14. Guerra-Guimarães, L.; Pinheiro, C.; Chaves, I.; Barros, D.; Ricardo, C. Protein dynamics in the plant extracellular space. *Proteomes* **2016**, *4*, 22. [CrossRef] [PubMed]
15. Hoffmann-Sommergruber, K. Pathogenesis-related (PR)-proteins identified as allergens. *Biochem. Soc. Trans.* **2002**, *30*, 930–935. [CrossRef]
16. Lee, O.R.; Pulla, R.K.; Kim, Y.-J.; Balusamy, S.R.D.; Yang, D.-C. Expression and stress tolerance of *PR10* genes from *Panax ginseng* C. A. Meyer. *Mol. Biol. Rep.* **2012**, *39*, 2365–2374. [CrossRef] [PubMed]
17. Moiseyev, G.P.; Fedoreyeva, L.I.; Zhuravlev, Y.N.; Yasnetskaya, E.; Jekel, P.A.; Beintema, J.J. Primary structures of two ribonucleases from ginseng calluses. New members of the PR-10 family of intracellular pathogenesis-related plant proteins. *FEBS Lett.* **1997**, *407*, 207–210. [CrossRef]
18. Kim, S.T.; Yu, S.; Kang, Y.H.; Kim, S.G.; Kim, J.-Y.; Kim, S.-H.; Kang, K.Y. The rice pathogen-related protein 10 (JIOsPR10) is induced by abiotic and biotic stresses and exhibits ribonuclease activity. *Plant Cell Rep.* **2008**, *27*, 593–603. [CrossRef]
19. Xie, X.-Z.; Xue, Y.-J.; Zhou, J.-J.; Zhang, B.; Chang, H.; Takano, M. Phytochromes regulate SA and JA signaling pathways in rice and are required for developmentally controlled resistance to *Magnaporthe grisea*. *Mol. Plant* **2011**, *4*, 688–696. [CrossRef]
20. Griebel, T.; Zeier, J. Light regulation and daytime dependency of inducible plant defenses in *Arabidopsis*, phytochrome signaling controls systemic acquired resistance rather than local defense. *Plant Physiol.* **2008**, *147*, 790–801. [CrossRef]
21. Yang, Y.; Li, Y.; Li, X.; Guo, X.; Xiao, X.; Tang, D.; Liu, X. Comparative proteomics analysis of light responses in cryptochrome1-304 and Columbia wild-type 4 of *Arabidopsis thaliana*. *Acta Biochem. Biophys. Sin. (Shanghai)* **2008**, *40*, 27–37. [CrossRef] [PubMed]
22. Wu, L.; Yang, H.-Q. CRYPTOCHROME 1 is implicated in promoting R protein-mediated plant resistance to *Pseudomonas syringae* in *Arabidopsis*. *Mol. Plant* **2010**, *3*, 539–548. [CrossRef] [PubMed]
23. Mayer, R.; Raventos, D.; Chua, N.H. Det1, Cop1, and Cop9 mutations cause inappropriate expression of several gene Sets. *Plant Cell* **1996**, *8*, 1951–1959. [CrossRef]
24. Kim, T.-H.; Kim, B.-H.; von Arnim, A.G. Repressors of photomorphogenesis. *Int. Rev. Cytol.* **2002**, *220*, 185–223. [CrossRef]
25. Purcell, E.B.; Crosson, S. Photoregulation in prokaryotes. *Curr. Opin. Microbiol.* **2008**, *11*, 168–178. [CrossRef]
26. Karniol, B.; Wagner, J.R.; Walker, J.M.; Vierstra, R.D. Phylogenetic analysis of the phytochrome superfamily reveals distinct microbial subfamilies of photoreceptors. *Biochem. J.* **2005**, *392*, 103–116. [CrossRef]
27. Fraikin, G.Y.; Strakhovskaya, M.G.; Belenikina, N.S.; Rubin, A.B. Bacterial photosensory proteins, regulatory functions and optogenetic applications. *Microbiology* **2015**, *84*, 461–472. [CrossRef]
28. Kraiselburd, I.; Moyano, L.; Carrau, A.; Tano, J.; Orellano, E.G. Bacterial photosensory proteins and their role in plant-pathogen interactions. *Photochem. Photobiol.* **2017**, *93*, 666–674. [CrossRef]
29. Wu, L.; McGrane, R.S.; Beattie, G.A. Light regulation of swarming motility in *Pseudomonas syringae* integrates signaling pathways mediated by a bacteriophytochrome and a LOV protein. *mBio* **2013**, *4*, e00334-13. [CrossRef]
30. Consiglieri, E.; Xu, Q.; Zhao, K.-H.; Gärtner, W.; Losi, A. The first molecular characterisation of Blue- and Red-light photoreceptors from *Methylobacterium radiotolerans*. *Phys. Chem. Chem. Phys.* **2020**, *22*, 12434–12446. [CrossRef]
31. Mukherjee, S.; Jemielita, M.; Stergioula, V.; Tikhonov, M.; Bassler, B.L. Photosensing and quorum sensing are integrated to control *Pseudomonas aeruginosa* collective behaviors. *PLoS Biol.* **2019**, *17*, e3000579. [CrossRef]
32. Hessling, M.; Spellerberg, B.; Hoenes, K. Photoinactivation of bacteria by endogenous photosensitizers and exposure to visible light of different wavelengths—A review on existing data. *FEMS Microbiol. Lett.* **2017**, *364*, fnw270. [CrossRef]
33. Wang, Y.; Wang, Y.; Wang, Y.; Murray, C.K.; Hamblin, M.R.; Hooper, D.C.; Dai, T. Antimicrobial Blue light inactivation of pathogenic microbes, state of the art. *Drug Resist. Updates* **2017**, *33–35*, 1–22. [CrossRef]
34. Hoenes, K.; Bauer, R.; Meurle, T.; Spellerberg, B.; Hessling, M. Inactivation effect of Violet and Blue light on ESKAPE pathogens and closely related non-pathogenic bacterial species—A promising tool against antibiotic-sensitive and antibiotic-resistant microorganisms. *Front. Microbiol.* **2020**, *11*, 612367. [CrossRef]

35. Abdollahi, H.; Luziatelli, F.; Cirilli, M.; Frioni, E.; Rugini, E.; Ruzzi, M.; Muleo, R. Involvement of a hypersensitive-like reaction in tolerance to fire blight in pear (*Pyrus Communis* L.). *Afr. J. Biotechnol.* **2014**, *13*, 2840–2849. [CrossRef]
36. de Waard, M.A.; Del Sorbo, G.; Hayashi, K.; Schoonbeek, H.; Vermeulen, T. ABC and MFS Transporters from Botrytis cinerea. In *Book of Abstracts XIIth International Botrytis Symposium*; Université de Reims: Reims, France, 3–7 July 2000; p. L40.
37. Saier, M.H. Families of transmembrane sugar transport proteins. *Mol. Microbiol.* **2000**, *35*, 699–710. [CrossRef] [PubMed]
38. Muleo, R.; Iacona, C. Light perception and timekeeping systems in plants, the biological value of the domain of time. *Riv. Biol.* **2007**, *100*, 16–21.
39. Genoud, T.; Millar, A.J.; Nishizawa, N.; Kay, S.A.; Schäfer, E.; Nagatani, A.; Chua, N.H. An *Arabidopsis* mutant hypersensitive to Red and Far-Red light signals. *Plant Cell* **1998**, *10*, 889–904. [CrossRef] [PubMed]
40. Karpinski, S.; Gabrys, H.; Mateo, A.; Karpinska, B.; Mullineaux, P.M. Light perception in plant disease defence signalling. *Curr. Opin. Plant Biol.* **2003**, *6*, 390–396. [CrossRef]
41. Goodspeed, D.; Chehab, E.W.; Min-Venditti, A.; Braam, J.; and Covington, M.F. *Arabidopsis* synchronizes jasmonate-mediated defense with insect circadian behavior. *Proc. Natl. Acad. Sci. USA* **2012**, *109*, 4674–4677. [CrossRef]
42. Zhou, M.; Wang, W.; Karapetyan, S.; Mwimba, M.; Marques, J.; Buchler, N.E.; Dong, X. Redox rhythm reinforces the circadian clock to gate immune response. *Nature* **2015**, *523*, 472–476. [CrossRef]
43. Pajerowska-Mukhtar, K.M.; Emerine, D.K.; Mukhtar, M.S. Tell me more, roles of NPRs in plant immunity. *Trends Plant Sci.* **2013**, *18*, 402–411. [CrossRef]
44. Ballaré, C.L. Light regulation of plant defense. *Annu. Rev. Plant Biol.* **2014**, *65*, 335–363. [CrossRef]
45. Janda, T.; Szalai, G.; Pál, M. Salicylic acid signalling in plants. *IJMS* **2020**, *21*, 2655. [CrossRef]
46. Perales, M.; Más, P. A functional link between rhythmic changes in chromatin structure and the *Arabidopsis* biological clock. *Plant Cell* **2007**, *19*, 2111–2123. [CrossRef]
47. Jin, H.; Choi, S.-M.; Kang, M.-J.; Yun, S.-H.; Kwon, D.-J.; Noh, Y.-S.; Noh, B. Salicylic acid-induced transcriptional reprogramming by the HAC–NPR1–TGA Histone Acetyltransferase complex in *Arabidopsis*. *Nucleic Acids Res.* **2018**, *46*, 11712–11725. [CrossRef] [PubMed]
48. Cirvilleri, G.; Spina, S.; Iacona, C.; Catara, A.; Muleo, R. Study of rhizosphere and phyllosphere bacterial community and resistance to bacterial canker in genetically engineered Phytochrome A cherry plants. *J. Plant Physiol.* **2008**, *165*, 1107–1119. [CrossRef] [PubMed]
49. Hashimoto, M.; Kisseleva, L.; Sawa, S.; Furukawa, T.; Komatsu, S.; Koshiba, T. A novel rice PR10 protein, RSOsPR10, specifically induced in roots by biotic and abiotic stresses, possibly via the Jasmonic acid signaling pathway. *Plant Cell Physiol.* **2004**, *45*, 550–559. [CrossRef]
50. Colditz, F.; Nyamsuren, O.; Niehaus, K.; Eubel, H.; Braun, H.-P.; Krajinski, F. Proteomic approach, identification of *Medicago truncatula* proteins induced in roots after infection with the pathogenic Oomycete *Aphanomyces euteiches*. *Plant Mol. Biol.* **2004**, *55*, 109–120. [CrossRef]
51. Poupard, P.; Brunel, N.; Leduc, N.; Viémont, J.-D.; Strullu, D.-G.; Simoneau, P. Expression of a Bet v 1 homologue gene encoding a PR10 protein in birch roots, induction by auxin and localization of the transcripts by in situ hybridization. *Funct. Plant Biol.* **2001**, *28*, 57. [CrossRef]
52. Losi, A.; Gärtner, W. A light life together, photosensing in the plant microbiota. *Photochem. Photobiol. Sci.* **2021**, *20*, 451–473. [CrossRef]
53. Akbudak, M.A.; Yildiz, S.; and Filiz, E. Pathogenesis related protein-1 (PR-1) genes in tomato (*Solanum lycopersicum* L.), bioinformatics analyses and expression profiles in response to drought stress. *Genomics* **2020**, *112*, 4089–4099. [CrossRef]
54. Sessa, G.; Yang, X.-Q.; Raz, V.; Eyal, Y.; Fluhr, R. Dark induction and subcellular localization of the pathogenesis-related PRB-1b protein. *Plant Mol. Biol.* **1995**, *28*, 537–547. [CrossRef]
55. Seo, P.J.; Lee, A.-K.; Xiang, F.; Park, C.-M. Molecular and functional profiling of *Arabidopsis* pathogenesis-related genes, insights into their roles in salt response of seed germination. *Plant Cell Physiol.* **2008**, *49*, 334–344. [CrossRef] [PubMed]
56. Staswick, P.E.; Tiryaki, I.; Rowe, M.L. Jasmonate response locus JAR1 and several related Arabidopsis genes encode enzymes of the firefly luciferase superfamily that show activity on jasmonic, salicylic, and indole-3-acetic acids in an as-say for adenylation. *Plant Cell* **2002**, *14*, 1405–1415. [CrossRef] [PubMed]
57. Chen, H.J.; Fu, T.Y.; Yang, S.L.; Hsieh, H.L. FIN219/JAR1 and cryptochrome1 antagonize each other to modulate photomorphogenesis under blue light in *Arabidopsis*. *PLoS Genet.* **2018**, *14*, e1007248. [CrossRef]
58. Hsieh, H.L.; Okamoto, H.; Wang, M.; Ang, L.H.; Matsui, M.; Goodman, H.; Deng, X.W. FIN219, an aux-in-regulated gene, defines a link between phytochrome A and the downstream regulator COP1 in light control of *Arabidopsis* development. *Genes Dev.* **2000**, *14*, 1958–1970. [PubMed]
59. Wang, J.G.; Chen, C.H.; Chien, C.T.; Hsieh, H.L. FAR-RED INSENSITIVE219 modulates CONSTITUTIVEPHOTOMORPHOGENIC1 activity via physical interaction to regulate hypocotyl elongation in Arabidopsis. *Plant Physiol.* **2011**, *156*, 631–646. [CrossRef]
60. Deng, X.-W.; Caspar, T.; Quail, P.H. Cop1, a regulatory locus involved in light-controlled development and gene expression in *Arabidopsis*. *Genes Dev.* **1991**, *5*, 1172–1182. [CrossRef]
61. Dechorgnat, J.; Patrit, O.; Krapp, A.; Fagard, M.; and Daniel-Vedele, F. Characterization of the *AtNRT2.6* gene is involved in the response of *Arabidopsis thaliana* to *Erwinia amylovora*. *PLoS ONE* **2012**, *7*, e42491. [CrossRef]

62. Farjad, M.; Clément, G.; Launay, A.; Jeridi, R.; Jo-livet, S.; Citerne, S.; Rigault, M.; Soulie, M.-C.; Dinant, S.; Fagard, M. Plant nitrate supply regulates *Erwinia amylovora* virulence gene expression in *Arabidopsis*. *Mol. Plant Pathol.* **2021**, 1–15. [CrossRef]
63. Deslandes, L.; Rivas, S. Catch me if you can, bacterial effectors and plant targets. *Trends Plant Sci.* **2012**, *17*, 644–655. [CrossRef] [PubMed]
64. Gust, A.A.; Pruitt, R.; Nürnberger, T. Sensing danger, key to activating plant immunity. *Trends Plant Sci.* **2017**, *22*, 779–791. [CrossRef] [PubMed]
65. Palmer, J.W. Light transmittance by apple leaves and canopies. *J. App. Ecol.* **1977**, *14*, 505–513. [CrossRef]
66. Grant, R.H. 1997. Partitioning of biologically active radiation in plant canopies. *Int. J. Biometeorol.* **1997**, *40*, 26–40. [CrossRef]
67. Bastías, R.M.; Corelli-Grappadelli, L. Light quality management in fruit orchards, physiological and technological aspects. *Chil. J. Agric. Res.* **2012**, *72*, 574–581. [CrossRef]
68. Radauer, C.; Breiteneder, H. Evolutionary biology of plant food allergens. *J. Allergy Clin. Immun.* **2007**, *120*, 518–525. [CrossRef] [PubMed]
69. Holefors, A.; Xue, Z.-T.; Zhu, L.-H.; Welander, M. The Arabidopsis phytochrome B gene influences growth of the apple rootstock M26. *Plant Cell Rep.* **2000**, *19*, 1049–1056. [CrossRef]
70. Sambrook, J.; Fritsch, E.; Maniatis, T. *Molecular Cloning, a Laboratory Manual*, 2nd ed.; Cold Spring Harbor Laboratory Press: New York, NY, USA, 1989; Volume 2, ISBN 978-0-87969-309-1.
71. Quoirin, M.; Lepoivre, P. Improved media for in vitro culture of *Prunus* sp. *Acta Hortic.* **1977**, *78*, 437–442. [CrossRef]
72. Sgamma, T.; Thomas, B.; and Muleo, R. Ethylene inhibitor silver nitrate enhances regeneration and genetic transformation of Prunus avium (L.) cv. Stella. *Plant Cell Tissue Organ Cult.* **2015**, *120*, 79–88. [CrossRef]
73. Livak, K.J.; Schmittgen, T.D. Analysis of relative gene expression data using real-time quantitative PCR and the 2(-Delta Delta C(T)) method. *Methods* **2001**, *25*, 402–408. [CrossRef] [PubMed]
74. Losi, A.; Gardner, K.H.; Möglich, A. Blue-light receptors for optogenetics. *Chem. Rev.* **2018**, *118*, 10659–10709. [CrossRef] [PubMed]

plants

MDPI

Article

Light Spectra and Root Stocks Affect Response of Greenhouse Tomatoes to Long Photoperiod of Supplemental Lighting

Jason Lanoue [1], Alyssa Thibodeau [1], Celeste Little [1], Jingming Zheng [1], Bernard Grodzinski [2] and Xiuming Hao [1,*]

[1] Harrow Research and Development Centre, Agriculture & Agri-Food Canada, Harrow, ON N0R1G0, Canada; Jason.Lanoue@agr.gc.ca (J.L.); Alyssa.Thibodeau@agr.gc.ca (A.T.); Celeste.Little@agr.gc.ca (C.L.); Jingming.Zheng@agr.gc.ca (J.Z.)

[2] Department of Plant Agriculture, University of Guelph, Guelph, ON N1G 2W1, Canada; bgrodzin@uoguelph.ca

* Correspondence: Xiuming.Hao@agr.gc.ca

Abstract: Plant biomass and yield are largely dictated by the total amount of light intercepted by the plant (daily light integral (DLI)—intensity × photoperiod). It is more economical to supply the desired DLI with a long photoperiod of low-intensity light because it uses fewer light fixtures, reducing capital costs. Furthermore, heat released by the light fixtures under a long photoperiod extended well into the night helps to meet the heating requirement during the night. However, extending the photoperiod beyond a critical length (>17 h) may be detrimental to production and lead to leaf chlorosis and a reduction in leaf growth and plant vigor in greenhouse tomato production. It is known that red light can increase leaf growth and plant vigor, as can certain rootstocks, which could compensate for the loss in plant vigor and leaf growth from long photoperiods. Therefore, this study investigated the response of tomatoes grafted onto different rootstocks to a long photoperiod of lighting under red and other light spectra. Tomato plants 'Trovanzo' grafted onto 'Emperator' or 'Kaiser' were subjected to two spectral compositions—100% red or a mix of red (75%), blue (20%), and green (5%) light for 17 h or 23 h. The four treatments supplied similar DLI. Leaf chlorosis appeared in all plants under 23 h lighting regardless of spectral compositions between 20 and 54 days into the treatment. The yield for 23 h mixed lighting treatment was lower than both 17 h lighting treatments. However, the 23 h red lighting treatment resulted in less leaf chlorosis and the plants grafted onto 'Emperator' produced a similar yield as both 17 h lighting treatments. Therefore, both spectral compositions and rootstocks affected the response of greenhouse tomatoes to long photoperiods of lighting. With red light and proper rootstock, the negative yield impact from long photoperiod lighting can be eliminated.

Keywords: light quality; photoperiod; tomato; light-emitting diode; greenhouse; photosynthesis; light spectra; root stock

Citation: Lanoue, J.; Thibodeau, A.; Little, C.; Zheng, J.; Grodzinski, B.; Hao, X. Light Spectra and Root Stocks Affect Response of Greenhouse Tomatoes to Long Photoperiod of Supplemental Lighting. *Plants* **2021**, *10*, 1674. https://doi.org/10.3390/plants10081674

Academic Editors: Valeria Cavallaro and Rosario Muleo

Received: 15 July 2021
Accepted: 12 August 2021
Published: 14 August 2021

1. Introduction

The daily light integral (DLI; light intensity x photoperiod duration) plays a vital role in plant biomass accumulation and yield. While the natural solar DLI is dictated by time of year, global location, and local weather, the DLI can be augmented by the introduction of supplemental lighting. Supplemental lighting can aid in the achievement of a desired/target DLI to increase plant growth and yield, specifically during low-light months [1]. The use of an extended photoperiod with supplemental light at a lower light intensity can have economic benefits by reducing the overall fixture need (i.e., capital cost) and by using electricity during the night, when electrical costs are low [2]. Furthermore, most of the input electricity in light fixtures is eventually converted into heat because plants only convert a small percentage of light into biomass. By utilizing LEDs during the subjective night period, the heat released from light fixtures can help to meet nighttime heating requirements. However, exceeding the tolerable limits of photoperiods, which are species-specific, can lead to diminished yield, photoperiod-related leaf injury, and an economic disadvantage for growers [3]. For tomatoes, photoperiods up to 17 h are associated with normal growth patterns [4]. Photoperiods beyond 17 h, which do not employ a drastic temperature dip, spectral change, or decrease in light intensity, have been shown to cause photoperiod-related injury characterized by leaf chlorosis, photosynthetic inhibition, and yield decrease [5–7]. However, prolonged photoperiods (>18 h) can theoretically lead to increased plant biomass and yield due to the added light available for photosynthesis, if photoperiod-related injury is not induced [8].

The underlying mechanism involved in photoperiod-related injury has yet to be determined. The *type III light harvesting chlorophyll a/b binding protein 13* (*CAB-13*) gene has been demonstrated to play an important role in photoperiod-related injury [9]. As the name suggests, *CAB-13* plays an important role in the light harvesting/photosystem II (PSII) super complex [10]. When plants were grown under continuous light (CL, 24 h), *CAB-13* expression was downregulated, leading to photoinhibition and photoperiod-related injury characterized by a decrease in the maximum quantum efficiency of PSII (F_v/F_m) [9]. Velez-Ramirez et al. [9] also suggest that photoperiod-related injury may be due to the unbalanced excitation between photosystem I (PSI) and PSII. Furthermore, Haque et al. [6] hypothesized that continuous lighting could cause damage to PSII, which ultimately reduces photosynthesis and yield. It has also been hypothesized that the restoration of proper leaf photochemistry, potentially through improved expression of genes such as *CAB-13*, as well as balancing source/sink strength may alleviate injury under extended photoperiods [6,9,11,12]. From this, we have deduced that the utilization of different spectral compositions during an extended photoperiod may play a key role in avoiding or lessening photoperiod-related leaf injury.

The ability to regulate gene expression can largely be traced back to the role of photoreceptors such as phytochrome and cryptochrome [13]. With the advancements in light-emitting diode (LED) technology, the interaction between photoperiod length and spectral composition has become of interest in optimizing growth conditions for high-value crops. With red and blue LEDs being the most efficient and these wavelengths being primarily absorbed by phytochrome and cryptochrome, respectively, much research relating to photoperiods and spectra has focused around these wavelengths [7,9,14]. Matsuda et al. [14] indicated that photoperiod-related injury was less severe when tomato seedlings were exposed to red or orange LEDs compared to blue or white during the subjective night period when grown under CL. Moreover, Velez-Ramirez et al. [15] determined that phytochrome A (PHY A) plays an important role in photoperiod-related injury. Together, these studies indicate that light spectral compositions/quality may play an important role in reducing tomato injury under extended photoperiods. However, both Matsuda et al. [14] and Velez-Ramirez et al. [15] performed experiments using controlled environment growth chambers, which would exclude any effect that the natural solar radiation would have. Furthermore, using such chambers would not facilitate adequate growth space for plants to reach maturity (both studies only used young plants up to the first flower stage) and thus did not allow for the assessment of yield during prolonged photoperiods.

Exposing tomatoes to extended photoperiods tends to lead to smaller leaf area [9,16]. It has been stated that even if tomatoes were genetically altered to be CL-tolerant, the overall leaf area would be low, resulting in reduced light capture and plant growth [9]. However, leaf expansion can be controlled by spectral compositions. Red supplemental light is generally thought of as a vegetative light, able to improve leaf expansion, whereas blue light is known to reduce leaf size and increase leaf thickness [17,18]. Therefore, utilizing red light during extended photoperiods may overcome the reduction in leaf size invoked by CL. It should also be noted that the rootstocks can affect plant growth. While grafting is traditionally done to invoke disease resistance, many studies have shown that proper selection of rootstock materials can increase leaf area as well as plant vigor, leading to improved yield [19–21]. Therefore, the use of wavelength-specific lighting such as red light and proper rootstock selection may alleviate photoperiod-related injury. For these reasons, we set out to test the response of large vining tomatoes to an extended photoperiod of lighting with different spectral compositions/quality and rootstocks to see if the negative impact of extended photoperiod lighting on tomato fruit production can be eliminated by proper selection of light spectral compositions and rootstocks.

2. Materials and Methods

2.1. Plant Material and Experimental Design

Tomato (*Solanum lycopersicum*) seedlings cv. 'Trovanzo' were grafted onto 'Emperator' (TE) or onto 'Kaiser' (TK) in a double-stemmed (twin-head) system. 'Emperator' has been observed to promote vegetative growth (or a vigorous rootstock) whereas 'Kaiser' has been shown to promote generative growth. In this way, we could observe the homeostatic balance between two different rootstock types. Twin-head transplants (5 weeks old) raised by a commercial propagator were placed into rockwool slabs on top of a raised growing trough (30 cm high) in a large glass greenhouse (200 m^2 growing area) at the Harrow Research and Development Centre, Agriculture and Agri-Food Canada, Harrow, Ontario, Canada (42.03° N, 82.9° W) on 11 November 2018 at a plant density of 2 plants m^{-2} (4 stems m^{-2}). The plants were planted into 6 rows, with the 2 outside rows serving as guard rows. The two stems/heads of each plant in the row were trained upward along the vertical strings into a 'V' system as in commercial production. The strings were hung onto the top wires (3.5 m high). Once the plant head reached the overhead wires, a few bottom leaves were removed and the plants were lowered twice every week. The plants were drip-irrigated using a complete nutrient solution [22]. The electrical conductivity and pH were set at 2.8 dS m^{-1} and 5.8, respectively. Plants were grown at an enriched CO_2 concentration of 800 $\mu L\ L^{-1}$ when the greenhouse was not ventilated. The average daytime temperature was held between 21 and 24 °C during the months of November, December, and January. Average daytime temperatures during the months of February and March were between 21 and 25 °C. Average daytime temperatures during the months of April and May were between 22 and 27 °C depending on the ambient solar radiation. Nighttime temperature was maintained at 20 ± 1 °C throughout the production period. Relative humidity of 70 ± 10% was maintained during both daytime and nighttime periods.

The 4 middle rows were divided into 16 plots via white curtains, which were impenetrable to light. There were 24 twin-head plants (48 stems) in each plot; 12 plants for each of the 2 rootstocks were planted. Four supplemental overhead lighting treatments were applied to the 16 plots in a Latin square design with 4 replications (one lighting treatment in each row or column of the 16 plots): 100% red from 23:00 to 16:00 (Red 17 h), 100% red from 17:00 to 16:00 (Red 23 h), mixed light from 23:00 to 16:00 (Mix 17 h), and mixed light from 17:00 to 16:00 (Red 23 h; Figure 1). The mixed light had 75% red (600–700 nm), 20% blue (400–499 nm), and 5% green light (500–599 nm), with green light being introduced via white diodes. The four lighting treatments provided similar DLIs (Table 1). All the lighting treatments were applied using Pro 325e smart LED fixtures from LumiGrow (Emeryville, California, USA). A 23 h monochromatic red photoperiod was utilized because red light tends to preferentially support vegetative growth. Furthermore, red diodes are highly efficient, and thus a pure red spectrum is able to achieve the highest photosynthetic photon efficacy (PPE) [23]. The mixed light spectrum was chosen as a control, which provides close to the recommended amounts of blue light [24] and some green light, which has also been shown to improve tomato growth and yield [25].

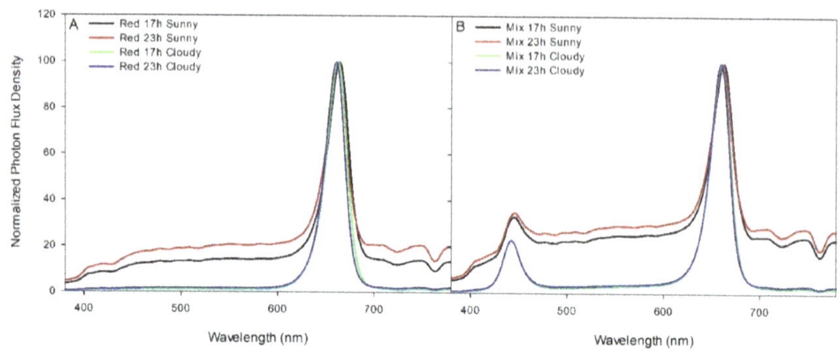

Figure 1. Normalized photon flux density (380–780 nm) from red and mixed LED lighting treatments during both sunny and cloudy days. The spectra were measured using a Li-180 Spectrometer (Li-COR Inc., Lincoln, NE, USA) at a distance of 80 cm from the light fixtures. Measurements for sunny days were done on 11 February 2019 and cloudy days on 14 February 2019 between 12:00 and 14:00. Panel (**A**) represents spectra from both red light treatments during sunny and cloudy days. Panel (**B**) represents spectra from both mixed light treatments during sunny and cloudy days.

Table 1. Photosynthetic photon flux density of supplemental lighting treatments (400–700 nm; 80 cm below the LED fixtures) as determined by using a Li-190R quantum line sensor at night. Red to far-red ratios (R:Fr) were estimated by dividing the total photons of red (600–700 nm) by the total photons of far-red (700–780 nm) during both sunny and cloudy days when the supplementary lighting fixtures were on. Phytochrome photostationary state (PSS) was calculated using Equation (1) [26].

Lighting Treatment	Supplemental Light Intensity (μmol m^{-2} s^{-1})	R:Fr—Sunny Day	R:Fr—Cloudy Day	PSS—Sunny Day	PSS—Cloudy Day	Supplemental DLI (mol m^{-2} d^{-1})
Red 17 h	176 ± 7	3.78	25.56	0.82	0.88	10.79 ± 0.41
Red 23 h	127 ± 4	2.81	17.77	0.80	0.87	10.54 ± 0.36
Mix 17 h	169 ± 4	2.55	15.29	0.79	0.86	10.32 ± 0.27
Mix 23 h	134 ± 5	2.30	12.01	0.78	0.86	11.09 ± 0.42

Application of the supplemental lighting treatments began on 16 November 2018. Supplemental light intensities as shown in Table 1 were determined at three different positions within a treatment at 80 cm from the light fixtures (just above the heads of the plants) using a Li-COR 190R (Li-COR Biosciences Inc., Lincoln, NE, USA) quantum line sensor during the nighttime period to exclude natural solar radiation. Red:far-red (R:Fr) were determined on both sunny and cloudy days (11 February and 14 February 2019, respectively) by dividing the total photons of red by the total photons of far-red (700–780 nm). Further to this, the phytochrome photostationary state (PSS) was determined using Equation (1) from Sager et al. [26], where N is the photon flux (mol m^{-2} s^{-1}) and σ_r and σ_{Fr} are the photochemical cross-section of phytochrome in the red absorbing state and far-red absorbing state, respectively.

$$PSS = \frac{\left(\sum_{380}^{780} N\sigma_r\right)}{\left(\sum_{380}^{780} N\sigma_r + \sum_{380}^{780} N\sigma_{Fr}\right)} \quad (1)$$

Spectral readings for these calculations were taken between 12:00 and 14:00 using a Li-COR Li-180 spectrometer on both sunny and cloudy days under their respective lighting treatments. The sunny day was one with no observable clouds in the sky and the cloudy day was fully overcast. Throughout the experiment, supplemental lighting remained on regardless of ambient light levels to ensure that all treatments received similar total DLIs

(Figure 2). The curtains were closed during cloudy days to prevent light contamination between treatments. During sunny days, the curtains were opened to prevent shading of natural light. On days which were partly cloudy/sunny, the forecast was used to determine the majority (i.e., sun or cloudy) and then curtains were opened or closed as appropriate. Because there was sunlight between 16:00 and 17:00 (i.e., during natural sunset), the actual photoperiods (including sunlight) were 18 h and 24 h (CL) for the 17 h and 23 h lighting, respectively. In commercial greenhouses, bumble bees are used as pollinators for fruit setting. The stopping of supplemental lighting between 16:00 and 17:00 was to facilitate the return of bees to their hives under natural dusk/sunset conditions; otherwise, the bees may get lost and significantly increase the number of bees (and associated cost) needed for pollination [27]. Dusk/sunset varied throughout the course of the experiment from 16:52 to 20:22, which was sufficient to allow for bees to return to their hives.

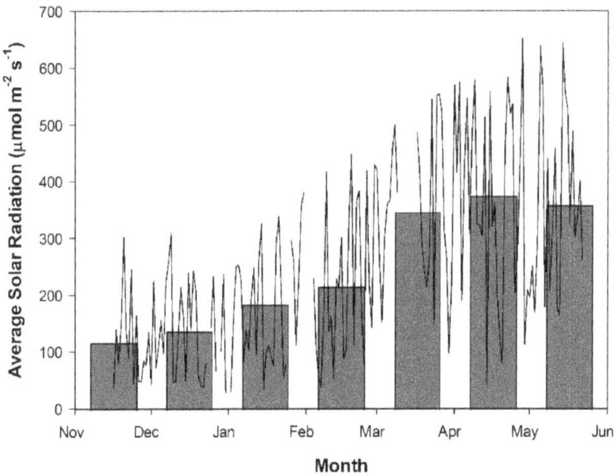

Figure 2. Daily average solar radiation as measured from 16 November 2018 to 22 May 2019 using a Li-COR LI-200R pyranometer converted from W m^{-2} to μmol m^{-2} s^{-1} using the conversion value of 2.1. Readings were taken above the greenhouse and then corrected for an approximate 50% transmissivity to account for shading from the greenhouse structure, lighting fixtures, and shade curtains. Measurements were taken every 2 h, beginning at 08:00 and concluding at 16:00, between the wavelengths of 400 and 1100 nm. Measurements during this period were averaged to provide an average solar radiation for each day (line plot). The bar graph indicates the average daily solar radiation throughout the month. Breaks in the line plot indicate periods of time which were not documented due to a technical malfunction.

2.2. Growth Measurements

Growth measurements were performed on 6 randomly selected plants from TE and TK at 31, 60, and 139 days into the treatment (DIT), corresponding with 16 December 2018, 14 January 2019, and 3 April 2019, respectively. Growth measurements included leaf length, leaf width, and chlorophyll content of the 5th, 10th, and 15th leaf when applicable. Leaf chlorophyll was measured using a SPAD meter (model 502, Konica Minota, Osaka, Japan) and values were converted to chlorophyll content using correction equations generated by spectrophotometric pigment analysis. Chlorophyll correction curves were generated by extracting leaf punches in 95% ethanol at 78 °C for approximately 3 h until the tissue was cleared. Samples were then analyzed at 664.2 nm, 648.6 nm, and 470 nm wavelengths using a spectrophotometer (Beckman DU-640 UV–Vis, Indianapolis, IN, USA). Concentrations of *chlorophyll a*, *b*, and carotenoids were determined via equations from [28].

2.3. Leaf Gas Exchange: Day and Night Measurements

The 5th leaves from TE plants were placed in the chamber of a Li-COR 6400 (Li-COR Inc., Lincoln, NE, USA), which was fitted with a 2 cm × 3 cm clear-top chamber. The leaf temperature was set to 24 °C, with a relative humidity of 55–65% and a CO_2 level held at 800 μL L^{-1}. Three leaves from separate plants under each treatment were used at 21 DIT (6 December 2018) and 55 DIT (9 January 2019) for both daytime and nighttime measurements. Measurements were taken during the day on cloudy days to maximize the effect of supplemental lighting while minimizing the effect of natural light. Nighttime measurements were taken between 18:00 and 20:00, which was at least 1 h after sunset on each respective day. This was done to ensure that plants had sufficient time to reach a steady-state photosynthetic or respiratory rate depending on treatment. Leaves were kept in the chamber until steady-state photosynthesis rates were obtained; then, the average from a 2-min period was taken.

2.4. Leaf Gas Exchange: Light Response Curves

The 5th leaves from TE plants were placed in the chamber of a Li-COR 6400, which was fitted with a 2 cm × 3 cm red/blue LED Li-COR standard light source (88Red/12%Blue). The leaf temperature was set to 24 °C, with a relative humidity of 55–65% and a CO_2 level held at 800 μL L^{-1}. Three leaves from separate plants under each treatment were used at 20, 54, and 134 DIT, corresponding with 5 December 2018, 8 January 2019, and 29 March 2019, respectively. Measurements were performed on cloudy days. Light curves began at a high light intensity and decreased gradually following the procedure from Lanoue et al. [29]. At each light level, the photosynthetic rate was allowed to reach a steady state; then, a measurement was taken for that light level. Photosynthetic rates were plotted against light intensity and fitted to a regression line following the equation $y = y_o + a(1-e^{(-b*x)})$, using SigmaPlot 10.0 to determine the photosynthetic maximum. A linear regression (y = mx + b) using the photosynthetic rates at the light levels of 0–100 μmol m^{-2} s^{-1} was used to calculate both the light compensation point (LCP) and quantum yield (QY).

2.5. Leaf Gas Exchange: CO_2 Response Curves

The 5th leaves from TE plants were placed in the chamber of a Li-COR 6400, which was fitted with a 2 cm × 3 cm red/blue LED Li-COR standard light source (88%R/12%B). The leaf temperature was set to 24 °C, with a relative humidity of 55–65% and a light level of 300 μmol m^{-2} s^{-1}. CO_2 response curves were specifically preformed at a light level of 300 μmol m^{-2} s^{-1} to assess the leaf photosynthetic capacity at, or near, an average growth condition. Thus, the maximum rate of Rubisco carboxylation (V_{cmax}) and the maximum rate of electron transport (J_{max}) are associated with a non-saturated light level. Three leaves from separate plants under each treatment were used at 20, 54, and 134 DIT, corresponding with 5 December 2018, 8 January 2019, and 29 March 2019, respectively. Measurements were performed on cloudy days. CO_2 response curves began at the ambient CO_2 concentration (800 μL L^{-1}) and reduced gradually to 50 μL L^{-1}. After the 50 μL L^{-1} measurement, the CO_2 concentration was set to 800 μL L^{-1} and was held steady until plant photosynthetic parameters returned to levels established during the beginning of the experiment. The CO_2 level was then increased incrementally to 2000 μL L^{-1}, at which point the CO_2 response curve was terminated. At each CO_2 concentration, the photosynthetic rate was allowed to reach a steady state; then, a measurement was taken to produce values for that CO_2 concentration. Photosynthetic rates were plotted against internal CO_2 concentration (C_i) and fitted to the FvCB model [30] and temperature-corrected [31,32] to determine the maximum rate of photosynthesis under Rubisco-limited and RuBP-limited conditions.

2.6. Chlorophyll Fluorescence Imaging

Intact leaflets from TE plants were dark-adapted using aluminum foil for 10 min. After the dark adaptation period, leaflets were detached and immediately used for chlorophyll imaging using a closed FluorCam model FC 800-C with FluorCam v.7.0 software (FluorCam,

Photon System Instruments, Brno, Czech Republic). The minimum fluorescence in a dark-adapted state (F_o) was acquired during a dark period of 5 s, after which an 800 ms saturating light pulse (2400 μmol m^{-2} s^{-1}) from a blue LED (peak emission of 449 nm) was used to measure maximum fluorescence in a dark-adapted state (F_m). From F_o and F_m, the variable fluorescence in a dark-adapted state (F_v) was calculated ($F_v = F_m - F_o$), which was used to determine the maximum photosystem II (PSII) quantum yield (F_v/F_m). In general, the lower the value of F_v/F_m is, the more severe the photoinhibition and thus the leaf injury [33]. By calculating F_v/F_m using chlorophyll fluorescence imaging, we were able to assess not only the prevalence of injury but also the spatial heterogeneity of F_v/F_m from a leaflet. Eight leaflets from the 5th leaf were used for each lighting treatment when plants were 23, 62, and 138 DIT, corresponding with 8 December 2018, 16 January 2019, and 2 April 2019, respectively.

2.7. Fruit Yield

First harvest began on 1 February 2019, 78 DIT. Clusters of tomato were harvested when 4 out of the 5 fruits in the cluster had become red, twice a week. The harvested fruit was graded according to commercial grading standards [22]. The fruit number and weight for marketable and unmarketable fruit were recorded from 1 February 2019 to 17 May 2019. On 22 May 2019, the plants were strip-harvested. During the strip harvest, all tomatoes regardless of ripeness stage were harvested and used during calculations. The plants were topped (growing head removed) 4 weeks before the strip harvest. By the time of strip harvest, all the clusters of fruit on the plants had reached full size but had still not reached commercial harvest stage yet (80% red). Throughout the manuscript, yield data from 1 February 2019 to 28 February 2019 are designated as February; yield data from 1 March 2019 to 31 March 2019 are designated March; yield data from 1 April 2019 to 30 April 2019 are designated April; and yield data from 1 May 2019 to the strip harvest on 22 May 2019 are designated May.

2.8. Statistical Analysis

All statistics were performed using SAS Studio 3.5. Means comparisons between the red 17 h, red 23 h, mix 17 h, and mix 23 h lighting treatments were done using a two-way ANOVA assessing the effects of photoperiod length, spectral quality, and the interaction with a Tukey–Kramer adjustment and a $p < 0.05$ indicating a significant difference.

3. Results

3.1. Chlorophyll Fluorescence and Photosynthesis

Assessing the maximum efficiency of PSII (F_v/F_m) via chlorophyll fluorescence measurements is often used as a proxy measurement to assess the health of a plant. Here, it was used to assess injury related to photoperiod extension. During the initial stage of growth (23 DIT, 8 December 2018), leaves from TE tomato plants had similar maximum efficiency of PSII (Figure 3A). At 62 DIT (16 January 2019), leaves exposed to both red 23 h and mix 23 h lighting treatments produced lower F_v/F_m values compared to leaves exposed to red 17 h and mix 17 h treatments (Figure 3B). During the late stage of growth (138 DIT, 2 April 2019), leaves under all lighting treatments again produced similar F_v/F_m values (Figure 3C). Of note, leaves under both red 23 h and mix 23 h treatments showed an increase in F_v/F_m values from 62 DIT to 138 DIT to values similar to those at the beginning of the experiment, indicating full recovery. As shown in Figure 4, leaves exposed to red 23 h and mix 23 h lighting displayed photoinhibition patterns characteristic of interveinal chlorosis. These patterns were not apparent on the 5th leaf of any treatments at 138 DIT (Figure 4). Leaves under the mix 23 h lighting treatment tended to have more interveinal chlorosis than leaves under the red 23 h lighting treatment at 62 DIT (Figure 4), which is also seen by a slightly higher F_v/F_m in Figure 3B.

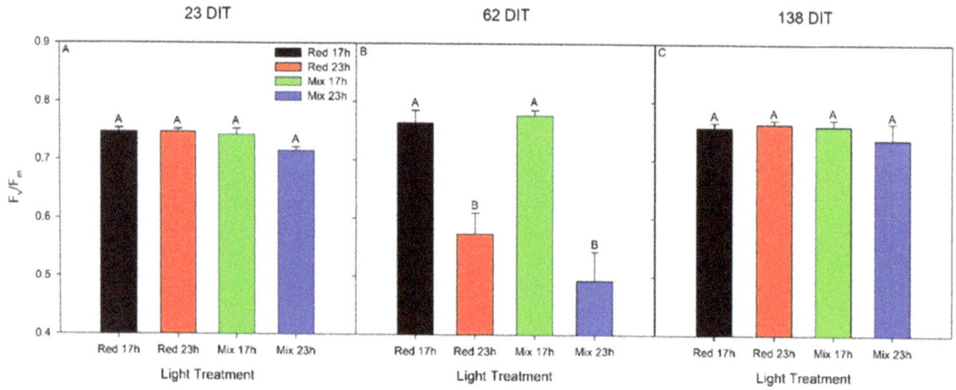

Figure 3. Maximum efficiency of PSII (F_v/F_m) from the 5th leaf of TE tomatoes grown under red 17 h, red 23 h, mix 17 h, or mix 23 h lighting treatments at 23 DIT (8 December 2018, panel **A**), 62 DIT (January 16, panel **B**), and 138 DIT (2 April 2019, panel **C**). Error bars represent the standard error of the mean of n = 8. Letter groups (A, B) represent significant differences between the lighting treatments at a specific time point and leaf position at $p < 0.05$.

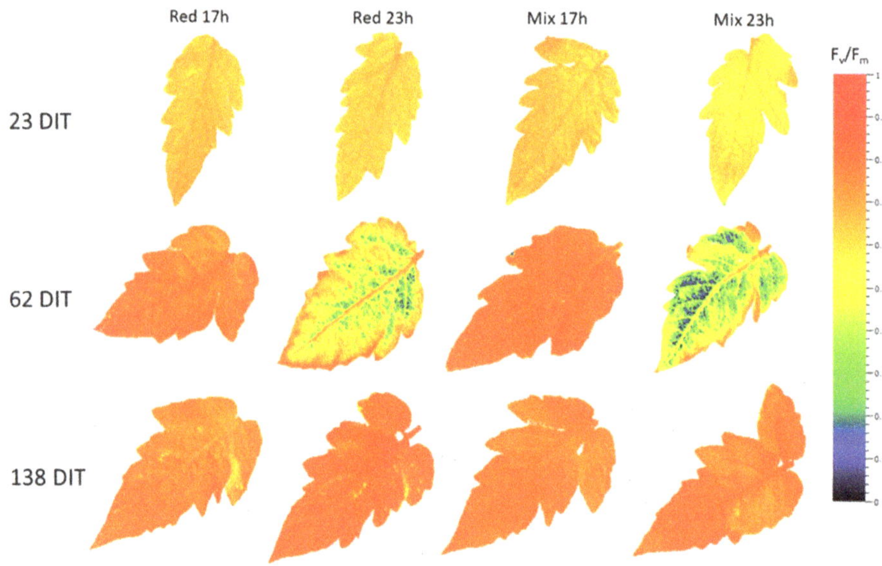

Figure 4. Spatial response of F_v/F_m from the 5th leaf of TE tomatoes grown under either red 17 h, red 23 h, mix 17 h, or mix 23 h lighting treatments at 23 DIT (8 December 2018), 62 DIT (16 January 2019), and 138 DIT (2 April 2019).

At 21 DIT (6 December 2018), leaves under the mix 17 h treatment produced the highest daytime net carbon exchange rate (NCER; Figure 5A). Both red 23 h and mix 23 h as well as the red 17 h treatment produced statistically lower daytime NCER values compared to leaves exposed to mix 17 h. Respiration rates, indicated by a negative NCER, were the highest under both 17 h lighting treatments (Figure 5A). In both 23 h lighting treatments, an increase in NCER was observed as there was light from the LEDs during this subjective nighttime period (Figure 5A). It should be noted that the red 23 h lighting treatment had a higher NCER than the mix 23 h lighting treatment during the subjective nighttime period, which may indicate the first signs of the alleviation of photoperiod-related injury by red light (Figure 5A). At 55 DIT (9 January 2019), leaves exposed to mix 23 h produced drastically reduced daytime NCER compared to leaves exposed to mix 17 h lighting treatment (Figure 5B), indicating severe damage caused by the long photoperiod with a mixed light spectrum. However, leaves exposed to red 23 h produced statistically similar daytime NCER values as both 17 h lighting treatments. Leaves exposed to both 17 h lighting treatments produced the highest respiration rates during the nighttime period, as expected (Figure 5B). Surprisingly, leaves under both 23 h lighting treatments also produced negative NCER values (indicating respiration), even during a period with an appreciable amount of supplemental light (Figure 5B). However, it should be noted that the respiratory rate in both 23 h treatments was lower than those in the 17 h treatment. While light was present in the night, it might not have been utilized to the full extent due to the leaf chlorosis (Figures 3B and 4).

Daytime transpiration rates at both 21 DIT and 55 DIT were similar among all lighting treatments (Figure 5C,D). At 21 DIT, nighttime transpiration rates were also similar among all lighting treatments (Figure 5C). However, at 55 DIT, the transpiration rate was the highest in leaves exposed to the mix 23 h lighting treatment and the lowest under the mix 17 h treatment (Figure 5D). Water use efficiency (WUE) indicates the rate of CO_2 and H_2O exchange through stomata, with a positive rate indicating photosynthesis and a negative rate indicating respiration. At 21 DIT and 55 DIT, the daytime WUE of leaves exposed to the mix 17 h lighting treatment was higher than leaves exposed to the mix 23 h lighting treatment (Figure 5E,F). At 55 DIT, WUE from leaves exposed to the red 17 h treatment was also higher than leaves under the mix 23 h treatment. However, WUE was not different between 17 h and 23 h with red light. Therefore, light spectral compositions did affect the response of WUE to the long photoperiod (23 h). Nighttime WUE at 21 DIT was the lowest under the red 17 h treatment and the highest under both red 23 h and mix 17 h (Figure 5E). At 55 DIT, nighttime WUE was the lowest in leaves exposed to the mix 17 h lighting treatment but the highest under the mix 23 h lighting treatment (Figure 5F).

Light use efficiency (LUE) is the calculation of how much CO_2 is fixed per incoming unit of photons. In this way, it provides a metric which allows for the assessment of light capture and carbon fixation. All leaves at 21 DIT produced similar LUE values (Figure 5G). At 55 DIT, leaves grown under both 17 h lighting treatments produced higher LUE values than leaves under the mix 23 h lighting treatment (Figure 5H). At both time periods, nighttime LUE for the 17 h treatments was non-resultant due to a light intensity of 0 μmol m^{-2} s^{-1} (Figure 5G,H). At 21 DIT, the nighttime LUE was higher under the red 23 h treatment than the mix 23 h lighting treatments (Figure 5G). Similar results were obtained during measurements at 55 DIT (Figure 5H). This indicates that leaves grown under an extended red photoperiod were better able to utilize the light during the subjective nighttime period than those under the mixed spectrum, likely due to less photoperiod-related injury.

At 20 DIT (5 December 2018), photosynthetic light response curves were generated for the fifth leaf of TE tomatoes (Figure 6). All photosynthetic parameters (i.e., respiration rate, light compensation point (LCP), quantum yield (QY), and maximum photosynthetic rate (Pn_{max})) were similar among all treatments (Table 2). At 54 DIT (8 January 2019), leaves exposed to the mix 17 h lighting treatment produced the lowest respiration rate and leaves exposed to the mix 23 h lighting treatment produced the highest (Table 2). Both 23 h lighting treatments produced drastically higher LCP than leaves exposed to the 17 h treatments, showing an inability to utilize light well (Figure 6F; Table 2). At 54 DIT, leaves exposed to the mix 17 h treatment produced the highest QY out of all lighting treatments. Furthermore, leaves exposed to the red 17 h treatment produced higher QY than leaves exposed to either red 23 h or mix 23 h treatments (Table 2). At 54 DIT, Pn_{max} was greatly reduced in both 23 h lighting treatments compared to both red 17 h and mix 17 h lighting treatments (Table 2). At 134 DIT (29 March 2019), respiration rate, LCP, and QY were similar between all lighting treatments. However, Pn_{max} was higher in leaves exposed to the mix 23 h lighting treatment than the red 17 h lighting treatment (Table 2).

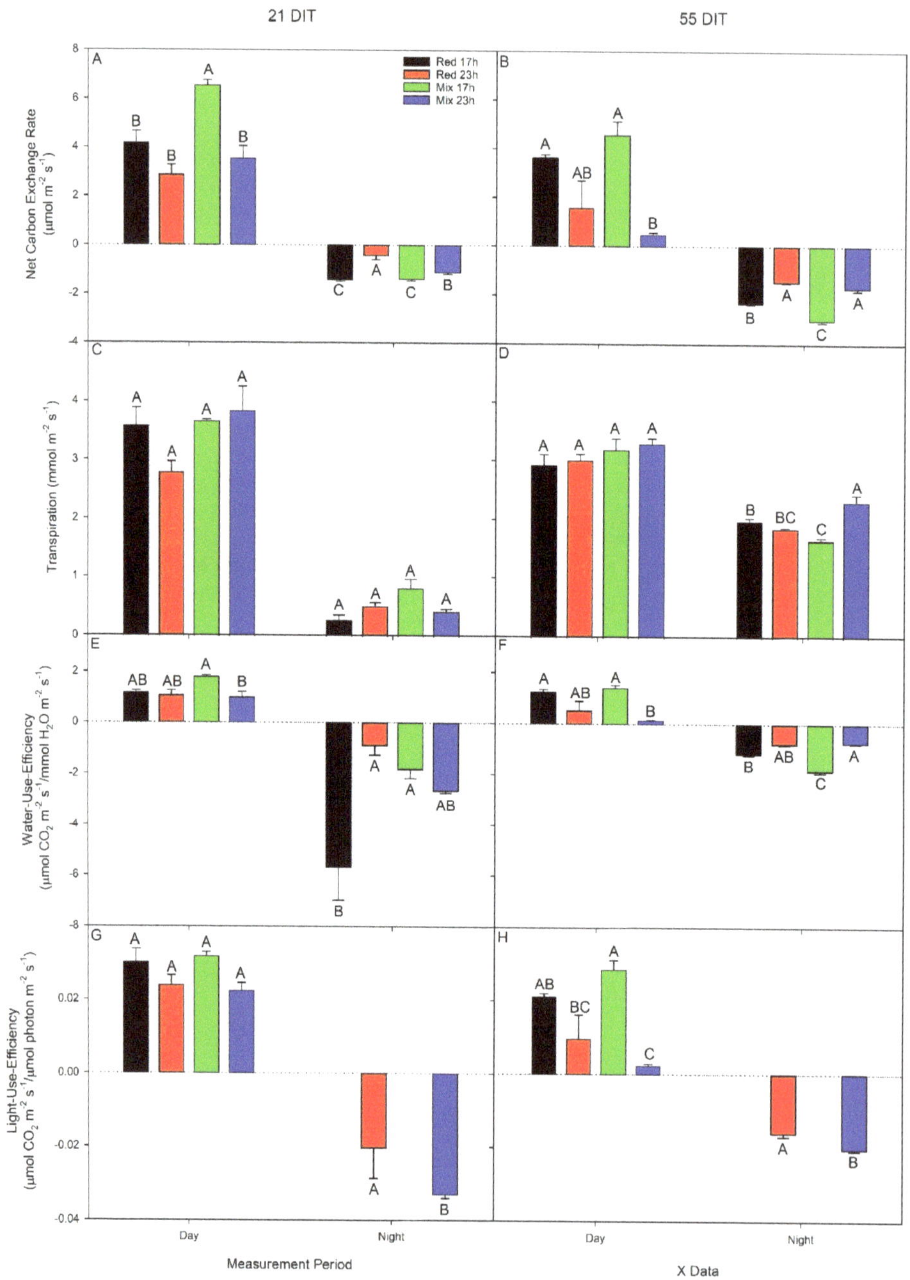

Figure 5. Net carbon exchange rate (NCER; panel **A**,**B**), transpiration (panel **C**,**D**), water use efficiency (panel **E**,**F**), light use efficiency (panel **G**,**H**) of the 5th leaf from TE tomato plants grown under either red 17 h, red 23 h, mix 17 h, or mix 23 h lighting treatments at 21 DIT (6 December 2018, panels **A**,**C**,**E**,**G**) or 55 DIT (9 January 2019, panels **B**,**D**,**F**,**H**) during the daytime and nighttime. Measurements were performed using a Li-COR 6400 fitted with a clear-top chamber on a cloudy day or night and thus represent the NCER driven mostly by the supplemental lighting. Error bars represent the standard error of the mean of n = 3. Letter groups (A, B, C) represent significant differences within a panel between the lighting treatments at a specific data collection period at $p < 0.05$.

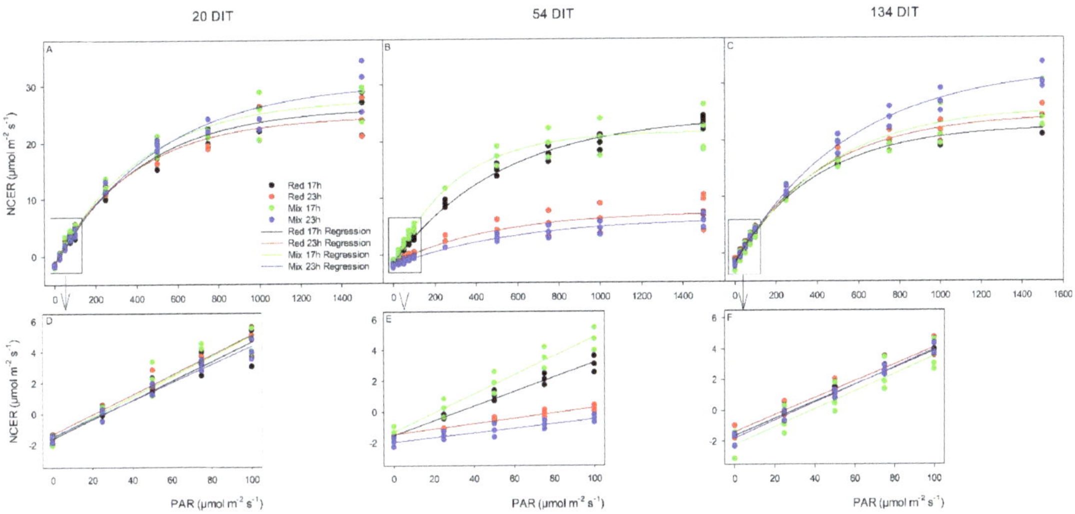

Figure 6. Photosynthetic light response curves from TE leaves grown under either red 17 h, red 23 h, mix 17 h, or mix 23 h lighting treatments at 20 DIT (5 December 2018, panel **A**), 54 DIT (8 January 2019, panel **B**), and 134 DIT (29 March 2019, panel **C**) as determined using a Li-COR 6400 with a red/blue standard Li-COR light source. Measurements were performed at a CO_2 concentration of 800µL L^{-1}, leaf temperature of 24 °C, and a relative humidity of 55–65%. Regression lines were fit to $y = y_o + a(1 - e^{(-b*x)})$ for each lighting treatment. Panels (**D–F**) are magnifications of 0–100 µmol m^{-2} s^{-1} PAR regions fit to the regression line y = mx + b.

Table 2. Summary of the major physiological traits as determined by leaf light response curves (Figure 6) from tomatoes grown under red 17 h, red 23 h, mix 17 h, and mix 23 h at 20 (5 December 2018), 54 (8 January 2019), and 134 DIT (29 March 2019). Respiration values were the averages of NCER when the light level was 0 µmol m^{-2} s^{-1}. The light compensation point (LCP) and quantum yield (QY) were calculated from a regression line (y = mx + b) fitted to the values between the PAR values of 0–100 µmol m^{-2} s^{-1}. The photosynthetic maximum (Pn_{max}) was calculated from $y = y_o + a(1 - e^{(-b*x)})$. Values ± the standard error of the mean are representative of n = 3. Within each parameter and time of measurement, letter groups (A, B, C, D) represent a statistical difference as determined by a two-way ANOVA with a Tukey–Kramer adjustment ($p < 0.05$).

Lighting Treatment	Respiration (µmol CO_2 m^{-2} s^{-1})	LCP (µmol m^{-2} s^{-1})	QY (µmol CO_2 m^{-2} s^{-1}/ µmol m^{-2} s^{-1})	Pn_{max} (µmol CO_2 m^{-2} s^{-1})
		20 DIT		
Red 17 h	−1.61 ± 0.24 A	26.23 ± 3.71 A	0.062 ± 0.006 A	28.34 ± 2.13 A
Red 23 h	−1.31 ± 0.13 A	20.79 ± 2.33 A	0.064 ± 0.005 A	25.97 ± 2.67 A
Mix 17 h	−1.52 ± 0.38 A	23.07 ± 5.64 A	0.065 ± 0.005 A	29.59 ± 2.16 A
Mix 23 h	−1.47 ± 0.22 A	25.12 ± 3.08 A	0.058 ± 0.004 A	29.68 ± 1.23 A
		54 DIT		
Red 17 h	−1.65 ± 0.05 B,C,D	32.48 ± 2.59 B	0.047 ± 0.003 B	25.80 ± 0.23 A
Red 23 h	−1.45 ± 0.04 A,B,C	86.14 ± 7.37 A	0.017 ± 0.002 C	8.88 ± 1.88 B
Mix 17 h	−1.38 ± 0.24 A,B	21.45 ± 5.13 B	0.062 ± 0.002 A	23.87 ± 2.76 A
Mix 23 h	−1.98 ± 0.18 C,D	131.60 ± 10.21 A	0.015 ± 0.0003 C	8.24 ± 1.10 B
		134 DIT		
Red 17 h	−1.71 ± 0.08 A	29.58 ± 0.93 A	0.055 ± 0.002 A	24.19 ± 1.24 B
Red 23 h	−1.45 ± 0.24 A	25.44 ± 4.78 A	0.055 ± 0.002 A	25.93 ± 1.03 AB
Mix 17 h	−2.31 ± 0.49 A	39.00 ± 9.12 A	0.058 ± 0.002 A	28.15 ± 2.36 AB
Mix 23 h	−1.83 ± 0.24 A	31.08 ± 3.30 A	0.058 ± 0.002 A	35.00 ± 1.36 A

At 20 DIT (5 December 2018), all parameters related to photosynthetic performance (V_{cmax} and J_{max}) were similar among all lighting treatments (Figure 7; Table 3). At 54 DIT (8 January 2019), leaves exposed to the mix 17 h lighting treatment produced the highest V_{cmax}, J_{max}, and Pn_{max} compared to the other lighting treatments (Table 3). During the same time period, leaves exposed to the red 17 h lighting treatment produced higher values of V_{cmax}, J_{max}, and Pn_{max} than leaves exposed to either 23 h lighting treatment (Table 3). At 134 DIT (29 March 2019), V_{cmax}, J_{max}, and Pn_{max} were similar among all lighting treatments, returning to levels observed at the beginning of the experiment (Table 3).

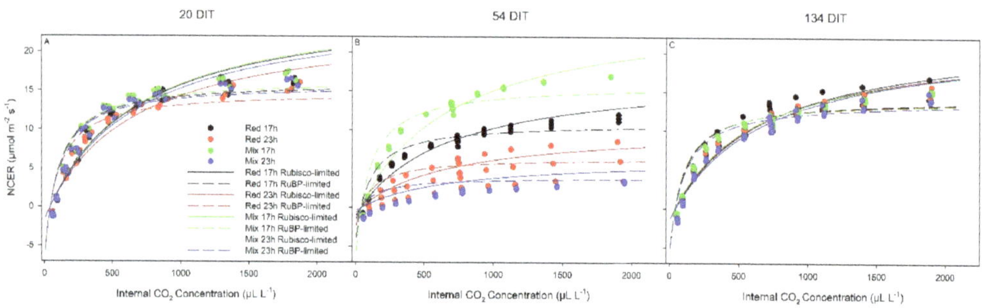

Figure 7. Photosynthetic CO_2 response curve from leaves grown under red 17 h, red 23 h, mix 17 h, and mix 23 h lighting treatments at 20 DIT (5 December 2018, panel **A**), 54 DIT (8 January 2019, panel **B**), and 134 DIT (29 March 2019, panel **C**). As determined using a Li-COR 6400 with a red/blue standard Li-COR light source. Measurements were performed at 300 µmol m^{-2} s^{-1} PAR, a temperature of 24 °C, and relative humidity of 55–65%. Rubisco- and RuBP-limited fit lines were determined using temperature corrections from [31,32].

Table 3. Summary of the major physiological traits as determined by leaf CO_2 response curves (Figure 7) from tomatoes grown under red 17 h, red 23 h, mix 17 h, and mix 23 h at 20 (5 December 2018), 54 (8 January 2019), and 134 DIT (29 March 2019). The maximum rate of Rubisco carboxylation (V_{cmax}) and the maximum rate of electron transport (J_{max}) were determined using equations from [31,32]. Pn_{max} was calculated from $y = y_o + a(1-e^{(-b*x)})$ and indicates the maximum rate of photosynthesis at a light level of 300 µmol m^{-2} s^{-1} at a saturating CO_2 level. Values ± the standard error of the mean are representative of n = 3. Within each parameter and time of measurement, letter groups (A, B, C) represent a statistical difference as determined by a two-way ANOVA with a Tukey–Kramer adjustment ($p < 0.05$).

Lighting Treatment	V_{cmax} (µmol CO_2 m^{-2} s^{-1})	J_{max} (µmol e^- m^{-2} s^{-1})	Pn_{max} (µmol CO_2 m^{-2} s^{-1})
		20 DIT	
Red 17 h	26.98 ± 1.24 A	100.16 ± 8.26 A	19.71 ± 0.82 A
Red 23 h	24.48 ± 0.66 A	85.61 ± 4.22 A	18.26 ± 0.53 A
Mix 17 h	27.19 ± 0.45 A	103.18 ± 3.04 A	20.13 ± 0.31 A
Mix 23 h	26.23 ± 0.29 A	95.80 ± 1.64 A	19.35 ± 0.20 A
		54 DIT	
Red 17 h	17.71 ± 0.46 B	55.61 ± 1.67 B	13.07 ± 0.31 B
Red 23 h	10.56 ± 2.11 C	30.84 ± 6.66 C	7.80 ± 1.52 C
Mix 17 h	26.05 ± 0.20 A	96.30 ± 1.58 A	19.08 ± 0.14 A
Mix 23 h	6.54 ± 0.14 C	18.26 ± 0.39 C	4.97 ± 0.12 C
		134 DIT	
Red 17 h	23.17 ± 1.38 A	80.98 ± 7.66 A	17.00 ± 1.03 A
Red 23 h	22.33 ± 0.51 A	79.51 ± 2.48 A	17.02 ± 0.39 A
Mix 17 h	21.76 ± 0.10 A	78.50 ± 0.64 A	16.72 ± 0.11 A
Mix 23 h	22.77 ± 0.90 A	77.00 ± 3.36 A	16.69 ± 0.53 A

3.2. Plant Parameters

Leaf length and width were measured in TE and TK plants at 31 DIT (16 December 2018) and resulted in similar values between two rootstocks for both metrics (Table 4). At 60 DIT (14 January 2019), the 5th and the 10th leaf from TE produced similar lengths and widths, respectively, under each lighting treatment (Table 4). However, the fifth leaves from TK exposed to the red 17 h lighting treatment were longer than those same leaves exposed to the mix 23 h treatment (Table 4), indicating that the rootstock can affect the response to the lighting treatments. This was not observed with leaf width, where all treatments produced the same width for the fifth leaves of TK plants. The interaction between lighting treatments and rootstocks demonstrates a competitive advantage during the use of extended photoperiods. This highlights the importance of selecting lighting treatments and rootstocks which combat the leaf area reduction typically observed under CL to improve light capture. Both the leaf length and leaf width of the 10th leaf at 60 DIT were similar in leaves exposed to all lighting treatments of TK plants (Table 4). At 139 DIT (3 April 2019), similar leaf lengths and widths were observed under all lighting treatments for both TE and TK (Table 4).

Table 4. Leaf parameters of plants grown under red 17 h, red 23 h, mix 17 h, and mix 23 h at 31 (16 December 2018), 60 (14 January 2019), and 139 DIT (3 April 2019). Values ± the standard error of the mean are representative of n = 6 for TE and TK plants under all lighting treatments. TE values are under white columns while TK values are under shaded columns. Different letter groups (A, B) represent a statistical difference between rootstocks and lighting treatments within a time point, leaf rank, and leaf parameter at $p < 0.05$.

		Leaf Length (cm)		**Leaf Width (cm)**	
Cultivar		TE	TK	TE	TK
Lighting Treatment	Leaf Rank				
		31 DIT			
Red 17 h	5th	46.33 ± 0.80 A	47.83 ± 1.25 A	36.00 ± 1.51 A	42.50 ± 2.17 A
Red 23 h	5th	46.17 ± 1.14 A	43.83 ± 1.89 A	39.83 ± 0.98 A	35.67 ± 3.69 A
Mix 17 h	5th	45.83 ± 0.95 A	47.67 ± 1.69 A	45.83 ± 0.95 A	38.67 ± 1.74 A
Mix 23 h	5th	42.67 ± 1.43 A	44.50 ± 1.18 A	42.67 ± 1.43 A	38.67 ± 2.08 A
		60 DIT			
Red 17 h	5th	38.67 ± 1.09 A	42.33 ± 1.56 A	36.67 ± 3.16 A	40.50 ± 1.34 A
	10th	46.50 ± 1.33 A	49.67 ± 1.61 A	61.50 ± 3.28 A	66.33 ± 4.24 A
Red 23 h	5th	41.67 ± 0.84 A	41.67 ± 1.05 AB	40.50 ± 1.12 A	41.67 ± 2.12 A
	10th	50.17 ± 0.83 A	48.50 ± 1.45 A	67.00 ± 2.80 A	61.50 ± 3.91 A
Mix 17 h	5th	42.17 ± 1.14 A	42.67 ± 0.95 AB	42.17 ± 2.84 A	41.17 ± 0.95 A
	10th	50.33 ± 1.26 A	51.50 ± 1.11 A	67.00 ± 3.09 A	67.50 ± 4.03 A
Mix 23 h	5th	39.67 ± 0.99 A	38.00 ± 0.73 B	39.67 ± 2.85 A	36.67 ± 1.99 A
	10th	47.33 ± 2.16 A	50.00 ± 1.73 A	62.33 ± 3.82 A	62.50 ± 3.54 A
		139 DIT			
Red 17 h	5th	39.17 ± 1.19 A	38.83 ± 0.91 A	35.00 ± 2.41 A	33.67 ± 1.87 A
	10th	45.17 ± 0.83 A	46.83 ± 1.19 A	56.67 ± 1.86 A	55.50 ± 2.49 A
	15th	47.17 ± 2.09 A	45.17 ± 1.01 A	56.50 ± 3.66 A	57.50 ± 3.02 A
Red 23 h	5th	37.50 ± 0.92 A	40.33 ± 1.02 A	35.17 ± 3.17 A	38.00 ± 2.28 A
	10th	44.17 ± 2.47 A	45.00 ± 1.75 A	56.00 ± 2.77 A	54.83 ± 2.41 A
	15th	46.17 ± 1.22 A	45.83 ± 1.38 A	56.83 ± 2.56 A	54.00 ± 1.29 A
Mix 17 h	5th	37.67 ± 1.31 A	39.33 ± 1.20 A	34.50 ± 1.34 A	35.17 ± 2.47 A
	10th	42.17 ± 2.26 A	43.33 ± 1.73 A	53.83 ± 2.76 A	49.67 ± 3.02 A
	15th	44.33 ± 1.41 A	49.17 ± 1.38 A	50.83 ± 4.07 A	55.50 ± 2.32 A
Mix 23 h	5th	40.33 ± 1.54 A	39.50 ± 0.89 A	38.17 ± 1.42 A	40.00 ± 1.73 A
	10th	46.67 ± 1.76 A	42.17 ± 1.19 A	59.17 ± 2.47 A	53.83 ± 2.50 A
	15th	43.33 ± 2.30 A	45.50 ± 1.71 A	58.17 ± 2.89 A	57.83 ± 2.23 A

At 31 DIT (16 December 2018), leaves from both rootstocks produced similar total chlorophyll and carotenoid concentrations when exposed to all lighting treatments (Table 5). At 60 DIT (14 January 2019), TE leaves grown under the mix 17 h lighting treatment produced the highest total chlorophyll and carotenoid concentrations of any lighting treatments at both the 5th and 10th leaf position (Table 5). TK leaves grown under either the red 17 h or mix 17 h lighting treatments produced higher total chlorophyll concentrations than leaves grown under the mix 23 h lighting treatment at the fifth leaf position (Table 5), indicating that mix 23 h caused leaf chlorophyll reduction when grafted onto 'Kaiser' (TK). At the 10th leaf position, leaves grown under the mix 17 h lighting treatment produced a higher total chlorophyll concentration than did leaves grown under the mix 23 h lighting treatment at the same position (Table 5). At the fifth leaf position of TK leaves at 60 DIT, those grown under the mix 23 h lighting treatment produced the lowest carotenoid concentration of any lighting treatment (Table 5). At 139 DIT (3 April 2019), leaves from TE plants produced similar total chlorophyll and carotenoid concentrations at each leaf position under each lighting treatment (Table 5). Leaves at the fifth leaf position from TK plants had higher total chlorophyll and carotenoid concentrations when grown under the mix 17 h treatment than either the red 17 h or red 23 h lighting treatment at 139 DIT (Table 5). Overall, plants grown under the mix 17 h lighting treatment produced high levels of pigments related to light capture and photosynthesis.

3.3. Fruit Yield

During the month of February, the average fruit weight (size, g fruit^{-1}) from TE plants was similar between red 17 h, red 23 h, and mix 17 h (Figure 8A). Notably, plants grown under the mix 23 h treatment produced a lower fruit weight than plants under the mix 17 h treatment during February ($p = 0.0046$; Figure 8A). During the month of March, the average fruit weights for TE plants grown under the red 17 h and mix 17 h lighting treatments were observed to be higher than plants grown under the mix 23 h lighting treatment ($p = 0.0022$; Figure 8A). In April, TE plants produced a similar average fruit weight among all lighting treatments (Figure 8A). During May, plants grown under both red 23 h and mix 23 h supplemental lighting treatments produced a higher average fruit weight than plants grown under the red 17 h treatment ($p = 0.0022$; Figure 8A).

In February, TK plants under the mix 17 h lighting treatment produced a higher average fruit weight than did plants grown under either red 23 h and mix 23 h treatments ($p = 0.003$; Figure 8B). During the month of March, plants grown under the red 17 h and mix 17 h lighting treatments produced similar average fruit weight (Figure 8B). Plants under both 17 h lighting treatments produced higher average fruit weights than did the 23 h lighting treatments ($p < 0.0001$; Figure 8B). Notably, plants grown under the mix 23 h lighting treatment produced the lowest fruit weight of any treatments (Figure 8B). During the month of April, plants grown under the red 17 h treatment produced higher average fruit weight than plants grown under the red 23 h lighting treatment ($p = 0.0036$; Figure 8B). Furthermore, plants grown under both 17 h lighting treatments produced higher fruit weights than did plants under the mix 23 h lighting treatment ($p = 0.0025$; Figure 8B). In May, the average fruit weight was similar among all lighting treatments (Figure 8B).

Throughout the harvest period (i.e., from 1 February 2019 to 22 May 2019), within a rootstock and lighting treatment, fruit production increased each month (Table 6) as more sunlight became available. In the month of February, within each rootstock, all lighting treatments produced similar values of fruit number per stem (Table 6). TE plants grown under both 23 h lighting treatments produced the lowest total fruit weight per stem during the month of February ($p = 0.0091$; Table 6). During the month of March, TE plants produced the same number of fruits per stem under all lighting treatments (Table 6). However, TK plants grown under both 23 h lighting treatments produced low numbers of fruits per stem, with plants grown under the mix 23 h treatment being the lowest of all treatments ($p = 0.0006$; Table 6), indicating that 23 h lighting caused more damage with TK than TE. TE plants grown under the mix 17 h lighting treatment produced higher fruit weight per stem than plants grown under the mix 23 h lighting treatment ($p = 0.0018$; Table 6). TK plants grown under both 17 h lighting treatments had higher fruit weight per stem than both 23 h lighting treatments, with plants grown under the mix 23 h lighting treatment having the lowest overall ($p < 0.0001$; Table 6). Importantly, during the month of March, both TE and TK plants grown under the mix 23 h lighting treatment had the lowest total fruit weight per stem (Table 6).

Table 5. Pigment analysis of plants grown under red 17 h, red 23 h, mix 17 h, and mix 23 h at 31 (16 December 2018), 60 (14 January 2019), and 139 DIT (3 April 2019). Values ± the standard error of the mean are representative of n = 6 for TE and TK plants under all lighting treatments. TE values are under white columns while TK values are under shaded columns. Different letter groups (A, B) represent a statistical difference between lighting treatments within a cultivar, time point, leaf rank, and pigment at $p < 0.05$.

		Chlorophyll $a + b$ (μg cm^{-2})		Carotenoids (μg cm^{-2})	
Cultivar		**TE**	**TK**	**TE**	**TK**
Lighting Treatment	Leaf Rank				
		31 DIT			
Red 17 h	5th	39.98 ± 0.90 ^{A}	38.05 ± 2.79 ^{A}	7.93 ± 0.15 ^{A}	7.59 ± 0.47 ^{A}
Red 23 h	5th	38.19 ± 1.68 ^{A}	36.62 ± 1.21 ^{A}	7.63 ± 0.28 ^{A}	7.37 ± 0.20 ^{A}
Mix 17 h	5th	38.61 ± 1.72 ^{A}	38.80 ± 1.74 ^{A}	7.70 ± 0.29 ^{A}	7.73 ± 0.29 ^{A}
Mix 23 h	5th	40.11 ± 1.81 ^{A}	40.44 ± 1.13 ^{A}	7.94 ± 0.39 ^{A}	8.00 ± 0.19 ^{A}
		60 DIT			
Red 17 h	5th	26.77 ± 1.36 ^{B}	31.47 ± 1.28 ^{A}	5.68 ± 0.24 ^{B}	6.50 ± 0.22 ^{A}
	10th	34.30 ± 1.42 ^{B}	36.61 ± 1.02 ^{AB}	6.98 ± 0.24 ^{B}	7.37 ± 0.29 ^{A}
Red 23 h	5th	29.31 ± 2.31 ^{B}	29.91 ± 1.91 ^{AB}	6.21 ± 0.51 ^{B}	6.22 ± 0.34 ^{AB}
	10th	34.49 ± 2.17 ^{B}	33.75 ± 0.96 ^{AB}	7.00 ± 0.37 ^{B}	6.89 ± 0.21 ^{A}
Mix 17 h	5th	39.47 ± 3.49 ^{A}	38.21 ± 2.16 ^{A}	8.45 ± 0.77 ^{A}	7.63 ± 0.36 ^{A}
	10th	41.46 ± 2.02 ^{A}	38.62 ± 1.36 ^{A}	8.17 ± 0.33 ^{A}	8.26 ± 0.39 ^{A}
Mix 23 h	5th	28.47 ± 2.73 ^{B}	22.75 ± 2.42 ^{B}	5.98 ± 0.60 ^{B}	4.95 ± 0.44 ^{B}
	10th	34.63 ± 1.07 ^{AB}	32.36 ± 1.34 ^{B}	7.04 ± 0.18 ^{B}	6.88 ± 0.38 ^{A}
		139 DIT			
Red 17 h	5th	40.94 ± 1.89 ^{A}	41.94 ± 1.25 ^{B}	8.08 ± 0.31 ^{A}	8.25 ± 0.20 ^{B}
	10th	43.86 ± 1.54 ^{A}	44.16 ± 1.25 ^{A}	8.56 ± 0.25 ^{A}	8.61 ± 0.20 ^{A}
	15th	34.10 ± 2.20 ^{A}	37.82 ± 2.02 ^{A}	6.94 ± 0.37 ^{A}	7.57 ± 0.34 ^{A}
Red 23 h	5th	42.59 ± 0.54 ^{A}	42.78 ± 0.86 ^{B}	8.36 ± 0.09 ^{A}	8.39 ± 0.14 ^{B}
	10th	44.80 ± 1.76 ^{A}	45.43 ± 0.94 ^{A}	8.71 ± 0.29 ^{A}	8.82 ± 0.15 ^{A}
	15th	37.11 ± 1.16 ^{A}	32.57 ± 1.55 ^{A}	7.45 ± 0.19 ^{A}	6.68 ± 0.26 ^{A}
Mix 17 h	5th	43.26 ± 1.17 ^{A}	46.36 ± 1.13 ^{A}	8.46 ± 0.19 ^{A}	8.97 ± 0.18 ^{A}
	10th	45.19 ± 2.26 ^{A}	44.43 ± 1.09 ^{A}	8.77 ± 0.36 ^{A}	8.65 ± 0.18 ^{A}
	15th	35.89 ± 3.57 ^{A}	38.33 ± 1.94 ^{A}	7.22 ± 0.61 ^{A}	7.65 ± 0.32 ^{A}
Mix 23 h	5th	45.65 ± 2.23 ^{A}	44.30 ± 1.31 ^{AB}	8.85 ± 0.36 ^{A}	8.63 ± 0.21 ^{AB}
	10th	42.53 ± 1.06 ^{A}	45.38 ± 0.87 ^{A}	8.35 ± 0.17 ^{A}	8.81 ± 0.14 ^{A}
	15th	34.38 ± 2.25 ^{A}	31.98 ± 1.89 ^{A}	6.98 ± 0.38 ^{A}	6.58 ± 0.33 ^{A}

During the month of April, both TE plants ($p = 0.0019$) and TK plants ($p = 0.0003$) grown under the mix 23 h lighting treatment produced the lowest total fruit weight per stem (Table 6). However, total fruit weight per stem from TE under the red 23 h was similar to both 17 h lighting treatments while the total fruit weight per stem from TK under the red 23 h lighting treatment was still lower than the 17 h red lighting treatment. The above results indicate that both light spectra and rootstocks played a role in photoperiod-related injury as an improved yield recovery time was observed from TE compared to TK. During the month of May, both TE and TK plants produced similar values of both total fruit number per stem and fruit weight per stem, indicating the full recovery of fruit production under the 23 h lighting treatments (Table 6).

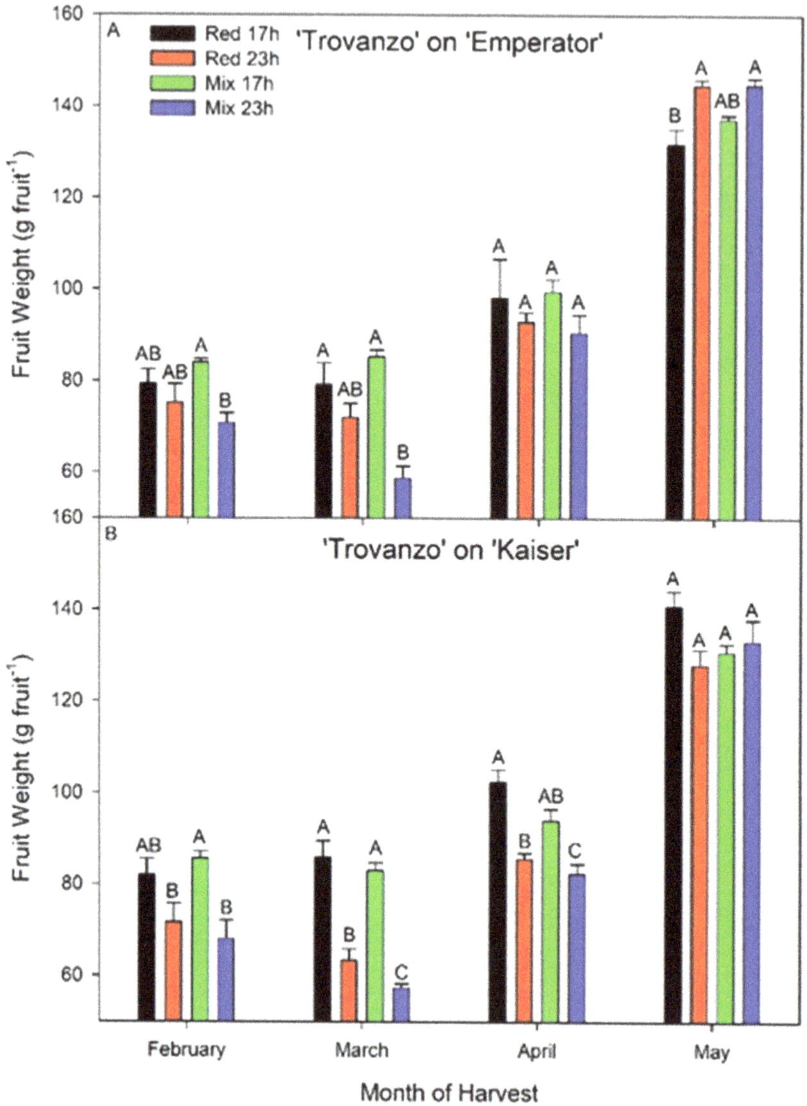

Figure 8. Average fruit weight (size) for cv. 'Trovanzo' grafted onto 'Emperator' (TE; panel **A**) or onto 'Kaiser' (TK; panel **B**), grown under red 17 h, red 23 h, mix 17 h, and mix 23 h lighting treatments. Monthly fruit number includes the 1st of each month to the last day of each respective month. Values represent the mean ± the standard error of the mean, where n = 4 for TE and TK. Of note, values representing May yield include a strip harvest on 22 May 2019. Within the month and cultivar, letter groups (A, B, C) represent a statistical difference in fruit weight as determined by a two-way ANOVA with a Tukey–Kramer adjustment ($p < 0.05$).

For both TE ($p = 0.0029$) and TK ($p = 0.0001$) plants, growth under the mix 23 h lighting treatment was associated with the lowest cumulative number of fruit per stem throughout the entire production period (Table 6). Similarly, for TE ($p = 0.0017$) and TK ($p < 0.0001$), plants grown under the mix 23 h lighting treatment produced the lowest cumulative fruit weight per stem (Table 6). It should be noted that TE plants grown under the red 23 h lighting treatment produced similarly high cumulative fruit number and weight per stem to both 17 h treatments, whereas TK plants under the red 23 h still produced a lower weight per stem than the 17 h treatments (Table 6). This again shows that TK plants tended to be more affected by the extended photoperiod than TE plants, indicating an effect of rootstock material on photoperiod-related injury. Taken together, these results suggest that the use of a broad (mix) spectrum lighting treatment during an extended photoperiod tends to have a more negative effect on fruit yield than a monochromatic red spectrum.

Table 6. Monthly yield analysis for cv. 'Trovanzo' grafted onto cv. 'Emperator' (TE) or onto cv. 'Kaiser' (TK), grown under red 17 h, red 23 h, mix 17 h, and mix 23 h lighting treatments. Values represent the mean ± the standard error of the mean, where n = 4 for TE and TK. TE values are under white columns while TK values are under shaded columns. Of note, values representing May yield include a strip harvest on 22 May 2019. Total yield is the summation of the harvest period. Within each yield parameter, month (or total), and rootstock, letter groups (A, B, C) represent a statistical difference as determined by a two-way ANOVA with a Tukey–Kramer adjustment ($p < 0.05$). p values at the bottom of the table are representative of the cumulative data after a two-way ANOVA.

	Fruit per Stem		Fruit Weight per Stem (kg stem^{-1})	
Cultivar	**TE**	**TK**	**TE**	**TK**
		February		
Lighting Treatment				
Red 17 h	9 ± 1 ^{A}	8 ± 2 ^{A}	0.68 ± 0.06 ^{AB}	0.63 ± 0.07 ^{A}
Red 23 h	7 ± 2 ^{A}	10 ± 1 ^{A}	0.52 ± 0.10 ^{B}	0.67 ± 0.05 ^{A}
Mix 17 h	11 ± 2 ^{A}	8 ± 2 ^{A}	0.88 ± 0.12 ^{A}	0.65 ± 0.15 ^{A}
Mix 23 h	7 ± 1 ^{A}	8 ± 1 ^{A}	0.50 ± 0.04 ^{B}	0.52 ± 0.02 ^{A}
		March		
Red 17 h	21 ± 2 ^{A}	22 ± 2 ^{A}	1.66 ± 0.20 ^{AB}	1.90 ± 0.14 ^{A}
Red 23 h	16 ± 2 ^{A}	17 ± 1 ^{BC}	1.13 ± 0.16 ^{AB}	1.07 ± 0.08 ^{B}
Mix 17 h	20 ± 2 ^{A}	21 ± 2 ^{AB}	1.70 ± 0.10 ^{A}	1.72 ± 0.12 ^{A}
Mix 23 h	17 ± 2 ^{A}	12 ± 1 ^{C}	0.97 ± 0.11 ^{B}	0.70 ± 0.03 ^{C}
		April		
Red 17 h	20 ± 1 ^{A}	22 ± 1 ^{A}	1.97 ± 0.26 ^{A}	2.23 ± 0.13 ^{A}
Red 23 h	22 ± 2 ^{A}	21 ± 1 ^{A}	2.06 ± 0.14 ^{A}	1.77 ± 0.07 ^{B}
Mix 17 h	19 ± 1 ^{A}	22 ± 1 ^{A}	1.93 ± 0.15 ^{A}	2.05 ± 0.10 ^{AB}
Mix 23 h	16 ± 1 ^{B}	12 ± 1 ^{B}	1.42 ± 0.06 ^{A}	0.95 ± 0.05 ^{C}
		May		
Red 17 h	26 ± 1 ^{A}	24 ± 1 ^{A}	3.47 ± 0.09 ^{A}	3.32 ± 0.17 ^{A}
Red 23 h	24 ± 1 ^{A}	25 ± 2 ^{A}	3.50 ± 0.10 ^{A}	3.22 ± 0.10 ^{A}
Mix 17 h	25 ± 1 ^{A}	27 ± 1 ^{A}	3.47 ± 0.10 ^{A}	3.45 ± 0.13 ^{A}
Mix 23 h	25 ± 1 ^{A}	26 ± 1 ^{A}	3.66 ± 0.14 ^{A}	3.37 ± 0.21 ^{A}
		Cumulative		
Red 17 h	74 ± 1 ^{A}	75 ± 3 ^{A}	7.29 ± 0.17 ^{AB}	8.08 ± 0.36 ^{A}
Red 23 h	72 ± 1 ^{A}	72 ± 1 ^{A}	7.51 ± 0.08 ^{A}	6.75 ± 0.11 ^{B}
Mix 17 h	73 ± 1 ^{A}	77 ± 2 ^{A}	7.73 ± 0.15 ^{A}	7.87 ± 0.20 ^{A}
Mix 23 h	64 ± 1 ^{B}	57 ± 2 ^{B}	6.54 ± 0.16 ^{B}	5.54 ± 0.20 ^{C}

4. Discussion

4.1. Effect of Photoperiod Extension on Leaf Physiology

Extending the natural solar photoperiod via the implementation of supplemental lighting has proven to be a beneficial lighting strategy in terms of plant growth and yield during greenhouse production of tomatoes [1]. However, decades of research have shown that photoperiod extension beyond a critical length causes photoperiod-related injury characterized by leaf chlorosis [9,34–36]. Indeed, this negative result observed under photoperiods greater than 17 h has nullified any theoretical advantages they may have [8].

The increased use of wavelength-specific LED fixtures during controlled environment plant production has the potential to play a pivotal role in reducing injury related to photoperiod extension. During CL, Matsuda et al. [14] indicated that young tomato plants (23 days after planting and 13 days into the treatment) grown under white light during the day and either red or orange light during the subjective nighttime period produced a lower degree of injury than plants grown under white light during the day and either white or blue light at night. A recent study by Lanoue et al. [7] produced injury-free tomatoes during a 6-month production period grown under CL by using an alternating spectrum of red during the day and blue during the subjective nighttime period. However, a reduction in the light level also occurred during the spectral shift from 200 μmol m^{-2} s^{-1} of red light to 50 μmol m^{-2} s^{-1} of blue light (close to light compensation point), which may have confounded any effects of the spectral shift [9]. In this study, the light intensity was maintained at the same level for the two lighting treatments, allowing for the comparison of spectral compositions/quality without any confounding effects.

During the initial stages of plant growth, plants grown under all lighting treatments were observed to be injury-free (Figures 3A and 4). However, as determined by measurements between 54 and 62 DIT (8–16 January 2019), plants grown under either red 23 h or mix 23 h treatments developed photoperiod-related injury, characterized by leaf chlorosis, consistent with previous research (Figures 3B and 4) [6,35–37]. At this period (around 2–4 weeks before the starting of fruit harvest—78 DIT; it usually takes 2 weeks for a cluster of fruit from mature green (full size) to reach harvesting stage—4 red fruits out of the 5 fruits in each cluster), the clusters of fruits were in the fastest growing stage (fastest increase in size). The plants had heavy fruit sink and weak leaf source due to low natural sunlight (Figure 2, 8–16 January 2019). During this period of measurements, parameters related to photosynthesis, such as LCP, QY, Pn_{max}, V_{cmax}, and J_{max}, were all reduced from leaves that were exposed to both 23 h lighting treatments compared to those exposed to either 17 h lighting treatment (Tables 2 and 3). The downregulation in parameters related to photosynthesis may be related to either excess carbohydrate accumulation or a downregulation of *CAB-13*, affecting PSII function [9,38]. Interestingly, during this period of production, which was characterized by photoperiod-related injury and a downregulation in parameters related to photosynthesis, both stomatal conductance and transpiration rates were unaffected (Figure 5). It should be noted that Lanoue et al. [7] did not observe issues with carbon metabolism during CL. Therefore, the major driver of injury during extended photoperiods cannot simply be attributed to improper stomatal function or overaccumulation of carbohydrates. Furthermore, as a decrease in the chlorophyll content was observed (Table 5), photoperiod-related injury seems to be more closely related to light capture.

From 54 DIT (8 January 2019) onwards, plants and leaves grown under both 23 h treatments were observed to recover from the photoperiod-related injury (Figures 3C and 6C). The photosynthetic capacity and F_v/F_m values from leaves under 23 h treatments returned to levels similar to the 17 h treatments, and fruit yield during the months of April and May also increased (Table 6). However, no changes to the experimental settings had occurred to invoke such changes. What did change was the peak intensity of natural solar radiation (Figure 2) and the natural photoperiod from January to April. In fact, from the period of peak photoperiod injury (62 DIT, 16 January 2019) to when photoperiod injury was completely alleviated (138 DIT, 2 April 2019), the solar radiation nearly doubled in average daily intensity from 88 W m^{-2} to 165 W m^{-2} (Figure 2). Moreover, during this period, the plant growth was more balanced—less fruit load due to fruit harvesting and more leaf growth due to strong sunlight. The changes in both solar radiation and plant growth balance (generative vs. vegetative or source vs. sink) may, in some way, account for the observed recovery of plants under both 23 h lighting treatments. Further specifically designed studies will be needed to separate the light intensity effect from that of fruit load/plant growth stage. Potentially, as the natural light intensity increased later in the study, the strong exogenous signaling (i.e., high peak light intensity) was able to override the endogenous signal causing injury [34,37,39]. However, an in-depth look at this hypothesis was beyond the scope of our study.

4.2. Effect of Photoperiod Extension on Fruit Production

Photoperiod-related leaf injury characterized by chlorosis and a downregulation of photosynthesis can drastically reduce fruit yield due to a decrease in carbon assimilation essential for plant growth. However, if photoperiod-related injury can be avoided, an increase in plant growth and yield is theoretically possible [8]. The use of wavelength-specific LEDs has shown promise in reducing photoperiod-related injury during the vegetative growth stage [14]; however, little is known about how the use of wavelength-specific LEDs will affect the yield of high-wire-fruiting vegetables during extended photoperiods [7,40]. Between the months of February and April, those which followed the photoperiod-related injury in leaves, the average fruit weight (size, g fruit^{-1}) was observed to decrease in both 23 h lighting treatments (Figures 5 and 8). These data indicate a clear correlation between the onset and persistence of injury and the reduction in fruit production. Notably, by April, all yield parameters for both 23 h lighting treatments were similar to those of the 17 h lighting treatments (Figure 8 and Table 6), again coinciding with increased natural solar radiation.

Throughout the harvest period in both TE and TK, those under the red 23 h lighting treatment tended to show less yield reduction than plants under the mix 23 h treatment when compared to the 17 h lighting treatments (Figure 8 and Table 6). In fact, our cumulative yield analysis indicates that growth under the red 23 h lighting treatment of both TE and TK plants produced higher yield than did plants grown under the mix 23 h lighting treatment (Table 6). Furthermore, TE and TK plants grown under the mix 23 h lighting treatment produced fewer fruits per stem than all other lighting treatments (Table 6). These results indicate that the use of supplemental red light during photoperiod extension may be more beneficial than a mixed lighting treatment including appreciable amounts of blue light (Figure 1).

Generally, under a traditional 16 h photoperiod, the addition of some blue light (6–12%) to a predominantly red supplemental lighting spectrum has shown increases in biomass and total fruit number in greenhouse tomatoes [24]. For this reason, in commercial production practices, the addition of blue light is typically thought of as generative light, while increasing the red light component is thought to promote vegetative growth (i.e., vegetative light). In this way, growers can steer the plant towards a more generative or vegetative growing pattern depending on current/future environmental and plant conditions. Administering light during an extended, nearly continuous photoperiod can also be described as a generative light environment as there is constant photon energy pressuring the plant during the subjective nighttime period—in effect, forcing growth. Thus, we hypothesize that the interaction of a mixed spectrum with an extended photoperiod may cause imbalances in plant growth patterns, while using a pure red, more vegetative spectrum during an extended photoperiod may allow for proper vegetative vs. generative homeostasis. We believe that because of the use of the more vegetative red 23 h lighting treatment, the plants grown under this treatment displayed better yield performance than did the more generative mix 23 h treatment (Figure 8 and Table 6). Therefore, the interaction between photoperiod and light spectrum needs to be taken into account during tomato production.

4.3. Interaction between Rootstocks and Photoperiod Length

While wild tomato species are generally tolerant to extended photoperiods, domesticated tomatoes have been determined to be sensitive to extended photoperiods (i.e., photoperiod-related leaf injury has been observed) [9]. Our study involved the use of one domesticated tomato cultivar as the scion ('Trovanzo') and two domesticated tomato cultivars as rootstocks ('Emperator' and 'Kasier'). Interestingly, the interaction between light spectra and photoperiod impacted the severity of injury—and, ultimately, yield—differently between rootstocks. The TE ('Trovanzo' grafted on 'Emperator') plants were observed to be more tolerant to photoperiod extension than the TK ('Trovanzo' grafted onto 'Kaiser') plants (Figure 8 and Table 6). It should also be noted that TE plants had statistically similar yield under the red 23 h lighting treatment as both 17 h lighting treatments; this was not the case for TK plants (Figure 8 and Table 6).

Rootstock material can have a large impact on the growing patterns of the plant as a whole [19–21]. Rahmatian et al. [20] showed that overall plant biomass and yield could be increased simply by grafting the same scion onto different rootstock materials, demonstrating the impact that proper rootstock selection can have. In muskmelon and orange trees, the use of different rootstocks also led to differences in fruit production and carbohydrate status in the fruit [41,42]. These studies indicate that the source vs. sink balance can also be impacted by the rootstock selection [20,21,41,42]. This then can have an impact on photoperiod-related injury due to the improper balance between overall vegetative and generative plant growth. In our study, we used two different rootstocks that have previously demonstrated generative ('Kasier') and vegetative ('Emperator') growing patterns. TE plants generally performed better in terms of fruit yield during extended photoperiods than TK plants, indicating that rootstocks were interacting with the light environment in some fashion. Velez-Ramirez et al. [43] showed that when a CL-sensitive scion was grafted onto a CL-tolerant rootstock, the CL-sensitive scion was less affected by CL. However, in Velez-Ramirez et al.'s study [43], the CL-tolerant rootstock was also allowed to grow a shoot along with the CL-sensitive scion. This allowed the CL-tolerant accession to have leaves exposed to the CL and thus the authors proposed that a "transferable injurious substance" or a signaling molecule could be transferred between the CL-tolerant and CL-sensitive accession [43]. In our study, no vegetative shoots from the rootstock were allowed to grow; thus, the rootstocks were not able to interact with the light environment directly. In a recent study, Paponov et al. [44] observed the modulation of phytohormones in the root zone due to varied canopy light environments. Thus, during growth under extended photoperiods, it is important to take into account the rootstocks being used and not simply the scions as there may be potential communication via hormone signaling involved in the regulation of photoperiod-related injury [45].

5. Conclusions

In conclusion, our data indicate that tomatoes grown under extended photoperiods under sole red supplemental light have improved fruit yield compared to those grown under a mixed lighting treatment. In fact, growth under a 23 h photoperiod with red supplemental light resulted in similar cumulative yield parameters to plants grown under a red or mixed light spectrum with 17 h lighting. By utilizing a low intensity and long photoperiod with energy-efficient red LEDs, a decrease in capital fixture cost and electrical costs can be realized. Under both red and mixed 23 h lighting treatments, physiological parameters indicated the presence of photoperiod-related injury at 54 DIT (8 January 2019) during the measurement period. However, during the subsequent production period, photoperiod-related injury was alleviated. Because no experimental parameters were altered, we propose that the increase in peak solar intensity or overall DLI from January to April could play a role in alleviating injury, but further study will be needed to confirm this hypothesis. Interestingly, the rootstocks used were observed to have an interaction with the light spectra and photoperiod length, which led to improved yield from TE plants compared to TK plants. These data suggest that both rootstocks and light spectral compositions/quality can impact photoperiod-related injury. With red light and proper rootstock, the negative yield impact from long photoperiod lighting can be eliminated.

Author Contributions: Conceptualization, J.L., B.G. and X.H.; methodology, J.L., B.G. and X.H.; validation, J.L., A.T., C.L. and J.Z.; formal analysis, J.L.; data curation, J.L., A.T., C.L. and J.Z.; writing—original draft preparation, J.L.; writing—review and editing, J.L., B.G. and X.H.; funding acquisition, X.H. All authors have read and agreed to the published version of the manuscript.

Funding: This research was funded by the Foundational Science Program of Agriculture and Agri-Food Canada, grant number J-002228.001.04, to X.H.

Institutional Review Board Statement: Not applicable.

Informed Consent Statement: Not applicable.

Data Availability Statement: All data reported can be found within the manuscript.

Acknowledgments: LumiGrow Inc. (Emeryville, CA 96408, USA) provided the Smart LED lighting system and helped to set up the lighting treatments.

Conflicts of Interest: The authors declare no conflict of interest.

Abbreviations

CAB-13: type III light harvesting chlorophyll a/b binding protein 13; CL: Continuous lighting; DIT: Days into the treatment; DLI: Daily light integral; F_o: Minimum fluorescence in a dark-adapted state; F_m: Maximum fluorescence in a dark-adapted state; F_v: Variable fluorescence in a dark-adapted state; F_v/F_m: Maximum quantum efficiency of PSII; J_{max}: Maximum rate of electron transport; LCP: Light compensation point; LED: Light-emitting diode; LUE: Light use efficiency; N: Photon flux; NCER: Net carbon exchange rate; PHY A: Phytochrome A; Pn_{max}: Photosynthetic maximum; PPE: Photosynthetic photon efficacy; PSI: Photosystem I; PSII: Photosystem II; PSS: Phytochrome photostationary state; QY; Quantum yield; R:Fr: Red to far-red ratio; TE: 'Trovanzo' grafted onto 'Emperator'; TK: 'Trovanzo' grafted onto 'Kaiser'; V_{cmax}: Maximum rate of Rubisco carboxylation; WUE: Water use efficiency; σ_r: Photochemical cross-section of phytochrome in the red absorbing state; σ_{Fr}: Photochemical cross-section of phytochrome in the red absorbing state.

References

1. McAvoy, R.J.; Janes, H.W.; Godfriaux, B.L.; Secks, M.; Duchai, D.; Wittman, W.K. The effect of total available photosynthetic photon flux on single truss tomato growth and production. *J. Hortic. Sci.* **1989**, *64*, 331–338. [CrossRef]
2. IESO. Independent Electricity System Operator. Available online: www.ieso.ca/power-data (accessed on 27 January 2020).
3. Hao, X.; Guo, X.; Lanoue, J.; Zhang, Y.; Cao, R.; Zheng, J.; Little, C.; Leonardos, D.; Kholsa, S.; Grodzinski, B.; et al. A review on smart application of supplemental lighting in greenhouse fruiting vegetable production. *Acta Hortic.* **2018**, *1227*, 499–506. [CrossRef]
4. Demers, D.A.; Gosselin, A. Growing greenhouse tomato and sweet pepper under supplemental lighting: Optimal photoperiod, negative effects of long photoperiod and their causes. *Acta Hortic.* **2002**, *580*, 83–88. [CrossRef]
5. Velez-Ramierez, A.I.; van Ieperen, W.; Vreugdenhil, D.; Millenaar, F.F. Plants under continuous light. *Trends Plant Sci.* **2011**, *16*, 310–318. [CrossRef] [PubMed]
6. Haque, M.S.; De Sousa, A.; Soares, C.; Kjaer, K.H.; Fidalgo, F.; Rosenqvist, E.; Ottosen, C.-O. Temperature Variation under Continuous Light Restores Tomato Leaf Photosynthesis and Maintains the Diurnal Pattern in Stomatal Conductance. *Front. Plant Sci.* **2017**, *8*, 1602. [CrossRef]
7. Lanoue, J.; Zheng, J.; Little, C.; Thibodeau, A.; Grodzinski, B.; Hao, X. Alternating Red and Blue Light-Emitting Diodes Allows for Injury-Free Tomato Production with Continuous Lighting. *Front. Plant Sci.* **2019**, *10*, 1114. [CrossRef] [PubMed]
8. Velez-Ramirez, A.; Van Ieperen, W.; Vreugdenhil, D.; Millenaar, F. Continuous light as a way to increase greenhouse tomato production: Expected challenges. *Acta Hortic.* **2012**, *956*, 51–57. [CrossRef]
9. Velez-Ramirez, A.I.; van Ieperen, W.; Vreugdenhil, D.; van Poppel, P.M.J.A.; Heuvelink, E.; Millenaar, F.F. A single locus confers tolerance to continuous light and allows substantial yield increase in tomato. *Nat. Commun.* **2014**, *5*, 4549. [CrossRef]
10. Kouřil, R.; Dekker, J.P.; Boekema, E.J. Supramolecular organization of photosystem II in green plants. *Biochim. Et Biophys. Acta Bioenerg.* **2012**, *1817*, 2–12. [CrossRef]
11. Heuvelink, E.; Dorais, M. Crop growth and yield. In *Tomatoes*; Heuvelink, E., Ed.; CABI Publishing: Wallingford, UK, 2005; pp. 85–144.
12. LaNoue, J.; Leonardos, E.D.; Grodzinski, B. Effects of Light Quality and Intensity on Diurnal Patterns and Rates of Photo-Assimilate Translocation and Transpiration in Tomato Leaves. *Front. Plant Sci.* **2018**, *9*, 756. [CrossRef] [PubMed]
13. Petrillo, E.; Herz, M.A.G.; Barta, A.; Kalyna, M.; Kornblihtt, A.R. Let there be light: Regulation of gene expression in plants. *RNA Biol.* **2014**, *11*, 1215–1220. [CrossRef]
14. Matsuda, R.; Yamano, T.; Murakami, K.; Fujiwara, K. Effects of spectral distribution and photosynthetic photon flux density for overnight LED light irradiation on tomato seedling growth and leaf injury. *Sci. Hortic.* **2016**, *198*, 363–369. [CrossRef]
15. Velez-Ramirez, A.I.; Vreugdenhil, D.; Millenaar, F.F.; Van Ieperen, W. Phytochrome A Protects Tomato Plants From Injuries Induced by Continuous Light. *Front. Plant Sci.* **2019**, *10*, 19. [CrossRef]

16. Hao, X.; Zhang, Y.; Guo, X.; Little, C.; Zheng, J. Dynamic temperature control strategy with a temperature drop improves responses of greenhouse tomatoes and sweet peppers to long photoperiods of supplemental lighting and saves energy. *Acta Hortic.* **2018**, *1227*, 291–298. [CrossRef]
17. Hogewoning, S.; Trouwborst, G.; Maljaars, H.; Poorter, H.; Van Ieperen, W.; Harbinson, J. Blue light dose-responses of leaf photosynthesis, morphology, and chemical composition of Cucumis sativus grown under different combinations of red and blue light. *J. Exp. Bot.* **2010**, *61*, 3107–3117. [CrossRef]
18. Hernández, R.; Kubota, C. Physiological responses of cucumber seedlings under different blue and red photon flux ratios using LEDs. *Environ. Exp. Bot.* **2016**, *121*, 66–74. [CrossRef]
19. Rahmatian, A.; Delshad, M.; Salehi, R. Effect of grafting on growth, yield and fruit quality of single and double stemmed tomato plants grown hydroponically. *Hortic. Environ. Biotechnol.* **2014**, *55*, 115–119. [CrossRef]
20. Khah, E.M.; Kakava, E.; Mavromatis, A.; Chachalis, D.; Goulas, C. Effect of grafting on growth and yield of tomato (*Lycopersicon esculentum* Mill.) in greenhouse and open-field. *J. Appl. Hortic.* **2006**, *8*, 3–7. [CrossRef]
21. Kyriacou, M.C.; Rouphael, Y.; Colla, G.; Zrenner, R.; Schwarz, D. Vegetable grafting: The implications of a growing agronomic imperative for vegetatble fruit quality and nutritive value. *Front. Plant Sci.* **2017**, *8*, 741. [CrossRef] [PubMed]
22. Ontario Ministry of Agriculture. *Food and Rural Affairs. Growing Greenhouse Vegetables in Ontario*; Queen's Printer: Toronto, ON, Canada, 2010.
23. Kusuma, P.; Pattison, P.M.; Bugbee, B. From physics to fixtures to food: Current and potential LED efficacy. *Hortic. Res.* **2020**, *7*, 1–9. [CrossRef] [PubMed]
24. Kaiser, E.; Ouzounis, T.; Giday, H.; Schipper, R.; Heuvelink, E.; Marcelis, L.F.M. Adding Blue to Red Supplemental Light Increases Biomass and Yield of Greenhouse-Grown Tomatoes, but Only to an Optimum. *Front. Plant Sci.* **2019**, *9*, 2002. [CrossRef]
25. Kaiser, E.; Weerheim, K.; Schipper, R.; Dieleman, J.A. Partial replacement of red and blue by green light increases biomass and yield in tomato. *Sci. Hortic.* **2019**, *249*, 271–279. [CrossRef]
26. Sager, J.C.; Smith, W.O.; Edwards, J.L.; Cyr, K.L. Photosynthetic efficiency and phytochrome photoequilibria determination using spectral data. *Trans. Am. Soc. Agric. Eng.* **1988**, *31*, 1882–1889. [CrossRef]
27. Bloch, G.; Bar-Shai, N.; Cytter, Y.; Green, R. Time is honey: Circadian clocks of bees and flowers and how their interactions may influence ecological communities. *Philos. Trans. R. Soc. B Biol. Sci.* **2017**, *372*, 20160256. [CrossRef] [PubMed]
28. Lichtenthaler, H.K. Chlorophylls and carotenoids: Pigments of photosynthetic biomembranes. *Methods Enzymol.* **1987**, *148*, 350–382.
29. LaNoue, J.; Leonardos, E.D.; Khosla, S.; Hao, X.; Grodzinski, B. Effect of elevated CO2 and spectral quality on whole plant gas exchange patterns in tomatoes. *PLoS ONE* **2018**, *13*, e0205861. [CrossRef] [PubMed]
30. Farquhar, G.D.; Von Caemmerer, S.; Berry, J.A. A biochemical model of photosynthetic CO2 assimilation in leaves of C3 species. *Planta* **1980**, *149*, 78–90. [CrossRef] [PubMed]
31. McMurtrie, R.E.; Wang, Y. Mathematical models of the photosynthetic response of tree stands to rising CO2 concentrations and temperatures. *Plant Cell Environ.* **1993**, *16*, 1–13. [CrossRef]
32. Bernacchi, C.J.; Singsaas, E.L.; Pimentel, C.; Portis, A.R., Jr.; Long, S.P. Improved temperature response functions for models of rubisco-limited photosynthesis. *Plant Cell Environ.* **2001**, *24*, 253–259. [CrossRef]
33. Baker, N.R. Chlorophyll fluorescence: A probe of photosynthesis in vivo. *Annu. Rev. Plant Biol.* **2008**, *59*, 89–113. [CrossRef]
34. Hillman, W.S. Injury of tomato plants by continuous light and unfavorable photoperiodic cycles. *Am. J. Bot.* **1956**, *43*, 89–96. [CrossRef]
35. Matsuda, R.; Ozawa, N.; Fujiwara, K. Leaf photosynthesis, plant growth, and carbohydrate accumulation of tomato under different photoperiods and diurnal temperature differences. *Sci. Hortic.* **2014**, *170*, 150–158. [CrossRef]
36. Haque, M.S.; Kjaer, K.H.; Rosenqvist, E.; Ottosen, C.-O. Continuous light increases growth, daily carbon gain, antioxidants, and alters carbohydrate metabolism in a cultivated and a wild tomato species. *Front. Plant Sci.* **2015**, *6*, 522. [CrossRef] [PubMed]
37. Velez-Ramirez, A.I.; Dünner-Planella, G.; Vreugdenhil, D.; Millenaar, F.F.; Van Ieperen, W. On the induction of injury in tomato under continuous light: Circadian asynchrony as the main triggering factor. *Funct. Plant Biol.* **2017**, *44*, 597. [CrossRef] [PubMed]
38. Azcón-Bieto, J.; Federico, R.; Giartosio, C.E. Inhibition of Photosynthesis by Carbohydrates in Wheat Leaves. *Plant Physiol.* **1983**, *73*, 681–686. [CrossRef] [PubMed]
39. Kristoffersen, T. Interactions of photoperiod and temperature in growth and development of young tomato plants (*Lycopersicon esculentum* Mill.). *Physiol. Plant.* **1963**, *16*, 1–98.
40. Lanoue, J.; Zheng, J.; Little, C.; Grodzinski, B.; Hao, X. Continuous Light Does Not Compromise Growth and Yield in Mini-Cucumber Greenhouse Production with Supplemental LED Light. *Plants* **2021**, *10*, 378. [CrossRef]
41. Liu, Y.; Li, T.; Qi, H.; Li, J.; Yin, X. Effects of grafting on carbohydrate accumulation and sugar-metbolic enzyme activities in muskmelon. *Afr. J. Biotechnol.* **2010**, *9*, 25–35.
42. Barry, H.G.; Castle, W.S.; Davies, F.S. Rootstocks and plant water relations affect sugar accumulation of citrus fruit via osmotic adjustment. *J. Am. Soc. Hort. Sci.* **2014**, *129*, 881–889. [CrossRef]
43. Velez-Ramierez, A.I.; van Ieperen, W.; Vreugdenhil, D.; Millenaar, F.F. Continuous-light tolerance in tomato is graft-transferable. *Planta* **2015**, *241*, 285–290. [CrossRef]

44. Paponov, M.; Kechasov, D.; Lacek, J.; Verheul, M.J.; Paponov, I. Supplemental Light-Emitting Diode Inter-Lighting Increases Tomato Fruit Growth Through Enhanced Photosynthetic Light Use Efficiency and Modulated Root Activity. *Front. Plant Sci.* **2020**, *10*, 1656. [CrossRef] [PubMed]
45. Albacete, A.; Ghanem, M.E.; Martínez-Andújar, C.; Acosta, M.; Sánchez-Bravo, J.; Martínez, V.; Lutts, S.; Dodd, I.C.; Pérez-Alfocea, F. Hormonal changes in relation to biomass partitioning and shoot growth impairment in salinized tomato (*Solanum lycopersicum* L.). *Plants J. Exp. Bot.* **2008**, *59*, 4119–4131. [CrossRef] [PubMed]

plants MDPI

Article

Effects of Different Light Spectra on Final Biomass Production and Nutritional Quality of Two Microgreens

Stefania Toscano [1], Valeria Cavallaro [2], Antonio Ferrante [3], Daniela Romano [1,*] and Cristina Patané [2]

[1] Department of Agriculture, Food and Environment (Di3A), Università degli Studi di Catania, 95123 Catania, Italy; stefania.toscano@unict.it
[2] IBE-Istituto di BioEconomia, Consiglio Nazionale delle Ricerche, 95126 Catania, Italy; valeria.cavallaro@cnr.it (V.C.); cristinamaria.patane@cnr.it (C.P.)
[3] Department of Agricultural and Environmental Sciences, Università degli Studi di Milano, 20133 Milan, Italy; antonio.ferrante@unimi.it
* Correspondence: dromano@unict.it

Abstract: To improve microgreen yield and nutritional quality, suitable light spectra can be used. Two species—amaranth (*Amaranthus tricolor* L.) and turnip greens (*Brassica rapa* L. subsp. *oleifera* (DC.) Metzg)—were studied. The experiment was performed in a controlled LED environment growth chamber (day/night temperatures of 24 ± 2 °C, 16 h photoperiod, and 50/60% relative humidity). Three emission wavelengths of a light-emitting diode (LED) were adopted for microgreen lighting: (1) white LED (W); (2) blue LED (B), and (3) red LED (R); the photosynthetic photon flux densities were 200 ± 5 μmol for all light spectra. The response to light spectra was often species-specific, and the interaction effects were significant. Morphobiometric parameters were influenced by species, light, and their interaction; at harvest, in both species, the fresh weight was significantly greater under B. In amaranth, Chl *a* was maximized in B, whereas it did not change with light in turnip greens. Sugar content varied with the species but not with the light spectra. Nitrate content of shoots greatly varied with the species; in amaranth, more nitrates were measured in R, while no difference in turnip greens was registered for the light spectrum effect. Polyphenols were maximized under B in both species, while R depressed the polyphenol content in amaranth.

Keywords: LED; light spectrum; ascorbic acid; chlorophylls; carotenoids

Citation: Toscano, S.; Cavallaro, V.; Ferrante, A.; Romano, D.; Patané, C. Effects of Different Light Spectra on Final Biomass Production and Nutritional Quality of Two Microgreens. *Plants* **2021**, *10*, 1584. https://doi.org/10.3390/plants10081584

Academic Editor: Eva Darko

Received: 6 July 2021
Accepted: 28 July 2021
Published: 31 July 2021

1. Introduction

Light is one of the major factors for growth. It represents the main signal perceived by plants, and it has been largely demonstrated that different light qualities, light intensity, and photoperiod have broad regulatory effects on the morphogenesis, physiological metabolism, growth and development, and nutritional quality of plants [1–4]. Plant morphogenesis and its related aspects are mainly regulated by various photoreceptors which are activated by photons in the blue, red, and far-red regions of the light spectrum [5]. Light-emitting diodes (LEDs) are an emerging source of light in protected and indoor cultivations. They have several advantages over conventional lighting systems (fluorescent light, halide metal, high-pressure solid, and incandescent), e.g., long operating lifetime, relatively lower heat emission, high photosynthetically active radiation efficiency, small size, and control of spectral composition. All these advantages make LED an ideal light source for the artificial regulation of plant growth and an easy disposal without any environmental hazards [6]. Moreover, LEDs offer the advantage to emit specific spectral patterns [7] and regulate the light intensities, in accordance with the needs of the plants, optimizing the production processes and/or the production of secondary metabolites [4,7]. For these reasons, LEDs are attracting increasing attention for indoor facilities, vertical farming, and greenhouse productions, especially with leafy vegetables, such as lettuce and rockets [8–12]. According to the manufacturers' indications and measured light fluence rates, LED lids would require about 32% less energy than fluorescent tubes, per $\mu mol \cdot m^2 \cdot s^{-1}$ delivered to the plants [13].

Approximately 90% of red and blue light that falls on plant leaves is absorbed. It is well known that those sections of the spectrum strongly influence plant development and physiology [14]. Blue and red light are absorbed by photosynthetic pigments (chlorophylls) and photomorphogenetic (cryptochromes, phytochromes) receptors [15].

Red light influences leaf expansion in red lettuce [16], as well as increases plant height in tomato [17] and in vitro grown chestnut seedlings [18]. Blue light suppresses hypocotyl elongation and induces biomass production [18]. In combination with red and blue light, green light increases plant and leaf growth, as well as early stem elongation [18–20].

Microgreens are young, tender greens of edible plants that are harvested at the first true leaf stage. Microgreens are much smaller than regular greens, even "baby" greens. They are harvested when plants are no taller than 5 cm, taking about 1–3 weeks after seeding. Microgreens have emerged on the market and become popular for their nutrient concentrations that are higher than those of their mature leaf counterparts [21–23]. The attention toward this category of products is confirmed by the very high number of items published about microgreens. Moreover, microgreens have an eye-catching appearance; they can be grown in small spaces and on indoor farms, thus representing a potentially useful addition to urban diets [24].

Microgreens are also frequently used to add color and flavor to meals. They have a double function as food and garnish on plates. Micro versions of basil, coriander, chard, beetroot, and red garnet amaranth were originally used to complement the flavor of dishes and as a garnish. Today, since their popularity has widened, people can even buy pots ready to grow your own.

The levels of nutrients in microgreens vary with the species. Nonetheless, they typically have higher levels of vitamin C, vitamin E, and carotenoids than mature plants [25]. Due to their adaptation to different cultivation environments, they can be cultivated in individual households, as well as on a large scale for commercial purposes [26]. Grown in a greenhouse with supplemental lighting and heating, microgreens can be produced throughout the entire year.

Numerous vegetables and crops can be used for microgreen production. Among these, the following are of considerable importance: turnip green and amaranth. Turnip green (*Brassica rapa* L. subsp. *oleifera*) is a member of the Brassicaceae family. The Brassicaceae microgreen effects on health are tied to their high levels of bioactive compounds such as ascorbic acid, carotenoids, tocopherols, and phenolic compounds in addition to glucosinolates and mineral nutrients [23].

Amaranth (*Amaranthus tricolor* L.) is one of the most preferable greens in terms of texture, flavor, appearance, and overall eating quality [25].

Recent studies highlighted the possibility of regulating seedling growth and increasing the content of important nutritional compounds (as glucosinolates in rocket and sugars, proteins, flavonoids, and vitamin C in lettuce) through appropriate regulation of the light spectrum used [9–12].

In recent years, spectral effects of red/blue/red–blue light have been investigated in microgreen species, belonging to different families, e.g., Brassicacceae, Lamiaceae, Apiaceae, Boraginaceae, and Chenopodiaceae [27–32]. However, for new and emerging microgreen species, information on plant secondary metabolites profiles and how these bioactive compounds respond to LED spectral quality is lacking. Instead, there is a need, as it is often a species-specific response, to investigate the mechanism of different light spectra on the phytochemical profiles of some microgreens [32].

With this in mind, a study was conducted to evaluate the effects of different LED spectra (white, red and blue), on the final biomass and nutritional traits, in two different microgreen species. The hypothesis of the work was to enhance the microgreen composition modulating the light quality.

2. Results

2.1. Seedling Height and Biomass

Seedling height was influenced by species, light, and their interaction (Table 1, Figure 1). Seedlings of amaranth were significantly smaller than those of turnip greens, under all lights ($p \leq 0.001$).

Table 1. Main effects of species (amaranth and turnip greens) and LED treatment (W = white, B = blue, R = red) on plant height, fresh biomass, and dry biomass percentage of microgreens.

		Seedling Height (H, cm)	Fresh Biomass (FW, mg·Plant^{-1})	Dry Biomass (DW, %)
Species (*S*)	Amaranth	3.9 ± 0.2^{b}	18.5 ± 1.8^{b}	5.4 ± 0.4
	Turnip greens	5.4 ± 0.2^{a}	66.4 ± 2.8^{a}	5.2 ± 0.3
LED treatments (*L*)	W	3.9 ± 0.4^{c}	37.1 ± 9.9^{b}	5.9 ± 0.3^{a}
	B	5.2 ± 0.4^{a}	50.1 ± 11.8^{a}	5.9 ± 0.3^{a}
	R	4.9 ± 0.2^{b}	40.2 ± 10.8^{b}	4.2 ± 0.1^{b}
Significance	*S*	***	***	ns
	L	***	***	***
	S × *L*	**	ns	ns

Values (mean ± *se*) within each column, followed by the same letter, do not significantly differ at $p \leq 0.05$ according to Tukey's test; ns = not significant; significant at $p \leq 0.01$ (**) and 0.001 (***). Three biological replicates were used for measurements ($n = 3$).

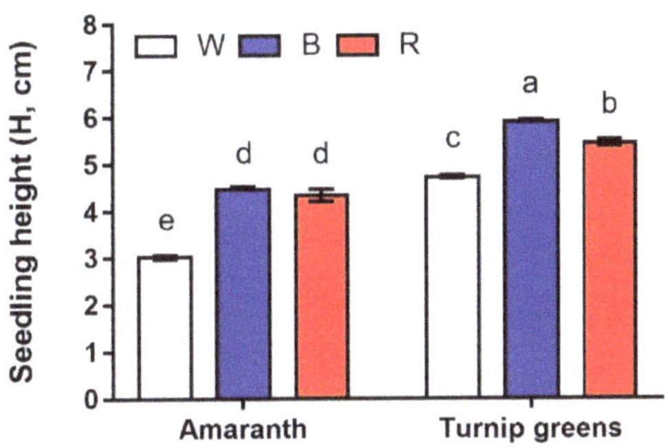

Figure 1. Interaction effect of *species* × *light treatment* (W = white, B = blue, R = red) on seedling height (H, cm) of microgreens. Data are means ± standard error ($n = 3$). Three biological replicates were used for the measurements. Different letters indicate significance at $p \leq 0.05$ according to Tukey's test.

However, light exerted a different effect, depending on species ($S \times L, p \leq 0.01$). In amaranth, seedlings were almost 3 cm tall in W, whilst their height exceeded 4 cm in B and R. In turnip greens, blue light (B) promoted plant growth, resulting in a final seedling height close to 6 cm; in this species, white light (W) adversely affected plant growth, leading to a final height < 5 cm (Figure 1).

Fresh biomass of the single plant varied with species and light but not with their interaction. According to seedling height, turnip greens produced a fresh biomass more than threefold greater than that of amaranth ($p \leq 0.001$) under the same experimental conditions, revealing a faster growth. Light also affected the biomass accumulation, and fresh weight of the single plant at harvest was significantly greater under blue light in both species ($L, p \leq 0.001$; $S \times L, p \geq 0.05$). Fresh biomass produced in W and R did not differ at ANOVA. However, dry biomass was the lowest (<4.3%) under red light, in both species, indicating a greater plant water content under these growing conditions.

When the height/dry biomass (cm·mg^{-1}) ratio was calculated, interesting results on plant morphology were obtained (Figure 2). While no differences among light treatments were observed in turnip for the ratio (<2 cm·mg^{-1}), the significantly higher value (7 cm·mg^{-1}) calculated for amaranth ($S \times L$, $p \leq 0.001$) in R with respect to W and B (3.7 cm·mg^{-1}, on average) indicates that the same dry matter was distributed over longer plants under red light, i.e., the hypocotyls were thinner under these experimental conditions, contributing to total plant biomass.

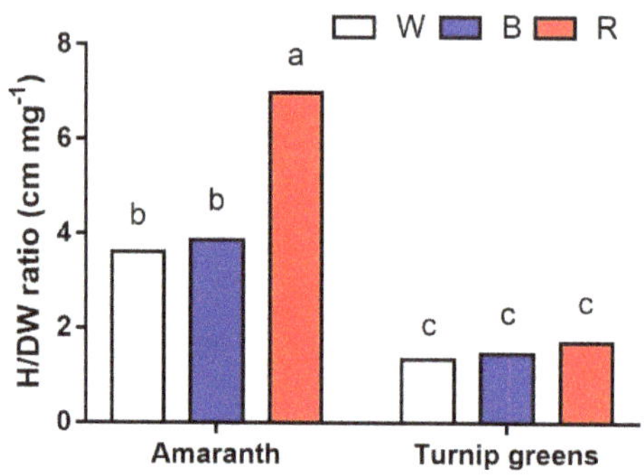

Figure 2. Interaction effect of *species* × *light treatment* (W = white, B = blue, R = red) on plant height/dry biomass ratio (H/DW, cm·mg^{-1}) of microgreens. Data are means ± standard error (n = 3). Three biological replicates were used for the measurements. Different letters indicate significance at $p \leq 0.05$ according to Tukey's test.

2.2. Chlorophyll (a, b, and Total) and Carotenoids

The content of Chl *a* exhibited a different pattern in relation to species and light. Overall, greater contents were measured in shoots of amaranth (Table 2, Figure 3a). In this species, Chl *a* was maximized in B, whereas, in turnip greens, Chl *a* content did not differ with light ($S \times L$, $p \leq 0.01$). No effect of species and light was observed according to ANOVA on Chl *b* content (S, L, $S \times L$, $p \geq 0.05$) (Table 2, Figure 3b). As a result, *total* Chl content followed the same pattern of Chl *a*, being higher in amaranth. In this species, as for Chl *a*, total Chl peaked under blue light, whereas it did not change with light in turnip greens ($S \times L$, $p \leq 0.05$) (Table 2, Figure 3c).

Table 2. Main effects of species (amaranth and turnip greens) and LED treatment (W = white, B = blue, R = red) on chlorophyll *a*, *b*, chlorophyll *a*/*b* ratio (Chl *a*/Chl *b*), total chlorophyll, carotenoids, and chlorophyll/carotenoid ratio (Chl/Car) of microgreens.

		Chl *a* (mg·g^{-1} FW)	Chl *b* (mg·g^{-1} FW)	Chl *a*/Chl *b* (mg·g^{-1} FW)	Total Chl (mg·g^{-1} FW)	Carotenoids (mg·g^{-1} FW)	Chl/Car (mg·g^{-1} FW)
Species (S)	Amaranth	0.41 ± 0.0 ^{a}	0.11 ± 0.0 ^{a}	3.79 ± 0.18 ^{a}	0.51 ± 0.0 ^{a}	0.10 ± 0.00 ^{a}	5.1 ± 0.1
	Turnip greens	0.32 ± 0.0 ^{b}	0.10 ± 0.0 ^{b}	3.20 ± 0.20 ^{b}	0.43 ± 0.0 ^{b}	0.08 ± 0.00 ^{b}	5.2 ± 0.1
LED treatment (L)	W	0.34 ± 0.1	0.11 ± 0.0	3.18 ± 0.11 ^{b}	0.45 ± 0.0 ^{b}	0.08 ± 0.00 ^{b}	5.3 ± 0.1
	B	0.40 ± 0.0	0.11 ± 0.0	3.74 ± 0.33 ^{a}	0.51 ± 0.0 ^{a}	0.10 ± 0.01 ^{a}	5.1 ± 0.2
	R	0.35 ± 0.0	0.10 ± 0.0	3.49 ± 0.08 ^{ab}	0.45 ± 0.0 ^{b}	0.09 ± 0.00 ^{b}	5.1 ± 0.1
Significance	S	***	***	***	***	***	ns
	L	ns	ns	**	**	**	ns
	$S \times L$	***	ns	ns	*	**	ns

Values (mean ± *se*) within each column, followed by the same letter, do not significantly differ at $p \leq 0.05$ according to Tukey's test; ns = not significant; significant at $p \leq 0.05$ (*), 0.01 (**), and 0.001 (***). Three biological replicates were used for the analysis (n = 3).

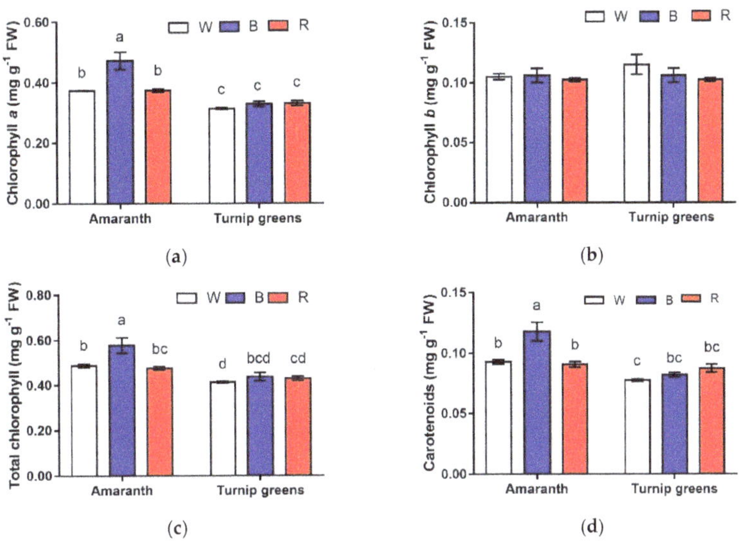

Figure 3. Interaction effect of *species* × *light treatment* (W = white, B = blue, R = red) on chlorophyll (*a* (**a**), *b* (**b**), and total (**c**)) and carotenoids (**d**) of microgreens. Data are means ± standard error (n = 3). Different letters indicate significance at $p \leq 0.05$ according to Tukey's test.

Carotenoids were accumulated in a larger amount in microgreens of amaranth (0.10 mg·g^{-1} FW, against 0.08 mg·g^{-1} in turnip greens, $p \leq 0.01$). They exhibited opposite trends in the two species, in response to light ($S \times L$, $p \leq 0.01$) (Table 2, Figure 3d). In amaranth, blue light (B) promoted the biosynthesis of carotenoids, whose content was almost the 30% higher than that in W and R. Such differences were not evidenced in turnip greens, where the content of carotenoids did not change with the light conditions of growth. The chlorophyll a/b ratio showed that W treatment provided a lower value while the highest ratio was found in the B light treatment. The relationship between total chlorophyll and carotenoids expressed as a ratio did not show any significant difference (Table 2).

2.3. Sugar Content

Sugar content varied with the species, being more than 40% greater in turnip greens (>1.3 mg·g^{-1} FW) than in amaranth (<0.7 mg·g^{-1} FW) (Table 3).

Table 3. Main effects of species (amaranth and turnip greens) and LED treatment (W = white, B = blue, R = red) on total sugars and nitrate content of microgreens.

		Total Sugars (mg·g^{-1} FW)	**Nitrate (mg·kg^{-1})**
Species (S)	Amaranth	0.7 ± 0.0 ^{b}	1990.9 ± 140.3 ^{a}
	Turnip greens	1.3 ± 0.0 ^{a}	704.9 ± 48.0 ^{b}
LED treatment (L)	W	1.0 ± 0.4	1137.1 ± 202.5 ^{b}
	B	1.0 ± 0.4	1247.1 ± 318.7 ^{b}
	R	0.9 ± 0.4	1659.5 ± 357.7 ^{a}
Significance	S	***	***
	L	ns	***
	$S \times L$	Ns	***

Values (mean ± *se*) within each column, followed by the same letter, do not significantly differ at $p \leq 0.05$ according to Tukey's test; ns = not significant; significant at $p \leq 0.001$ (***). Three biological replicates were used for the analysis (n = 3).

Light did not exert any clear effect on this trait (L, $p \geq 0.05$), while the interactive effect with species was not significant ($S \times L$, $p \geq 0.05$).

2.4. Nitrate Content

The nitrate content of shoots greatly varied with the species (S, $p \leq 0.001$) (Figure 4). Greater amounts were measured in amaranth, where nitrates peaked under red light (>2000 mg·kg^{-1}). Lower contents (1583 to 1946 mg·kg^{-1}) were measured in this species in W and B, showing no difference according to ANOVA. Unlike amaranth, very low nitrate contents (<705 mg·kg^{-1}) were detected in turnip greens, regardless of the light conditions of growth ($S \times L$, $p \geq 0.05$).

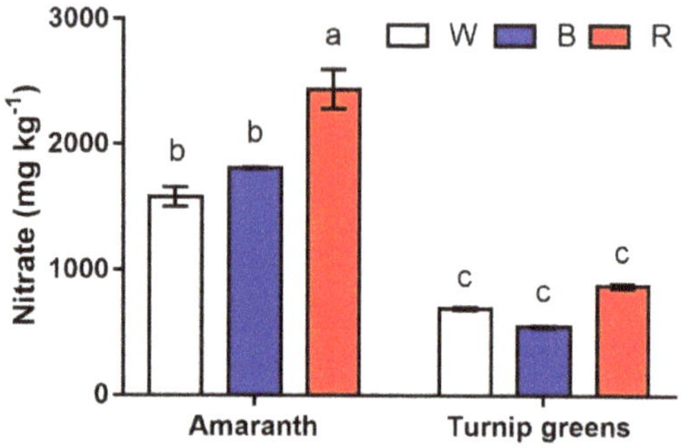

Figure 4. Interaction effect of *species* × *light treatment* (W = white, B = blue, R = red) on the nitrate content (ppm) of microgreens. Data are means ± standard error (n = 3). Three biological replicates were used for the analysis. Different letters indicate significance at $p \leq 0.05$ according to Tukey's test.

2.5. Antioxidants and Antioxidant Activity

The content of the main antioxidants (polyphenols, carotenoids, and ascorbic acid) was measured in this experiment. Significant differences were found for total polyphenols in relation to species, light, and their interaction (Table 4, Figure 5). On average across light conditions, the antioxidant content was slightly but significantly higher in turnip greens (>145 mg GAE·100 g^{-1} FW vs. 124 mg·g^{-1} in amaranth). Polyphenols were maximized under blue light (B, >165 mg·g^{-1} FW) in both species. Red light (R) somehow depressed the biosynthesis of polyphenols, leading to a final content that was overall the lowest in amaranth (<75 mg·g^{-1} FW), but did not differ from that in W (128.2 and 135.9 mg·g^{-1} FW in R and W, respectively) in turnip greens ($S \times L$, $p \leq 0.01$).

Table 4. Main effects of species (amaranth and turnip greens) and LED treatment (W = white, B = blue, R = red) on total phenolic content (TPC), ascorbic acid (Asc), and antioxidant activity (DPPH) of microgreens.

		TPC (mg GAE·100 g^{-1} FW)	Asc (mg·g^{-1} FW)	DPPH (mg TE·100 g^{-1} FW)
Species (S)	Amaranth	124.8 ± 13.5 [b]	0.20 ± 0.0 [b]	54.6 ± 5.2 [b]
	Turnip greens	145.6 ± 7.7 [a]	0.78 ± 0.2 [a]	180.5 ± 22.2 [a]
LED treatment (L)	W	135.6 ± 2.3 [b]	0.79 ± 0.2 [a]	102.4 ± 26.5 [b]
	B	168.6 ± 6.2 [a]	0.20 ± 0.0 [c]	168.5 ± 42.9 [a]
	R	104.4 ± 12.2 [c]	0.51 ± 0.1 [b]	82.3 ± 15.7 [c]
Significance	S	**	***	***
	L	***	***	***
	$S \times L$	**	***	**

Values (mean ± *se*) within each column, followed by the same letter, do not significantly differ at $p \leq 0.05$ according to Tukey's test; significant at $p \leq 0.01$ (**) and 0.001 (***). Three biological replicates were used for the analysis (n = 3).

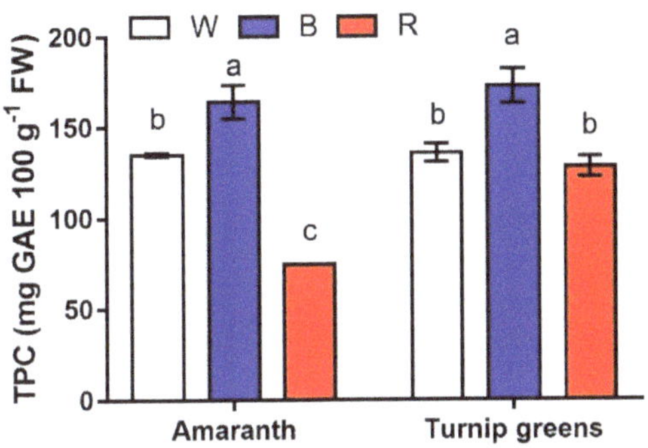

Figure 5. Interaction effect of *species* × *light treatment* (W = white, B = blue, R = red) on TPC (mg GAE·100 g^{-1} FW) of microgreens. Data are means ± standard error (n = 3). Three biological replicates were used for the analysis. Different letters indicate significance at $p \leq 0.05$ according to Tukey's test.

Ascorbic acid (Asc) is another antioxidant that was detected in the microgreens of the two species. Unlike carotenoids, much greater contents of Asc (up to 1.3 mg·g^{-1} FW) were found in turnip greens ($p \leq 0.001$) (Table 4, Figure 6). Light strongly affected (L, $p \leq 0.001$) the content of this antioxidant, being significantly higher under blue light (B) in both species. A significant $S \times L$ interaction ($p \leq 0.001$) was found according to ANOVA, indicating that, unlike amaranth, whose microgreens had the same Asc in W and R, white light (W) significantly reduced the accumulation of this metabolite in shoots of turnip greens, whose final content was <0.3 mg·g^{-1} FW.

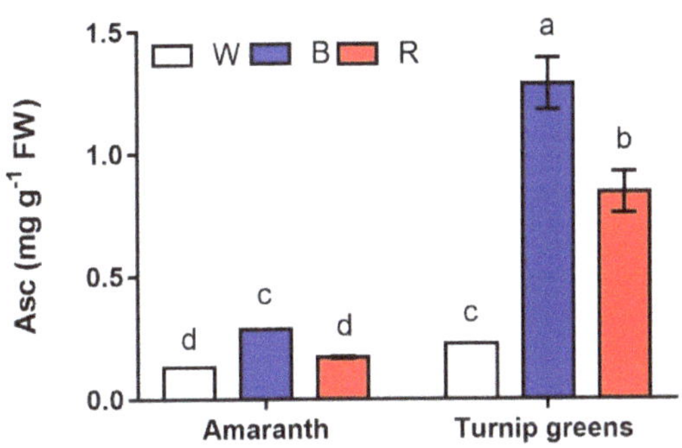

Figure 6. Interaction effect of *species* × *light treatment* (W = white, B = blue, R = red) on ascorbic acid (Asc, mg·100 g^{-1} FW) of microgreens. Data are means ± standard error (n = 3). Three biological replicates were used for the analysis. Different letters indicate significance at $p \leq 0.05$ according to Tukey's test.

Antioxidant activity (AA), expressed as DPPH free-radical scavenging activity, was seemingly correlated to Asc (r = 0.82*) more than to other antioxidants ($r = 0.54^{ns}$ vs. carotenoids, $r = 0.41^{ns}$ vs. TPC). As a result, on average across light conditions, higher AA corresponded to turnip greens (up to 260 mg TE·100 g^{-1} FW) with respect to microgreens of amaranth (AA <74%) (Figure 7). Light also exerted a significant effect on this trait, with AA being higher in microgreens grown under blue light (B). However, a significant $S \times L$ interaction ($p \leq 0.001$) revealed that, while no differences were observed for AA between W and R in amaranth (46 mg TE·100 g^{-1} FW, on average), red light (R) adversely affected the antioxidant activity in amaranth, which was the 55% and 27% lower than AA in B and W, respectively.

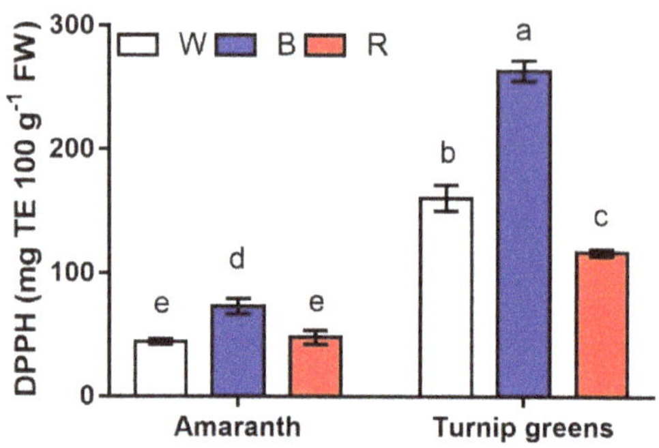

Figure 7. Interaction effect of *species* × *light treatment* (W = white, B = blue, R = red) on the antioxidant activity (DPPH, mg TE·100 g^{-1} FW) of microgreens. Data are means ± standard error (n = 3). Three biological replicates were used for the analysis. Different letters indicate significance at $p \leq 0.05$ according to Tukey's test.

2.6. Mineral Composition

Multifactorial ANOVA showed that the mineral contents were significantly affected by species and the LED treatments, as well as by their interaction (Table 5). Most of the mineral elements were different in the two species except for Fe and Ni. Amaranth showed higher concentrations of Mg, K, Cu, Zn, and P, but lower concentrations of Na, Ca, and Mn compared to turnip greens (Table 5).

Light treatments significantly influenced the concentration of Mg, Ca, Mn, Fe, Ni, and P. In particular, R light increased the concentrations of Mg, Mn, Fe, and Ni, while W light increased the concentrations of Ca and P (Table 5).

The effect of LED treatment was more pronounced for Mg and the microelements (Mn, Fe, and Cu), which were significantly higher in turnip greens under the red LED. Amaranth grown under the red and blue LED showed a high Fe concentration. No significant differences were observed for Ni, Zn, and P (Table 5).

2.7. RGB Color Analysis

The color analysis of microgreens showed that light significantly affected the RB components of amaranth, while no significant differences were observed for the RGB components of turnip greens (Figure 8). In amaranth the white light led to the highest R and B values, while the red light lowered the B value. In the comparison between species, the G component was significantly higher in turnip greens compared to amaranth (Figure 8).

When all the effects were summarized in a PCA score plot, differential reactions of amaranth and turnip greens to different light spectra were observed (Figure 9a,b). The first two PCs were related with eigen values >1 and explained more than 90% of the total variance, with PC1 and PC2 accounting for 56.4% and 43.6% for amaranth, and 54.1% and 45.9% for turnip greens. In amaranth, we identified four groups of positively correlated variables: (1) the group in the upper left quadrant, which included Chl *b*, sugars, and Na; (2) the group in the upper right quadrant, which included carotenoids, total Chl, Chl *a*, Asc, DPPH, TPC, Ca, and Cu; (3) the group clustered in the lower right quadrant, which included nitrates and most mineral elements (Fe, Mg, Zn, Mn, and Ni); (4) the group in the lower left quadrant, which included P, K, FW, and % DW (Figure 9a).

Table 5. Results of the multifactorial ANOVA for sodium (Na), magnesium (Mg), potassium (K), calcium (Ca), manganese (Mn), iron (Fe), nickel (Ni), copper (Cu), zinc (Zn), and phosphorus (P) concentrations of microgreens.

			Na ($g \cdot kg^{-1}$ DW)	Mg ($g \cdot kg^{-1}$ DW)	K ($g \cdot kg^{-1}$ DW)	Ca ($g \cdot kg^{-1}$ DW)	Mn ($mg \cdot kg^{-1}$ DW)	Fe ($mg \cdot kg^{-1}$ DW)	Ni ($mg \cdot kg^{-1}$ DW)	Cu ($mg \cdot kg^{-1}$ DW)	Zn ($mg \cdot kg^{-1}$ DW)	P ($g \cdot kg^{-1}$ DW)
Species (*S*)		Amaranth	2.5 ± 0.1^{b}	11.2 ± 0.3^{a}	94.2 ± 1.7^{a}	8.1 ± 0.2^{b}	48.8 ± 1.4^{b}	1.7 ± 0.2	9.5 ± 1.6	27.1 ± 1.3^{a}	102.7 ± 1.5^{a}	12.2 ± 0.2^{a}
		Turnip greens	4.8 ± 0.1^{a}	9.5 ± 1.1^{b}	72.0 ± 1.4^{b}	10.9 ± 0.2^{a}	54.9 ± 4.5^{a}	2.1 ± 0.5	8.2 ± 1.4	23.7 ± 1.7^{b}	77.5 ± 2.0^{b}	9.3 ± 0.1^{b}
			$p \leq 0.001$	$p \leq 0.05$	$p \leq 0.001$	$p \leq 0.001$	$p \leq 0.01$	$p \geq 0.05$	$p \geq 0.05$	$p \leq 0.05$	$p \leq 0.001$	$p \leq 0.001$
LED treatment (*L*)		W	3.6 ± 0.4	8.9 ± 0.6^{b}	83.9 ± 4.4	9.9 ± 0.7^{a}	47.5 ± 1.7^{b}	1.1 ± 0.0^{b}	5.2 ± 0.8^{b}	23.4 ± 1.4	91.7 ± 4.9	11.0 ± 0.7^{a}
		B	3.5 ± 0.5	9.9 ± 1.1^{b}	83.0 ± 3.7	9.4 ± 0.4^{ab}	47.6 ± 2.5^{b}	1.5 ± 0.2^{b}	7.3 ± 1.2^{b}	24.9 ± 2.5	88.0 ± 7.2	10.4 ± 0.5^{b}
		R	3.7 ± 0.6	12.2 ± 0.8^{a}	82.6 ± 7.2	9.2 ± 0.8^{b}	60.4 ± 5.3^{a}	3.0 ± 0.5^{a}	14.1 ± 0.8^{a}	27.8 ± 1.5	90.6 ± 5.7	10.8 ± 0.7^{ab}
			$p \geq 0.05$	$p \leq 0.01$	$p \geq 0.05$	$p \leq 0.05$	$p \leq 0.001$	$p \leq 0.001$	$p \leq 0.001$	$p \geq 0.05$	$p \geq 0.05$	$p \leq 0.05$
S × L	Amaranth	W	2.7 ± 0.0^{b}	10.2 ± 0.1^{ab}	93.6 ± 1.6^{a}	8.3 ± 0.3^{c}	43.8 ± 0.5^{b}	1.1 ± 0.1^{c}	4.6 ± 2.1	24.8 ± 4.2^{ab}	101.7 ± 6.5	12.6 ± 0.5
		B	2.4 ± 0.1^{b}	12.3 ± 0.3^{a}	90.8 ± 3.8^{a}	8.5 ± 0.4^{c}	52.7 ± 0.7^{b}	1.9 ± 0.1^{b}	9.2 ± 2.2	30.2 ± 0.2^{a}	103.7 ± 5.6	11.6 ± 0.3
		R	2.4 ± 0.3^{b}	11.1 ± 0.5^{a}	98.2 ± 6.7^{a}	7.5 ± 0.8^{c}	49.9 ± 2.9^{b}	2.1 ± 0.3^{b}	14.5 ± 2.7	26.2 ± 4.2^{ab}	102.8 ± 0.6	12.3 ± 0.5
	Turnip greens	W	4.6 ± 0.0^{a}	7.5 ± 0.2^{b}	74.1 ± 0.1^{b}	11.5 ± 0.5^{a}	51.2 ± 2.0^{b}	1.2 ± 0.1^{c}	5.7 ± 1.8	22.0 ± 2.6^{ab}	81.7 ± 4.4	9.4 ± 0.2
		B	4.7 ± 0.2^{a}	7.7 ± 1.2^{b}	75.1 ± 1.1^{b}	10.2 ± 0.4^{b}	42.6 ± 3.9^{b}	1.2 ± 0.1^{c}	5.3 ± 2.0	19.6 ± 3.4^{b}	72.3 ± 3.1	9.3 ± 0.2
		R	5.0 ± 0.1^{a}	13.3 ± 2.6^{a}	66.9 ± 2.3^{b}	10.9 ± 0.3^{ab}	70.9 ± 8.9^{a}	3.9 ± 1.1^{a}	13.6 ± 1.2	29.4 ± 3.0^{a}	78.4 ± 6.3	9.3 ± 0.3
			$p \leq 0.05$	$p \leq 0.01$	$p \leq 0.01$	$p \leq 0.05$	$p \leq 0.001$	$p \leq 0.01$	$p \geq 0.05$	$p \leq 0.05$	$p \geq 0.05$	$p \geq 0.05$

The mean values associated with the two factors and their interaction were evaluated according to Tukey's test. Means significantly different are indicated by different letters; ns not significant, significant at $p \leq 0.05$ (*), 0.01 (**), and 0.001 (***). Three biological replicates were used for the analysis ($n = 3$).

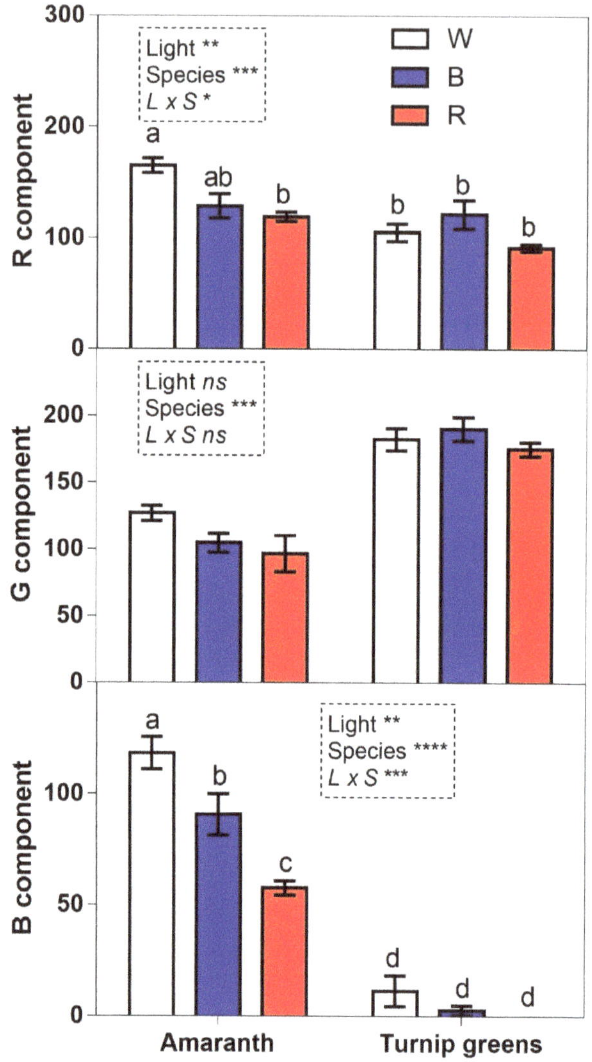

Figure 8. RGB component analysis of microgreen photos grown under different light conditions: white (W), blue (B), and red (R). Values are means with standard errors (n = 4). Four biological replicates were used for the analysis. Data were subjected to two-way ANOVA. Differences among means were determined using Tukey's test. Different letters highlight significant differences at $p \leq 0.05$; ns not significant, significant at $p \leq 0.05$ (*), 0.01 (**), and 0.001 (***).

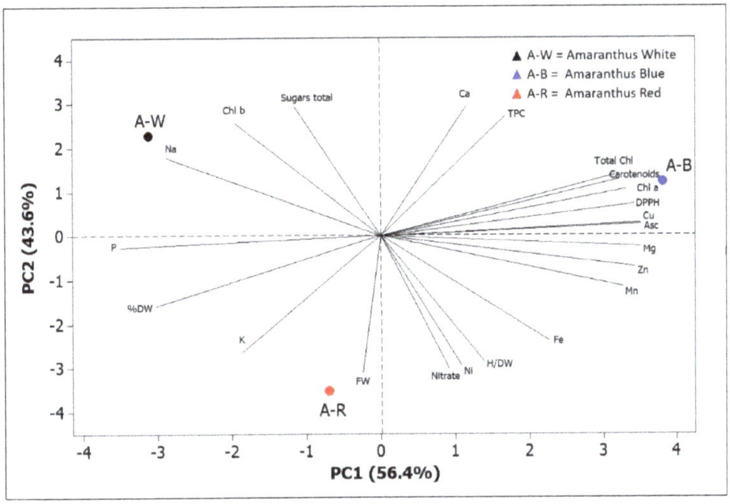

(a)

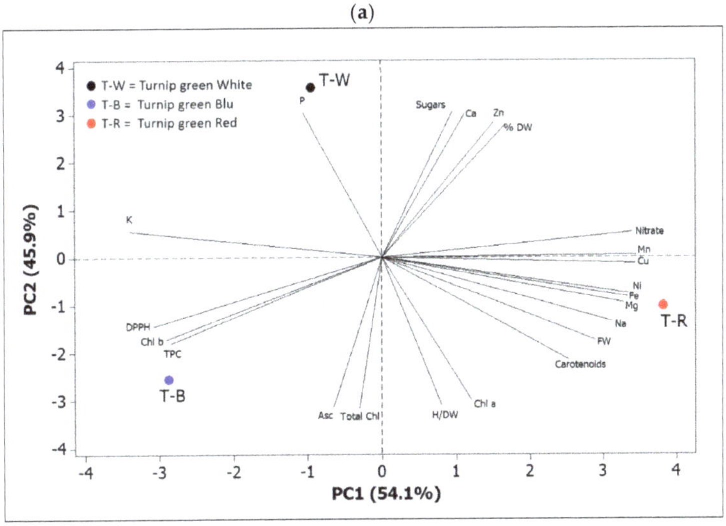

(b)

Figure 9. Principal component loading plot and scores of PCA fresh weight and dry biomass, H/DW, photosynthetic pigments (Chl *a*, Chl *b*, total Chl, and carotenoids), mineral concentrations (nitrate, Na, Mg, P, K, Ca, Mn, Fe, Ni, Cu, and Zn), DPPH, TPC, total sugars, Asc, and total phenolic concentrations for amaranth (**a**) and turnip greens (**b**) as modulated by LED treatments. W = white LED treatment; B = blue LED treatment; R = red LED treatment.

For turnip greens, we identified the following groups: (1) the group in the upper left quadrant, which included K and Na; (2) the group in the upper right quadrant, which included sugars, nitrates, % DW, and some mineral elements (Ca, Zn, and Mn); (3) the group clustered in the lower right quadrant, which included carotenoids, Chl *a*, H/DW, FW, and most mineral elements (Cu, Ni, Fe, Mg, and Na); (4) the group in the lower left quadrant, which included antioxidant activity (DPPH and TPC), Chl *b*, total Chl, and Asc (Figure 9b).

Plants of amaranth grown under red LED, positioned in the lower left quadrant of the PCA score plot, exhibited a higher concentration of P and K, whilst those grown under blue LED, positioned in the upper right quadrant, were characterized by higher total Chl, Chl *a*, and antioxidant activity. Plants of turnip greens, grown under blue LED, positioned in lower left quadrant, were characterized by higher antioxidant level (TPC and Asc), antioxidant activity (DPPH), and total Chl, whilst those grown under red LED, positioned in the lower right quadrant of the PCA score plot, showed a higher content of mineral elements (Cu, Ni, Fe, Mg, and Na).

The PCA analysis reported in the present study could, therefore, help to better understand the influence of LED treatments on morphological and nutraceutical characteristics of the two studied species.

3. Discussion

The results from this study revealed that the growth of hypocotyls in microgreens was affected by the quality of light. It has been reported in the literature how the hypocotyl growth may be influenced by artificial lights [33]; this aspect is relevant because the hypocotyls represent one of the main edible parts of sprouts and microgreens. To facilitate the machine harvest for labor savings, the height of microgreens needs to reach ~5 cm. The two studied microgreens exhibited a very different hypocotyl height. Among the three LED lights tested, blue and, to a lesser extent, red light seemed to be more effective than white light in promoting fresh biomass accumulation and hypocotyl growth. Similar significant increases in hypocotyl and shoot dry and fresh weight under monocromatic blue and red light were reported in microgreens of mustard and kale [34]. The blue LED, compared with the combined red and blue LED, was reported to increase the hypocotyl length of buckwheat sprouts [35]. Similarly, compared with the white LED, both blue and red LEDs were able to significantly increase the stem length of pea microgreens [36].

The leaf color of the two species, red (in amaranth) and green (in turnip greens), probably modifies the response to light spectra, as also observed in two cultivars of lettuce differing in leaf color (red and green) [37]. The PCA scatterplot clearly evidenced the differences between amaranth and turnip greens cultivated under different LED spectra (Figures S1 and S2).

Unlike fresh biomass, no significant difference between blue and white light was detected for dry biomass, revealing a higher water content in the plantlets grown under blue light. Our results partially differed from those obtained with other two leafy species, lettuce [38] and rocket [12], where better results in terms of fresh and dry biomass were obtained under red light, along with no difference in plant biomass between blue and white light. Microgreens are plants with a short growth period; therefore, the light spectrum influences more photomorphogenesis than photosynthesis. Photomorphogenic processes activated by the blue photomorphogenetic (cryptochromes, phytochromes) receptors constitute a default developmental process triggered by blue light in sprouts and microgreens during their development from seeds to edible vegetable products. The effects of blue light in microgreens may be different from those of mature plants [33,39].

In this study, a significant rise in Chl *a* and total Chl content under blue light was observed, although only in amaranth. Blue LEDs, used alone or in combination with red light, were reported to increase the chlorophyll ratio [40,41] and the chlorophyll content [42] in different leafy species and microgreens [15].

Sugar content significantly changed only with the species, being more than 40% greater in turnip greens than in amaranth.

Nitrate concentration in fresh vegetables is an important qualitative feature since its intake at high levels is associated with increased probability for carcinogenic nitrosamine formation in the stomach [32,43]. Approximately 80% of human dietary nitrates comes from vegetables; therefore, a low nitrate accumulation in vegetables is a primary concern [33]. In this study, nitrates were much greater in amaranth than in turnip greens, but lower than the maximum levels in European Commission (EC) Regulation No. 1258/2011 [44]. Moreover, the response to light spectra was species-specific, with the nitrate content significantly enhanced by red light in amaranth and unaffected by light spectra in turnip greens. This study confirms that nitrate accumulation capacity is a trait strongly associated with the genetic background of plants [21], even among genera within the same family. Contrasting results were reported in the literature for nitrate accumulation in response to red light; according to our results, an enhancement induced by monochromatic red light was found in mustard (*Brassica juncea* 'Red Lace') by Brazaitytė et al. [34]. Conversely, a reduced nitrate content under a red LED was reported in *Perilla frutescens* (L.) and radish microgreens.

The content in antioxidants is a very relevant quality index of sprouts and microgreens. The phenolic phytochemical accumulation can be stimulated by cultivation under different LEDs. In our study, blue light positively influenced total polyphenols, carotenoids, ascorbic acid (Asc), and antioxidant activity. As compared to white light, red light exerted similar or negative effects on the antioxidants with the only exception being Asc content. Previous studies showed that total phenolic content was significantly increased under a blue LED, as compared with white LEDs in Chinese kale and common buckwheat sprouts [45,46]. In Chinese kale sprouts, the highest antioxidant capacity was measured under a blue LED [33,46].

Increasing blue light dosage has been recognized to increase the level of phenolics in lamb lettuce [47], of phenols and antioxidant activity in pea sprouts [48], and of carotenoids in some microgreens [15]. Blue light is reported to stimulate the accumulation of carotenoids via cryptochromes [49,50]. Carotenoids play a relevant role in photoprotective efficiency in plants [51,52]. Carotenoids protect plants from photo-oxidative damage through thermal dissipation by means of the xanthophyll cycle (converting violaxanthin to zeaxanthin) [53]. β-Carotene directly participates in light absorption, absorbing light in the blue region at 448 and 454 nm [49].

Differences between red and green lettuce [28,54] and basil [55] in growth, antioxidant levels, and photosynthetic response to red LED parameters were reported, which highlighted that red (purple) cultivars are less sensitive to environmental impacts. Similarly, in our study, according the PCA analysis, the two species with red (amaranth) and green (turnip greens) leaves showed a distinct response under the same lighting conditions.

4. Materials and Methods

This study was conducted on two microgreens: amaranth (*Amaranthus tricolor* L.) and turnip greens (*Brassica rapa* L. subsp. *oleifera* (DC.) Metzg) (CN Seeds, Ltd., Pymoor, Ely, Cambridgeshire, UK).

The experiment was performed in a controlled-environment growth chamber. Day and night temperature was maintained at 24 ± 2 °C within a 16/8 h light/dark photoperiod and a relative humidity of 50/60% was maintained. During the experiments, the air temperature and relative humidity (RH%) were measured using a meteo station (Avidsen Italia). Plantlets were grown in sowing substrate ('Brill® Semina Bio', Agrochimica S.p.A., Bolzano, Italy) and vermiculite in containers (14×9 cm) for 10 days from sowing to harvest.

Three containers (i.e., three replicates) were used for each experimental treatment.

Light-emitting diode (LED)-based lighting units, consisting of commercially avail-able LEDs with emission wavelengths of (1) white LED (W) (LEDW—blue 21%; green 38%; red 35%; dark red 6%—Grow Light C65 NS12—Valoya Oy Helsinki, Finland), (2) blue LED (B) (LEDR/B, BS Biosystem, Catania), and (3) red LED (R) (100% BS Biosystem, Catania), were used for microgreen lighting. The measured photosynthetic photon flux density (PPFD) sources (i.e., at the pot top level) were 200 ± 5 μmol for all the sectors. Spectral outputs from the various LED lamps were verified using a calibrated spectroradiometer LI-190R (LiCor, Inc., Lincoln, NE, USA, LICOR Biosciences).

4.1. Chemicals and Reagents

Analytical reagent-grade chemicals and bi-distilled water were used throughout this experiment. Methanol used was of HPLC-grade, ≥99.9%, CHROMASOLV™ (Honeywell Riedel-de Haën™); KNO_3, acetone (Multisolvent® HPLC grade), NaOH, and H_2SO_4 were purchased from Merck KGaA, Darmstadt, Germany. Methyl viologen, oxalic acid anhydrous 99%, salicylic acid acs 99%, sodium carbonate (Na_2CO_3), glucose solution, anthrone 97%, L-ascorbic acid, Folin–Ciocâlteu reagent, DPPH• radical reagent, Trolox, and gallic acid were purchased from Merck KGaA, Darmstadt, Germany. Standard solutions were prepared with bi-distilled water.

4.2. Measurement and Data Collection of Growth Parameters

At harvest time, morphological parameters, seedling fresh biomass (g), seedling dry biomass (g), and seedling height (cm) were measured. The height (H), fresh weight (FW), and dry weight (DW) were determined on 15 seedlings, randomly selected within each container. The weight was expressed as micrograms per seedling. The dry biomass (DW) of the plants was obtained by putting weighed samples in a thermo-ventilated oven at 70 °C until they reached a constant weight. Stem and leaves were immersed in liquid nitrogen and kept at −80 °C for phytochemical analysis. The plant height/plant dry weight ratio (H/DW, cm·mg^{-1}) was also calculated. For all chemical analysis, three replicates were performed.

4.3. Chlorophyll and Carotenoid Pigments

The contents of chlorophyll (Chl *a*, Chl *b*, and total Chl) and carotenoids was analyzed using the spectrophotometric method. Samples of 150 mg were extracted using 99% methanol and incubated in dark room (4 °C for 24 h). The absorbance of samples was read at 665.2 nm, 652.4 nm, and 470 nm, respectively, for Chl *a*, Chl *b*, and carotenoids in a spectrophotometer (7315 Spectrophotometer, Jenway, Staffordshire, UK). Chlorophyll and carotenoid contents were calculated as described by Lichtenthaler et al. [56].

$\text{Chl } a = 16.75A_{665.2} - 9.16A_{652.4}$.

$\text{Chl } b = 34.09A_{652.4} - 15.28A_{665.2}$.

$\text{Carotenoids} = (1000A_{470} - 1.63\text{Chl } a - 104.96\text{Chl } b)/221$.

4.4. Total Sugars

The total sugars were determined spectrophotometrically following the anthrone method with slight modifications [57]. The anthrone reagent (10.3 mM) was prepared by dissolving anthrone in ice-cold 95% H_2SO_4. The reagent was left to stand for 30–40 min before use. Then, 1 g of fresh sample was extracted in 3 mL of distilled water and centrifuged at 3000× *g* for 15 min at room temperature (RT). Next, 0.5 mL of extract was placed on top of 2.5 mL of anthrone reagent incubated in ice for 5 min. The reactions were heated to 95 °C for 10 min and left to cool in ice. The absorbance was read at 620 nm. A calibration curve was generated using a glucose solution (0 to 0.05 mg·mL^{-1}) (R^2 = 0.9995).

4.5. Nitrate Concentrations

Nitrate concentrations were determined following the salicyl sulfuric acid method [58]. First, 1 g of fresh sample was homogenized in 3 mL of distilled water and then centrifuged (4000 rpm, 15 min), collecting the supernatant. Then, 20 µL of extract was added to 80 µL of 5% salicylic acid in sulfuric acid and to 3 mL of NaOH 1.5 N. The samples were cooled, and the spectrophotometer readings were read at 410 nm. A calibration curve was generated using a KNO_3 standard (0, 1, 2.5, 5, 7.5, and 10 mM KNO_3) (R^2 = 0.9918).

4.6. Ascorbic Acid Analysis

The ascorbic acid content was determined using a spectrophotometric method [59]. Fresh plant tissue (1 g) was homogenized in 10 mL of 5% oxalic acid and then centrifuged (5 min, 4000 rpm). The extract (1 mL) was added to 2 mL of 0.1% methyl viologen and 2 mL of 2 mol·L^{-1} NaOH. The colored radical ion was read at 600 nm against the radical blank. The concentration of ascorbic acid was calculated as a function of the values obtained from the L-ascorbic acid standard curve (100–500 µg·mL^{-1}) (R^2 = 0.9907). Results were expressed as mg·g^{-1} fresh weight.

4.7. Total Phenolic Compounds and 2,2-Diphenyl-1-picrylhydrazyl (DPPH) Radical-Scavenging Activity

First, 1 g FW of sample was homogenized in a solution containing 50% acetone and 50% water (1:10). The samples were vortexed and incubated for 15 h at 20 °C. Then, 100 µL of supernatant was mixed with 0.5 mL of Folin–Ciocâlteu reagent (Sigma-Aldrich, Italy) and 6 mL of distilled water. Next, 1.5 mL of Na_2CO_3 (20%) was added, before incubating at 20 °C for 2 h. The absorbance was read at 765 nm. The concentration of total phenolic compounds was calculated as a function of the values obtained from the gallic acid standard curve (0, 50, 100, 250, and 500 mg·L^{-1}) (R^2 = 0.9954). The total phenolic content was expressed as mg·100 g^{-1} gallic acid equivalent.

The antioxidant activity was determined using DPPH. About 1 g of fresh weight was mixed with 1.5 mL of methanol solution (80%), sonicated for 30 min, and centrifuged (10 min, 5 °C, 5000× *g*). Then, 0.01 mL of supernatant was mixed with 1.4 mL of 150 µM DPPH solution in methanol and water (95:5), before incubating for 30 min in the dark. The sample was read at 517 nm. The antioxidant activity was calculated as a function of the values obtained from the Trolox standard curve (0 to 0.5 mg·mL^{-1}) (R^2 = 0.9995). DPPH scavenging activity values were expressed as Trolox equivalent antioxidant activity (mg TE·100 g^{-1}).

4.8. Meso and Micro Elements

Meso and micro element (Na, Mg, K, Ca, Mn, Fe, Ni, Cu, Zn, and P) concentrations were determined on oven-dried samples (80 °C for 48 h). Samples of 300 mg of dry matter were mineralized at 120 °C in 5 mL of 14.4 M HNO_3, clarified with 1.5 mL of 33% H_2O_2, and dried at 80 °C. The mineralized material was solubilized in 5 mL of 1 M HNO_3 and filtered on a 0.45 µm nylon membrane. Mineral elements were measured using inductively coupled plasma mass spectroscopy (ICP-MS; Varian 820-MS, ICP Mass Spectrometer). Concentrations of mineral elements were expressed on a dry weight basis.

4.9. RGB Color Analysis

Photos of microgreens grown in different treatments were taken at 30 cm distance. The colors of the photos were analyzed using online tools (https://imagecolorpicker.com/, accessed on 8 June 2021) for the measurements the RGB components.

4.10. Statistical Analysis

The experiment was performed using a completely randomized design. Three biological replicates were used for the analysis. Data were subjected to two-way ANOVA, and differences among means were determined using Tukey's post hoc test ($p \leq 0.05$) All statistical analyses were performed using CoStat release 6.311 (CoHort Software, Monterey, CA, USA). The principal component loading plot and scores of PCA were obtained using Minitab 16, LLC. The data presented in the figures are the means ± *se* (Graphpad 7.0).

5. Conclusions

The results obtained with our trial indicate that the following:

- blue light was particularly effective in enhancing the growth and nutritional characteristics (particularly antioxidant activity) of the two studied microgreens as compared to the more traditionally used white light;
- red light seemed to be more effective than white light in promoting fresh biomass accumulation and hypocotyl growth. However, its effects on nutraceutical characteristics were quite different for the two genotypes, since it did not influence those of turnip greens but worsened those of amaranth (see nitrates, nickel, and total polyphenol contents) as compared to the other lights;
- the response to the spectral system is typically species-specific; for this reason, it is possible to adopt a specific light formula that allows maximizing both plant growth and nutritional quality, thereby enhancing the microgreen industry.

Supplementary Materials: The following are available online at https://www.mdpi.com/article/10.3390/plants10081584/s1. Figure S1: Amaranth and turnip greens 10 days after sowing under different LED treatments; Figure S2: Microgreens in growth chamber under different LED treatments (white and blue).

Author Contributions: Conceptualization, S.T. and D.R.; methodology, S.T.; software, S.T.; validation, D.R.; formal analysis, S.T. and A.F.; investigation, S.T. and D.R.; data curation, S.T. and C.P.; writing—original draft preparation, S.T., V.C., A.F., D.R. and C.P.; writing—review and editing, S.T., V.C., A.F., D.R. and C.P.; visualization, D.R.; supervision, V.C., A.F., D.R. and C.P. All authors have read and agreed to the published version of the manuscript.

Funding: This research received no external funding.

Institutional Review Board Statement: Not applicable.

Informed Consent Statement: Not applicable.

Data Availability Statement: Main data are contained within the article; further data presented in this study are available on request from the corresponding author.

Acknowledgments: The authors would like to thank Salvatore La Rosa and Alessandra Pellegrino for their technical assistance.

Conflicts of Interest: The authors declare no conflict of interest.

References

1. Ward, J.M.; Cufr, C.A.; Denzel, M.A.; Neff, M.M. The Dof transcription factor OBP3 modulates phytochrome and cryptochrome signaling in arabidopsis. *Plant Cell* **2005**, *17*, 475–485. [CrossRef]
2. Pérez-Balibrea, S.; A Moreno, D.; García-Viguera, C. Influence of light on health-promoting phytochemicals of broccoli sprouts. *J. Sci. Food Agric.* **2008**, *88*, 904–910. [CrossRef]
3. Liu, X.; Chang, T.; Guo, S.; Xu, Z.; Li, J. Effect of different light quality of led on growth and photosynthetic character in cherry tomato seedling. *Acta Hortic.* **2011**, 325–330. [CrossRef]
4. Batista, D.; Felipe, S.H.S.; Silva, T.D.; De Castro, K.M.; Mamedes-Rodrigues, T.C.; Miranda, N.; Ríos, A.M.R.; Faria, D.; Fortini, E.A.; Chagas, K.; et al. Light quality in plant tissue culture: Does it matter? *In Vitro Cell. Dev. Biol. Anim.* **2018**, *54*, 195–215. [CrossRef]
5. Gupta, S.D.; Agarwal, A.; Pradhan, S. Phytostimulatory effect of silver nanoparticles (AgNPs) on rice seedling growth: An insight from antioxidative enzyme activities and gene expression patterns. *Ecotoxicol. Environ. Saf.* **2018**, *161*, 624–633. [CrossRef]
6. Gupta, S.D. *Light Emitting Diodes for Agriculture*; Springer: Singapore, 2017; p. 334.
7. Gupta, S.D.; Jatothu, B. Fundamentals and applications of light-emitting diodes (LEDs) in in vitro plant growth and morphogenesis. *Plant Biotechnol. Rep.* **2013**, *7*, 211–220. [CrossRef]
8. Darko, E.; Heydarizadeh, P.; Schoefs, B.; Sabzalian, M.R. Photosynthesis under artificial light: The shift in primary and secondary metabolism. *Philos. Trans. R. Soc. Biol. Sci.* **2014**, *369*, 20130243. [CrossRef]
9. Bian, Z.-H.; Cheng, R.-F.; Yang, Q.-C.; Wang, J.; Lu, C. Continuous light from red, blue, and green light-emitting diodes reduces nitrate content and enhances phytochemical concentrations and antioxidant capacity in lettuce. *J. Am. Soc. Hortic. Sci.* **2016**, *141*, 186–195. [CrossRef]
10. Zhang, C.; Liu, J.; Zhang, Y.; Cai, X.; Gong, P.; Zhang, J.; Wang, T.; Li, H.; Ye, Z. Overexpression of SlGMEs leads to ascorbate accumulation with enhanced oxidative stress, cold, and salt tolerance in tomato. *Plant Cell Rep.* **2010**, *30*, 389–398. [CrossRef]
11. Ntagkas, N.; Woltering, E.J.; Marcelis, L.F. Light regulates ascorbate in plants: An integrated view on physiology and biochemistry. *Environ. Exp. Bot.* **2018**, *147*, 271–280. [CrossRef]
12. Signore, A.; Bell, L.; Santamaria, P.; Wagstaff, C.; Van Labeke, M.-C. Red light is effective in reducing nitrate concentration in rocket by increasing nitrate reductase activity, and contributes to increased total glucosinolates content. *Front. Plant Sci.* **2020**, *11*, 604. [CrossRef]
13. Shukla, M.R.; Singh, A.S.; Piunno, K.; Saxena, P.K.; Jones, A.M.P. Application of 3D printing to prototype and develop novel plant tissue culture systems. *Plant Methods* **2017**, *13*, 1–10. [CrossRef]
14. Terashima, I.; Fujita, T.; Inoue, T.; Chow, W.S.; Oguchi, R. Green light drives leaf photosynthesis more efficiently than red light in strong white light: Revisiting the enigmatic question of why leaves are green. *Plant Cell Physiol.* **2009**, *50*, 684–697. [CrossRef]
15. Girardi, F.M.; Barra, M.B.; Zettler, C.G. Papillary thyroid carcinoma: Does the association with Hashimoto's thyroiditis affect the clinicopathological characteristics of the disease? *Braz. J. Otorhinolaryngol.* **2015**, *81*, 283–287. [CrossRef]
16. Johkan, M.; Shoji, K.; Goto, F.; Hashida, S.-N.; Yoshihara, T. Blue light-emitting diode light irradiation of seedlings improves seedling quality and growth after transplanting in red leaf lettuce. *HortScience* **2010**, *45*, 1809–1814. [CrossRef]

17. Kalaitzoglou, P.; Van Ieperen, W.; Harbinson, J.; Van Der Meer, M.; Martinakos, S.; Weerheim, K.; Nicole, C.C.S.; Marcelis, L.F.M. Effects of continuous or end-of-day far-red light on tomato plant growth, morphology, light absorption, and fruit production. *Front. Plant Sci.* **2019**, *10*, 322. [CrossRef]
18. Park, J.-W.; Kang, P.; Park, H.; Oh, H.-Y.; Yang, J.-H.; Kim, Y.-H.; Kwon, S.-K. Synthesis and properties of blue-light-emitting anthracene derivative with diphenylamino-fluorene. *Dye Pigment* **2010**, *85*, 93–98. [CrossRef]
19. Johkan, M.; Shoji, K.; Goto, F.; Hahida, S.; Yoshihara, T. Effect of green light wavelength and intensity on photomorphogenesis and photosynthesis in *Lactuca sativa*. *Environ. Exp. Bot.* **2012**, *75*, 128–133. [CrossRef]
20. Kwon, Y.; Sunesh, C.D.; Choe, Y. Light-emitting properties of cationic iridium complexes containing phenanthroline based ancillary ligand with blue-green and green emission colors. *Opt. Mater.* **2015**, *39*, 40–45. [CrossRef]
21. Kyriacou, M.; Rouphael, Y.; Di Gioia, F.; Kyratzis, A.; Serio, F.; Renna, M.; De Pascale, S.; Santamaria, P. Micro-scale vegetable production and the rise of microgreens. *Trends Food Sci. Technol.* **2016**, *57*, 103–115. [CrossRef]
22. Kyriacou, M.; Soteriou, G.A.; Colla, G.; Rouphael, Y. The occurrence of nitrate and nitrite in Mediterranean fresh salad vegetables and its modulation by preharvest practices and postharvest conditions. *Food Chem.* **2019**, *285*, 468–477. [CrossRef]
23. Xiao, Z.; Rausch, S.R.; Luo, Y.; Sun, J.; Yu, L.; Wang, Q.; Chen, P.; Yu, L.; Stommel, J.R. Microgreens of Brassicaceae: Genetic diversity of phytochemical concentrations and antioxidant capacity. *LWT* **2019**, *101*, 731–737. [CrossRef]
24. Turner, E.R.; Luo, Y.; Buchanan, R.L. Microgreen nutrition, food safety, and shelf life: A review. *J. Food Sci.* **2020**, *85*, 870–882. [CrossRef]
25. Xiao, Z.; Lester, G.E.; Park, E.; Saftner, R.A.; Luo, Y.; Wang, Q. Evaluation and correlation of sensory attributes and chemical compositions of emerging fresh produce: Microgreens. *Postharvest Biol. Technol.* **2015**, *110*, 140–148. [CrossRef]
26. Ilakiya, T.; Parameswari, E.; Davamani, V.; Prakash, E. Microgreens combating malnutrition problem. *Biot. Res. Today* **2020**, *2*, 110–112.
27. Brazaitytė, A.; Sakalauskienė, S.; Samuolienė, G.; Jankauskienė, J.; Viršilė, A.; Novičkovas, A.; Sirtautas, R.; Miliauskienė, J.; Vaštakaitė-Kairienė, V.; Dabašinskas, L.; et al. The effects of LED illumination spectra and intensity on carotenoid content in Brassicaceae microgreens. *Food Chem.* **2015**, *173*, 600–606. [CrossRef]
28. Samuolienė, G.; Sirtautas, R.; Brazaitytė, A.; Duchovskis, P. LED lighting and seasonality effects antioxidant properties of baby leaf lettuce. *Food Chem.* **2012**, *134*, 1494–1499. [CrossRef]
29. Samuolienė, G.; Brazaitytė, A.; Viršilė, A.; Jankauskienė, J.; Sakalauskienė, S.; Duchovskis, P. Red light-dose or wave-length-dependent photoresponse of antioxidants in herb microgreens. *PLoS ONE* **2016**, *11*, e0163405. [CrossRef]
30. Craver, J.K.; Gerovac, J.R.; Lopez, R.G.; Kopsell, D.A. Light intensity and light quality from sole-source light-emitting diodes impact phytochemical concentrations within Brassica microgreens. *J. Am. Soc. Hortic. Sci.* **2017**, *142*, 3–12. [CrossRef]
31. Lobiuc, A.; Vasilache, V.; Oroian, M.; Stoleru, T.; Burducea, M.; Pintilie, O.; Zamfirache, M.-M. Blue and red LED illumination improves growth and bioactive compounds contents in *Acyanic* and *Cyanic Ocimum basilicum* L. microgreens. *Molecules* **2017**, *22*, 2111. [CrossRef]
32. Kyriacou, M.C.; El-Nakhel, C.; Pannico, A.; Graziani, G.; Soteriou, G.A.; Giordano, M.; Zarrelli, A.; Ritieni, A.; De Pascale, S.; Rouphael, Y. Genotype-specific modulatory effects of select spectral bandwidths on the nutritive and phytochemical composition of microgreens. *Front. Plant Sci.* **2019**, *10*, 1501. [CrossRef]
33. Zhang, X.; Bian, Z.; Yuan, X.; Chen, X.; Lu, C. A review on the effects of light-emitting diode (LED) light on the nutrients of sprouts and microgreens. *Trends Food Sci. Technol.* **2020**, *99*, 203–216. [CrossRef]
34. Brazaitytė, A.; Miliauskienė, J.; Vaštakaitė-Kairienė, V.; Sutulienė, R.; Laužikė, K.; Duchovskis, P.; Małek, S. Effect of different ratios of blue and red LED light on Brassicaceae microgreens under a controlled environment. *Plants* **2021**, *10*, 801. [CrossRef]
35. Lee, S.-W.; Seo, J.M.; Lee, M.-K.; Chun, J.-H.; Antonisamy, P.; Arasu, M.V.; Suzuki, T.; Al-Dhabi, N.A.; Kim, S.-J. Influence of different LED lamps on the production of phenolic compounds in common and Tartary buckwheat sprouts. *Ind. Crops Prod.* **2014**, *54*, 320–326. [CrossRef]
36. Wu, M.-C.; Hou, C.-Y.; Jiang, C.-M.; Wang, Y.-T.; Wang, C.-Y.; Chen, H.-H.; Chang, H.-M. A novel approach of LED light radiation improves the antioxidant activity of pea seedlings. *Food Chem.* **2007**, *101*, 1753–1758. [CrossRef]
37. Viršilė, A.; Brazaitytė, A.; Vaštakaitė-Kairienė, V.; Miliauskienė, J.; Jankauskienė, J.; Novičkovas, A.; Laužikė, K.; Samuolienė, G. The distinct impact of multi-color LED light on nitrate, amino acid, soluble sugar and organic acid contents in red and green leaf lettuce cultivated in controlled environment. *Food Chem.* **2020**, *310*, 125799. [CrossRef]
38. Zhang, T.; Shi, Y.; Piao, F.; Sun, Z. Effects of different LED sources on the growth and nitrogen metabolism of lettuce. *Plant Cell Tissue Organ Cult. (PCTOC)* **2018**, *134*, 231–240. [CrossRef]
39. Ying, Q.; Kong, Y.; Jones-Baumgardt, C.; Zheng, Y. Responses of yield and appearance quality of four Brassicaceae microgreens to varied blue light proportion in red and blue light-emitting diodes lighting. *Sci. Hortic.* **2020**, *259*, 108857. [CrossRef]
40. Li, H.; Tang, C.; Xu, Z.; Liu, X.; Han, X. Effects of different light sources on the growth of non-heading Chinese cabbage (*Brassica campestris* L.). *J. Agric. Sci.* **2012**, *4*, 262. [CrossRef]
41. Mizuno, T.; Amaki, W.; Watanabe, H. Effects of monochromatic light irradiation by led on the growth and anthocyanin contents in leaves of cabbage seedlings. *Acta Hortic.* **2011**, 179–184. [CrossRef]
42. Li, Q.; Kubota, C. Effects of supplemental light quality on growth and phytochemicals of baby leaf lettuce. *Environ. Exp. Bot.* **2009**, *67*, 59–64. [CrossRef]

43. European Food Safety Authority. Opinion of the scientific panel on contaminants in the food chain on a request for the European Commission to perform a scientific risk assessment on nitrate in vegetables. *EFSA J.* **2008**, *689*, 1–79. Available online: https://seguridadalimentaria.elika.eus/wp-content/uploads/articulos/Archivo291/CONTAM_NitratosVeg08.pdf (accessed on 15 June 2021).
44. European Commission Commission. Regulation (EU) No 1258/2011 of 2 December 2011 amending regulation (EC) No. 1881/2006 as regards maximum levels for nitrates in foodstuffs. *Off. J. Eur. Union* **2011**, *320*, 15–17.
45. Nam, T.G.; Kim, D.-O.; Eom, S.H. Effects of light sources on major flavonoids and antioxidant activity in common buckwheat sprouts. *Food Sci. Biotechnol.* **2017**, *27*, 169–176. [CrossRef] [PubMed]
46. Qian, H.; Liu, T.; Deng, M.; Miao, H.; Cai, C.; Shen, W.; Wang, Q. Effects of light quality on main health-promoting compounds and antioxidant capacity of Chinese kale sprouts. *Food Chem.* **2016**, *196*, 1232–1238. [CrossRef]
47. Długosz-Grochowska, O.; Kołton, A.; Wojciechowska, R. Modifying folate and polyphenol concentrations in Lamb's lettuce by the use of LED supplemental lighting during cultivation in greenhouses. *J. Funct. Foods* **2016**, *26*, 228–237. [CrossRef]
48. Liu, H.; Chen, Y.; Hu, T.; Zhang, S.; Zhang, Y.; Zhao, T.; Yu, H.; Kang, Y. The influence of light-emitting diodes on the phenolic compounds and antioxidant activities in pea sprouts. *J. Funct. Foods* **2016**, *25*, 459–465. [CrossRef]
49. Lefsrud, M.G.; Kopsell, D.; Sams, C.E. Irradiance from distinct wavelength light-emitting diodes affect secondary metabolites in kale. *HortScience* **2008**, *43*, 2243–2244. [CrossRef]
50. Kopsell, D.A.; Sams, C.E. Increases in shoot tissue pigments, glucosinolates, and mineral elements in sprouting broccoli after exposure to short-duration blue light from light emitting diodes. *J. Am. Soc. Hortic. Sci.* **2013**, *138*, 31–37. [CrossRef]
51. Cazzaniga, S.; Li, Z.; Niyogi, K.K.; Bassi, R.; Dall'Osto, L. The Arabidopsis szl1 mutant reveals a critical role of b-carotene in photosystem I photoprotection. *Plant Physiol.* **2012**, *159*, 1745–1758. [CrossRef]
52. Samuolienė, G.; Brazaitytė, A.; Viršilė, A.; Miliauskienė, J.; Vaštakaitė-Kairienė, V.; Duchovskis, P. Nutrient levels in Brassicaceae Microgreens increase under tailored light-emitting diode spectra. *Front. Plant Sci.* **2019**, *10*, 1475. [CrossRef] [PubMed]
53. Stange, C.; Flores, C. Carotenoids and photosynthesis regulation of carotenoid biosyntesis by photoreceptors. In *Advances in Photosynthesis: Fundamental Aspects*; InTech: Rijekia, Croatia, 2012. [CrossRef]
54. Son, K.-H.; Oh, M.-M. Growth, photosynthetic and antioxidant parameters of two lettuce cultivars as affected by red, green, and blue light-emitting diodes. *Hortic. Environ. Biotechnol.* **2015**, *56*, 639–653. [CrossRef]
55. Hosseini, A.; Mehrjerdi, M.Z.; Aliniaeifard, S.; Seif, M. Photosynthetic and growth responses of green and purple basil plants under different spectral compositions. *Physiol. Mol. Biol. Plants* **2019**, *25*, 741–752. [CrossRef] [PubMed]
56. Lichtenthaler, H.K. Chlorophylls and carotenoids: Pigments of photosynthetic biomembranes. *Method Enzymol.* **1987**, *148*, 350–382. [CrossRef]
57. Cocetta, G.; Rossoni, M.; Gardana, C.; Mignani, I.; Ferrante, A.; Spinardi, A. Methyl jasmonate affects phenolic metabolism and gene expression in blueberry (*Vaccinium corymbosum*). *Physiol. Plant.* **2014**, *153*, 269–283. [CrossRef]
58. Cataldo, D.A.; Maroon, M.; Schrader, L.E.; Youngs, V.L. Rapid colorimetric determination of nitrate in plant tissue by nitration of salicylic acid. *Commun. Soil Sci. Plant Anal.* **1975**, *6*, 71–80. [CrossRef]
59. Janghel, E.; Gupta, V.; Rai, M.; Rai, J. Micro determination of ascorbic acid using methyl viologen. *Talanta* **2007**, *72*, 1013–1016. [CrossRef]

plants

MDPI

Review

Comparative Effects of Different Light Sources on the Production of Key Secondary Metabolites in Plants In Vitro Cultures

Mariam Hashim [1], Bushra Ahmad [2], Samantha Drouet [3], Christophe Hano [3], Bilal Haider Abbasi [4] and Sumaira Anjum [1,*]

1 Department of Biotechnology, Kinnaird College for Women, Jail Road, Lahore 54000, Pakistan; mariamhashim07@gmail.com
2 Shaheed Benazir Bhutto Women University, Peshawar 25000, Pakistan; bushraahmad@sbbwu.edu.pk
3 Laboratoire de Biologie des Ligneux et des Grandes Cultures, INRAE USC1328, Eure & Loir Campus, University of Orleans, 28000 Chartres, France; samantha.drouet@univ-orleans.fr (S.D.); hano@univ-orleans.fr (C.H.)
4 Department of Biotechnology, Quaid-i-Azam University, Islamabad 15320, Pakistan; bhabbasi@qau.edu.pk
* Correspondence: sumaira.anjum@kinnaird.edu.pk; Tel.: +92-30-0695-7038

Abstract: Plant secondary metabolites are known to have a variety of biological activities beneficial to human health. They are becoming more popular as a result of their unique features and account for a major portion of the pharmacological industry. However, obtaining secondary metabolites directly from wild plants has substantial drawbacks, such as taking a long time, posing a risk of species extinction owing to over-exploitation, and producing a limited quantity. Thus, there is a paradigm shift towards the employment of plant tissue culture techniques for the production of key secondary metabolites in vitro. Elicitation appears to be a viable method for increasing phytochemical content and improving the quality of medicinal plants and fruits and vegetables. In vitro culture elicitation activates the plant's defense response and increases the synthesis of secondary metabolites in larger proportions, which are helpful for therapeutic purposes. In this respect, light has emerged as a unique and efficient elicitor for enhancing the in vitro production of pharmacologically important secondary metabolites. Various types of light (UV, fluorescent, and LEDs) have been found as elicitors of secondary metabolites, which are described in this review.

Keywords: LED light; fluorescent light; UV light; elicitation; plant secondary metabolites; plant in vitro cultures

Citation: Hashim, M.; Ahmad, B.; Drouet, S.; Hano, C.; Abbasi, B.H.; Anjum, S. Comparative Effects of Different Light Sources on the Production of Key Secondary Metabolites in Plants In Vitro Cultures. *Plants* **2021**, *10*, 1521. https://doi.org/10.3390/plants10081521

Academic Editors: Valeria Cavallaro and Rosario Muleo

Received: 7 July 2021
Accepted: 23 July 2021
Published: 26 July 2021

1. Introduction

Plants are complex species and have gained importance due to their nutritional and pharmaceutical values. Apart from the production of primary metabolites such as carbohydrates, lipids, and proteins that plants need for their growth and development, low molecular weight organic compounds involved in defense against stress conditions called secondary metabolites are also synthesized by higher plants [1]. Secondary metabolites are involved in the production of pharmaceuticals, industrially important biochemicals, food additives, and flavors [2]. The production of secondary metabolites in the wild is limited to some re-gional and environmental constraints, which limit the production of compounds commercially [3]. Traditional cultivation of certain types of plants is often difficult and may take several years for their growth [4].Recent trends have focused on developing in vitro culture techniques as a convenient alternative to cope with the demand for medicinal plants, as more than 60% of anti-cancer drugs are manufactured directly or indirectly from plants [5,6]. In vitro cultures are an efficient means of production of biomass, leading to rapid growth and consistent metabolite productivity [7]. In addition, elicitation has proved beneficial in the production of in vitro cultures [2]. Usually, under stress conditions or environment variability, the output of these compounds is improved to cope with se-vere stress effects. For scaling up the development of these phytochemicals, in vitro techniques can prove beneficial.

Elicitation, where certain pathways are activated by introducing agents (elicitors) for triggering a plant's defense mechanisms, is amongst the most relevant and effec-tive techniques [3,8]. To increase the production of secondary metabolites, many biotic and abiotic elicitors are used. Light is, however, an influential abiotic elicitor that af-fects the growth, development, and morphogenesis in plants [9,10]. Light also plays a critical role in controlling primary and secondary metabolism in order to achieve op-timum growth in plants [11–13]. Light stress has been designed to increase secondary metabolite production from various in vitro cultures of medicinally im-portant plants [12,14]. The signaling, regulatory, and metabolic mechanisms involved in eliciting secondary metabolites, as well as the mechanism of light precipitation, are not thoroughly characterized in the literature. However, it is reasonable to speculate that oxidative stress, in addition to other mechanisms, plays a significant role in light perception and signaling. Oxidative damage produced as a result of environmental stress leads to the production of highly reactive free radicals that halts the growth and development of plants [15–17]. To counteract the effect of these radicals, plants have natural antioxi-dant defense mechanisms that are involved in producing a wide range of secondary metabolites [18–20].

The improved production of various valuable secondary metabolites through light elicitation has unlocked a new area of research that could have significant economic benefits for the pharmaceutical and nutraceutical industry. To date, different sources of light such as ultraviolet (UV), light-emitting diodes (LED) and fluorescent lights have been reported as efficient elicitors of pharmacologically important secondary metabolites, as summarized in Figure 1 [21–23]. These light sources have been used either alone or in combination with each other in order to maximize the production of valuable metabolites in in vitro cultures of plants. In this review, in-depth literature on the role of light as an elicitor of valuable secondary metabolites has been critically reviewed. Furthermore, the mechanistic aspects of various sources of light as elicitors of secondary metabolites through activation/regulation of various genes are also discussed.

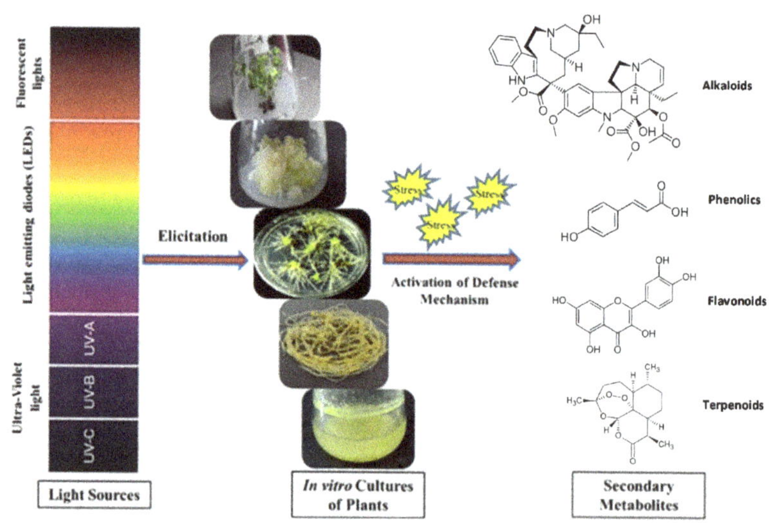

Figure 1. An overview of light's function as an elicitor of important secondary metabolites in various in vitro plant cultures maintained under controlled conditions, including shoot, callus hairy root,

adventitious root, and cell suspension cultures (from top to bottom). Different light sources, including UV light but also excessive light, can cause stress and activate the defense response, resulting in the production of a variety of bioactive plant secondary metabolites such as alkaloids (e.g., vinblastine), phenolics (e.g., *p*-coumaric acid), flavonoids (e.g., quercetin), or terpenoids (e.g., artemisinin).

2. Light as an Elicitor

In vitro cultures of several plant species have been documented to elicit secondary metabolites using a variety of light sources in the past. These sources have been classified into three main categories in this review as UV lights, LED lights, and fluorescent lights, compared and discussed herein in detail.

2.1. UV Lights

A significant abiotic elicitor used in the past to boost the production of secondary metabolites in a variety of plant cultures is UV [24]. The wavelength of UV (400–200 nm) accounts for only a small por-tion of the solar radiation that reaches the earth's surface, yet it has a significant bio-logical impact on living species, including plants. UV radiation is categorized into three parts: UV-A (320 to 390 nm), UV-B (280 to 320 nm), and UV-C (280 nm and be-low). The intensity of UV-A is greater than UV-B, but this difference is not biologically significant [25]. UV-B photons are perhaps the most intense wave that hits the surface of the earth, and even tiny changes in their quantity can have a huge impact on vital processes and properties at all scales, from species to ecosystems [26]. In addition to the rest of the UV groups, UV-C radiation has also proved to be the most effective in stimulating the production of plant secondary metabolites such as phenolics, alkaloids, or flavonoids [27–29] (Table 1).

Table 1. Effects of UV light as an elicitor of key secondary metabolites in controlled and in vitro plant culture systems.

Light Sources	Intensity	Exposure Time	Plant Species	Culture System	Secondary Metabolite	Yield Increase	References
UV-A/B	4–5 Wm^{-2}/10–14 Wm^{-2}	3 h per day for 16 days	*Capsicum annum*	CGC (leaf)	Cynaroside	-	[30]
UV-B	73.08 $kJ/m^2/day$	7 h per day for 13 days	*Nymphoides humboldtiana*	CGC (leaf)	Flavonoids	-	[31]
UV-B	40 J/cm^2	8 h	*Alternanthera Sessilis* *Alternanthera Brasiliana*	CGC (shoot)	Flavonoids	51% 62%	[32]
UV-B	1.14 $kJ/m^2/day$	4 h per day for 14 days	*Capsicum annuum*	CGC (leaf)	Flavonoids	-	[33]
UV-B	313 nm	7–14 days	*Ginkgo biloba*	CGC (leaf)	Quercetin, kaempferol, and isorhamnetin	2.05- to 2.4-fold and 16.67- to −42-fold, respectively	[34]
UV-B	1.26 $\mu W/cm^2$	5 min	*Catharanthus roseus*	Cell suspension culture	Catharanthine and vindoline	3-fold and 12-fold	[35]
UV-B	224 $\mu mol\ m^{-2}\ s^{-1}$	1 h for 2 days and 2 h for 2 days	*Ocimum basilicum* Green basil	CGC (leaf)	Anthocyanin, phenolics, and flavonoids	9–23%, 28–126% and 80–169%, respectively	[21]
		2 h for 2 days and 2 h for five days	*Ocimum basilicum* Purple Basil		Phenolics and flavonoids	29–63% and 37–79%	
UV-B	102 $kJ/m^2/day$	3 days	*Ocimum basilicum*	CGC (plantlet)	Phenolics		[36]
UV-B	20 $\mu W/cm^2$	4 days	*Triticum aestivium*	CGC (seedling)	Phenolics	26.3%	[37]
UV-C	254 nm	15 min	*Vitis vinifera*	Callus culture	*trans*-resveratrol	26-fold	[38]
UV-C	254 nm	5 min after 24 h 10 min after 48 h	*Vitis vinifera*	Callus culture	*trans*-resveratrol Catechin	8-fold -	[39]
UV-C	3 W/m^2	60 min	*Lepidium sativum*	Callus culture	Chlorogenic acid, kaemferol, and quercetin	2.5-fold	[40]

Table 1. *Cont.*

Light Sources	Intensity	Exposure Time	Plant Species	Culture System	Secondary Metabolite	Yield Increase	References
UV-C + Photoperiod UV-C + Dark	3.6 kJ/m^2 + 16/8 h 1.8 kJ/m^2 + 24 h dark	10–60 min	*Linum usitatissimum*	Callus culture	Secoisolariciresinol diglucoside (SDG)	1.86-fold	[41]
					Lariciresinol diglucoside (LDG)	2.25-fold	
					Guaiacylglycerol-β-coniferyl alcohol ether glucoside (GGCG)	1.33-fold	
					Total phenolic production	2.82-fold	
					Total flavonoid production	2.94-fold	
					Dehydrodiconiferyl alcohol glucoside (DCG)	1.36-fold	
UV-C	3 W/m^2	10 min	*Ocimum basilicum*	Callus culture	Rosmarinic acid	2.3-fold	[42]
					Chichoric acid and cyanide	4.1-fold	
					Peonidin	2.7-fold	
UV-C	254 nm	60 min	*Echinacea purpurea*	Callus culture Cell suspension culture	Phenolics	-	[43]

CGC = controlled growth chamber.

2.1.1. UV-A and UV-B

Since UV light acts as an elicitor, it is basically involved in activating the defense mechanisms of plants which in turn produce secondary metabolites useful for humans for therapeutic purposes as they are not required by plants for their growth [28]. UV-A light can act as a potential elicitor to stimulate the production of secondary metabolites in plants grown under controlled conditions and/or in vitro cultures (Table 1).

Flavonols, also known as 3-hydroxylavones, are the most common flavonoids found in food. They are structurally similar to flavones, but they differ in that they have a hydroxyl group at the 3-position on the C-ring, while flavones have a ketone group with an unsaturated carbon–carbon bond [44]. Cynaroside (luteo-lin-7-glucoside), a flavone, is used for a variety of medical purposes; it may protect heart cells from apoptosis caused by reactive oxygen species (ROS). Cynaroside also reduces kidney damage caused by the chemotherapeutic drug cisplatin, which is used to treat cancer [45]. The production of cynaroside has been increased in in vitro culture of Capsicum annum (aka bell pepper plant) by elicitation of UVA/B light for a period of 16 days [30].

UV-B has also been reported as a potential elicitor candidate to induce various changes in the metabolism of plants [46]. This leads to the activation of plant protec-tion mechanism by the formation of secondary metabolites such as alkaloids and fla-vonoids [47,48]. Physical responses in plant tissues, such as increased amounts of spe-cific phenolics, make plants more resistant to UV-B radiation than other species, and improved levels of pigmentation are induced [49,50]. Plant secondary metabolites containing polyphenolic structure, i.e., flavonoids, can be found in a wide range of foods, including fruits and vegetables. They have antioxidant and biochemical properties that can help with disorders like cancer, Alzheimer, atherosclerosis, and many others [51–53]. In a recent study, elicitation of *Nymphoides humboldtiana* with UV-B reported the production of pharmaceutically important flavonoids such as phloroglucinol, chlorogenic acid, epicatechin, quercetin, and ferulic acid [31]. Similarly, shoots of *Alternanthera* species, i.e., *A. sessilis* and *A. brasiliana*, on elicitation with UV-B for 8 h showed a 51% and 62% increase in flavonoid content, respectively, in comparison with control [32]. Likewise, an increase in the production of flavonoids was also reported in *Capsicum annum* L. with the elicitation of UV- B [33]. Under UV-B exposure, the production of flavonols, particularly quercetin, kaempferol, and isorhamnetin, were also vastly improved in *Ginkgo biloba* leaves [34]. Vinblastine and vincristine are chemotherapy drugs that are made by linking the alkaloids catharanthine and vindoline and are used to treat a variety of cancers [54]. The impact of UV-B on cell suspension cultures of the *Catharanthus roseus* plant was investigated in which the production of these important alkaloids, catharanthine and vindoline, was improved to

3 and 12 fold, respectively [35]. *Ocimum basilicum*, also known as sweet basil, possesses a wide range of potent activities due to the presence of precious secondary metabolites. It is used in traditional medicine as a result of its bioactive dary metabolites. It is used in traditional medicine as a result of its bioactive metabo-lites [55,56]. Elicitation with UV-B irradiation at an intensity of 224 μmol m^{-2} s^{-1} dramatically raised the production of anthocyanin, phenolics, and flavonoids in leaves of *Ocimum basilicum* [21].

When aerobic or photosynthetic metabolism is disrupted by various environmental stresses, phenolic compounds can play an important role in plant development by functioning as protective substances and signal molecules in plants, as well as safeguarding them from ROS. According to several studies, whenever a plant is infected with a disease, phenolic compounds are produced in response to that infection [57]. Most phenolics were accumulated when UV-B radiation was applied at a rate of 20 W/cm^2 to wheat seedlings. On day 4, total phenolics, DPPH, and ABTS levels increased by 26.3, 25.1, and 12.0%, respectively, as compared to un-irradiated wheat seedlings [37]. Likewise, in another study, seeds of *Ocimum basilicum* on irradiation of UV-B for three days showed enhanced production of phenolics [36].

2.1.2. UV-C

UV-C region of the spectrum includes wavelengths below 280 nm. These intense wavelengths are absorbed by ozone and do not reach the earth's surface [58]. Thus, the application of UV-C on various plant cultures using artificial lamps may be a promising strategy in raising the production of valuable secondary metabolites. Trans-resveratrol (TR) and resveratrol are non-flavonoid phenolics of stilbenes and are classified under biologically active isomers found majorly in red grapes and berries. They are mainly employed in pharmaceutical industries due to their anti-carcinogenic, anti-diabetic, anti-acne, antioxidant and anti-inflammatory activities. These properties make them a promising candidate for novel drugs [59]. The elicited callus culture of *Vitis vinifera* L. showed 8–10 times higher production of resveratrol levels in the 48 h after UV-C application than in the first 24 h. The highest resveratrol accumulation was recorded to be 62.23 µg/g of FW in a 12-day old culture [38]. Likewise, in another study, calli that had been exposed to UV-C for 5 min showed higher content of TR (8.43 µg g^{-1}) when harvested after 24 h. Catechin accumulation (8.89 mg/g) was also higher in calli exposed to UV-C for 10 min when harvested after 48 h. In conclusion, UV-C radiation aided the accumulation of secondary metabolites in the calli of the Okuzgozu grape cultivar [39].

Polyphenols found in *Lepidium sativum* L. have a diverse variety of medical and pharmaceutical purposes. Essential oils are also abundant in this therapeutic plant and exhibit exceptional anti-cancer properties in animal models as well as in various cell lines [60–62]. UV-C was irradiated on the callus of *L. sativum* for varying periods of time and different concentrations of melatonin. Phytochemical investigations revealed the production of three significant compounds: chlorogenic acid, kaempferol, and quercetin in callus culture. Cultures treated with melatonin (20 µM) showed three times higher production (36.36 mg/g DW), whereas cultures exposed to UV-C (60 min) showed a 2.5-fold increase (32.33 mg/g DW) in production in comparison with control (13.94 mg/g DW). This study compares both the elicitors and their effects on the initiation of physiological pathways in *L. sativum* for the formation of secondary metabolites [40].

Linum usitatissimum L. is considered a functional food and is widely used as oilseed crops in Europe [63,64]. It is a nutrient-dense diet that can be used as a substitute for food because it contains all dietary components [65,66]. When lignans and neolignans were quantified using reverse phase-high performance liquid chromatography (RP-HPLC), cultures of *L. usitatissimum* exposed to UV-C + photoperiod (16 h light/8 h dark) showed the presence of secoisolariciresinol diglucoside (SDG), dehydrodiconiferyl alcohol glucoside (DCG), guaiacylglycerol-β-coniferyl alcohol ether glucoside (GGCG), and lariciresinol diglucoside (LDG). UV-C radiation of 3.6 kJ/m^2 resulted in a higher accumulation of SDG, LDG, and GGCG by 1.86-fold, followed by 2.25-fold and 1.33-fold in cell cultures grown under UV + photoperiod. Furthermore, the accumulation of DCG was raised by 1.36-fold in cell cultures grown under UV + dark under an intensity of 1.8 kJ/m^2. Moreover, total phenolics, flavonoids, and antioxidants were also enhanced by 2.82-fold, 2.94-fold, and 1.04-fold, respectively, in cell cultures maintained under UV + photoperiod at a dosage of 3.6 kJ/m^2 of UV-C radiation. These findings broadened the scope of what can be done in *L. usitatasimum* cell cultures to produce biologically active lignans and neolignans [41].

Ocimum basilicum L. var purpurascens (Lamiaceae) has gained popularity due to its ornamental and aromatic qualities as well as the presence of important volatile secondary metabolites such as rosmarinic acid, flavonoids, and anthocyanin s [67]. In a recent report, the effect of UV-C on the synthesis of phenylpropanoid metabolites was examined in in vitro culture of *Ocimum basilicum* L. When compared to the control, UV-C (10 min) exposure resulted in a 2.3-fold increase in rosmarinic acid (134.5 mg/g DW). Chichoric acid (51.52 mg/g DW) and anthocyanin (cyanide 0.50 mg/g DW) were about 4.1-fold greater after 50 min of UV-C exposure, whereas peonidin was 2.7-fold higher. Overall, UV-C proved to be efficient elicitors, as there was a positive association between induced phenolic compound synthesis and antioxidant activity of basil callus extracts [42]. Similarly, in another study formation of secondary metabolites in *Echinacea purpurea* callus and cell suspension cultures were observed after 24, 48, and 72 h of elicitation. Callus and cell suspension cultures were exposed to UV-C radiation for 15, 30, and 60 min, respectively. UV-C irradiation for 60 min was the most successful in promoting the aggregation of most secondary metabolites with varying incubation times depending on tissue culture meth ods [43]. This suggests that UV-C could be a potential elicitor of secondary metabolite synthesis in plants in vitro cultures.

2.2. Fluorescent Lights

When plants are grown under sole-source electric lighting, lamp spectral customization can be an approach for achieving desired plant characteristics [68]. Fluorescent lights are a major source of light energy for stimulating the production of secondary metabolites in in vitro cultures [9]. Previous studies have confirmed that, depending on plant species, light quality has a direct impact on morphological and physiological responses [9,69]. For the management of controlled-environment agriculture facilities, it is vital to conserve electrical energy expenditures. In this regard, fluorescent lamps are considered much cheaper than LEDs [70]. They are particularly appealing for a variety of applications due to their high efficiency, outstanding color rendering, and extended life [71]. As a result, their use could be beneficial in triggering various in vitro cultures for increased secondary metabolite production (Table 2).

Table 2. Role of fluorescent light as an elicitor of key secondary metabolites in various in vitro systems of plants.

Light Type	Light Characteristic	Exposure Time	Plant Species	Culture System	Secondary Metabolite	Yield Increase	References
Blue light	380–560 nm	30 days	*Stevia rebaudiana*	Callus culture	Phenolics and Flavonoids	-	[22]
Blue light	40–50 µmol m^{-2} s^{-1}	3 weeks	*Prunella vulgaris*	Callus culture	Phenolics and Flavonoids	-	[14]
Blue light	60 µmol m^{-2} s^{-1}	6 weeks	*Scutellaria lateriflora* L.	Shoot culture	Glucuronides, Baicalin, Wogonoside, Verbascoside	1.54-, 1.49-, 2.05- and 1.86-fold, respectively	[72]
Red light	660 nm	5 weeks	*Hypericum perforatum*	Root culture	Hypericins, Flavonoids	-	[73]
Blue light	470 nm	1 week			Hypericins, Phenolics	52% and 26%	

Table 2. *Cont.*

Light Type	Light Characteristic	Exposure Time	Plant Species	Culture System	Secondary Metabolite	Yield Increase	References
Blue light White light	50 μmol m^{-2} s^{-1}	30 days	*Hyptis marrubioides*	Micro propagation	Rutin	-	[74]
Fluores cent light	50 μmol m^{-2} s^{-1}	4 weeks	*Panax ginseng*	Hairy root culture	Ginsenoside	-	[75]
Green light	40–50 μmol m^{-2} s^{-1}	3 weeks	*Artemisia absinthium*	Callus culture	Phenolics and Flavonoids	-	[9]

In a study, enhanced production of secondary metabolites and biomass accumulation was observed in *Stevia rebaudiana* callus cultures on exposure to different fluorescent spectral lights. Maximum capability in enhancing total phenolic content (102.32 μg/g of DW), total antioxidant potential (11.63 μg/g DW), and total flavonoids content (22.07 μg/g DW) was shown under blue light [22]. Similarly, blue light increased total phenolic (23.9 mg/g DW) and flavonoids content (1.65 mg/g DW) in callus cultures of *Prunella vulgaris* [14]. Likewise, in another study, blue light promoted aggregation of metabolites to the greatest degree in shoot culture of *Scutellaria lateriflora*. The major flavonoids that showed the highest concentrations were baicalin, verbascoside, glucuronides, and wogonoside. Their levels were 1.49, 1.86, 1.54, and 2.05 times higher than the control under white light, respectiv ely [72]. Based on the findings of the above investigations, it can be hypothesized that blue light can operate as a potential elicitor in diverse in vitro cultures, promoting the formation of secondary metabolites.

Hypericin and pseudohypericin, which are derivates of naphtodianthrones, are structurally similar phenolic compounds that have gained importance commercially due to their unique activities [76]. However, conventional cultivation of these metabolites is unable to fulfill the fierce competition in the pharmaceutical industry in terms of both quantity and quality. Therefore, elicitation of root culture of *Hypericum perforatum* with red light showed the highest production of total hypericins (i.e., hyperin + pseudohypericin) (9.61 ± 0.3 μg/g), whereas the lowest on exposure to fluorescent light (7.12 ± 0.26 μg/g). Roots grown under red light also showed the highest content of flavonoids (41.17 ± 7.21 mg/g). In this study, the impact of blue light was also evaluated on the production of key secondary metabolites. A considerable increase in the production of metabolites was observed after one week of exposure; however, an inhibitory effect was detected after five weeks of incubation. Results also showed that the production of hypericin and total phenolic content were increased to 52% and 26%, respectively, in root culture after one-week exposure to blue ligh t [73].

Hyptis marrubioides has been traditionally used to cure infections related to gastrointestines, cramps, discomfort, and skin infections [77] in many regions. Effect of various fluorescent lights was observed on the production of important phenolic compounds in seed culture of *H. marrubioides*. White (0.308 mg/g of DW) and blue (0.298 mg/g of DW) light was shown to accumulate the highest amount of rutin, while red light improved plant development and increased dry weight and leaf number in in vitro-cultivated seeds of *H. marrubioides* [74]. Ginsenosides (saponins) found in the root extract of *Panax ginseng* are the most active components known to exhibit immunomodulatory properties and provide protection against heart and liver diseases [78]. Rb and Rg are formed from the structures of 20(S) protopanaxadiol and 20(S) protopanaxatriol, and are two important groups of ginsenosides [79]. The Rg group of ginsenosides (5.3–0.1 mg/g DW) accumulated more than the Rb group (3.7–0.7 mg/g DW) in hairy roots grown under fluorescent light. These findings imply that growing hairy roots in dim or bright settings can affect the Rb and Rg ginsenoside productions in in vitro cultures [75].

Artemisia absinthium L., often known as "Wormwood", is referred to as a "universal treatment for all ailments" due to its therapeutic medical characteristics [80,81]. This plant has traditionally been used to treat diarrhea, cough, and common cold due [82] to its insecticidal, bitter [83,84], vermifuge, trematocidal [85], diuretic, and antispasmodic properties [86]. Total phenolics, total flavonoids, and antioxidant activity were found to be more supported by the green spectrum grown for three weeks under a photosynthetic photon flux density of 40–50 μmol m^{-2} s^{-1} in callus culture of *A. absinthium* [9]. As a result of the preceding studies, it is concluded that light elicitation (fluorescent, blue, red, green, and white light) has a positive influence and that different light regimes can aid in optimizing plant growth and developmental changes for the formation of commercially significant secondary metabolites in vitro.

2.3. LED Lights

The amount of light a plant receives has a big impact on its development, growth, and production. Agriculture employs traditional artificial light sources such as high-pressure sodium lamps (HPSLs), metal-halide lamps (MHLs), and fluorescent lamps (FLs) to provide a controlled atmosphere. Fluorescent lamps have risen in favor among these. However, the wavelengths of these lightning sources span from 350 to 750 nm, and for plant growth and development, it is considered of low quality. They possess a limited lifetime of activity and a low photosynthetic flux, limiting their use in plant illumination systems that require a large agricultural production. LEDs technology has been able to be employed in a rising variety of new sectors, including plant growth and development, due to the implementation of new types of semiconductor materials. As a substitute to conventional lighting systems, LEDs have proved to be a smarter source of artificial lighting to provide controlled conditions in agriculture and in vitro systems. Since LEDs can emit over specified spectral areas, they may be utilized to manage the amount of photosynthetically active and photomorphogenic radiation required for plant growth and development. Matching LED wavelengths to photoreceptors in plants can allow for optimal output while also altering plant shape and metabolism. As a result, these solid-state light sources can be used in developing lighting lamps for sustainable production and photo-morphogenesis research [87].

Plant growth, production, and secondary metabolism are all influenced by light in general and light quality in particular. LEDs exist in various colors, i.e., white, blue, green, red, yellow, violet, and far-red. Many scientists believe that red (600–700 nm) and blue (400–500 nm) light are more crucial for stimulating photosynthesis than other light wavelengths since they have the highest photosynthetic photon efficacy values. Green light (500–600 nm) can, on the other hand, penetrate deep inside the leaf due to its high transmittance and reflectance [88]. LED lights have been widely used to elicit the production of key secondary metabolites in various plant culture systems (Table 3).

Table 3. Role of LED lights as an elicitor of key secondary metabolites in plant culture systems.

Light Types	Light Characteristic	Exposure Time	Plant Species	Culture System	Secondary Metabolite	Yield Increase	References
Monochromatic Blue LED	456 nm	27 days	*Solanum lycopersicum*	Closed-type plant production system (seedling)	Phenolics and flavonoids	-	[23]
Blue LED	200 μmol m^{-2} s^{-1}	14/10 h	Lettuce	CGC (leaf)	Flavonoids	2.07-fold	[89]
Blue LED	50 μmol m^{-2} s^{-1}	28 days	*Curculigo Orchioides*	CGC (shoot bud)	Phenolics and flavonoids	-	[90]
Blue LED	30 ±1 μmol m^{-2} s^{-1}	8 h per day 40 days	*Anoectochilus roxburghii*	CGC (leaf)	Flavonoids and polyphenols	24.2%	[91]
Blue (+B) and Blue-violet (+BV) LED	450 nm and 420–440 nm	10 days	*Ocimum basilicum* *Eruca sativa*	CGC (leaf)	Phenolics Flavonoids		[92]
Red LED	35 μmol m^{-2} s^{-1}	6 weeks	*Myrtus communis*	In vitro shoot culture	Gallic acid and Myricetin	-	[93]
Red LED	40–50 μmol m^{-2} s^{-1}	24 h	*Silybum marianum*	Callus culture	Silymarin	2-fold	[94]

Table 3. *Cont.*

Light Types	Light Characteristic	Exposure Time	Plant Species	Culture System	Secondary Metabolite	Yield Increase	References
Blue LED	50 μmol m^{-2} s^{-1}	4 weeks	*Rehmannia glutinosa*	In vitro shoot culture	Phenolics and flavonoids	39.3%, 33.6%	[95]
Red LED	450 nm			CGC (root and leaf)	Phenolics, Flavonoids	33.6%, 61.7%	[96]
Blue LED	40–50 μmol m^{-2} s^{-1}	4 weeks	*Ocimum basilicum*	Callus culture	Rosmarinic acid and eugenol	2.46- and 2.25-fold	[97]
Red LED					Peonidin and cyanidin	3.5- and 4.53-fold	
White LED					Chicoric acid	4.52-fold	
White LED	380 nm	4 days	*Triticum aestivium*	CGC (seedling)	Epicatechin	1.3- to 1.46-fold	[98]
Blue LED	470 nm	8 days			Gallic acid and quercetin	1.2–1.5 fold	
Red LED	660 nm	8 days			Ferulic acid	-	
		12 days			*p*-coumaric acid	1.27- to 1.77-fold	
Red Blue light	137 μmol m^{-2} s^{-1}	96 h	*Fagopyrum* sp.	CGC (seedling)	Flavonoids	-	[99]
Blue LED	177 μmol m^{-2} s^{-1}				Anthocyanins	-	
Red and Blue 2R:1B	120 μmol m^{-2} s^{-1}	12 h each day for 17 days	*Ocimum basilicum*	CGC (seedling)	Phenolics and flavonoids	1.63–1.87-fold and 2.06-fold	[100]
Blue–red (1:1) LED	44.80 μmol/s	30 days for 16 h	*Dendrobium*	In vitro protocorn-like body culture	Flavonoids	-	[101]
Red: Blue LED	33 μmol m^{-2} s^{-1}	4 weeks	*Cnidium officinale*	Callus culture	Phenolics and flavonoids	-	[10]
Red LED Blue LED	660 nm 460 nm	24 h	*Eclipta alba*	Callus culture	Phenolics and flavonoids	2-fold	[102]

2.3.1. Blue LED

Traditional lighting systems require color filters, but LED lighting systems may create the light of any desired color without them. LEDs are versatile enough to produce only the type of light that plants require. Plants require specific spectrums or colors for distinct morphogenic responses, and LED systems can be fine-tuned to provide only those. Blue light controls stomatal opening and transpiration, as well as preventing "red light syndrome" [103]. Thus, blue LEDs can play an efficient role in stimulating the production of pharmacologically active secondary metabolites, as demonstrated in Table 3.

Tomatoes (*Solanum lycopersicum* L.) are the world's seventh most produced crop species and one of the year-round value crops grown in greenhouses. Tomato fruits are nutritionally dense because they include vital nutrients as well as phytochemicals that promote health [104]. LED was used as a source of artificial light to evaluate its effect on *S. lycopersicum* L. 'Cuty' (tomato seedlings). Results showed raised production of phenolics, flavonoids, and antioxidants under blue light when compared to control. This concludes that manipulating light quality using LED's could stimulate the production of bioactive compounds and antioxidants [23].

The use of nutritional or therapeutic plant-based natural substances to treat disease has become a novel paradigm in clinical science. Flavonoids, particularly phenolic compounds, can be found in nearly all plants. Antioxidant, anti-cancer, anti-diabetic, and cardiovascular effects are observed in phenolics and flavonoids [105]. A study reported the highest concentrations of soluble protein and flavonoid in lettuce when exposed to blue light [89]. Another study found that leaves of synseed grown under blue LED had higher amounts of chlorophyll a, flavonoids, phenolics, and carotenoids than those grown under fluorescent or blue-red LED. Anti-oxidant activity was similarly boosted in synseed-derived seedlings grown in blue LED light [90]. Likewise, the overall polyphenol content in the blue LED treatment was also considerably higher than in the control treatment in *Anoectochilus roxburghii* [91,101]. Similarly, both blue and blue–violet light supplements boosted phenolic acid production in *Ocimum basilicum*, while in *Eruca sativa*, higher flavonoid synthesis was seen in response to both light supplements, but greater production was observed under blue– violet [92]. The above study concludes the potential of using blue LED as a promising elicitor in raising the production dramatically.

2.3.2. Red LED

Red LEDs has been widely employed as an alternative source of illumination for in vitro survival and improved production of metabolites in medicinally important plants. Red light is considered the most efficient spectral light in driving photosynthesis, and many studies reported it as an efficient elicitor in elevating the production of pharmacologically active constituents in in vitro systems of various plants [106]. *Myrtus communis* has long been used in medicine and possesses therapeutic characteristics [107,108]. *M. communis* (leaves, flowers, and fruits) contain various components that are critical to the pharmaceutical, food, liqueur, and cosmetic industries; hence, large-scale manufacturing of this plant is necessary [109]. *M. communis* leaf extracts are enriched sources of phenolic acids and flavonoids. In a recent study, the impact of red LED was evaluated on in vitro culture of *M. communis*. Results showed that out of all flavonoids, myricetin (347.02–1118.69 mg 100 g^{-1} DW) was the main constituent with the highest concentration and 6-benzyladenine (BA) level, whereas gallic acid had the highest concentrations of the phenolic acids (95.58 mg 100 g^{-1} DW on average) at a photosynthetic photon flux density of 35 μmol m^{-2} s^{-1} [93].

Milk thistle (*Silybum marianum* L.) is a well-known hepato-protective medicinal herb that has been extensively researched. The effects of various LEDs lights were investigated, and it was discovered that red light greatly increased phenolics, flavonoids, and superoxide dismutase (SOD) activities in this plant. Under red light, HPLC analysis revealed a substantially doubled total silymarin concentration (18.67 mg g^{-1} DW) when compared to the control (9.17 mg g^{-1} DW). When exposed to red light, the levels of isosilychristin, silybin A, silybin B, silychristin, and silydianin were found to be highest. This demonstrates that the quality of light has a significant impact on the callus culture of *S. marianum* morphological and biochemical properties [94].

2.3.3. Blue and Red LED in Combination

Several studies have reported the use of blue and red LEDs in combination on one or more plants simultaneously. Many reports suggested that the use of different LEDs in combination can enhance the production of bioactive compounds to several folds. Thus, their combination can be employed in eliciting various cultures that have gained importance pharmacologically [95,97,98]. Since ancient times, root extracts of *Rehmannia glutinosa* Libosch (Chinese foxglove) have been used to treat many disorders such as anemia and hypertension [96]. The therapeutic use of *R. glutinosa* roots is well known due to the presence of bioactive compounds such as catalpol, aucubin, and rehmaglutin [96,110]. In a study, plant development as well as phytochemicals with both defensive and prospective medicinal characteristics, such as phenolics and flavonoids, were influenced by spectral characteristics in both leaf and root tissues of *R. glutinosa*. On irradiation with blue LED, leaf extracts were found to have the highest total phenol concentration (35 ± 0.05 μg GAE/mg) as compared to root extracts. On the other hand, red LED exposure also increased the overall flavonoid content of the leaf extract by 33.6% and the root extracts by 61.7%. As a result, blue and red LEDs may be the most enticing light sources for *R. glutinosa* proliferation in in vitro conditions [95].

Bioactive compounds like phenolics and flavonoids are of significant importance in therapeutic plants because they can act as free radical scavengers [111]. The effects of red and blue LEDs on the accumulation of phenolic and flavonoids in *Ocimum basilicum* callus cultures were investigated in a recent study. Rosmarinic acid and eugenol were considerably enhanced under blue light (2.46 and 2.25 times greater than control). While under red light, the highest amounts of cyanidin (0.1216 mg/g DW) and peonidin (0.127 mg/g DW) were detected. Chicoric acid accumulation was about 4.52 times higher than callus grown under the control of continuous white light (81.40 mg/g DW) [97]. This suggests that employing LEDs to accelerate the production of physiologically active chemicals in vitro could be beneficial.

Fruits, nuts, medicinal herbs, and vegetables all contain phenylpropanoid, a type of secondary metabolite that is involved in reducing the risk of diabetes and heart disease through inhibiting carcinogenesis [112,113]. Plants are protected from bacteria by the antimicrobial activities of some phenylpropanoids during their interactions [114]. The generation of phenylpropanoid metabolites was examined using LEDs as an elicitor in wheat sprouts, and qRT-PCR and HPLC results revealed that white light (380 nm) was the best wavelength for epicatechin biosynthesis in wheat sprouts. Blue light (470 nm) was involved in the increased accumulation of gallic acid and quercetin, whereas red light (660 nm) increased the aggregation of ferulic acid on the 8th day and p-coumaric acid on the 12th day [98]. Similarly, flavonoids and anthocyanins, which also belong to a class of phenylpropanoids, were also enhanced in buckwheat, *Fagopyrum* sp. under red + blue (combination), and blue LED, respectively [99]. Likewise, phenolics and flavonoids were enhanced in seeds of *Ocimum basilicum* under red–blue (2R:1B) LED. In addition, the highest flavonoid content of 16.79 mmol/g FW was also achieved for protocorm-like bodies pretreated with white LED for more than three cultures cycles under blue–red (1:1) LED [101]. Thus, red and blue LED ratios can be altered to generate improved growth and phenolic content in both red and green basil microgreens as a practical technique for generating superior quality foods [100]. Furthermore, in the callus culture of *Cnidium officinale*, mixed (red–blue) LED illumination enhanced the synthesis of phenolics and flavonoids [10]. In another study, in vitro generated *Eclipta alba* callus culture was subjected to multispectral lighting under regulated aseptic conditions. Results showed that the red light enhanced the production of phenolics (57.8 mg/g) and flavonoids (11.1 mg/g). whereas, on exposure to blue light, the production of four major compounds coumarin (1.26 mg/g), eclalbatin (5.00 mg/g), wedelolactone (32.54 mg/g), and demethylwedelolactone (23.67 mg/g), as well as two minor compounds β-amyrin (0.38 mg/g) and luteolin (0.39 mg/g) were increased [102].

3. Mechanistic Considerations and Future Prospects

Light is a key driver of plant secondary metabolite biosynthesis and accumulation [115]. When photons activate photoreceptors, they activate signaling pathways and cause changes in gene expression. The light-absorbing properties are defined by the interaction of a photoreceptor protein and a chromophore [116]. Figure 2 depicts the absorption bands of essential photoreceptors, as well as the quantities and processes that they affect. In comparison to photosynthesis, only a small fraction of photons absorbed are used to activate photoreceptors [117]. Plants have four major photoreceptors, absorbing photons not only in the photosynthetically active radiation (PAR) domain but also in the UV and far-red regions. They interact with many signal transduction factors and are in charge of initiating various physiological changes and adaptations, including secondary metabolite production, triggered by light.

To address secondary metabolites in plants, photosynthetically active radiation (PAR) is insufficient, and a broader wavelength range (physiologically active radiation) is required.

The protein cryptochrome absorption is comprised between the UV-A and blue light domains (340−520 nm). Phytochrome absorbs light primarily in the red/far-red area around 665 nm (Pr) and 730 nm (Pfr), but also in the blue/near-UV region. Phototropin light absorption is comprised between 350 and 520 nm, whereas UVR8 shows maximum absorption in the range between 280 and 350 nm.

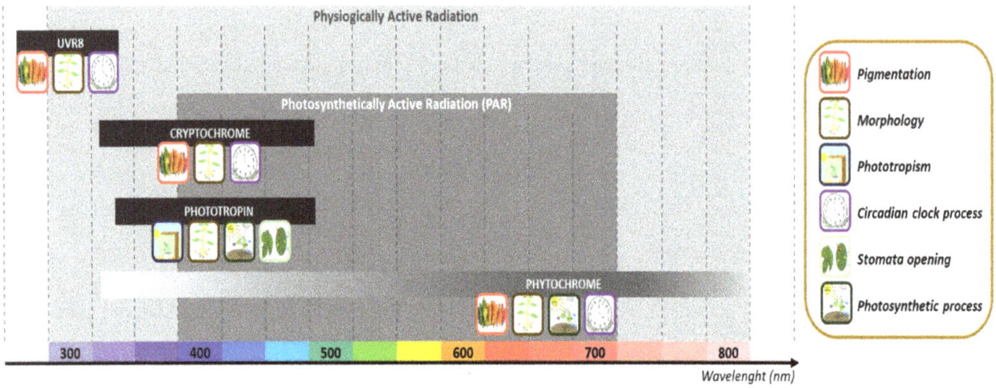

Figure 2. Photoreceptor absorption bands in plants and their processes (adapted from [115]). Absorbing photons activate different types of photoreceptors not just in the PAR domain but also in the UV and far-red regions. Each absorption band and photoreceptor couple are engaged in a different physiological response that may be essential for biomass and secondary metabolite synthesis, such as pigmentation, morphology, phototropism, circadian clock process, stomata opening, or photosynthetic activity.

Many significant players have been identified in the literature to explain how light affects secondary metabolite biosynthesis. The first is ROS; when exposed to large levels of UV-B radiation, plants preferentially produce a wide range of phenylpropanoid structures, including flavonoids, supporting the theory that flavonoids are principally engaged in photoprotection by absorbing the shortest solar wavelengths [118]. There is, however, substantial evidence that flavonoid production is similarly enhanced in response to high photosynthetically active radiation either in the presence or absence of UV radiation. For instance, the presence of phytochromes, at a modest level, as well as UVR8 and cryptochrome, are important for anthocyanin accumulation; therefore, wavelengths beyond UV also show impacts [115]. This lends credence to the idea that flavonoids can act as scavengers of ROS produced by excessive light [118].

Phytohormonal control is also an important regulation that could be linked to light. Phytohormones have been linked to the regulation of secondary metabolite synthesis under both developmental and stress circumstances [119,120]. Many phytochrome proteins have been connected to variations in phytohormone levels, including gibberellins and auxins, ethylene, jasmonates, and abscisic acid (ABA) [117,121], all of which are, for example, significant players in the control of plant secondary metabolism, including pigments [115].

Light may also regulate the expression of transcription factors, which are essential regulators of secondary metabolite biosynthesis. For instance, cryptochrome has been shown to regulate the accumulation of anthocyanins [122], whereas green light has been shown to reverse the effects of blue light in anthocyanin accumulation [123]. The expression of R2R3MYB transcription factors, including V1MYBA1-2 and VIMYBA2, which regulate the expression of anthocyanin synthesis-related genes like *VvUFGT*, increased in grape berries irradiated with blue light, which is consistent with this observation [124].

It all demonstrates how light can regulate the formation of secondary metabolites at various levels and via various signaling pathways. Yet, in vitro systems have received far too little attention at these levels of regulation, while in vitro systems could be appealing models for conducting this type of research since, unlike the whole plant, they allow for uniformity, accessibility, and reduced complexity [125].

Light allows for the manipulation of growth conditions and secondary metabolite synthesis in both controlled and in vitro cultures, but several elements must be addressed before it can be completely utilized. In recent years, LED systems appear to have a lot of promise, as light influences plant growth and development at every step. It affects morphogenesis, differentiation, and proliferation rates in the plant cell, tissue, and organ cultures, all of which are important for the biosynthesis of secondary metabolites. Conscious manipulation of light quality, along with the advantages of LED technology, will lead to increased biomass production with high secondary metabolite content under regulated in vitro settings. It will also ensure that homogenous material is obtained in a reasonably short amount of time without the need to cultivate the whole plant to the fully mature stage in the field. However, according to a review of the literature, the response appears to be depending on the plant species and the culture systems. Biosynthesis of plant secondary metabolites is regulated effectively by various signaling events, as the production of secondary metabolites depends upon the plant species and their respective phytochemical classes. As a result, proposing a general mechanism of light elicitation on the production of different secondary metabolites is complicated, and further research is needed to gain a comprehensive understanding. To uncover the molecular and cellular mechanism of light elicitation, an omics-based analysis could be a useful approach. Therefore, in conclusion, it may be inferred that different light regimes can aid in the improvement of growth and developmental alterations for the in vitro generation of safe secondary metabolites.

4. Conclusions

The current review deciphers the role of light as a promising elicitor in enhancing the production of key metabolites in different in vitro systems of plants. Among various abiotic elicitors, light has gained attention in enhancing secondary metabolite production due to its specified wavelengths, cost-effectiveness, and durability. A plethora of studies have reported the effect and quality of different types of light sources on enhanced secondary metabolite accumulation in several plant species in vitro. These findings pave the way for a more thorough examination of light as a plant elicitor. The fact that most studies have focused on plant growth and development, including the generation of primary metabolites, highlights the paucity of information about the different factors regulating light elicitation mechanisms. Thus, many studies still lack at depicting the mechanisms of elicitation promoting the production of pharmacologically important secondary metabolites since it may differ depending upon the type of plant species, culture conditions, and light source used.

Author Contributions: Conceptualization, S.A.; methodology, M.H., B.A., and S.D.; validation, C.H., B.H.A., and S.A.; formal analysis, S.A.; investigation, M.H., B.A., and S.D.; resources, C.H., B.H.A., and S.A.; data curation, C.H., B.H.A., and S.A.; writing—original draft preparation, M.H., B.A., and S.A.; writing—review and editing, C.H., B.H.A., and S.A.; supervision, C.H., B.H.A., and S.A.; project administration, S.A.; funding acquisition, C.H., B.H.A., and S.A. All authors have read and agreed to the published version of the manuscript.

Funding: This research was supported by Cosmetosciences, a global training and research program dedicated to the cosmetic industry. Located in the heart of the Cosmetic Valley, this program led by the University of Orleans is funded by the Region Centre-Val de Loire (VALBIOCOSM and INNOCOSM).

Institutional Review Board Statement: Not applicable.

Informed Consent Statement: Not applicable.

Data Availability Statement: All data are included in the present study.

Acknowledgments: B.H.A. and C.H. acknowledge the research fellowship of Le Studium-Institute for Advanced Studies, Loire Valley, Orléans, France. S.D. acknowledges Cosmetosciences and Région Centre Val de Loire for research fellowships. The authors would like to acknowledge networking support by Le Studium COSMENOVIC consortium and CNRS GDR3711 COSM'ACTIFS. The authors gratefully acknowledge the Région Centre Val de Loire for funding the project VALBIOCOSM and INNOCOSM. C.H. and B.H.A. gratefully acknowledge the support of Campus France through the PHC PERIDOT.

Conflicts of Interest: The authors declare no conflict of interest.

References

1. Wang, J.W.; Wu, J.Y. Effective elicitors and process strategies for enhancement of secondary metabolite production in hairy root cultures. In *Biotechnology of Hairy Root Systems*; Springer: Berlin, Germany, 2013; pp. 55–89.
2. Murthy, H.N.; Lee, E.-J.; Paek, K.-Y. Production of secondary metabolites from cell and organ cultures: Strategies and approaches for biomass improvement and metabolite accumulation. *Plant Cell Tissue Organ. Cult.* **2014**, *118*, 1–16. [CrossRef]
3. Yue, W.; Ming, Q.-l.; Lin, B.; Rahman, K.; Zheng, C.-J.; Han, T.; Qin, L.-P. Medicinal plant cell suspension cultures: Pharmaceutical applications and high-yielding strategies for the desired secondary metabolites. *Crit. Rev. Biotechnol.* **2016**, *36*, 215–232. [CrossRef]
4. Gonçalves, S.; Romano, A. Production of plant secondary metabolites by using biotechnological tools. In *Secondary Metabolites, Sources and Applications*; IntechOpen: London, UK, 2018; pp. 81–99.
5. Hussain, M.S.; Fareed, S.; Saba Ansari, M.; Rahman, A.; Ahmad, I.Z.; Saeed, M. Current approaches toward production of secondary plant metabolites. *J. Pharm. Bioall. Sci.* **2012**, *4*, 10. [CrossRef] [PubMed]
6. Cragg, G.M.; Newman, D.J. Plants as a source of anti-cancer agents. *J. Ethnopharmacol.* **2005**, *100*, 72–79. [CrossRef] [PubMed]
7. Cui, X.-H.; Chakrabarty, D.; Lee, E.-J.; Paek, K.-Y. Production of adventitious roots and secondary metabolites by *Hypericum perforatum* L. in a bioreactor. *Bioresour. Technol.* **2010**, *101*, 4708–4716. [CrossRef]
8. Yang, L.; Stöckigt, J. Trends for diverse production strategies of plant medicinal alkaloids. *Nat. Prod. Rep.* **2010**, *27*, 1469–1479. [CrossRef]
9. Tariq, U.; Ali, M.; Abbasi, B.H. Morphogenic and biochemical variations under different spectral lights in callus cultures of *Artemisia absinthium* L. *J. Photochem. Photobiol. B Biol.* **2014**, *130*, 264–271. [CrossRef]
10. Adil, M.; Ren, X.; Jeong, B.R. Light elicited growth, antioxidant enzymes activities and production of medicinal compounds in callus culture of Cnidium officinale Makino. *J. Photochem. Photobiol. B Biol.* **2019**, *196*, 111509. [CrossRef]
11. Abbasi, B.H.; Tian, C.-L.; Murch, S.J.; Saxena, P.K.; Liu, C.-Z. Light-enhanced caffeic acid derivatives biosynthesis in hairy root cultures of *Echinacea purpurea*. *Plant Cell Rep.* **2007**, *26*, 1367–1372. [CrossRef] [PubMed]
12. Shohael, A.; Ali, M.; Yu, K.; Hahn, E.; Islam, R.; Paek, K. Effect of light on oxidative stress, secondary metabolites and induction of antioxidant enzymes in *Eleutherococcus senticosus* somatic embryos in bioreactor. *Process Biochem.* **2006**, *41*, 1179–1185. [CrossRef]
13. Liu, C.; Zhao, Y.; Wang, Y. Artemisinin: Current state and perspectives for biotechnological production of an antimalarial drug. *Appl. Microbiol. Biotechnol.* **2006**, *72*, 11–20. [CrossRef]
14. Fazal, H.; Abbasi, B.H.; Ahmad, N.; Ali, S.S.; Akbar, F.; Kanwal, F. Correlation of different spectral lights with biomass accumulation and production of antioxidant secondary metabolites in callus cultures of medicinally important *Prunella vulgaris* L. *J. Photochem. Photobiol. B Biol.* **2016**, *159*, 1–7. [CrossRef] [PubMed]
15. Moran, J.F.; Becana, M.; Iturbe-Ormaetxe, I.; Frechilla, S.; Klucas, R.V.; Aparicio-Tejo, P. Drought induces oxidative stress in pea plants. *Planta* **1994**, *194*, 346–352. [CrossRef]
16. Alexieva, V.; Sergiev, I.; Mapelli, S.; Karanov, E. The effect of drought and ultraviolet radiation on growth and stress markers in pea and wheat. *Plant Cell Environ.* **2001**, *24*, 1337–1344. [CrossRef]
17. Yu, T.-W.; Anderson, D. Reactive oxygen species-induced DNA damage and its modification: A chemical investigation. *Mutat. Res. Fundam. Mol. Mech. Mutagenesis* **1997**, *379*, 201–210. [CrossRef]
18. Mittler, R. Oxidative stress, antioxidants and stress tolerance. *Trends Plant Sci.* **2002**, *7*, 405–410. [CrossRef]
19. Ashry, N.A.; Mohamed, H.I. Impact of secondary metabolites and related enzymes in flax resistance and or susceptibility to powdery mildew. *World J. Agric. Sci* **2011**, *7*, 78–85.
20. Samuolienė, G.; Brazaitytė, A.; Urbonavičiūtė, A.; Šabajevienė, G.; Duchovskis, P. The effect of red and blue light component on the growth and development of frigo strawberries. *Zemdirbyste Agric.* **2010**, *97*, 99–104.
21. Dou, H.; Niu, G.; Gu, M. Pre-harvest UV-B radiation and photosynthetic photon flux density interactively affect plant photosynthesis, growth, and secondary metabolites accumulation in basil (*Ocimum basilicum*) plants. *Agronomy* **2019**, *9*, 434. [CrossRef]
22. Ahmad, N.; Rab, A.; Ahmad, N. Light-induced biochemical variations in secondary metabolite production and antioxidant activity in callus cultures of *Stevia rebaudiana* (Bert). *J. Photochem. Photobiol. B Biol.* **2016**, *154*, 51–56. [CrossRef] [PubMed]
23. Kim, E.-Y.; Park, S.-A.; Park, B.-J.; Lee, Y.; Oh, M.-M. Growth and antioxidant phenolic compounds in cherry tomato seedlings grown under monochromatic light-emitting diodes. *Hortic. Environ. Biotechnol.* **2014**, *55*, 506–513. [CrossRef]
24. Xuan, T.D.; Khanh, T.D.; Khang, D.T.; Quan, N.T.; Elzaawely, A.A. Changes in chemical composition, total phenolics and antioxidant activity of Alpinia (*Alpinia zerumbet*) leaves exposed to UV. *Int. Lett. Nat. Sci.* **2016**, *55*, 25–34. [CrossRef]
25. Stapleton, A.E. Ultraviolet radiation and plants: Burning questions. *Plant Cell* **1992**, *4*, 1353. [CrossRef] [PubMed]
26. Parihar, P.; Singh, S.; Singh, R.; Singh, V.P.; Prasad, S.M. Changing scenario in plant UV-B research: UV-B from a generic stressor to a specific regulator. *J. Photochem. Photobiol. B Biol.* **2015**, *153*, 334–343. [CrossRef]
27. Freitas, A.; Moldão-Martins, M.; Costa, H.S.; Albuquerque, T.G.; Valente, A.; Sanches-Silva, A. Effect of UV-C radiation on bioactive compounds of pineapple (*Ananas comosus* L. Merr.) by-products. *J. Sci. Food Agric.* **2015**, *95*, 44–52. [CrossRef]
28. Yin, X.; Singer, S.D.; Qiao, H.; Liu, Y.; Jiao, C.; Wang, H.; Li, Z.; Fei, Z.; Wang, Y.; Fan, C. Insights into the mechanisms underlying ultraviolet-C induced resveratrol metabolism in grapevine (*V. amurensis* Rupr.) cv."Tonghua-3". *Front. Plant Sci.* **2016**, *7*, 503. [CrossRef] [PubMed]
29. Abbasi, B.H.; Khan, T.; Khurshid, R.; Nadeem, M.; Drouet, S.; Hano, C. UV-C mediated accumulation of pharmacologically significant phytochemicals under light regimes in in vitro culture of *Fagonia indica* (L.). *Sci. Rep.* **2021**, *11*, 1–16. [CrossRef] [PubMed]

30. Ellenberger, J.; Siefen, N.; Krefting, P.; Schulze Lutum, J.-B.; Pfarr, D.; Remmel, M.; Schröder, L.; Röhlen-Schmittgen, S. Effect of UV Radiation and Salt Stress on the Accumulation of Economically Relevant Secondary Metabolites in Bell Pepper Plants. *Agronomy* **2020**, *10*, 142. [CrossRef]
31. Nocchi, N.; Duarte, H.M.; Pereira, R.C.; Konno, T.U.P.; Soares, A.R. Effects of UV-B radiation on secondary metabolite production, antioxidant activity, photosynthesis and herbivory interactions in *Nymphoides humboldtiana* (Menyanthaceae). *J. Photochem. Photobiol. B Biol.* **2020**, *212*, 112021. [CrossRef] [PubMed]
32. Klein, F.R.S.; Reis, A.; Kleinowski, A.M.; Telles, R.T.; Amarante, L.d.; Peters, J.A.; Braga, E.J.B. UV-B radiation as an elicitor of secondary metabolite production in plants of the genus Alternanthera. *Acta Bot. Bras.* **2018**, *32*, 615–623. [CrossRef]
33. Rodríguez-Calzada, T.; Qian, M.; Strid, Å.; Neugart, S.; Schreiner, M.; Torres-Pacheco, I.; Guevara-González, R.G. Effect of UV-B radiation on morphology, phenolic compound production, gene expression, and subsequent drought stress responses in chili pepper (*Capsicum annuum* L.). *Plant Physiol. Biochem.* **2019**, *134*, 94–102. [CrossRef] [PubMed]
34. Zhao, B.; Wang, L.; Pang, S.; Jia, Z.; Wang, L.; Li, W.; Jin, B. UV-B promotes flavonoid synthesis in *Ginkgo biloba* leaves. *Ind. Crops Prod.* **2020**, *151*, 112483. [CrossRef]
35. Ramani, S.; Jayabaskaran, C. Enhanced catharanthine and vindoline production in suspension cultures of *Catharanthus roseus* by ultraviolet-B light. *J. Mol. Signal.* **2008**, *3*, 1–6. [CrossRef]
36. Mosadegh, H.; Trivellini, A.; Ferrante, A.; Lucchesini, M.; Vernieri, P.; Mensuali, A. Applications of UV-B lighting to enhance phenolic accumulation of sweet basil. *Sci. Hortic.* **2018**, *229*, 107–116. [CrossRef]
37. Chen, Z.; Ma, Y.; Weng, Y.; Yang, R.; Gu, Z.; Wang, P. Effects of UV-B radiation on phenolic accumulation, antioxidant activity and physiological changes in wheat (*Triticum aestivum* L.) seedlings. *Food Biosci.* **2019**, *30*, 100409. [CrossRef]
38. Keskin, N.; Kunter, B. Production of *trans*-resveratrol in callus tissue of Öküzgözü (*Vitis vinifera* L.) in response to ultraviolet-C irradiation. *J. Anim Plant Sci.* **2010**, *20*, 197–200.
39. Cetin, E.S. Induction of secondary metabolite production by UV-C radiation in *Vitis vinifera* L. Öküzgözü callus cultures. *Biol. Res.* **2014**, *47*, 1–7. [CrossRef]
40. Ullah, M.A.; Tungmunnithum, D.; Garros, L.; Drouet, S.; Hano, C.; Abbasi, B.H. Effect of ultraviolet-C radiation and melatonin stress on biosynthesis of antioxidant and antidiabetic metabolites produced in in vitro callus cultures of *Lepidium sativum* L. *Int. J. Mol. Sci.* **2019**, *20*, 1787. [CrossRef]
41. Anjum, S.; Abbasi, B.H.; Doussot, J.; Favre-Réguillon, A.; Hano, C. Effects of photoperiod regimes and ultraviolet-C radiations on biosynthesis of industrially important lignans and neolignans in cell cultures of *Linum usitatissimum* L. (Flax). *J. Photochem. Photobiol. B Biol.* **2017**, *167*, 216–227. [CrossRef]
42. Nazir, M.; Asad Ullah, M.; Mumtaz, S.; Siddiquah, A.; Shah, M.; Drouet, S.; Hano, C.; Abbasi, B.H. Interactive effect of melatonin and UV-C on phenylpropanoid metabolite production and antioxidant potential in callus cultures of purple basil (*Ocimum basilicum* L. var. *purpurascens*). *Molecules* **2020**, *25*, 1072. [CrossRef]
43. Abd El-Aal, M.S.; Rabie, K.A.; Manaf, H.H. The effect of UV-C on secondary metabolites production of *Echinacea purpurea* culture in vitro. *J. Biol. Chem. Environ. Sci.* **2016**, *11*, 465–483.
44. Zuiter, A. *Proanthocyanidin: Chemistry and Biology: From Phenolic Compounds to Proanthocyanidins*; Elsevier: Amsterdam, The Netherlands, 2014.
45. Sun, X.; Sun, G.B.; Wang, M.; Xiao, J.; Sun, X.B. Protective effects of cynaroside against H_2O_2-induced apoptosis in H9c2 cardiomyoblasts. *J. Cell. Biochem.* **2011**, *112*, 2019–2029. [CrossRef]
46. Schreiner, M.; Krumbein, A.; Mewis, I.; Ulrichs, C.; Huyskens-Keil, S. Short-term and moderate UV-B radiation effects on secondary plant metabolism in different organs of nasturtium (*Tropaeolum majus* L.). *Innov. Food Sci. Emerg. Technol.* **2009**, *10*, 93–96. [CrossRef]
47. Zhang, W.J.; Björn, L.O. The effect of ultraviolet radiation on the accumulation of medicinal compounds in plants. *Fitoterapia* **2009**, *80*, 207–218. [CrossRef]
48. Schreiner, M.; Mewis, I.; Huyskens-Keil, S.; Jansen, M.; Zrenner, R.; Winkler, J.; O'brien, N.; Krumbein, A. UV-B-induced secondary plant metabolites-potential benefits for plant and human health. *Crit. Rev. Plant Sci.* **2012**, *31*, 229–240. [CrossRef]
49. Flint, S.D.; Ryel, R.J.; Caldwell, M.M. Ecosystem UV-B experiments in terrestrial communities: A review of recent findings and methodologies. *Agric. For. Meteorol.* **2003**, *120*, 177–189. [CrossRef]
50. Caldwell, M.M.; Bornman, J.; Ballaré, C.; Flint, S.D.; Kulandaivelu, G. Terrestrial ecosystems, increased solar ultraviolet radiation, and interactions with other climate change factors. *Photochem. Photobiol. Sci.* **2007**, *6*, 252–266. [CrossRef]
51. Burak, M.; Imen, Y. Flavonoids and their antioxidant properties. *Turkiye Klin. Tip. Bil. Derg.* **1999**, *19*, 296–304.
52. Castañeda-Ovando, A.; de Lourdes Pacheco-Hernández, M.; Páez-Hernández, M.E.; Rodríguez, J.A.; Galán-Vidal, C.A. Chemical studies of anthocyanins: A review. *Food Chem.* **2009**, *113*, 859–871. [CrossRef]
53. Lee, Y.K.; Yuk, D.Y.; Lee, J.W.; Lee, S.Y.; Ha, T.Y.; Oh, K.W.; Yun, Y.P.; Hong, J.T. (−)-Epigallocatechin-3-gallate prevents lipopolysaccharide-induced elevation of beta-amyloid generation and memory deficiency. *Brain Res.* **2009**, *1250*, 164–174. [CrossRef] [PubMed]
54. Nisar, A.; Mamat, A.S.; Hatim, M.I.; Aslam, M.S.; Ahmad, M.S. An updated review on *Catharanthus roseus*: Phytochemical and pharmacological analysis. *Indian Res. J. Pharm. Sci.* **2016**, *3*, 631–653.
55. Bora, K.S.; Arora, S.; Shri, R. Role of *Ocimum basilicum* L. in prevention of ischemia and reperfusion-induced cerebral damage, and motor dysfunctions in mice brain. *J. Ethnopharmacol.* **2011**, *137*, 1360–1365. [CrossRef]

56. Loughrin, J.H.; Kasperbauer, M.J. Light reflected from colored mulches affects aroma and phenol content of sweet basil (*Ocimum basilicum* L.) leaves. *J. Agric. Food Chem.* **2001**, *49*, 1331–1335. [CrossRef]
57. Petkovšek, M.M.; Štampar, F.; Veberič, R. Accumulation of phenolic compounds in apple in response to infection by the scab pathogen, *Venturia inaequalis*. *Physiol. Mol. Plant Pathol.* **2009**, *74*, 60–67. [CrossRef]
58. Bridgen, M. Using ultraviolet-C (UV-C) irradiation on greenhouse ornamental plants for growth regulation. In Proceedings of the VIII International Symposium on Light in Horticulture 1134, East Lansing, MI, USA, 22–26 May 2016; pp. 49–56.
59. Fiod Riccio, B.V.; Fonseca-Santos, B.; Colerato Ferrari, P.; Chorilli, M. Characteristics, biological properties and analytical methods of trans-resveratrol: A review. *Crit. Rev. Anal. Chem.* **2020**, *50*, 339–358. [CrossRef] [PubMed]
60. Hardman, W.E.; Avula, C.R.; Fernandes, G.; Cameron, I.L. Three percent dietary fish oil concentrate increased efficacy of doxorubicin against MDA-MB 231 breast cancer xenografts. *Clin. Cancer Res.* **2001**, *7*, 2041–2049. [PubMed]
61. Kassie, F.; Pool-Zobel, B.; Parzefall, W.; Knasmüller, S. Genotoxic effects of benzyl isothiocyanate, a natural chemopreventive agent. *Mutagenesis* **1999**, *14*, 595–604. [CrossRef]
62. Kassie, F.; Laky, B.; Gminski, R.; Mersch-Sundermann, V.; Scharf, G.; Lhoste, E.; Kansmüller, S. Effects of garden and water cress juices and their constituents, benzyl and phenethyl isothiocyanates, towards benzo (a) pyrene-induced DNA damage: A model study with the single cell gel electrophoresis/Hep G2 assay. *Chem. Biol. Interact.* **2003**, *142*, 285–296. [CrossRef]
63. Schmidt, T.J.; Klaes, M.; Sendker, J. Lignans in seeds of Linum species. *Phytochemistry* **2012**, *82*, 89–99. [CrossRef] [PubMed]
64. Wang, H.; Wang, J.; Qiu, C.; Ye, Y.; Guo, X.; Chen, G.; Li, T.; Wang, Y.; Fu, X.; Liu, R.H. Comparison of phytochemical profiles and health benefits in fiber and oil flaxseeds (*Linum usitatissimum* L.). *Food Chem.* **2017**, *214*, 227–233. [CrossRef]
65. Omar, K.A.; Shan, L.; Zou, X.; Song, Z.; Wang, X. Effects of two emulsifiers on yield and storage of flaxseed oil powder by response surface methodology. *Pak. J. Nutr.* **2009**, *8*, 1316–1324. [CrossRef]
66. Waszkowiak, K.; Gliszczyńska-Świgło, A. Binary ethanol–water solvents affect phenolic profile and antioxidant capacity of flaxseed extracts. *Eur. Food Res. Technol.* **2016**, *242*, 777–786. [CrossRef]
67. Zheng, W.; Wang, S.Y. Antioxidant activity and phenolic compounds in selected herbs. *J. Agric. Food Chem.* **2001**, *49*, 5165–5170. [CrossRef] [PubMed]
68. Hernández, R.; Eguchi, T.; Deveci, M.; Kubota, C. Tomato seedling physiological responses under different percentages of blue and red photon flux ratios using LEDs and cool white fluorescent lamps. *Sci. Hortic.* **2016**, *213*, 270–280. [CrossRef]
69. Ali, M.; Abbasi, B.H. Light-induced fluctuations in biomass accumulation, secondary metabolites production and antioxidant activity in cell suspension cultures of *Artemisia absinthium* L. *J. Photochem. Photobiol. B Biol.* **2014**, *140*, 223–227. [CrossRef]
70. Ohashi-Kaneko, K.; Takase, M.; Kon, N.; Fujiwara, K.; Kurata, K. Effect of light quality on growth and vegetable quality in leaf lettuce, spinach and komatsuna. *Environ. Control. Biol.* **2007**, *45*, 189–198. [CrossRef]
71. Jack, A.; Vrenken, L. Fluorescent lamps and low pressure sodium lamps. *IEE Proc. A* **1980**, *127*, 149–157. [CrossRef]
72. Kawka, B.; Kwiecień, I.; Ekiert, H. Influence of culture medium composition and light conditions on the accumulation of bioactive compounds in shoot cultures of *Scutellaria lateriflora* L. (American Skullcap) grown in vitro. *Appl. Biochem. Biotechnol.* **2017**, *183*, 1414–1425. [CrossRef] [PubMed]
73. Sobhani Najafabadi, A.; Khanahmadi, M.; Ebrahimi, M.; Moradi, K.; Behroozi, P.; Noormohammadi, N. Effect of different quality of light on growth and production of secondary metabolites in adventitious root cultivation of *Hypericum perforatum*. *Plant Signal. Behav.* **2019**, *14*, 1640561. [CrossRef]
74. Pedroso, R.C.N.; Branquinho, N.A.A.; Hara, A.C.; Costa, A.C.; Silva, F.G.; Pimenta, L.P.; Silva, M.L.A.; Cunha, W.R.; Pauletti, P.M.; Januario, A.H. Impact of light quality on flavonoid production and growth of *Hyptis marrubioides* seedlings cultivated in vitro. *Rev. Bras. Farmacogn.* **2017**, *27*, 466–470. [CrossRef]
75. Yu, K.-W.; Murthy, H.N.; Hahn, E.-J.; Paek, K.-Y. Ginsenoside production by hairy root cultures of *Panax ginseng*: Influence of temperature and light quality. *Biochem. Eng. J.* **2005**, *23*, 53–56. [CrossRef]
76. Murthy, H.N.; Kim, Y.-S.; Park, S.-Y.; Paek, K.-Y. Hypericins: Biotechnological production from cell and organ cultures. *Appl. Microbiol. Biotechnol.* **2014**, *98*, 9187–9198. [CrossRef] [PubMed]
77. McNeil, M.; Facey, P.; Porter, R. Essential oils from the *Hyptis* genus—A review (1909–2009). *Nat. Prod. Commun.* **2011**, *6*, 1934578X1100601149. [CrossRef]
78. Coon, J.T.; Ernst, E. Panax ginseng. *Drug Saf.* **2002**, *25*, 323–344. [CrossRef] [PubMed]
79. Palazón, J.; Cusidó, R.M.; Bonfill, M.; Mallol, A.; Moyano, E.; Morales, C.; Piñol, M.T. Elicitation of different *Panax ginseng* transformed root phenotypes for an improved ginsenoside production. *Plant Physiol. Biochem.* **2003**, *41*, 1019–1025. [CrossRef]
80. Sharopov, F.S.; Sulaimonova, V.A.; Setzer, W.N. Composition of the Essential oil of *Artemisia absinthium* from Tajikistan. *Rec. Nat. Prod.* **2012**, *6*, 127–134.
81. Baker, P. *The Book of Absinthe: A Cultural History*; Grove Press: Greenwich Village, NY, USA, 2001.
82. Hayat, M.Q.; Khan, M.A.; Ashraf, M.; Jabeen, S. *Ethnobotany of the Genus Artemisia L. (Asteraceae) in Pakistan*; University of Hawaii at Manoa: Honolulu, HI, USA, 2009.
83. Anderson, F.J. *An Illustrated History of the Herbals*; iUniverse: Bloomington, IN, USA, 1997.
84. Smith, A.; Secoy, D. Plants used for agricultural pest control in western Europe before 1850. *Chem. Ind.* **1981**, *70*, 12–17.
85. Ferreira, J.F.; Peaden, P.; Keiser, J. In vitro trematocidal effects of crude alcoholic extracts of *Artemisia annua*, *A. absinthium*, *Asimina triloba*, and *Fumaria officinalis*. *Parasitol. Res.* **2011**, *109*, 1585–1592. [CrossRef]

86. Mohamed, A.E.-H.H.; El-Sayed, M.; Hegazy, M.E.; Helaly, S.E.; Esmail, A.M.; Mohamed, N.S. Chemical constituents and biological activities of *Artemisia herba-alba*. *Rec. Nat. Prod.* **2010**, *4*, 1–25.
87. Gupta, S.D.; Agarwal, A. *Light Emitting Diodes for Agriculture*; Springer: Berlin, Germany, 2017.
88. Terashima, I.; Fujita, T.; Inoue, T.; Chow, W.S.; Oguchi, R. Green light drives leaf photosynthesis more efficiently than red light in strong white light: Revisiting the enigmatic question of why leaves are green. *Plant Cell Physiol.* **2009**, *50*, 684–697. [CrossRef]
89. Zhang, T.; Shi, Y.; Piao, F.; Sun, Z. Effects of different LED sources on the growth and nitrogen metabolism of lettuce. *Plant Cell Tissue Organ. Cult.* **2018**, *134*, 231–240. [CrossRef]
90. Gupta, S.D.; Kumar, A.; Agarwal, A. Impact of light-emitting diodes (LEDs) on the growth and morphogenesis of encapsulated shoot buds of *Curculigo orchioides* Gaertn., an endangered medicinal herb. *Acta Physiol. Plant* **2019**, *41*, 50. [CrossRef]
91. Wang, W.; Su, M.; Li, H.; Zeng, B.; Chang, Q.; Lai, Z. Effects of supplemental lighting with different light qualities on growth and secondary metabolite content of *Anoectochilus roxburghii*. *PeerJ* **2018**, *6*, e5274. [CrossRef] [PubMed]
92. Taulavuori, K.; Pyysalo, A.; Taulavuori, E.; Julkunen-, R. Responses of phenolic acid and flavonoid synthesis to blue and blue-violet light depends on plant species. *Environ. Exp. Bot.* **2018**, *150*, 183–187. [CrossRef]
93. Cioć, M.; Szewczyk, A.; Żupnik, M.; Kalisz, A.; Pawłowska, B. LED lighting affects plant growth, morphogenesis and phytochemical contents of *Myrtus communis* L. in vitro. *Plant Cell Tissue Organ. Cult.* **2018**, *132*, 433–447. [CrossRef]
94. Younas, M.; Drouet, S.; Nadeem, M.; Giglioli-Guivarc'h, N.; Hano, C.; Abbasi, B.H. Differential accumulation of silymarin induced by exposure of *Silybum marianum* L. callus cultures to several spectres of monochromatic lights. *J. Photochem. Photobiol. B Biol.* **2018**, *184*, 61–70. [CrossRef]
95. Manivannan, A.; Soundararajan, P.; Halimah, N.; Ko, C.H.; Jeong, B.R. Blue LED light enhances growth, phytochemical contents, and antioxidant enzyme activities of *Rehmannia glutinosa* cultured in vitro. *Hortic. Environ. Biotechnol.* **2015**, *56*, 105–113. [CrossRef]
96. Zhang, R.-X.; Li, M.-X.; Jia, Z.-P. *Rehmannia glutinosa*: Review of botany, chemistry and pharmacology. *J. Ethnopharmacol.* **2008**, *117*, 199–214. [CrossRef]
97. Nadeem, M.; Abbasi, B.H.; Younas, M.; Ahmad, W.; Zahir, A.; Hano, C. LED-enhanced biosynthesis of biologically active ingredients in callus cultures of *Ocimum basilicum*. *J. Photochem. Photobiol. B Biol.* **2019**, *190*, 172–178. [CrossRef] [PubMed]
98. Cuong, D.M.; Ha, T.W.; Park, C.H.; Kim, N.S.; Yeo, H.J.; Chun, S.W.; Kim, C.; Park, S.U. Effects of LED lights on expression of genes involved in phenylpropanoid biosynthesis and accumulation of phenylpropanoids in wheat sprout. *Agronomy* **2019**, *9*, 307. [CrossRef]
99. Seo, J.-M.; Arasu, M.V.; Kim, Y.-B.; Park, S.U.; Kim, S.-J. Phenylalanine and LED lights enhance phenolic compound production in Tartary buckwheat sprouts. *Food Chem.* **2015**, *177*, 204–213. [CrossRef]
100. Lobiuc, A.; Vasilache, V.; Oroian, M.; Stoleru, T.; Burducea, M.; Pintilie, O.; Zamfirache, M.-M. Blue and red LED illumination improves growth and bioactive compounds contents in acyanic and cyanic *Ocimum basilicum* L. microgreens. *Molecules* **2017**, *22*, 2111. [CrossRef] [PubMed]
101. Yeow, L.C.; Chew, B.L.; Sreeramanan, S. Elevation of secondary metabolites production through light-emitting diodes (LEDs) illumination in protocorm-like bodies (PLBs) of *Dendrobium* hybrid orchid rich in phytochemicals with therapeutic effects. *Biotechnol. Rep.* **2020**, *27*, e00497. [CrossRef]
102. Khurshid, R.; Ullah, M.A.; Tungmunnithum, D.; Drouet, S.; Shah, M.; Zaeem, A.; Hameed, S.; Hano, C.; Abbasi, B.H. Lights triggered differential accumulation of antioxidant and antidiabetic secondary metabolites in callus culture of *Eclipta alba* L. *PLoS ONE* **2020**, *15*, e0233963. [CrossRef] [PubMed]
103. Davis, P.A.; Burns, C. Photobiology in protected horticulture. *Food Energy Secur.* **2016**, *5*, 223–238. [CrossRef]
104. García-Closas, R.; Berenguer, A.; Tormo, M.J.; Sánchez, M.J.; Quiros, J.R.; Navarro, C.; Arnaud, R.; Dorronsoro, M.; Chirlaque, M.D.; Barricarte, A. Dietary sources of vitamin C, vitamin E and specific carotenoids in Spain. *Br. J. Nutr.* **2004**, *91*, 1005–1011. [CrossRef] [PubMed]
105. Al-Snafi, A.E. Phenolics and flavonoids contents of medicinal plants, as natural ingredients for many therapeutic purposes-A review. *IOSR J. Pharm.* **2020**, *10*, 42–81.
106. Hogewoning, S.W.; Trouwborst, G.; Maljaars, H.; Poorter, H.; van Ieperen, W.; Harbinson, J. Blue light dose–responses of leaf photosynthesis, morphology, and chemical composition of *Cucumis sativus* grown under different combinations of red and blue light. *J. Exp. Bot.* **2010**, *61*, 3107–3117. [CrossRef] [PubMed]
107. Yildirim, F.; Şan, B.; Yildirim, A.; Polat, M.; Ercişli, S. Mineral composition of leaves and fruit in some myrtle (*Myrtus communis* L.) genotypes. *Erwerbs Obstbau* **2015**, *57*, 149–152. [CrossRef]
108. Bouaziz, A.; Abdalla, S.; Baghiani, A.; Charef, N. Phytochemical analysis, hypotensive effect and antioxidant properties of *Myrtus communis* L. growing in Algeria. *Asian Pac. J. Trop. Biomed.* **2015**, *5*, 19–28. [CrossRef]
109. Shah, A.; Kazi, T.; Baig, J.; Afridi, H.; Kandhro, G.; Khan, S.; Kolachi, N.; Wadhwa, S. Determination of total mercury in chicken feed, its translocation to different tissues of chicken and their manure using cold vapour atomic absorption spectrometer. *Food Chem. Toxicol.* **2010**, *48*, 1550–1554. [CrossRef] [PubMed]
110. Chung, I.M.; Kim, J.J.; Lim, J.D.; Yu, C.Y.; Kim, S.H.; Hahn, S.J. Comparison of resveratrol, SOD activity, phenolic compounds and free amino acids in *Rehmannia glutinosa* under temperature and water stress. *Environ. Exp. Bot.* **2006**, *56*, 44–53. [CrossRef]
111. Machlin, L.J.; Bendich, A. Free radical tissue damage: Protective role of antioxidant nutrients 1. *FASEB J.* **1987**, *1*, 441–445. [CrossRef] [PubMed]

112. Yang, C.S.; Landau, J.M.; Huang, M.-T.; Newmark, H.L. Inhibition of carcinogenesis by dietary polyphenolic compounds. *Annu. Rev. Nutr.* **2001**, *21*, 381–406. [CrossRef]
113. Yao, L.H.; Jiang, Y.-M.; Shi, J.; Tomas-Barberan, F.; Datta, N.; Singanusong, R.; Chen, S. Flavonoids in food and their health benefits. *Plant Foods Hum. Nutr.* **2004**, *59*, 113–122. [CrossRef]
114. Naoumkina, M.A.; Zhao, Q.; Gallego-Giraldo, L.; Dai, X.; Zhao, P.X.; Dixon, R.A. Genome-wide analysis of phenylpropanoid defence pathways. *Mol. Plant Pathol.* **2010**, *11*, 829–846. [CrossRef]
115. Thoma, F.; Somborn-Schulz, A.; Schlehuber, D.; Keuter, V.; Deerberg, G. Effects of light on secondary metabolites in selected leafy greens: A review. *Front. Plant Sci.* **2020**, *11*, 497. [CrossRef]
116. Folta, K.M.; Carvalho, S.D. Photoreceptors and control of horticultural plant traits. *HortScience* **2015**, *50*, 1274–1280. [CrossRef]
117. Brennicke, A.; Schopfer, P. *Pflanzenphysiologie*; Spektrum Akademischer Verlag: Berlin, Germany, 2010.
118. Brunetti, C.; Fini, A.; Sebastiani, F.; Gori, A.; Tattini, M. Modulation of phytohormone signaling: A primary function of flavonoids in plant–environment interactions. *Front. Plant Sci.* **2018**, *9*, 1042. [CrossRef]
119. Renouard, S.; Corbin, C.; Lopez, T.; Montguillon, J.; Gutierrez, L.; Lamblin, F.; Lainé, E.; Hano, C. Abscisic acid regulates pinoresinol–lariciresinol reductase gene expression and secoisolariciresinol accumulation in developing flax (*Linum usitatissimum* L.) seeds. *Planta* **2012**, *235*, 85–98. [CrossRef]
120. Markulin, L.; Corbin, C.; Renouard, S.; Drouet, S.; Durpoix, C.; Mathieu, C.; Lopez, T.; Auguin, D.; Hano, C.; Lainé, É. Characterization of LuWRKY36, a flax transcription factor promoting secoisolariciresinol biosynthesis in response to *Fusarium oxysporum* elicitors in *Linum usitatissimum* L. hairy roots. *Planta* **2019**, *250*, 347–366. [CrossRef] [PubMed]
121. Casal, J.J. Phytochromes, cryptochromes, phototropin: Photoreceptor interactions in plants. *Photochem. Photobiol.* **2000**, *71*, 1–11. [CrossRef]
122. Fox, A.R.; Soto, G.C.; Jones, A.M.; Casal, J.J.; Muschietti, J.P.; Mazzella, M.A. cry1 and GPA1 signaling genetically interact in hook opening and anthocyanin synthesis in Arabidopsis. *Plant Mol. Biol.* **2012**, *80*, 315–324. [CrossRef] [PubMed]
123. Folta, K.M. Green light stimulates early stem elongation, antagonizing light-mediated growth inhibition. *Plant Physiol.* **2004**, *135*, 1407–1416. [CrossRef] [PubMed]
124. Rodyoung, A.; Masuda, Y.; Tomiyama, H.; Saito, T.; Okawa, K.; Ohara, H.; Kondo, S. Effects of light emitting diode irradiation at night on abscisic acid metabolism and anthocyanin synthesis in grapes in different growing seasons. *Plant Growth Regul.* **2016**, *79*, 39–46. [CrossRef]
125. Hano, C.; Addi, M.; Fliniaux, O.; Bensaddek, L.; Duverger, E.; Mesnard, F.; Lamblin, F.; Lainé, E. Molecular characterization of cell death induced by a compatible interaction between *Fusarium oxysporum* f. sp. linii and flax (*Linum usitatissimum*) cells. *Plant Physiol. Biochem.* **2008**, *46*, 590–600. [CrossRef]

plants

MDPI

Article

Single Wavelengths of LED Light Supplement Promote the Biosynthesis of Major Cyclic Monoterpenes in Japanese Mint

Takahiro Ueda, Miki Murata and Ken Yokawa *

Faculty of Engineering, Kitami Institute of Technology, Hokkaido 090-8507, Japan; t.ueda7991@gmail.com (T.U.); muratamk@mail.kitami-it.ac.jp (M.M.)

* Correspondence: yokawaken@mail.kitami-it.ac.jp; Tel.: +81-157-26-9434

Abstract: Environmental light conditions influence the biosynthesis of monoterpenes in the mint plant. Cyclic terpenes, such as menthol, menthone, pulegone, and menthofuran, are major odor components synthesized in mint leaves. However, it is unclear how light for cultivation affects the contents of these compounds. Artificial lighting using light-emitting diodes (LEDs) for plant cultivation has the advantage of preferential wavelength control. Here, we monitored monoterpene contents in hydroponically cultivated Japanese mint leaves under blue, red, or far-red wavelengths of LED light supplements. Volatile cyclic monoterpenes, pulegone, menthone, menthol, and menthofuran were quantified using the head-space solid phase microextraction method. As a result, all light wavelengths promoted the biosynthesis of the compounds. Remarkably, two weeks of blue-light supplement increased all compounds: pulegone (362% increase compared to the control), menthofuran (285%), menthone (223%), and menthol (389%). Red light slightly promoted pulegone (256%), menthofuran (178%), and menthol (197%). Interestingly, the accumulation of menthone (229%) or menthofuran (339%) was observed with far-red light treatment. The quantification of glandular trichomes density revealed that no increase under light supplement was confirmed. Blue light treatment even suppressed the glandular trichome formation. No promotion of photosynthesis was observed by pulse-amplitude-modulation (PAM) fluorometry. The present result indicates that light supplements directly promoted the biosynthetic pathways of cyclic monoterpenes.

Keywords: mint; monoterpenes; solid phase microextraction (SPME); hydroponics; LED supplement

Citation: Ueda, T.; Murata, M.; Yokawa, K. Single Wavelengths of LED Light Supplement Promote the Biosynthesis of Major Cyclic Monoterpenes in Japanese Mint. *Plants* **2021**, *10*, 1420. https://doi.org/10.3390/plants10071420

Academic Editors: Valeria Cavallaro and Rosario Muleo

Received: 21 May 2021
Accepted: 6 July 2021
Published: 12 July 2021

1. Introduction

Secondary metabolites are chemical products enzymatically converted from primary metabolites in plants. Plants use these compounds to adapt to their environment, for example, defense against pathogens or insect attacks, or other stresses. In human history, secondary metabolites have also been an essential source of medicines [1]. Although the development of modern chemistry enables us to synthesize a broad range of chemical compounds, many plant secondary metabolites, especially terpenes, are still considered important pharmaceutical materials. This is because it is convenient to use plant-derived compounds as synthetic starting materials to obtain the desired molecules. Thus, significant efforts have been made to acquire valuable secondary metabolites [2,3].

A glandular trichome (GT) is a plant-specific storage organ distributed in the aerial part of the plant body. In mint plants, GTs play an essential role as a tiny cell factory to synthesize and accumulate secondary metabolites. In terms of biotechnological or pharmacological interests, the number of studies on plant GTs has increased recently [4,5]. Transcriptomic analysis was conducted to elucidate the regulation of GT-specific terpene biosynthesis, for example, in spearmint [6] and *Artemisia* plants [7]. An essential oil generated catalytically in GTs is susceptible to environmental fluctuations or stresses [8]. Particularly, an increase in the terpene contents after various abiotic stress treatments in medicinal plants was confirmed [9].

Mint plants belonging to the *Mentha* genus are the most famous herbs that have been used in the past. A major terpene produced by the plant is menthol. Menthol causes a sensation of coolness through the direct reaction with the transient receptor potential melastatin 8 (TRPM8) channel [10]. Menthol-containing essential oil is used extensively for many purposes. Menthol has many biological actions, and its antifungal activity is a well-known function [11]. In addition to menthol, mint plants produce cyclic monoterpenes, for example, pulegone, menthone (the intermediates in menthol biosynthesis), and menthofuran (Figure 1). The balance of these contents features an odor of the mint leaves, and the growth environment influences the biosynthesis of the compounds.

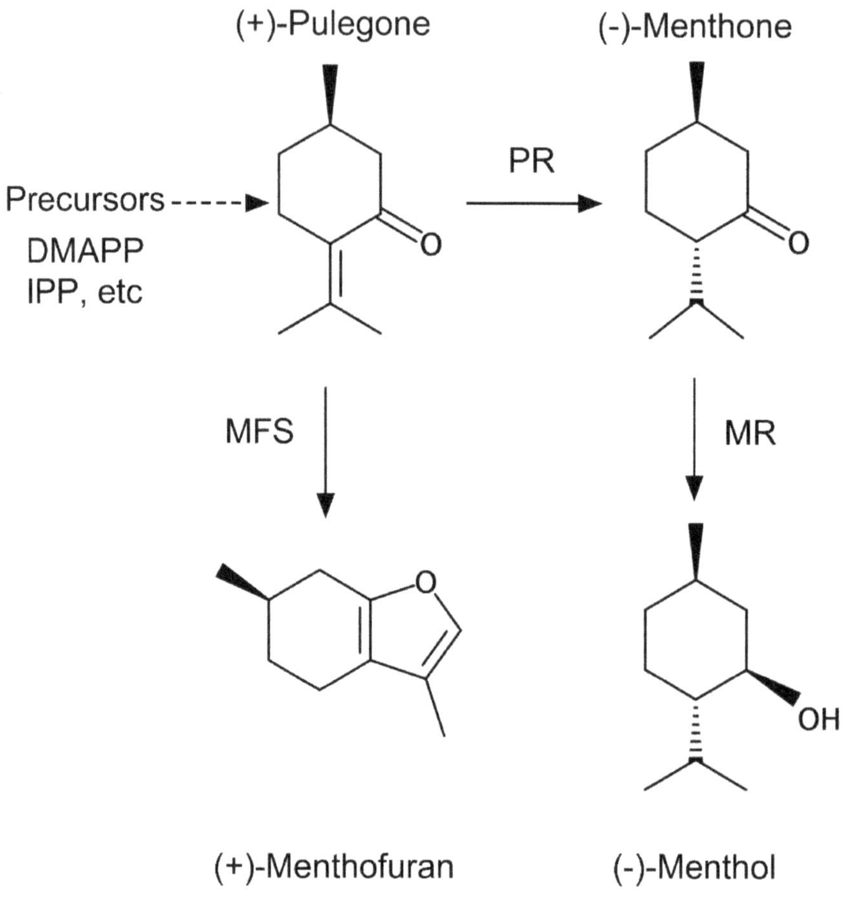

Figure 1. Cyclic monoterpene 12 synthesized in glandular trichomes in Japanese mint. These four molecules are the major components of Japanese mint flavor. PR: pulegone reductase; MR: (−)-menthol reductase; MFS: menthofuran synthase.

Light is one of the physical factors necessary for plant growth, development, and metabolism. Many studies have focused on light and changes in secondary metabolites, such as anthocyanins, carotenoids, and flavanols, under the control of photoreceptors [12]. The light effect on monoterpene metabolism of mint plants has long been studied [13]. It was assessed that a short-day photoperiod treatment for three mint species significantly increased the oil content [14]. In a controlled light environment using light-emitting diodes (LED), red LED was highly effective in increasing the oil content in *M. piperita*; blue and white LEDs were also effective [15]. However, it is still unclear which wavelength of light affects both the contents and composition of monoterpenes produced by mint plants.

Here, we analyzed the contents of four major cyclic monoterpenes in a Japanese mint plant cultivated under single-wavelength supplementation of blue, red, or far-red LED light. The number of GTs on the growing mint leaves was counted and compared between the treatments. The possible physiological mechanism of light supplementation on terpene biosynthesis was also discussed.

2. Materials and Methods

2.1. Plant Preparation and Growth Conditions

A rootstock of the major Japanese mint cultivar "HOKKAI JM 23" (*Mentha canadensis* L.) was obtained from the National Agriculture and Food Research Organization Genebank, Tsukuba, Japan (collection ID: JP176265). The plants were first recovered from the rootstock and grown in commercial soil. Fourteen young shoots were harvested from a well-grown mint plant by stem cutting at the position below the fourth leaf. The cuttings were incubated until adventitious roots were generated in distilled water for nine days at room temperature under a 16 h/8 h light/dark cycle at 23 °C. The young plants were then placed in a commercially available hydroponic cultivating system, as shown in Figure 2A (Green Farm, U-ing, Osaka, Japan). The box-shaped system was equipped with programmable white LED lighting, a ventilator, and automatic water flow/aeration. For acclimation to the new environment, the plants were pre-grown for one week with distilled water. At the beginning of the experiment, the hydroponic water tank was filled with 4 L of standard Hoagland's cultivating solution (2.5 mM KNO_3, 1.25 mM $Ca(NO_3)_2 \cdot 4H_2O$, 0.5 mM NH_4NO_3, 1 mM $MgSo_4 \cdot 7H_2O$, 0.25 mM KH_2PO_4, 25 mM NaFe(III)EDTA, 23 mM H_3BO_3, 4.55 mM $MnCl_2 \cdot 4H_2O$, 0.39 mM $ZnSO_4 \cdot 7H_2O$, 0.1 mM $CuSO_4 \cdot 5H_2O$, and 0.25 mM $Na_2MoO_4 \cdot 2H_2O$; FUJIFILM Wako Chemicals, Osaka, Japan), and the pH of the solution was measured daily using a portable pH meter (LAQUAtwin pH-33B, HORIBA, Kyoto, Japan). The pH was adjusted to 6.2 using 1 M 2-morpholinoethanesulfonic acid (Dojindo, Kumamoto, Japan) buffer, and 5 M NaOH was used every two days [16].

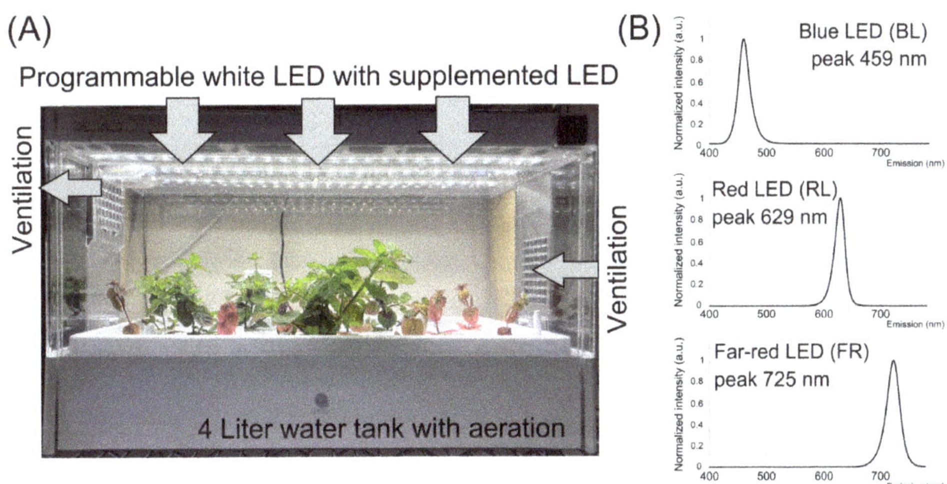

Figure 2. Hydroponic cultivation of Japanese mint and LED supplement. (**A**) A commercially available hydroponic system was modified for the light-supplementation experiment. The system is semi-closed and automatic. (**B**) Three light spectra were emitted from three LED sources used in the study: blue light (BL), red light (RL), and far-red (FR).

2.2. LED Lighting Conditions

The photosynthetic photon flux density (PPFD) of all light sources used in the study was measured using a Light Analyzer LA-105 (NK Systems, Tokyo, Japan). In the cultivating system, the average PPFD value of white light (WL) LED at the foliar position was at 161 μmol/m^2/s. An arrayed LED source of blue light (BL), red light (RL), or far-red (FR) supplements was placed on the ceiling of the cultivating box. All light spectra are shown in Figure 2B. PPFD values of BL, RL, and FR were at 6.7, 7.1, and 3.7 μmol/m^2/s, respectively. The photocycle of the basal WL was 16 h light/8 h dark. The daily light supplement of each wavelength began at the same time when the WL was on and lasted 6 h. In the hydroponic culture system used in the study, 16 young mint plants were simultaneously cultivated and treated with LED light for each independent experiment. The duration of the experiment of each light supplement was two weeks.

2.3. HS-SPME and GC–MS Analysis for Cyclic Monoterpenes

After two weeks of light treatment, the second and third leaves (total four leaves) were harvested by cutting the petiole and weighed. The first leaves newly emerged during the 2-week LED treatment, and the size of the leaves was small and immature. To observe the effect of the LED supplements on expanded leaves, we decided to use the second and third leaves for further analyses. The analysis of a volatile compound with the head-space solid phase microextraction (HS-SPME) method was adapted from a previous report [17]. The leaves were then frozen with liquid nitrogen and ground finely with a pre-cooled pestle and mortar. The ground powder was put into 2 mL of 2 M $CaCl_2 \cdot H_2O$ solution in a head-space 20-mL glass vial (Supelco/Sigma-Aldrich, Tokyo, Japan) to prevent unnecessary enzymatic reactions. The tightly sealed vials were incubated in a hot bath at a temperature of 40 °C for 5 min. An SPME fiber coated with 60-μm-thick PDMS/DVB (polydimethylsiloxane/divinylbenzene, Supleco) was inserted into the head-space through the vial septum, and the volatile released from the warmed solution was absorbed. The fiber was removed from the vial and inserted quickly into the inlet of a gas chromatography–mass spectroscopy (GC–MS) system (GC-17A, Shimadzu, Kyoto, Japan) with a polyethylene glycol column (60 m long, 0.25 mm internal diameter, 0.25 μm film thickness; GL Sciences, Tokyo, Japan). The temperature of the inlet was held at 180 °C with a 9:1 split. Helium was used as the carrier gas at a 2 mL/min flow rate, 245.9 kPa. The starting oven temperature was 100 °C, followed by an 8 °C/min ramp until 180 °C was reached and held for 2 min. The interface to the mass spectrometer was held at 250 °C. The mass spectrometer was run in the scan mode with electron impact ionization at 1 keV, from m/z 40 to m/z 300. Compounds were identified using the National Institute of Standard and Technology Mass Spectral Database (NIST12). The GC–MS analyses of terpenes were conducted independently six times in different LED lightning treatments.

2.4. Preparation of Standards

Standard solutions of pure chemicals of menthol and menthone (FUJIFILM Wako Chemicals, Osaka, Japan), pulegone, and menthofuran (Sigma-Aldrich, Tokyo, Japan) were dissolved in pure methanol. The dilution series of each solution was prepared with methanol. A volume of 2 μL of the standard solution was added to the glass vial with 2 mL $CaCl_2$, and the compound was measured using GC–MS with the same method described above. Three to four concentrations of each solution were used to make standard curves. Peak areas of each compound obtained both from samples and standards were used for calculations.

2.5. Quantification of GTs

All experiments were conducted using plant samples grown under different light supplements for two weeks. At the bottom, middle, or top of the leaf on both the abaxial and adaxial sides, six locations were imaged using a stereomicroscope (WRAYMER 820T, WRAYMER, Osaka, Japan). In the same manner, the number of GTs on the second and third leaves was compared. An area of 4 mm^2 was chosen randomly from each picture, and the total number of GTs was counted. The data were analyzed independently from sixteen plants as biological replications.

2.6. Chlorophyll Quantification

Chlorophyll was extracted and quantified according to a previous method [18]. The chlorophyll and anthocyanin extractions were made using leaves different from the ones used for the terpene quantifications. The leaf position (e.g., lighting) alters the chlorophyll contents. Therefore, to minimize the errors among the analyses, we carefully harvested and chose the second and third leaves with similar sizes for terpene or chlorophyll measurement. Briefly, weighed leaf samples were placed into 2 mL of pure N,N-dimethylformamide (DMF; FUJIFILM Wako Chemicals, Osaka, Japan) at 4 °C in darkness overnight. The absorbance values at 647 and 664 nm of the aliquot of the extracted solution were measured using a spectrophotometer (U-5100, Hitachi, Tokyo, Japan). DMF was used as the blank control. The total amount of chlorophyll was calculated based on the following equation [18]: $Chl\ a + b = 17.67 * A_{647} + 7.12 * A_{664}$.

2.7. Anthocyanin Quantification

For anthocyanin extraction and quantification, a method reported previously was adapted to this study [19,20]. Weighed leaf samples were ground with liquid nitrogen with a pre-cooled pestle and mortar. The powder was put into 2 mL of 1% HCl–methanol and incubated overnight at 4 °C in darkness. A volume of 300 μL of the aliquot was collected into a new 1.5 mL tube, and 200 μL of distilled water and 500 μL of chloroform were added. After mixing, the tube was centrifuged for 5 min at 13,000× *g* at 4 °C. A volume of 400 μL of the aliquot was collected carefully into a new tube, and 400 μL of 60% methanol + 1% HCl solution was added. The absorbance of the solution containing anthocyanin was measured at 530 nm and 657 nm.

2.8. Measurement of Photosynthetic Efficiency

The efficiency of photosynthesis was non-invasively monitored using a pulse-amplitude-modulation (PAM) chlorophyll fluorometer in each plant grown under different light wavelengths. Plants were placed in a dark environment for 30 min before the PAM measurement to obtain the maximum quantum efficiency of PSII (Fv/Fm). Then, PAM measurements were conducted under dim light using Junior-PAM (WALZ, Effeltrich, Germany).

2.9. Statistical Analyses

All numerical data were analyzed using Tukey's honestly significant difference (HSD) for parametric analysis and the Mann–Whitney *U* test for non-parametric analyses with R software version 4.0.2. (https://www.r-project.org/ accessed on 23 July 2020). Differences were considered significant when p-values were smaller than 0.05.

3. Results and Discussion

3.1. Promotion of Terpene Biosynthesis by Light Supplements

After two weeks of light supplementation, the increase of the cyclic monoterpene content in the treated leaves was observed. Especially, BL increased all four monoterpenes (Figure 3): pulegone (362% increase compared to the WL control), menthofuran (285%), menthone (223%), and menthol (389%). RL slightly promoted pulegone (256%), menthofuran (178%), and menthol (197%). These results are consistent with a previous study that showed an increase in fresh weight and essential oils in several mint species (*Mentha piperita, Mentha spicata*, etc.) cultivated in a red-blue LED incubator [15]. Another previous report also observed the increase of terpene contents in *Cannabis* plants under RL-BL sub-canopy LED lighting [21]. The precursors of all higher plant monoterpene biosynthesis, isopentenyl diphosphate (IPP) or dimethylallyl pyrophosphate (DMAPP), were proposed to be increased by the RL-BL supplement [21]. It is likely to explain our result showing the increase of cyclic monoterpene with either BL or RL supplement in Japanese mint (Figure 3). Interestingly, the accumulation of menthone (229%) or menthofuran (339%) was observed with FR light treatment. Our result is the first report that showed that FR light treatment increased terpenes in mint plants. In natural conditions, long-day treatment to mint plants showed the accumulation of menthofuran [14]. It indicates that the content of menthofuran is associated with the life cycle of the mint plant. Our result showed that the FR treatment disturbed the plant response to the photoperiod, affecting the metabolism of menthofuran. Furthermore, the reaction to FR indicates that the biosynthesis of these compounds is possibly under the control of phytochromes. In our study, no significant promotion of plant growth was observed during two weeks of LED light supplementation at the fluence rate used. Also, the contents of cyclic monoterpenes shown in Figure 3 were normalized with the weight of fresh leaves. This suggests that the increase of cyclic monoterpenes was not due to the leaf area's expansion. In many *Lamiaceae* plants, GTs in the aerial part of the plant body influence the biosynthesis of secondary metabolites, including terpenes, and store them in a cavity space surrounded by a cuticle. Here, we hypothesized why the terpenes increased after LED treatments as follows: (1) BL treatment increased the density of GTs, and (2) light stimulated the processes of terpene biosynthesis.

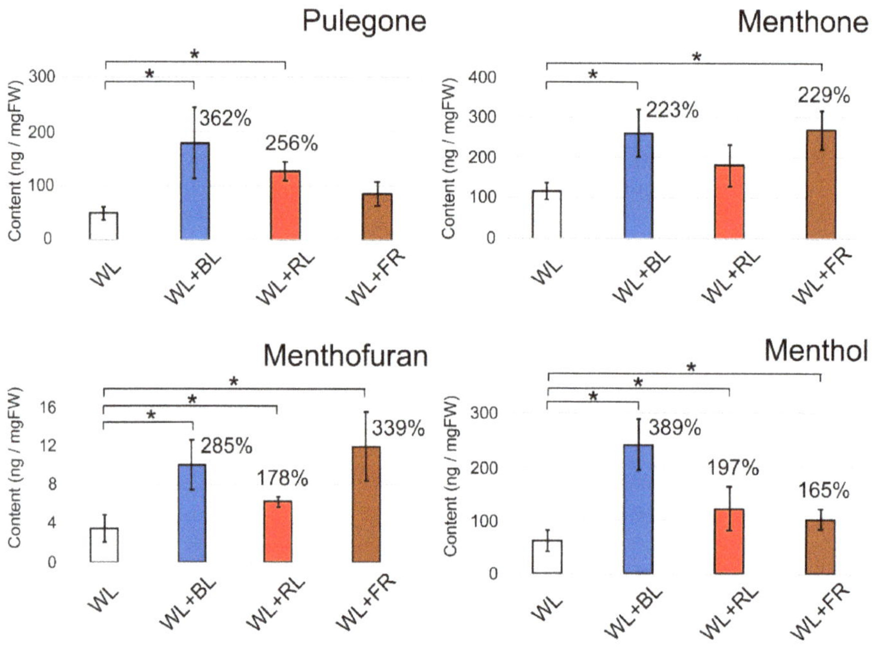

Figure 3. Modulation of the contents of cyclic monoterpenes in two-week light-supplemented Japanese mint leaves. The contents of pulegone, menthone, menthofuran, and menthol are presented. The values in the graphs indicate the increasing rates compared to the WL control. The measurements were conducted independently six times for biological replicates. Error bars indicate standard deviations from the mean. * Asterisks represent a significant difference from the WL control in the Mann–Whitney *U* test ($p < 0.05$).

3.2. Alteration of GT Density

We next observed whether the light treatment promoted the emergence of GTs. After two weeks of LED light treatment, the second and third leaves from the top of mint cuttings were harvested. The number of GTs in three regions of the leaf (apex, middle and basal regions) was counted separately and averaged. Overall, the density of GTs in the abaxial side of the leaves was slightly higher than that on the adaxial side (Figure 4A,B and Figure 5A,B). On both sides of the second leaf, except the apex region of the abaxial side, the GT density was decreased by the BL treatment (Figure 4A,B). There was a tendency for an increase in GT density by the FR light treatment. On the third leaf, a BL-dependent GT increase was similarly observed, whereas FR showed no promotion of GT emergence (Figure 5A,B). We observed three regions of the second and third leaves to determine whether GT was newly generated as the leaf expands (apex region) or light directly regulated GT density regardless of the leaf regions. Here, we confirmed that environmental factors, including light conditions, could flexibly control GT density. A previous study showed that two light conditions (full solar radiation vs. shade) had no impact on GT density in a medicinal plant, *Ocimum campechianum* [22]. In cultivated tomato plants, a density of type VI leaf GTs increased under high light conditions (approximately 300 μmol/m^2/s) [23]. Thus, light supplementation is likely to affect the density of GTs. The emergence of GTs is under the regulation of the well-known MYB transcription factor. Overexpression of AaMYB17 in sweet wormwood (*Artemisia annua* L.) increased the density of GTs [24]. Homeodomain-leucine zipper IV transcription factor was also shown to be involved in the control of GT density [25,26]. Our results show that the modulation of GT density by BL and FR is possible because of the regulation of transcription factors by light perception. However, the reduction of GT density with BL treatment was inconsistent with the increase of terpene contents, as shown in Figure 3. This suggests that biological processes of terpene biosynthesis were promoted directly by single wavelengths of LED light supplements.

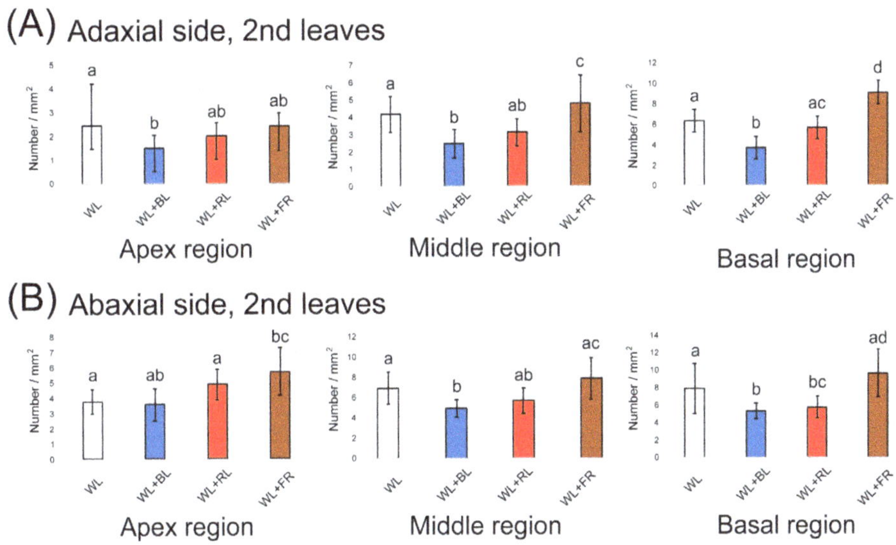

Figure 4. The number of glandular trichomes generated on the second leaf under different light treatments for two weeks. (**A**) Quantification on the adaxial side in the three leaf regions ($n = 16$). (**B**) Quantification on the abaxial side in the three leaf regions ($n = 16$). The data were obtained from four plants in four independent experiments as biological replications. Error bars indicate standard deviations from the mean. Different alphabets indicate significant differences according to Tukey's HSD test; $p < 0.05$.

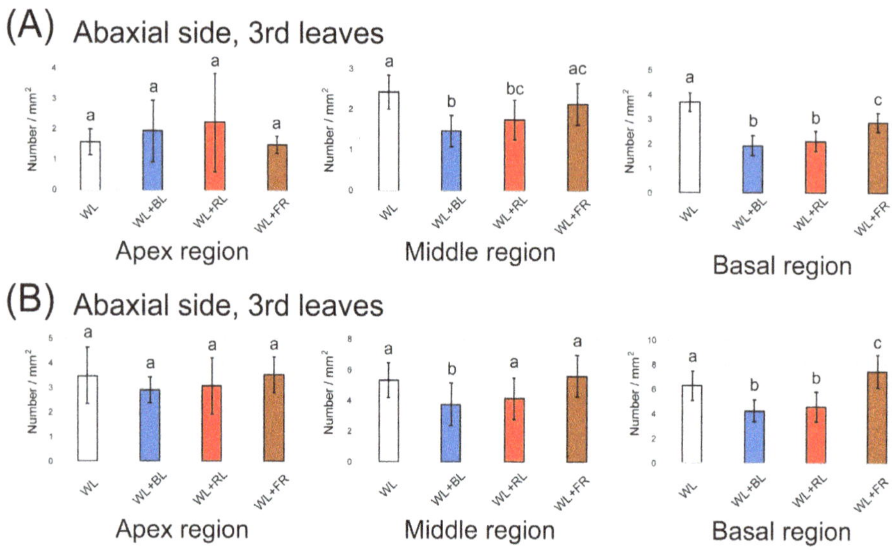

Figure 5. The number of glandular trichomes generated on the third leaf under different light treatments for two weeks. (**A**) Quantification on the adaxial side in the three leaf regions ($n = 16$). (**B**) Quantification on the abaxial side in the three leaf regions ($n = 16$). The data were obtained from four plants in four independent experiments as biological replications. Error bars indicate standard deviations from the mean. Different alphabets indicate significant differences according to Tukey's HSD test; $p < 0.05$.

3.3. Photosynthesis Was Not Boosted by LED Light Supplements

In mint plants, monoterpenes are biosynthesized from isopentenyl pyrophosphate and dimethylallyl pyrophosphate, which are provided by the plastidial methylerythritol phosphate (MEP) pathway [27]. The initial compound of the MEP pathway is pyruvic acid, which is supplied from sugars as a photosynthetic by-product. Next, we confirmed whether the photosynthetic activity was the source of monoterpene biosynthesis under light supplementation. We measured chlorophyll contents in the two-week light-treated mint leaves and compared them among the different light wavelengths. Similar to GT density results, BL and FR treatments significantly increased chlorophyll but not anthocyanin (Figure 6A). Anthocyanin biosynthesis was induced by BL treatment to plants. In *Arabidopsis thaliana*, anthocyanin biosynthesis was stimulated by 2.5 W/m^2 of BL (approximately 11 μmol/m^2/s) [28], which is higher than the intensity we used in this study (6.7 μmol/m^2/s). We considered that the increase in the chlorophyll contents induced by either BL or FR supplements could help the biosynthesis of cyclic monoterpenes in GTs. To check the contribution of chlorophyll to photosynthesis, we quantified the quantum yield of photosystem II based on chlorophyll fluorescence.

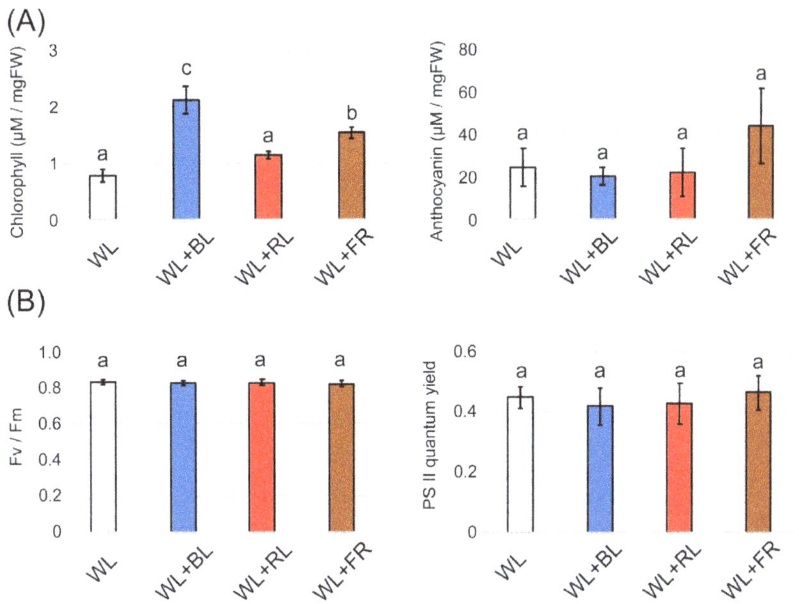

Figure 6. Effects of light supplementation on leaf pigment and photosynthesis in two-week light-supplemented Japanese mint leaves. (**A**) Total chlorophyll (n = 4) and anthocyanin contents (n = 4) were quantified using a spectrophotometer. (**B**) Photosynthetic parameters, Fv/Fm, and PSII quantum yield on leaves were monitored using a PAM chlorophyll fluorometer (n = 8–13). The data were obtained from independent experiments as biological replications. Error bars indicate standard deviations from the mean. Different alphabets indicate significant differences according to Tukey's HSD test; $p < 0.05$.

Contrary to our expectations, both Fv/Fm and the quantum yield showed no change with any wavelength of light treatment (Figure 6B). This suggests that the increase in cyclic monoterpenes shown in Figure 3 was not due to the promotion of photosynthesis by the light supplements. Evans and Terashima reported no correlation between the chlorophyll content in spinach leaves and photosynthetic activity [29]. In *Arabidopsis*, red and far-red wavelengths supplemented with WL enhanced chlorophyll biosynthesis [30]. A BL supplement was also shown to increase the chlorophyll content in seven plant species [31]. In conclusion, chlorophyll generation observed in light-supplementation experiments might be the result of the transient activation of photoreceptors and does not influence photosynthesis and terpene biosynthesis.

4. Conclusions

We propose a model based on the results as shown in Figure 7. The background WL illumination was still important for photosynthesis to provide sugars for terpene biosynthesis in GTs. We then speculate that the increase in cyclic monoterpenes in the GTs was probably due to the direct activation of processes of terpene biosynthesis. Studies have revealed that light modulated the biosynthesis of terpenes in many plant species. For example, a RL receptor, phytochrome, was shown to promote monoterpene production in thyme seedlings [32]. In cannabis plants, the content of a meroterpenoid, cannabinoid, was increased by BL treatment with a short photoperiod (12 h light/12 h dark) [33]. This indicates that photoreceptors or related factors, such as transcription factors, might directly or indirectly facilitate the enzymatic processes of terpene biosynthesis in secretory cells in GTs.

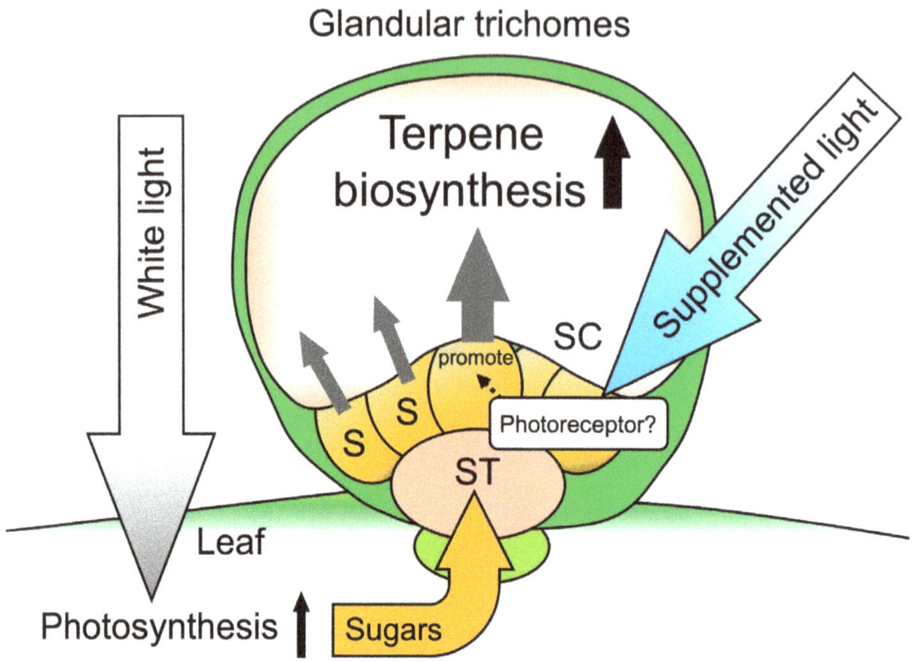

Figure 7. Schematic of glandular trichomes on Japanese mint leaves. Background WL provides energy for photosynthesis and provides a sugar source, which is necessary for terpene biosynthesis in glandular trichomes. LED light supplementation may promote the terpene biosynthetic pathway in glandular trichome cells through photoreceptor activation. S: secretory cells; ST: stalk cells; SC: storage cavity.

Recently, GT research is spotlighted because certain secondary metabolites produced by plants have high pharmaceutical value [4,5]. A lot of chemical compounds for producing medicines still rely on plant-derived starting materials. For example, *Artemisia annua* is recognized as a precious plant resource of artemisinin, an antimalarial agent, stored in GTs. Since a net synthesis of artemisinin has great difficulty and costly, the cultivation of *Artemisia* plants is necessary to obtain the compound. In addition to the plant, many other medicinal plants store valuable second metabolites, including terpenes in GTs. It is known that the contents of the compounds in GTs are dependent on the growing environment [8]. Therefore, a considerable effort is being paid to cultivating medicinal plants in an artificially modified environment to maximize the contents of desired compounds in GTs. Results obtained from Japanese mint as a model of GTs metabolism showed that light, which is one of the critical environmental factors, promotes monoterpene synthesis, and it will be helpful for GT research. Although we first speculated that the BL affects specific terpene biosynthesis, all four monoterpenes (pulegone, menthofuran, menthone, and menthol) were significantly increased. It suggests that the BL might stimulate the biosynthesis and accumulation of terpene precursors such as DMAPP or IPP in GTs (Figures 1 and 7). These compounds are the necessary starting materials for all higher plant monoterpene biosynthesis, including *Artemisia* plants, as mentioned above. Thus, the information on supplement lighting obtained from the Japanese mint study will be helpful to boost the contents of valuable terpene compounds in medicinal plants for industrial cultivating conditions. As no genomic information on the Japanese mint used in the study was provided, genetic verification was not conducted in this study. Therefore, in future research, further details on physiological regulation by the light perception in the Japanese mint need clarification.

Author Contributions: Conceptualization, K.Y. and M.M.; methodology, K.Y.; formal analysis, K.Y.; investigation, T.U.; writing and editing manuscript, K.Y., M.M. and T.U. All authors have read and agreed to the published version of the manuscript.

Funding: This research was funded by the Lotte Research Promotion Grant (K.Y.) and in part by the Leading Initiative for Excellent Young Researchers of MEXT, Japan (K.Y.).

Institutional Review Board Statement: Not applicable.

Informed Consent Statement: Not applicable.

Data Availability Statement: The data that support the findings of this study are available from the corresponding author, K.Y., upon reasonable request.

Acknowledgments: We thank research collaboration between Kitami Hakka Tsusho and HakkaLab, Kitami Institute of Technology.

Conflicts of Interest: The authors declare no conflict of interest.

References

1. Bourgaud, F.; Gravot, A.; Milesi, S.; Gontier, E. Production of plant secondary metabolites: A historical perspective. *Plant Sci.* **2001**, *161*, 839–851. [CrossRef]
2. Courdavault, V.; O'Connor, S.E.; Oudin, A.; Besseau, S.; Papon, N. Towards the microbial production of plant-derived anticancer drugs. *Trends Cancer* **2020**, *6*, 444–448. [CrossRef] [PubMed]
3. Oksman-Caldentey, K.M.; Inzé, D. Plant cell factories in the post-genomic era: New ways to produce designer secondary metabolites. *Trends Plant Sci.* **2004**, *9*, 433–440. [CrossRef] [PubMed]
4. Huchelmann, A.; Boutry, M.; Hachez, C. Plant glandular trichomes: Natural cell factories of high biotechnological interest. *Plant Physiol.* **2017**, *175*, 6–22. [CrossRef]
5. Schuurink, R.; Tissier, A. Glandular trichomes: Micro-organs with model status? *New Phytol.* **2020**, *225*, 2251–2266. [CrossRef]
6. Jin, J.; Panicker, D.; Wang, Q.; Kim, M.J.; Liu, J.; Yin, J.L.; Wong, L.; Jang, I.C.; Chua, N.H.; Sarojam, R. Next generation sequencing unravels the biosynthetic ability of spearmint (*Mentha spicata*) peltate glandular trichomes through comparative transcriptomics. *BMC Plant Biol.* **2014**, *14*, 292. [CrossRef] [PubMed]
7. Soetaert, S.S.; Van Neste, C.M.; Vandewoestyne, M.L.; Head, S.R.; Goossens, A.; Van Nieuwerburgh, F.C.; Deforce, D.L. Differential transcriptome analysis of glandular and filamentous trichomes in *Artemisia annua*. *BMC Plant Biol.* **2013**, *13*, 220. [CrossRef]
8. Selmar, D.; Kleinwächter, M. Stress enhances the synthesis of secondary plant products: The impact of stress-related over-reduction on the accumulation of natural products. *Plant Cell Physiol.* **2013**, *54*, 817–826. [CrossRef]
9. Ramakrishna, A.; Ravishankar, G.A. Influence of abiotic stress signals on secondary metabolites in plants. *Plant Signal Behav.* **2011**, *6*, 1720–1731.
10. Xu, L.; Han, Y.; Chen, X.; Aierken, A.; Wen, H.; Zheng, W.; Wang, H.; Lu, X.; Zhao, Z.; Ma, C.; et al. Molecular mechanisms underlying menthol binding and activation of TRPM8 ion channel. *Nat. Commun.* **2020**, *11*, 3790. [CrossRef]
11. Suchodolski, J.; Feder-Kubis, J.; Krasowska, A. Antifungal activity of ionic liquids based on (-)-menthol: A mechanism study. *Microbiol. Res.* **2017**, *197*, 56–64. [CrossRef] [PubMed]
12. Thoma, F.; Somborn-Schulz, A.; Schlehuber, D.; Keuter, V.; Deerberg, G. Effects of light on secondary metabolites in selected leafy greens: A review. *Front. Plant Sci.* **2020**, *11*, 497. [CrossRef] [PubMed]
13. Burbott, A.J.; Loomis, W.D. Effects of light and temperature on the monoterpenes of peppermint. *Plant Physiol.* **1967**, *42*, 20–28. [CrossRef] [PubMed]
14. Farooqi, A.H.A.; Sangwan, N.S.; Sangwan, R.S. Effect of different photoperiodic regimes-on growth, flowering and essential oil in Mentha species. *Plant Growth Regul.* **1999**, *29*, 181–187. [CrossRef]
15. Sabzalian, M.R.; Heydarizadeh, P.; Zahedi, M.; Boroomand, A.; Agharokh, M.; Sahba, M.R.; Schoefs, B. High performance of vegetables, flowers, and medicinal plants in a red-blue LED incubator for indoor plant production. *Agron. Sustain. Dev.* **2014**, *34*, 879–886. [CrossRef]
16. Zeng, H.; Xia, C.; Zhang, C.; Chen, L. A simplified hydroponic culture of *Arabidopsis*. *Bio-Protocol* **2018**, *101*, e3121. [CrossRef]
17. Johnson, A.J.; Meyerson, E.; de la Parra, J.; Savas, T.L.; Miikkulainen, R.; Harper, C.B. Flavor-cyber-agriculture: Optimization of plant metabolites in an open-source control environment through surrogate modeling. *PLoS ONE* **2019**, *14*, e0213918. [CrossRef] [PubMed]
18. Porra, R.J.; Thompson, W.A.; Kriedemann, P.E. Determination of accurate extinction coefficients and simultaneous equations for assaying chlorophylls a and b extracted with four different solvents: Verification of the concentration of chlorophyll standards by atomic absorption spectroscopy. *Biochim. Biophys. Acta Bioenerg.* **1989**, *975*, 384–394. [CrossRef]
19. Neff, M.M.; Chory, J. Genetic interactions between phytochrome A, phytochrome B, and cryptochrome 1 during *Arabidopsis* development. *Plant Physiol.* **1998**, *118*, 27–35. [CrossRef]
20. Sims, D.A.; Gamon, J.A. Relationships between leaf pigment content and spectral reflectance across a wide range of species, leaf structures and developmental stages. *Remote Sens. Environ.* **2002**, *81*, 337–354. [CrossRef]

21. Hawley, D.; Graham, T.; Stasiak, M.; Dixon, M. Improving cannabis bud quality and yield with subcanopy lighting. *HortScience* **2018**, *53*, 1593–1599. [CrossRef]
22. Martínez-Natarén, D.A.; Villalobos-Perera, P.A.; Munguía-Rosas, M.A. Morphology and density of glandular trichomes of *Ocimum campechianum* and *Ruellia nudiflora* in contrasting light environments: A scanning electron microscopy study. *Flora* **2018**, *248*, 28–33. [CrossRef]
23. Escobar-Bravo, R.; Ruijgrok, J.; Kim, H.K.; Grosser, K.; Van Dam, N.M.; Klinkhamer, P.G.L.; Leiss, K.A. Light intensity-mediated induction of trichome-associated allelochemicals increases resistance against thrips in tomato. *Plant Cell Physiol.* **2018**, *59*, 2462–2475. [CrossRef]
24. Qin, W.; Xie, L.; Li, Y.; Liu, H.; Fu, X.; Chen, T.; Hassani, D.; Li, L.; Sun, X.; Tang, K. An R2R3-MYB transcription factor positively regulates the glandular secretory trichome initiation in *Artemisia annua* L. *Front. Plant Sci.* **2021**, *12*, 657156. [CrossRef]
25. Yan, T.; Chen, M.; Shen, Q.; Li, L.; Fu, X.; Pan, Q.; Tang, Y.; Shi, P.; Lv, Z.; Jiang, W.; et al. Homeodomain Protein 1 is required for jasmonate-mediated glandular trichome initiation in *Artemisia annua*. *New Phytol.* **2017**, *213*, 1145–1155. [CrossRef] [PubMed]
26. Yan, T.; Li, L.; Xie, L.; Chen, M.; Shen, Q.; Pan, Q.; Fu, X.; Shi, P.; Tang, Y.; Huang, H.; et al. A novel HD-ZIP IV/MIXTA complex promotes glandular trichome initiation and cuticle development in *Artemisia annua*. *New Phytol.* **2018**, *218*, 567–578. [CrossRef] [PubMed]
27. Croteau, R.B.; Davis, E.M.; Ringer, K.L.; Wildung, M.R. (-)-Menthol biosynthesis and molecular genetics. *Naturwissenschaften* **2005**, *92*, 562–577. [CrossRef] [PubMed]
28. Yang, H.Q.; Wu, Y.J.; Tang, R.H.; Liu, D.; Liu, Y.; Cashmore, A.R. The C termini of *Arabidopsis* cryptochromes mediate a constitutive light response. *Cell* **2000**, *103*, 815–827. [CrossRef]
29. Evans, J.R.; Terashima, I. Effects of nitrogen nutrition on electron transport components and photosynthesis in spinach. *Funct. Plant Biol.* **1987**, *14*, 59–68. [CrossRef]
30. Price, L.; Klein, W.H. Red, far-red response & chlorophyll synthesis. *Plant Physiol.* **1961**, *36*, 733–735.
31. Snowden, M.C.; Cope, K.R.; Bugbee, B. Sensitivity of seven diverse species to blue and green light: Interactions with photon flux. *PLoS ONE* **2016**, *11*, e0163121. [CrossRef] [PubMed]
32. Tanaka, S.; Yamaura, T.; Shigemoto, R.; Tabata, M. Phytochrome-mediated production of monoterpenes in thyme seedlings. *Phytochemistry* **1989**, *28*, 2955–2957. [CrossRef]
33. Magagnini, G.; Grassi, G.; Kotiranta, S. The effect of light spectrum on the morphology and cannabinoid content of *Cannabis sativa* L. *Med. Cannabis Cannabinoids* **2018**, *1*, 19–27. [CrossRef]

plants

MDPI

Article

Morpho-Physiological Responses of *Arabidopsis thaliana* L. to the LED-Sourced CoeLux® System

Peter Beatrice [1,*], Mattia Terzaghi [2], Donato Chiatante [1], Gabriella Stefania Scippa [3] and Antonio Montagnoli [1]

[1] Department of Biotechnology and Life Sciences, University of Insubria, 21100 Varese (VA), Italy; donato.chiatante@uninsubria.it (D.C.); antonio.montagnoli@uninsubria.it (A.M.)
[2] Department of Chemistry and biology 'A. Zambelli', University of Salerno, 84084 Fisciano (SA), Italy; mterzaghi@unisa.it
[3] Department of Biosciences and Territory, University of Molise, 86090 Pesche (IS), Italy; scippa@unimol.it
[*] Correspondence: p.beatrice@uninsubria.it

Abstract: The CoeLux® lighting system reproduces the true effect of natural sunlight entering through an opening in the ceiling, with a realistic sun perceived at an infinite distance surrounded by a clear blue sky. It has already been demonstrated that this new lighting system generates long-term positive effects on human beings; however, there are no investigations so far concerning the plant responses to CoeLux® lighting. To fill this gap, the model plant *Arabidopsis thaliana* L. was grown at four different distances from the light source, corresponding to four different light intensities (120, 70, 30, 20 μmol m^{-2} s^{-1}). High-pressure sodium lamps were used as control light. Plant phenology and morpho-physiological traits were monitored to assess for the first time the ability of plants to grow and develop under the light spectrum and intensity of the CoeLux® system. Plants grown at the lower light intensities showed a delayed life cycle and were significantly smaller than plants grown with more light. Furthermore, plants grown under the CoeLux® light type showed an additional deficit when compared to control plants. Overall, our results show that both the light spectrum and intensity of the CoeLux® system had a strong impact on *A. thaliana* growth performance.

Keywords: CoeLux®; LEDs; light intensity; light spectrum; *Arabidopsis thaliana*; photomorphogenesis; growth and development; confined environment; low light

Citation: Beatrice, P.; Terzaghi, M.; Chiatante, D.; Scippa, G.S.; Montagnoli, A. Morpho-Physiological Responses of *Arabidopsis thaliana* L. to the LED-Sourced CoeLux® System. *Plants* **2021**, *10*, 1310. https://doi.org/10.3390/plants10071310

Academic Editors: Petronia Carillo, Valeria Cavallaro and Rosario Muleo

Received: 28 April 2021
Accepted: 24 June 2021
Published: 28 June 2021

1. Introduction

Historically, several lighting systems have been used for indoor plant growth, among them fluorescent lamps, metal-halide, high-pressure sodium (HPS), and incandescent lamps [1]. These different light types share common negative features like huge energy consumption, short lifetime, and unwanted heat generation [2]. Recently, the lighting industry has seen rapid growth and the introduction of several new lighting systems. One of the most interesting and quickly developing are light-emitting diodes (LEDs), which show high efficiency, long lifetime, and negligible heat emission [3]. Furthermore, LEDs allow an enormous variety of lighting effects to be produced, among these, the CoeLux® lighting system is one of the last arrivals on the market [4]. CoeLux® system is an innovative LED-based technology for indoor lighting that uses nanostructured materials and optical systems to reproduce Rayleigh scattering effect that occurs when light crosses the earth's atmosphere [5]. Furthermore, CoeLux® is able to simulate the visual effect of the sun in a blue sky and project realistic shadows in the room. The key difference with other artificial lighting systems is that CoeLux® provides a real impression of natural sunlight together with all its properties [6]. Thus far, the numerous applications of the CoeLux® system include the lighting of hospital wards, subway systems, underground rooms and offices, and, in general, all those spaces that are not naturally illuminated. Furthermore, there is an increasing interest in the possible effects of the CoeLux® lighting systems on human health, in particular on human mood, cognition, and physiological reactions. It has already been demonstrated that this artificial skylight generates positive long-term psycho-physiological effects on human beings comparable to the real counterpart [7].

On the other hand, there are no investigations so far concerning plant responses to CoeLux® lighting. The great suitability of CoeLux® technology for closed or underground environments raises the question of whether this lighting system could be appropriate to grow crop plants for human subsistence [8] or ornamental plants for human well-being [9]. In this context, it must be taken into account that both the quality [10] and quantity [11] of visible light received by plants are crucial for their growth and development. Terrestrial green plants absorb photons unevenly across the electromagnetic spectrum, and only photosynthetically active radiation (PAR) is used to carry out photosynthesis [12]. The photosynthetic pigments in the chloroplasts respond mainly to blue (400–490 nm) and red light (590–700 nm), whereas green and yellow light (490–590 nm) is considerably less efficient in driving photosynthesis [13]. Moreover, in the natural environment, every species of plants is adapted to manage a certain variety of light intensity [14], as in the sunbeam the radiation can easily reach values of 1000–2000 μmol m^{-2}s^{-1}, whereas in the shade, radiation intensity can lower down to 10–20 μmol m^{-2}s^{-1} [15]. Several features of plant form, physiology, and resource allocation vary with the level of irradiance to which plants are acclimated [16]. Plant species adapted to live at a high light intensity show a shade avoidance response when they grow at low light intensity [17].

The interest for further development of the CoeLux® technology continues to grow due to its application in a wide range of artificially illuminated environments. In this context, it is crucial to understand how plants react to this peculiar lighting system and assess if this artificial skylight could sustain plant growth in underground or confined environments. We hypothesized that the low light intensity of these systems could be the principal limit for their use for plant growth, while the light spectrum might affect plant growth only marginally. To test our hypothesis, *Arabidopsis thaliana* plants were grown at four different distances from the CoeLux® system light source, each of them corresponding to different light intensities (120, 70, 30, 20 μmol m^{-2} s^{-1}). High-pressure sodium (HPS) lamps, historically considered as an ideal light source for indoor plant growth [18,19], were used to provide a control light type in our study.

2. Results

2.1. Phenological Analysis

Arabidopsis thaliana plants grown with 120 μmol m^{-2} s^{-1} under the HPS light type (control) completed their life cycle, from sowing to the fruit ripening and senescence phenological stage, in 57 days (dark green solid line in Figure 1). Plants delayed their life cycle completion when growing with lower light intensity (Figure 1). In particular, life cycle duration was inversely related to light intensity (dashed lines). This delay was even wider during the reproductive phase (bolting to ripening stage). Although a similar delay can be observed between control (HPS) and treated plants (CoeLux®), the latter plants showed a higher magnitude for all light intensities considered. Significant differences between plants grown under the two different light types increased with the lowering of the light intensity, showing the smaller delay at 120 μmol m^{-2} s^{-1} and the highest delay at 20 μmol m^{-2}s^{-1}. Furthermore, at the lowest light condition (20 μmol m^{-2} s^{-1}), *A. thaliana* plants were unable to complete their life cycle with the production of ripe seeds, both under the HPS light type and the CoeLux® system's light type. In particular, under the CoeLux® light type, seeds were produced only at the highest light intensities (70 and 120 μmol m^{-2} s^{-1}). These seeds were viable and germinated regularly at 98% when sown (data not shown).

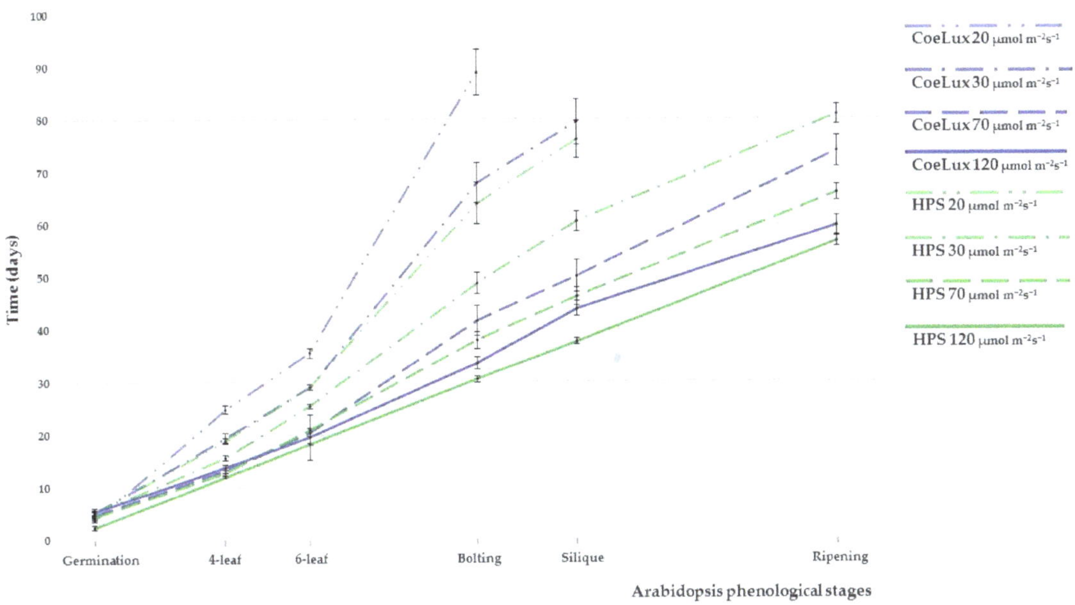

Figure 1. Phenological stages observation were recorded both under CoeLux® light (blue) and under HPS light (green) at four different light intensities, namely 20, 30, 70, and 120 µmol $m^{-2}s^{-1}$. Error bars represent the 95% confidence interval.

2.2. Morphological Traits

The biomass of both leaves and roots was found to increase with the increase of the light intensity (Figure 2a,b). The highest biomass values were measured for plants growing at 120 µmol $m^{-2}s^{-1}$, for both leaves (Figure 2a) and roots (Figure 2b) organs. For both leaves and roots biomass, significant differences between plants grown under the CoeLux® light type and plants grown under the HPS light type were measured at 20, 30, and 120 µmol $m^{-2}s^{-1}$. The root biomass of plants grown at 20 µmol $m^{-2}s^{-1}$ and 30 µmol $m^{-2}s^{-1}$ was not measured due to the low weight, which was lower than the limit of the scale range (0.0001 g). The shoot/root ratio data (Figure 2c) were significantly higher in plants grown with 70 µmol $m^{-2}s^{-1}$. Moreover, at 120 µmol $m^{-2}s^{-1}$, plants grown under the CoeLux® light type showed a significantly lower shoot/root ratio (Figure 2c).

The Projected Rosette Area (PRA) increased with the increase in light intensity independently of the light type analyzed (Figure 3a). The only exception was found for the CoeLux® light type, with no differences in PRA between 70 and 120 µmol m^{-2} s^{-1} (Figure 3a). Plants grown under the HPS light type had significantly higher PRA values than plants grown under the CoeLux® light type, with the only exception at 70 µmol m^{-2} s^{-1} (Figure 3a).

The diameter of the rosette (RD) increased with the increase in the light intensity independently of the light type analyzed (Figure 3b). Plants grown under both light types did not show significant differences in RD between 70 and 120 µmol m^{-2} s^{-1} (Figure 3b). Plants grown under the HPS light type had significantly higher values of RD than plants grown under the CoeLux® light type with the only exception at 70 µmol m^{-2} s^{-1} (Figure 3b).

The lamina to petiole ratio (L/P) increased with the increase in light intensity independently of the light type analyzed (Figure 3c). In the case of plants grown under the CoeLux® light type, similar values were measured between 20 and 30 µmol $m^{-2}s^{-1}$ and between 70 and 120 µmol $m^{-2}s^{-1}$ (Figure 3c). In the case of plants grown under the HPS light type, L/P values were similar between 70 and 120 µmol $m^{-2}s^{-1}$. Plants grown under the HPS light type had significantly higher L/P values than plants grown under the CoeLux® light type only at 30 µmol $m^{-2}s^{-1}$ (Figure 3c).

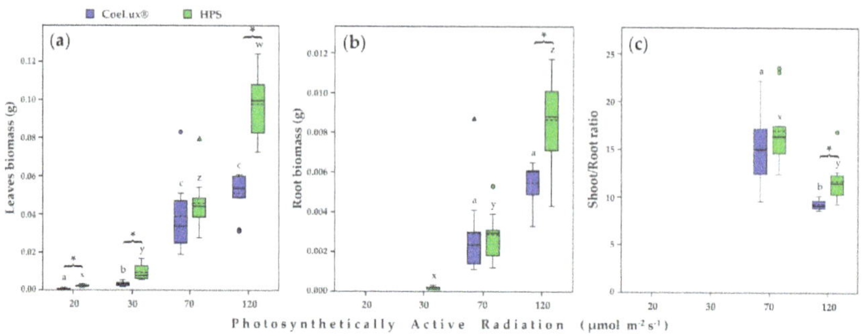

Figure 2. (**a**) Leaves biomass (g), (**b**) root biomass (g), and (**c**) shoot-to-root ratio for different light intensities. Blue and green bars indicate data of plants grown under the CoeLux® and the HPS light type, respectively. Black asterisks indicate statistically significant differences ($p < 0.05$) between plants grown under the CoeLux® and the HPS light type within the same light intensity. Letters indicate statistically significant differences ($p < 0.05$) between plants grown under different light intensities within the same light type. Vertical boxes represent approximately 50% of the observations, and lines extending from each box are the upper and lower 25% of the distribution. Within each box, the solid horizontal line is the median value, whereas the dotted horizontal line is the mean.

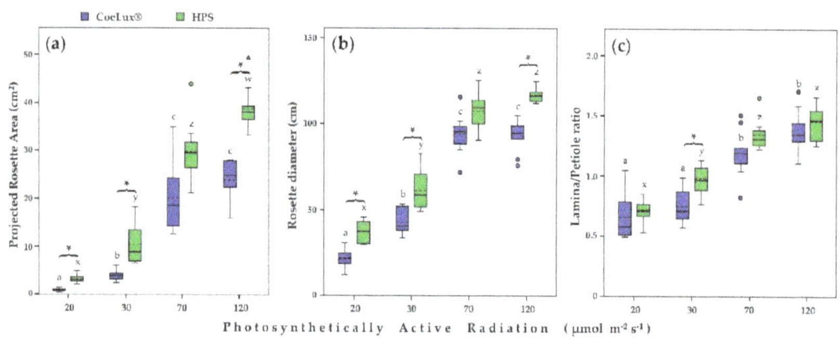

Figure 3. (**a**) Projected rosette area (cm^2), (**b**) rosette diameter (cm), and (**c**) lamina-to-petiole length ratio for different light intensities. Blue and green bars indicate data of plants grown under the CoeLux® and the HPS light types, respectively. The lamina-to-petiole length ratio is the mean of three leaves for each of the ten replicates. Black asterisks indicate statistically significant differences ($p < 0.05$) between plants grown under the CoeLux® and the HPS light type within the same light intensity. Letters indicate statistically significant differences ($p < 0.05$) between plants grown under different light intensities within the same light type. Vertical boxes represent approximately 50% of the observations and lines extending from each box are the upper and lower 25% of the distribution. Within each box, the solid horizontal line is the median value, whereas the dotted horizontal line is the mean.

2.3. *Physiological Measurements*

The chlorophyll content increased with the increase in the light intensity, and no significant differences were detected between CoeLux® and control light (Figure 4a). In the case of plants grown under the CoeLux® light type, the highest chlorophyll concentrations were found in plants grown with a light intensity of 70 and 120 µmol m^{-2}, while the lowest concentrations were found for 20 and 30 µmol m^{-2} s^{-1}, which did not differ from each other (Figure 4a). In the case of plants grown under the HPS light type, the highest and the lowest values were found for 120 and 20 µmol m^{-2} s^{-1}, respectively (Figure 4a).

The flavonoid content also increased with the increase in light intensity (Figure 4b). In the case of plants grown under the CoeLux® light type, the highest values were found for 70 and 120 µmol m^{-2} s^{-1} and the lowest values for 20 µmol m^{-2} s^{-1} (Figure 4b). In the case of plants grown under the HPS light type, the highest and the lowest values were found for 120 and 20 μmol m^{-2} s^{-1}, respectively (Figure 4b). A significant difference between plants grown under the CoeLux® and the HPS light type was observed only at 20 μmol m^{-2} s^{-1}.

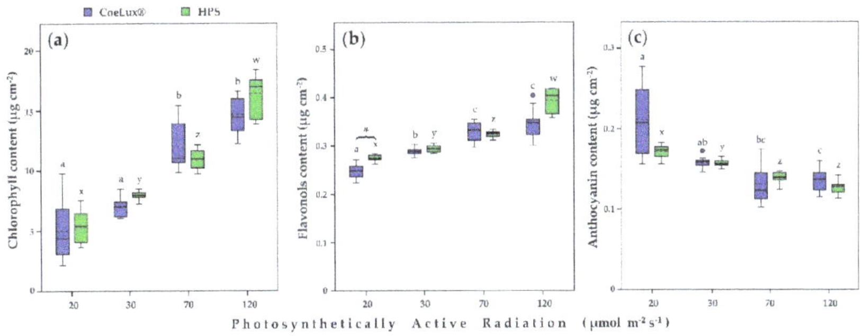

Figure 4. (**a**) Chlorophyll content (μg cm^{-2}), (**b**) flavonols content (μg cm^{-2}), and (**c**) anthocyanin content (μg cm^{-2}) for different light intensities. Blue and green bars indicate data of plants grown under the CoeLux® and the HPS light types, respectively. Black asterisks indicate statistically significant differences ($p < 0.05$) between plants grown under the CoeLux® and the HPS light type within the same light intensity. Letters indicate statistically significant differences ($p < 0.05$) between plants grown under different light intensities within the same light type. Vertical boxes represent approximately 50% of the observations and lines extending from each box are the upper and lower 25% of the distribution. Within each box, the solid horizontal line is the median value, whereas the dotted horizontal line is the mean.

The anthocyanin concentration decreased with the increase in light intensity independently of the light type considered (Figure 4c). In the case of plants grown under the CoeLux® light type, the highest and the lowest values were found for 20 and 30 μmol m^{-2} s^{-1} and 70 and 120 μmol m^{-2} s^{-1}, respectively. In the case of plants grown under the HPS light type, the highest values were found for 20 μmol m^{-2} s^{-1} and the lowest values for 70 and 120 μmol m^{-2} s^{-1}. No significant difference was observed between plants grown under the CoeLux® and the HPS light type, independently of the light intensity considered (Figure 4c).

The maximum quantum efficiency of PSII photochemistry (Fv/Fm) increased with the increase in the light intensity independently of the light type considered (Figure 5a). In the case of plants grown under the CoeLux® light type, Fv/Fm values were similar at 70 and 120 μmol m^{-2} s^{-1} (Figure 5a). Plants grown under the HPS light type had similar values of Fv/Fm at 20 and 30 μmol m^{-2} s^{-1}. Plants grown under the CoeLux® light type had significantly higher Fv/Fm values than plants grown under the HPS light type at 30 and 70 μmol m^{-2}s^{-1} (Figure 5a).

The PSII operating efficiency in the light (ΦPSII) was not different among different light intensities for plants grown under the CoeLux® light type (Figure 5b). In the case of plants grown under the HPS light type, ΦPSII slightly increased with the increase in the light intensity, with the highest and lowest values measured at 20 and 120 μmol m^{-2}s^{-1}, respectively (Figure 5b). Plants grown under the CoeLux® light type at 20 and 30 μmol m^{-2}s^{-1} had significantly higher ΦPSII values than plants grown under the HPS light type (Figure 5b).

The Non-Photochemical Quenching (NPQ) for plants grown under the CoeLux® light type was not different among different light intensities at 30, 70, and 120 μmol m^{-2}s^{-1}, while a significantly lower value was observed at 20 μmol m^{-2}s^{-1}. In the case of plants grown under the HPS light type, NPQ increased with the increase in light intensity, with the only exception of 30 μmol m^{-2}s^{-1}, which was the lower value, while the highest value was observed at 120 μmol m^{-2}s^{-1}. Plants grown under the CoeLux® light type at 20 and 70 μmol m^{-2}s^{-1} had significantly lower NPQ values than plants grown under HPS light type.

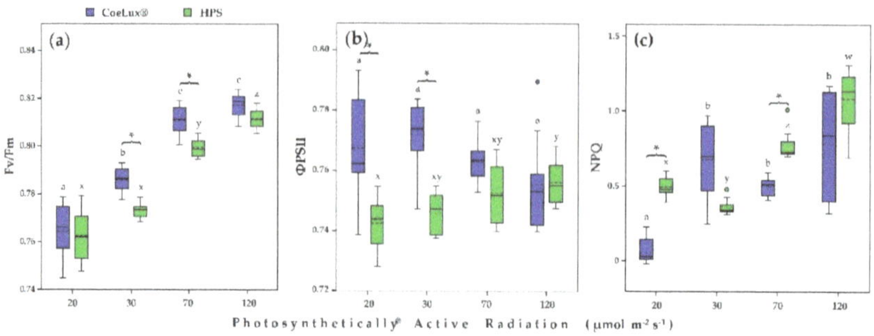

Figure 5. (**a**) Maximum quantum efficiency of PSII photochemistry (*Fv*/*Fm*), (**b**) PSII operating efficiency in the light (ΦPSII), and (**c**) non-photochemical quenching (NPQ) for different light intensities. Blue and green bars indicate data of plants grown under the CoeLux® and the HPS light types, respectively. Black asterisks indicate statistically significant differences ($p < 0.05$) between plants grown under the CoeLux® and the HPS light type within the same light intensity. Letters indicate statistically significant differences ($p < 0.05$) between plants grown under different light intensities within the same light type. Vertical boxes represent approximately 50% of the observations and lines extending from each box are the upper and lower 25% of the distribution. Within each box, the solid horizontal line is the median value, whereas the dotted horizontal line is the mean.

The net photosynthetic rate (Pn) increased with the increase of the light intensity independently of the light type considered (Figure 6a). For plants grown under both CoeLux® and HPS light type, the highest and lowest Pn values were measured at 120 and 20 µmol $m^{-2}s^{-1}$, respectively (Figure 6a). At 20 and 30 µmol $m^{-2}s^{-1}$, negative photosynthetic values were measured due to the glass delimiting the instrument cuvette chamber, which lowered the incident light received by the encapsulated leaf of 52.9 ± 7.3 µmol $m^{-2}s^{-1}$. Plants grown under the CoeLux® light type had significantly lower Pn values than plants grown under the HPS light type at 20, 70, and 120 µmol $m^{-2}s^{-1}$.

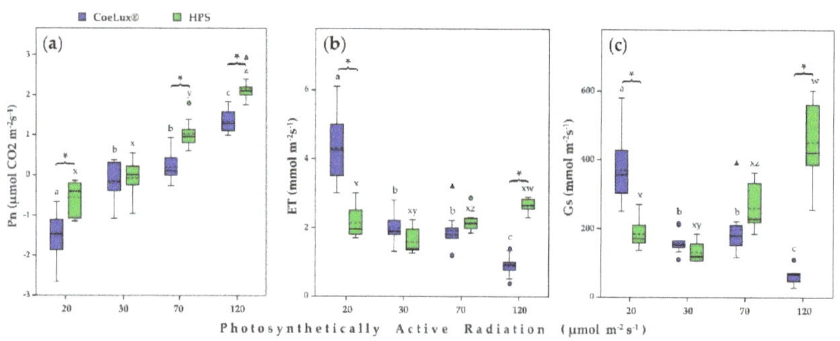

Figure 6. (**a**) Pn: net photosynthetic rate (µmol CO_2 $m^{-2}s^{-1}$), (**b**) ET: evapotranspiration (mmol $m^{-2}s^{-1}$), and (**c**) Gs: stomatal conductance (mmol $m^{-2}s^{-1}$) for different light intensities. Blue and green bars indicate data of plants grown under the CoeLux® and the HPS light types, respectively. Black asterisks indicate statistically significant differences ($p < 0.05$) between plants grown under the CoeLux® and the HPS light type within the same light intensity. Letters indicate statistically significant differences ($p < 0.05$) between plants grown under different light intensities within the same light type. Vertical boxes represent approximately 50% of the observations and lines extending from each box are the upper and lower 25% of the distribution. Within each box, the solid horizontal line is the median value, whereas the dotted horizontal line is the mean.

The evapotranspiration rate (ET) for plants grown under the CoeLux® light type decreased with the increase in the light intensity (Figure 6b). The highest and lowest values were found for plants grown, respectively, at 20 and 120 µmol $m^{-2}s^{-1}$, while intermediate values were found at 30 and 70 μmol $m^{-2}s^{-1}$. In the case of plants grown under the HPS light type, ET values did not differ among different light intensities (Figure 6b). The ET values measured at 20 μmol $m^{-2}s^{-1}$ for plants grown under the CoeLux® light type were significantly higher than the values measured for plants grown under the HPS light type (Figure 6b). The ET values measured at 120 μmol $m^{-2}s^{-1}$ were significantly higher for plants grown under the HPS light type than plants grown under the CoeLux® light type (Figure 6b).

The stomatal conductance (Gs) decreased with the increase in the light intensity in the case of plants grown under the CoeLux® light type (Figure 6c). The highest and lowest Gs values were measured, respectively, at 20 and 120 μmol $m^{-2}s^{-1}$, while intermediate values were found at 30 and 70 μmol $m^{-2}s^{-1}$. In the case of plants grown under the HPS light type, the Gs values were not different among light intensities, with the only exception of 120 μmol $m^{-2}s^{-1}$, which showed the highest values (Figure 6c). At 20 μmol $m^{-2}s^{-1}$, the Gs value measured for plants grown under the CoeLux® light type was significantly higher than the value measured for plants grown under the HPS light type (Figure 6c). On the contrary, at 120 μmol $m^{-2}s^{-1}$, the Gs value measured for plants grown under the HPS light type was significantly higher than the value measured for plants grown under the CoeLux® light type (Figure 6c).

3. Discussion

In our study, we used the model plant *Arabidopsis thaliana* to assess for the first time if the light spectrum and intensity of the CoeLux® 45HC lighting system could be suitable for plant growth in controlled environments. Both light quantity and quality are fundamental for plant growth and development [20]. In this context, LEDs show unrivaled advantages, since LED bulbs can be assembled in countless ways to obtain exactly the light characteristics needed for optimal plant growth [10]. However, the CoeLux systems have peculiar constraints due to the physical effects involved in the setting up of their characteristic visual effects [4,6]. Thus, light quantity and quality cannot be adjusted like with other LED-based lighting systems currently used for plant growth [10]. We observed that the light emitted by the 45HC CoeLux® system, even inside the sunbeam, was characterized by low levels of photosynthetic active radiation (PAR). The registered values were similar to those that can be normally found in shaded environments, for example, under a dense forest canopy [21]. Consequently, even if natural sunlight's visual effects were perfectly reproduced, this artificial skylight cannot be compared to its natural counterpart in terms of light intensity. Shade-adapted plants are certainly the most suitable to grow under this lighting system, as photosynthesis is directly influenced by the amount of light reaching the plant's leaves [16].

With the phenological analysis, we observed that *A. thaliana* plants grown under the CoeLux® light type showed a significant delay with respect to plants grown under the HPS light type, and this delay was independent of the light intensities considered. Moreover, this plant development delay was particularly evident at the last growth stages such as *Bolting* and *Silique* set, and it was of higher magnitude at the lowest light intensities. In particular, plants grown at 20 and 30 μmol $m^{-2}s^{-1}$ could not reach the seed maturity stage during the 100-days period analyzed in our study. Other studies also reported a 2-week flowering delay in *A. thaliana* plants grown under reduced light intensity and lowered R/FR [22]. Morphological data are in line with the phenological observations, highlighting the negative influence of the CoeLux® light type on *A. thaliana* growth. In fact, for all morphological parameters analyzed, we observed a similar trend that grows with the increase in the light intensity but was always slightly lower with the CoeLux® light type than with the HPS light type.

Plants have to balance the biomass allocation to leaves, stems, and roots in a way that matches the physiological functions performed by these organs. In stress situations, plants allocate relatively more biomass to roots if the limiting factor for growth is below ground (e.g., nutrients or water), whereas they will allocate relatively more biomass to shoots if the limiting factor is above ground (e.g., light or CO_2) [23]. That is, plants that received a lower irradiance showed increased allocation to the shoots in an attempt to enhance the uptake to the most limiting factor, light. Surprisingly, plants grown under CoeLux® light type showed slightly lower shoot–root ratios relative to control plants.

In addition to biomass, also the PRA and the RD showed a clear detrimental effect of the CoeLux® light type with respect to the HPS light type, demonstrating that this light type is less appropriate than the control one for *A. thaliana* plants growth. This effect could be explained by the different fractions of blue and red light radiated by the two light types, as the blue and red components represent 59% of the total irradiation under the HPS light type and only 55% under the CoeLux® light type (Figure 9). The CoeLux® light type showed a higher yellow component (+4%); however, yellow light is less efficient in driving photosynthesis, as plant's photosystems respond mainly to red and blue light.

In *A. thaliana*, the lamina to petiole ratio is one of the principal indicators of shade avoidance syndrome (SAS) [24]. In low light conditions, plants grew a longer petiole and a shorter lamina in an effort to collect more light, consequently decreasing the L/P ratio below 1.0. Furthermore, plants that were grown under the CoeLux® light type showed slightly lower L/P ratios than control plants, indicating the onset of a more severe shade avoidance syndrome (SAS) caused by the light quality. Specifically, the CoeLux® light type is characterized by a lower blue component and a lower B/G ratio (Figure 9), which could trigger an SAS via the cryptochrome pathway [24,25]. In natural environments, light reflected or transmitted through photosynthetic tissues of plants in close proximity is depleted in blue, red, and UV-B wavelengths. Therefore, the reflected or transmitted light is enriched in green and far-red spectral regions, resulting in lowered R/FR and B/G ratios. Plants perceive these differences through multiple photoreceptors to regulate shade avoidance responses and tune the plant growth under suboptimal light environments [17].

During shade avoidance responses, many aspects of leaf development are modified, including pigment production [26]. The chlorophyll content is known to decrease at low light intensities [27,28]. A pattern of this nature was also observed in our experiment, with no significant differences between the two different light types analyzed. Flavonoids, such as flavonols and anthocyanins, are also involved in plant's responses to light stress, as they were proposed to protect against high irradiance, both UV and visible [29]. Furthermore, flavonoids are also antioxidants that can scavenge reactive oxygen species (ROS) and can be observed frequently when plants are exposed to other physiological stresses such as extreme temperatures, drought, or nutritional stresses, in addition to high light and UV radiation [30]. Thus, the biosynthesis of these compounds is regulated by the interplay of multiple factors. Furthermore, the pigment content varies in leaves of different ages [31], and young leaves of many plants have transiently high concentrations of anthocyanins, disappearing as leaves mature [32]. *A. thaliana* plants growing at lower light intensities displayed a strong growth delay (Figure 1); consequently, the pigment concentration measurements were taken on younger leaves with the lowering of light intensity, explaining the unexpected reduction in anthocyanins content observed with the increase in light intensity in *A. thaliana* leaves.

The Fv/Fm ratio gives a robust indicator of the maximum quantum yield of PSII chemistry and is commonly used to detect plant stress in leaves [33]. Plants grown at lower light intensities showed lower Fv/Fm, suggesting a stress condition related to light quantity. However, the CoeLux® light type appears to have a positive effect on PSII photochemistry, as we found slightly higher Fv/Fm values compared to control plants. This observation is probably related to the higher photoinhibition of control plants grown under the HPS light type (Figure 5c), as Murchie et al. reported lowered values of Fv/Fm in leaves in a quenched state [34]. Nonetheless, an equal reduction in Fv/Fm was not observed in response to the increased NPQ with the increase in light intensity, suggesting the involvement of multiple factors. The use of leaf samples with different pigment contents may also be a source of inaccuracies [33]. The quantum yield of PSII (Φ_{PSII}) showed only minimal differences between the different light intensities, both with the CoeLux® and the HPS light type (Figure 5b).

The drop in light intensity resulted in a lowered net photosynthetic rate in *A. thaliana* plants. Furthermore, the CoeLux® light type negatively influenced the Pn at three of the four light intensities tested, explaining the patterns observed in Figure 2a,b. The lower CO_2 assimilation under the CoeLux® light type causes a lack of essential building blocks and, consequently, an impaired biomass production. Evapotranspiration rate and stomatal conductance showed similar patterns but no clear differences between the two light types were detected (Figure 5b,c).

Overall, our results showed that the intensity of the light, both under control and CoeLux® light types, had a strong impact on plant growth performance, demonstrating that the light intensity could be the major limiting factor for plants growing under this led-sourced artificial skylight. Furthermore, the light quality of the CoeLux® system showed a negative impact on *A. thaliana* growth, independently of the light intensity considered, demonstrating that light quality could be an additional limiting factor for plants growing under this light source. Further research is needed to assess if shade-tolerant plant species could perform better than *A. thaliana* under this peculiar lighting system, while the comprehension of the molecular mechanisms underlying the observed phenomena could provide significant starting points for the development of CoeLux-adapted plant strains.

4. Materials and Methods

4.1. Light Characterization

The CoeLux® growth room (University of Insubria) is composed of two standard 1 TEU containers assembled one above the other. The upper container hosts the two CoeLux® 45HC lighting systems, while the lower one is insulated and equipped with an air conditioner for temperature control to function as a growth room. The lighting system is sourced by full-spectrum white LEDs with a color temperature of 6500 K. This light is subsequently filtered to obtain the desired skylight effect [6], modifying both spectra and intensity of the original light. Therefore, both light quality and intensity were characterized within a representative section of the growth room (Figure 7).

The HD 2302.0 Light Meter (Delta Ohm) was used to measure the photosynthetically active radiation (PAR) along the central section of the growth room (566 cm $\times$ 256 cm). A custom-made rail was designed to guide the instrument sensor along the selected section and perform measurements exactly every 10 cm across the whole surface. The resulting data were analyzed to obtain a color-scale map (Figure 7). The light radiated by the CoeLux® systems is not uniformly diffused inside the growth room, being concentrated within the sunbeam ray of light with a fixed angle of 45°. Within the sunbeam, the highest PAR intensity, measured at 10 cm from the lighting system, was 140 μmol $m^{-2}s^{-1}$, while at a further distance, it drops rapidly to around 20 μmol $m^{-2}s^{-1}$. These values are even of a lower magnitude when measured within the shade, ranging from 26 μmol $m^{-2}s^{-1}$ under the blue sky from the system to less than 1 μmol $m^{-2}s^{-1}$ in the most shaded parts of the growth room. Increased light intensity was observed in some shade areas due to light reflection on the walls of the growth room and the frames of the CoeLux® systems skylight (Figure 7).

Spectra measurements every 4 nm in the range between 380 nm and 780 nm were taken on a horizontal white reflector using the SpectraScan PR655 (Photo Research), both inside the CoeLux® growth room and under the HPS lamps that we used as control. Inside the CoeLux® growth room, a total of 23 measurements were performed: 17 of them along the central section of the growth-room at five different heights from the ground floor (0, 50, 100, 150, 200 cm), inside the sunbeam of the CoeLux® system, outside the sunbeam but under the blue panel of the lighting system (sky), and in the deep shade part of the container (Figure 7). The other 6 measurements were taken near the lateral and bottom walls of the growth room to investigate the influence on the light spectra of light reflecting on the grey walls of the growth room. Within the same measurement, the instrument also provides a light intensity value in the form of luminance (cd/m^2), which was used to normalize the spectra measurements. The spectra were divided into color components: blue light is the integral between 400 and 490 nm, green light is the integral between 490 and 560 nm, yellow light is the integral between 560 and 590 nm, red light is the integral between 590 and 700 nm, and far-red light is the integral between 700 and 780 nm. The red-to-far-red ratio (R/FR) and the blue-to-green ratio (B/R) were calculated according to Sellaro et al. [25].

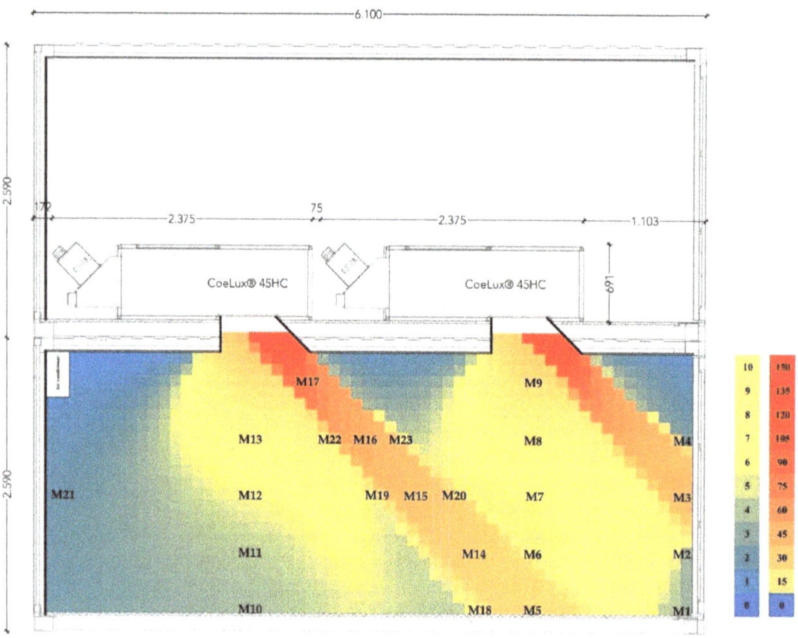

Figure 7. The photosynthetically active radiation (PAR) color-scale map of the double container showroom at Insubria University Campus (Varese, IT). The PAR values in the figure are given in μmol m^{-2}s^{-1}. Spectra measurements (M) were taken at positions M1 to M23.

We observed only small differences between the light spectrum measured within the sunbeam (red lines in Figure 8) and that measured within the shade (light blue lines in Figure 8), independently of the distance from the light source or room walls. In both cases, the spectra covered almost the entire visible wavebands; however, the total irradiance was differently distributed. Within the sunbeam, the spectra presented a broad peak between 490 and 700 nm and a sharp peak of irradiance of comparable height in the blue region (400–490 nm), representing 14% of the entire irradiance. Within the shade, the spectrum had a similar pattern but with a higher peak at 450 nm (representing 26% of the entire irradiance) and lower values in the red component of light between 590 and 700 nm (30% vs. 41%). Thus, at an equal light intensity, plants placed in a shade position received more blue and green light while plants placed inside the sunbeam receive more red and far-red light. In the small frontier positions between sun and shade, we found spectra with an intermediate shape (yellow lines in Figure 8).

High-pressure sodium (HPS) lamps (Philips MasterColour CDM-T MW eco 230W/842) were used to provide a control light type in our study. To characterize this light spectrum, a total of 12 spectra measurements were performed at different positions in the range between 120 and 20 μmol m^{-2}s^{-1}. Data were normalized on luminance to compare the spectra generated by the CoeLux® systems within the sunbeam with those of HPS lamps

(Figure 9). The HPS light type were shown to have a higher blue component (24% vs. 14%), while the CoeLux® light type had more yellow (15% vs. 11%) and red (41% vs. 35%) components. The green light component was almost identical even if a statistically significant difference was detected, while the far-red component showed no significant difference between the two, as it represented 6% of the total radiation for both types of light (Figure 9).

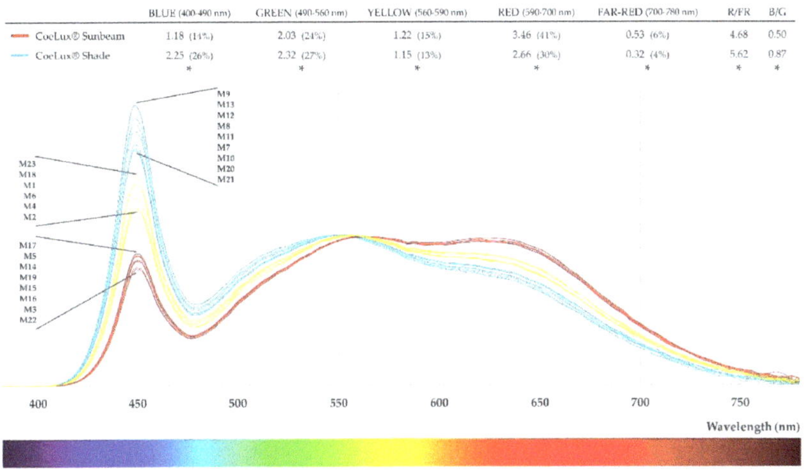

Figure 8. Light spectra measured within the sunbeam (red lines), in shade areas (light blue lines), and in frontier positions between sun and shade (yellow lines). Measurements (M) are reported in order of appearance in the graph. Color components mean values are reported and statistically significant differences are marked with an asterisk (*). R/FR: red/far-red ratio; B/G: blue/green ratio.

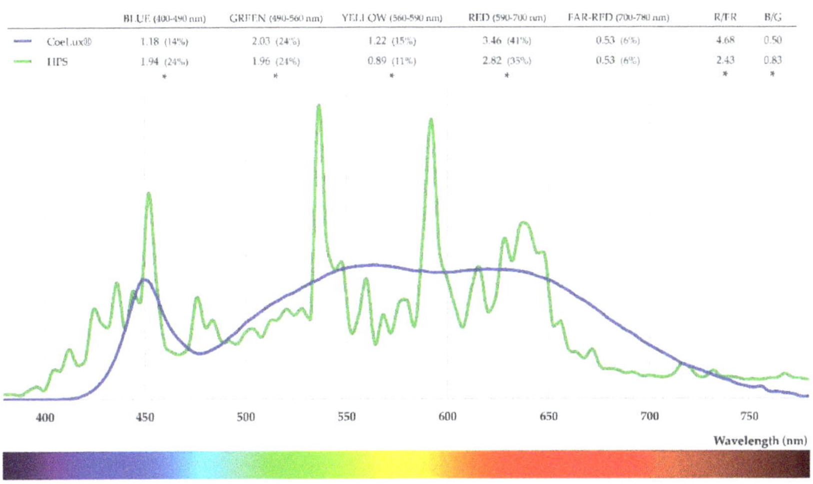

Figure 9. Mean light spectra curves collected within the CoeLux® sunbeam (blue) and under HPS light (green). Color components mean values are reported and statistically significant differences are marked with an asterisk (*). R/FR: red-to-far-red ratio; B/G: blue-to-green ratio.

4.2. Plant Material and Growth Conditions

Arabidopsis thaliana wild-type (WT) seeds were stratified at 4 °C for 5 days on 1% agar gel and subsequently transferred to pot flats (Araflats; Arasystem; Ghent/Belgium) composed of 51 individual pot cavities with a 5 cm diameter, filled with sterilized commercial soil-less substrate. Plants were grown at a temperature as close as possible to 22 °C, with an air humidity ranging between 50% and 70%, and a photoperiod of 14 h. A constant 1-cm water layer was maintained in the tray and 1mL liquid fertilizer (NPK 7.5–3−6) was supplied weekly. In the CoeLux® growth room, full pot flats were located at four different positions at progressive distances from the light source (20, 85, 205, 365 cm) inside the system's sunbeam, each corresponding to a different value of light intensity, respectively 120, 70, 30, 20 μmol $m^{-2}s^{-1}$. In our CoeLux® facility, 120 μmol $m^{-2}s^{-1}$ is the position suitable for plant growth with the highest light intensity achievable. In a separate growth room, with the same environmental parameters of the CoeLux® growth room, plants were illuminated with HPS lamps as reference light (control), recreating the same light intensity of each of the four positions under the CoeLux® light.

4.3. Plant Analysis

Phenological analysis [35] was performed through the recording of different developmental stages for a period of 100 days after sowing, considered as day 0 (Table 1). A total of 10 plants were monitored for each growth condition.

Table 1. *A. thaliana* growth stages recorded in this work.

Stage	Description
Germination	Plants with fully expanded cotyledons
4-leaf stage	Plants with the first two rosette leaves bigger than the cotyledons
6-leaf stage	Plants with the second couple of rosette leaves bigger than the first one
Bolting and flowering	Plants with a floral stalk taller than 1 cm
Silique formation	Plants with at least one fully developed silique
Ripening and senescence	Plants with at least one silique totally brown or open with ripe seeds

For morphological measurements, a total of 10 *A. thaliana* plants for each growth condition were sampled after 33 days from sowing. The leaves (complete rosette) of the plants were scanned at 800 dpi with the Epson Expression 12000XL instrument and then oven-dried at 70 °C until constant weight. Plant roots were freed from soil media by carefully washing them under running water and subsequently oven-dried until constant weight. The dry organs were then weighed on an analytical balance (Orma AL220S) and the shoot-to-root ratio (S/R) was calculated. The scanned images were processed with WinRhizo (Regent Instrument) to measure the projected rosette area (PRA) and with ImageJ (NIH, USA) to measure the rosette diameter (RD) as well as lamina and petiole length of three leaves for each plant. The lamina-to-petiole ratio (L/P) was then calculated.

Physiological measurements were conducted as follows on 10 different plants for each growth condition.

(A) Chlorophyll fluorescence analysis was performed with a modulated chlorophyll fluorometer (OS1-FL; Opti-Sciences) after 35 days from sowing. The maximum quantum efficiency of photosystem II (PSII) photochemistry (Fv/Fm) was measured pre-dawn, while the PSII operating efficiency (ΦPSII) was measured after at least 3 h of plant exposure to light. Non-photochemical quenching (NPQ) was subsequently calculated using Fm and Fm′ values [33,36].

(B) The leaf pigment content was measured on the upper face of completely expanded young leaves with the Dualex Scientific Instrument (Force-A) 36 days after sowing. The concentrations of chlorophyll, anthocyanins, and flavonoids are reported by the instrument in μg/cm^2.

(C) The Ciras 2 instrument (PP Systems) was used to measure the net photosynthetic rate (Pn), the stomatal conductance (Gs), and the evapotranspiration rate (ET) 48 days after sowing. The single leaf gas exchange measurements were taken under ambient light with the 25 × 7 mm leaf cuvette oriented perpendicularly to the light source. At least three measurements for each leaf were taken on completely expanded young leaves. For leaves smaller than 25 × 7 mm, digital pictures were made to determine the projected leaf area inside the cuvette and properly scale the measurement [37].

4.4. Statistical Analysis

Statistical analysis was carried out with SPSS 20.0 (SPSS Inc). The post hoc Dunnett's test was used for multiple comparisons, while the Student's t-test was used when only two means were compared. Statistically significant differences ($p < 0.05$) between the means were marked with the letters a, b, c, d for the CoeLux® light type, with the letters x, y, z, w for the HPS light type, and with a black asterisk for comparisons between the two types of light. In boxplots, colored circles and triangles represent respectively outliers (outside the 3rdQ + 1.5 × IQR and the 1stQ − 1.5 × IQR) and extreme outliers (outside the 3rdQ + 3 × IQR and the 1stQ − 3 × IQR). Microsoft Excel functions were used to show the 95% confidence interval error bars in Figure 1.

Author Contributions: Conceptualization, A.M., G.S.S. and D.C.; methodology, A.M. and P.B.; software, M.T.; validation, A.M.; formal analysis, P.B. and M.T.; investigation, P.B.; resources, A.M.; data curation, P.B. and M.T.; writing—original draft preparation, P.B.; writing—review and editing, P.B. and A.M.; visualization, P.B.; supervision, A.M.; project administration, A.M.; funding acquisition, A.M., D.C. and G.S.S. All authors have read and agreed to the published version of the manuscript.

Funding: This research was funded by the University of Insubria [FAR 2019–2020; PhD scholarship of Peter Beatrice], and by the EC FP7 [ZEPHYR, grant number 308313, 2012–2015].

Institutional Review Board Statement: Not applicable.

Informed Consent Statement: Not applicable.

Data Availability Statement: The data presented in this study are available on request from the corresponding author.

Acknowledgments: Authors acknowledge Paolo di Trapani, Paolo Ragazzi, and Antonio Lotti (Coelux® S.r.l.) for providing lighting systems and light spectrum analysis and Massimiliano Marchica and Alice Tettamanti (Coelux® S.r.l.) for helping with the growth room design. We are also grateful to Silvia Baronti (CNR-IBIMET, Florence) for providing Dualex and Ciras instruments, to Franco Miglietta (CNR-IBIMET, Florence) for sharing *A. thaliana* seeds, and to Alessio Miali (University of Insubria) for his valuable help with the laboratory work. P.B. is a PhD student of the "Life Sciences and Biotechnology" course at Università degli Studi dell'Insubria.

Conflicts of Interest: The authors declare no conflict of interest. The funders had no role in the design of the study; in the collection, analyses, or interpretation of data; in the writing of the manuscript; or in the decision to publish the results.

References

1. Dutta Gupta, S.; Agarwal, A. Artificial Lighting System for Plant Growth and Development: Chronological Advancement, Working Principles, and Comparative Assessment. In *Light Emitting Diodes for Agriculture: Smart Lighting*; Springer: Singapore, 2017; pp. 1–334. ISBN 9789811058073.
2. Seiler, F.; Soll, J.; Bölter, B. Comparative phenotypical and molecular analyses of arabidopsis grown under fluorescent and LED Light. *Plants* **2017**, *6*, 24. [CrossRef]
3. Tetri, E.; Sarvaranta, A.; Syri, S. Potential of new lighting technologies in reducing household lighting energy use and CO_2 emissions in Finland. *Energy Effic.* **2014**, *7*, 559–570. [CrossRef]
4. Di Trapani, P.; Magatti, D. Artificial Lighting System for Simulating Natural Lighting. US Patent 2014/01,331,125 A1, 15 May 2014.
5. Rayleigh, J.W. On the instability of jets. *Proc. Lond. Math. Soc.* **1878**, *10*, 4–13. [CrossRef]
6. Di-Trapani, P.; Magatti, D. Artificial Illumination Device. US Patent 9,709,245 B2, 18 July 2017.
7. Canazei, M.; Laner, M.; Staggl, S.; Pohl, W.; Ragazzi, P.; Magatti, D.; Martinelli, E.; Di Trapani, P. Room- and illumination-related effects of an artificial skylight. *Light. Res. Technol.* **2016**, *48*, 539–558. [CrossRef]
8. Dong, C.; Fu, Y.; Liu, G.; Liu, H. Low light intensity effects on the growth, photosynthetic characteristics, antioxidant capacity, yield and quality of wheat (*Triticum aestivum* L.) at different growth stages in BLSS. *Adv. Sp. Res.* **2014**, *53*, 1557–1566. [CrossRef]

9. Bringslimark, T.; Hartig, T.; Patil, G.G. The psychological benefits of indoor plants: A critical review of the experimental literature. *J. Environ. Psychol.* **2009**, *29*, 422–433. [CrossRef]
10. Montagnoli, A.; Dumroese, R.K.; Terzaghi, M.; Pinto, J.R.; Fulgaro, N.; Scippa, G.S.; Chiatante, D. Tree seedling response to LED spectra: Implications for forest restoration. *Plant Biosyst.* **2018**, *152*, 515–523. [CrossRef]
11. Chiang, C.; Bånkestad, D.; Hoch, G. Reaching natural growth: The significance of light and temperature fluctuations in plant performance in indoor growth facilities. *Plants* **2020**, *9*, 1312. [CrossRef]
12. Clementson, L.A.; Wojtasiewicz, B. Dataset on the absorption characteristics of extracted phytoplankton pigments. *Data Br.* **2019**, *24*, 103875. [CrossRef]
13. Wahid, A.; Rasul, E. *Handbook of Photosynthesis*; CRC Press: Boca Raton, FL, USA, 2005; ISBN 9781420027877.
14. Valladares, F.; Niinemets, Ü. Shade tolerance, a key plant feature of complex nature and consequences. *Annu. Rev. Ecol. Evol. Syst.* **2008**, *39*, 237–257. [CrossRef]
15. Gommers, C.M.M.; Visser, E.J.W.; Onge, K.R.S.; Voesenek, L.A.C.J.; Pierik, R. Shade tolerance: When growing tall is not an option. *Trends Plant Sci.* **2013**, *18*, 65–71. [CrossRef]
16. Givnish, T.J. Adaptation to sun and shade: A whole-plant perspective. *Aust. J. Plant Physiol.* **1988**, 139–143. [CrossRef]
17. Sessa, G.; Carabelli, M.; Possenti, M.; Morelli, G.; Ruberti, I. Multiple pathways in the control of the shade avoidance response. *Plants* **2018**, *7*, 102. [CrossRef]
18. Islam, M.A.; Kuwar, G.; Clarke, J.L.; Blystad, D.R.; Gislerød, H.R.; Olsen, J.E.; Torre, S. Artificial light from light emitting diodes (LEDs) with a high portion of blue light results in shorter poinsettias compared to high pressure sodium (HPS) lamps. *Sci. Hortic.* **2012**, *147*, 136–143. [CrossRef]
19. Pinho, P.; Jokinen, K.; Halonen, L. Horticultural lighting—Present and future challenges. *Light. Res. Technol.* **2012**, *44*, 427–437. [CrossRef]
20. Shirley, H.L. The Influence of Light Intensity and Light Quality Upon the Growth of Plants. *Am. J. Bot.* **1929**, *16*, 354–390. [CrossRef]
21. Horton, J.L.; Neufeld, H.S. Photosynthetic responses of Microstegium vimineum (Trin.) A. Camus, a shade-tolerant, C4 grass, to variable light environments. *Oecologia* **1998**, *114*, 11–19. [CrossRef]
22. Cipollini, D. Interactive effects of lateral shading and jasmonic acid on morphology, phenology, seed production, and defense traits in Arabidopsis thaliana. *Int. J. Plant Sci.* **2005**, *166*, 955–959. [CrossRef]
23. Poorter, H.; Niklas, K.J.; Reich, P.B.; Oleksyn, J.; Poot, P.; Mommer, L. Biomass allocation to leaves, stems and roots: Meta-analyses of interspecific variation and environmental control. *New Phytol.* **2012**, *193*, 30–50. [CrossRef]
24. Keller, M.M.; Jaillais, Y.; Pedmale, U.V.; Moreno, J.E.; Chory, J.; Ballaré, C.L. Cryptochrome 1 and phytochrome B control shade-avoidance responses in Arabidopsis via partially independent hormonal cascades. *Plant J.* **2011**, *67*, 195–207. [CrossRef]
25. Sellaro, R.; Crepy, M.; Trupkin, S.A.; Karayekov, E.; Buchovsky, A.S.; Rossi, C.; Casal, J.J. Cryptochrome as a sensor of the blue/green ratio of natural radiation in Arabidopsis. *Plant Physiol.* **2010**, *154*, 401–409. [CrossRef] [PubMed]
26. Smith, H.; Whitelam, G.C. The shade avoidance syndrome: Multiple responses mediated by multiple phytochromes. *Plant Cell Environ.* **1997**, *20*, 840–844. [CrossRef]
27. Brouwer, B.; Gardeström, P.; Keech, O. In response to partial plant shading, the lack of phytochrome A does not directly induce leaf senescence but alters the fine-tuning of chlorophyll biosynthesis. *J. Exp. Bot.* **2014**, *65*, 4037–4049. [CrossRef] [PubMed]
28. Lichtenthaler, H.K.; Ač, A.; Marek, M.V.; Kalina, J.; Urban, O. Differences in pigment composition, photosynthetic rates and chlorophyll fluorescence images of sun and shade leaves of four tree species. *Plant Physiol. Biochem.* **2007**, *45*, 577–588. [CrossRef]
29. Owens, D.K.; Alerding, A.B.; Crosby, K.C.; Bandara, A.B.; Westwood, J.H.; Winkel, B.S.J. Functional analysis of a predicted flavonol synthase gene family in arabidopsis. *Plant Physiol.* **2008**, *147*, 1046–1061. [CrossRef]
30. Zeng, X.Q.; Chow, W.S.; Su, L.J.; Peng, X.X.; Peng, C.L. Protective effect of supplemental anthocyanins on Arabidopsis leaves under high light. *Physiol. Plant.* **2010**, *138*, 215–225. [CrossRef]
31. Bailey, S.; Horton, P.; Walters, R.G. Acclimation of Arabidopsis thaliana to the light environment: The relationship between photosynthetic function and chloroplast composition. *Planta* **2004**, *218*, 793–802. [CrossRef]
32. Karageorgou, P.; Manetas, Y. The importance of being red when young: Anthocyanins and the protection of young leaves of Quercus coccifera from insect herbivory and excess light. *Tree Physiol.* **2006**, *26*, 613–621. [CrossRef]
33. Murchie, E.H.; Lawson, T. Chlorophyll fluorescence analysis: A guide to good practice and understanding some new applications. *J. Exp. Bot.* **2013**, *64*, 3983–3998. [CrossRef]
34. Murchie, E.H.; Horton, P. Toward C4 rice: Learning from the acclimation of photosynthesis in the C3 leaf. *Charting New Pathw. C4 Rice* **2008**, 333–350. [CrossRef]
35. Boyes, D.C.; Zayed, A.M.; Ascenzi, R.; McCaskill, A.J.; Hoffman, N.E.; Davis, K.R.; Görlach, J. Growth stage-based phenotypic analysis of Arabidopsis: A model for high throughput functional genomics in plants. *Plant Cell* **2001**, *13*, 1499–1510. [CrossRef]
36. Maxwell, K.; Johnson, G.N. Chlorophyll fluorescence—A practical guide. *J. Exp. Bot.* **2000**, *51*, 659–668. [CrossRef]
37. Kölling, K.; George, G.M.; Künzli, R.; Flütsch, P.; Zeeman, S.C. A whole-plant chamber system for parallel gas exchange measurements of Arabidopsis and other herbaceous species. *Plant Methods* **2015**, *11*, 1–12. [CrossRef]

plants

Article

The Effect of Plant Growth Compensation by Adding Silicon-Containing Fertilizer under Light Stress Conditions

Natalya A. Semenova [1], Alexandr A. Smirnov [1], Andrey A. Grishin [1], Roman Y. Pishchalnikov [2,*], Denis D. Chesalin [2], Sergey V. Gudkov [2,3], Narek O. Chilingaryan [1], Anastasia N. Skorokhodova [4], Alexey S. Dorokhov [1] and Andrey Y. Izmailov [1]

1 Federal State Budgetary Scientific Institution "Federal Scientific Agroengineering Center VIM" (FSAC VIM), 109428 Moscow, Russia; natalia.86@inbox.ru (N.A.S.); as984788@gmail.com (A.A.S.); 5145412@mail.ru (A.A.G.); narek-s@list.ru (N.O.C.); dorokhov.vim@yandex.ru (A.S.D.); vim@vim.ru (A.Y.I.)

2 Prokhorov General Physics Institute of the Russian Academy of Sciences, 119991 Moscow, Russia; genoa-and-pittsburgh@mail.ru (D.D.C.); s_makariy@rambler.ru (S.V.G.)

3 Institute of Biology and Biomedicine, Lobachevsky State University of Nizhni Novgorod, 603022 Nizhni Novgorod, Russia

4 Moscow Timiryazev Agricultural Academy, Russian State Agrarian University, 127550 Moscow, Russia; red-green216@mail.ru

* Correspondence: rpishchal@kapella.gpi.ru; Tel.: +7-916-518-7076

Abstract: The effects of different spectral compositions of light-emitting diode (LED) sources and fertilizer containing biologically active silicon (Si) in the nutrient solution on morphological and physiological plant response were studied. Qualitative indicators and the productivity of plants of a red-leaved and a green-leaved lettuce were estimated. Lettuce was grown applying low-volume hydroponics in closed artificial agroecosystems. The positive effect of Si fertilizer used as a microadditive in the nutrient solution on the freshly harvested biomass was established on the thirtieth day of vegetation under LEDs. Increase in productivity of the red-leaved lettuce for freshly harvested biomass was 26.6%, while for the green-leaved lettuce no loss of dry matter was observed. However, being grown under sodium lamps, a negative impact of Si fertilizer on productivity of both types of plants was observed: the amount of harvested biomass decreased by 22.6% and 30.3% for the green- and red-leaved lettuces, respectively. The effect of using Si fertilizer dramatically changed during the total growing period: up to the fifteenth day of cultivation, a sharp inhibition of the growth of both types of lettuce was observed; then, by the thirtieth day of LED lighting, Si fertilizer showed a stress-protective effect and had a positive influence on the plants. However, by the period of ripening there was no effect of using the fertilizer. Therefore, we can conclude that the use of Si fertilizers is preferable only when LED irradiation is applied throughout the active plant growth period.

Keywords: light-emitting diode; sodium lamps; plants cultivation; silicon fertilizer; red-leaved lettuce; green-leaved lettuce

Citation: Semenova, N.A.; Smirnov, A.A.; Grishin, A.A.; Pishchalnikov, R.Y.; Chesalin, D.D.; Gudkov, S.V.; Chilingaryan, N.O.; Skorokhodova, A.N.; Dorokhov, A.S.; Izmailov, A.Y. The Effect of Plant Growth Compensation by Adding Silicon-Containing Fertilizer under Light Stress Conditions. *Plants* **2021**, *10*, 1287. https://doi.org/10.3390/plants10071287

Academic Editors: Valeria Cavallaro and Rosario Muleo

Received: 29 April 2021
Accepted: 19 June 2021
Published: 24 June 2021

1. Introduction

Considering environmental factors in agriculture, light is one of the most important ingredients influencing plant growth, development, and production. It is known that most of Russia's territory, particularly its northern regions, suffers from a lack of sunlight, which is needed to maintain a high level of plant production during winter periods. For this reason, the deficiency of natural sunlight is usually compensated by using supplementary assimilation lamps in greenhouses. For instance, high-pressure sodium lamps are a very common additional lighting source. Despite well-established greenhouse technologies, the development of LED lighting systems has proved to be a subject of considerable attention over the last decade [1,2]. Above all, the efficiency of LEDs is higher compared to the sodium lamps; moreover, the potential design and optimization of LED lighting systems are rather flexible. Due to low energy emission, a light source can be located in close proximity to or within the lampshade [3]. Since LEDs emit in a narrow spectral range, any combination of diodes of different colors can be applied to control plant growth and development. Nevertheless, to consider LEDs as a valuable light source for greenhouse technology and horticulture, it would be helpful to conduct quantitative studies on plant responses to LEDs of different spectral ranges [4].

The absorption spectrum of a green leaf is characterized by three pronounced frequency ranges: the 300–400 nm high-energy range corresponds to the Soret band of chlorophylls; 400–550 nm is the region of carotenoid absorption; and 600–800 nm is a region of intense absorption of the Qy electronic transition of chlorophylls [5]. Light quanta of other spectral regions are also absorbed by plants through photoreceptors that stimulate specific developmental processes [6]. Combinations of incident lights from the 300–800 nm range affect plant morphology and can cause some changes in flowering and flower color [7]. Absorption in the red region drives basic photosynthesis processes, which is why in horticulture the most commercially used light sources are red ones. Generally, red light stimulates the growth of branches and bud outcome. Green light corresponds to the low-energy part of the green leaf spectrum, and its possible influence on photomorphogenesis is still under debate. It is assumed that green light can penetrate deeper into the leaf, increasing the light absorption in lower leaf layers, and, therefore, the intensity of photosynthetic processes. It has been reported that, with an increase in the proportion of green light, the dry mass of lettuce is also increased [8]. On the other hand, there are some studies that report no pronounced effects of green light or unconvincing results [9,10]. Blue light is essential for normal functioning of plants. Only about 10% of blue light is needed to prevent any photosynthetic dysfunction caused by its lack in lighting. Blue light sources can be used additionally to improve growth and prevent unwanted effects such as excessive stem elongation. Thus, it is obvious that variations of intensities of the irradiation spectrum can control the photomorphogenic response of plants and might significantly enhance crop production [11–13].

Besides variations in light conditions, different types of fertilizers can be used to improve plant growth and production, particularly silicon-containing fertilizers [14–20]. Si is one of the most abundant elements on Earth; its concentration corresponds to 14–20 mg Si/L [21,22], which does not go beyond the concentrations of other inorganic elements. Since plants take up Si and transfer it from roots to shoots in the form of H_4SiO_4 [23], any soil can be classified by the availability of soluble Si. Many studies have revealed that Si actively moderates morphological and physiological responses in plants [21,24–26]. It has been shown that by using Si fertilizers the number of foliar and soilborne diseases can be significantly decreased in many agricultural crops [27–30]. Moreover, the efficiency of some photosynthetic processes involved in the regulation of antioxidant mechanisms is improved for plants grown with Si fertilizers [21].

In this study we assess the combined effect of different spectral ranges of irradiation and using Si fertilizers during the growing period of red- and green-leaved lettuces [31–37]. Considering various light types used in our study as sources of stress conditions, Si nutrient solution was used to explore a possible effect of stress compensation. Working with two climatic chambers allowed us to grow plants simultaneously under sodium lamps and under LEDs. Two regimes of lighting in the LED climatic chamber were employed: one was for cultivation of lettuce before the massive appearance of shoots (first five days), and the other was for the subsequent cultivation of lettuce up to the ripeness of the product. Varying the amount of Si fertilizer throughout cultivation, we estimated the effect it has on the cultivation of the plants.

2. Materials and Methods

2.1. Climate and Light Conditions

All experiments were carried out in two climatic chambers developed in the Federal Scientific Agroengineering Center VIM, Russia (Figure 1A,B). The chambers are of the same size–2500 × 1600 × 1700 mm. Each chamber has only one level for plant cultivation. The maximum plant height is 1500 mm. The total usable area for growing was 3.8 m^2. The total power consumption was 3 kW. The intensity of light was measured by a TKA-VD spectrocolorimeter (TKA Scientific Instruments, Saint-Petersburg, Russia) at 15-cm height in each climatic chamber.

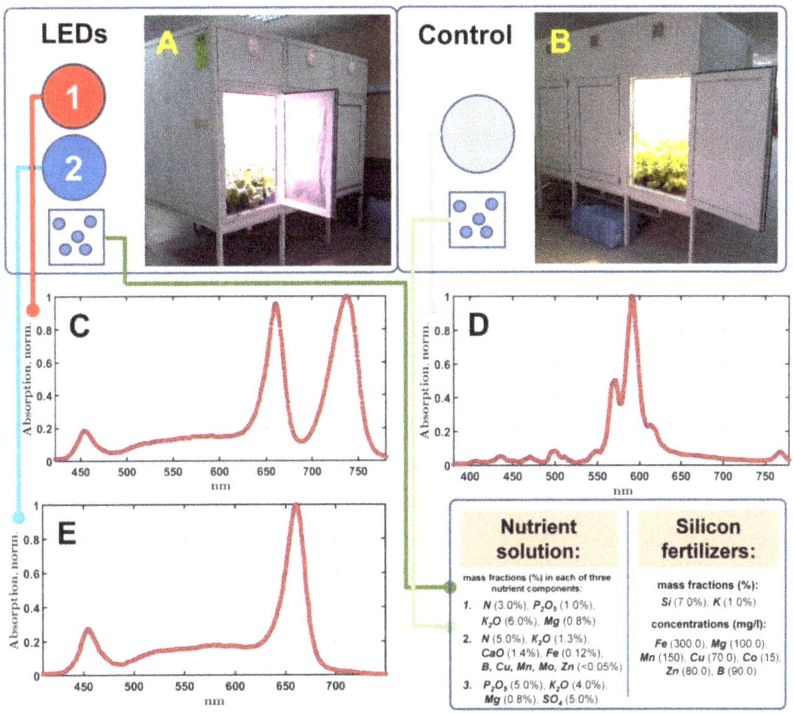

Figure 1. Scheme of the experimental setup. Two climatic chambers equipped with LEDs (**A**) or sodium lamps (**B**) and the corresponding spectra of the lighting regimes are shown in plots (**C–E**).

Climatic chamber 1 (Figure 1A) is equipped with a system of lighting based on LEDs, and it has two spectral regimes of irradiation. The first regime of lighting is characterized by the visible spectrum shown in Figure 1C. It was applied at the earlier stage of plant growth (the first 5 days) and has the distribution of photosynthetic phonon flux density (PFD) over the frequency ranges as shown in Table 1. The second regime of lighting was applied after 5 days of cultivation and was intended to stimulate the reverse response of phytochromes. This is achieved by reducing the intensity in the red region of the LED spectrum (Figure 1E). The PFD distribution for this regime is shown in Table 1.

Lighting in the second climatic chamber (Figure 1B) was provided by two tubular sodium lamps (yellow color) and one metal halide lamp (white color). The averaged PFD of the chamber is 132.3 µmol m^{-2} s^{-1}. The temperature settings in both climatic chambers at the first stage of cultivation were 19–20 °C during the day and 16–18 °C during the night; those of at the second stage of cultivation were 20–22 °C and 18–20 °C, respectively. The relative humidity in the chambers was maintained at 65–70% throughout the experiments. The temperature of the nutrient solution was 18–20 °C.

Preparation of the climatic chambers for sowing the seeds was as follows: (1) sanitization of the inner and outer surfaces of the chambers; (2) equipping the chambers with containers of mineral wool used as a substrate for the plants; (3) setting up the spectral composition of LEDs corresponding to the initial stage of cultivation; (4) regulation of the automatic watering system for the plants.

Table 1. Averaged biometric parameters of the aerial part of green and red lettuce (*Lactuca sativa* L.) under conditions of a regulated agroecosystem on the 15th day of cultivation. Values represent mean ± SE (n = 12). The different letters indicate significant differences among treatments mean non-significant differences among treatments according to Duncan's test ($p \leq 0.05$).

Lettuce	Lighting	Si	Plant Mass, g		Number of Leaves	Length of a Leaf, cm	Moisture, %	Leaf area, cm^2
			Fresh	Dry				
Green	Sodium lamps	No	3.4 ± 0.8a	0.20 ± 0.02a	6.2 ± 0.4b	11.7 ± 0.9b	94.2 ± 0.2a	122.5 ± 8.6b
		Yes	2.7 ± 0.7c	0.17 ± 0.03	5.5 ± 0.4e	10.0 ± 1.2b	93.7 ± 0.4b	74.8 ± 4.4c
	LEDs	No	1.9 ± 0.4d	0.11 ± 0.01d	5.0 ± 0.4d	9.5 ± 1.1c	94.3 ± 0.5a	69.6 ± 7.2c
		Yes	1.4 ± 0.4e	0.08 ± 0.01e	4.7 ± 0.5	7.8 ± 0.8	94.1 ± 0.1a	58.6 ± 8.0c
Red	Sodium lamps	No	3.5 ± 0.9a	0.20 ± 0.03a	6.6 ± 0.5a	12.8 ± 1.1a	94.3 ± 0.2a	162.6 ± 10.9a
		Yes	3.1 ± 0.6b	0.20 ± 0.02b	6.2 ± 0.6b	12.1 ± 1.5a	93.8 ± 0.3b	99.6 ± 5.9bc
	LEDs	No	1.9 ± 0.4d	0.12 ± 0.01d	5.3 ± 0.7c	9.2 ± 1.1c	93.7 ± 0.5b	76.1 ± 3.0c
		Yes	1.7 ± 0.6d	0.11 ± 0.01d	5.2 ± 0.4c	9.1 ± 1.6c	93.6 ± 0.1b	67.2 ± 6.9c

2.2. Cultivation of the Plants

Ten containers filled with the chemically inert substrate (mineral wool), which is highly permeable to water and has high water-holding capacity, were installed in each climatic chamber. The mineral wool in the container was divided into two rows, each of which contained five square nests. In each nest, five small depressions were made in which one seed of the red- or green-leaved lettuce (*Lactuca sativa* L.) was sown [38]. Such types of salad lettuces are commonly used to perform diverse studies on plant cultivation under different conditions [39–42], We experimented with two varieties of lettuce: the green-leaved lettuce called Azart (Prestizh Semena, www.pr-semena.ru, accessed on 1 May 2020, Russia), and the red-leaved lettuce called Robin (MoravoSeed, 1 May 2020, Czech Republic).

Before sowing the seeds, the mineral wool was prepared by completely moistening it with the nutrient solution, the concentration of which was 10% lower than that of the main solution. The nutrient solution is based on the FloraSeries fertilizer kit (GHE), which provides all the necessary macro and micronutrients. The kit contains FloraGro, FloraMicro SW, and FloraBloom components. We used the following ratio of the components: 2.5:1.6:1. The corresponding mixture of hydroponic kits was chosen according to the recommendations of the fertilizer manufacturer (generalhydroponics.com, 4 December 2019), taking into account the specificity of our research (focusing on green mass). Particularly, at the phase of seed germination the concentration of hydroponic kits in the solution was less than at the phase of vegetative growth. FloraGro supplement promotes structural growth and gain of active green mass. This supplement provides plants with a sufficient amount of nitrogen and potassium, as well as secondary minerals: 1.0% of ammonia nitrogen, 2.0% of nitrate nitrogen, 1.0% of phosphorus P_2O_5, 1.0% of soluble potassium K_2O, and 0.8% of magnesium. FloraMicro helps to stabilize the pH level of the nutrient solution. It contains 1.5% of ammonia nitrogen, 3.5% of nitrate nitrogen, 1.3% of soluble potassium, and the following EDTA-chelated microelements: Cu (0.01%), Mn (0.05%), Zn (0.015%); EDDHA and OPTA chelated Fe (0.12%) and Mo (0.004%). FloraBloom supplies plants with an adequate amount of phosphorus and potassium, which are involved in bud and fruit formation. The content of microelements in this supply is the following: P_2O_5 (5.0%), K_2O (4.0%), Mg (3.0%), SO_4 (5.0%).

2.3. Silicon Fertilizer

A liquid silicon-containing fertilizer was used in the experiments. The preparation is composed of silicon and potassium with mass fractions Si (7.0%) and K (1.0%) and trace elements in easily available chelate forms for plants (mg/L): Fe (300.0), Mg (100.0), Cu (70.0), Zn (80.0), Mn (150.0), Co (15.0), B (90.0).

2.4. Biometric and Biochemical Analysis

To determine the impact of different light conditions and fertilizer treatment, careful selection and preparation of plant samples were made. Leaves of the selected samples (for plants that have not reached the stage of forming flower stalks) were fresh, healthy, and undamaged. Their shape, color, and smell corresponded to their botanical type and variety. The leaf surface was not damaged by pests or their waste products. The assimilating leaf surface showed no sign of excessive external moisture.

The lettuce was sampled by taking one plant from each container and then cutting a rosette at the base of the plant. All samples were placed in hermetically sealed bags for further study. The total number of samples was 25 plants for each type of experiment (variety, lighting, fertilizer). Twelve plants were used to determine morphological parameters and 13 plants to determine their photosynthetic parameters. Morphological parameters (weight, number, and area of leaves) were determined for each selected plant. The biometric parameters of the morphological organs of the plants by the phases of their development were assessed 15, 30, and 45 days after the lettuce shoots appeared.

To determine the mass of dry matter in the plants, a sample was crushed manually using a hand cutting tool until the fragment size of no more than 1 cm. After grinding, the sample was mixed to avoid the inhomogeneity of fragments. Then, two weighed portions with mass of at least 5 g were isolated in two replicates using an analytical balance. The samples were dried in an oven for either 3 h at 60–70 °C or 1 h at 105 °C.

To estimate the moisture in fresh green leafy mass, the test sample was crushed and mixed, and two portions of 25–50 g were weighed on a balance with an accuracy of 0.001 g. Then the portions were placed in weighing bottles pre-dried to a constant weight. The containers with weighed portions were placed for 20–30 minutes in an oven heated to 120–130 °C to inactivate enzymes and then dried at 105 °C to constant weight.

The quantitative analysis of pigments (chlorophylls, carotenoids) included their extraction from the plant tissues using acetone, separation of the mixture into individual components, and spectrophotometry.

The concentrations of pigments in 100% acetone were calculated according to Holm–Wettstein as follows:

$$C_{Chl_a} = 9.784 \times D_{662} - 0.990 \times D_{644}$$

$$C_{Chl_b} = 21.426 \times D_{644} - 4.650 \times D_{662}$$

$$C_{Chl_a+Chl_b} = 5.134 \times D_{662} + 20.436 \times D_{644}$$

$$C_{car} = 4.695 \times D_{440.5} - 0.268 \times C_{Chl_a+Chl_b}$$

Here C_{Chl_a}, C_{Chl_b}, and C_{car} are the concentrations of chlorophylls *a* and *b* and the carotenoids. D_ω is the optical density of the extract at the corresponding wavelength ω in nm.

To determine nitrate ion concentration, a 1% solution of aluminum-potassium alum was used as the extraction solution. The concentration of nitrate ions was measured using an NO_3-selective electrode connected to an Ekspert-001 ionometric station (Econiks, Russia). The station was calibrated using solutions containing a known concentration of KNO_3 (Sigma-Aldrich, SL, USA).

Vitamin C in the leaves of lettuce plants was determined by a spectrophotometric method with a dye solution of 2,6-dichlorophenol indophenol [43]. The sucrose concentration was determined by refractometry [44].

2.5. Statistical Analysis

The biometric and biochemical parameters were processed by applying ANOVA. To estimate the statistical significance of the considered parameters, the F-test and the least significant difference test were applied.

3. Results

3.1. Effect of Fertilizers and Lighting on the Productivity of Lettuces 15 Days after the Emergence of Mass Shoots

At the initial stage of growth (up to 15 days), lettuce exposed to light stress (chamber 1) lagged behind in growth, and the addition of Si fertilizer to the nutrient solution had a depressing effect in all variants of the experiment for both lettuce varieties (Table 1).

Analysis of variance of the data showed a significant effect of factors such as lighting and the use of Si fertilizer on the fresh weight of both lettuce varieties. Under the sodium lamps, lettuce of both varieties developed faster, and by the 15th day of cultivation the increase in fresh weight was about 53%, LSD = 0.26 (Table S1). This effect is caused by light stress during the germination phase.

The use of Si fertilizer oppressed the plants regardless of the variety and the lighting. The average loss of fresh weight was 18.5% (Table 1). All three factors (lighting, variety of lettuce, Si fertilizer) had a significant impact on the number of lettuce leaves (Table S1). Under sodium lamps, the number of leaves of both varieties is on average 1 pc. more (by 21%) than under LED irradiation, LSD = 0.21 (Table S1). Also, Si fertilizer negatively influenced the number of leaves in the first stage of development of both varieties of lettuce regardless of the type of lighting. Independent of other factors, the red lettuce produced more leaves than the green lettuce.

A significant effect of lighting and the use of Si fertilizer on the dry weight of lettuce of both varieties was revealed. The dry weight of the lettuce changed significantly depending on lighting, and the reaction of the plants was variety specific. For red lettuce (Robin), the best results on the accumulation of dry matter were observed under LED irradiation, while for the green lettuce (Azart) it was under that of sodium lighting.

The area of the leaf surface when illuminated with sodium lamps was 69% greater than under LED lighting. Generally, the use of Si fertilizer caused a decrease in the photosynthetic surface of plants by about 43.5%. However, the largest significant difference of the leaf surface area was observed under sodium illumination without the addition of Si fertilizer.

Analysis of pigment concentrations showed that the red lettuce, regardless of the influence of other factors, accumulated 10% more chlorophyll *a* than the green lettuce; moreover, the addition of Si fertilizer increased the concentration of chlorophyll *a* in the lettuce leaves by 37.5% (Table 2).

It should be stressed that the green lettuce accumulated more chlorophyll *a* when using Si fertilizer (71% more with sodium lighting), and the highest concentration of chlorophyll *a* was observed with LED lighting using 0.15% Si fertilizer (95% more than control under sodium lamps without the use of Si fertilizer).

For the red lettuce under sodium lighting, the addition of Si fertilizer to the solution caused an increase in the concentration of chlorophyll *a* (the best variant of the experiment), while under LED lighting the fertilizer contributed to a decrease in the concentration of the pigment. The data analysis showed that the specificity of the influence of factors on the concentration of chlorophyll *b* and carotenoids is similar to that of chlorophyll *a*, LSD = 0.2 (Table S2).

Table 2. Average concentrations of photosynthetic pigments in green and red lettuce (*Lactuca sativa* L.) under conditions of a regulated agroecosystem on the 15th day of cultivation. Values represent mean ± SE (n = 12). The different letters indicate significant differences among treatments means non-significant differences among treatments according to Duncan's test ($p \leq 0.05$).

Lettuce	Lighting	Si	Chlorophyll *a*, mg/g	Chlorophyll *b*, mg/g	Carotenoids, mg/g
Green	Sodium lamps	No	1.04 ± 0.21e	0.34 ± 0.04e	0.38 ± 0.07f
		Yes	2.00 ± 0.04cd	0.59 ± 0.04d	0.69 ± 0.03b
	LEDs	No	1.65 ± 0.01d	0.59 ± 0.02d	0.52 ± 0.01cd
		Yes	2.27 ± 0.11c	0.67 ± 0.04c	0.63 ± 0.02c
Red	Sodium lamps	No	1.53 ± 0.01d	0.48 ± 0.03de	0.48 ± 0.02d
		Yes	2.98 ± 0.02a	0.96 ± 0.08a	1.04 ± 0.02a
	LEDs	No	2.45 ± 0.05b	0.80 ± 0.03b	0.75 ± 0.02b
		Yes	1.39 ± 0.34de	0.52 ± 0.09de	0.37 ± 0.15f

3.2. Effect of Fertilizer and Lighting on the Productivity of Lettuces after 30 Days of the Emergence of Mass Shoots

A significant positive effect of Si fertilizer was found on the 30th day of cultivation: the increase in fresh weight was 13.8%, LSD = 9.28 (Table S3), which indicates a delayed stress-protective effect of Si fertilizer even under light stress (Table 3).

Table 3. Averaged biometric parameters of the aerial part of green and red lettuce (*Lactuca sativa* L.) under the conditions of a regulated agroecosystem on the 30th day of cultivation. Values represent mean ± SE (n = 12). The different letters indicate significant differences among treatments and ns means non-significant differences among treatments according to Duncan's test ($p \leq 0.05$).

Lettuce	Lighting	Si	Plant Mass, g		Number of Leaves	Length of a Leaf, cm	Moisture, %	Leaf Area, cm^2	Photo Productivity, g/m^2 Per Day
			Fresh	Dry					
Green	Sodium lamps	No	33.8 ± 15.0a	9.9 ± 0.9a	9.9 ± 2.1ab	22.6 ± 4.5ns	90.2 ± 0.1ns	1847.7 ± 7.6ns	5.5 ± 0.9a
		Yes	26.5 ± 11.8a	6.5 ± 0.5a	9.5 ± 1.4ab	20.0 ± 2.2ns	91.9 ± 1.0ns	1492.8 ± 8.1ns	6.3 ± 0.7a
	LEDs	No	25.6 ± 13.9a	6.0 ± 0.4a	8.1 ± 1.2b	21.1 ± 4.3ns	91.4 ± 0.9ns	1500.1 ± 5.6ns	6.2 ± 0.3a
		Yes	27.2 ± 8.8a	7.1 ± 0.5a	9.1 ± 1.8ab	22.8 ± 4.9ns	91.3 ± 1.2ns	1614.0 ± 9.1ns	6.1 ± 0.7a
Red	Sodium lamps	No	27.8 ± 11.9a	8.1 ± 0.3a	11.2 ± 2.6a	23.8 ± 5.1ns	90.3 ± 0.4ns	1871.2 ± 9.6ns	4.4 ± 0.2b
		Yes	19.3 ± 8.2b	5.3 ± 0.8b	9.9 ± 1.7ab	22.8 ± 2.3ns	90.9 ± 1.1ns	1402.3 ± 7.4ns	5.5 ± 0.8a
	LEDs	No	23.3 ± 13.4b	6.9 ± 0.7b	10.2 ± 2.3a	21.7 ± 3.3ns	90.1 ± 1.2ns	1428.4 ± 6.9ns	4.4 ± 0.9b
		Yes	29.4 ± 9.2a	8.8 ± 0.7a	11.1 ± 1.8a	23.2 ± 2.4ns	90.0 ± 0.9ns	1675.6 ± 8.4ns	6.1 ± 0.3a

Under the sodium lamps, the Si fertilizer caused a decrease in the fresh weight of lettuce of both varieties by 25.6%. The factors of lighting and nutrition did not have a significant effect on the dry mass, as well as on the leaf surface area, which means there is no loss in product quality. After light stress in chamber No. 1, by the 30th day, the plants had already adapted and significantly increased in terms of fresh and dry weight in a relatively short period of time. The same tendency was observed in the number of leaves in lettuce of both varieties (Table 3). The data show that the red lettuce is characterized by more intensive (14.2%) leaf formation than the green lettuce, which is a drumhead kind of lettuce, LSD = 1.51 (Table S3).

The best values of the sugar content for green lettuce were obtained under the sodium light with the addition of 0.15% of the Si fertilizer (12.1% higher sugar); however, for the red lettuce, the maximum sugar content was obtained under LED lighting also with the addition of Si fertilizer (by 8.2% compared to the control). In general, it appeared that the green lettuce accumulates more sugars by 9.9% (Table 4). The minimum concentration of nitrates for both varieties was observed under sodium lamps with Si fertilizer (14.8%

lower); the concentrations in all other cases were within the allowed limits for this type of chemical compound (up to 4500 mg/kg). The use of LED and sodium lamps, as well as Si fertilizer, did not have a significant effect on the content of vitamin C and the basic pigments (Table S4).

Table 4. Averaged concentrations of sucrose, vitamin C, and nitrates in green and red lettuce (*Lactuca sativa* L.) under the conditions of a regulated agroecosystem on the 30th day of cultivation. Values represent mean ± SE (n = 12). The different letters indicate significant differences among treatments and ns means non-significant differences among treatments according to Duncan's test ($p \leq 0.05$).

Lettuce	Lighting	Si	Vitamin C, mg/100 g	Nitrate, mg/kg	Chlorophyll *a*, mg/g	Chlorophyll *b*, mg/g	Carotenoids, mg/g
Green	Sodium lamps	No	15.1 ± 0.1ns	5.1 ± 0.1ab	2.2 ± 0.5ns	0.6 ± 0.1ns	0.7 ± 0.2ns
		Yes	14.9 ± 0.2ns	5.0 ± 0.4ab	2.6 ± 0.8ns	0.8 ± 0.2ns	0.9 ± 0.3ns
	LEDs	No	15.0 ± 0.1ns	5.3 ± 0.1a	2.5 ± 0.6ns	0.8 ± 0.2ns	0.8 ± 0.2ns
		Yes	14.9 ± 0.1ns	5.3 ± 0.1a	2.7 ± 0.5ns	0.9 ± 0.2ns	0.8 ± 0.2ns
Red	Sodium lamps	No	14.9 ± 0.03ns	5.7 ± 0.4a	2.1 ± 0.9ns	0.7 ± 0.3ns	0.7 ± 0.3ns
		Yes	15.0 ± 0.03ns	4.3 ± 0.1b	2.5 ± 0.4ns	0.8 ± 0.1ns	0.7 ± 0.1ns
	LEDs	No	14.9 ± 0.1ns	4.9 ± 0.2b	2.2 ± 0.6ns	0.7 ± 0.2ns	0.7 ± 0.2ns
		Yes	14.9 ± 0.1ns	5.4 ± 0.1a	2.1 ± 0.1ns	0.6 ± 0.0ns	0.7 ± 0.04ns

3.3. Effect of Fertilizer and Lighting on the Productivity of Lettuces after 45 Days of the Emergence of Mass Shoots

The last sampling of the lettuce was done on the 45th day of cultivation when the plants correspond to marketable products (Figure 2). Considering the type of lighting and the variety of the lettuce as factors of statistical analysis, it was found that these factors had a significant effect on the fresh weight of the plants (Tables 5 and 6). The distribution of the lettuce fresh weight group average of both varieties shows the advantage (about 25%) of sodium lighting over LEDs, LSD = 8.64 (Table S5). The green lettuce accumulates fresh mass more intensively than the red lettuce (by 22.3%).

Table 5. Averaged biometric parameters of the aerial part of green and red lettuce (*Lactuca sativa* L.) under conditions of a regulated agroecosystem on the 45th day of cultivation. Values represent mean ± SE (n = 12). The different letters indicate significant differences among treatments and ns means non-significant differences among treatments according to Duncan's test ($p \leq 0.05$).

Lettuce	Lighting	Si	Plant Mass, g		Number of Leaves	Length of a leaf, cm	Moisture, %	Leaf Area, cm^2	Photo Productivity, g/m^2 per day
			Fresh	Dry					
Green	Sodium lamps	No	81.1 ± 21.7a	3.3 ± 0.1a	13.8 ± 1.5ab	30.5 ± 3.5a	95.9 ± 0.1a	4661.2 ± 6.5a	1.0 ± 0.7ns
		Yes	74.3 ± 19.3a	3.0 ± 0.1a	12.8 ± 1.4ab	27.7 ± 2.9b	95.9 ± 0.1a	3135.6 ± 5.2a	0.4 ± 0.2ns
	LEDs	No	57.6 ± 21.2b	2.3 ± 0.1b	11.9 ± 2.3b	27.3 ± 3.7b	96.0 ± 0.1a	2525.5 ± 5,7b	1.1 ± 0.4ns
		Yes	57.6 ± 18.8b	2.6 ± 0.1b	11.6 ± 1.1b	26.8 ± 3.0b	95.5 ± 0.2b	2748.4 ± 4.5b	1.2 ± 0.2ns
Red	Sodium lamps	No	65.5 ± 25.8b	2.8 ± 0.1b	17.2 ± 5.1a	31.4 ± 3.7a	95.7 ± 0.1b	3253.4 ± 5.7a	0.8 ± 0.2ns
		Yes	55.7 ± 14.9b	2.4 ± 0.1b	15.2 ± 2.7a	29.0 ± 2.8a	95.6 ± 0.2b	2825.9 ± 8.1a	0.9 ± 0.2ns
	LEDs	No	41.1 ± 19.8c	1.9 ± 0.04c	14.7 ± 3.7ab	25.9 ± 4.1c	95.3 ± 0.1c	2413.3 ± 6.6b	1.5 ± 0.2ns
		Yes	48.2 ± 20.7b	2.3 ± 0.1b	14.8 ± 3.6ab	24.3 ± 3.0c	95.2 ± 0.1c	2299.1 ± 7.0b	1.3 ± 0.3ns

Table 6. Average concentrations of chlorophyll *a* and *b*, carotenoids, and nitrates in green and red lettuce (*Lactuca sativa* L.) under conditions of a regulated agroecosystem on the 45th day of cultivation. Values represent mean ± SE (n = 12). The different letters indicate significant differences among treatments means non-significant differences among treatments according to Duncan's test ($p \leq 0.05$).

Lettuce	Lighting	Si	Nitrate, mg/kg	Chlorophyll *a*, mg/g	Chlorophyll *b*, mg/g	Carotenoids, mg/g
Green	Sodium lamps	No	5.13 ± 0.08b	2.09 ± 0.07b	0.71 ± 0.05c	0.74 ± 0.12b
		Yes	5.16 ± 0.43b	2.50 ± 0.21a	1.27 ± 0.15a	0.92 ± 0.01ab
	LEDs	No	5.25 ± 0.02b	2.18 ± 0.41b	1.33 ± 0.22a	1.01 ± 0.04a
		Yes	5.5 ± 0.12ab	2.33 ± 0.20a	1.23 ± 0.05a	0.97 ± 0.07a
Red	Sodium lamps	No	5.58 ± 0.16a	2.07 ±0.20b	1.06 ± 0.11b	0.84 ± 0.02ab
		Yes	4.55 ± 0.02c	2.54 ± 0.09a	1.12 ± 0.04b	0.70 ± 0.06b
	LEDs	No	5.11 ± 0.07b	1.99 ± 0.11b	1.00 ± 0.02b	0.88 ± 0.01ab
		Yes	5.78 ± 0.06a	2.08 ± 0.05b	0.74 ± 0.03c	0.79 ± 0.02b

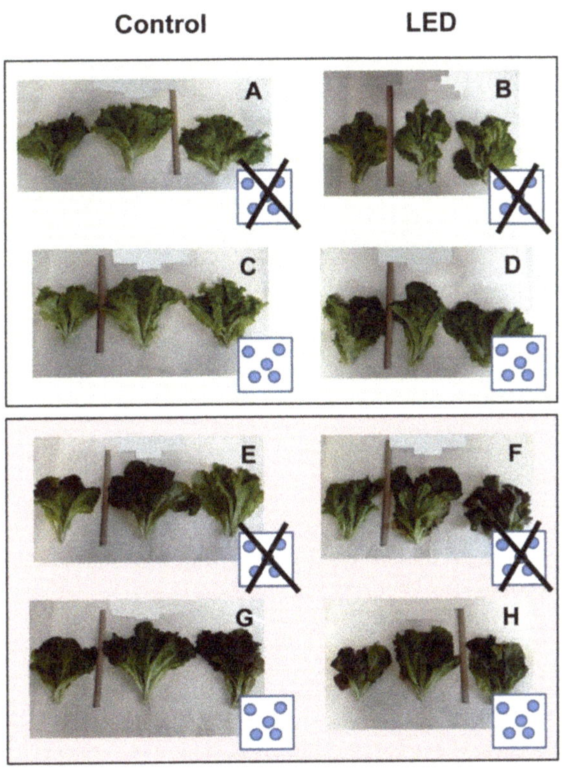

Figure 2. Comparison of the appearance of green (**A–D**) and red (**E–H**) lettuce on the 45th day of cultivation. Plants after sodium lamp illumination are on the left, after LED illumination on the right. (**C**,**D**,**G**,**H**) are the samples grown with Si fertilizer. (**A**,**B**,**E**,**F**) were grown without Si fertilizer.

The same advantage of sodium over LED lighting, about 12.2%, was revealed comparing the number of leaves of both varieties, LSD = 1.26 (Table S5). The red lettuce showed a tendency for more intensive leaf formation (19.4%).

Silicon fertilizer has had a significant effect on the content of chlorophyll *a*. The distribution of group averages of chlorophyll *a* concentration shows that the use of Si fertilizer increases the content of this pigment by 8.3%. It turned out that all three factors (type of lettuce, lighting, and Si fertilizer) have a significant effect on the content of chlorophyll *b*. The green lettuce accumulates 14% more chlorophyll *b* than that of the red lettuce. The largest accumulation of chlorophyll *b* in the green lettuce was observed under the LED illumination (21.9%), and for the red lettuce it was observed under sodium lamps (+18.5%). Addition of Si fertilizer increased the chlorophyll *b* concentration by 26.7% under the sodium lamp lighting, while under the LED lighting it was decreased by 15.5%. Finally, the use of Si fertilizer contributed to the accumulation of total chlorophyll in both varieties grown under sodium lamps by 26% (green) and 14.5% (red).

The range of the chlorophyll *a* to chlorophyll *b* ratio (which is about 2–3) for different combinations of the studied factors is due to the averaged samples taken from various tiers of plants (for heliophilous crops, this ratio is about three).

All three factors also significantly affect the concentration of carotenoids: the green lettuce accumulates 13% more of this pigment than the red lettuce (Table 6). Under the LED illumination, the carotenoids were accumulated by 13%, but the ratio of total chlorophyll to carotenoids is within the physiological norm (from three to eight), which means the plants adapted to the light conditions and do not have stress. The use of Si fertilizer caused an increase in the concentration of carotenoids for the green lettuce and a decrease for the red lettuce. In general, the concentration of nitrates in lettuces was affected by all factors, LSD = 0.95 (Table S6). The accumulation of nitrates was more than 5.5% under the LEDs. The addition of Si fertilizer reduced the concentration of nitrates by 9.3% under the sodium lamp lighting, but it was increased by 8.2% under the LEDs. For all cases, the total nitrates concentration was hundreds of times less than the threshold limit value on this product.

4. Discussion

Analysis of the results of this three-factor study allowed us to assess the influence of the biologically active Si fertilizer as well as the effect of different spectral compositions of LEDs in comparison with sodium lamps by using two climatic chambers (No. 1 and No. 2) on the biomass of freshly harvested lettuce of different varieties, red and green. Comparison of the biometric and biochemical data made it possible to draw certain conclusions on plant cultivation and development.

Some earlier works claimed [45] that application of Si-containing fertilizer did not increase the lettuce yield. However, in our current and several recent studies [1,46], increased yield was observed as the positive effect of Si fertilizer, used as a microadditive to the nutrient solution, on the freshly harvested biomass of the green and red lettuces, grown under LED illumination on the 30th day of vegetation was established (Figure 3B). Table 7 contains the ranks for each combination of the three factors. One can see that for both types of lettuce and LED illumination on the 30th day, the second and the third ranks have samples grown with Si fertilizer. The increase in productivity of red lettuce for freshly harvested biomass was 26.6%, and for green lettuce 6.3%, while no loss of dry matter was observed.

It is interesting to note that when using hydroponic nutrient solution with Si [47], the lettuce plants showed a lower level of shoot dry matter, whereas the reverse effect [48] has been demonstrated too: the application of Si fertilizer via fertigation favored an increase in dry matter of lettuce. However, in our study a negative effect of Si fertilizer on the productivity of lettuce plants of both varieties on the 30th day of cultivation in chamber No. 2, under sodium lamp illumination, was noted (the 5th and the 7th ranks after the 30th day in Table 7). The productivity of green lettuce by estimating freshly harvested biomass decreased by 22.6%, while the red one decreased by 30.3% (without loss in dry matter percentage). Moreover, we observed an increase in the dry weight of lettuce (4.4%) when using the Si fertilizer in chamber No. 1 with LEDs' irradiators on the 45 day.

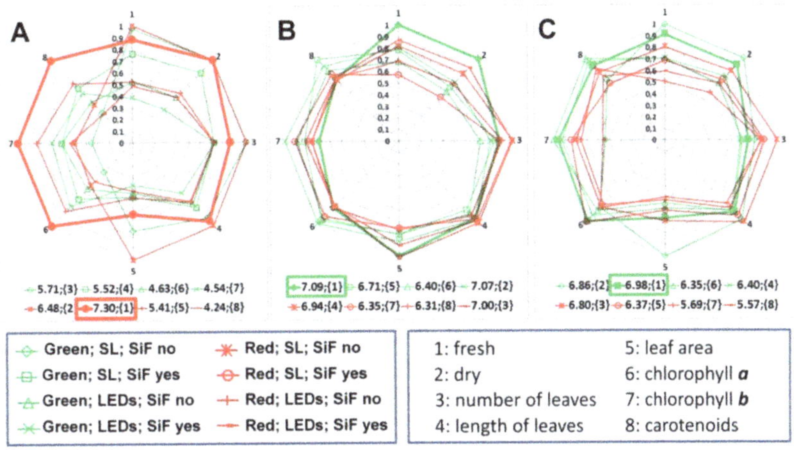

Figure 3. Comparison of the results of biometric and biochemical analysis for red and green lettuce after the 15th (**A**), 30th (**B**), and 45th (**C**) days of cultivation. The diagrams display the normalized parameters presented in Tables 1–6 (eight parameters shown in the right panel at the bottom of the diagrams). The sets of parameters corresponding to a combination of three factors—type of lettuce, type of lighting, and the use of Si fertilizer—are connected by red or green lines with markers (the left panel at the bottom of the diagrams). The rank of each set of measured parameters is shown in curly brackets. Rectangular frames indicate the best combination of parameters obtained in the experiments.

Table 7. Rank of each combination of factors (plants, lighting, Si fertilizer (SiF)) after the 15th, 30th, and 45th days of cultivation and the total ranks (Σ).

Factors	Rank after			Σ
	the 15th day	the 30th day	the 45th day	
Green; SL; SiF no	3	1	2	6
Green; SL; SiF yes	4	5	1	10
Green; LEDs; SiF no	6	6	6	18
Green; LEDs; SiF yes	7	2	4	13
Red; SL; SiF no	2	4	3	9
Red; SL; SiF yes	1	7	5	13
Red; LEDs; SiF no	5	8	7	20
Red; LEDs; SiF yes	8	3	8	19

We found that Si fertilizer helps to increase the leaf surface area and improve the marketable appearance of plants of both varieties (Figure 2). In addition, it increased the content of total chlorophyll, the maximum amount of which was observed when it was illuminated under sodium light.

There is a study [49] in which Si fertilizer was used in pots where lettuce was grown. In this case it did not cause significant differences in the leaf contents of nitrogen but increasing doses of silicate caused nitrogen deficiency. Our data and analysis revealed that the addition of Si fertilizer under sodium illumination reduced the concentration of nitrates by 9.3%; under the LEDs it was increased by 8.2%. Being grown under LED illumination, a more intense accumulation of nitrates was observed (about 5.5% more), which indicates that the synthesis of nitrogen-containing substances was intensified.

The effect of using Si fertilizer dramatically changed during the growing time: up to the 15th day of cultivation (Figure 3A) there was a sharp inhibition of the growth of lettuce plants of both varieties (the 7th and 8th ranks after the 15th day in Table 7); then, by the 30th day under LED lighting, the addition of Si fertilizer showed a stress-protective effect and had a positive influence in general. By the time of ripeness, the effect of using Si fertilizer was no longer observed. Therefore, the use of such a type of supplementary nutrition is advisable only when using LED illumination during the period of active plant growth (from the 10th day after germination).

It was recently reported [50] that the LED spectrum provided by the combination of far-red, deep-red, and blue LEDs is more favorable than sodium lamps for promoting the growth and nutrient uptake of plants. However, we found that the total productivity (3271.7 g) under the entire set of lettuce growth conditions (regardless of the lettuce variety) did not depend on the composition of the nutrient solution, but growth in chamber No. 2 (when illuminated with sodium lamps) was 35.5% higher than the plant productivity (2414.71 g) of the same lettuce varieties grown in chamber No. 1 (under the LED lighting).

5. Conclusions

Thus, considering the biometric and biochemical analysis, we can conclude that it makes sense to add Si fertilizer when growing red lettuce under LED lighting, while for the green lettuce the addition of such fertilizer is not useful. In the case of sodium lamp illumination, the best option for growing both varieties is without Si fertilizer.

Supplementary Materials: The following are available online at https://www.mdpi.com/article/10.3390/plants10071287/s1, Table S1. LCD values for biometric parameters of the aerial part of green and red lettuce (*Lactuca sativa* L.) under conditions of a regulated agroecosystem on the 15th day of cultivation; Table S2. LCD values for concentrations of photosynthetic pigments in green and red lettuce (*Lactuca sativa* L.) under conditions of a regulated agroecosystem on the 15th day of cultivation; Table S3. LCD values for biometric parameters of the aerial part of green and red lettuce (*Lactuca sativa* L.) under the conditions of a regulated agroecosystem on the 30th day of cultivation; Table S4. LCD values for concentrations of sucrose, vitamin C, and nitrates in green and red lettuce (*Lactuca sativa* L.) under the conditions of a regulated agroecosystem on the 30th day of cultivation; Table S5. LCD values for biometric parameters of the aerial part of green and red lettuce (*Lactuca sativa* L.) under the conditions of a regulated agroecosystem on the 45th day of cultivation; Table S6. LCD values for concentrations of chlorophyll a and b, carotenoids, and nitrates in green and red lettuce (*Lactuca sativa* L.) under the conditions of a regulated agroecosystem on the 45th day of cultivation.

Author Contributions: Conceptualization and methodology, S.V.G., N.O.C.; validation, N.A.S.; formal analysis, A.N.S., N.A.S.; investigation and resources, A.A.S., A.A.G.; writing—original draft preparation, R.Y.P., D.D.C.; writing—review and editing, R.Y.P., visualization, N.A.S.; supervision and funding acquisition, A.Y.I.; project administration, A.S.D. All authors have read and agreed to the published version of the manuscript.

Funding: This research was funded by a grant of the Ministry of Science and Higher Education of the Russian Federation for large scientific projects in priority areas of scientific and technological development (subsidy identifier 075-15-2020-774).

Institutional Review Board Statement: Not applicable.

Informed Consent Statement: Not applicable.

Data Availability Statement: No additional data available.

Conflicts of Interest: The authors declare no conflict of interest. The funders had no role in the design of the study, in the collection, analyses, or interpretation of the data, in the writing of the manuscript, or in the decision to publish the results.

References

1. Virsile, A.; Brazaityte, A.; Vastakaite-Kairiene, V.; Miliauskiene, J.; Jankauskiene, J.; Novickovas, A.; Lauzike, K.; Samuoliene, G. The distinct impact of multi-color LED light on nitrate, amino acid, soluble sugar and organic acid contents in red and green leaf lettuce cultivated in controlled environment. *Food Chem.* **2020**, *310*, 125799. [CrossRef] [PubMed]
2. Camejo, D.; Frutos, A.; Mestre, T.C.; Piñero, M.D.; Rivero, R.M.; Martínez, V. Artificial light impacts the physical and nutritional quality of lettuce plants. *Hortic. Environ. Biotechnol.* **2020**, *61*, 69–82. [CrossRef]
3. Davis, P.A.; Burns, C. Photobiology in protected horticulture. *Food Energy Secur.* **2016**, *5*, 223–238. [CrossRef]
4. Najera, C.; Urrestarazu, M. Effect of the Intensity and Spectral Quality of LED Light on Yield and Nitrate Accumulation in Vegetables. *Hortscience* **2019**, *54*, 1745–1750. [CrossRef]
5. Shevela, D.; Pishchainikov, R.Y.; Eichacker, L.A.; Govindjee. Stress Biology of Cyanobacteria: Molecular Mechanism to Cellular Responses. In *Oxygenic Photosynthesis in Cyanobacteria*; CRC Press: Boca Raton, FL, USA, 2013; pp. 3–40.
6. Pishchalnikov, R.Y.; Razjivin, A.P. From localized excited states to excitons: Changing of conceptions of primary photosynthetic processes in the twentieth century. *Biochem. Mosc.* **2014**, *79*, 242–250. [CrossRef] [PubMed]
7. Gudkov, S.V.; Andreev, S.N.; Barmina, E.V.; Bunkin, N.F.; Kartabaeva, B.B.; Nesvat, A.P.; Stepanov, E.V.; Taranda, N.I.; Khramov, R.N.; Glinushkin, A.P. Effect of visible light on biological objects: Physiological and pathophysiological aspects. *Phys. Wave Phenom.* **2017**, *25*, 207–213. [CrossRef]
8. Kim, H.H.; Goins, G.D.; Wheeler, R.M.; Sager, J.C. Green-light supplementation for enhanced lettuce growth under red- and blue-light-emitting diodes. *Hortscience* **2004**, *39*, 1617–1622. [CrossRef]
9. Hernandez, R.; Kubota, C. Physiological responses of cucumber seedlings under different blue and red photon flux ratios using LEDs. *Environ. Exp. Bot.* **2016**, *121*, 66–74. [CrossRef]
10. Johkan, M.; Shoji, K.; Goto, F.; Hashida, S.; Yoshihara, T. Blue Light-emitting Diode Light Irradiation of Seedlings Improves Seedling Quality and Growth after Transplanting in Red Leaf Lettuce. *Hortscience* **2010**, *45*, 1809–1814. [CrossRef]
11. Huche-Thelier, L.; Crespel, L.; Le Gourrierec, J.; Morel, P.; Sakr, S.; Leduc, N. Light signaling and plant responses to blue and UV radiations-Perspectives for applications in horticulture. *Environ. Exp. Bot.* **2016**, *121*, 22–38. [CrossRef]
12. Bhuiyan, R.; van Iersel, M.W. Only Extreme Fluctuations in Light Levels Reduce Lettuce Growth Under Sole Source Lighting. *Front. Plant Sci.* **2021**, *12*, 24. [CrossRef]
13. Gherghina, E.; Luta, G.; Dobrin, E.; Draghici, E.M.; Balan, D.; Sanmartin, A.M. Biochemical changes under artificial led lighting in some *Lactuca sativa* L. varieties. *Agrolife Sci. J.* **2020**, *9*, 141–148.
14. Rizwan, M.; Rehman, M.Z.U.; Ali, S.; Abbas, T.; Maqbool, A.; Bashir, A. *Biochar Is a Potential Source of Silicon Fertilizer: An Overview*; Elsevier Science Bv: Amsterdam, The Netherlands, 2019; pp. 225–238. [CrossRef]
15. Tripathi, P.; Na, C.I.; Kim, Y. Effect of silicon fertilizer treatment on nodule formation and yield in soybean (*Glycine max* L.). *Eur. J. Agron.* **2021**, *122*. [CrossRef]
16. Stephano, M.F.; Geng, Y.H.; Cao, G.J.; Wang, L.C.; Meng, W.; Zhang, M.L. Effect of Silicon Fertilizer and Straw Return on the Maize Yield and Phosphorus Efficiency in Northeast China. *Commun. Soil Sci. Plant Anal.* **2021**, *52*, 116–127. [CrossRef]
17. Huang, H.L.; Rizwan, M.; Li, M.; Song, F.R.; Zhou, S.J.; He, X.; Ding, R.; Dai, Z.H.; Yuan, Y.; Cao, M.H.; et al. Comparative efficacy of organic and inorganic silicon fertilizers on antioxidant response, Cd/Pb accumulation and health risk assessment in wheat (*Triticum aestivum* L.). *Environ. Pollut.* **2019**, *255*, 113146. [CrossRef]
18. Fan, Y.; Shen, W.Y.; Cheng, F.Q. Reclamation of two saline-sodic soils by the combined use of vinegar residue and silicon-potash fertiliser. *Soil Res.* **2018**, *56*, 801–809. [CrossRef]
19. Franca, A.A.; Schultz, J.; Borges, R.; Wypych, F.; Mangrich, A.S. Rice Husk Ash as Raw Material for the Synthesis of Silicon and Potassium Slow-Release Fertilizer. *J. Braz. Chem. Soc.* **2017**, *28*, 2211–2217. [CrossRef]
20. Eneji, A.E.; Inanaga, S.; Muranaka, S.; Li, J.; Hattori, T.; An, P.; Tsuji, W. Growth and nutrient use in four grasses under drought stress as mediated by silicon fertilizers. *J. Plant Nutr.* **2008**, *31*, 355–365. [CrossRef]
21. Debona, D.; Rodrigues, F.A.; Datnoff, L.E. Silicon's Role in Abiotic and Biotic Plant Stresses. In *Annual Review of Phytopathology*; Leach, J.E., Lindow, S.E., Eds.; Annual Reviews: Palo Alto, CA, USA, 2017; Volume 55, pp. 85–107.
22. Houben, D.; Sonnet, P.; Cornelis, J.T. Biochar from Miscanthus: A potential silicon fertilizer. *Plant Soil* **2014**, *374*, 871–882. [CrossRef]
23. Ma, J.F.; Yamaji, N.; Mitani-Ueno, N. Transport of silicon from roots to panicles in plants. *Proc. Jpn. Acad. Ser. B-Phys. Biol. Sci.* **2011**, *87*, 377–385. [CrossRef]
24. Huang, C.P.; Wang, L.; Gong, X.Q.; Huang, Z.T.; Zhou, M.R.; Li, J.; Wu, J.S.; Chang, S.X.; Jiang, P.K. Silicon fertilizer and biochar effects on plant and soil PhytOC concentration and soil PhytOC stability and fractionation in subtropical bamboo plantations. *Sci. Total. Environ.* **2020**, *715*, 136846. [CrossRef]
25. Currie, H.A.; Perry, C.C. Silica in plants: Biological, biochemical and chemical studies. *Ann. Bot.* **2007**, *100*, 1383–1389. [CrossRef]
26. Cuong, T.X.; Ullah, H.; Datta, A.; Hanh, T.C. Effects of Silicon-Based Fertilizer on Growth, Yield and Nutrient Uptake of Rice in Tropical Zone of Vietnam. *Rice Sci.* **2017**, *24*, 283–290. [CrossRef]
27. Miyake, Y.; Takahashi, E. Effect of silicon on the growth of solution-cultured cucumber plant. *Soil Sci. Plant Nutr.* **1983**, *29*, 71–83. [CrossRef]
28. Seebold, K.W.; Kucharek, T.A.; Datnoff, L.E.; Correa-Victoria, F.J.; Marchetti, M.A. The influence of silicon on components of resistance to blast in susceptible, partially resistant, and resistant cultivars of rice. *Phytopathology* **2001**, *91*, 63–69. [CrossRef]

29. Stevens, W.; Rhine, M.; Vories, E. Effect of Irrigation and Silicon Fertilizer on Total Rice Grain Arsenic Content and Yield. *Crop. Forage Turfgrass Manag.* **2017**, *3*, 1–6. [CrossRef]
30. Yu, T.H.; Peng, Y.Y.; Lin, C.X.; Qin, J.H.; Li, H.S. Application of iron and silicon fertilizers reduces arsenic accumulation by two Ipomoea aquatica varities. *J. Integr. Agric.* **2016**, *15*, 2613–2619. [CrossRef]
31. Bian, Z.H.; Lei, B.; Cheng, R.F.; Wang, Y.; Li, T.; Yang, Q.C. Selenium distribution and nitrate metabolism in hydroponic lettuce (*Lactuca sativa* L.): Effects of selenium forms and light spectra. *J. Integr. Agric.* **2020**, *19*, 133–144. [CrossRef]
32. Yan, Z.N.; He, D.X.; Niu, G.H.; Zhou, Q.; Qu, Y.H. Growth, Nutritional Quality, and Energy Use Efficiency of Hydroponic Lettuce as Influenced by Daily Light Integrals Exposed to White versus White Plus Red Light-emitting Diodes. *Hortscience* **2019**, *54*, 1737–1744. [CrossRef]
33. Yan, Z.N.; He, D.X.; Niu, G.H.; Zhai, H. Evaluation of growth and quality of hydroponic lettuce at harvest as affected by the light intensity, photoperiod and light quality at seedling stage. *Sci. Hortic.* **2019**, *248*, 138–144. [CrossRef]
34. Virsile, A.; Brazaityte, A.; Vastakaite-Kairiene, V.; Miliauskiene, J.; Jankauskiene, J.; Novickovas, A.; Samuoliene, G. Lighting intensity and photoperiod serves tailoring nitrate assimilation indices in red and green baby leaf lettuce. *J. Sci. Food Agric.* **2019**, *99*, 6608–6619. [CrossRef] [PubMed]
35. Naznin, M.T.; Lefsrud, M.; Gravel, V.; Azad, M.O.K. Blue Light added with Red LEDs Enhance Growth Characteristics, Pigments Content, and Antioxidant Capacity in Lettuce, Spinach, Kale, Basil, and Sweet Pepper in a Controlled Environment. *Plants* **2019**, *8*, 93. [CrossRef] [PubMed]
36. Meng, Q.W.; Kelly, N.; Runkle, E.S. Substituting green or far-red radiation for blue radiation induces shade avoidance and promotes growth in lettuce and kale. *Environ. Exp. Bot.* **2019**, *162*, 383–391. [CrossRef]
37. Shevtsova, L.P.; Shyurova, N.A.; Bashinskaya, O.S.; Toigildin, A.L.; Toigildina, I.A. Practices of Raising the Cropping Power of Green Large Seed Lentil in the Volga Region Steppe. *Res. J. Pharm. Biol. Chem. Sci.* **2016**, *7*, 113–119.
38. Chung, H.Y.; Chang, M.Y.; Wu, C.C.; Fang, W. Quantitative Evaluation of Electric Light Recipes for Red Leaf Lettuce Cultivation in Plant Factories. *Horttechnology* **2018**, *28*, 755–763. [CrossRef]
39. Spadafora, N.D.; Cocetta, G.; Ferrante, A.; Herbert, R.J.; Dimitrova, S.; Davoli, D.; Fernandez, M.; Patterson, V.; Vozel, T.; Amarysti, C.; et al. Short-Term Post-Harvest Stress that Affects Profiles of Volatile Organic Compounds and Gene Expression in Rocket Salad during Early Post-Harvest Senescence. *Plants* **2020**, *9*, 4. [CrossRef]
40. Lee, H.J.; Chun, J.H.; Kim, S.J. Effects of Pre Harvest Light Treatments (LEDs, Fluorescent Lamp, UV-C) on Glucosinolate Contents in Rocket Salad (*Eruca sativa*). *Hortic. Sci. Technol.* **2017**, *35*, 178–187. [CrossRef]
41. Jin, J.; Koroleva, O.A.; Gibson, T.; Swanston, J.; Magan, J.; Zhang, Y.; Rowland, I.R.; Wagstaff, C. Analysis of Phytochemical Composition and Chemoprotective Capacity of Rocket (Eruca sativa and Diplotaxis tenuifolia) Leafy Salad Following Cultivation in Different Environments. *J. Agric. Food Chem.* **2009**, *57*, 5227–5234. [CrossRef]
42. Santamaria, P.; Gonnella, M.; Elia, A.; Parente, A.; Serio, F. Ways of reducing rocket salad nitrate content. In *Proceedings of the International Symposium on Growing Media and Hydroponics*; Maloupa, E., Gerasopoulos, D., Eds.; International Society for Horticultural Science: Leuven, Belgium, 2001; pp. 529–536. [CrossRef]
43. Karayannis, M.I. Kinetic determination of ascorbic acid by the 2,6-dichlorophenolindophenol reaction with a stopped-flow technique. *Anal. Chim. Acta* **1975**, *76*, 121–130. [CrossRef]
44. Turhan, A.; Kuscu, H.; Ozmen, N.; Serbeci, M.S.; Demir, A.O. Effect of different concentrations of diluted seawater on yield and quality of lettuce. *Chil. J. Agric. Res.* **2014**, *74*, 111–116. [CrossRef]
45. Ferreira, R.L.F.; Souza, R.J.; de Carvalho, J.G.; Neto, S.E.D.; Yuri, J.E. Evaluation lettuce cultivars fertilizer with silifertil (R). *Rev. Caatinga* **2009**, *22*, 5–10.
46. Esmaili, M.; Mashal, M.; Aliniaeifard, S.; Urrestarazu, M.; Carrillo, F.F. Impact of Silicon on Chemical Properties of Drainage Water from Lettuce Following Determination of Proper Cultivar and Light Spectrum. *Commun. Soil Sci. Plant Anal.* **2021**, *52*, 756–768. [CrossRef]
47. Luz, J.; Guimaraes, S.T.M.R.; Korndörfer, G.H. Produção hidropônica de alface em solução nutritiva com e sem silício. *Hortic. Bras.* **2006**, *24*, 295–300. [CrossRef]
48. De Souza, R.S.; Rezende, R.; de Freitas, P.S.L.; Goncalves, A.C.A.; Rezende, G.S. Dry matter production and macronutrient leaf composition in lettuce under fertigation with nitrogen, potassium and silicon. *Rev. Bras. Eng. Agric. Ambient.* **2015**, *19*, 1166–1171. [CrossRef]
49. Ferreira, R.L.F.; de Souza, R.J.; de Carvalho, J.G.; de Araujo, S.E.D.; Mendonca, V.; Wadt, P.G.S. Evaluation of lettuce cultivars fertilized with calcium silicate in greenhouse. *Cienc. Agrotecnol.* **2010**, *34*, 1093–1101. [CrossRef]
50. Pinho, P.; Jokinen, K.; Halonen, L. The influence of the LED light spectrum on the growth and nutrient uptake of hydroponically grown lettuce. *Lighting Res. Technol.* **2017**, *49*, 866–881. [CrossRef]

plants

MDPI

Article

Increased Plant Quality, Greenhouse Productivity and Energy Efficiency with Broad-Spectrum LED Systems: A Case Study for Thyme (*Thymus vulgaris* L.)

Jenny Manuela Tabbert [1,2,*], Hartwig Schulz [1,3] and Andrea Krähmer [1,*]

1 Plant Analysis and Storage Product Protection, Institute for Ecological Chemistry, Julius Kühn Institute—Federal Research Centre for Cultivated Plants, Königin-Luise-Str. 19, 14195 Berlin, Germany; hs.consulting.map@t-online.de

2 Institute of Pharmacy, Freie Universität Berlin, Königin-Luise-Str. 2–4, 14195 Berlin, Germany

3 Consulting & Project Management for Medicinal and Aromatic Plants, Waltraudstraße 4, 14532 Stahnsdorf, Germany

* Correspondence: jenny.tabbert@julius-kuehn.de (J.M.T.); andrea.kraehmer@julius-kuehn.de (A.K.); Tel.: +49-308-304-2210 (A.K.)

Abstract: A light-emitting diode (LED) system covering plant-receptive wavebands from ultraviolet to far-red radiation (360 to 760 nm, "white" light spectrum) was investigated for greenhouse productions of *Thymus vulgaris* L. Biomass yields and amounts of terpenoids were examined, and the lights' productivity and electrical efficiency were determined. All results were compared to two conventionally used light fixture types (high-pressure sodium lamps (HPS) and fluorescent lights (FL)) under naturally low irradiation conditions during fall and winter in Berlin, Germany. Under LED, development of *Thymus vulgaris* L. was highly accelerated resulting in distinct fresh yield increases per square meter by 43% and 82.4% compared to HPS and FL, respectively. Dry yields per square meter also increased by 43.1% and 88.6% under LED compared to the HPS and FL lighting systems. While composition of terpenoids remained unaffected, their quantity per gram of leaf dry matter significantly increased under LED and HPS as compared to FL. Further, the power consumption calculations revealed energy savings of 31.3% and 20.1% for LED and FL, respectively, compared to HPS. In conclusion, the implementation of a broad-spectrum LED system has tremendous potential for increasing quantity and quality of *Thymus vulgaris* L. during naturally insufficient light conditions while significantly reducing energy consumption.

Keywords: light-emitting diode; daily light integral; volatile organic compounds; energy consumption; plant morphology; biomass efficacy

Citation: Tabbert, J.M.; Schulz, H.; Krähmer, A. Increased Plant Quality, Greenhouse Productivity and Energy Efficiency with Broad-Spectrum LED Systems: A Case Study for Thyme (*Thymus vulgaris* L.). *Plants* **2021**, *10*, 960. https://doi.org/10.3390/plants10050960

Academic Editors: Valeria Cavallaro and Rosario Muleo

Received: 26 April 2021
Accepted: 8 May 2021
Published: 12 May 2021

1. Introduction

Insufficient natural light intensities and short photoperiods drastically limit plant development during winter months in northern regions. Although most common horticultural crops depend on daily light integrals (DLIs) of 6 to 50 mol m^{-2} d^{-1} [1], outdoor solar DLIs often do not exceed 10 mol m^{-2} d^{-1} in higher latitudes during light-limited winter months [2] and are further reduced by up to 60% inside greenhouses [3–5]. Therefore, greenhouse industries and research facilities seasonally apply supplemental light sources to prolong cultivation periods and optimize plant growths. However, potentials for (year-round) horticultural productions remain under-utilized, as traditional light sources consume unfeasible amounts of energy [6] and are not tailored to the plants' photoreceptors [7]. Hence, new technology, which significantly reduces electricity consumption while improving crop value, is of great interest to greenhouse industry and research facilities [8].

Today, light-emitting diodes (LEDs) have the potential to replace traditional light sources such as high-pressure sodium lamps (HPS) [9,10] and fluorescent lights (FL) [11]. They show important technical advantages such as high energy efficiency, small size, durability, long operating lifetime, low thermal emission, and adjustable spectral wavelength range (reviewed in [12–14]). Consequently, the utilization of LED technology as horticultural lighting increases [15].

However, the majority of LED radiation studies on plant development have only included narrow wavebands of red (R) and blue (B) light, as these wavelengths are maximally absorbed by the plant's light-capturing chlorophylls [16]. Initial LED plant-lighting research proved that plants could complete their life cycle with R light alone [17], but the plants' morphogenesis including compact growth and leaf expansion, as well as plants' flowering, were significantly improved when differing proportions of B light were included [18–22]. Additionally, specific B light proportions positively influence physiological plant responses such as stomatal opening, chlorophyll contents, and secondary metabolism [8,22].

Recently, studies have suggested further photosynthetic improvements by adding far-red (FR) wavelengths to R spectra, for example, increasing FR radiations promoted growth of seedlings by increasing leaf expansion and whole-plant net assimilation, decreased anthocyanins and carotenoids, and reduced antioxidant potentials [23–25].

As recent studies confirm, green (G) light can also contribute to plant development and growth [26–28]. Enhanced lettuce growth under RB illumination complemented with G light and improved cucumber growth under HPS supplemented with G light have been reported [29–31]. However, G light stimulates early stem elongation and stomatal closure, antagonizing the typical blue-light mediated growth inhibition and stomatal opening [32–34].

Due to the multitude of photobiological studies conducted, it is now well established that wavelengths between ~ 360 and 760 nm influence plants' photosynthesis, physiology, morphogenesis, and phytochemical contents [7] and that specific spectral regions can be used to induce specific plant traits of interest.

Nevertheless, negative side effects resulting from narrow waveband LED applications, such as unwanted photomorphogenic and physiological disorders, pest and disease pressures, as well as difficult visual assessment of plant-status absent under (natural) broad light spectra, have to be further minimized [17,35].

In consequence, LED fixtures with broader spectral quality covering the range of the photosynthetically active radiation (PAR) between 400 and 700 nm (perceived as white light) sometimes including the flanking regions of UV (~360–400 nm) and FR (~600–760 nm) radiation are emerging recently [36] and are becoming popular as sole-source lighting for horticulture [37,38].

For example, Spalholz et al. (2020) compared the response of two lettuce cultivars to a sun-simulated spectrum and other commonly applied B:R spectra, providing a biologically active radiation between 300–800 nm of 200 μmol m^{-2} s^{-1} [39]. The study elucidated unique responses including greatest fresh-to-dry mass ratio, greater leaf area, excessive stem extension, and flower initiation under the sun-simulated spectrum despite a 36% greater photosynthetic photon flux density (PPFD) in B:R treatments. Coinciding results were published by Gao and coworkers (2020), who tested the effects of white and different monochromatic (B, G, Y, R) LEDs on Welsh onions [40]. In addition to increased plant yield, net photosynthetic rate and photosynthetic efficiency were significantly higher under white light than under those of the monochromatic light treatments. Matysiak and Kowalski (2019) observed greatest fresh weights under W and R light treatment for lamb's lettuce and garden rocket, whereas for two sweet basil cultivars, no differences in fresh weight were detected under all tested light treatments [5]. However, supplementation with B resulted in more compact growth of green-leaved basil. For red pak choi, a white light including UV and FR was evaluated as ideal for best overall yield performance [41], and the importance of white light on shoot and root fresh weights of lettuce was demonstrated [42].

Thus, it has been found that broad LED spectra, covering a wider plant-receptive spectral range rather than single narrow bands, and at best including flanking regions in the FR and UV, can lead to greater plant development. So far, however, such a broad LED spectrum has not been tested under insufficient light conditions in greenhouses. Therefore, our aim was to evaluate the advantages and disadvantages of broadband LED lighting during the winter season in northern central Europe (Berlin, Germany, 52.5° N, 1.33° E) in a practical case study and to compare the results with the common HPS and FL setups found in the greenhouse industry and research facilities today.

As a model plant, we chose moderately light-dependent *Thymus vulgaris* L., which belongs to the Lamiaceae family rich in other genera such as *Salvia* and *Organum* [43] and which is widely used in European cuisine and folk medicine for its expectorant, antitussive, antibroncholitic, antispasmodic, antimicrobial, antioxidant, anti-inflammatory, anthelmintic, carminative, and diuretic properties. The major bio-active metabolite responsible for the therapeutic properties of aromatic *Thymus vulgaris* L. is the monoterpene thymol [44].

The aim of this study was to conduct a greenhouse experiment during winter in order to assess the development, biomass, and health-promoting terpenoid yields of Thymus under a prototype broad-spectrum LED, as well as to obtain the prototypes' power consumption and efficacy. To further evaluate the practical applicability and potential for greenhouse businesses and research facilities, we aimed at comparing the broad-spectrum LED results with results assessed under HPS and FL fixtures under their common setup conditions.

2. Results and Discussion

2.1. Biomass Yield, Partitioning, and Morphology

The LED system resulted in distinct yield increases, biomass partitioning, and a differentiated morphological appearance of *Thymus vulgaris* L. in comparison to the HPS and FL systems (Figures 1 and 2). While the LEDs produced a fresh biomass of averagely 28.1 $\pm$ 2.0 g plant^{-1}, the HPS systems accounted for a fresh biomass of 15.9 $\pm$ 2.3 g plant^{-1} within the same cultivation period. The lowest fresh biomass of 4.9 $\pm$ 0.4 g plant^{-1} was produced under FL (Figure 1A). Accordingly, dry matter yields of *Thymus vulgaris* L. were significantly enhanced by the LED system (5.6 $\pm$ 0.8 g plant^{-1}) in comparison to HPS (3.2 $\pm$ 0.5 g plant^{-1}) and FL (0.6 $\pm$ 0.1 g plant^{-1}), representing an increase of 1.75- and eight-fold, respectively (Figure 1B). Thereby, the weight proportion of dry leaves did not differ from the (mostly lignified) weight proportion of stems in thyme plants cultivated under the LED and HPS systems, respectively. Under FL, however, the majority of dry yield consisted of leaves (83.3%) and only 16.7% consisted of (unwooded) shoots (Figure 1C). With a Pearson correlation coefficient of r = 0.97 and R^2 = 0.95 (p < 0.001), dry mass yields under the differing supplemental lighting systems were highly related to the individual daily light integrals (DLI).

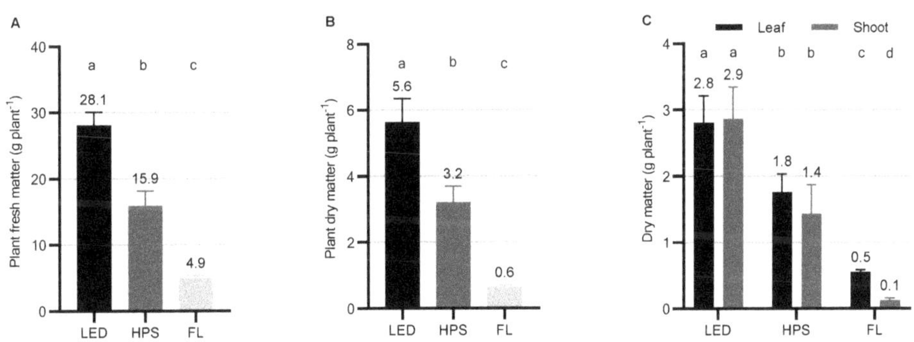

Figure 1. Biomass yields and partitioning of *Thymus vulgaris* L. cultivated under different supplemental lighting systems during fall and winter in Berlin, Germany. LED = light-emitting diode, HPS = high-pressure sodium lamp, FL = fluorescent

light. (**A**) Fresh matter yields in gram per plant *, (**B**) Dry matter yields in gram per plant *, (**C**) Leaf and shoot dry matter partitioning in gram per plant **. * Presented are mean plant yields of four independent spatial replications per light treatment ($n = 4$) ± standard deviation (SD) of 32 harvested plants per spatial replication and light treatment (N = 384, $n = 128$ plants per supplemental light treatment, $n = 32$ plants per spatial replication). Significant differences ($p \leq 0.01$) were determined according to Dunnett's T3 multiple comparisons test after Brown-Forsythe and Welch ANOVA test ($p \leq 0.001$). Different letters indicate significant differences. ** Presented are mean dry leaf and shoot matter yields of four independent spatial replications per light treatment ($n = 4$) ± SD (standard deviation) of 16 harvested plants per spatial replication and light treatment (N = 192, $n = 64$ plants per supplemental light treatment, $n = 16$ plants per spatial replication). Significant differences ($p \leq 0.05$) were determined according to Dunnett's T3 multiple comparisons test after Brown-Forsythe and Welch ANOVA test ($p \leq 0.001$). Different letters indicate significant differences.

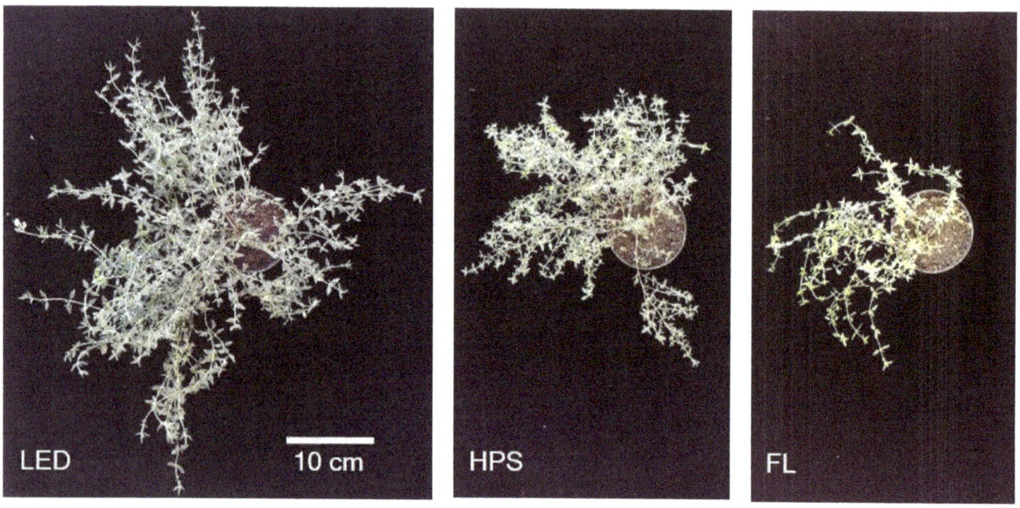

Figure 2. Visual appearance of *Thymus vulgaris* L. at harvest cultivated under different supplemental lighting systems during fall and winter of Berlin, Germany. LED = light-emitting diode, HPS = high-pressure sodium lamp, FL = fluorescent light.

As indicated in Figure 2, the stem biomass of *Thymus vulgaris* L. was greatly increased under the LED system at the end of the experimental period and led to a profoundly different visual appearance in comparison to thyme plants grown under the other two supplemental lighting fixtures. Despite the lowest corresponding leaf-to-shoot ratio, which was 0.9 for LED, 1.3 for HPS, and 5 for FL, the leaf dry matter (L_{DM}) of thyme was significantly increased and highest under LED (Figure 1C).

The reason for the outstanding biomass accumulations and the concomitant rapid thyme development under the LED system is clearly found in the heightened DLI between 400 and 700 nm, as shown by the correlation coefficient of $r = 0.97$ ($R^2 = 0.95$). That increasing DLIs accelerate the development and growth of plants up to a certain point is well established [45,46]. The correlation of DLIs and plant growth is known to be linear between each species-specific light compensation point and light saturation point [7].

Faust stated that optimal DLIs vary from 6 to 50 mol m^{-2} d^{-1} for various crops, and moderately light-dependent thyme requires a DLI of at least 18 mol m^{-2} d^{-1} [46]. The natural average DLI in greenhouses during winter in northern latitudes however is often as low as 1 to 5 mol m^{-2} d^{-1} and reached approximately 3.9 mol m^{-2} d^{-1} during our greenhouse trial [3–5]. Hence, supplemental lighting is essential for winter greenhouse productions. Since FLs raised the total DLI (natural DLI 3.9 mol m^{-2} d^{-1} + supplemental DLI 3 mol m^{-2} d^{-1}) only to approximately 7 mol m^{-2} d^{-1} during winter production, the FLs are neither suitable for the production of thyme nor presumably for the majority of greenhouse crops under the given cultivation conditions. HPS elevated the total DLI (natural DLI 3.9 mol m^{-2} d^{-1} + supplemental DLI 7 mol m^{-2} d^{-1}) to an estimated level of 11 mol m^{-2} d^{-1}. Therewith, the biomass accumulation of *Thymus vulgaris* L. increased significantly in comparison to FL; however, the DLI remains insufficient for an optimal thyme production during winter. With a total DLI of approximately 16 mol m^{-2} d^{-1} (natural DLI 3.9 mol m^{-2} d^{-1} + supplemental DLI 11 mol m^{-2} d^{-1}) during the low light season, the tested LED system achieved the highest DLI and approached the recommended DLI of $\geq$ 18 mol m^{-2} d^{-1} the most (Table 1, Section 3. Material and Methods). Further, only the LED system would be able to achieve the recommended DLI of the moderately light-dependent thyme by simply extending the photoperiod from 14 to 16 h per day during winter.

Table 1. Spectral composition of the supplemental lighting fixtures used in the greenhouse for the cultivation of thyme (*Thymus vulgaris* L.).

Parameter *	Supplemental Light Fixtures **					
	LED		HPS		FL	
	µmol m^{-2} s^{-1}	% ***	µmol m^{-2} s^{-1}	% ***	µmol m^{-2} s^{-1}	% ***
PPFD (400–700 nm)	212	91.2	132	92.5	57	95
PFD (360–760 nm)	232	100	143	100	60	100
PFD-Ultraviolet (360–399 nm)	1.7	0.7	1.6	1.1	0.6	1.0
PFD-Blue (400–519 nm)	65.7	28.4	16.9	11.9	8.2	13.6
PFD-Green (520–559 nm)	33.5	14.5	7.0	4.9	13.5	22.4
PFD-Yellow (560–624 nm)	56.9	24.5	83.6	58.7	27.9	46.4
PFD-Red (625–700 nm)	55.3	23.9	24.3	17.1	7.6	12.7
PFD-Far Red (701–760 nm)	18.7	8.1	9.0	6.4	2.4	3.9
R/FR ratio (660/730 nm) ‡	2.8		2.4		0.1	
DLI (mol m^{-2} d^{-1}) ±	10.6/11.7		6.6/7.2		2.9/3.0	

* *PPFD* = photosynthetic photon flux density, *PFD* = photon flux density, *R/FR* ratio = red to far-red ratio, DLI = daily light integral. ** LED = light-emitting diode, HPS = high-pressure sodium lamp, FL = fluorescent light. *** Values represent percentages of total *PFD*. ‡ R/FR ratio is based on the absorption maxima of phytochromes at 660 and 730 nm [47]. ± DLI based on *PPFD/PFD*.

2.2. Content and Composition of Volatile Organic Compounds (VOCs)

Applying GC-MS analysis, 12 monoterpenes and one sesquiterpene were identified in the leaf extracts of *Thymus vulgaris* L., representing $\geq$ 94% of all detected volatile constituents. Major identified volatile organic compounds (VOCs) in the leaf extracts of *Thymus vulgaris* L. under the supplemental lighting systems were thymol, γ-terpinene, and *p*-cymene, respectively, which is consistent with the results of former research [48,49]. Thereby, the chemical makeup remained unaffected by the different lighting systems. The total content of VOCs per g of L_{DM} is highly enhanced by LED (2.7%) and HPS (2.3%) as compared to by FL (1.1%). The difference in quantity of VOCs per g of L_{DM} between thyme plants cultivated under LED and HPS is not significant (p = 0.088). The LED considerably increased the amounts of all 13 evaluated terpenoids in the leaves of *Thymus vulgaris* L. in contrast to the FL system. The HPS system also enabled considerable increases in comparison to the FL system, even though the differences between the amounts of γ-terpinene and borneol are less profound, with p = 0.099 and p = 0.075, respectively. Differences between LED and HPS treatments were only detected for α-pinene, while myrcene (p = 0.077) as well as limonene (p = 0.057) differed only in tendency. All results are summarized in Table 2.

Table 2. Effect of three different supplemental lighting systems on the chemical composition of 13 main volatile organic compounds (VOCs) of *Thymus vulgaris* L. cultivated in the greenhouse during fall and winter of Berlin, Germany.

Compound	RI *	Light-Emitting Diode (LED)		High-Pressure Sodium Lamp (HPS)		Fluorescent Light (FL)	
		% **	μg 100 mg^{-1} LDM ***	%	μg 100 mg^{-1} LDM	%	μg 100 mg^{-1} LDM
monoterpene hydrocarbons							
α-pinene	938.0 ± 0.4	0.8 ± 0.1	14.03 ± 0.8 ^{a}	0.8 ± 0.1	11.41 ± 1.1 ^{b}	1.0 ± 0.1	7.22 ± 1.1 ^{c}
sabinene	977.8 ± 0.4	1.4 ± 0.4	34.21 ± 8.5 ^{a}	1.5 ± 0.6	33.57 ± 6.4 ^{a}	1.3 ± 0.5	13.95 ± 1.5 ^{b}
myrcene	991.8 ± 0.3	1.6 ± 0.2	35.16 ± 1.8 ^{a}	1.5 ± 0.2	28.50 ± 3.6 ^{a}	1.8 ± 0.2	17.19 ± 1.6 ^{b}
α-terpinene	1020.8 ± 0.3	1.9 ± 0.4	39.47 ± 4.3 ^{a}	2.1 ± 0.7	33.43 ± 3.2 ^{a}	2.2 ± 0.8	19.72 ± 2.1 ^{b}
p-cymene	1029.1 ± 0.5	8.5 ± 2.2	157.70 ± 27.4 ^{a}	8.3 ± 2.5	123.90 ± 8.0 ^{a}	6.9 ± 2.5	55.16 ± 4.7 ^{b}
limonene	1033.1 ± 0.3	0.6 ± 0.1	9.81 ± 0.7 ^{a}	0.6 ± 0.1	8.22 ± 0.8 ^{a}	0.6 ± 0.1	4.59 ± 0.5 ^{b}
γ-terpinene	1064.4 ± 0.8	15.2 ± 5.4	323.20 ± 56.7 ^{a}	16.0 ± 4.7	259.90 ± 49.0 ^{ab}	20.1 ± 6.3	179.50 ± 22.6 ^{b}
oxygenated monoterpenes							
cis-sabinene hydrate	1071.7 ± 0.4	1.3 ± 0.1	28.41 ± 2.3 ^{a}	1.4 ± 0.2	25.38 ± 1.6 ^{a}	1.3 ± 0.3	12.31 ± 0.8 ^{b}
linalool	1100.9 ± 0.5	2.6 ± 0.7	61.92 ± 11.3 ^{a}	2.7 ± 0.8	52.22 ± 11.2 ^{a}	2.3 ± 0.9	21.25 ± 1.9 ^{b}
borneol	1173.6 ± 0.3	0.9 ± 0.4	44.44 ± 2.6 ^{a}	1.0 ± 0.6	46.50 ± 5.9 ^{ab}	1.5 ± 0.7	31.75 ± 3.7 ^{b}
thymol	1297.6 ± 1.8	54.6 ± 6.9	1134.00 ± 86.3 ^{a}	52.9 ± 6.4	917.10 ± 142.9 ^{a}	50.2 ± 7.9	429.90 ± 57.3 ^{b}
carvacrol	1304.6 ± 1.1	2.4 ± 0.4	60.82 ± 4.9 ^{a}	2.3 ± 0.5	48.82 ± 8.2 ^{a}	2.0 ± 0.6	21.42 ± 2.9 ^{b}
sesquiterpene hydrocarbons							
β-caryophyllene	1436.55 ± 0.5	3.1 ± 1.3	47.69 ± 7.1 ^{a}	3.67 ± 1.1	49.50 ± 7.0 ^{a}	3.32 ± 1.1	21.91 ± 3.4 ^{b}
% of total extract **		94.93 ± 1.50		94.68 ± 2.22		94.53 ± 1.95	
Total VOCs [% g^{-1} LDM] ****		2.7 ± 0.22 ^{a}		2.3 ± 0.25 ^{a}		1.1 ± 0.10 ^{b}	

* Retention indices (RI) relative to C_6-C_{24} n-alkanes on a HP-5MS column for compound identification. Indices are presented as means ± *SD* with $n = 192$. ** Percentages were calculated from GC-FID TIC data after weight correction and presented as means ± *SD* with $n = 64$. *** Amounts of major compounds were calculated based on density corrected calibration functions obtained from reference standards analyzed under the same GC-FID conditions as the samples. Presented are mean amounts of volatile compounds (μg 100 mg^{-1} L_{DM} (=leaf dry matter)) of four independent spatial replications per light treatment ($n = 4$) ± *SD* of 16 collected dried leaf samples per spatial replication and light treatment ($N = 192$, $n = 64$ dry leaf samples per supplemental light treatment, $n = 16$ dry leaf samples per spatial replication). Significant differences ($p \leq 0.05$) were determined according to Dunnett's T3 multiple comparisons test after Brown-Forsythe and Welch ANOVA test ($p \leq 0.02$). Different letters within a row indicate significant differences at $p \leq 0.05$, and bold amounts indicate significant differences at $p \leq 0.1$. **** Percentage of total VOCs (volatile organic compounds) was calculated based on the results of the internal standard (6-methyl-5-penten-2-one), which was co-analyzed in each sample. Presented are mean percentages per g L_{DM} (% g^{-1} L_{DM} (=leaf dry matter)) of four independent spatial replications per light treatment ($n = 4$) ± *SD* of 16 collected dried leaf samples per spatial replication and light treatment ($N = 192$, $n = 64$ dry leaf samples per supplemental light treatment, $n = 16$ dry leaf samples per spatial replication). Significant differences ($p \leq 0.05$) were determined according to Dunnett's T3 multiple comparisons test after Brown-Forsythe and Welch ANOVA test ($p \leq 0.001$). Different letters within a row indicate significant differences at $p \leq 0.05$ and bold amounts indicate significant differences at $p \leq 0.1$.

Gouinguene and Turlings (2002) showed in their study that young corn plants (*Zea mays* L.) significantly increased their emissions of volatiles as light intensity increased up to 10,000 lm [50]. However, beyond 10,000 lm volatile emissions in *Zea mays* L. did not enhance any further, suggesting a kind of saturation or limitation was reached. The authors of [51,52] also detected this proposed light quantity-dependency. Multiple studies also suggest that terpene synthesis involves phytochromes [51–54], red and far-red light-sensing photoreceptors (reviewed by [55]), making the production of terpenes also dependent on light quality, specifically on the *R/FR* ratio [56]. For example, in thyme seedlings, red light strongly promoted the production of mono- and sesquiterpenes (thymol, γ-terpinene, *p*-cymene and carvacrol, β-caryophyllene) and the number of essential oil-containing trichomes per cotyledone, two stimulatory effects that proved to be completely reversible by a subsequent exposure to far-red irradiation [53,54]. Later, a partial reduction of volatile emissions was detected in *Arabidopsis thaliana* (L.) Heynh exposed to a low *R/FR* ratio of 0.2 as compared to plants exposed to a high *R/FR* ratio of 2.2 when controlling for light intensity [56]. These findings could explain the comparatively low VOC contents detected in thyme leaves grown under FL, as both their light intensity (*PFD* 60 µmol m^{-2} s^{-1}, *PFD*-R 7.6 µmol m^{-2} s^{-1}) as well as their *R/FR* ratio (0.1) were significantly reduced as compared to LED and HPS under our experimental conditions. It would also suggest that the similar contents of VOCs per gram of thyme leaves found under LED and HPS are the result of their similar high *R/FR* ratios of 2.8 and 2.4, respectively. However, *R/FR* ratios dramatically decline under vegetational canopies. As described by Franklin (2008), a single leaf reduces a given *R/FR* ratio of 1.2 to 0.2 for the leaves growing underneath, and the ratio reduces further to 0.1 underneath a second leaf [55]. As leaf and shoot yields of *Thymus vulgaris* L. significantly increased under the LED system, it is reasonable to believe that the actual *R/FR* ratio underneath the more densely stands of thyme plants grown under the LED system was much lower than under the less dense canopy of thyme plants grown under HPS. This idea coincides with the result from Kegge et al. (2013), who detected a reduction of volatile emissions in plants grown in high density stands [56]. Another explanation for the similar contents of VOCs per gram of thyme leaves found under LED and HPS may be found in the high B light proportion found under the broad LED light spectrum, as it was recently shown that essential oil contents of *Thymus vulgaris* L. decrease with increasing proportions of blue light [57]. The associated suppressions of terpene synthesis under low *R/FR* ratios as well as under low *R/B* ratios may have been partially compensated by the LEDs' elevated light intensity (*PFD* 232 µmol m^{-2} s^{-1}, *PFD*-R 55.3 µmol m^{-2} s^{-1}) compared to the intensity of the HPS system (*PFD* 143 µmol m^{-2} s^{-1}, *PFD*-R 24.3 µmol m^{-2} s^{-1}) in our study. Additionally, though air temperatures under the given experimental conditions did not differ between the supplemental lighting systems, it is known that leaf temperature increases under HPS lights as comparted to other lighting systems [58]. As elevated temperatures evidently increase the emission of volatiles [50], a greater leaf temperature under HPS may have been present and contributed to the terpene synthesis in HPS-grown thyme plants. Further, as we did not adjust fertilization, though the LED system yielded much greater biomasses than FL and HPS, it is plausible that a reduced nutrient availability for LED-grown thyme plants limited their production of VOCs, as demonstrated by Gouinguene, and Turlings (2002), who showed that fertilization rate positively effects volatile emissions [50].

Nevertheless, as the LED lights were able to increase the production of volatiles in thyme leaves significantly compared to the HPS lights, the LEDs' volatile productivity per square meter doubled in absolute terms (2.5 vs. 1.3 g m^{-2}) with $p < 0.06$ (Table 3).

2.3. Productivity

The broad-spectrum LED system enabled a highly significant increase in leaf and stem production of fresh thyme per square meter, representing increases of 43.3% and 82.4% in comparison to the HPS and FL system, respectively. Additionally, the dry matter productions of HPS and FL were highly reduced by 43.1% and 88.6% in comparison to the LED system. Further, the LED system enabled an increase in production of VOCs per square meter under the given greenhouse conditions in comparison to the conventionally used HPS system (at *p*-value of 0.051). Both systems (LED and HPS) considerably promoted the VOC production in comparison to the FL system. Table 3 summarizes the results. Despite the lower leaf-to-shoot ratio for LED lightning of 0.9 as compared to both HPS (1.3) and FL (5, see Section 2.1), absolute L_{DM} and overall quantity of VOCs were highest for LED. Therefore, LED lightning offers an attractive alternative for thyme cultivation, both for essential oil production and delivery to fresh market.

Table 3. Fresh and dry plant production as well as content of volatile fraction of thyme (*Thymus vulgaris* L.) per m^2 under three supplemental lighting systems.

Light Fixture *	FM ** per Square Meter [g m^{-2}] ***	DM ** per Square Meter [g m^{-2}] ***	VOC ** per Square Meter [mg m^{-2}] ****
LED	897.9 ± 64.65^a	180.2 ± 24.69 ^{a}	**2472 ± 626.4 ^{a}**
HPS	509.4 ± 72.88 ^{b}	102.6 ± 16.87 ^{b}	**1273 ± 334.0 ^{a}**
FL	158.0 ± 6.73 ^{c}	20.62 ± 2.06 ^{c}	199.1 ± 30.98 ^{b}

* LED = light-emitting diode, HPS = high-pressure sodium lamp, FL = fluorescent light. ** FM = total fresh matter, DM = total dry matter, VOC = total content of volatile organic compounds of total leaf dry matter. *** Presented data are means of cumulated fresh and dry matter productions of four independent spatial replications per light treatment (n = 4) ± SD of 32 harvested plants per spatial replication and light treatment (N = 384, n = 128 plants per supplemental light treatment, n = 32 plants per spatial replication). Significant differences ($p \leq 0.01$) were determined according to Dunnett's T3 multiple comparisons test after Brown-Forsythe and Welch ANOVA test ($p \leq 0.001$). Different letters indicate significant differences. **** Presented data are means of cumulated volatile productions in thyme leaves of four independent spatial replications per light treatment (n = 4) ± SD of 16 harvested plants per spatial replication and light treatment (N = 192, n = 64 dry leaf samples per supplemental light treatment, n = 16 dry leaf samples per spatial replication). Significant differences were determined according to Dunnett's T3 multiple comparisons test after Brown-Forsythe and Welch ANOVA test ($p \leq 0.002$). Different letters within the column indicate significant differences at $p \leq 0.02$, and bold amounts indicate a difference by trend at $p < 0.06$.

2.4. Power Consumption and Biomass Efficiacy

The LEDs consumed the least electricity with 257.7 W m^{-2}, followed by the FLs with the use of 299.4 W m^{-2}, whereas the HPS lamp consumed the highest amount of electricity with 374.9 W m^{-2}. At the end of the cultivation period, the power consumptions per m^2 of LED and FL lighting system resulted in high energy savings of 31.3% and 20.1%, respectively, when compared to the consumption of the HPS system. While each LED system enabled ± 1.92 g of fresh thyme per kWh and square meter, the HPS and FL enabled only ± 40% and ± 16% of these yields per kWh and square meter, respectively. Accordingly, the dry thyme production per kWh and square meter under LED (±396.3 mg) was significantly higher than the dry thyme production under HPS (±155.2 mg) and FL (±39.1 mg). Further, the production of VOCs per kWh and square meter was significantly elevated underneath the LEDs (±5.4 mg) as compared to HPS (±1.9 mg) and FL (±0.4 mg). Results and calculations are combined in Table 4.

Table 4. Power consumption per square meter of the supplemental lighting fixtures for the production of thyme (*Thymus vulgaris* L.) grown in a greenhouse during fall and winter of Berlin, Germany.

Light Fixture *	Power Consumption per Meter^2 [W m^{-2}]	Power Consumption for Thyme Cultivation [kWh m^{-2}]	Power Savings Compared to HPS [%]	Fresh Thyme Production ** [g kWh^{-1} m^{-2}]	Dry Thyme Production ** [mg kWh^{-1} m^{-2}]	VOC Production *** [mg kWh^{-1} m^{-2}]
LED	257.7	454.6	31.3	1.92 ± 0.15 ^{a}	396.3 ± 54.31 ^{a}	5.4 ± 1.4^a
HPS	374.9	661.3	*na*	0.77 ± 0.11 ^{b}	155.2 ± 25.5 ^{b}	1.9 ± 0.5 ^{b}
FL	299.4	528.1	20.1	0.30 ± 0.03 ^{c}	39.1 ± 3.9 ^{c}	0.4 ± 0.1 ^{c}

* LED = light-emitting diode, HPS = high-pressure sodium lamp, FL = fluorescent light, *na* = not applicable. ** Presented are calculated average fresh and dry thyme productions per power consumption of each light fixture type within a square meter during the cultivation period (g or mg per kWh and m^2) of four independent spatial replications per light treatment (n = 4) ± SD of 32 harvested plants per spatial replication and light treatment (N = 384, n = 128 plants per supplemental light treatment, n = 32 plants per spatial replication). Significant differences ($p \leq 0.01$) were determined according to Dunnett's T3 multiple comparisons test after Brown-Forsythe and Welch ANOVA test ($p \leq 0.001$). Different letters indicate significant differences. *** Presented are calculated average productions of volatile organic compounds per power consumption of each light fixture type within a square meter during the cultivation period (mg per kWh and m^2) of four independent spatial replications per light treatment (n = 4) ± SD of 16 harvested plants per spatial replication and light treatment (N = 192, n = 64 plants per supplemental light treatment, n = 16 plants per spatial replication). Significant differences ($p \leq 0.05$) were determined according to Dunnett's T3 multiple comparisons test after Brown-Forsythe and Welch ANOVA test ($p \leq 0.004$). Different letters indicate significant differences.

Our power consumption results and thus the potential of LEDs for reducing energy costs coincide with numerous studies and reviews [6,9,59], stating energy reductions up to 70% compared to traditional light sources while producing similar crop yields at equal light intensities, and confirm the current trend of LEDs' increasing photon efficiencies: While HPS and LED fixtures had nearly identical photon efficiencies until ~2015 [6,35], the best evaluated LED fixture was 40% more photon-efficient than HPS due to technological improvements of LEDs within the PAR region soon after [35,59]. A current study by Hernandez et al. (2020) confirms the corresponding increase in biomass efficacy of LEDs, as their LED treatment led to a 2.4 to 3.1 times greater biomass efficacy than HPS, which matches our findings [60]. Another study in which LED and FL treatments were compared, reported a biomass efficacy three to five times higher under LED than under FL lighting [61]. In contrast, the LED system used in this current study greatly exceeds their findings, as the LED enabled a biomass efficacy 6 to 10 times higher than the FL system (Table 4) under our experimental conditions.

Further, in our study, plant growth may have been limited by nutrient availability, and an adjustment of fertilization based on the differing thyme growth rates may further increase biomass efficacies under HPS and especially under the broad-spectrum LED system. Nevertheless, when using the broad-spectrum LED lighting system, the significantly more inhomogeneous light intensity distribution compared to HPS and FL lamps (Figure 3) must be taken into account when light uniformity is necessary for the greenhouse application as it demands more LED light fixtures per area.

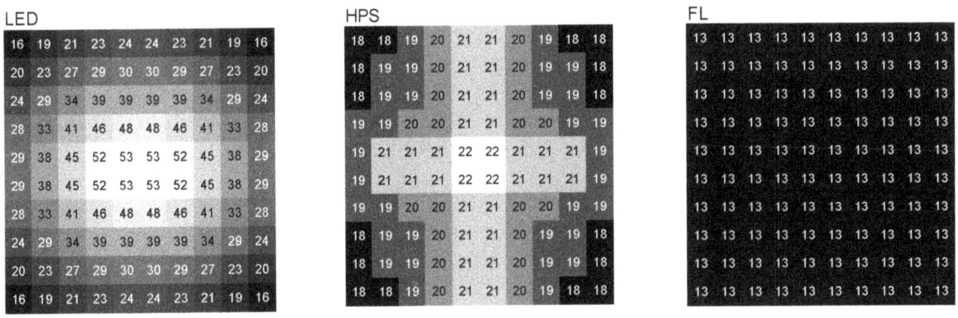

Figure 3. Irradiance profiles (W m^{-2} nm^{-1}) of the experimental plots (1 m^2) underneath each supplemental lighting system. (LED = light-emitting diode, HPS = high-pressure sodium lamp, FL = fluorescent light).

Nevertheless, at their edges, where the lowest light intensities occur, the LEDs achieve values of 16 W m^{-2} nm^{-1}, which are sufficient for high-quality thyme production. Therefore, if homogeneous plant development is not necessarily required, plants of marketable quality are also available with the LED setup used here without additional lamps.

3. Materials and Methods

3.1. Experimental Design

To investigate biomass yields and contents of volatile organic compounds (VOCs) of *Thymus vulgaris* L. grown under a broad-spectrum LED system, and to compare the lights' productivity as well as electrical efficiency with conventionally used lighting fixtures for the cultivation of thyme under naturally low irradiated greenhouse conditions during fall and winter in Berlin, Germany, a one-factorial experiment with a randomized block design with three different supplemental light sources and four spatially independent replications (N = 384; n = 32 thyme plants per replication) was conducted.

3.2. Lighting Systems and Illumination Conditions

Three different supplemental light sources ((1) fan-cooled light-emitting diode (LED) (SUNtec Technology, FUTURELED®, Berlin, Germany, dimensioning 47.5 × 21.5 × 19.5 cm^3), (2) high-pressure sodium (HPS) lamps (bulb: SON GreenPower CG T 400 W E40 1SL, PHILIPS, Hamburg, Germany; ballast: HST, SILL Leuchten®, Berlin, Germany, dimensioning 50 × 30 × 19 cm^3), and (3) fluorescent lamps (FL) (VENEDIG, Pracht®, Berlin, Germany, dimensioning 50 × 50 × 16 cm^3)) were horizontally mounted onto given steel frames 1.40 m above greenhouse benches, resulting in distances between the bottom of the LED, HPS, and FL light sources and the greenhouse benches of 1.14, 1.13, and 1.09 m, respectively. Based on weather recordings from WetterKontor [62], plants were exposed to an average of 2.5 h of sunshine per day during the experiment. In addition to the natural sunlight, plants were subject to supplemental lighting from 6:00 a.m. to 8:00 p.m. for a photoperiod of 14 h per day during the greenhouse experiment. Plastic sheeting extending from above the light fixtures to below the greenhouse benches eliminated neighboring light pollution.

3.3. Irradiance Profile Measurements

Irradiance measurements of the light fixtures were taken prior to the experiment using a spectral PAR meter (PG200N, UPRtek, Aachen, Germany) at night. Light intensity, spectral composition, and irradiance profiles (light distribution patterns) were measured and recorded at bench level under experimental conditions. The software package of the spectrometer (uSpectrum PC laboratory software) automatically calculated all electromagnetic parameters including photon flux density (PFD in $\mu mol\ m^{-2}\ s^{-1}$) and spectral irradiance ($W\ m^{-2}\ nm^{-1}$) between 360 and 760 nm and photosynthetic photon flux density (PPFD in $\mu mol\ m^{-2}\ s^{-1}$) between 400 and 700 nm. During a sunny day, light transmission between 350 and 800 nm of the natural irradiance through the greenhouse glass was determined to be 28% (±5%) by comparing the output of the spectrometer inside the greenhouse with the output outside the greenhouse and resulted in a photon flux density of ~434 $\mu mol\ m^{-2}\ s^{-1}$ at bench level, which amounts to an approximate natural daily light integral of 3.9 $mol\ m^{-2}\ d^{-1}$ when combined with the weather recordings (Section 3.2). Light spectra and detailed spectral compositions of the supplemental lighting systems are depicted in Figure 4 and summarized in Table 1, respectively. Under each lighting system, the irradiance profiles within one square meter (representing the replicated experimental plots during the greenhouse experiment) were measured over a flat plane below the fixtures at intervals of 10 cm between 360 and 760 nm. Each measurement represents the spectral irradiance in $W\ m^{-2}\ nm^{-1}$ and was replicated three times and averaged, leading to a total dataset of 100 measurements per square meter. These irradiance profiles are depicted in Figure 3.

3.4. Plant Material and Growth Conditions

Seeds of *Thymus vulgaris* L. (Rühlemann's Kräuter- und Duftpflanzen, Horstedt, Germany) were sown in 128-cell plug trays (Ø 4 cm) filled with potting substrate (Fruhstorfer Einheitserde Typ P, HAWITA, Vechta, Germany) on 9 October 2018 and placed under the differing lighting fixtures in a NS-orientated greenhouse located in Berlin, Germany (52.5° N, 13.3° E). After six weeks on 20 November 2018, 32 representative seedlings were transplanted into pots (Ø 9 cm) containing substrate with an elevated nutrient composition (Fruhstorfer Einheitserde Typ T, HAWITA, Vechta, Germany) and evenly placed (quadratic) within 1 m^2 under each light fixture. Each treatment was replicated four times, positioned in a randomized block design, and surrounded by 28 border plants to avoid boundary effects. Starting 27 December 2018, plants were fertilized weekly with 100 mL of a 0.2% (v/v) nutrient solution (Hakaphos® Blau, COMPO EXPERT®, Muenster, Germany) containing 15% N, 10% P_2O_5, 15% K_2O, 2% MgO, 0.01% B, 0.02% Cu, 0.075% Fe, 0.05% Mn, 0.001% Mo, and 0.015% Zn. However, due to the low biomass accumulation of the plants grown under the fluorescent lamp system, the plant fertilization was started two weeks later under the FL treatment. All thyme plants grew for a total period of 18 weeks until harvest on 12 February 2019. Climatic conditions including temperature and relative humidity of the greenhouse air were continuously monitored at canopy level via data loggers (EL-USB−2, Lascar, CONRAD, Hirschau, Germany). Average temperatures (°C ± SD) under LED, HPS, and FL lighting were 20.4 ± 1.4, 21.1 ± 2.0, and 20.9 ± 1.2, with a measuring accuracy of 1 °C. Average humidities (%rh ± SD) under LED, HPS, and FL lighting were 47.2 ± 7.0, 44.2 ± 7.1, and 38.8 ± 7.1 with a measuring accuracy of 2.25%rh. Both climatic conditions did not differ between treatments.

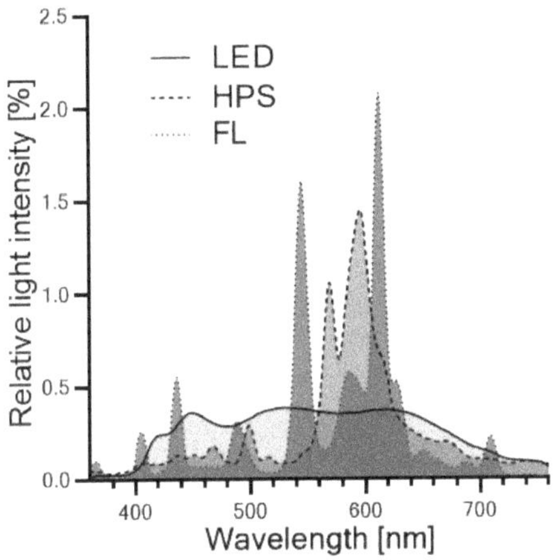

Figure 4. Light spectra of the three artificial light sources (light-emitting diode (LED) = solid line, high-pressure sodium lamp (HPS) = dashed line, fluorescent light (FL) = dotted line) used during the greenhouse experiment.

3.5. Harvest and Crop Managements

To analyze the effect of the supplemental lighting systems on the yield, all 32 experimental thyme plants were harvested separately from each treatment condition and replication. Fresh matter (FM) of the above-ground plant parts was individually recorded at harvest on 12 February 2019. Total dry matter (DM) was measured after drying the samples in a circulated drying oven at 30 °C until stable mass was attained (≤seven days). Leaf dry matter (L_{DM}) was determined for 16 plants selected from each treatment and replication, and the corresponding shoot dry matter (S_{DM}) was calculated by subtracting the LDM from DM. All dried leaf samples were vacuum-sealed (V.300®, Landig + Lava GmbH & Co KG, Bad Saulgau, Germany) and stored in the dark at 4 °C until further processing.

3.6. Energy Measurements

The power draw of current (I) and voltage (U) as well as electrical characteristics including real power (P) and apparent power (S) from representative lamps of each lighting treatment were measured using a power meter (ENERGY MONITOR 3000, VOLTCRAFT®, Wernberg-Köblitz, Germany) in order to estimate energy consumptions and biomass efficacies of the light fixtures. To correct for the detected difference between P and S due to heat dissipation of the HPS system, the measured *cos phi* of 0.93 was incorporated into the HPS' power consumption calculations. According to the manufacturer's specifications, the HPS system allows a homogeneously illuminated area of 1.56 m^2 (1.2 × 1.3 m). Thus, the measured power consumptions were adjusted to the power consumption per square meter (W m^{-2}) via rule of three. No adjustments were necessary for the LED and FL system.

3.7. Chemicals

Pure standard substances (α-pinene, α- and γ-terpinene, β-caryophyllene, borneol, carvacrol, limonene, linalool, myrcene, *p*-cymene, sabinene, thymol, and 6-methyl-5-penten-2-one) were purchased as analytical standards with a purity of at least 95% for GC reference analysis from Alfa Aesar (Kandel, Germany), Carl Roth (Karlsruhe, Germany), Merck KgaA (Darmstadt, Germany), and Fluka (Seelze, Germany). Isooctane (> 99%, HPLC grade) for solvent extraction of volatile organic compounds was obtained from Th. Geyer (Renningen, Germany).

3.8. Extraction of Volatile Organic Compounds

The volatile organic compounds (VOCs) of 16 plants per treatment and replication (n = 64) were extracted according to the following procedure: 100 mg ($\pm$2%) of gently oven-dried and powdered (3 intervals of 10 s at 15,000 rpm via Tube Mill control, IKA®, Staufen, Germany) thyme leaves were transferred into 2 mL screw cap micro tubes (SARSTEDT AG & Co. KG, Nümbrecht, Germany) including two steal grinding balls (Ø 2 mm). The plant material was homogenized in 1.0 mL of isooctane (containing 1:40,000 (v/v) 6-methyl-5-penten-2-one as internal standard) for 10 min at 30 rps with a ball mill (MM400, Retsch®, Haan, Germany). After 10 min of ultra-sonication (Sonorex RK 106, BANDELIN electronic GmbH & Co. KG, Berlin, Germany) and 10 min of centrifugation at 13,000 rpm (Heraeus™ Labofuge™ 400 R, Thermo Scientific™, Osterode, Germany) at 22 °C respectively, the supernatants were transferred into GC-vials and stored at −70 °C until analysis.

3.9. GC-FID and GC-MS Analysis

A total of 1 µL of the obtained extracts of volatiles was analyzed by GC–FID using an Agilent gas chromatograph 6890N fitted with an HP-5MS column (30 m $\times$ 250 µm $\times$ 0.5 µm) in splitless mode. Detector and injector temperatures were set to 250 °C. The following oven temperature program was used: 50 °C for 2 min, heating from 50 to 320 °C at a rate of 5 K min^{-1}. The final temperature was held for 6 min. Hydrogen was used as carrier gas with a constant flow rate of 1.2 mL min^{-1}. GC-MS was performed using an Agilent mass spectrometer 5975B, on an HP-5MS column (see GC), operating at 70 eV ionization energy, using the same temperature program as above. Helium was used as carrier gas with a constant flow rate of 1.2 mL min^{-1}. Retention indices were calculated by using retention times of C_6-C_{24}-alkanes that were injected under the same chromatographic conditions.

3.10. Identification and Quantification of Volatile Organic Compounds

All main organic compounds of the volatile extracts were identified by comparing their mass spectra with those of internal reference libraries (Adams, NIST). Additionally, the identification of α-pinene, α- and γ-terpinene, β-caryophyllene, borneol, carvacrol, limonene, linalool, myrcene, *p*-cymene, sabinene, and thymol was confirmed by authentic reference standards by comparing their individual retention indices. Quantitative data of each main compound were obtained with serial dilutions of external standard solutions using at least six known concentrations (β-caryophyllene, *cis*-sabinene hydrate and linalool: 0.5, 1, 2, 5, 10, and 50 ng µL^{-1}; α-pinene, α-terpinene, borneol, limonene, myrcene, and sabinene: 1, 2, 5, 10, 50, and 100 ng µL^{-1}; *p*-cymene: 1, 2, 5, 10, 50, 100, 200, and 400 ng µL^{-1}; γ-terpinene: 1, 2, 5, 10, 50, 100, 200, 400, and 600 ng µL^{-1}; carvacrol: 1, 2, 5 10, 50, 100, 200, 400, 600, and 1200 ng µL^{-1}; thymol: 5, 10, 50, 100, 200, 400, 600, 1200, and 1800 ng µL^{-1}), which covered concentration ranges detected for each compound in all samples.

3.11. Statistical Analysis and Calculations

Statistical analysis was performed using GraphPad Prism 8.4.2.679 (San Diego, MO, USA). Data of each spatial replication were tested for normality via Anderson-Darling, D'Agostino and Pearson, Shapiro–Wilk test, and Kolmogorov–Smirnov test. If normality test failed, outliers were identified via ROUT method (Q = 10%) and removed to establish normality of all data sets. The means of each spatial replication (n = 4) were tested for normality with the Shapiro–Wilk test, and all data sets passed the normality test at $\alpha = 0.05$. Finally, all data sets were statistically analyzed using a Brown-Forsythe and Welch ANOVA test due to unequal variances between treatments. Multiple comparisons were conducted via Dunnett T3. Two-tailed Pearson correlation between biomasses and DLIs was calculated after first computing means of 18 replicates per spatially independent light treatment and then analyzing those means (n = 12). Biomass efficacies (g or mg kWh^{-1} m^2) were calculated based on electricity consumed within a square meter during the cultivation period.

4. Conclusions

With its outstanding biomass as well as terpenoid efficacy, the broad-spectrum LED system represents a strong competitor to the conventionally used HPS and FL lighting systems in greenhouses under naturally insufficient light conditions as investigated in this study. Marketable *Thymus vulgaris* L. can be achieved faster and thus more often when replacing HPS and FL light sources with the tested LED system, ultimately resulting in greater revenues at simultaneously highly reduced production costs for greenhouse growers. The comparatively high initial capital costs of LEDs which have delayed their establishment in the past are decreasing [59]. Based on our results, combined with the typically low maintenance and long operating lifetime [13,15,63,64], the initial investment into LEDs should quickly become a source of profit for greenhouse growers. Additionally, different adaptive control approaches making use of the dimmability of LEDs [35,65] can further decrease the power consumptions and help to achieve consistent growth rates at a daily and seasonal level as shown by [60,66]. Our results suggest that an implementation of a broad-spectrum LED system in greenhouses could provide the possibility to cultivate a greater variety of crops with greater DLI-requirements under naturally insufficient light levels as conventional lighting systems are capable of today. Further, the broad-spectrum LED system could extend greenhouse production seasons, which are currently constrained by low supplemental DLIs, and allow a year-round production of a wider variety of selected greenhouse crops then HPS and FL systems are able to at present. However, further trails with a variety of greenhouse crops need to be investigated to confirm the suggested applicability for a range of crops. Therefore, in prospective broad-spectrum LED studies, the crops' individual DLI requirements need to be incorporated and compared to commonly applied mono- and dichromatic LED light spectra at equal light intensities for advancing our knowledge on the impact of LED light spectra on morphological, physiological, and metabolic plant responses. In addition, more studies examining the impact of light qualities on terpenoid biosynthesis, content, and composition are needed to optimize the quality of aromatic plant species in the future.

Author Contributions: Conceptualization and project administration, J.M.T. and H.S.; methodology, H.S.; formal analysis and validation, J.M.T., H.S., and A.K.; investigation, J.M.T.; data curation, A.K.; Visualization, J.M.T.; writing—original draft preparation, J.M.T.; writing—review and editing, H.S. and A.K.; supervision, H.S.; funding acquisition and resources, H.S. All authors have read and agreed to the published version of the manuscript.

Funding: This work was supported by the European Innovation Partnership for Improvement of Agricultural Productivity and Sustainability (grant number: 204016000016/80168353) via the European Agricultural Fund for Rural Development.

Institutional Review Board Statement: Not applicable.

Informed Consent Statement: Not applicable.

Data Availability Statement: The data that support the findings of this study are available from the corresponding author, A.K., upon reasonable request.

Acknowledgments: We thank Roland Buchhorn, Claudia Könecke, Heike Bäumer, Sabrina Pilz, and Fabian Hinze for cultivation assistance, René Grünwald for analytical assistance, and Marcus Müller from Humboldt University of Berlin for technical assistance. We further thank Torsten Meiners and Christoph Böttcher for their critical revision of the manuscript, and Matthias Melzig from Free University Berlin for his support.

Conflicts of Interest: The authors declare no conflict of interest.

References

1. Faust, J.E.; Logan, J. Daily light integral: A research review and high-resolution maps of the United States. *HortScience* **2018**, *53*, 1250–1257. [CrossRef]
2. Korczynski, P.C.; Logan, J.; Faust, J.E. Mapping monthly distribution of daily light integrals across the contiguous United States. *HortTechnology* **2002**, *12*, 12–16. [CrossRef]
3. Pramuk, L.A.; Runkle, E.S. Photosynthetic daily light integral during the seedling stage influences subsequent growth and flowering of *Celosia*, *Impatiens*, *Salvia*, *Tagetes* and *Viola*. *HortScience* **2005**, *40*, 1336–1339. [CrossRef]
4. Kong, Y.; Llewellyn, D.; Zheng, Y. Response of growth, yield, and quality of pea shoots to supplemental light-emitting diode lighting during winter greenhouse production. *Can. J. Plant Sci.* **2017**, *98*, 732–740. [CrossRef]
5. Matysiak, B.; Kowalski, A. White, blue and red LED lighting on growth, morphology and accumulation of flavonoid compounds in leafy greens. *Zemdirbyste* **2019**, *106*, 281–286. [CrossRef]
6. Nelson, J.A.; Bugbee, B. Economic analysis of greenhouse lighting: Light emitting diodes vs. high intensity discharge fixtures. *PLoS ONE* **2014**, *9*, e99010. [CrossRef] [PubMed]
7. Eichhorn-Bilodeau, S.; Wu, B.S.; Rufyikiri, A.S.; MacPherson, S.; Lefsrud, M. An update on plant photobiology and implications for cannabis production. *Front. Plant Sci.* **2019**, *10*, 296. [CrossRef]
8. Darko, E.; Heydarizadeh, P.; Schoefs, B.; Sabzalian, M.R. Photosynthesis under artificial light: The shift in primary and secondary metabolism. *Philos. Trans. R. Soc. B* **2014**, *369*, 20130243. [CrossRef]
9. Martineau, V.; Lesrud, M.; Naznin, M.T.; Kopsell, D.A. Comparison of light-emitting diode and high-pressure sodium light treatments for hydroponics growth of boston lettuce. *HortScience* **2012**, *47*, 477–482. [CrossRef]
10. Wojciechowska, R.; Dlugosz-Grochowska, O.; Kolton, A.; Żupnik, M. Effects of LED supplemental lighting on yield and some quality parameters of lamb's lettuce grown in two winter cycles. *Sci. Hortic.* **2015**, *187*, 80–86. [CrossRef]
11. Park, J.E.; Park, Y.G.; Jeong, B.R.; Hwang, S.J. Growth and Anthocyanin content of lettuce as affected by artificial light source and photoperiod in a closed-type plant production system. *Korean J. Hortic. Sci. Technol.* **2012**, *30*, 673–679. [CrossRef]
12. Massa, G.D.; Kim, H.-H.; Wheeler, R.M.; Mitchell, C.A. Plant productivity in response to LED lighting. *HortScience* **2008**, *43*, 1951–1956. [CrossRef]
13. Morrow, R.C. LED lighting in horticulture. *HortScience* **2008**, *43*, 1947–1950. [CrossRef]
14. Bourget, C.M. An introduction to light-emitting diodes. *HortScience* **2008**, *43*, 1944–1946. [CrossRef]
15. Mitchell, C.A.; Dzakovich, M.P.; Gomez, C.; Lopez, R.; Burr, J.F.; Hernández, R.; Kubota, C.; Currey, C.J.; Meng, Q.; Runkle, E.S.; et al. Light-emitting diodes in horticulture. In *Horticultural Reviews*, 1st ed.; Janick, J., Ed.; Wiley-Blackwell: Hoboken, NJ, USA, 2015; Volume 43.
16. McCree, K.J. The action spectrum absorptance and quantum yield of photosynthesis in crop plants. *Agric. Meteorol.* **1972**, *9*, 191–216. [CrossRef]
17. Goins, G.D.; Yorio, N.C.; Sanwo-Lewandowski, M.M.; Brown, C.S. Life cycle experiments with *Arabidopsis* grown under red light-emitting diodes (LEDs). *Life Support Biosph. Sci.* **1998**, *5*, 143–149. [PubMed]
18. Hoenecke, M.E.; Bula, R.J.; Tibbitts, T.W. Importance of 'blue' photon levels for lettuce seedlings grown under red-light-emitting diodes. *HortScience* **1992**, *27*, 427–430. [CrossRef]
19. Ahmad, M.; Grancher, N.; Heil, M.; Black, R.C.; Giovani, B.; Galland, P.; Lardemer, D. Action spectrum for cryptochrome-dependent hypocotyl growth inhibition in Arabidopsis. *Plant Physiol.* **2002**, *129*, 774–785. [CrossRef] [PubMed]
20. Folta, K.M.; Spalding, E.P. Unexpected roles for cryptochrome 2 and phototropin revealed by high-resolution analysis of blue light-mediated hypocotyl growth inhibition. *Plant J.* **2001**, *26*, 471–478. [CrossRef] [PubMed]
21. Yorio, N.C.; Goins, G.D.; Kagie, H.R.; Wheeler, R.M.; Sager, J.C. Improving spinach, radish, and lettuce growth under red light-emitting diodes (LEDs) with blue light supplementation. *HortScience* **2001**, *36*, 380–383. [CrossRef]
22. Hernandez, R.; Kubota, C. Physiological responses of cucumber seedlings under different blue and red photon flux ratios using LEDs. *Environ. Exp. Bot.* **2016**, *121*, 66–74. [CrossRef]
23. Park, Y.; Runkle, E.S. Far-red radiation promotes growth of seedlings by increasing leaf expansion and whole-plant net assimilation. *Environ. Exp. Bot.* **2017**, *136*, 41–49. [CrossRef]
24. Stutte, G.W.; Edney, S. Photoregulation of bioprotectant content of red leaf lettuce with light-emitting diodes. *HortScience* **2009**, *44*, 79–82. [CrossRef]
25. Li, Q.; Kubota, C. Effects of supplemental light quality on growth and phytochemicals of baby leaf lettuce. *Environ. Exp. Bot.* **2009**, *67*, 59–64. [CrossRef]
26. Schenkels, L.; Saeys, W.; Lauwers, A.; De Proft, M.P. Green light induces shade avoidance to alter plant morphology and increases biomass production in *Ocimum basilicum* L. *Sci. Hortic.* **2020**, *261*, 109002. [CrossRef]
27. Smith, H.L.; McAusland, L.; Murchie, E.H. Don't ignore the green light: Exploring diverse roles in plant processes. *J. Exp. Bot.* **2017**, *68*, 2099–2110. [CrossRef]

28. Johkan, M.; Shoji, K.; Goto, F.; Hahida, S.; Yoshihara, T. Effect of green light wavelength and intensity on photomorphogenesis and photosynthesis in *Lactuca sativa*. *Environ. Exp. Bot.* **2012**, *75*, 128–133. [CrossRef]
29. Kim, H.-H.; Goins, G.D.; Wheeler, R.M.; Sager, J.C. Green-light supplementation for enhanced lettuce growth under red- and blue-light-emitting diodes. *HortScience* **2004**, *39*, 1617–1622. [CrossRef] [PubMed]
30. Kong, S.-W.; Chung, H.-Y.; Chang, M.-Y.; Fang, W. The contribution of different spectral sections to increase fresh weight of boston lettuce. *HortScience* **2015**, *50*, 1006–1010. [CrossRef]
31. Novičkovas, A.; Brazaitytė, A.; Duchovskis, P.; Jankauskienė, J.; Samuolienė, G.; Virŝilė, A.; Sirtautas, R.; Bliznikas, Z.; Žukauskas, A. Solid-state lamps (LEDs) for the short-wavelength supplementary lighting in greenhouses: Experimental results with cucumber. *Acta Hortic.* **2012**, *927*, 723–730. [CrossRef]
32. Folta, K.M. Green light stimulates early stem elongation, antagonizing light-mediated growth inhibition. *Plant Physiol.* **2004**, *135*, 1407–1416. [CrossRef] [PubMed]
33. Frechilla, S.; Talbott, L.D.; Bogomolni, R.A.; Zeiger, E. Reversal of blue light-stimulated stomatal opening by green light. *Plant Cell Physiol.* **2000**, *41*, 171–176. [CrossRef]
34. Talbott, L.D.; Nikolova, G.; Ortiz, A.; Shayevich, I.; Zeiger, E. Green light reversal of blue-light-stimulated stomatal opening is found in a diversity of plant species. *Am. J. Bot.* **2002**, *89*, 366–368. [CrossRef]
35. Gomez, C.; Izzo, L.G. Increasing efficiency of crop production with LEDs. *AIMS* **2018**, *3*, 135–153. [CrossRef]
36. Ouzounis, T.; Rosenqvist, E.; Ottosen, C.-O. Spectral effects of artificial light on plant physiology and secondary metabolism: A review. *HortScience* **2015**, *50*, 1128–1135. [CrossRef]
37. Cope, K.R.; Bugbee, B. Spectral effects of three types of white light-emitting diodes on plant growth and development: Absolute versus relative amounts of blue light. *HortScience* **2013**, *48*, 504–509. [CrossRef]
38. Burattini, C.; Mattoni, B.; Bisegna, F. The impact of spectral composition of white LEDs on spinach (*Spinacia oleracea*) growth and development. *Energies* **2017**, *10*, 1383. [CrossRef]
39. Spalholz, H.; Perkins-Veazie, P.; Hernández, R. Impact of sun-simulated white light and varied blue:red spectrums on the growth, morphology, development, and phytochemical content of green- and red-leaf lettuce at different growth stages. *Sci. Hortic.* **2020**, *264*, 109195. [CrossRef]
40. Gao, S.; Liu, X.; Liu, Y.; Cao, B.; Chen, Z.; Xu, K. Photosynthetic characteristics and chloroplast ultrastructure of welsh onion (*Allium fistulosum* L.) grown under different LED wavelengths. *BMC Plant Biol.* **2020**, *20*, 78. [CrossRef] [PubMed]
41. Mickens, M.A.; Torralba, M.; Robinson, S.A.; Spencer, L.E.; Romeyn, W.M.; Massa, G.D.; Wheeler, R.M. Growth of red pak choi under red and blue, supplemented white, and artificial sunlight provided by LEDs. *Sci. Hortic.* **2019**, *245*, 200–209. [CrossRef]
42. Lin, K.-H.; Huang, M.-Y.; Huang, W.-D.; Hsu, M.-H.; Yang, Z.-W.; Yang, C.-M. The effects of red, blue, and white light-emitting diodes on the growth, development, and edible quality of hydroponically grown lettuce (*Lactuca sativa* L. var. capitata). *Sci. Hortic.* **2013**, *150*, 86–91. [CrossRef]
43. Perrino, E.V.; Valerio, F.; Gannouchi, A.; Trani, A.; Mezzapesa, G. Ecological and plant community implication on essential oils composition in useful wild officinal species: A pilot case study in Apulia (Italy). *Plants* **2021**, *10*, 574. [CrossRef] [PubMed]
44. Salehi, B.; Mishra, A.P.; Shukla, I.; Sharifi-Rad, M.; del Mar Contreras, M.; Segura-Carretero, A.; Fathi, H.; Nasrabadi, N.N.; Kobarfard, F.; Sharifi-Rad, J. Thymol, thyme, and other plant sources: Health and potential uses. *Phytother. Res.* **2018**, *32*, 1688–1706. [CrossRef] [PubMed]
45. Craver, J.K.; Boldt, J.K.; Lopez, R.G. Comparison of supplemental lighting provided by high-pressure sodium lamps or light-emitting diodes for the propagation and finishing of bedding plants in a commercial greenhouse. *HortScience* **2019**, *54*, 52–59. [CrossRef]
46. Faust, J.E. First Research Report. Light Management in Greenhouses. I. Daily light Integral: A Useful Tool for the U.S. Floriculture Industry. Available online: https://www.specmeters.com/assets/1/7/A051.pdf (accessed on 17 April 2020).
47. Mathew, S. Phytochrome-mediated development in land plants: Red light sensing evolves to meet the challenges of changing light environments. *Mol. Ecol.* **2006**, *15*, 3482–3503. [CrossRef] [PubMed]
48. Thompson, J.D.; Chalchat, J.C.; Michet, A.; Linhart, Y.B.; Ehlers, B. Qualitative and quantitative variation in monoterpene co-occurrence and composition in the essential oil of *Thymus vulgaris* chemotypes. *J. Chem. Ecol.* **2003**, *29*, 859–880. [CrossRef]
49. Rota, M.C.; Herrera, A.; Martínez, R.M.; Sotomayor, J.A.; Jordán, M.J. Antimicrobial activity and chemical composition of *Thymus vulgaris*, *Thymus zygis* and *Thymus hyemalis* essential oils. *Food Control* **2008**, *19*, 681–687. [CrossRef]
50. Gouinguene, S.P.; Turlings, T.C.J. The effects of abiotic factors on induced volatile emissions in corn plants. *Plant Physiol.* **2002**, *129*, 1296–1307. [CrossRef]
51. Spring, O.; Priester, T.; Hager, A. Light-induced accumulation of sesquiterpenes lactones in sunflower seedlings. *J. Plant Physiol.* **1986**, *123*, 79–89. [CrossRef]
52. Gleizes, M.; Pauly, G.; Bernard-Dagan, C.; Jacques, P. Effects of light on terpene hydrocarbon synthesis in *Pinus pinaster*. *Physiol. Plant.* **1980**, *50*, 16–20. [CrossRef]
53. Yamaura, T.; Tanaka, S.; Tabata, M. Participation of Phytochrome in the Photoregulation of Terpenoid Synthesis in Thyme Seedlings. *Plant Cell Physiol.* **1991**, *32*, 6. [CrossRef]
54. Tanaka, S.; Yamaura, T.; Shigemoto, R.; Tabata, M. Phytochrome-mediated production of monoterpenes in thyme seedlings. *Phytochemistry* **1989**, *28*, 2955–2957. [CrossRef]
55. Franklin, K.A. Shade avoidance. *New Phytol.* **2008**, *179*, 930–944. [CrossRef]

56. Kegge, W.; Weldegergis, B.T.; Soler, R.; Vergeer-Van Eijk, M.; Dicke, M.; Voesenek, L.A.C.J.; Pierik, R. Canopy light cues affect emission of constitutive and methyl jasmonate-induced volatile organic compounds in *Arabidopsis thaliana*. *New Phytol.* **2013**, *200*, 861–874. [CrossRef]
57. Tohidi, B.; Rahimmalek, M.; Arzani, A.; Sabzalian, M.R. Thymol, carvacrol, and antioxidant accumulation in *Thymus* species in response to different light spectra emitted by light-emitting diodes. *Food Chem.* **2020**, *307*, 125521. [CrossRef] [PubMed]
58. Poel, B.R.; Runkle, E.S. Seedling growth is similar under supplemental greenhouse lighting from high-pressure sodium lamps or light-emitting diodes. *HortScience* **2017**, *52*, 388–394. [CrossRef]
59. Wallace, C.; Both, A.J. Evaluating operating characteristics of light sources for horticultural applications. *Acta Hortic.* **2016**, *1134*, 435–443. [CrossRef]
60. Hernandez, E.; Hernandez, M.B.; Mattson, N.S. Quality, yield, and biomass efficacy of several hydroponic lettuce (*Lactuca sativa* L.) cultivars in response to high pressure sodium lights or light emitting diodes for greenhouse supplemental lighting. *Horticulturae* **2020**, *6*, 7. [CrossRef]
61. Piovene, C.; Orsini, F.; Bosi, S.; Sanoubar, R.; Bregola, V.; Dinelli, G.; Gianquinto, G. Optimal red:blue ratio in LED lighting for nutraceutical indoor horticulture. *Sci. Hortic.* **2015**, *193*, 202–208. [CrossRef]
62. WetterKontor. Available online: https://www.wetterkontor.de/de/wetter/deutschland/rueckblick.asp?id=20&datum0=09.10.2018&datum1=31.12.2018&jr=2021&mo=3&datum=08.03.2021&t=8&part=0 (accessed on 10 March 2021).
63. Yeh, N.; Chung, J.P. High-brightness LEDs—Energy efficient lighting sources and their potential in indoor plant cultivation. *Renew. Sustain. Energy Rev.* **2009**, *13*, 2175–2180. [CrossRef]
64. Barta, D.J.; Tibbitts, T.W.; Bula, R.J.; Morrow, R.C. Evaluation of light emitting diode characteristics for a space-based plant irradiation source. *Adv. Space Res.* **1992**, *12*, 141–149. [CrossRef]
65. Van Iersel, M.W.; Gianino, D. An adaptive control approach for light-emitting diode lights can reduce the energy costs of supplemental lighting in greenhouses. *HortScience* **2017**, *52*, 72–77. [CrossRef]
66. Harbick, K.; Albright, L.D.; Mattson, N.S. Electrical savings comparison of supplemental lighting control systems in greenhouse environments. In Proceedings of the American Society of Agricultural and Biological Engineers (ASABE) Annual International Meeting, Orlando, FL, USA, 17–20 July 2016. Paper Number 162460478. [CrossRef]

plants

MDPI

Communication

Metabolic Analysis of Root, Stem, and Leaf of *Scutellaria baicalensis* Plantlets Treated with Different LED Lights

Hyeon-Ji Yeo [1,†], Chang-Ha Park [1,†], Soo-Yun Park [2], Sun-Ok Chung [3,4], Jae-Kwang Kim [5,*] and Sang-Un Park [1,4,*]

1 Department of Crop Science, Chungnam National University, 99 Daehak-Ro, Yuseong-gu, Daejeon 34134, Korea; guswl7627@gmail.com (H.-J.Y.); parkch804@gmail.com (C.-H.P.)
2 National Institute of Agricultural Sciences, Rural Development Administration, Wanju-gun, Jeonbuk 55365, Korea; psy22@korea.kr
3 Department of Agricultural Machinery Engineering, Graduate School, Chungnam National University, 99 Daehak-Ro, Yuseong-gu, Daejeon 34134, Korea; sochung@cnu.ac.kr
4 Department of Smart Agriculture Systems, Chungnam National University, 99 Daehak-Ro, Yuseong-gu, Daejeon 34134, Korea
5 Division of Life Sciences, College of Life Sciences and Bioengineering, Incheon National University, Yeonsugu, Incheon 22012, Korea
* Correspondence: kjkpj@inu.ac.kr (J.-K.K.); supark@cnu.ac.kr (S.-U.P.); Tel.: +82-32-835-8241 (J.-K.K.); +82-42-821-5730 (S.-U.P.)
† Hyeon Ji Yeo and Chang Ha Park contributed equally to this work.

Citation: Yeo, H.-J.; Park, C.-H.; Park, S.-Y.; Chung, S.-O.; Kim, J.-K.; Park, S.-U. Metabolic Analysis of Root, Stem, and Leaf of *Scutellaria baicalensis* Plantlets Treated with Different LED Lights. *Plants* **2021**, *10*, 940. https://doi.org/10.3390/plants10050940

Academic Editors: Valeria Cavallaro and Rosario Muleo

Received: 6 April 2021
Accepted: 4 May 2021
Published: 8 May 2021

Abstract: Light emitting diodes (LEDs) have recently been considered an efficient artificial light source in plant factories for enhancing plant growth and nutritional quality. Accordingly, this study aimed to review blue, red, and white LED light sources for efficiency and length of the growing period to produce seedlings of *Scutellaria baicalensis* with high nutritional value. The roots, stems, and leaves of *S. baicalensis* seedlings were grown under different LED lights and harvested after two and four weeks, and analyzed using high-performance liquid chromatography and gas chromatography time-of-flight mass spectrometry to identify and quantify primary and secondary metabolites. Roots, particularly in the seedlings treated with white LEDs were determined to contain the greatest concentrations of the representative compounds present in *S. baicalensis*: baicalin, baicalein, and wogonin, which show highly strong biological properties compared to the other plant organs. A total of 50 metabolites (amino acids, sugars, sugar alcohols, organic acids, phenolic acids, and amines) were detected in the roots, stems, and leaves of *S. baicalensis* seedlings, and the concentrations of primary and secondary metabolites were generally decreased with the increasing duration of LED illumination. Therefore, this study suggests that white LED light and a 2-week growing period are the most efficient conditions for the production of baicalin, baicalein, and wogonin.

Keywords: LED lights; medicinal plant; *Scutellaria baicalensis*; flavones; metabolites

1. Introduction

Scutellaria baicalensis Georgi, known as Huang Qin in Chinese medicine, has been used as a conventional herbal remedy in East Asia and is formally listed in the Chinese Pharmacopeia [1]. According to previous research, the root extract of *S. baicalensis* causes apoptosis of hepatocellular, prostatic, pancreatic, urothelial carcinoma, and breast cells, and suppresses the growth of cancer cells in vitro, and it is often used in conjunction with other medicinal plants [2].

Flavonoids are found in vegetables, seeds, nuts, flowers and stems, wine, tea [3], honey, and propolis [4], and the roots of *S. baicalensis* contain flavonoids such as baicalin, baicalein, wogonoside, and wogonin [5]. Baicalin is biosynthesized using several enzymes, including phenylalanine ammonia-lyase (PAL), cinnamate 4-hydroxylase (C4H), 4-coumarate: CoA ligase (4CL), chalcone synthase (CHS), and chalcone isomerase (CHI). It is catalyzed to baicalein through β-glucuronidase (GUS) or vice versa with UDPglucuronate: baicalein 7-O-glucuronosyltransferase (UBGAT) [6]. Similarly, baicalein has in vitro antioxidative, anti-inflammatory, lipoxygenase inhibitory, antiviral, and anti-allergic activities [7]. Wogonin, one of the main chemical components of *S. baicalensis*, is a flavanone derivative containing the nucleus of a phenylbenzopyrone [8] that suppresses tumor growth and angiogenesis in vitro [9].

Artificial light has been known to improve plant development, growth, and phytochemical production; in plant factories that require strong light to grow vegetables, light-emitting diodes (LEDs) are a promising source due to their durability, cool temperature, long life, diverse wavelengths, and small diode size [10]. According to previous studies, LEDs have positive effects on the accumulation of various secondary metabolites, such as glucosinolate, phenylpropanoid, and carotenoid, in *Brassica juncea* sprouts, wheat sprouts, and the callus of *Scutellaria baicalensis* [11–13].

However, there are no studies on the effects of LED lights and their duration on metabolites in *S. baicalensis* sprouts. Therefore, this study aimed to investigate the effects of different LED light sources (white, blue, and red) and their duration on metabolic changes in *S. baicalensis* sprouts and to optimize the most efficient qualities for the production of flavones (baicalin, baicalein, and wogonin).

2. Results

2.1. HPLC Analysis of Baicalin, Baicalein, and Wogonin in Root, Stem, and Leaf of S. baicalensis Treated with Different LED Light Sources

The three LED lights (red, blue, and white) and their treatment duration caused variations in flavones (baicalin, baicalein, and wogonin) in the roots, leaves, and stems of *S. baicalensis*. Baicalin and baicalein were detected in all plant parts, whereas wogonin was only found in the roots (Figure 1). Roots showed the greatest concentrations of the flavones compared with leaves and stems, and the most abundant was baicalin, followed by baicalein and wogonin. After two weeks under white LED light treatment the roots of *S. baicalensis* seedlings produced the highest levels of baicalin (100.42 ± 0.32 mg/g dry weight (dw)) and wogonin (4.51 ± 0.09 mg/g dw); whereas levels of these compounds decreased in the roots under all three LED colors after four weeks. Similarly, roots under white and blue LED lights, contained slightly higher levels of baicalein than those under red LED light. In stems, baicalin began accumulating after four weeks regardless of light color and those treated with red LED light contained the greatest amounts of baicalin (0.17 ± 0.05 mg/g dw). In contrast, baicalein concentrations showed a slightly increasing accumulation pattern under red LED illumination, whereas stems treated with white and blue LED light revealed decreasing levels with increasing duration. Baicalin and baicalein were also present in leaves, and those treated with red LED light showed increasing patterns of baicalin and baicalein accumulation with increasing illumination duration.

2.2. Metabolite-Specific Profiling of Root, Stem, and Leaf of S. baicalensis Treated with Different LED Light Sources

GC-TOFMS was used to detect 50 metabolites (amino acids, sugars, sugar alcohols, organic acids, phenolic acids, and amines) in the roots, stems, and leaves of *S. baicalensis* seedlings treated with different LED light sources (red, blue, and white). In leaves and stems treated with blue LED light, a greater number of metabolites were detected than in those treated with red and white LED light. The majority of the amino acids, organic acids, and TCA cycle intermediates showed decreasing patterns in leaves and stems treated with increasing durations of LED light regardless of the source. In contrast, the levels of most sugars and sugar alcohols had slightly increasing patterns in both plant parts. Similarly, roots of seedlings treated with blue or red LED lights contained greater concentrations of metabolites and displayed decreasing patterns of most amino acids, organic acids, and TCA cycle intermediates over time regardless of the light source. White LED light induced slightly increasing levels of sugars and sugar alcohols in the roots, whereas blue and red LED lights revealed decreasing accumulations of these metabolites. Additionally, roots under white LED light for two weeks contained lower levels of sugars and sugar alcohols than those under blue and red LED lights. Partial least-squares discriminant analysis (PLS-DA) was performed with the data derived from GC-TOFMS and HPLC to investigate the metabolic changes in the roots, stems, and leaves of *S. baicalensis* seedlings under various LED light treatments and their durations (Figure 2). The PLS-DA results showed a separation between the leaf group at 2 weeks from that at 4 weeks. This separation was attributable to changes in amino acids, organic acids, sugars, and sugar alcohols as related previously.

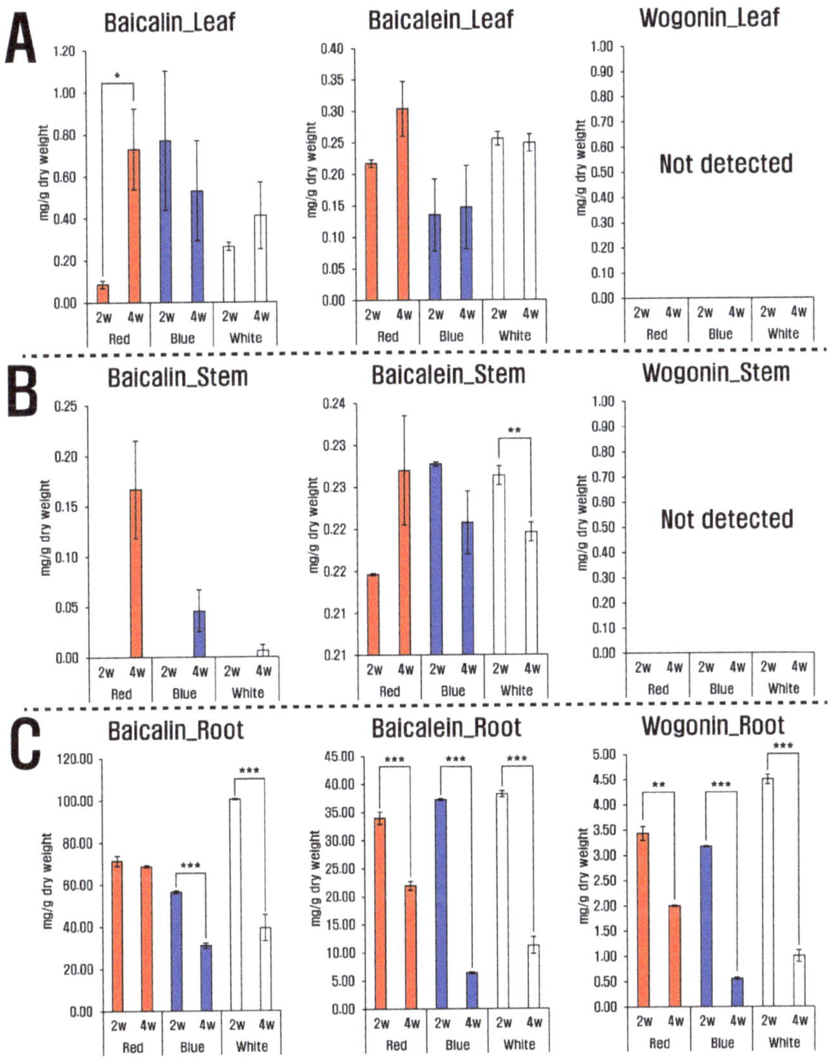

Figure 1. Flavone contents of the leaf (**A**), stem (**B**), and root (**C**) of *S. baicalensis* seedlings grown under LED treatment of varying duration. 2 w and 4 w indicate 2 weeks and 4 weeks, respectively (*t*-test, * $p < 0.05$, ** $p < 0.01$, *** $p < 0.005$).

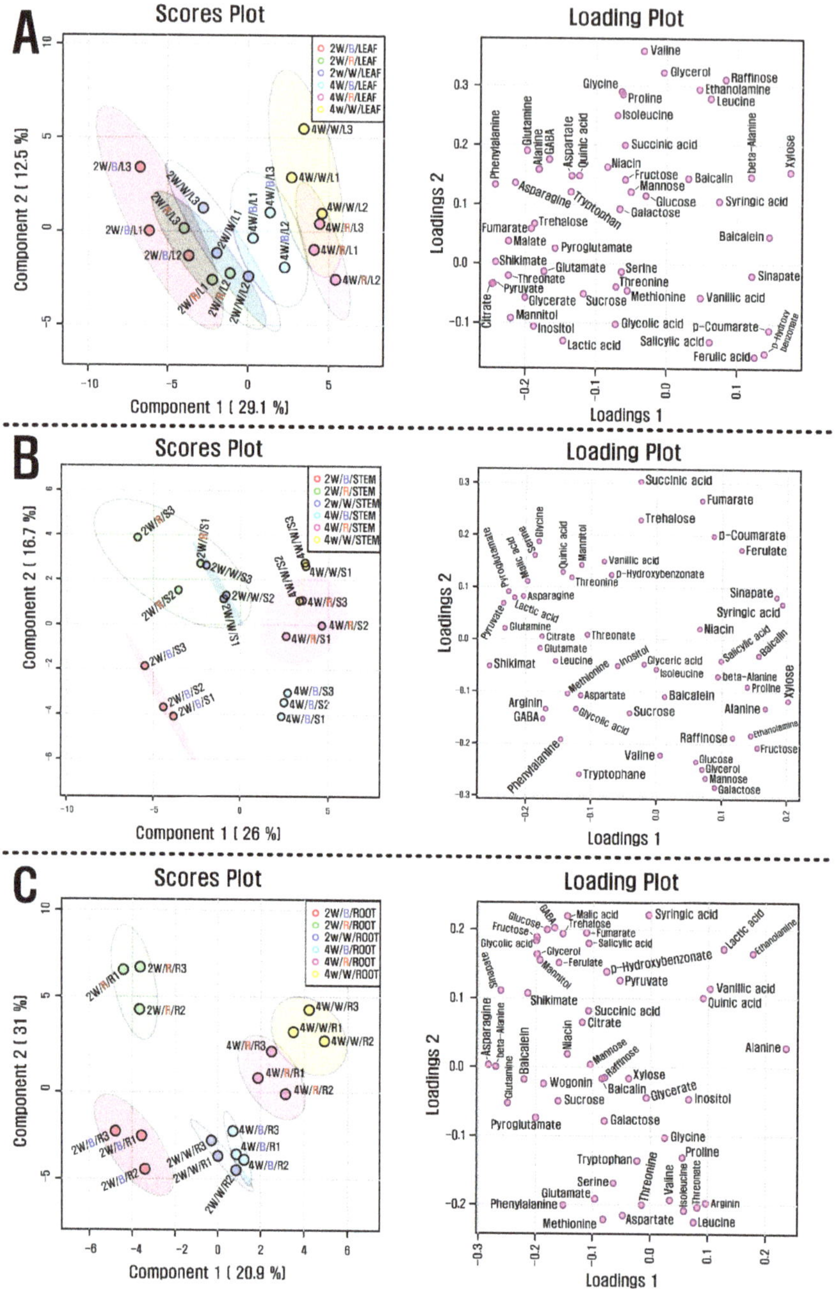

Figure 2. (**A**) Scores and loading plots of the PCA model using metabolites from the leaves of *S. baicalensis* seedlings grown under the LED treatment of varying duration, (**B**) Scores and loading plots of the PCA model using metabolites from the stem of *S. baicalensis* seedlings grown under the LED treatment of varying duration, and (**C**) Scores and loading plots of the PCA model using metabolites from the root of *S. baicalensis* seedlings grown under the LED treatment of varying duration. 2 w and 4 w indicate 2 weeks and 4 weeks, respectively, as well as B, R, and W indicate blue, red, and white, respectively.

To measure the relationship between different metabolites quantified in the roots, stems, and leaves of *S. baicalensis* seedlings treated with different LED lights, an HCA was performed using Pearson's correlation results (Figure 3). Compounds involved in nitrogen metabolism into amino acids (glutamine, glutamic acid, aspartic acid, and asparagine) and other nitrogen-containing compounds, were positively correlated, and these amino acids and their derivatives were also positively correlated in the roots, stems, and leaves of *S. baicalensis* seedlings treated with different LED lights. Phenylalanine and tryptophan, arising from the shikimate biosynthesis pathway, had a positive relationship with shikimate. The carbohydrates sucrose, galactose, mannose, and raffinose also returned positive correlations. Phenylalanine is a precursor of phenolic acid and flavonoid biosynthesis, and it showed a negative correlation with most phenolic acids and flavonoids. Similarly, most carbohydrates, which act as energy sources, were negatively correlated with most phenolics detected.

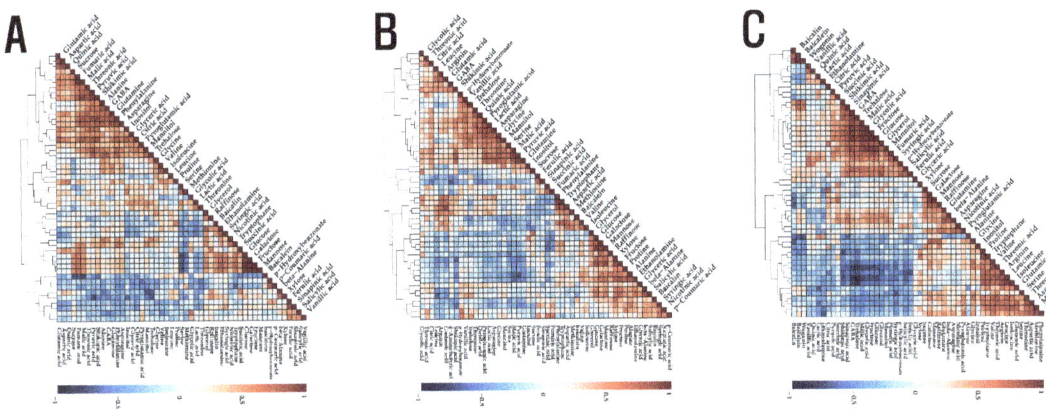

Figure 3. Correlation matrix (**A**) of metabolites obtained from the leaf of *S. baicalensis* seedlings grown under the LED treatment of varying duration, correlation matrix (**B**) of metabolites obtained from the stem of *S. baicalensis* seedlings grown under the LED treatment of varying duration, and correlation matrix (**C**) of metabolites obtained from the root of *S. baicalensis* seedlings grown under the LED treatment of varying duration. Each square indicates the Pearson's correlation coefficient for a pair of compounds, and the value of the correlation coefficient is represented by the intensity of the deep blue or deep red color, as indicated on the color scale.

3. Discussion

In this study, the roots of *S. baicalensis* seedlings treated with white LED light contained the highest levels of baicalin, baicalein, and wogonin and lower levels of most sugars than the other plant parts, suggesting the need for energy to enhance the biosynthesis of phenolic compounds, including the three described here. These results agree with previous studies showing that sugar concentrations for anthocyanin accumulation were lower in purple kohlrabi than in green kohlrabi [14] and that a fungal elicitor allowed for the more rapid depletion of sugar pools to promote alkaloid biosynthesis in cell cultures of *Papaver somniferum* [15].

Numerous previous studies have reported that LED illumination can enhance secondary metabolite production in vegetables and medicinal plants. White LED illumination has been shown to increase the accumulation of phenolics in *Agastache rugosa* seedlings [16], carotenoids in *Fagopyrum tataricum* sprouts [17], and glucosinolates in *Brassica juncea* sprouts [11], compared with other colored LED lights, consistent with the findings of this study. Blue LED light has been reported to increase accumulations of phenolics in *Brassica napus* [18] and *Glycine max* sprouts [19], and red LED light has been shown to enhance both phenolic compounds in the leaves of *Myrtus communis* in vitro [20] and carotenoid production in the outer peel layer of citrus fruit [21].

The metabolic networks of glutamine, glutamate, aspartate, and asparagine are involved in various nitrogen-related processes, including nitrogen assimilation by plants, metabolism into amino acids and other nitrogen-containing compounds, transport between source and sink, stress-associated metabolism, and carbon-nitrogen partitioning. Glutamine is derived from ammonium assimilation and can be converted into glutamic acid with α-ketoglutarate, a TCA cycle intermediate. This glutamic acid is further metabolized into aspartic acid, which is converted to asparagine. The first three compounds can be used for the synthesis of proteinogenic and non-proteinogenic amino acids, amides, and other nitrogenous compounds. Asparagine is a prominent nitrogen transport agent as well as a proteinogenic amino acid [22–24].

In this study, glutamate, glutamine, asparagine, and aspartate showed decreasing concentration levels in the roots, leaves, and stems of *S. baicalensis* seedlings treated with different LED lights, with a related reduction of their derivatives. This result was supported by the positive correlations between these four compounds and their derivatives. Furthermore, since shikimate and phenylalanine, which are derived from the shikimate pathway, were negatively correlated with most phenolic compounds, the biosynthesis of these compounds, including phenolic acids and flavones must have been assisted by intermediates or precursors. These findings corroborate a previous study reporting that the internal pool of phenylalanine was lower in purple kohlrabi, which contained a high amount of phenolic compounds, reflecting a precursor supply to produce phenolic acids and anthocyanins [14].

Artificial LED source is important to regulate the lighting systems in a plant factory to produce high-quality plant materials. Therefore, this study suggests that *S. baicalensis* seedlings, containing a high number of health-beneficial compounds, can be produced under LED lights in limited space since *S. baicalensis* was generally cultivated in the field and indicates that the optimal light was white LED for flavone accumulation in *S. baicalensis* seedlings.

4. Materials and Methods

4.1. Preparation of Plant Materials

S. baicalensis seeds were purchased from Aram Seed Co. (Seoul, Korea). Seeds for germination were soaked overnight in water. To produce seedlings, 50 seeds were placed in each pot (diameter: 12 cm, height: 11 cm) containing vermiculite and grown in a growth chamber equipped with fluorescent light with a flux rate of 35 $\mu mol \cdot m^{-2} \cdot s^{-1}$ at 25 °C. After 2 weeks, the seedlings in six pots were moved to a room in a growth chamber equipped with each blue, white, and red LED light with a flux rate of 90 $\mu mol \cdot m^{-2} \cdot s^{-1}$ at 25 °C with an 8 h dark/16 h light cycle. The leaves, stems, and roots from seedlings were harvested with liquid nitrogen and then freeze-dried for further metabolite analysis after two and four weeks of LED light treatment. The LED light sources, and their specific information are described in Appendix A Table A1 and a previous study [18]. Seedlings from three pots were used as independent replicates for each LED light for each duration.

4.2. High-Performance Liquid Chromatography (HPLC) Analysis for Flavones

We detected three flavones (baicalin, baicalein, and wogonin) using a slightly modified method of Park et al., [25]. The freeze-dried samples were ground into a powder using a grinder (Wonder blender WB-1, SANPLATEC CORP, Osaka, Japan). The *S. baicalensis* root, stem, and leaf powders (0.1 g each) from seedlings treated with various LED lights were extracted with 2 mL of 80% (v/v) aqueous MeOH and vortexed for 30 s. Following sonication for 1 h, the samples were centrifuged at 10,000× g at 4 °C for 20 min, and the crude extracts were syringe-filtered to a vial for analysis. The HPLC system and analysis conditions were the same as those used in the method reported by Park et al. [25] (Table A2). The three different flavones were identified by retention time and spike tests, and the equation of calibration curves for each flavone was obtained to quantify the compounds in the roots, stems, and leaves of *S. baicalensis* seedlings treated with the different LED lights.

4.3. Gas Chromatography Time-of-Flight Mass Spectrometry (GC-TOFMS) Analysis

Hydrophilic metabolites were detected using the method reported by Park et al. [26]. The root, stem, and leaf powders (0.1 g each) of *S. baicalensis* seedlings treated with different LED lights were extracted with 2 mL of 80% (*v*/*v*) aqueous MeOH and vortexed for 30 s. After sonication for 1 h, each sample was centrifuged at 10,000× *g* at 4 °C for 20 min, and then the crude extracts were syringe-filtered into a vial for analysis. The system and analysis conditions were reported by Park et al., 2021. Retention time comparison and spike test were conducted to identify the three different flavones, and the equation of calibration curves for each flavone was obtained to quantify the compounds in the roots, stems, and leaves of the *S. baicalensis* seedlings. The tissue powders (0.01 g each) were placed in a 2 mL tube along with 1 mL of a water/chloroform/methanol mixture (1:1:2.5 *v*/*v*/*v*) and 60 µL of ribitol (0.2 g/L; Sigma, St. Louis, MO, USA) as an internal standard. The extracts were mixed at 1200× *g* using a thermomixer, followed by centrifugation at 10,000× *g* for 5 min. The polar phase (0.8 mL) was transferred to a fresh tube containing water for chromatography (0.4 mL) and evaporated for 3 h. The dried residues were derived by adding 0.08 mL of methoxyamine hydrochloride/pyridine (20 g/L), followed by shaking at 37 °C for 2 h. After the addition of 0.08 mL of *N*-methyl-*N*-(trimethylsilyl)trifluoroacetamide, each tube was heated at 37 °C for 30 min. The final extract was placed in a vial for GC analysis. The analysis system, condition, and program of GC-TOFMS were used to identify and quantify metabolites in the roots, stems, and leaves of *S. baicalensis* seedlings treated with different LED lights according to the previous studies [26,27].

4.4. Statistical Analysis

SPSS (version 24.0;(IBM, Chicago, IL, USA)) was used to perform a t-test and MetaboAnalyst 5.0 (http://www.metaboanalyst.ca/, accessed on 5 March 2021) was used for principal component analysis (PCA) and hierarchical cluster analysis (HCA) using Pearson correlations for the metabolites detected in roots, stems, and leaves of *S. baicalensis* seedlings treated with different LED lights. The resolution of the resulting figures was improved using Adobe Illustrator.

5. Conclusions

Considering flavone content, white LED light for 2 weeks was the most efficient for the production of the three different flavones in the roots, stems, and leaves of *S. baicalensis* seedlings. Based on the results from the current and previous studies, it appears that the effect of different LED lights on the accumulation of secondary metabolites may depend on plant species, and this study reports that white LED lights are the most optimal for flavone accumulation in *S. baicalensis* seedlings.

Author Contributions: S.-U.P. and J.-K.K. conceived and designed the experiments. H.-J.Y., C.-H.P., S.-Y.P. and S.-O.C. performed the experiments and analyzed the data. H.-J.Y. and C.-H.P. wrote, reviewed, and edited the manuscript. All authors have read and agreed to the published version of the manuscript.

Funding: This work was supported by Incheon National University Research Concentration Professors Grant in 2021.

Conflicts of Interest: The authors declare no conflict of interest.

Appendix A

Table A1. Plant growth method using LED lights.

LED Plant Growth Chamber	
Product	**Multi-Room Chamber HB-302S-4 (Hanbaek Scientific Co.,)**
Picture, which is taken from a previous study [10]	
Dimension of each room (L × W × H)	136 cm × 78 cm × 168 cm
LED lights	The white (450–660 nm), blue (450 nm), or red (660 nm) LED lights (PGL-PFL series) were manufactured from PARUS LED Co., Cheonan, Korea

Table A2. HPLC analysis method.

HPLC Analysis Performed Using Our Previous Study [25]	
Equipment	NS-4000 HPLC apparatus (Futecs, Daejeon, Korea)
Detector	UV-Vis
Column	optimapak C18 column (250 mm × 4.6 mm, 5 μm; RStech, Daejon, Korea)
Detector wavelength	275 nm
Oven temperature	30 °C
Flow rate	1.0 mL/min
Mobile phase	Acetonitrile, solvent A and 0.2% (*v*/*v*) acetic acid, solvent B
Gradient program	Solvent B 90%; 0 min, solvent B 80%; 10 min, solvent B 80%; 15 min, solvent B 75%; 20 min, solvent B 75%; 25 min, solvent B 40%; 50 min, solvent B 90%; 50.1–60 min

References

1. Li, H.-B.; Jiang, Y.; Chen, F. Separation methods used for *Scutellaria baicalensis* active components. *J. Chromatogr. B* **2004**, *812*, 277–290. [CrossRef]
2. Zhang, D.Y.; Wu, J.; Ye, F.; Xue, L.; Jiang, S.; Yi, J.; Zhang, W.; Wei, H.; Sung, M.; Wang, W. Inhibition of cancer cell proliferation and prostaglandin E2 synthesis by *Scutellaria baicalensis*. *Cancer Res.* **2003**, *63*, 4037–4043.
3. Middleton, E., Jr. The impact of plant flavonoids on mammalian biology: Implications for immunity, inflammation and cancer. In *The Flavonoids: Advances in Research Since 1986*; Harborne, J.B., Ed.; Chapman & Hall: London, UK, 1993; pp. 619–652.
4. Grange, J.M.; Davey, R.W. Antibacterial properties of propolis (bee glue). *J. R. Soc. Med.* **1990**, *83*, 159–160. [CrossRef]

5. Lee, W.H.; Ku, S.-K.; Bae, J.-S. Anti-inflammatory effects of Baicalin, Baicalein, and Wogonin in vitro and in vivo. *Inflammation* **2015**, *38*, 110–125. [CrossRef]
6. Yuan, Y.; Wu, C.; Liu, Y.; Yang, J.; Huang, L. The *Scutellaria baicalensis* R2R3-MYB transcription factors modulates flavonoid biosynthesis by regulating GA metabolism in transgenic tobacco plants. *PLoS ONE* **2013**, *8*, e77275. [CrossRef]
7. Chen, C.-H.; Huang, T.-S.; Wong, C.-H.; Hong, C.-L.; Tsai, Y.-H.; Liang, C.-C.; Lu, F.-J.; Chang, W.-H. Synergistic anti-cancer effect of baicalein and silymarin on human hepatoma HepG2 Cells. *Food Chem. Toxicol.* **2009**, *47*, 638–644. [CrossRef] [PubMed]
8. Hui, K.M.; Huen, M.S.; Wang, H.Y.; Zheng, H.; Sigel, E.; Baur, R.; Ren, H.; Li, Z.W.; Wong, J.T.-F.; Xue, H. Anxiolytic effect of wogonin, a benzodiazepine receptor ligand isolated from *Scutellaria baicalensis* Georgi. *Biochem. Pharmacol.* **2002**, *64*, 1415–1424. [CrossRef]
9. Kimura, Y.; Sumiyoshi, M. Anti-tumor and anti-metastatic actions of wogonin isolated from *Scutellaria baicalensis* roots through anti-lymphangiogenesis. *Phytomedicine* **2013**, *20*, 328–336.
10. Yeo, H.J.; Park, C.H.; Lee, K.B.; Kim, J.K.; Park, J.S.; Lee, J.-W.; Park, S.U. Metabolic analysis of *Vigna unguiculata* sprouts exposed to different light-emitting diodes. *Nat. Prod. Commun.* **2018**, *13*, 1934578X1801301029.
11. Park, C.H.; Park, Y.E.; Yeo, H.J.; Kim, J.K.; Park, S.U. Effects of Light-Emitting Diodes on the Accumulation of Phenolic Compounds and Glucosinolates in *Brassica juncea* Sprouts. *Horticulturae* **2020**, *6*, 77. [CrossRef]
12. Cuong, D.M.; Ha, T.W.; Park, C.H.; Kim, N.S.; Yeo, H.J.; Chun, S.W.; Kim, C.S.; Park, S.U. Effects of LED lights on expression of genes involved in phenylpropanoid biosynthesis and accumulation of phenylpropanoid in wheat sprout. *Agronomy* **2019**, *9*, 307. [CrossRef]
13. Tuan, P.A.; Park, C.H.; Park, W.T.; Kim, Y.B.; Kim, Y.J.; Chung, S.O.; Kim, J.K.; Park, S.U. Expression levels of carotenoid biosynthetic genes and carotenoid production in the callus of scutellaria baicalensis exposed to white, blue, and red light-emitting diodes. *Appl. Biol. Chem.* **2017**, *60*, 591–596. [CrossRef]
14. Park, C.H.; Yeo, H.J.; Kim, N.S.; Eun, P.Y.; Kim, S.-J.; Arasu, M.V.; Al-Dhabi, N.A.; Park, S.-Y.; Kim, J.K.; Park, S.U. Metabolic profiling of pale green and purple kohlrabi (*Brassica oleracea* var. *gongylodes*). *Appl. Biol. Chem.* **2017**, *60*, 249–257. [CrossRef]
15. Zulak, K.G.; Weljie, A.M.; Vogel, H.J.; Facchini, P.J. Quantitative 1 H NMR metabolomics reveals extensive metabolic reprogramming of primary and secondary metabolism in elicitor-treated opium poppy cell cultures. *BMC Plant Biol.* **2008**, *8*, 1–19. [CrossRef] [PubMed]
16. Park, W.T.; Yeo, S.K.; Sathasivam, R.; Park, J.S.; Kim, J.K.; Park, S.U. Influence of light-emitting diodes on phenylpropanoid biosynthetic gene expression and phenylpropanoid accumulation in *Agastache rugosa*. *Appl. Biol. Chem.* **2020**, *63*, 1–9. [CrossRef]
17. Tuan, P.A.; Thwe, A.A.; Kim, Y.B.; Kim, J.K.; Kim, S.-J.; Lee, S.; Chung, S.-O.; Park, S.U. Effects of white, blue, and red light-emitting diodes on carotenoid biosynthetic gene expression levels and carotenoid accumulation in sprouts of tartary buckwheat (*Fagopyrum tataricum* Gaertn.). *J. Agric. Food Chem.* **2013**, *61*, 12356–12361. [CrossRef] [PubMed]
18. Park, C.H.; Kim, N.S.; Park, J.S.; Lee, S.Y.; Lee, J.-W.; Park, S.U. Effects of light-emitting diodes on the accumulation of glucosinolates and phenolic compounds in sprouting canola (*Brassica napus* L.). *Foods* **2019**, *8*, 76. [CrossRef]
19. Azad, M.O.K.; Kim, W.W.; Park, C.H.; Cho, D.H. Effect of artificial LED light and far infrared irradiation on phenolic compound, isoflavones and antioxidant capacity in soybean (*Glycine max* L.) sprout. *Foods* **2018**, *7*, 174. [CrossRef]
20. Ma, G.; Zhang, L.; Kato, M.; Yamawaki, K.; Kiriiwa, Y.; Yahata, M.; Ikoma, Y.; Matsumoto, H. Effect of blue and red LED light irradiation on β-cryptoxanthin accumulation in the flavedo of citrus fruits. *J. Agric. Food Chem.* **2012**, *60*, 197–201. [CrossRef]
21. Cioć, M.; Szewczyk, A.; Żupnik, M.; Kalisz, A.; Pawłowska, B. LED lighting affects plant growth, morphogenesis and phytochemical contents of *Myrtus communis* L. in vitro. *Plant Cell Tiss. Organ Cult.* **2018**, *132*, 433–447. [CrossRef]
22. Galili, S.; Amir, R.; Galili, G. Genetic engineering of amino acid metabolism in plants. *Adv. Plant Biochem. Mol. Biol.* **2008**, *1*, 49–80.
23. Miflin, B.J.; Habash, D.Z. The role of glutamine synthetase and glutamate dehydrogenase in nitrogen assimilation and possibilities for improvement in the nitrogen utilization of crops. *J. Exp. Bot.* **2002**, *53*, 979–987. [CrossRef]
24. Stitt, M.; Muller, C.; Matt, P.; Gibon, Y.; Carillo, P.; Morcuende, R.; Scheible, W.R.; Krapp, A. Steps towards an integrated view of nitrogen metabolism. *J. Exp. Bot.* **2002**, *53*, 959–970. [CrossRef]
25. Park, C.H.; Xu, H.; Yeo, H.J.; Park, Y.E.; Hwang, G.-S.; Park, N.I.; Park, S.U. Enhancement of the flavone contents of *Scutellaria baicalensis* hairy roots via metabolic engineering using maize Lc and *Arabidopsis* PAP1 transcription factors. *Metab. Eng.* **2021**, *64*, 64–73. [CrossRef]
26. Park, C.H.; Park, S.-Y.; Park, Y.J.; Kim, J.K.; Park, S.U. Metabolite Profiling and Comparative Analysis of Secondary Metabolites in Chinese Cabbage, Radish, and Hybrid ×*Brassicoraphanus*. *J. Agric. Food Chem.* **2020**, *68*, 13711–13719. [CrossRef] [PubMed]
27. Zhao, S.; Park, C.H.; Yang, J.; Yeo, H.J.; Kim, T.J.; Kim, J.K.; Park, S.U. Molecular characterization of anthocyanin and betulinic acid biosynthesis in red and white mulberry fruits using high-throughput sequencing. *Food Chem.* **2019**, *279*, 364–372. [CrossRef] [PubMed]

plants

MDPI

Article

Green Light Improves Photosystem Stoichiometry in Cucumber Seedlings (*Cucumis sativus*) Compared to Monochromatic Red Light

Nicholas B. Claypool and J. Heinrich Lieth *

Plant Sciences, University of California, Davis, CA 95616, USA; nbclaypool@gmail.com
* Correspondence: jhlieth@ucdavis.edu

Abstract: It has been shown that monochromatic red and blue light influence photosynthesis and morphology in cucumber. It is less clear how green light impacts photosynthetic performance or morphology, either alone or in concert with other wavelengths. In this study, cucumber (*Cucumis sativus*) was grown under monochromatic blue, green, and red light, dichromatic blue–green, red–blue, and red–green light, as well as light containing red, green, and blue wavelengths, with or without supplemental far-red light. Photosynthetic data collected under treatment spectra at light-limiting conditions showed that both red and green light enhance photosynthesis. However, photosynthetic data collected with a 90% red, 10% blue, 1000 μmol photons m^{-2} s^{-1}, saturating light show significantly lower photosynthesis in the green, red, and red–green treatments, indicating a blue light enhancement due to photosystem stoichiometric differences. The red–green and green light treatments show improved photosynthetic capacity relative to red light, indicating partial remediation by green light. Despite a lower quantum efficiency and the lowest ambient photosynthesis levels, the monochromatic blue treatment produced among the tallest, most massive plants with the greatest leaf area and thickest stems.

Keywords: blue light; *Cucumis sativus L.* (cucumber); green light; light-emitting diode (LED); morphology; photosynthesis; red light; intrinsic water use efficiency (iWUE); photostationary state of phytochrome (PSS); photosynthetic photon flux density (PPFD); yield photon flux (YPF)

Citation: Claypool, N.B.; Lieth, J.H. Green Light Improves Photosystem Stoichiometry in Cucumber Seedlings (*Cucumis sativus*) Compared to Monochromatic Red Light. *Plants* **2021**, *10*, 824. https://doi.org/10.3390/plants10050824

Academic Editor: Valeria Cavallaro

Received: 29 March 2021
Accepted: 18 April 2021
Published: 21 April 2021

1. Introduction

While light provides energy for photosynthesis, it also directs how plants grow through the use of photoreceptors, such as phytochrome and cryptochrome, which allow the plant to respond to changes in spectral quality ranging from ultraviolet to far-red wavelengths [1]. These responses have implications for plant growth in natural conditions, from the forest floor to field conditions, as well as artificial environments such as indoor agriculture illuminated entirely by electrical lighting [2–5].

Already in the 1970s, research showed that various wavelengths of light had differing effects on photosynthesis on a quantum yield basis [6–9]. In particular, red and blue wavelengths were shown to result in greater rates of photosynthesis than green wavelengths. More recently, studies have examined chemical and structural changes to photosystem stoichiometry and function as they relate to photosynthesis [10,11]. It has been found that monochromatic red light results in poor growth characterized by a low photosynthetic capacity, unresponsive stomatal conductance, low specific leaf weight (leaf mass divided by leaf area), and low maximum quantum efficiency of photosystem II [10–13]. However, the addition of blue light can ameliorate these negative responses, restoring photosynthetic and physiological characteristics comparable to plants grown under white light [13]. In addition to photosynthetic responses, there is widespread interest in how spectral quality changes other aspects of physiology and development.

One commonly reported morphological response is specific leaf weight (SLW), also called leaf mass area (LMA), which is the mass of a leaf divided by its area. This is because SLW represents an investment by the plant per unit of leaf area created, so that plants with the same plant-level net photosynthesis could have very different leaf area due to differences in SLW and different net photosynthesis rates per unit leaf area. Previous studies have found that SLW tends to increase with increasing proportion of blue light [10,14–16].

Many studies have focused on the role of red, blue, and combinations of red and blue light [10,11,15,17–19]. Comparatively little research has been done on green light [20–22]. Nevertheless, it is important to include green light in spectral quality studies, as physiological responses can be the result of interactions between different wavelengths as well as other environmental variables. Green light pulses inhibited blue light-induced phototropism in dark-grown seedlings while enhancing blue light-induced phototropism in light-grown seedlings [23]. Earlier studies showed that green light reversed blue light-induced stomatal opening [24,25].

The present study used cucumber as a model plant for several reasons. First, cucumber has been documented to have high sensitivity to light quality [14–16,26,27]. Second, cucumber is one of the most produced crops under protected cultivation globally under artificial and supplemental lighting systems [28]. Third, while responses to red and blue light have been studied somewhat extensively in cucumber, to our knowledge, the response to green light and interactions between blue, green, and red light is less well understood [11,15,16,29].

Our objective was to characterize cucumber photosynthetic adaptation to diverse spectra containing combinations of red, green, and blue light to determine how light signals in complex spectra interact to influence photosynthesis. Additionally, we sought to understand how photosynthetic differences influence biomass accumulation and morphology.

2. Results

2.1. Photosynthesis

Fitting net photosynthesis (A) vs. cellular CO_2 concentration (C_c) curves to net photosynthesis over a range of CO_2 concentrations allows for the estimation of parameters that relate to leaf-level photosynthesis and the underlying biochemistry limiting photosynthetic assimilation of CO_2 (Figure 1), with estimates for the potential electron transport rate (J) and maximum RuBP carboxylation rate (V_{cmax}) in Table 1. A C_c value of ~270 ppm corresponded to ambient concentrations of CO_2 of 400 ppm under measurement conditions. At that level, photosynthesis was highest in RB and GB (23.9 and 23.6 μmol CO_2 m^{-2} s^{-1}, respectively) followed by RGB, RGB + FR, and B (21.0, 20.5, and 19.6 μmol CO_2 m^{-2} s^{-1}, respectively). Considerably lower photosynthesis values are found for G and RG (11.3 and 10.2 μmol CO_2 m^{-2} s^{-1}, respectively) with R having the lowest photosynthesis of all groups at 5.8 μmol CO_2 m^{-2} s^{-1}.

The estimated maximum rate of Rubisco carboxylation is significantly higher in RB and RGB + FR than all other treatments except GB. The G, R, and RG treatments have significantly lower estimates for V_{cmax} and J than treatments containing blue light—the B, GB, RB, RGB, and RGB + FR treatments (Table 1).

The photosynthesis measurements under 1000 μmol photons m^{-2} s^{-1}, saturating light differ substantially from the photosynthesis measurements under ambient, treatment light at 170 μmol photons m^{-2} s^{-1} (Figure 2). Under ambient conditions, net photosynthesis was highest in RGB and RG, followed by RB and RGB + FR which had significantly higher net photosynthesis than GB or R. The B and G treatments had the lowest net photosynthesis under ambient conditions.

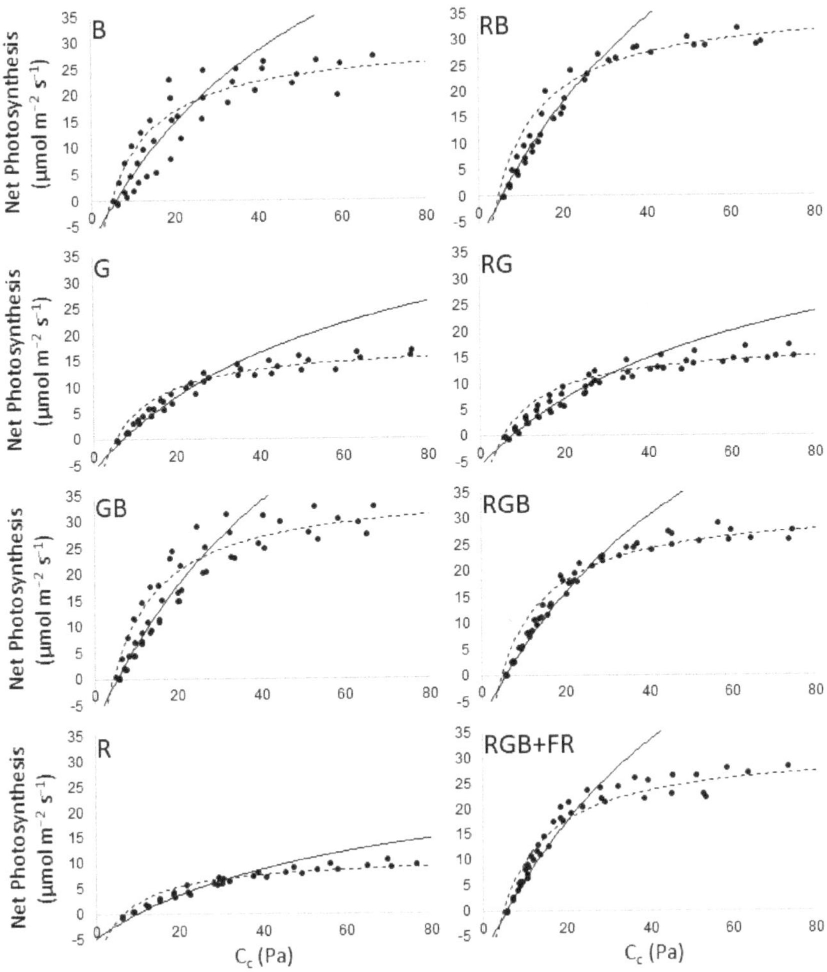

Figure 1. Net photosynthesis (A) vs. cellular CO_2 concentration (C_c) curve fitting for each light treatment. Filled circles represent observed net photosynthesis (A) relative to calculated Cc values. The solid line shows Rubisco limitation, while the dotted line fits RuBP limitation. Triose-phosphate utilization (TPU) limitation was not apparent.

Table 1. Maximum carboxylation rate of Rubisco (Vcmax) and potential rate of photosynthetic electron transport (J) estimated for each light treatment. Different letters indicate significant differences ($p \leq 0.05$; n = 3 or 4).

Treatment	J	V_{cmax}
	μmol m^{-2} s^{-1}	μmol m^{-2} s^{-1}
B	132.0 ± 5.3 c	87.0 ± 4.6 b
G	83.4 ± 1.2 d	53.6 ± 1.2 c
GB	155.5 ± 4.3 ab	102.1 ± 3.6 ab
R	54.3 ± 0.7 e	32.2 ± 0.8 d
RB	157.3 ± 2.8 a	103.0 ± 2.3 a
RG	82.1 ± 1.3 d	49.0 ± 1.2 c
RGB	140.5 ± 1.6 bc	93.2 ± 1.2 b
RGB + FR	137.7 ± 2.7 c	101.9 ± 2.1 a

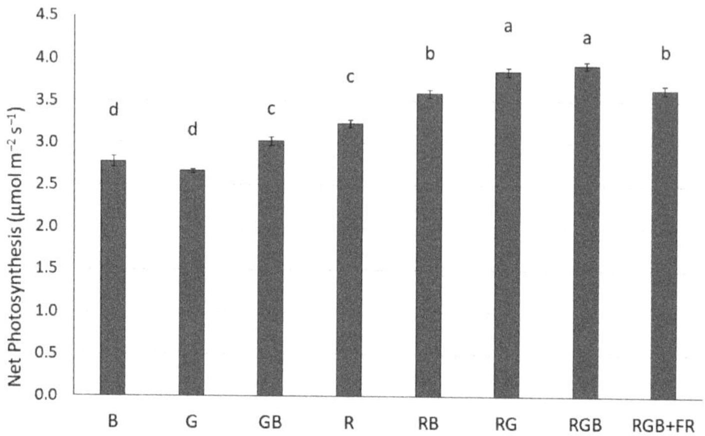

Figure 2. Net photosynthesis (A) under ambient treatment lighting. Different lowercase letters indicate significant differences ($p \leq 0.05$; $n = 6$). Error bars are the standard error. Uppercase letters indicate light treatments.

Overall, there was no correlation between net photosynthesis under ambient, treatment lighting at 170 μmol photons m^{-2} s^{-1} and those observed at the same CO_2 concentration under saturating 90% red, 10% blue light at 1000 μmol photons m^{-2} s^{-1}.

2.2. Chlorophyll Fluorescence

The relative operating efficiency of PSII was highest in the GB treatment and lowest in the R treatment (Table 2). The R and RG treatments had significantly higher ΦPSII values than the R treatment, but significantly lower than all treatments containing blue light. The maximum quantum efficiency of PSII photochemistry (F_v/F_m) was slightly under 0.83—indicating mild stress—for all treatments with no significant differences observed for any treatments (Table 2). Both light-induced and non-light-induced nonphotochemical quenching were higher in treatments lacking blue light (R, G, and RG) compared to treatments containing blue light (B, RB, GB, RGB, and RGB + FR).

Table 2. Maximum quantum efficiency of PSII (Fv/Fm), relative PSII operating efficiency (ΦPSII), coefficient of photochemical quenching (qp), the quantum yield of non-light-induced nonphotochemical quenching (ΦNPQ), the quantum yield of light-induced nonphotochemical quenching (ΦNO), and the fraction of oxidized plastoquinone (q_L) calculated using measurements under saturating (1000 μmol photons m^{-2} s^{-1}) 90% red, 10% blue light. Different letters indicate significant differences ($p \leq 0.05$; $n = 3$ or 4).

Treatment	ΦPSII	F_v/F_m	ΦNPQ	ΦNO	q_P	q_L
B	0.26 ± 0.01 b	0.81 ± 0.01 a	0.48 ± 0.01 c	0.26 ± 0.00 f	0.43 ± 0.01 b	0.23 ± 0.01 ab
G	0.14 ± 0.00 c	0.80 ± 0.00 a	0.54 ± 0.00 a	0.32 ± 0.00 c	0.24 ± 0.01 c	0.12 ± 0.00 d
GB	0.29 ± 0.01 a	0.82 ± 0.00 a	0.44 ± 0.01 e	0.26 ± 0.00 f	0.46 ± 0.01 a	0.24 ± 0.01 a
R	0.09 ± 0.00 e	0.79 ± 0.01 a	0.53 ± 0.00 a	0.38 ± 0.00 a	0.14 ± 0.00 e	0.06 ± 0.00 f
RB	0.26 ± 0.01 b	0.82 ± 0.01 a	0.47 ± 0.01 cd	0.26 ± 0.00 f	0.42 ± 0.01 b	0.22 ± 0.01 bc
RG	0.12 ± 0.00 d	0.80 ± 0.01 a	0.51 ± 0.00 b	0.37 ± 0.00 b	0.19 ± 0.01 d	0.08 ± 0.00 e
RGB	0.27 ± 0.01 b	0.82 ± 0.00 a	0.43 ± 0.01 e	0.29 ± 0.00 d	0.42 ± 0.01 b	0.20 ± 0.01 c
RGB + FR	0.27 ± 0.01 b	0.81 ± 0.01 a	0.45 ± 0.01 de	0.28 ± 0.00 e	0.43 ± 0.01 b	0.23 ± 0.01 b

Under light-saturating conditions, net photosynthesis was significantly lower in the G, R, and RG treatments than all other treatments (Figure 1). However, this had no apparent effect on photosynthesis under ambient conditions (Figure 2). While ambient photosynthesis was lowest in the G treatment, net photosynthesis in R was comparable with GB, and significantly higher than B or G. Finally, ambient photosynthesis in RG was significantly higher than all other treatments save RGB (Figure 2).

2.3. Shoot Characteristics

Qualitative differences between treatments can be seen in Figure 3, which shows exemplar plants (those closest to treatment average in height and mass) from the replication experiment for each treatment. Shoot dry weight showed no clear trends, except that far-red light increased shoot dry weight, with an average of 2.53 g per plant for the RGB + FR treatment and only 1.63 g for the RGB treatment (Figure 4). The RGB + FR and B treatments had significantly higher dry weight than the GB, R, and RB treatments, while the RGB + FR treatment also had significantly higher dry weight than the RG and RGB treatments (Figure 4).

Figure 3. Profile and overhead images of representative plants, chosen by selecting the plant that was closest in dry weight and height to the treatment average. This image is a composite to allow for visual comparison between treatments. Cucumber seedlings were grown under blue (B), green (G), green–blue (GB), red (R), red–blue (RB), red–green (RG), red–green–blue (RGB), and red–green–blue with far-red (RGB + FR) light. The white bar in the upper middle is 10 cm for profile images, while the yellow bar in the lower middle is 10 cm for overhead images.

Far-red light also increased plant height, with the RGB + FR treatment being significantly taller than the RGB treatment (Figure 4). Conversely, supplemental blue light decreased plant height, with shorter plants in RB than R, GB than G, and RGB than RG. However, plants grown in the B treatment were taller than all other treatments except RGB + FR. The RGB + FR and B treatments, in addition to being the tallest, also had the lowest leaf dry weight fraction (leaf dry weight divided by shoot dry weight) (Figure 4).

There were no clear trends for stem diameter, except that far-red light enhanced stem diameter, with plants in the RGB + FR treatment having significantly greater stem diameter than plants in the RGB treatment (Figure 4).

The RGB + FR treatment also resulted in significantly greater leaf area than the RGB treatment (Figure 4). Due to high within groups variability, there were no other significant trends in leaf area, although with a larger sample size, a trend of decreasing leaf area with supplemental blue light may be observed.

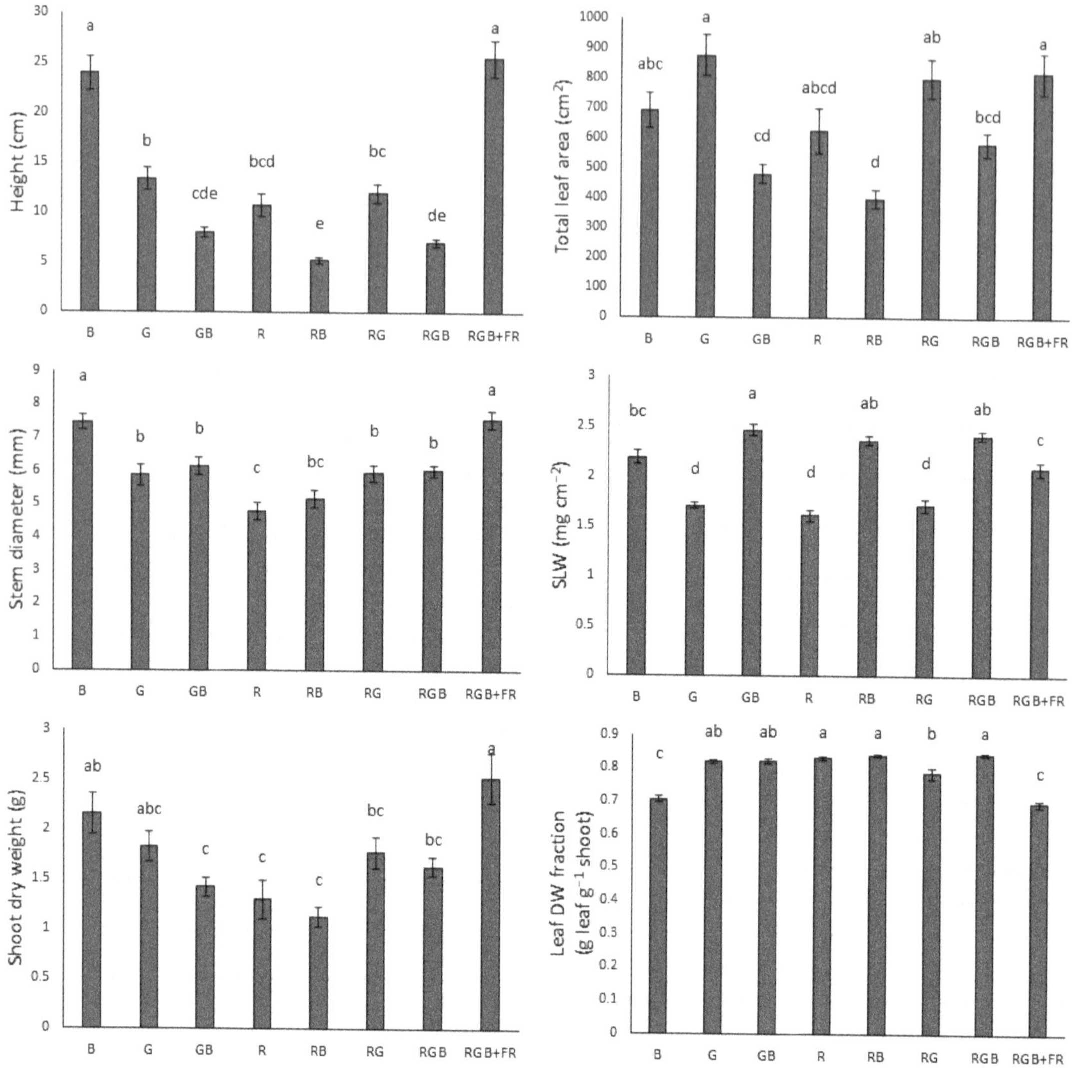

Figure 4. Shoot characteristics of cucumber under diverse spectra. Different lowercase letters indicate significant differences ($p \leq 0.05$; n is between 25 and 30 for each treatment). Uppercase letters indicate light treatments.

Specific leaf weight (SLW), the dry weight of a leaf divided by its area, does show a clear trend with blue light significantly increasing SLW. Higher specific leaf weights were observed in the GB treatment relative to G, RB relative to R, and RGB relative to RG (Figure 4). Far-red light decreased SLW, with the RGB + FR treatment having significantly lower SLW than the RGB treatment.

2.4. Stomatal Characteristics

Stomatal conductance under ambient lighting was significantly higher in B (0.24 mol m^{-2} s^{-1}) relative to R (0.09 mol m^{-2} s^{-1}) or G (0.09 mol m^{-2} s^{-1}) and significantly higher in GB (0.28 mol m^{-2} s^{-1}) relative to G, in RB (0.19 mol m^{-2} s^{-1}) relative to R, and in RGB (0.27 mol m^{-2} s^{-1}) relative to RG (0.09 mol m^{-2} s^{-1}), demonstrating a blue light-mediated increase in stomatal conductance (Figure 5A).

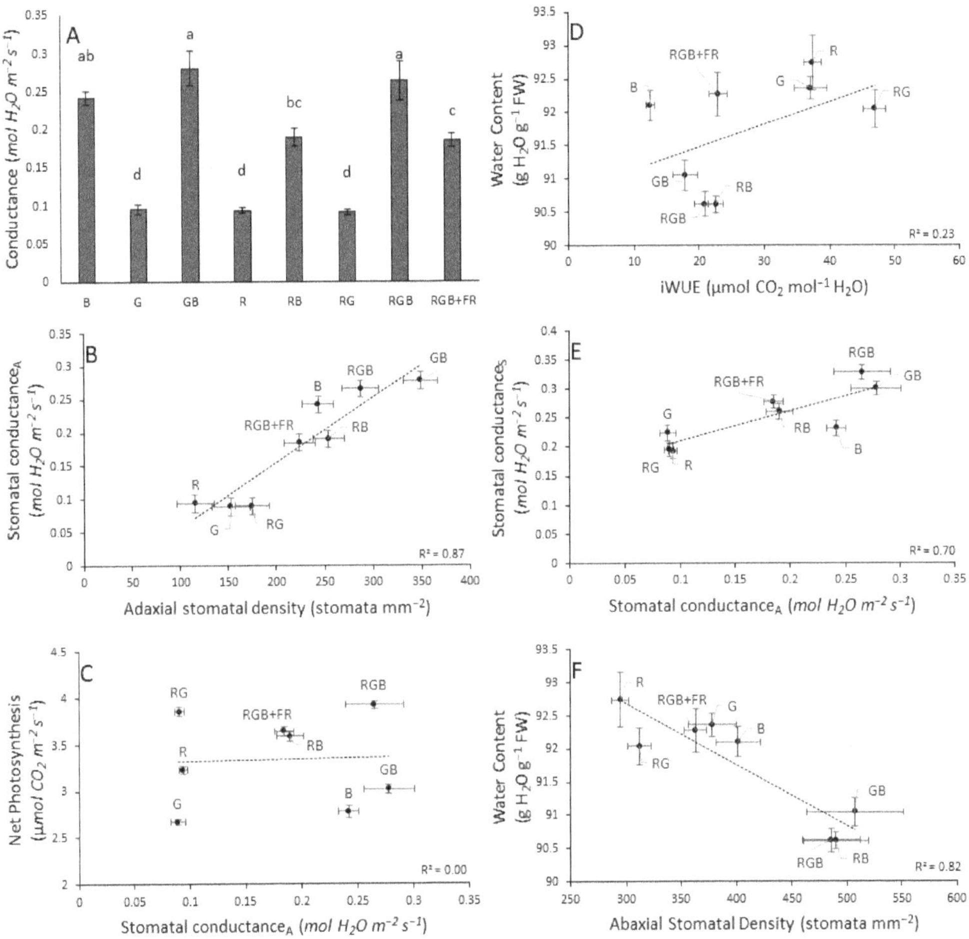

Figure 5. (**A**): Stomatal conductance under saturating light by light treatment. Different lowercase letters indicate significant differences ($p \leq 0.05$; n is between 25 and 30 for each treatment). (**B**): Stomatal conductance under ambient, treatment lighting (Stomatal conductance$_A$) vs. adaxial stomatal density. (**C**): Net photosynthesis under ambient, treatment lighting vs. stomatal conductance under ambient, treatment lighting. (**D**): Water content vs. instantaneous water use efficiency. (**E**): Stomatal conductance under saturating light vs. stomatal conductance under ambient, treatment lighting. (**F**): Water content vs. abaxial stomatal density. Uppercase letters indicate light treatments.

We found a significant increase in conductance from RB to RGB, but there was no difference in stomatal conductance between G, R, and the RG treatments (Figure 5A). Far-red light decreased stomatal conductance, with conductance significantly lower in RGB + FR (0.18 mol m^{-2} s^{-1}) compared to RGB. Stomatal conductance under ambient, treatment lighting was highly correlated with adaxial stomatal density (Figure 5B, R^2 = 0.87). However, stomatal conductance was not correlated to net photosynthesis under ambient, treatment lighting (Figure 5C, R^2 = 0.00).

When measuring the A vs. C_c curves, all plants were subjected to saturating levels of 90% red, 10% blue light at 1000 µmol photons m^{-2} s^{-1}. Despite being illuminated with the same spectrum, conductance trends were similar to those obtained when illuminated by treatment spectra. Overall, average conductance values under saturating light were higher in all treatments compared to ambient lighting conditions except the B treatment, with R^2 = 0.70 ($p < 0.01$) (Figure 5E).

Like conductance, blue light resulted in an increased stomatal density, with abaxial stomatal density higher in B relative to R (although not different from G), and higher abaxial density in GB relative to G, RB relative to R, and RGB relative to RG. The same trends were found for adaxial stomatal density, except that adaxial stomatal density in B was significantly higher than G (Table 3).

Table 3. Stomatal density, abaxial (AB) to adaxial (AD) stomatal density ratio, intrinsic water use efficiency, and water content of cucumber.

Treatment	Abaxial	Adaxial	AB:AD	iWUE	Water Content
	Stomata/mm^2	Stomata/mm^2		μmol CO_2 mol^{-1} H_2O	g H_2O 100 g^{-1} FW
B	402 ± 22 bcd	243 ± 16 bc	1.66 ± 0.11 b	12.5 ± 1.4 d	92.1 ± 0.2 a
G	378 ± 26 cde	153 ± 20 de	2.67 ± 0.13 a	37.1 ± 1.5 b	92.4 ± 0.2 a
GB	507 ± 25 a	349 ± 18 a	1.55 ± 0.12 b	17.9 ± 1.4 cd	91.0 ± 0.2 b
R	295 ± 25 e	116 ± 19 e	2.76 ± 0.13 a	37.4 ± 1.4 b	92.7 ± 0.4 a
RB	490 ± 22 ab	254 ± 16 b	1.99 ± 0.11 b	22.6 ± 1.4 c	90.6 ± 0.1 b
RG	312 ± 24 de	175 ± 18 cde	1.82 ± 0.12 b	47.0 ± 1.3 a	92.0 ± 0.3 a
RGB	486 ± 25 abc	287 ± 19 ab	1.73 ± 0.13 b	20.9 ± 1.4 c	90.6 ± 0.2 b
RGB + FR	363 ± 21 de	224 ± 16 bcd	1.66 ± 0.11 b	23.0 ± 1.4 c	92.3 ± 0.3 a

Abaxial stomatal density was also lower in RGB + FR than RGB, though there was no difference in adaxial stomatal density (Table 3). We observed significantly higher abaxial:adaxial ratios for the G and R treatments relative to all other treatments.

Intrinsic water use efficiency (iWUE) under ambient conditions, which is calculated by dividing the net photosynthesis by stomatal conductance, was highest in RG, followed by G and R, while iWUE was lowest in the B treatment. Like [30], we found no difference in iWUE between RB and RGB; however, [31] did find a significant increase in iWUE in a low R:FR treatment compared to a high R:FR treatment, while we found no difference between the RGB and RGB + FR treatment.

Intrinsic water use efficiency had only a weak correlation with water content at harvest (Figure 5D, R^2 = 0.23). Water content, the percentage of fresh weight from water, was lower in GB, RB, and RGB than all other treatments. Interestingly, while stomatal conductance was best explained by adaxial stomatal density, water content at harvest was best explained by abaxial stomatal density (R^2 = 0.82, Figure 5F).

3. Discussion

Previous experiments found that cucumber measured under saturating light and grown under monochromatic red light showed lower photosynthesis than cucumber plants grown under red–blue light, consistent with our findings [10].

The J and V_{cmax} values calculated suggest that blue light significantly enhances photosynthetic capacity relative to treatments lacking blue light. Since these values are lowest in the R treatment, and significantly higher in the RG and G treatments, we can also conclude that green light improves photosynthetic capacity relative to monochromatic red light. However, because the RG and G treatments have significantly lower J and V_{cmax} values than treatments containing blue light, the effect of green light must be lesser than that of blue light.

Previously, calculations of V_{cmax} and J were found to be significantly higher in B than R; however, they found no difference in V_{cmax} between R and RB, while estimates for V_{cmax} were significantly higher in RB than R in our experiment [12]. Others calculated V_{cmax} for a low R:FR treatment as significantly lower than for a high R:FR treatment, while we found V_{cmax} to be significantly higher in the RGB + FR treatment compared to the RGB treatment [31].

Our findings suggest that green light enhanced net photosynthesis, since values for V_{cmax}, J, and net photosynthesis were significantly greater in GB relative to B, RG relative to R, and RGB relative to RB (Figure 3). Photosynthesis was also significantly higher in RB relative to B, RG relative to G, and RGB relative to GB, suggesting a red light enhancement. Others have found that a broader spectrum resulted in higher fixation than a red–blue light treatment for tomato and poinsettia but saw no difference in cucumber at ambient CO_2 concentrations [32]. They did observe a difference at elevated CO_2, similar to the enhancement we saw for photosynthesis in RGB relative to RB at ambient CO_2 and light levels. Others have found higher fixation in B than R for cucumber while we observed the opposite under ambient conditions [11,12,15,33]. During A vs. C_c measurements, plants from the B treatment had a higher net photosynthesis level than plants in the R treatment, indicating that the choice of spectral composition, intensity, or both is critically important to comparing net photosynthesis levels between treatments, even when the same light source is used for each treatment.

It is possible that the red and green light enhancement can be described in part by the 'enhancement effect' or 'Emerson effect' which refers to the phenomenon where photosynthesis from combined spectra can be greater than the sum of its parts due to excitation energy distribution between photosystem I and photosystem II [34–36].

3.1. Chlorophyll Fluorescence

The maximum quantum efficiency of PSII photochemistry (F_v/F_m) indicates how effectively PSII uses absorbed light energy to reduce the primary quinone acceptor of PSII (Q_A) [37]. In practice, this measure can be used to assess stress in plants, as a value of ~ 0.83 is very consistent across species in non-stressed leaves [38]. Values below 0.83 indicate stress and a reduced maximum photosynthetic capacity; however, photosynthesis may not be reduced under ambient conditions as the quantum yield of PSII (ΦPSII) is generally considerably lower than F_v/F_m, especially under high light intensity. A low F_v/F_m is one of the symptoms of red light syndrome [11–13].

F_v/F_m was qualitatively lower in G, R, and RG than all other treatments, indicating a reduced maximum photosynthetic efficiency with values suggesting mild stress (Table 4). These qualitative differences are supported by previous findings that a comparatively higher level of blue light in LED treatments increased F_v/F_m relative to high pressure sodium treatments [39]. Others have also concluded that blue light enhances PSII photochemistry relative to red light [11–13].

PSII operating efficiency decreases with increasing light intensity, primarily due to a reduced ability to oxidize Q_A rather than an increase in non-photochemical quenching (NPQ) [37,40]. Therefore, it is not surprising that under saturating light the PSII operating efficiencies observed were much lower than F_v/F_m. PSII operating efficiency was significantly lower in G, R, and RG than all other treatments (with ΦPSII in R significantly lower than RG, which was significantly lower than ΦPSII in G) (Table 2). It is not possible to estimate electron transport rate or the quantum yield of CO_2, since we cannot account for alternative electron sinks to PSII because these measurements were taken under atmospheric O_2 concentrations. Nevertheless, ΦPSII gives an estimate on the upper limit of possible photosynthetic carbon assimilation under a given condition, and the trend observed is very similar to the trend in net photosynthesis observed.

Despite the lower ΦPSII values, ΦNPQ, the quantum yield of light-induced quenching, and ΦNO, non-light-induced quenching, are both significantly higher in the G, R, and RG treatments than all other treatments (Table 2). Together, these data suggest that electron acceptors downstream of PSII are insufficient in the G, R, and RG treatments compared to the other treatments, and that G, R, and RG treatments are compensating by increasing nonphotochemical quenching to reduce photo-induced damage. Since the spectral quality and intensity used to excite the photosystems were identical across treatments during light-saturated measurements, one would expect differences in net photosynthesis and chlorophyll fluorescence to be related to adaptive differences between light treatments.

Table 4. Color breakdown for light treatment spectra as a percentage of total photosynthetic photon flux density (PPFD). Wavelength ranges for the traditional method indicate the typically defined range for each color, while the wavelength range for the bar method indicates the range in which >99% of the light is emitted from a given bar color.

	Treatment	Red (600–700 nm)	Green (500–600 nm)	Blue (400–500 nm)
Traditional method	B	0	1	99
	G	0	93	6
	GB	0	37	63
	R	100	0	0
	RB	47	1	52
	RG	65	31	3
	RGB	38	22	40
	RGB + FR	40	20	40
	Treatment	**Red (623–684 nm)**	**Green (486–582 nm)**	**Blue (432–500 nm)**
Bar method	B	0	0	100
	G	0	100	0
	GB	0	38	62
	R	100	0	0
	RB	47	0	53
	RG	65	35	0
	RGB	38	23	39
	RGB + FR	40	21	39

Plants have a variety of mechanisms to respond to changes in light quality. In the short term, light-harvesting complex II (LHC-II) can be transferred from PSII to PSI to help balance excitation energy between the two systems to improve electron transport efficiency [41]. In the long term, algae, cyanobacteria, and higher plants adjust the stoichiometry of photosystem I (PSI) and photosystem II (PSII) in response to light quality to improve photosynthetic efficiency [42–45] as well as their pigment composition [42,46] to more efficiently absorb ambient light.

PSII is primarily excited by wavelengths at ~450–640 nm while PSI uses light above 680 nm much more efficiently than PSII [47]. Since the blue LEDs are the only source of photons at 450 nm in our light treatments, treatments lacking these wavelengths (G, R, and RG) likely have adjusted stoichiometry to decrease the number of PSI complexes relative to PSII to improve electron transport efficiency due to less efficient excitation of PSII relative to PSI. As the green LEDs supply light within the range that PSII can use effectively, this stoichiometric adjustment would be expected to be most pronounced in the R treatment, and less so in the G and RG treatments.

When the plants were exposed to the novel light treatment (90% red, 10% blue) during A vs. C_c measurements they could use transient LHC-II to improve the balance of excitation between PSI and PSII, but the capacity to balance in the G, R, and RG treatments may have been limited by extreme stoichiometric differences not seen in the other treatments. This would explain the much poorer performance of these three treatments relative to the other treatments and the poorer performance of R relative to G and RG during A vs. C_c measurements, while the same long-term adaptations may have allowed for the trends seen in Figure 1 during measurement under ambient lighting.

In any case, neither the photosynthesis measurements under saturating 90% red, 10% blue light at 1000 μmol photons m^{-2} s^{-1} nor ambient light at 170 μmol photons m^{-2} s^{-1} correlate well with shoot dry weight. There are many potential reasons for the lack of correlation between shoot biomass and photosynthesis measurements. First, the dry weight data include only shoot biomass, not root biomass. It is possible that with root biomass, the whole-plant biomass values would correlate well with the net photosynthesis measurements observed. The photosynthesis values presented are on a per-area basis. Leaf area and specific leaf weight (leaf area divided by leaf dry weight) vary between treatments.

It is therefore possible that plants with equivalent net photosynthesis rates could have very different whole plant growth rates due to differences in leaf area [48].

3.2. Stomatal Conductance

Others have also noted a trend of increasing stomatal conductance in cucumber with increasing blue light [11,15]. Significantly higher stomatal conductance in cucumber grown under monochromatic blue light compared to cucumber grown under monochromatic red light was found here and previously [33]. Another study also measured significantly higher stomatal conductance in cucumber seedlings under a monochromatic blue light treatment relative to a monochromatic red light treatment and found no difference between the B treatment and RB treatment, which our results support [12]. However, we found conductance in RB to be significantly greater compared to R, while they found no difference in conductance between the R and RB treatment. Our results differ from a finding that there was no difference in stomatal conductance of cucumber between a white light LED treatment and a red–blue light treatment, while we found conductance to be significantly higher in RGB compared to RB [30]. It is possible that this is because our RGB treatment was roughly 2:1:2 B:G:R light while the white light treatment in their experiment was roughly 1:2:1 B:G:R light. In Arabidopsis, red–blue light increased stomatal aperture more than red light alone, while red–green–blue light showed no increase in aperture relative to red light alone [24].

Our findings on blue and green light effects on stomatal conductance are similar to those findings in Arabidopsis, as they observed no difference in aperture between R and RG. However, they found a significant decrease in stomatal aperture for monochromatic G compared to R or RG while we found no difference. It is possible that this is because our plants had time to form long-term adaptations to light quality, while the Arabidopsis leaves were being exposed to a novel lighting condition and therefore only had short-term responses.

Another study found that a decreased ratio of R:FR resulted in a decrease in stomatal conductance in cucumber relative to a treatment with high R:FR light [31]. This is similar to our findings between the RGB and RGB + FR treatments, where RGB + FR had significantly lower conductance than the RGB treatment.

The fact that similar conductance trends were observed despite illumination under very different spectral quality suggests that the results are driven more by physical differences in the leaves than transient chemical expression induced by light signaling.

This is supported by the stomatal density data shown in Table 3 and the correlation between stomatal conductance under ambient lighting and adaxial stomatal density (Figure 5).

The stomatal density trends we observed agree with previous findings of significantly higher average stomatal density in B and RB light treatments relative to R [29]. Others found that a decreased R:FR ratio resulted in a decreased stomatal density, although they found a significant decrease in adaxial rather than abaxial stomatal density [31]. They also found a significant difference in the abaxial to adaxial stomatal ratio (AB:AD) while we found no difference between RGB and RGB + FR.

Stomatal density also explains more than half of the variation found in intrinsic water use efficiency (iWUE) under ambient conditions, which is calculated by dividing the net photosynthesis by stomatal conductance ($R^2 = 0.58$, Figure 5).

3.3. Morphology

In red–blue light treatments, increasing proportions of blue light led to reduced leaf area [15]. Previously, it was found that monochromatic blue light resulted in the highest leaf area, followed by white light, which had significantly higher leaf area than monochromatic red light [33]. We did not find any statistically significant differences between these treatments, but that may be due to the high within-group variation and relatively low sample size, as the mean leaf areas in our treatments follow the same trend.

The same study found no difference between R, B, and RB treatments, while we found RB to have significantly lower leaf area than R or B [33]. The effect of decreasing the R:FR ratio on leaf area is species specific, with some showing decreased leaf area, while others like petunia show an increase in leaf area [49]. In our experiment, far-red light mediated increased leaf area, which has been found by others [50].

Cucumber grown under monochromatic red light was taller than cucumber grown under a 1:1 ratio of red to blue, with both treatments being shorter than a monochromatic blue treatment, which is consistent with our findings [15]. Far-red-regulated increases in plant height are well documented [50–53]. As the height of the light fixtures was not adjusted during the experiment, it is possible that plants which grew taller received more irradiation than shorter plants, potentially affecting total shoot biomass.

The leaf area and specific leaf weights observed may also help to explain the average shoot biomass for each treatment. For example, the G treatment had the lowest net photosynthesis per unit leaf area along with the B treatment (Figure 2). However, both the G and B treatments had high leaf area, potentially allowing for the same or greater total photosynthesis as a treatment with lower leaf area but higher net photosynthesis per unit area, such as the RB treatment which had higher net photosynthesis per unit area, but lower leaf area and shoot dry weight.

In the case of the G treatment, a low SLW meant that more leaf area could be produced using the same amount of photosynthate compared to the RB treatment. The SLWs for the B and RB treatments were not significantly different, but the B treatment plants were much taller, potentially resulting in higher average light intensity for the duration of the experiment.

4. Materials and Methods

The experimental work consisted of 2 replications over time. Cucumber (*Cucumis sativus* cv. Diva) seeds were germinated in total darkness at 32 °C. Once germinated, seeds were transplanted into 4 in. pots containing UC mix ($^1/_3$ peat moss, $^1/_3$ redwood sawdust, $^1/_3$ fine sand), covered with an additional 100 cm^3 of UC mix, and randomly distributed into their light treatment chambers. Plants were irrigated with $\frac{1}{2}$ strength Hoagland's solution every third day for the first two weeks, then daily thereafter [54]. Leaf photosynthetic rate, stomatal conductance, and fluorescence measurements were obtained during both replications. Morphological measurements were made four weeks after transplant.

In both replications, plants were grown in chambers 61 cm wide, 122 cm long, and 90 cm tall. An 8 in. duct fan exhausted air from the chambers so that the average temperature was 23.0 ± 0.2 °C when the lights were on and 20.9 ± 0.2 °C when the lights were off.

Each chamber was illuminated with lamps consisting of various light-emitting diode (LED) bars (Demegrow, Inc., Sacramento, CA, USA) specifically designed to provide a custom spectrum in each chamber. Fixtures consisted of different combinations of diodes emitting far-red, red, green, or blue light, with peak intensities at wavelengths of 744 nm, 661 nm, 521 nm, and 460 nm, respectively. Spectra of the resulting lamp systems were measured with a JAZ spectrometer (model: JAZ spectrometer, Ocean Optics, Largo, FL, USA). The full width at half maximum for each peak was 21.8, 20.7, 33.6, and 21.6 nm for far-red, red, green, and blue peaks, respectively (Figure 6). Green light in the RGB and RGB + FR treatments came from 15,000 K white LEDs, which is why the green peak is broader in these treatments than other treatments containing green light. Each fixture installation was configured so that all had comparable photosynthetic photon flux density (PPFD) levels of 120 µmol photons m^{-2} s^{-1} in each chamber. This was achieved by raising or lowering the lamp array in each chamber and averaging measurements over a 45-point grid.

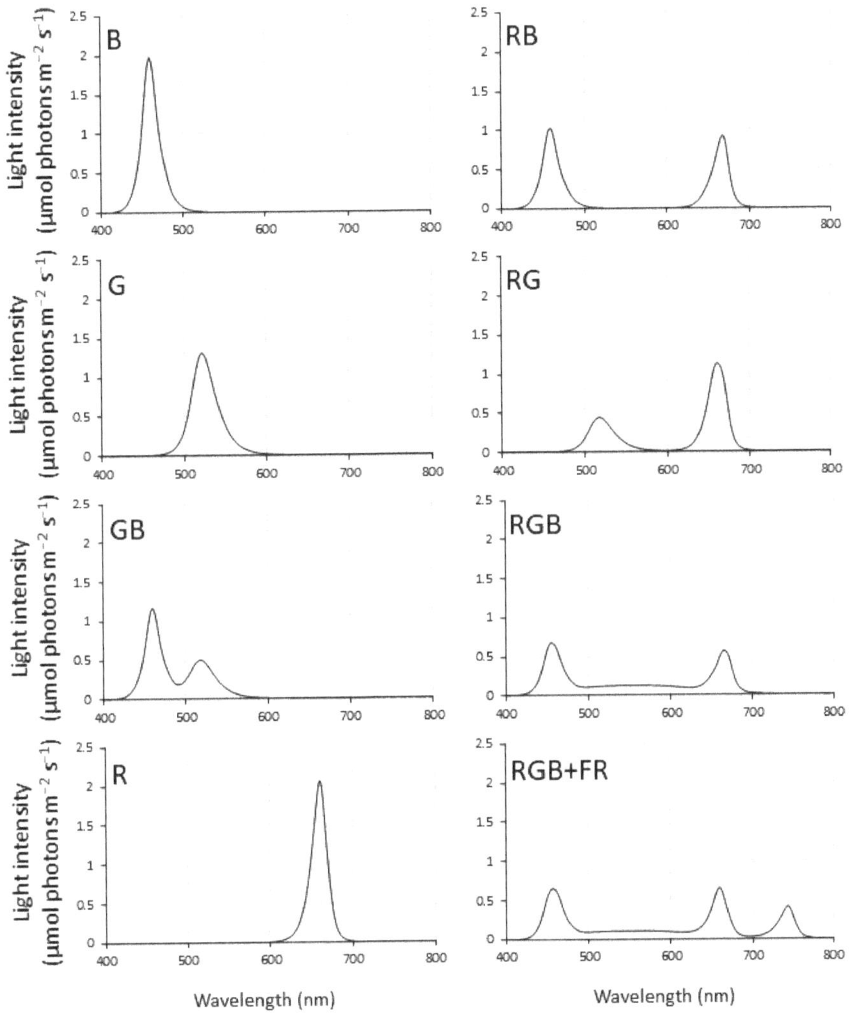

Figure 6. Light treatment spectra: monochromatic blue (B), monochromatic green (G), monochromatic red (R), green–blue (GB), red–blue (RB), red–green (RG), red–green–blue (RGB), and red–green–blue with far red (RGB + FR).

Where both colors were present, the intensity of blue and red are roughly 1:1, blue and green are 2:1, and red and green are 2:1; actual percentages of total light as in Table 4. Since the energy of far-red light does not contribute to photosynthetic photon flux density (PPFD), the RGB and RGB + FR treatments have roughly the same PPFD and light ratios between 400 and 700 nm, but 18% of all incident irradiation between 400 to 800 nm in the RGB + FR treatment was in the far-red (700 to 800 nm) region. One percent of incident photons were in the far-red region in the RGB treatment, while all other treatments had negligible levels of far-red light. In addition to the traditional color quantification (red 600–700 nm, green 500–600 nm, and blue 400–500 nm), the light is reported based on the quantity from each 'color' of LED bar. This was determined by only powering LED bars of a given light color and measuring PPFD, then calculating the percentage of total PPFD from that bar color (Table 4).

Yield photon flux (YPF) was calculated for each light treatment according to [9] by multiplying relative quantum efficiency at a given wavelength with the photon flux at that wavelength, then integrating from 300 to 800 nm (Table 5). The YPF model adjusts PPFD based on the likelihood that a photon of a given wavelength will be absorbed and the likelihood that the energy will be used for photosynthesis once absorbed.

Table 5. Yield photon flux (YPF) and photostationary state of phytochrome (PSS) for each light treatment.

Treatment	YPF	PSS
	$\mu mol\ m^{-2}\ s^{-1}$	P_{fr}:P_{total}
B	88	0.51
G	94	0.83
GB	90	0.62
R	114	0.89
RB	98	0.86
RG	106	0.88
RGB	102	0.86
RGB + FR	104	0.76

Finally, photostationary state of phytochrome (PSS), an estimate of active phytochrome as a portion of total phytochrome, was calculated using

$$PSS = \left(\sum_{300}^{800} N_\lambda \sigma_{r\ \lambda}\right) / \left(\sum_{300}^{800} N_\lambda \sigma_{r\ \lambda} + \sum_{300}^{800} N_\lambda \sigma_{fr\ \lambda}\right) \tag{1}$$

as reported in Table 2 [9].

Equation (1) gives PSS where N is incident photon flux at a given wavelength (λ), σ_r is the photochemical cross section of P_r (the red-absorbing, inactive form of phytochrome) at λ, and σ_{fr} is the photochemical cross section of P_{fr} (the far-red-absorbing, active form of phytochrome) at λ.

Photosynthesis was measured in two ways. First, using a LI-6400 with a clear-top chamber (LI-COR Biosciences, Lincoln, NE, USA), net photosynthesis (A) and stomatal conductance (g_s) were measured at an ambient CO_2 concentration (C_a) of 400 ppm and a leaf temperature of 25 °C, illuminated by treatment light spectra at ambient intensity.

Second, A vs. C_c response curves, the net photosynthesis rate obtained under varying concentrations of CO_2 in the chloroplast (C_c) under saturating light, were measured using the LI-6400 portable photosynthesis system with a 6400-40 leaf chamber fluorometer attachment (LI-COR Biosciences, Lincoln, NE, USA) in order to gain insight about possible molecular adaptations to the light environment. Measurements were taken at external CO_2 concentrations of 400, 300, 250, 200, 150, 100, 50, 400, 500, 600, 850, and 1000 ppm in that order. Following initial fluorescence measurements, the plants had half an hour to adapt to light at 1000 $\mu mol\ m^{-2}\ s^{-1}$. Each CO_2 concentration was held for two to four minutes at a flowrate of 300 $\mu mol\ s^{-1}$ with leaf temperature set to 25 °C while the plant was at room temperature (23 to 26 °C). The first true leaf, unshaded by neighboring leaves, was measured.

Plants were first dark adapted for half an hour before initial measurements. Fluorescence measurements were taken on the dark-adapted leaves, before acclimating to the light for a half hour. The minimum chlorophyll fluorescence for dark-adapted leaves (F_o), maximum light- and dark-adapted chlorophyll fluorescence (F_m' and F_m, respectively), and steady state light-adapted chlorophyll fluorescence (F′) were measured. The maximum quantum efficiency of PSII (F_v/F_m), the relative PSII operating efficiency (ΦPSII), the coefficient of photochemical quenching (q_p), the quantum yield of non-light-induced nonphotochemical quenching (ΦNPQ), and the quantum yield of light-induced nonphotochemical quenching (ΦNO) were calculated according to [37]. The fraction of oxidized plastoquinone, q_L, was calculated according to [55]. Due to the difficulties of measuring F_o', the minimal fluorescence of a light-adapted leaf, it was calculated using the equation $F_o' = F_o/[(F_v/F_m) + (F_o/F_m')]$ where F_o is the minimal fluorescence of a dark-adapted leaf,

F_m is the maximal fluorescence from a dark-adapted leaf, $F_m{}'$ is the maximal fluorescence from a light-adapted leaf, and F_v is the difference between F_m and F_o [40].

A vs. C_c curve fitting was done using SAS Studio 3.8 software via the NLIN procedure, a procedure for fitting nonlinear models, using Equations (2)–(4) [56,57]. Typically, these model fittings involve 3 segments representing photosynthesis as limited either by the maximum ribulose-1,5-bisphosphate (RuBP) carboxylation rate Equation (2), the RuBP regeneration rate Equation (3), or the triose phosphate utilization (TPU) rate. However, our data suggest that TPU was not a limiting factor and so we fitted to only the Rubisco-limiting (Equation (2)) and the RuBP-limiting curves (Equation (3)). The equation for calculating the concentration of CO_2 at Rubisco, C_c, has also been included (Equation (4)).

$$A = V_{cmax}\left[\frac{C_c - \Gamma^*}{C_c + K_c\left(1 + \frac{O}{K_o}\right)}\right] - R_d \tag{2}$$

$$A = J\left[\frac{C_c - \Gamma^*}{4C_c + 8\Gamma^*}\right] - R_d \tag{3}$$

$$C_c = C_i - \frac{A}{g_m P_{atm}} \tag{4}$$

where C_i is the intercellular concentration of CO_2, C_c is the concentration of CO_2 at Rubisco, A is net CO_2 assimilation, V_{cmax} is maximum carboxylation rate of Rubisco, Γ^* is the point at which oxygenation is twice the rate of carboxylation (CO_2 uptake equals CO_2 photorespiratory release), K_o is the inhibition constant of Rubisco for oxygen, K_c is the Michaelis–Menten constant of Rubisco for CO_2, O is the partial pressure of O_2 at Rubisco, R_d is non-photorespiratory CO_2 release, J is the rate of electron transport, P_{atm} is atmospheric pressure, and g_m is mesophyll conductance.

Due to the difficulty of accurately determining g_m due to the method of data collection and initial fittings determining that g_m did not significantly differ between any treatments, the overall average value of 2.12 μmol m^{-2} s^{-1} Pa^{-1} was used [58]. Likewise, since estimates of R_d did not significantly differ between treatments, an average value of 2.71 μmol CO_2 m^{-2} s^{-1} was used.

All plants' shoots were severed at the substrate surface and weighed for fresh weight, separated into leaf blades and all other material (stem, petioles, cotyledons, and leaves < 2 cm^2), oven dried for 72 h at 60 °C, and weighed to obtain dry weights. Stem diameter was measured with an electronic caliper just below the cotyledons with the caliper arm held parallel to the cotyledons to give a consistent measurement for seedlings with non-circular stem cross-sections. Stem height was measured from the point at which the shoot was severed to the base of the apical meristem to the nearest millimeter. The two largest leaves on each plant had length, width, petiole length, and leaf blade area measured. Additionally, total leaf area was measured using a LI-COR 3100 leaf area meter (LI-COR Biosciences, Lincoln, NE, USA). Finally, stomatal density was measured by taking a 1 cm by 2 cm section of leaf tissue adjacent to the midrib approximately halfway between leaf tip and leaf blade base and applying clear nail polish [59].

All means separations were determined using SAS Studio software 3.8 (SAS Institute Inc., Cary, NC, USA). Data from the two replications were treated as separate blocks with means separation analyzed by a Tukey–Kramer HSD (p = 0.05).

5. Conclusions

Plants adapt to light signals by adjusting photosystem stoichiometry. In monochromatic red light, this reduces photosynthetic capacity of the plant under broad spectra and saturating light conditions. However, this stoichiometric imbalance is not seen in spectra containing blue light and is partially remediated by spectra containing green light. Despite this observance under saturating light conditions, monochromatic green light had lower net photosynthesis rates than monochromatic red light under ambient conditions. Nevertheless, other factors, such as morphological adaptations like height, leaf area, and SLW seem to drive biomass accumulation as much or more than net photosynthesis per unit leaf area, given that plants grown under monochromatic blue light were tied for lowest ambient net photosynthetic rate with monochromatic green light, but the B treatment produced more massive plants than the GB, R, and RB treatments which all had higher net photosynthesis under ambient conditions than the B treatment plants.

Author Contributions: Conceptualization, N.B.C.; methodology, N.B.C.; validation, N.B.C.; formal analysis, N.B.C.; investigation, N.B.C.; resources, J.H.L.; data curation, N.B.C.; writing—original draft preparation, N.B.C.; writing—review and editing, N.B.C. and J.H.L.; visualization, N.B.C.; supervision, N.B.C.; project administration, N.B.C.; funding acquisition, J.H.L. All authors have read and agreed to the published version of the manuscript.

Funding: Undergraduate interns were funded by The Robert and Lien Chen Family Foundation.

Institutional Review Board Statement: Not applicable.

Informed Consent Statement: Not applicable.

Data Availability Statement: Data are available upon request.

Acknowledgments: We would like to thank student assistants Hannah Vahldick, Howard Tan, and Hala Alsaid for their help setting up the experiment and collecting data, and Sophia Belvoir, Makenzie Salyer, Jorge Rodriguez, and Shaokun Meng for their help collecting data.

Conflicts of Interest: The authors declare no conflict of interest. The funders had no role in the design of the study; in the collection, analyses, or interpretation of data; in the writing of the manuscript, or in the decision to publish the results.

References

1. Huché-Thélier, L.; Crespel, L.; Le Gourrierec, J.; Morel, P.; Sakr, S.; Leduc, N. Light signaling and plant responses to blue and UV radiations—Perspectives for applications in horticulture. *Environ. Exp. Bot.* **2016**, *121*, 22–38. [CrossRef]
2. Sellaro, R.; Crepy, M.; Trupkin, S.A.; Karayekov, E.; Buckovsky, A.S.; Rossi, C.; Casal, J.J. Cryptochrome as a Sensor of the Blue/Green Ratio of Natural Radiation in Arabidopsis. *Plant Physiol.* **2010**, *154*, 401–409. [CrossRef] [PubMed]
3. Kasperbauer, M.J. Far-red light reflection from green leaves and effects on phytochrome-mediated assimilate partitioning under field conditions. *Plant Physiol.* **1987**, *85*, 350–354. [CrossRef] [PubMed]
4. Ballaré, C.L. Light regulation of plant defense. *Annu. Rev. Plant Biol.* **2014**, *65*, 335–363. [CrossRef] [PubMed]
5. Massa, G.; Graham, T.; Haire, T.; Flemming II, C.; Gerard, N.; Wheeler, R. Light-emitting diode light transmission through leaf tissue of seven different crops. *HortScience* **2015**, *50*, 501–506. [CrossRef]
6. McCree, K.J. The action spectrum, absorptance and quantum yield of photosynthesis in crop plants. *Agric. Meteorol.* **1972**, *9*, 191–216. [CrossRef]
7. Inada, K. Action spectra for photosynthesis in higher plants. *Plant Cell Physiol.* **1976**, *17*, 355–365.
8. Evans, J.R. The dependence of quantum yield on wavelength and growth irradiance. *Aust. J. Plant Physiol.* **1987**, *14*, 69–79. [CrossRef]
9. Sager, J.C.; Smith, W.O.; Edwards, J.L.; Cyr, K.L. Photosynthetic efficiency and phytochrome photoequilibria determination using spectral data. *ASABE* **1988**, *31*, 1882–1889. [CrossRef]
10. Trouwborst, G.; Hogewoning, S.W.; van Kooten, O.; Harbinson, J.; van Ieperen, W. Plasticity of photosynthesis after the 'red light syndrome' in cucumber. *Environ. Exp. Bot.* **2016**, *121*, 75–82. [CrossRef]
11. Hogewoning, S.W.; Trouwborst, G.; Maljaars, H.; Poorter, H.; van Ieperen, W.; Harbinson, J. Blue light dose-response of leaf photosynthesis, morphology, and chemical composition of Cucumis sativus grown under different combinations of red and blue light. *J. Exp. Bot.* **2010**, *61*, 3107–3117. [CrossRef]
12. Miao, Y.; Wang, X.; Gao, L.; Chen, Q.; Mei, Q. Blue light is more essential than red light for maintaining the activities of photosystem II and I and photosynthetic electron transport capacity in cucumber leaves. *J. Integr. Agric.* **2016**, *15*, 87–100. [CrossRef]
13. Miao, Y.; Chen, Q.; Qu, M.; Gao, L.; Hou, L. Blue light alleviates 'red light syndrome' by regulating chloroplast ultrastructure, photosynthetic traits and nutrient accumulation in cucumber plants. *Sci. Hortic.* **2019**, *257*, 108680. [CrossRef]
14. Trouwborst, G.; Oosterkamp, J.; Hogewoning, S.W.; Harbinson, J.; van Ieperen, W. The responses of light interception, photosynthesis and fruit yield of cucumber to LED-lighting within the canopy. *Physiol. Plant.* **2010**, *138*, 289–300. [CrossRef] [PubMed]
15. Hernández, R.; Kubota, C. Physiological responses of cucumber seedlings under different blue and red photon flux ratios using LEDs. *Environ. Exp. Bot.* **2016**, *121*, 66–74. [CrossRef]

16. Hernández, R.; Kubota, C. Growth and morphological response of cucumber seedlings to supplemental red and blue photon flux ratios under varied solar daily light integrals. *Sci. Hortic.* **2014**, *173*, 92–99. [CrossRef]
17. Brown, C.S.; Schuerger, A.C.; Sager, J.C. Growth and photomorphogenesis of pepper plants under red light-emitting diodes with supplemental blue or far-red lighting. *J. Am. Soc. Hortic.* **1995**, *120*, 808–813. [CrossRef]
18. Yorio, N.C.; Goins, G.D.; Kagie, H.R.; Wheeler, R.M.; Sager, J.C. Improving spinach, radish, and lettuce growth under red light-emitting diodes (LEDs) with blue light supplementation. *HortScience* **2001**, *36*, 380–383. [CrossRef]
19. Matsuda, R.; Ohashi-Kaneko, K.; Fujiwara, K.; Goto, E.; Kurata, K. Photosynthetic characteristics of rice leaves grown under red light with or without supplemental blue light. *Plant Cell Physiol.* **2004**, *45*, 1870–1874. [CrossRef]
20. Wang, Y.; Zhang, T.; Folta, K.M. Green light augments far-red-light-induced shade response. *Plant Growth Regul.* **2013**, *77*, 145–155. [CrossRef]
21. Golovatskaya, I.F.; Karnachuk, R.A. Role of green light in physiological activity of plants. *Russ. J. Plant Physiol.* **2014**, *62*, 727–740. [CrossRef]
22. Smith, H.L.; McAusland, L.; Murchie, E.H. Don't ignore the green light: Exploring diverse roles in plant processes. *J. Exp. Bot.* **2017**, *68*, 2099–2110. [CrossRef] [PubMed]
23. McCoshum, S.; Kiss, J.Z. Green light affects blue-light-based phototropism in hypocotyls of Arabidopsis thaliana. *J. Torrey Bot. Soc.* **2011**, *138*, 409–417. [CrossRef]
24. Frechilla, S.; Talbott, L.D.; Bogomolni, R.A.; Zeiger, E. Reversal of blue light-stimulated stomatal opening by green light. *Plant Cell Physiol.* **2000**, *41*, 171–176. [CrossRef] [PubMed]
25. Talbott, L.D.; Nikolova, G.; Ortix, A.; Shmayevich, I.; Zeiger, E. Green light reversal of blue-light-stimulated stomatal opening is found in a diversity of plant species. *Am. J. Bot.* **2002**, *89*, 366–368. [CrossRef] [PubMed]
26. Hemming, S.; Mohammadkhani, V.; Dueck, T. Diffuse greenhouse covering materials—Material technology, measurements and evaluation of optical properties. *Acta Hortic.* **2008**, *797*, 469–476. [CrossRef]
27. Hernández, R.; Kubota, C. Tomato seedling growth and morphological responses to supplemental LED lighting red:Blue ratios under varied daily solar light integrals. *Acta Hortic.* **2012**, *956*, 187–194. [CrossRef]
28. Sabir, N.; Singh, B. Protected cultivation of vegetables in global arena: A review. *Indian J. Agric. Sci.* **2013**, *83*, 123–135.
29. Savvides, A.; Fanourakis, D.; van Ieperen, W. Co-ordination of hydraulic and stomatal conductances across light qualities in cucumber leaves. *J. Exp. Bot.* **2012**, *63*, 1135–1143. [CrossRef]
30. Jinxiu, S.; Qingwu, M.; Weifan, D.; Dongxian, H. Effects of light quality on growth and development of cucumber seedlings in controlled environment. *Int. J. Agric. Biol.* **2017**, *10*, 312–318.
31. Shibuya, T.; Endo, R.; Yuba, T.; Kitaya, Y. The photosynthetic parameters of cucumber as affected by irradiances with different red:far-red ratios. *Biol. Plant.* **2015**, *59*, 198–200. [CrossRef]
32. Bergstrand, K.J.; Suthaparan, A.; Mortensen, L.M.; Gislerød, H.G. Photosynthesis in horticultural plants in relation to light quality and CO2 concentration. *Eur. J. Hortic. Sci.* **2016**, *81*, 237–242. [CrossRef]
33. Wang, X.Y.; Xu, X.M.; Cui, J. The importance of blue light for leaf area expansion, development of photosynthetic apparatus, and chloroplast ultrastructure of Cucumis sativus grown under weak light. *Photosynthetica* **2015**, *53*, 213–222. [CrossRef]
34. Emerson, R.; Chalmers, R.; Cederstrand, C. Some factors influencing the long-wave limit of photosynthesis. *Proc. Natl. Acad. Sci. USA* **1957**, *43*, 133–143. [CrossRef] [PubMed]
35. Hogewoning, S.W.; Wientjes, E.; Douwstra, P.; Trouwborst, G.; van Ieperen, W.; Croce, R. Photosynthetic quantum yield dynamics: From photosystems to leaves. *Plant Cell* **2012**, *24*, 1921–1935. [CrossRef]
36. Murakami, K.; Matsuda, R.; Fujiwara, K. A mathematical model of photosynthetic electron transport in response to the light spectrum based on excitation energy distributed to photosystems. *Plant Cell Physiol.* **2018**, *59*, 1643–1651. [CrossRef] [PubMed]
37. Baker, N.R. Chlorophyll fluorescence: A probe of photosynthesis in vivo. *Annu. Rev. Plant Biol.* **2008**, *59*, 89–113. [CrossRef]
38. Björkman, O.; Demmig, B. Photon yield of O2 evolution and chlorophyll fluorescence characteristics at 77 K among vascular plants of diverse origins. *Planta* **1987**, *170*, 489–504. [CrossRef] [PubMed]
39. Bergstrand, K.J.; Mortensen, L.M.; Suthaparan, A.; Gislerød, H.G. Acclimatisation of greenhouse crops to differing light quality. *Sci. Hortic.* **2016**, *204*, 1–7. [CrossRef]
40. Oxborough, K.; Baker, N.R. Resolving chlorophyll a fluorescence images of photosynthetic efficiency into photochemical and nonphotochemical components—Calculation of qP and Fv'/Fm' without measuring Fo'. *Photosynth. Res.* **1997**, *54*, 135–142. [CrossRef]
41. Galka, P.; Santabarbara, S.; Khuong, T.T.H.; Degand, H.; Morsomme, P.; Jennings, R.C.; Boekema, E.J.; Caffarri, S. Functional analyses of the plant photosystem I-light-harvesting complex II supercomplex reveal that light-harvesting complex II loosely bound to photosystem II is a very efficient antenna for photosystem I in state II. *Plant Cell* **2012**, *24*, 2963–2978. [CrossRef]
42. Fujita, Y. A study on the dynamic features of photosystem stoichiometry: Accomplishments and problems for future studies. *Photosynth. Res.* **1997**, *53*, 83–93. [CrossRef]
43. Yamazaki, J.; Suzuki, T.; Maruta, E.; Kamimura, Y. The stoichiometry and antenna size of the two photosystems in marine green algae, Bryopsis maxima and Ulva pertusa, in relation to the light environment of their natural habitat. *J. Exp. Bot.* **2005**, *56*, 1517–1523. [CrossRef]

44. Dietzel, L.; Bräutigam, K.; Pfannschmidt, T. Photosynthetic acclimation: State transitions and adjustment of photosystem stoichiometry—Functional relationships between short-term and long-term light quality acclimation in plants. *FEBS J.* **2008**, *275*, 1080–1088. [CrossRef]
45. Schöttler, M.A.; Tóth, S.Z. Photosynthetic complex stoichiometry dynamics in higher plants: Environmental acclimation and photosynthetic flux control. *Front. Plant Sci.* **2014**, *5*, 188. [PubMed]
46. Croce, R.; van Amerongen, H. Natural strategies for photosynthetic light harvesting. *Nat. Chem. Biol.* **2014**, *10*, 492–501. [CrossRef]
47. Schöttler, M.A.; Tóth, S.Z.; Boulouis, A.; Kahlau, S. Photosynthetic complex stoichiometry dynamics in higher plants: Biogenesis, function, and turnover of ATP synthase and the cytochrome B6f complex. *J. Exp. Bot.* **2015**, *66*, 2373–2400. [CrossRef] [PubMed]
48. Weraduwage, S.M.; Chen, J.; Anozie, F.C.; Morales, A.; Weise, S.E.; Sharkey, T.D. The relationship between leaf area growth and biomass accumulation in Arabidopsis thaliana. *Front. Plant Sci.* **2015**, *6*, 167. [CrossRef] [PubMed]
49. Casal, J.J.; Aphalo, P.J.; Sanchez, R.A. Phytochrome effects on leaf growth and chlorophyll content in *Petunia axilaris*. *Plant Cell Environ.* **1987**, *10*, 509–514. [CrossRef]
50. Jeong, H.W.; Lee, H.R.; Kim, H.M.; Hwang, H.S.; Hwand, S.J. Using light quality for growth control of cucumber seedlings in closed-type plant production system. *Plants* **2020**, *9*, 639. [CrossRef]
51. Paul, L.J.; Kurana, J.P. Phytochrome-mediated light signaling in plants: Emerging trends. *Physiol. Mol. Biol. Plants* **2008**, *14*, 9–22. [CrossRef] [PubMed]
52. Kahlen, K.; Stützel, H. Simplification of a light-based model for estimating final internode length in greenhouse cucumber canopies. *Ann. Bot.* **2011**, *108*, 1055–1063. [CrossRef]
53. Demotes-Mainard, S.; Péron, T.; Corot, A.; Bertheloot, J.; Le Gourrierec, J.; Pelleschi-Travier, S.; Crespel, L.; Morel, P.; Huché-Thélier, L.; Boumaza, R.; et al. Plant responses to red and far-red lights, applications in horticulture. *Environ. Exp. Bot.* **2016**, *121*, 4–21. [CrossRef]
54. Arnon, D.I.; Hoagland, D.R. Crop production in artificial culture solutions and in soils with special reference to factors influencing yields and absorption of inorganic nutrients. *Soil Sci.* **1940**, *50*, 463–485.
55. Kramer, D.M.; Johnson, G.; Kiirats, O.; Edwards, G.E. New fluorescence parameters for the determination of QA redox state and excitation energy fluxes. *Photosynth. Res.* **2004**, *79*, 2019–2218. [CrossRef]
56. SAS Institute Inc. *SAS Studio 3.8;* SAS Institute Inc.: Cary, NC, USA, 2018.
57. Sharkey, T.D.; Bernacchi, C.J.; Farquhar, G.D.; Singsaas, E.L. Fitting photosynthetic carbon dioxide response curves for C3 leaves. *Plant Cell Environ.* **2007**, *30*, 1035–1040. [CrossRef]
58. Flexas, J.; Ribas-Carbo, M.; Diaz-Espejo, A.; Galmés, J.; Medrano, H. Mesophyll conductance to CO2: Current knowledge and future prospects. *Plant Cell Environ.* **2008**, *31*, 602–621. [CrossRef] [PubMed]
59. Jones, H.G. *Plants and Microclimate*, 2nd ed.; Cambridge University Press: Cambridge, UK, 1992; pp. 137–140.

plants

MDPI

Article

Canopy Size and Light Use Efficiency Explain Growth Differences between Lettuce and Mizuna in Vertical Farms

Theekshana C. Jayalath and Marc W. van Iersel *

Department of Horticulture, University of Georgia, Athens, GA 30602, USA; tcjay@uga.edu
* Correspondence: mvanier@uga.edu; Tel.: +1-706-583-0284

Abstract: Vertical farming is increasingly popular due to high yields obtained from a small land area. However, the energy cost associated with lighting of vertical farms is high. To reduce this cost, more energy efficient (biomass/energy use) crops are required. To understand how efficiently crops use light energy to produce biomass, we determined the morphological and physiological differences between mizuna (*Brassica rapa* var. *japonica*) and lettuce (*Lactuca sativa* 'Green Salad Bowl'). To do so, we measured the projected canopy size (PCS, a morphological measure) of the plants throughout the growing cycle to determine the total amount of incident light the plants received. Total incident light was used together with the final dry weight to calculate the light use efficiency (LUE, g of dry weight/mol of incident light), a physiological measure. Plants were grown under six photosynthetic photon flux densities (PPFD), from 50 to 425 μmol m^{-2} s^{-1}, for 16 h d^{-1}. Mizuna and lettuce were harvested 27 and 28 days after seeding, respectively. Mizuna had greater dry weight than lettuce ($p < 0.0001$), especially at higher PPFDs (PPFD $\geq$ 125 μmol m^{-2} s^{-1}), partly because of differences in the projected canopy size (PCS). Mizuna had greater PCS than lettuce at PPFDs $\geq$ 125 μmol m^{-2} s^{-1} and therefore, the total incident light over the growing period was also greater. Mizuna also had a higher LUE than lettuce at all six PPFDs. This difference in LUE was associated with higher chlorophyll content index and higher quantum yield of photosystem II in mizuna. The combined effects of these two factors resulted in higher photosynthetic rates in mizuna than in lettuce ($p = 0.01$). In conclusion, the faster growth of mizuna is the result of both a larger PCS and higher LUE compared to lettuce. Understanding the basic determinants of crop growth is important when screening for rapidly growing crops and increasing the efficiency of vertical farms.

Keywords: canopy size; incident light; light interception; light use efficiency; mizuna; projected canopy size; quantum yield of photosystem II

Citation: Jayalath, T.C.; van Iersel, M.W. Canopy Size and Light Use Efficiency Explain Growth Differences between Lettuce and Mizuna in Vertical Farms. *Plants* **2021**, *10*, 704. https://doi.org/10.3390/plants10040704

Academic Editors: Rosario Muleo and Valeria Cavallaro

Received: 28 February 2021
Accepted: 2 April 2021
Published: 6 April 2021

Publisher's Note: MDPI stays neutral with regard to jurisdictional claims in published maps and institutional affiliations.

1. Introduction

Vertical farming refers to hydroponic crop production in buildings with precise environmental control. Vertical farms do not require arable land and can obtain high crop yields from a small land area. Therefore, vertical farms are becoming popular in urban areas. However, the energy costs associated with the required electric lighting and environmental control, especially cooling and dehumidification, in vertical farming are high [1]. The cost of electric lighting in controlled environment agriculture in just the United States has been estimated at ~$600 million annually [2]. To bring this cost down, research into more efficient production techniques [3] and more energy-efficient crops (more biomass gain per unit of energy use) are required.

Overall, crop growth is a function of the amount of incident light reaching the canopy, which depends on projected canopy size (PCS) [4,5], and light use efficiency (LUE, grams of biomass produced per mol of incident light) [6]. To screen for crops with rapid growth, quantifying PCS development and LUE is essential.

Plants that produce a larger canopy can achieve faster growth by increasing the amount of incident light reaching the canopy compared to plants with a smaller canopy [4,5].

As incident light increases, canopy photosynthesis and biomass accumulation of plants also increase [5], as long as canopy photosynthesis is not light-saturated. This in turn helps plants to produce additional canopy faster than plants with a smaller canopy [4]. Therefore, quantifying the PCS and determining how this affects the amount of incident light, is important. Non-destructive digital imaging has been used in many crops, including tomatoes [7], soybean [8], and lettuce [5,9]. Periodic PCS measurements can be used to estimate the daily PCS [6]. Combining those data with PPFD data allows for estimation of the total incident light over the course of the growing cycle. Projected canopy size is also valuable to make crop growth predictions, since PCS early in the growing cycle may be correlated with the final dry weight of the crop, as we have previously shown in lettuce (*Lactuca sativa*) [6,10] and black-eyed susan (*Rudbeckia fulgida*) [11].

The LUE of a crop describes how efficiently plants use the incident light for growth. The LUE can be calculated by dividing the dry weight of a plant by the total incident light that the plant received throughout the growing period [6]. This provides a physiological measure of how efficiently crops use light, in contrast to calculating LUE based on the amount of light provided to the growing space, which provides insight into production efficiency, but not underlying physiological mechanisms [6]. Factors such as chlorophyll content, the quantum yield of photosystem II (Φ_{PSII}, the fraction of absorbed photons used to drive photochemistry), and CO_2 assimilation are all important in determining the LUE of a crop.

Higher chlorophyll content increases light absorption [12]. The energy of the absorbed photons can then be used in the light reactions of photosynthesis to drive photochemistry (electron transport), while some light energy is dissipated as heat (NPQ, non-photochemical quenching), or re-emitted as fluorescence. Photochemistry competes with NPQ and chlorophyll fluorescence for excitation energy from photons [13,14]. To understand how different lighting strategies affect plant growth, it is important to determine Φ_{PSII}, since this partly determines how efficiently plants use the absorbed light to drive photosynthesis and produce biomass. Previous studies have shown that increasing PPFD decreases Φ_{PSII}, due to the partial closing of photosystem II reaction centers and upregulation of NPQ [15,16]. Although Φ_{PSII} decreases, the electron transport rate (ETR) increases with increasing PPFD [15–17]. The Φ_{PSII} of different species responds differently to increasing PPFDs [16]. In addition, plants acclimated to high light have higher Φ_{PSII} and ETR than those acclimated to low light. Therefore, to increase the production efficiency in plant factories, identifying crops with both high Φ_{PSII} and high ETR is important, because a higher ETR will result in the production of more ATP and NADPH for use in CO_2 assimilation. Both high Φ_{PSII} and high chlorophyll content index (CCI) may increase the ETR, CO_2 assimilation rate, and potentially the growth rate of plants.

A previous indoor study with mizuna (*Brassica rapa* var. *japonica*) and oakleaf lettuce (*Lactuca sativa* 'Green Salad Bowl') found much faster growth of mizuna compared to lettuce [18]. This growth difference must be the result of morphological and/or physiological differences between the two crops, but those underlying reasons were not explored in that study. A better understanding of the underlying reasons for growth differences among crops will facilitate screening for rapid-growing crops and cultivars that are well-suited for vertical farming production systems. It can also enable breeding efforts by providing selection criteria for crops that are well-suited for production in vertical farms. Therefore, our objective was to determine the underlying mechanisms for the growth differences between mizuna and lettuce. We hypothesize that the faster growth of mizuna is the result of a greater PCS, increased canopy incident light, and higher LUE. Since crop growth is affected by PPFD, we also determined how different PPFDs affect the morphological and physiological factors underlying crop growth. Comparing two species, grown at different PPFDs, allowed us to determine how useful PCS and LUE are in explaining crop growth.

2. Results

2.1. Experiment

To determine the underlying morphological and/or physiological reasons for growth differences between mizuna and lettuce at different PPFDs, plants were grown at six PPFD levels (~50, 125, 200, 275, 350, and 425 μmol m^{-2} s^{-1} at the center of each section) for a 16-hr photoperiod. Projected canopy size was measured twice a week throughout the growing period, and those images were used to estimate the daily PCS and to calculate the total incident light per plant over the growing period. In addition, leaf chlorophyll content index (CCI), anthocyanin content index (ACI), Φ_{PSII}, and net CO_2 assimilation of both crops were measured during the study. Mizuna and lettuce were harvested 27 and 28 days after seeding, respectively. The total leaf area and shoot dry weight were measured. Finally, the LUE was calculated by dividing shoot dry weight by the total incident light over the growing period.

2.2. Projected Canopy Size

Projected canopy size of both crops increased sigmoidally over time. A PPFD of 50 μmol m^{-2} s^{-1} resulted in a much lower PCS of both crops compared to PPFD levels ≥ 125 μmol m^{-2} s^{-1}. The PCS of mizuna at the end of the growing cycle was ~340 cm^2/plant at PPFDs ≥ 200 μmol m^{-2} s^{-1} (Figures 1 and 2). For lettuce, PCS was similar (~240 cm^2/plant) at all PPFDs ≥ 125 μmol m^{-2} s^{-1} (Figures 1 and 2). Mizuna had a larger PCS than lettuce starting from the early growth stages at PPFDs > 50 μmol m^{-2} s^{-1} (Figure 1). The difference in PCS between the two species increased over time, especially at higher PPFDs (≥ 200 μmol m^{-2} s^{-1}) (Figure 1).

The PCS of both crops at harvest was low at a PPFD of 50 μmol m^{-2} s^{-1} and increased asymptotically with increasing PPFD ($p < 0.0001$, Figure 2). The PCS of both crops was similar at PPFDs of 50 and 125 μmol m^{-2} s^{-1}. However, the PCS of lettuce did not increase further at PPFDs ≥ 125 μmol m^{-2} s^{-1}, while mizuna PCS was similar at PPFDs ≥ 200 μmol m^{-2} s^{-1}. At PPFDs ≥ 200 μmol m^{-2} s^{-1}, mizuna had a greater PCS than lettuce (Figure 2).

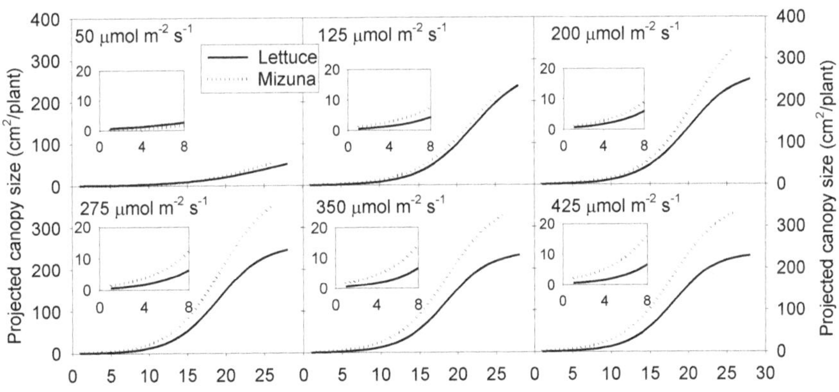

Figure 1. Sigmoidal regression curves fitted through the projected canopy size (PCS) data of mizuna (*Brassica rapa* var. *japonica*) and lettuce (*Lactuca sativa* 'Green Salad Bowl') over the course of the growing cycle for plants grown at six different photosynthetic photon flux densities (PPFD, upper left corner of each graph) ($R^2 \geq 0.99$ for all curves). Inserts show the PCS during the first eight days of the study.

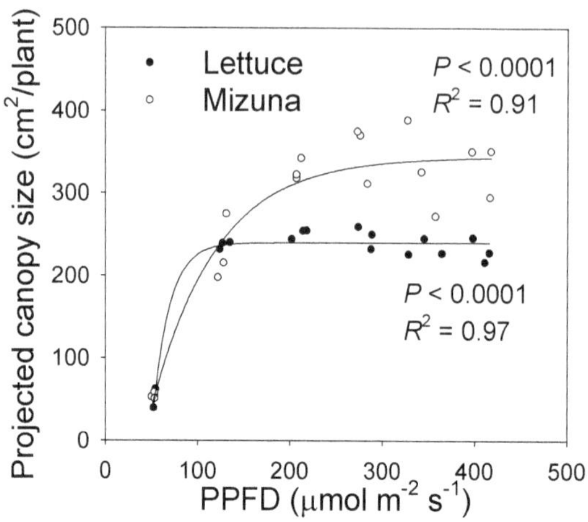

Figure 2. The projected canopy size (PCS) of mizuna (*Brassica rapa* var. *japonica*) and lettuce (*Lactuca sativa* 'Green Salad Bowl') plants grown at six different photosynthetic photon flux densities (PPFDs). Data were collected at the end of the growing cycle (27 and 28 days for mizuna and lettuce, respectively). Each data point represents the mean of nine plants.

2.3. Incident Light

The incident light integrated over the entire crop cycle increased with higher PPFD levels for both crops, but this increase was more pronounced for mizuna than for lettuce (Figure 3, $p < 0.0001$).

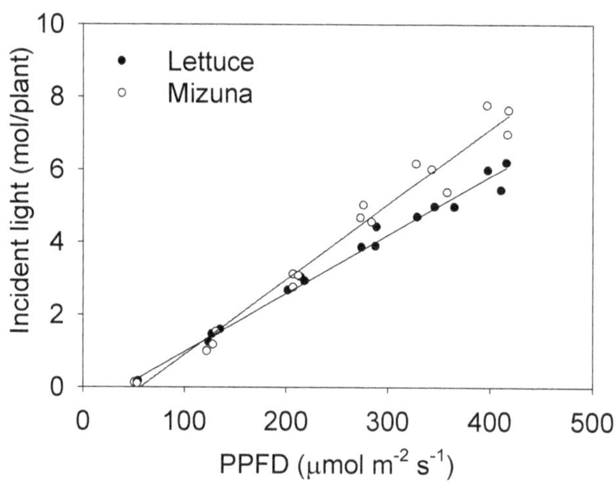

Figure 3. The total incident light on the plant canopy throughout the growing period of mizuna (*Brassica rapa* var. *japonica*) and lettuce (*Lactuca sativa* 'Green Salad Bowl') plants grown at six different photosynthetic photon flux densities (PPFDs) for 27 and 28 days, respectively. Lines show the results from multiple regression analysis ($R^2 = 0.98$), which indicated a significant species × PPFD interaction ($p < 0.0001$). Each data point represents the mean of nine plants.

2.4. Chlorophyll Content Index and Anthocyanin Content Index

Increasing the PPFD increased both CCI and anthocyanin content index (ACI) of both crops ($p \leq 0.003$) (Figure 4). This was more pronounced in mizuna than in lettuce; CCI of mizuna increased by 0.08 for each μmol m^{-2} s^{-1} increase in PPFD, compared to an increase of 0.01 per μmol m^{-2} s^{-1} in lettuce. As the PPFD increased from 50 to 400 μmol m^{-2} s^{-1}, the CCI of mizuna increased from ~7 to ~40, while that of lettuce increased from ~2 to about ~10 (Figure 4A).

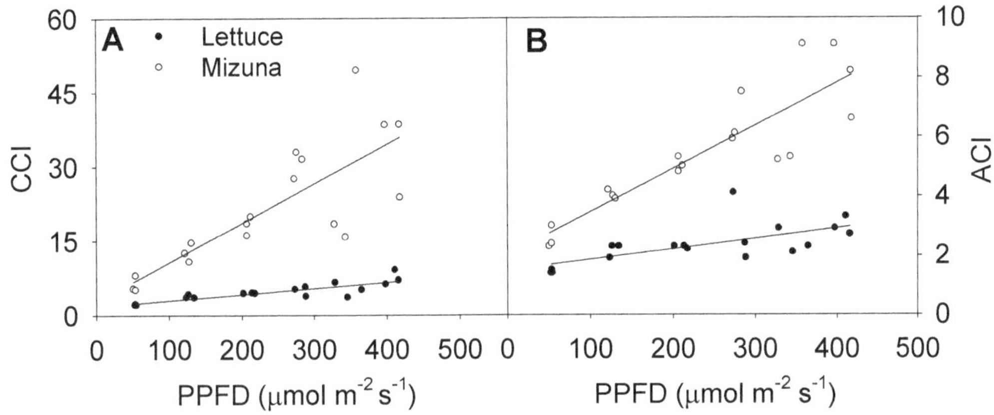

Figure 4. (**A**) Chlorophyll content index (CCI) and (**B**) anthocyanin content index (ACI) of mizuna (*Brassica rapa* var. *japonica*) and lettuce (*Lactuca sativa* 'Green Salad Bowl') plants grown at six different photosynthetic photon flux densities (PPFD). Data were collected a day before the harvesting (26 and 27 days for mizuna and lettuce, respectively). Lines show the results from multiple regression analysis, which indicated a significant species × PPFD interaction for both CCI (R^2 = 0.82, interaction $p < 0.0001$) and ACI (R^2 = 0.88, interaction $p < 0.0001$).

The ACI showed a similar pattern as CCI; increasing PPFD increased ACI in both crops ($p \leq 0.003$) (Figure 4B). Mizuna had a higher ACI than lettuce at all PPFD levels ($p < 0.0001$) and mizuna ACI increased more rapidly with increasing PPFD than that of lettuce. For each 1 μmol m^{-2} s^{-1} increase in PPFD, the ACI of mizuna increased by 0.015 and that of lettuce by 0.004. As the PPFD increased from 50 to 400 μmol m^{-2} s^{-1}, the ACI of mizuna increased from ~3 to ~8, while that of lettuce only increased from ~2 to about ~3.

2.5. *Quantum Yield of Photosystem II and CO_2 Assimilation*

The quantum yield of photosystem II (Φ_{PSII}) of both crops decreased linearly with increasing PPFD ($p = 0.0008$) (Figure 5A). Increasing PPFD by 1 μmol m^{-2} s^{-1} reduced Φ_{PSII} of lettuce and mizuna by 0.0003 mol mol^{-1}. Mizuna always had a higher Φ_{PSII} (~0.05 mol mol^{-1}) than lettuce regardless of PPFD ($p < 0.0001$). The net CO_2 assimilation rate of both crops increased with increasing PPFD, but this tended to be more pronounced in mizuna than in lettuce ($p = 0.08$). Both crops had a CO_2 assimilation rate of ~1 μmol m^{-2} s^{-1} at a PPFD of 50 μmol m^{-2} s^{-1}, but at PPFD of 425 μmol m^{-2} s^{-1} mizuna had a CO_2 assimilation rate of ~18 μmol m^{-2} s^{-1} while that of lettuce was only ~13 μmol m^{-2} s^{-1}. The assimilation rate of mizuna and lettuce increased by 0.044 and 0.035 μmol m^{-2} s^{-1} per 1 μmol m^{-2} s^{-1} increase in PPFD, respectively.

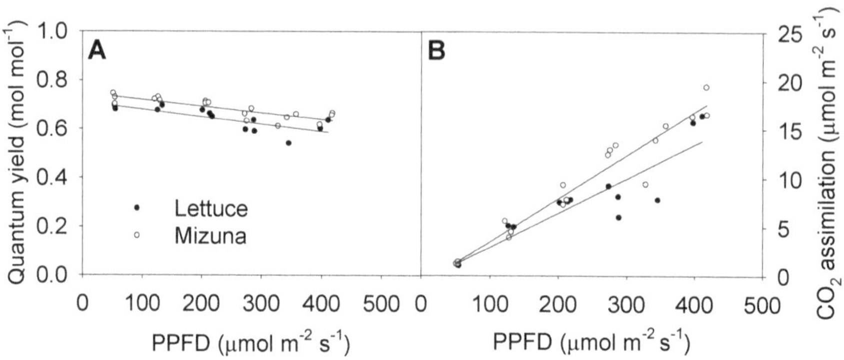

Figure 5. (**A**) Quantum yield of photosystem II and (**B**) net CO_2 assimilation rate of mizuna (*Brassica rapa* var. *japonica*) and lettuce (*Lactuca sativa* 'Green Salad Bowl') plants grown and measured at six photosynthetic photon flux densities (PPFDs). Data were collected a day before the harvesting (26 and 27 days for mizuna and lettuce, respectively). Lines show the results from multiple regression analysis, which indicated no significant species × PPFD interaction for quantum yield ($R^2 = 0.72$, interaction $p = 0.62$), but significant effects of PPFD and species (both $p < 0.0001$). For CO_2 assimilation rate, there was a weak species × PPFD interaction effect ($R^2 = 0.91$, interaction $p = 0.08$).

2.6. Final Leaf Area and Canopy Overlap Ratio

The final leaf area of both mizuna and lettuce increased asymptotically with increasing PPFD ($p < 0.0001$) (Figure 6A). Both crops had the highest leaf area at PPFDs ≥ 275 μmol m^{-2} s^{-1}, but lettuce leaf area increased faster with increasing PPFD than mizuna leaf area ($p < 0.0001$). At a PPFD of ≥ 275 μmol m^{-2} s^{-1} lettuce had a leaf area of ~1200 cm^2 per plant, while that of mizuna was only ~800 cm^2. However, lettuce had a lower PCS at harvest compared to mizuna at PPFDs ≥ 200 μmol m^{-2} s^{-1} (Figure 2). This apparent contradiction between a greater leaf area and a lower PCS of lettuce can be explained by the canopy overlap ratio (leaf area/PCS); lettuce had a much higher canopy overlap ratio than mizuna at PPFDs ≥ 125 μmol m^{-2} s^{-1} (Figure 6B). The canopy overlap ratio of lettuce increased more rapidly, from 1.2 to 5.2 with increasing PPFD compared to that of mizuna (increasing from 1.1 to 2.3) ($p < 0.0001$).

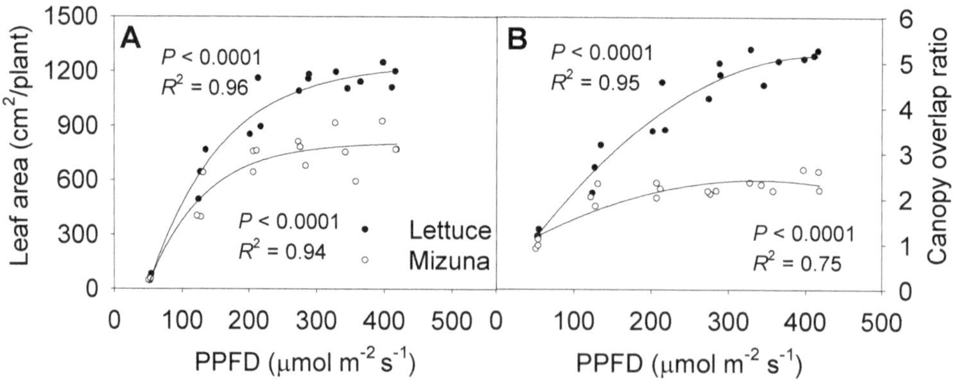

Figure 6. (**A**) Leaf area per plant and (**B**) canopy overlap ratio of mizuna (*Brassica rapa* var. *japonica*) and lettuce (*Lactuca sativa* 'Green Salad Bowl') plants grown at six different photosynthetic photon flux densities (PPFDs) for 27 and 28 days, respectively. The canopy overlap ratio is the ratio between the leaf area and the projected canopy size at harvest. Each data point represents the mean of nine plants.

2.7. Shoot Dry Weight and Specific Leaf Area

The shoot dry weight at 425 μmol m^{-2} s^{-1} was ~50 times higher than at 50 μmol m^{-2} s^{-1} for both lettuce and mizuna, although PPFD increased only ~9 × (Figure 7A, Figure S4). Lettuce and mizuna had a dry weight of 0.08 and 0.11 g/plant at a PPFD of 50 μmol m^{-2} s^{-1} and 3.85 and 6.02 g/plant at a PPFD of 425 μmol m^{-2} s^{-1}, respectively. A 1 μmol m^{-2} s^{-1} increase in PPFD increased the dry weight of lettuce by 10.0 mg/plant and that of mizuna by 15.8 mg/plant. With increasing PPFD, the specific leaf area (SLA, leaf area per gram of dry weight) of both crops decreased ($p < 0.0001$) (Figure 7B). However, due to the higher leaf area and lower dry weight of lettuce compared to mizuna, the SLA of lettuce was greater than that of mizuna at all PPFDs. The difference in SLA between the two species decreased with increasing PPFD (Figure 7B).

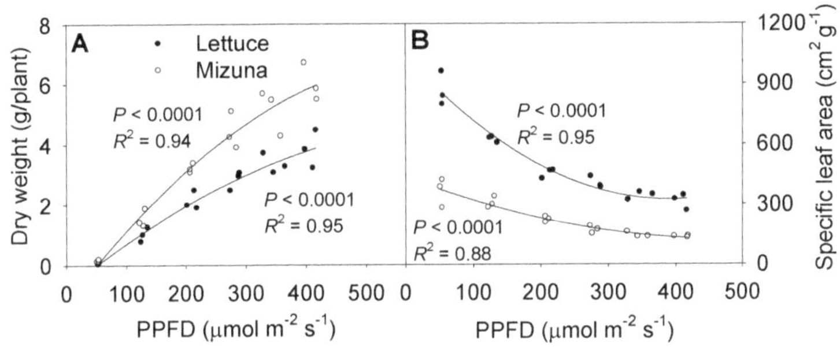

Figure 7. (**A**) Shoot dry weight and (**B**) specific leaf area of mizuna (*Brassica rapa* var. *japonica*) and lettuce (*Lactuca sativa* 'Green Salad Bowl') plants grown at six different photosynthetic photon flux densities (PPFDs) for 27 and 28 days, respectively. Specific leaf area is the ratio between leaf area and dry weight. Each data point represents the mean of nine plants.

2.8. Light Use Efficiency

Mizuna had a higher LUE than lettuce ($p < 0.0001$) (Figure 8). The LUE of mizuna was ~1.1 g mol^{-1} at PPFDs up to 200 μmol m^{-2} s^{-1} and decreased to ~0.75 g mol^{-1} at a PPFD of 425 μmol m^{-2} s^{-1}. In contrast, lettuce LUE was greatest at PPFDs of 125 to 350 μmol m^{-2} s^{-1} (~0.8 g mol^{-1}) and ~0.6 g mol^{-1} at PPFDs of 50 and 425 μmol m^{-2} s^{-1}.

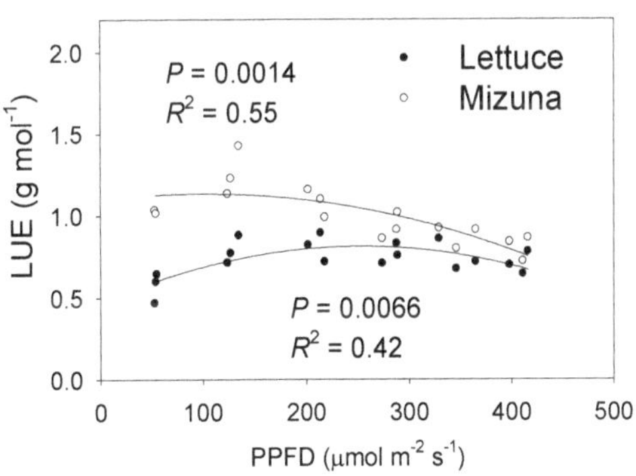

Figure 8. Light use efficiency (LUE, grams of shoot dry weight per mol of incident light) of mizuna (*Brassica rapa* var. japonica) and lettuce (*Lactuca sativa* 'Green Salad Bowl') plants grown at six different photosynthetic photon flux densities (PPFDs) for 27 and 28 days, respectively. Each data point represents the mean of nine plants.

3. Discussion

Mizuna grew faster than lettuce at PPFDs $\geq$ 125 µmol m^{-2} s^{-1} (Figure 7A). A previous study conducted to identify the effects of photoperiod on leafy greens found much faster growth of mizuna compared to lettuce [18]. The difference in biomass between the two crops increased with increasing PPFD (and shorter photoperiods). The goal of the current study was to determine the underlying reasons for the growth differences between these two crops.

3.1. Projected Canopy Size and Incident Light

Mizuna had a larger PCS than lettuce, starting from the early growth stages (Figure 1; Figure S1). One reason for this early difference in PCS is the faster germination and larger cotyledons of mizuna compared to lettuce. Due to their small canopy size, seedlings capture only a small fraction of the provided light. Therefore, increased PCS at early stages can increase light capture and growth of seedlings [11]. A previous study compared different lettuce cultivars and found that early PCS was a good predictor of final shoot biomass [10]. We observed the same pattern in our study, with a strong positive correlation ($R = 0.91$ for lettuce and $R = 0.89$ for mizuna, $p < 0.0001$) between early PCS (lettuce 10 d and mizuna 8 d after seeding) and the dry weight of both lettuce and mizuna (Figure S2).

This higher PCS of mizuna during the early part of the growing cycle (at PPFDs $\geq$ 125 µmol m^{-2} s^{-1}) may have helped it to capture more light and grow faster than lettuce. A previous greenhouse study conducted with 'Little Gem' lettuce observed the same trend; plants with larger PCS in early growth stages absorbed more light, grew faster, and produced additional canopy faster than plants with a smaller PCS [4]. In that prior study, plants were grown with different photoperiods, but the same DLI. In contrast, plants in the current study were grown with the same photoperiod, but different PPFDs, thus resulting in different DLIs.

At a PPFD of 50 µmol m^{-2} s^{-1}, both mizuna and lettuce grew slowly and had a low PCS at harvest compared to plants grown at higher PPFDs (Figure 2). At PPFDs $\geq$ 125 µmol m^{-2} s^{-1}, the PCS of both crops was much greater than at a PPFD of 50 µmol m^{-2} s^{-} However, at PPFD $\geq$ 200 µmol m^{-2} s^{-1}, the PCS of mizuna was greater than that of lettuce (Figure 2) ($p < 0.0001$). The growth difference between the two crops is at least partly because of the observed differences in PCS and its impact on incident light. The incident light of mizuna increased more rapidly with increasing PPFD than that of lettuce, consistent with the increasing difference in dry weight between the two crops as PPFD increased (Figures 3 and 7). Other studies also mentioned a positive correlation between PCS and incident light [5,11,19]. Additionally, several other studies on leafy greens show a positive correlation between biomass gain and incident light integrated over the entire crop cycle [5,20,21].

3.2. Leaf Area, Specific Leaf Area, and Canopy Overlap Ratio

Even though lettuce had a lower PCS than mizuna, its total leaf area was larger than that of mizuna at PPFDs $\geq$ 125 µmol m^{-2} s^{-1} (Figure 6A). This was due to the differences in leaf arrangement and morphology between the two species. The canopy overlap ratio of lettuce was higher than that of mizuna at PPFDs $\geq$ 125 µmol m^{-2} s^{-1} (Figure 6B), resulting in more intra-canopy shading in lettuce. With increasing PPFD, the differences in leaf area and canopy overlap ratio between the two species increased.

The SLA of lettuce was greater compared to mizuna, due to a larger total leaf area and lower shoot dry weight than mizuna at PPFDs $\geq$ 125 µmol m^{-2} s^{-1} (Figure 7B).

3.3. Chlorophyll Content Index and Anthocyanin Content Index

We observed a higher CCI in mizuna leaves compared to lettuce at all PPFDs (Figure 4A). Leaf light absorptance is positively associated with the CCI [12]. Therefore, mizuna likely had a higher leaf light absorptance in lettuce. However, we do not have the actual light absorptance data for this study.

With increasing PPFD, CCI increased in both species (Figure 4A). In previous studies, CCI increased with lower PPFD [4,18]. In those studies, lower PPFDs were combined with longer photoperiods to maintain the same DLI. Chlorophyll production is a light-regulated process [22]. Therefore, the CCI increase in low PPFD treatments in prior studies was associated with longer photoperiods, which increases the amount of time available for plants to produce chlorophyll [18]. In our study, photoperiod was the same in all treatments and therefore the daily light integral was higher at higher PPFDs.

In response to increasing PPFD, the CCI increase in mizuna was about four times greater than in lettuce, indicating that mizuna acclimates more strongly to different PPFDs than lettuce (Figure 4A). The greater CCI increase in mizuna may be partly the result of a lower SLA at PPFDs $\geq$ 125 μmol m^{-2} s^{-1}, compared to lettuce (Figure 7B). A higher SLA is associated with thinner leaves, which typically have low chlorophyll content per unit leaf area [23]. In our study, increasing PPFD decreased the SLA of both crops (Figure 7B). Such a decrease in SLA can increase the CCI, due to increased leaf thickness. We indeed observed strong negative correlations between the SLA and CCI of mizuna ($R = -0.76$, $p = 0.0003$) and lettuce ($R = -0.79$, $p = 0.0001$; Figure S3), but this relationship differed greatly between the two species. Mizuna's CCI decreased much more quickly with increasing SLA than that of lettuce. At mizuna's highest SLA (~375 cm^2 g^{-1}), its CCI was similar (~5) to the lettuce CCI at its lowest SLA (~325 cm^2 g^{-1})

The anthocyanin content index (ACI) was measured to identify whether the two crops differ in anthocyanin accumulation in response to increasing PPFD. Anthocyanins in leaves have a protective role against intense light and help dissipate excess excitation energy [24,25]. The ACI of both species did indeed increase with increasing PPFD (Figure 4B). However, the ACI of both crops was much lower than those previously reported for red leaf basil (ACI of 28–81) [26] and pak choi (ACI of 35–65) [27]. Consistent with the low ACI, we did not observe any red coloration on leaves of mizuna or lettuce, with increasing PPFD.

3.4. Quantum Yield of Photosystem II and CO_2 Assimilation

With increasing PPFD, Φ_{PSII} of both crops decreased at a similar rate (Figure 5A). At higher PPFD, a larger proportion of the PSII reaction centers are closed and more of the absorbed light energy is dissipated as heat to minimize photoinhibition [15,28]. This rise in heat dissipation results in a smaller fraction of the excitation energy being directed towards the PSII reaction centers, reducing Φ_{PSII} [17,29]. Many previous studies have observed a decrease in Φ_{PSII}, but increasing ETR, in response to increasing PPFD [15–17]. We observed the same pattern in Φ_{PSII} with increasing PPFD. We did not calculate the ETR because leaf absorptance was not measured. However, the higher CCI of mizuna suggests higher leaf absorptance compared to lettuce [12], in which case differences in electron transport rate between the two species would have been larger than the differences in Φ_{PSII}. Even though we observed a reduction in Φ_{PSII} of both crops with increasing PPFD, the Φ_{PSII} of mizuna was always higher (~0.05 mol mol^{-1}) than that of lettuce (Figure 5A) ($p < 0.0001$). A study conducted to understand the effects of different PPFDs on the photochemistry of three species adapted to different light levels, found a higher Φ_{PSII} and higher ETR in high-light adapted species compared to the species adapted to moderate or low-light [16]. This suggests that mizuna is better adapted to high light levels than lettuce, and therefore has a higher Φ_{PSII}.

With increasing PPFD, the net CO_2 assimilation rate of mizuna tended to increase more rapidly than that of lettuce (Figure 5B, $p = 0.08$). This is consistent with the higher CCI and Φ_{PSII} of mizuna. The lower SLA of mizuna suggests thicker leaves compared to lettuce. The higher CCI associated with thicker leaves can increase the light absorptance and CO_2 assimilation rate per unit leaf area [23].

3.5. Light Use Efficiency

Mizuna had a higher maximum (1.26 g mol^{-1}) LUE than lettuce (0.74 g mol^{-1}) ($p < 0.0001$) (Figure 8). Mizuna had its maximum LUE at PPFDs from 50 to 200 μmol m^{-2} s^{-1}, while lettuce LUE was maximal at PPFDs of 200 to 275 μmol m^{-2} s^{-1}. This indicates that mizuna can convert incident light into biomass more efficiently than lettuce, especially at lower PPFDs. The maximum lettuce LUE we observed (0.74 g mol^{-1}) is slightly higher than the LUE previously reported (0.61–0.65 g mol^{-1}) for 'Green Salad Bowl' lettuce [6]. Legendre and van Iersel [6] used the same method to calculate LUE, but, did not use CO_2 enrichment which may have reduced leaf photosynthesis and thus LUE.

Other studies reported substantially lower LUE values (<0.6 g mol^{-1}) for lettuce [23,30], but in those studies, LUE calculations were based on the amount of light provided to the growing space, rather than light reaching the canopy of the crop. Their LUE values are thus heavily dependent on the plant density of the growing spaces. Light use efficiency can also vary among cultivars [31,32]. Therefore, it is hard to compare our LUE values of lettuce with those studies. There are no prior reports for mizuna LUE.

Multiple factors contributed to the higher LUE of mizuna. The combined effects of more chlorophyll and higher Φ_{PSII} of mizuna likely resulted in higher ETR and thus more photosynthesis than in lettuce. High photosynthetic rates can increase the relative growth rate, which in turn reduces the fraction of carbohydrates allocated to maintenance respiration and increases carbon use efficiency [33]. That in turn can increase LUE.

The LUE decreased at high PPFDs for both mizuna (PPFD ≥ 200 μmol m^{-1} s^{-1}) and lettuce (PPFD ≥ 350 μmol m^{-1} s^{-1}) (Figure 8). This reduction of LUE at high PPFD may be due to the decrease of Φ_{PSII} with increasing PPFD (Figure 5).

3.6. Conclusions

The Asian leafy green mizuna (*Brassica rapa* var. *japonica*) grows faster than oakleaf lettuce (*Lactuca sativa* 'Green Salad Bowl') when PPFD ≥ 125 μmol m^{-2} s^{-1}. This faster growth of mizuna is the result of a higher CCI and larger PCS (at PPFDs ≥ 125 μmol m^{-2} s^{-1}), allowing mizuna to capture more light, a higher Φ_{PSII} and net CO_2 assimilation, and a higher LUE than lettuce. This study provides a framework for determining underlying morphological and physiological reasons for growth differences among crops. Understanding the basic determinants of crop growth is important to increase crop productivity and energy efficiency in vertical farms. Canopy imaging can be used to select crops that will grow well in vertical farms. Although we looked at a multitude of morphological and physiological factors, quantifying PCS and LUE would be adequate for the selection of crops with fast growth. Perhaps most intriguingly, our results confirm that early differences in PCS (8–10 days after seeding) are a good predictor of final biomass. Therefore, it may be possible to simply use early PCS to screen crops for rapid growth in controlled environment agriculture. This would allow for rapid throughput phenotyping and greatly accelerate the selection of promising genotypes.

4. Materials and Methods

4.1. Growth Chamber Setup

The study was conducted in a 4.4 m wide and 4.1 m long walk-in growth chamber. Cooling was provided using a top-mount refrigeration system and a dehumidifier maintained the relative humidity inside the growth chamber. The CO_2 level inside the growth chamber was measured and maintained by triggering a solenoid valve to open and release CO_2 from a compressed gas cylinder for 1-second intervals, whenever the CO_2 concentration dropped below 800 μmol mol^{-1}, using a CO_2 transmitter (GMC20; Vaisala, Helsinki, Finland) and a datalogger (CR6, Campbell Scientific, Logan, UT, USA). Temperature and relative humidity measurements were collected every ten seconds with a probe (HMP50; Vaisala) connected to the datalogger. Using those temperature and relative humidity values, vapor pressure deficit (VPD) was calculated. Average temperature, CO_2 level, and VPD inside the growth chamber were 24.2 ± 0.2 °C, 825 ± 38 μmol mol^{-1}, and 1.4 ± 0.12 kPa (mean ± SD), respectively.

The growth chamber contained three 2.4 m long × 0.6 m wide × 2.2 m high metal shelving racks with 0.9 m distance between the racks. Each rack had three shelves with a 0.6 m × 2.4 m ebb-and-flow tray on each shelf. Each tray had an individual irrigation tube connected to a submersible pump. Those pumps were submerged in a fertigation tank located under the bottom shelf of each of the three metal racks. Three pumps were submerged in each fertigation tank. Each ebb-and-flow tray was divided into two 1.2 m long sections. Therefore, each rack had six 1.2 m long × 0.6 m wide × 0.6 m high sections, for a total of 18 growing sections. Each growing section had two 1.1 m-long white LED lights (RAY series with Physiospec indoor spectrum; Fluence Bioengineering, Austin, TX, USA) hanging 0.4 m above the bottom of the ebb & flow tray. Air circulation was provided by four 4 × 4 cm^2 fans in each growing section.

4.2. Seeding and Plant Management

We placed two groups of nine 10-cm square pots in each growing section. Those pots were filled with a soilless substrate [80% peat: 20% perlite (v/v) (Fafard 1P; SunGro Horticulture, Agawam, MA, USA)]. Nine pots were seeded with mizuna (*Brassica rapa* var. *japonica*) and nine pots with lettuce (*Lactuca sativa* 'Green Salad Bowl'). To prevent algae growth on the surface of the substrate, the top 1 cm of each pot was filled with calcined clay (Turface® Pro League Elite, Profile Products LLC, Buffalo Grove, IL). Plants were sub-irrigated daily for 5 minutes with a nutrient solution containing 100 mg L^{-1} N made with a water-soluble fertilizer (15N–2.2P–12.45K, Peters Excel 15–5–15 Cal-Mag Special; Everris NA Inc, Dublin, OH, USA). Algaecide (ZeroTol 2.0, BioSafe Systems LLC, East Hartford, CT, USA) was applied to the surface of the substrate twice during the study, at a ratio of 1:400 (ZeroTol: water) as an algae preventative. Plants were grown under six treatments with different photosynthetic photon flux densities (PPFD) (~50, 125, 200, 275, 350, and 425 μmol m^{-2} s^{-1} at the center of each section) (Table 1). Treatments were randomly allocated to one of the six sections of each metal rack. The PPFD was controlled by sending a pulse width modulation signal from the datalogger to the dimmable drivers powering the light fixtures. The LED lights were on for 16 hours per day. Therefore, plants received daily light integrals of ~3.1, 7.4, 12.1, 16.2, 19.9, and 23.6 mol m^{-2} d^{-1} at the center of each section in the six treatments (Table 1).

Table 1. Photosynthetic photon flux densities (PPFD), and the average daily light integral (DLI) at the center of the tray of each treatment. The photoperiod was 16 hours for all the treatments. Values show the mean ± standard deviation.

PPFD (μmol m^{-2} s^{-1})	DLI (mol m^{-2} d^{-1})
53 ± 2	3.1 ± 0.1
128 ± 6	7.4 ± 0.3
210 ± 9	12.1 ± 0.5
281 ± 14	16.2 ± 0.8
345 ± 15	19.9 ± 0.8
410 ± 16	23.6 ± 1.0

4.3. Data Collection and Calculations

Mizuna and lettuce canopy images were captured twice a week throughout the growing period, using a chlorophyll fluorescence imaging setup. For the fluorescence imaging, we used a monochrome camera (CM3-U3-31S4M-CS, Chameleon3 USB3 camera, FLIR Systems, Inc., Arlington, VA, USA) with a 665 nm longpass filter (LP665 Dark Red Longpass Filter; Midopt Midwest Optical Systems, Inc., Palatine, IL, USA) attached to the lens. The camera was mounted facing downward inside of a 1.2 m × 0.6 m × 1.5 m grow tent. A blue LED panel was mounted inside the tent next to the camera to excite chlorophyll and induce fluorescence. Reemitted light from chlorophyll fluorescence was captured by the camera. Canopy images were taken biweekly on groups of nine plants. Those images were then analyzed with ImageJ software (ImageJ 1.52a, National Institute of Health, Bethesda, MD, USA) to determine the PCS. These PCS data were divided by nine to determine the PCS per plant. Those values were plotted against time and sigmoidal curves [f = a/(1 + exp(−(x−x0)/b))] were fitted (SigmaPlot 11.0, Systat software, Inc., San Jose, CA, USA). This was done for all individual treatments and replicates ($R^2 \geq 0.99$). Using the coefficients for the sigmoidal equation, the daily PCS was estimated (Microsoft Excel 365, Microsoft Corporation, Redmond, WA, USA). The daily PCS data were multiplied by the DLI received in each corresponding treatment to calculate the daily incident light per plant. By adding those daily incident light values, the total incident light on the canopy throughout the growing period was calculated.

One day before the harvest, leaf CCI and leaf anthocyanin content index (ACI) were measured using chlorophyll and anthocyanin meters (CCM-200 plus and ACM-200 plus; Apogee Instruments, Logan, UT, USA) on uppermost fully-expanded leaves. Then, the quantum yield of photosystem II (Φ_{PSII}) and CO_2 assimilation were measured using a leaf gas exchange system, equipped with a chlorophyll fluorometer (CIRAS-3 Portable Photosynthesis System: PP Systems, Amesbury, MA, USA). The corresponding PPFD level of each treatment was provided using the white LED light in the leaf cuvette during the measurements. The average leaf temperature, vapor pressure deficit, and CO_2 concentration inside the cuvette were 24.5 °C, 1.3 kPa, and 781 μmol mol^{-1}, respectively, to mimic the conditions inside the growth chamber at the time of the data collection.

At the end of the study, lettuce plants were harvested at 28 days after seeding and mizuna plants were harvested at 27 days after seeding. The total leaf area of each group of plants was measured using a leaf area meter (LI-3100 leaf area meter; LI-COR Biosciences, Lincoln, NE, USA). After drying them at 80 °C for 7 days, shoot dry weight of each group of plants was also measured. Both leaf area and shoot dry weight value were divided by nine to calculate the per plant leaf area and shoot dry weight. By dividing the leaf area by the associated PCS at harvest, the canopy overlap ratio was calculated. Specific leaf area (SLA) was calculated by dividing the leaf area of a plant by the shoot dry weight. To calculate LUE, shoot dry weight was divided by the total incident light that the plant received over the growing period.

4.4. Experimental Design and Statistical Analysis

The experiment design was a randomized complete block with a split-plot with one metal shelving rack as a block, 3 blocks, six PPFD levels as the main treatment, and the two crops as the split plot. The experimental unit was a group of nine plants. Regression analyses (linear, quadratic, sigmoidal, and exponential rise to maximum) were conducted with time or average PPFD of each treatment as the independent variable (SigmaPlot 11.0, Systat software, Inc., San Jose, CA, USA). Finally, multiple regression analysis ($\alpha < 0.05$) was performed using SAS (SAS University edition; SAS Institute, Cary, NC, USA) with PPFD as a continuous and species as a class variable to test for PPFD and species effects on PCS, incident light over the growing period, CCI, ACI, Φ_{PSII}, CO_2 assimilation, leaf area, canopy overlap ratio, dry weight, SLA, and LUE. When there was no significant interaction between species and PPFD, the interaction term was removed from the model and the main effects of species and PPFD were used to describe the treatment effects.

Supplementary Materials: The following supplementary figures are available online at https://www.mdpi.com/article/10.3390/plants10040704/s1, Figure S1. The projected canopy images captured during the growing period of mizuna (*Brassica rapa* var. *japonica*) (A–D) and lettuce (*Lactuca sativa* 'Green Salad Bowl') (E–H) grown at 425 μmol m^{-2} s^{-1}. The first three images of mizuna (A–C) and lettuce (E–G) were captured at 5 (A,E), 15 (B,F), and 19 days (C,G) after seeding. The last images of mizuna (D) and lettuce (H) were captured during the harvest (27 and 28 days, respectively); Figure S2. Correlation between early projected canopy size (lettuce at 10 days and mizuna at 8 days, after seeding) and final shoot dry weight of lettuce (*Lactuca sativa* 'Green Salad Bowl') and mizuna (*Brassica rapa* var. *japonica*); Figure S3. Correlation between specific leaf area (leaf area/shoot dry weight) and chlorophyll content index (CCI) of lettuce (*Lactuca sativa* 'Green Salad Bowl') and mizuna (*Brassica rapa* var. *japonica*); Figure S4. The appearance of lettuce (*Lactuca sativa* 'Green Salad Bowl', top) and mizuna (*Brassica rapa* var. *japonica*, bottom) at the end of the growing cycle for plants grown at six different photosynthetic photon flux densities (upper left corner of each picture).

Author Contributions: Conceptualization, T.C.J., and M.W.v.I.; methodology, T.C.J. and M.W.v.I.; formal analysis, T.C.J. and M.W.v.I.; investigation, T.C.J. and M.W.v.I.; data curation, T.C.J.; writing—original draft preparation, T.C.J.; writing—review and editing, T.C.J. and M.W.v.I.; visualization, T.C.J.; supervision, M.W.v.I.; funding acquisition, M.W.v.I. Both authors have read and agreed to the published version of the manuscript.

Funding: This research was funded by the U.S. Department of Agriculture–National Institute of Food and Agriculture–Specialty Crop Research Initiative Award No. 2018-51181-28365 (Project "Lighting Approaches to Maximize Profits". www.hortlamp.org).

Institutional Review Board Statement: Not applicable.

Informed Consent Statement: Not applicable.

Data Availability Statement: Data used in this study is available at https://drive.google.com/drive/folders/1BgP771XRRsbZHCttrHB-SueqV4nU0JMf.

Acknowledgments: We thank Eleanor Jane Rager for her assistance with data collection and Fluence Bioengineering for donating the light fixtures.

Conflicts of Interest: The authors declare no conflict of interest. The funders had no role in the design of the study; in the collection, analyses, or interpretation of data; in the writing of the manuscript, or in the decision to publish the results.

References

1. Birkby, J. Vertical farming. In *ATTRA Sustainable Agriculture*; National Center for Appropriate Technology: Butte, MT, USA, 2016; pp. 1–12.
2. Stober, K.; Lee, K.; Yamada, M.; Pattison, M. *Energy Savings Potential of SSL in Horticultural Applications*; Office of Energy Efficiency and Renewable Energy, U.S. Dept. of Energy: Washington, DC, USA, 2017.
3. Banerjee, C.; Adenaeuer, L. Up, up and away! The economics of vertical farming. *J. Agric. Stud.* **2014**, *2*, 40–60. [CrossRef]
4. Weaver, G.; van Iersel, M.W. Longer Photoperiods with Adaptive Lighting Control Can Improve Growth of Greenhouse-grown 'Little Gem' Lettuce (*Lactuca sativa*). *HortScience* **2020**, *55*, 573–580. [CrossRef]
5. Klassen, S.P.; Ritchie, G.L.; Frantz, J.M.; Pinnock, D.R.; Bugbee, B. Real-Time Imaging of Ground Cover: Relationships with Radiation Capture, Canopy Photosynthesis, and Daily Growth Rate. In *Digital Imaging and Spectral Techniques: Applications to Precision Agriculture and Crop Physiology*; American Society of Agronomy: Minneapolis, MN, USA, 2003; pp. 3–14.
6. Legendre, R.; van Iersel, M.W. Supplemental Far-Red Light Stimulates Lettuce Growth: Disentangling Morphological and Physiological Effects. *Plants* **2021**, *10*, 166. [CrossRef] [PubMed]
7. Campillo, C.; Prieto, M.H.; Daza, C.; Moñino, M.J.; García, M.I. Using Digital Images to Characterize Canopy Coverage and Light Interception in a Processing Tomato Crop. *HortScience* **2008**, *43*, 1780. [CrossRef]
8. Purcell, L. Soybean Canopy Coverage and Light Interception Measurements Using Digital Imagery. *Crop Sci.* **2000**, *40*, 834–837. [CrossRef]
9. Cometti, N.N.; Frantz, J.; Bugbee, B. *Imaging Lettuce Growth: A Comparison between % Ground Coverage and % PPF Absorption by Lettuce Grown in Hydroponics*; Crop Physiology Laboratory, USU: Logan, UT, USA, 2003.
10. Kim, C.; van Iersel, M.W. Morphological and physiological screening for growth differences among 11 lettuce cultivars. In Proceedings of the ASHS 2019 Annual Conference, Las Vegas, NV, USA, 21–25 July 2019; pp. S89–S90.
11. Elkins, C.; van Iersel, M.W. Longer Photoperiods with the Same Daily Light Integral Improve Growth of Rudbeckia Seedlings in a Greenhouse. *HortScience* **2020**, *55*, 1676–1682. [CrossRef]
12. Bauerle, W.L.; Weston, D.J.; Bowden, J.D.; Dudley, J.B.; Toler, J.E. Leaf absorptance of photosynthetically active radiation in relation to chlorophyll meter estimates among woody plant species. *Sci. Hortic.* **2004**, *101*, 169–178. [CrossRef]
13. Baker, N.R. Chlorophyll fluorescence: A probe of photosynthesis in vivo. *Annu. Rev. Plant Biol.* **2008**, *59*, 89–113. [CrossRef]
14. Maxwell, K.; Johnson, G.N. Chlorophyll fluorescence—A practical guide. *J. Exp. Bot.* **2000**, *51*, 659–668. [CrossRef]
15. Weaver, G.; van Iersel, M.W. Photochemical Characterization of Greenhouse-grown Lettuce (*Lactuca sativa* L. 'Green Towers') with Applications for Supplemental Lighting Control. *HortScience* **2019**, *54*, 317–322. [CrossRef]
16. Zhen, S.; van Iersel, M.W. Photochemical Acclimation of Three Contrasting Species to Different Light Levels: Implications for Optimizing Supplemental Lighting. *J. Am. Soc. Hortic. Sci.* **2017**, *142*, 346–354. [CrossRef]
17. Demmig-Adams, B.; Cohu, C.M.; Muller, O.; Adams, W.W. Modulation of photosynthetic energy conversion efficiency in nature: From seconds to seasons. *Photosynth Res.* **2012**, *113*, 75–88. [CrossRef]
18. Palmer, S.; van Iersel, M.W. Increasing Growth of Lettuce and Mizuna under Sole-Source LED Lighting Using Longer Photoperiods with the Same Daily Light Integral. *Agronomy* **2020**, *10*, 1659. [CrossRef]

19. Bhuiyan, R.; van Iersel, M.W. Only Extreme Fluctuations in Light Levels Reduce Lettuce Growth Under Sole Source Lighting. *Front. Plant Sci.* **2021**, *12*, 24. [CrossRef]
20. Tei, F.; Scaife, A.; Aikman, D.P. Growth of Lettuce, Onion, and Red Beet. 1. Growth Analysis, Light Interception, and Radiation Use Efficiency. *Ann. Bot.* **1996**, *78*, 633–643. [CrossRef]
21. Wurr, D.C.E.; Fellows, J.R. The influence of solar-radiation and temperature on the head weight of crisp lettuce. *J. Hortic. Sci.* **2015**, *66*, 183–190. [CrossRef]
22. Reinbothe, S.; Reinbothe, C. The regulation of enzymes involved in chlorophyll biosynthesis. *Eur. J. Biochem.* **1996**, *237*, 323–343. [CrossRef]
23. Zou, J.; Zhang, Y.; Zhang, Y.; Bian, Z.; Fanourakis, D.; Yang, Q.; Li, T. Morphological and physiological properties of indoor cultivated lettuce in response to additional far-red light. *Sci. Hortic.* **2019**, *257*, 108725. [CrossRef]
24. Manetas, Y.; Drinia, A.; Petropoulou, Y. High Contents of Anthocyanins in Young Leaves are Correlated with Low Pools of Xanthophyll Cycle Components and Low Risk of Photoinhibition. *Photosynthetica* **2002**, *40*, 349–354. [CrossRef]
25. Karageorgou, P.; Manetas, Y. The importance of being red when young: Anthocyanins and the protection of young leaves of *Quercus coccifera* from insect herbivory and excess light. *Tree Physiol.* **2006**, *26*, 613–621. [CrossRef]
26. Ferrarezi, R.S.; Bailey, D.S. Basil Performance Evaluation in Aquaponics. *HortTechnology* **2019**, *29*, 85. [CrossRef]
27. Mickens, M.A.; Torralba, M.; Robinson, S.A.; Spencer, L.E.; Romeyn, M.W.; Massa, G.D.; Wheeler, R.M. Growth of red pak choi under red and blue, supplemented white, and artificial sunlight provided by LEDs. *Sci. Hortic.* **2019**, *245*, 200–209. [CrossRef]
28. Ruban, A.V. Quantifying the efficiency of photoprotection. *Philos. Trans. R. Soc. Lond. B Biol. Sci.* **2017**, *372*, 20160393. [CrossRef]
29. Ruban, A.V. Evolution under the sun: Optimizing light harvesting in photosynthesis. *J. Exp. Bot.* **2015**, *66*, 7–23. [CrossRef] [PubMed]
30. Nemali, K.S.; van Iersel, M.W. Acclimation of wax begonia to light intensity: Changes in photosynthesis, respiration, and chlorophyll concentration. *J. Am. Soc. Hortic. Sci.* **2004**, *129*, 745–751. [CrossRef]
31. Muurinen, S.; Peltonen-Sainio, P. Radiation-use efficiency of modern and old spring cereal cultivars and its response to nitrogen in northern growing conditions. *Field Crop. Res.* **2006**, *96*, 363–373. [CrossRef]
32. Rosenthal, W.; Gerik, T. Radiation use efficiency among cotton cultivars. *Agron. J.* **1991**, *83*, 655–658. [CrossRef]
33. Van Iersel, M. Carbon use efficiency depends on growth respiration, maintenance respiration, and relative growth rate. A case study with lettuce. *Plant Cell Environ.* **2003**, *26*, 1441–1449. [CrossRef]

plants

MDPI

Article

LED Illumination Spectrum Manipulation for Increasing the Yield of Sweet Basil (*Ocimum basilicum* L.)

Md Momtazur Rahman *, **Mikhail Vasiliev** and **Kamal Alameh**

School of Science, Edith Cowan University, 270 Joondalup Drive, Joondalup, WA 6027, Australia; vasiliev.mikhail@gmail.com (M.V.); kalameh@bigpond.net.au (K.A.)
* Correspondence: mohammad.rahman@ecu.edu.au

Abstract: Manipulation of the LED illumination spectrum can enhance plant growth rate and development in grow tents. We report on the identification of the illumination spectrum required to significantly enhance the growth rate of sweet basil (*Ocimum basilicum* L.) plants in grow tent environments by controlling the LED wavebands illuminating the plants. Since the optimal illumination spectrum depends on the plant type, this work focuses on identifying the illumination spectrum that achieves significant basil biomass improvement compared to improvements reported in prior studies. To be able to optimize the illumination spectrum, several steps must be achieved, namely, understanding plant biology, conducting several trial-and-error experiments, iteratively refining experimental conditions, and undertaking accurate statistical analyses. In this study, basil plants are grown in three grow tents with three LED illumination treatments, namely, only white LED illumination (denoted W*), the combination of red (R) and blue (B) LED illumination (denoted BR*) (relative red (R) and blue (B) intensities are 84% and 16%, respectively) and a combination of red (R), blue (B) and far-red (F) LED illumination (denoted BRF*) (relative red (R), blue (B) and far-red (F) intensities are 79%, 11%, and 10%, respectively). The photosynthetic photon flux density (PPFD) was set at 155 μmol m^{-2} s^{-1} for all illumination treatments, and the photoperiod was 20 h per day. Experimental results show that a combination of blue (B), red (R), and far-red (F) LED illumination leads to a one-fold increase in the yield of a sweet basil plant in comparison with only white LED illumination (W*). On the other hand, the use of blue (B) and red (R) LED illumination results in a half-fold increase in plant yield. Understanding the effects of LED illumination spectrum on the growth of plant sweet basil plants through basic horticulture research enables farmers to significantly improve their production yield, thus food security and profitability.

Keywords: artificial lighting; energy use efficiency; protected horticulture; light exposure; far-red illumination; medicinal plants; water use efficiency; growth analysis

Citation: Rahman, M.M.; Vasiliev, M.; Alameh, K. LED Illumination Spectrum Manipulation for Increasing the Yield of Sweet Basil (*Ocimum basilicum* L.). *Plants* **2021**, *10*, 344. https://doi.org/10.3390/plants10020344

Academic Editor: Valeria Cavallaro

Received: 17 January 2021
Accepted: 9 February 2021
Published: 11 February 2021

1. Introduction

Global food demand is expected to increase by approximately 70 percent by 2050 due to increasing population growth [1]. The use of energy-efficient light-emitting diode (LED) sources in a protected-crop environment is an attractive approach that enables high-quality crops [2–5] to be produced cost-effectively, meeting human food demands. While white LED-generated photons can stimulate the plant photosynthesis process, the entire spectral components of white LED light would not equally participate in the photosynthesis process [6], and the impact of far-red wavelengths on the growth of sweet basil is not fully understood yet.

According to the Australian Department of Agriculture, horticulture production has the biggest market share in the Australian agriculture market, estimated to increase by 3% to $11.7 billion in 2020 [7]. Due to recurrent devastating natural calamities, such as bushfires, rainfall deficiency, Australian crop production has severely been affected, and this has encouraged farmers to adopt protected cropping practices to offer high-quality crops to consumers while maintaining a high level of food security by increasing consumer access to non-seasonal foods all over the year [8]. Hence, protected cropping systems used in urban areas have become one of the fastest-growing and dominant food-producing sector in Australian Horticulture, valued at around $1.8 billion per year in 2019, which is equivalent to 20% of the value of leafy vegetable food production in Australia [9]. Note that protected cropping offers several advantages, including diverse production structures (e.g., optical illumination in dark tents, greenhouses, vertical farms, etc.) [10], ambient temperature control, which maintains a high crop yield [11], better control of the CO_2 [12] concentration, water, minerals, and fertilizers, which improve the photosynthesis process. The heating/cooling and lighting costs together typically stand for 25~35% of the total cost in a greenhouse environment [13]. Due to higher operation costs, conventional greenhouses are still unable to provide agricultural products to consumers cost-effectively. For example, sweet basil, parsley, coriander, and kale species are relatively expensive to produce commercially and require accurate adjustment of the growth conditions. These high costs have recently driven the market of LED-based indoor farming, mainly because of the high-efficiency, durability (~50,000 h), low-heat-generation and low-cost of light-emitting diode (LED) technology and the wide range of LED wavelength bands availability. LEDs are 40~70% more efficient than high-pressure sodium (HPS) lights or metal halide (MH) lamps (most common light sources used in indoor farming [13]).

Figure 1 shows the prominent wavelength bands that contribute to the development of sweet basil.

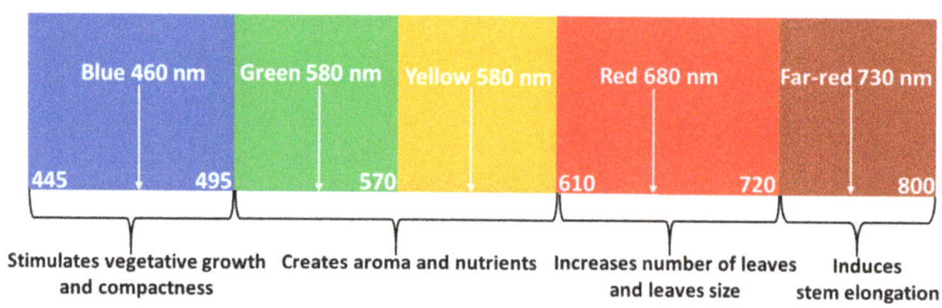

Figure 1. Effective wavelength bands that affect the growth of sweet basil plants.

In addition, with the absence of sunlight at night times, LEDs can be used to illuminate the crop for a more extended period, thus shortening the crop cultivation cycle and improving the crop yield. In addition, LED-based high-insulation grow tents are relatively cheap and portable structures that can maintain ideal growing temperature and provide sufficient lighting at any time of day or night, in comparison with a greenhouse environment, where natural outdoor weather conditions dictate the cooling and heating requirements with a high degree of unpredictability. The grow tents were silver-coated on the inside and black on the outside. The grow tents were considerably remained closed for most of the time, except for panels with getting to permit airflow in, on three of four sides towards the tent's base for ventilation. The fourth side was the tent entrance, and the zip was left open towards the bottom further to increase airflow and ventilation [14]. The tents were opened daily to water each plant; otherwise, the tents remained nearly closed. The visible light is typically the major contributor to the photosynthesis process for sweet basil plants [15]. According to the earlier publication reports, the red and blue spectral components are the major contributors to crop growth, such as promotes vegetative growth [16] and compactness [17,18] and creates aroma and nutritional value [19], as shown in Figure 1. While blue light has a short wavelength, it helps the plant adjust its growth [20] with the environmental interaction at a different stage and promotes early vegetative growth. Moreover, the blue component significantly affects the shoot architecture, resulting in a compact and dense plant [21], and increases the vegetative growth (leaves). In contrast, red light, promotes leaf elongation [22,23]. These wavelengths correlate to the five photosensory systems of a plant biosystem, namely, phytochromes (PHY), phototropins (PHOTO), cryptochrome (CRY), Zeitlupe (ZLT), and UVB-resistance locus 8 (UVR8). Each one of these photosensory systems triggers different morphological changes in plants. For example, phytochromes (PHY), whose unique photosensory properties can profoundly have a major role in governing plant elongation, flowering time, and leaf expansion [24], perceives light strongly in the red (660~700 nm) and far-red regions (700~750 nm) [25]. In contrast, the phototropin (PHOTO), cryptochrome (CRY), and ZLT system absorb light actively in the blue (400~495 nm) and UV-A (315~400 nm) regions [26], predominantly regulates plants hypocotyl elongation, and play an indispensable role in blue light facilitated stomatal opening [27,28] and controls the prosperity of an effective photoperiodic blossoming inducer [28], while the UVB-resistance locus 8 (UVR8) system perceives light intensely in the UV-B (280~315 nm) regions [29], and controls the biosynthesis-related genes expressions [30]. On the other hand, the far-red and UV ranges have secondary impacts on specific plants' growth. In contrast, the green-yellow spectral components (sometimes called tertiary light due to their minimal impact) have a marginal role in the photosynthesis process. It is essential to mention that less research has been conducted to investigate the optimum LED illumination and its effect on water and electricity uses efficiency and the morphological development of sweet basil plants.

In addition to that, to increase the further crop yield, greenhouse designers are currently investigating the use of smart glass that enables effective light management by transmitting the solar spectral components that effectively contribute to the photosynthesis process while blocking the ineffective radiations, which typically lead to plant transpiration and photoinhibition.

The market for industrial sweet basil (*Ocimum basilicum* L.) for pesto, frozen, and dried markets are developing. Besides being popular as a spice, basil contains essential oils used in the medicine and chemical industry. Note that relatively little information exists regarding the effects of narrow-band lighting technologies on the physiological and morphological development of sweet basil and its resource use activity. However, several research groups have investigated the light spectrum's effects on the yield of different greenhouse-grown vegetables and herbs [31,32]. Nevertheless, relatively few research groups have reported the impact of lighting outside of the well-known photosynthetically active radiation (PAR) wavelengths (400 nm to 700 nm), especially the effect of far-red wavelengths (~735 nm) on the physiological development of sweet basil plants. Therefore, this work aims to investigate the illumination spectrum that achieves significant biomass improvement of sweet basil plants through prior iterative refining of experimental circumstances and after a careful statistical analysis.

2. Materials and Methods

2.1. Plant Materials and Growing Conditions

The experiments were carried out during March 2020 in three separate grow tents at the Electron Science Research Institute (ESRI), Edith Cowan University (ECU), Australia.

A Heliospectra™ LX602C tunable LED light source was hung inside in each grows tent above the plants at 1.26 m from the top of each pot. All pots were placed on the floor. The tunable LED sources are designed to cover a 2 m × 1.80 m area at a mounting height of 2 m. Therefore, they were adequate for illuminating the species inside the three 1.5 m × 1.5 m × 2.0 m grow tents. We used 60 plants for the experiments (20 plants per light treatment). After seedling transplanting, the pots were randomly distributed around the center of each tent over a circular area of a diameter ~0.75 m, and hence, the LED illumination was almost uniform for all pots. The sweet basil (*Ocimum basilicum* L.) seedlings were obtained from Bunnings warehouse, Joondalup, Western Australia, and transplanted individually into 125 mm (diameter) plastic pots with 650 gm of potting mix from Scotts Osmocote® Plus Organics Vegetable and Herb Mix, Australia. The plants were watered regularly to maintain optimum soil moisture. We used a conventional soil moisture meter that indicates the amount of water needed to keep the soil moisture at the same level. After transplanting the seedlings, each pot was watered according to needs and kept as it is for one day to allow the water to soak in properly. To supply nutrients to the plants, a 10 mL of diluted liquid nutrient (Scotts Osmocote®, Melbourne, Australia) was mixed with 1490 mL of water. The resultant 1500 mL solution was split equally into 15,100 mL units, and each unit was supplied to a pot after every seven days from the start of LED illumination. Data (canopy temperature, air temperature, soil temperature, soil pH, humidity, LED consumption power, LED illumination uniformity) were collected at night.

2.2. Lighting Treatments

Three different LED illumination spectra were applied in the grow tents, namely, W*, BR*, and BRF*. The power consumption was measured for each tent using an intelligent power meter (Electus Distribution Pty Ltd, Sydney, Australia). Note that the Heliospectra™ LX602C tunable LED source enables the output power level of the individual LED spectral components to be systematically controlled between 0% and 100%, allowing a wide range of LED illumination spectra to be generated. The Heliospectra™ LX602C LED sources were switched on every day from the afternoon (3:00 p.m.) to morning (11:00 a.m.). Thus, the illumination period was 20 h per day. Figure 2 shows the measured photosynthetic photon flux density (PPFD) at different distances from the LED source for all tents. As shown in Figure 2, the measured photosynthetic photon flux densities (PPFD) at 1.26 m from the LED sources (i.e., at the pot top-level) were 155 ± 1 μmol m^{-2} s^{-1} for all the tents. Note that the photosynthetic photon flux density is the sum of the flux densities corresponding to the individual LED colors used to generate the LED illumination spectrum [33]. The measured air temperature and relative humidity were in the 22 °C to 26 °C range and the 60~75% range, respectively, for all tents.

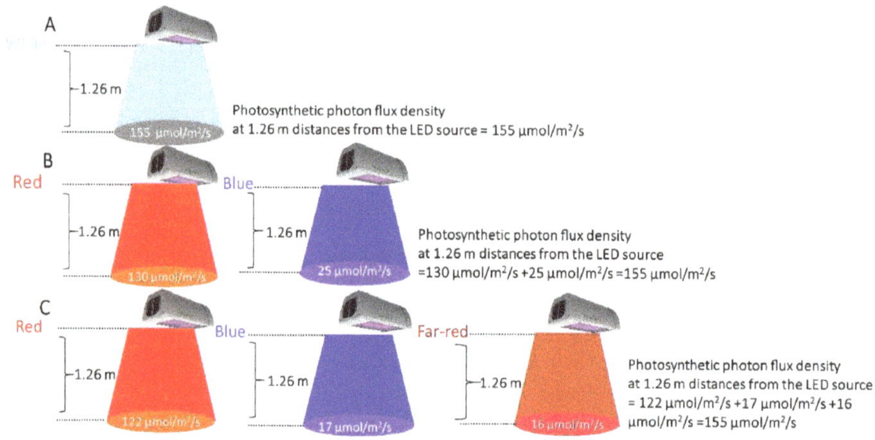

Figure 2. The photosynthetic photon flux densities (PPFD) of the Heliospectra™ grow lights measured using a LaserCheck optical power meter, at different distances from the light-emitting diode (LED) source, for all tents.

In order to accurately measure the LED intensity, we calibrated the LED power levels with a hand-held optical power meter (LaserCheck, Coherent Inc., Santa Clara, CA, USA), which has a power range of 10 μW to 1000 mW and an active aperture diameter of 8.0 mm and operates over the wavelength range 400 nm to 1064 nm. The power density (W/m^2) at a specific distance from each light source was calculated by measuring the power output and dividing it by the active aperture area of the optical power meter. The average fresh mass (FW) of plants was measured using a high-sensitivity scientific laboratory balance (Westlab, Mitchell Park, Australia) with an accuracy of 0.01 g. Then, the leaves were placed in Kraft paper envelopes (27 cm × 22 cm) and heated in an oven (Furnace Technologies

Pty Ltd, Jandakot, Australia) at 60 °C for 120 h until moisture was fully evaporated and a constant dry mass state was attained [34].

2.3. Measurements and Data Collection

After 28 days of LED illumination, morphological parameters of 24 out of 36 plants, namely, plant fresh mass (g), plant dry mass (g), plant height (cm), plant stem diameter (mm), leaf fresh mass (g), leaf dry mass (g), energy use efficiency (EUE) and water use efficiency (WUE) were measured. In order to measure the improvement in cultivation time for the BRF* illumination, the remaining 12 plants were kept in the white tent for two more weeks (14 days), and their morphological parameters were measured (i.e., on day 42). Note that the EUE, expressed as g FW kW^{-1}, is defined as the ratio of the fresh mass of the sweet basil plant and the electricity consumption of LEDs, and the WUE, expressed in g FW L^{-1} H_2O, is defined as the ratio between the leaf fresh mass and the volume of water used. Generally, WUE increases the plant's fresh mass [35] and fruit yield [36]. During the experiments, the air temperature (T_{air} in °C) and relative humidity (RH%) inside each tent were measured using a thermometer (Green May International Ltd, Shenzhen, China), and data were manually logged on every day at 8:00 pm. The leaf temperatures were measured manually every day at 8:00 pm using an infrared thermometer whose laser beam probe was applied to the leaf surface area (Wiltronics Research Pty Ltd, Alfredton, Australia). At the end of the experiments, the average height of the sweet basil plants was measured and recorded using a tape measure (Stanley Tools, Melbourne, Australia) with 0.01 m precision. The average stem diameter of the sweet basil plants was measured using a digital Vernier caliper with an accuracy of 0.01 mm (Kincrome Australia Pty Ltd, Melbourn, Australia). The Heliospectra LED light sources are programmed to emit wavelengths of light at the broader spectrum white LED light (5700 K white visible light having peaked at approximately 446 nm, 534 nm, and 625 nm). As shown in Figure 3, the white LED spectrum contains 24% blue light, 58% green light, and 18% red light.

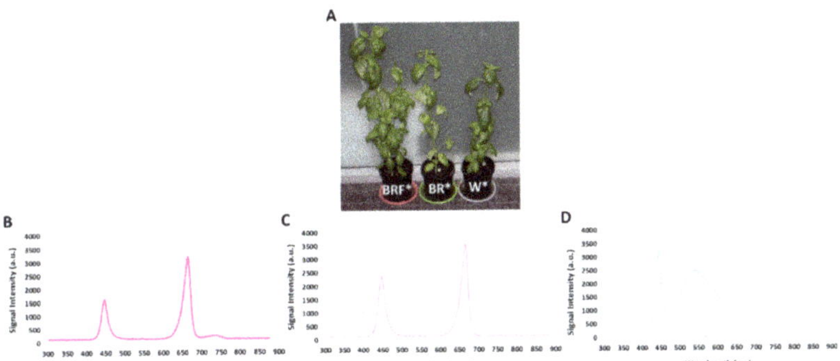

Figure 3. (**A**) Photos of typical sweet basil pots from the BRF*, BR*, and W* grow tents after 28 days of LED illumination. (**B**) BRF* illumination spectrum; (**C**) BR* illumination spectrum; and (**D**) W* illumination spectrum. W*—only white LED illumination; BR*— combination of red (R) and blue (B) LED illumination with relative R and B intensities of 84% and 16%, respectively); BRF*— combination of R, B and far-red (F) LED illumination, with relative R, B and far-red (F) intensities of 79%, 11%, and 10%, respectively.

The soil moisture (dry to wet range only) and pH were measured using a ZD-07 4 in 1 digital soil pH meter (NDI Instrument and Hand Tools, Cheltenham, Australia) by inserting the probe of the instrument as vertically as possible and down halfway between the plant stem and the edge of the pot and taking several readings for averaging. Table 1 shows the measured PPFD, relative humidity, soil temperature, pH level, leaf temperature, and energy consumption for all grow tents.

Table 1. The power being consumed (kWh, total per entire growth period per growth tent) and growing conditions applied to the cultivation of sweet basil plants in the grow tent.

Light Types	PPFD (μmol m^{-2} s^{-1})	Relative Humidity (%)	Soil Temperature (°C)	pH	Leaf Temperature (°C)	Energy Consumption (kWh)	Plant Height (cm)	Plant Stem Diameter (mm)	Leaf Fresh Mass (g)	Leaf Dry Mass (g)
W*	155 ± 1	64 ± 2	24.88 ± 1	6.20	22.78 ± 1	133 ± 1	383.75	3.79	11.04	1.45
BR*	155 ± 1	64 ± 2	24.98 ± 1	6.35	22.57 ± 1	133 ± 1	416.87	3.77	15.83	2.97
BRF*	155 ± 1	64 ± 2	25.19 ± 1	6.30	23.34 ± 1	133 ± 1	464.37	4.09	19.25	4.62

2.4. Statistics

The data sets collected from the experiment were analyzed using the one-way analysis of variance (ANOVA) to find out if our experimental results are significant. Once we conducted the ANOVA test and found the test results statistically significant, at least two groups have significantly different means. Then, we intended to locate the specific difference and wanted to find out where our differences indeed came from. Therefore, to determine which groups are different from each other, we had to conduct a post hoc test. Descriptive statistical parameters, such as mean, standard deviation, and standard error (SE), were calculated using the IBM SPSS Statistics software package, version 27.

3. Results

3.1. Effect of Light Quality on Plant Growth and Morphology

In this study, the effects of blue, red, and far-red light spectra are mainly investigated. Results show that the plants' growth is substantially different under the grow-tent environment, such as post hoc analyses revealed that the W*-illuminated tent produced the lowest average fresh biomass, 22.08 g, which was considerably lower than that produced by the BRF*-illuminated tent, 40.58 g, and the BR*-illuminated tent, 31.58 g, as shown in Figure 4A.

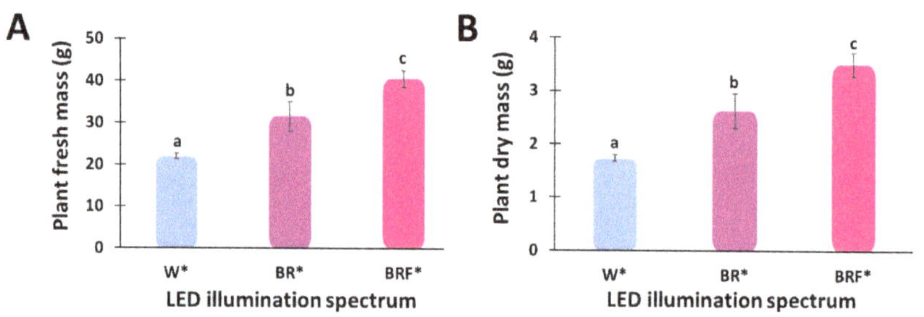

Figure 4. Effect of different LED illumination spectra on the growth, and development parameters, such as (**A**) fresh mass (g); and (**B**) dry mass of sweet basil (*Ocimum basilicum* L.) plants grown in 2 m^2 tents. Different letters indicate a significant difference among the treatments, $p \leq 0.05$.

We have identified the best LED wavelength bands required to significantly improve the growth rate of sweet basil (*Ocimum basilicum* L.) plants in grow tent environments by optimizing the LED spectrum illuminating the plants. Post hoc comparisons using the Tukey's–Kramer HSD tests indicated that the W*-illuminated tent produces the lowest mean dry biomass (M = 1.73 g; standard deviation, SD = 0.20 g), and this is substantially lower than that produced by the BRF*-illuminated tent (M = 3.49 g; SD = 0.62 g), and the BR*-illuminated tent (M = 2.62 g; SD = 0.93 g), as shown in Figure 4B.

3.2. Effect of Light Quality on Resource Usage Efficiency

Figure 5A shows the WUE for the three LED illumination spectra. The WUE for the W*-, BR*- and BRF*-illuminated tents are 13 g FW L^{-1} H_2O, 18 g FW L^{-1} H_2O, and 24 g FW L^{-1} H_2O.

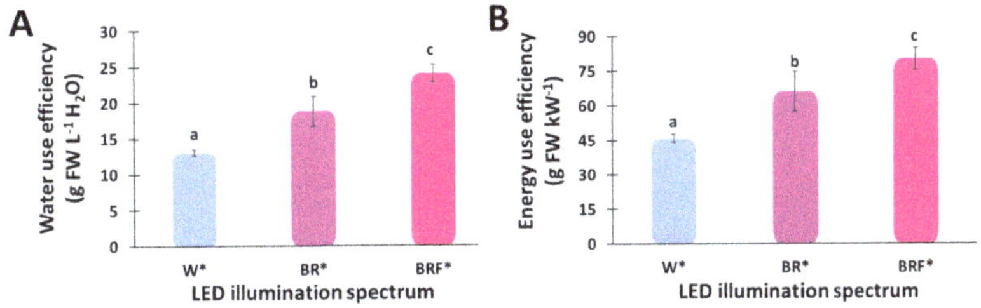

Figure 5. Effect of the different LED illumination spectra on the mean values of (**A**) water use efficiency (WUE) and (**B**) energy use efficiency (EUE) of sweet basil (*Ocimum basilicum* L.) plant grown in 2 m^2 tents. Different letters indicate a significant difference among the treatments, $p \leq 0.05$.

Note that the uniformity of photosynthetic photon flux (PPF) is crucial for indoor plant growth. This prevents photons from being spread over, reducing energy consumption by improving their utilization inside the grow tent. The energy use efficiencies (EUE) is defined as the average biomass produced per unit of electrical energy used for plant illumination.

The energy use efficiencies (EUE) for the three LED illumination spectra are shown in Figure 5B, revealing that the EUE for the W*-, BR*- and BRF*-illuminated tents are 46 ± 1.7 g FW kW^{-1}, 65 ± 8.6 g FW kW^{-1}, and 80 ± 4.8 g FW kW^{-1}. Here the BRF* spectrum yielded a significantly higher EUE than that of the W* spectrum.

3.3. Cultivation Cycle Improvement

Figure 6 shows the biomasses of basil plants after 6 weeks in the W*-illuminated tent and after 4 weeks in the BRF*-illuminated tents. Note that, after 6 weeks, the average dry mass of the basil plants grows in the W*-illuminated tent was around 3.2455 g per plant, and that was 93% of the average dry mass per plant grown over a period of 4 weeks in the BRF*-illuminated tent. Note that more than 60% of the basil plants grown in the W*-illuminated tent bolted after 6 weeks, whereas none of the basic plants showed bolting after 4 weeks. These results demonstrate the ability of the BRF* illumination spectrum to (i) achieve high crop yield and (ii) high crop quality (no bolting).

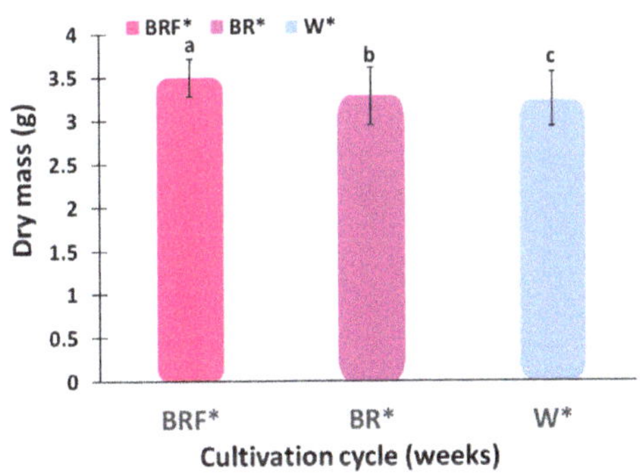

Figure 6. Biomasses of basil plants after 6 weeks in the W* illuminated tent and after 4 weeks in the BRF*-illuminated tents. Different letters indicate a significant difference among the treatments, $p \leq 0.05$.

4. Discussion

The spectral components of LED illumination significantly affect both plant physiology and growth morphology [37,38]. The visible light is typically the major contributor to the photosynthesis process for sweet basil plants, and by incorporating LED supplemental lighting in indoor farming, the plant growth rate can be increased significantly [39,40]. In the present study, crop yield enhancement was associated with a higher fraction of far-red light, a phenomenon that both Lee et al. [41] and Murchie et al. [42] have reported. According to the Emerson effect, both the red and far-red bands are significant contributors to the photosynthetic process for plants [43]. Thus, adding far-red LED illumination typically increases morphological parameters of indoor-grown plants. For example, increasing the ratio of the red intensity to far-red intensity increases the leaf length [44] and yield [45]. Sometimes, when the plant sizes become large, the degree of shading in the greenhouse becomes high, and plants tend to modulate their growth to resume a light-seeking strategy [2]. This is referred to as the shade avoidance syndrome and is, in effect, partial etiolation. Shade avoidance enables a plant to anticipate future competition for light by reducing reliance on resources for branching and capitalizing more on height growth [3]. Shade avoidance also causes an altered partitioning of photosynthate in favor of vegetative tissues, which can decrease yield in seed-producing crops [4]. The shade-avoidance response can be induced under other conditions such as crowding and under different conditions that cause a reduced ratio of red to far-red light, indicating phytochromes' involvement [4].

Note that, in order to validate these results, another set of experiments was conducted in September 2020, and the variances in plant yield, water use efficiency, and energy use efficiency were less than $\pm 2.5\%$, as shown in Figure 7.

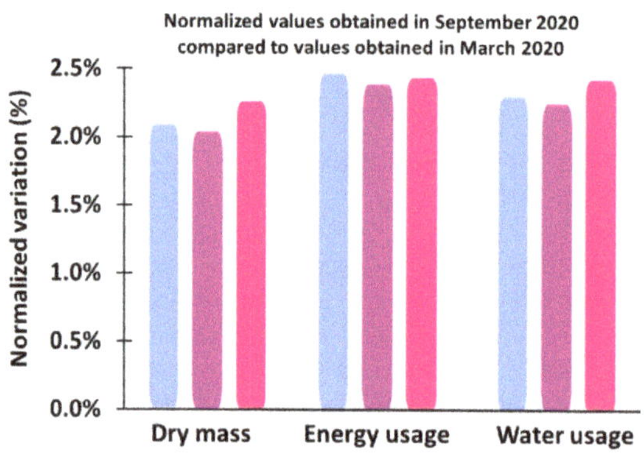

Figure 7. The normalized values obtained in the first experiment in September 2020 compared to values obtained in the second experiment in March 2020.

The inside walls of each grow tent used in the experiments were coated with silver reflective coatings to prevent shading. Shading typically results in plant "bolting" and reduces the oil content, and hence, the biomass, in the basil leaves [46,47]. Note that too much light would also reduce the plant biomass as a result of plant tip burns. Sweet Basil (*Ocimum basilicum* L.) plants were specially selected for this study since they can be grown in a protected environment with higher temperatures in a long-day growth mode. Note that, after 6 weeks, the average biomass of the sweet basil plants grown in the W*-illuminated tent was around 3.2455 g per plant, and that was 93% of the average biomass per plant produced over a period of 4 weeks in the BRF*-illuminated tent. Also note that, in this experiment, more than 60% of the basil plants grown in the W*-illuminated tent bolted after 6 weeks, whereas none of the basic plants showed bolting after 4 weeks. Therefore, by better understanding the effects of lighting conditions (including the effect of the length of the day and night illumination) on the vegetative growth and reproductive growth (bolting, also known as preliminary flowering), the plant yields can be improved significantly.

Furthermore, in the experiments, it was observed that increasing the red LED intensity inhibits the transition to flowering (i.e., bolting) in basil plants, and this also has been observed recently [48]. It was shown that compared to 100% red LED illumination, the combination of red and blue LED illumination sources (91% red + 9% blue, by the photon counts) has a positive impact on sweet basil, spinach, and lettuce, in terms of biomass, plant height and leaf size [49]. Moreover, it was reported that the highest shoot dry mass of sweet basil (*Ocimum basilicum* L.) plants is attained with R70B30 LED illumination (70% red + 30% blue light with 250 ± 10 μmol m^{-2} s^{-1}) [50]. Note that the phytochrome (PHY) photoreceptors of sweet basil absorb light strongly in the red (660~700 nm) and far-red regions (700~750 nm) [25], and this enhances the vegetative development (biomass) as well as architectural development of the plant [51]. In the experiments, a similar kind of results was achieved. For example, BRF*-illuminated tent produced a higher dry mass, 3.49 g, after 4 weeks, which was considerably higher than the average dry mass of 1.73 g produced by the W*-illuminated tent at the same time.

The basil's WUE values are 3 g FW L^{-1} H_2O in open field cultivation, while 20 to 22 g FW L^{-1} H_2O are observed in potted grown basil in European climate conditions [52]. Similar results were also found in the present experiments, such as higher biomass were produced with lower water consumption, i.e., the WUE increased to above 24 g FW L^{-1} H_2O in BRF*-illuminated tent (Figure 5A) [53]. Note that, in the present study, the WUE in the BRF*-illuminated tent was increased by 83% and 27%, respectively, compared to the W*-illumination (Figure 4A) and the BR*-illumination. In contrast, the WUE for BR*-illuminated tent was improved by 47% compared to the W*-illuminated tent. This is due to the higher red portion of the LED spectrum. While the red part of the LED spectrum increased, the quantum efficiency of the photosynthesis process decreased; however, the transpiration decreased more rapidly, resulting in increased water usage efficiency [54]. This increased water usage efficiency could also be associated with changes in the stomatal behavior of the plant. Typically, the soil temperature, which also affects basil plants' growth rate, depends on the ratio of the energy absorbed by the soil to the energy lost from the soil. Note that the soil temperature affects the soil moisture because high soil temperature leads to water evaporation and crop transpiration. Hence, to maintain a high plant growth rate, the amount of water supplied to the plants must be continuously monitored and optimized. The basil plants illuminated by the W* spectrum exhibited a higher leaf surface temperature than the plants illuminated by the BR* and BRF* spectra (Table 1). This could be attributed to the non-photosynthetic spectral components being absorbed by the crop and converted to heat. Note that the leaf surface temperature is typically affected by illumination, relative humidity and ambient temperature. When a photon of light hits the plant leaf, it can either be reflected, transmitted, or absorbed. The photons that participate in the photosynthesis process (e.g., blue, red, and far-red photons) typically have less impact on the leaf temperature than the photons absorbed by the plant but do not contribute to photosynthesis (e.g., UV, green, and IR photons). Therefore, measuring the leaf surface temperature under light illumination is an indirect indication of the effectiveness of the illumination spectrum on the photosynthesis process, thus energy usage efficiency. In the experiments, based on the basil yield, it should be noted that the EUE was greater when higher spectral portions were allocated to the red region as for the case of the BRF* illumination (80 ± 4.8 g FW kW^{-1}) and the BR* illumination (65 ± 8.6 g FW kW^{-1}) (Figure 4B), due to a larger yield increase observed in these treatments, compared to the W*-illumination (46 ± 1.7 g FW kW^{-1}).

5. Conclusions

We have experimentally investigated the effect of LED illumination spectra on the growth of sweet basil plants. Specifically, the plant fresh mass (g), plant dry mass (g), energy use efficiency (EUE), water use efficiency (WUE), and plant cultivation cycle were measured for sweet basil plants grown in three different grow tents illuminated with (i) white (W*), (ii) blue (B) and red (R); and (iii) blue (B), red (R) and far-red (F) LED spectra. Post hoc analyses have revealed that the BRF*- and BR*-illuminated tents produced, respectively, 83% and 42% higher average fresh biomass than that produced in the W*-illuminated tent. For the average dry mass, results have shown that the BRF*- and BR*-illuminated tents produced, respectively, 100% and 51% higher average dry biomass than that produced in the W*-illuminated tent. Results have also shown that, after 6 weeks, the average biomass of the basil plants grown in the W*-illuminated tent is 93% of the average biomass per plant grown over a period of 4 weeks in the BRF*-illuminated tent, and that, more than 60% of the basil plants grown in the W*-illuminated tent bolted after 6 weeks, whereas the basil plants were bolting-free after 4 weeks. These results have demonstrated that the BRF*-illumination treatment enables higher crop yield and quality (no bolting) in comparison with the W*-illumination treatment.

In addition, experimental results have shown that the water usage efficiency with the BRF* spectrum was 24 g FW L^{-1} H_2O, compared to 13 g FW L^{-1} H_2O, and 18 g FW L^{-1} H_2O for the W*- and BR*-spectra, respectively. Moreover, results have revealed that the electricity usage efficiency with the BRF* spectrum was 80 $\pm$ 4.8 g FW kW^{-1}, compared to 46 $\pm$ 1.7 g FW kW^{-1} and 65 $\pm$ 8.6 g FW kW^{-1} for the W* and BR* spectra, respectively. Therefore, the results of this study have demonstrated the commercial viability of both BRF*-, and BR*-illuminated grow tents compared to the commonly used W*-illuminated counterparts.

The protected environment results presented in this work paves the way towards the development of glass greenhouses employing spectrally selective optical coatings that pass the optimum solar spectral components through, which significantly increase the yield of sweet basil.

Author Contributions: Conceptualization, M.M.R., M.V. and K.A.; data curation, M.M.R., M.V. and K.A.; formal analysis, M.M.R., M.V. and K.A.; funding acquisition, K.A.; investigation, M.M.R., M.V. and K.A.; methodology, M.M.R., M.V. and K.A.; project administration, M.M.R. and K.A.; resources, M.M.R., M.V. and K.A.; software, M.M.R. and M.V.; supervision, M.V. and K.A.; validation, M.M.R., M.V. and K.A.; visualization, M.M.R. and K.A.; writing—original draft, M.M.R., M.V. and K.A.; writing—review and editing, M.M.R., M.V. and K.A. All authors have read and agreed to the published version of the manuscript.

Funding: This research received no external funding.

Data Availability Statement: The datasets generated and/or analysed during this study are available from the corresponding author on reasonable request.

Acknowledgments: The research was supported by Edith Cowan University (ECU), Australia. The authors wish to thank Jacqualine Anne Thomas, Mohammad Nur E Alam, Sretan Askraba, Paul Roach, our colleagues at Edith Cowan University for their suggestions on preparing the manuscript. We are also grateful for the insightful comments offered by the anonymous peer reviewers.

Conflicts of Interest: The authors declare no conflict of interest.

References

1. Linehan, V.; Thorpe, S.; Andrews, N.; Kim, Y.; Beaini, F. *Food Demand to 2050: Opportunities for Australian Agriculture*; Paper presented at the 42nd ABARES Outlook conference; ABARES: Canberra, Australia, 2012; pp. 2–7.
2. Zhang, X.; Bian, Z.; Yuan, X.; Chen, X.; Lu, C. A review on the effects of light-emitting diode (LED) light on the nutrients of sprouts and microgreens. *Trends Food Sci. Technol.* **2020**, *99*, 203–216. [CrossRef]
3. De Keyser, E.; Dhooghe, E.; Christiaens, A.; Van Labeke, M.-C.; Van Huylenbroeck, J. LED light quality intensifies leaf pigmentation in ornamental pot plants. *Sci. Hortic.* **2019**, *253*, 270–275. [CrossRef]
4. Paucek, I.; Pennisi, G.; Pistillo, A.; Appolloni, E.; Crepaldi, A.; Calegari, B.; Spinelli, F.; Cellini, A.; Gabarrell, X.; Orsini, F.; et al. Supplementary LED Interlighting Improves Yield and Precocity of Greenhouse Tomatoes in the Mediterranean. *Agronomy* **2020**, *10*, 1002. [CrossRef]
5. Gao, W.; He, D.; Ji, F.; Zhang, S.; Zheng, J. Effects of Daily Light Integral and LED Spectrum on Growth and Nutritional Quality of Hydroponic Spinach. *Agronomy* **2020**, *10*, 1082. [CrossRef]
6. Bolton, J.R.; Hall, D.O. The Maximum Efficiency of Photosynthesis *. *Photochem. Photobiol.* **1991**, *53*, 545–548. [CrossRef]
7. Jackson, T.; Zammit, K.; Hatfield-Dodds, S. Snapshot of Australian agriculture 2020. *Dep. Agric. Water Environ.* **2020**, *1*, 2–7.
8. Brown, C.; Miller, S. The Impacts of Local Markets: A Review of Research on Farmers Markets and Community Supported Agriculture (CSA). *Am. J. Agric. Econ.* **2008**, *90*, 1298–1302. [CrossRef]
9. Smith, G. *Overview of the Australian Protected Cropping Industry*; PCA: The Hague, The Netherlands, 2011.
10. Walters, K.J.; Currey, C.J. Hydroponic Greenhouse Basil Production: Comparing Systems and Cultivars. *HortTechnology* **2015**, *25*, 645–650. [CrossRef]
11. Choi, H.; Moon, T.; Jung, D.H.; Son, J.E. Prediction of Air Temperature and Relative Humidity in Greenhouse via a Multilayer Perceptron Using Environmental Factors. *Prot. Hortic. Plant. Fact.* **2019**, *28*, 95–103. [CrossRef]

12. Ramirez, T.; Kläring, H.-P.; Körner, O. Does interruption of electricity supply for supplementary lighting affect the long-term carbon dioxide exchange of greenhouse tomato crops? *Biosyst. Eng.* **2019**, *187*, 69–80. [CrossRef]
13. Kacira, M. Greenhouse production in US: Status, challenges, and opportunities. In Proceedings of the CIGR 2011 Conference on Sustainable Bioproduction WEF, Tokyo, Japan, 19–23 September 2011; pp. 19–23.
14. Thomas, A.J.; Vasiliev, M.; Nur-E-Alam, M.; Alameh, K. Increasing the Yield of Lactuca sativa, L. in Glass Greenhouses through Illumination Spectral Filtering and Development of an Optical Thin Film Filter. *Sustainability* **2020**, *12*, 3740. [CrossRef]
15. Dou, H.; Niu, G.; Gu, M.; Masabni, J.G. Effects of Light Quality on Growth and Phytonutrient Accumulation of Herbs under Controlled Environments. *Horticulturae* **2017**, *3*, 36. [CrossRef]
16. Bantis, F.; Koukounaras, A.; Siomos, A.S.; Fotelli, M.; Kintzonidis, D. Bichromatic red and blue LEDs during healing enhance the vegetative growth and quality of grafted watermelon seedlings. *Sci. Hortic.* **2020**, *261*, 109000. [CrossRef]
17. Lee, S.-W.; Seo, J.M.; Lee, M.-K.; Chun, J.-H.; Antonisamy, P.; Arasu, M.V.; Suzuki, T.; Al-Dhabi, N.A.; Kim, S.-J. Influence of different LED lamps on the production of phenolic compounds in common and Tartary buckwheat sprouts. *Ind. Crop. Prod.* **2014**, *54*, 320–326. [CrossRef]
18. Han, T.; Vaganov, V.; Cao, S.; Li, Q.; Ling, L.; Cheng, X.; Peng, L.; Zhang, C.; Yakovlev, A.N.; Zhong, Y. Improving "color rendering" of LED lighting for the growth of lettuce. *Sci. Rep.* **2017**, *7*, 45944. [CrossRef] [PubMed]
19. Watcharatanon, K.; Ingkaninan, K.; Putalun, W. Improved triterpenoid saponin glycosides accumulation in in vitro culture of Bacopa monnieri (L.) Wettst with precursor feeding and LED light exposure. *Ind. Crop. Prod.* **2019**, *134*, 303–308. [CrossRef]
20. Huche-Thelier, L.; Crespel, L.; Gourrierec, J.G.-L.; Morel, P.; Sakr, S.; LeDuc, N. Light signaling and plant responses to blue and UV radiations—Perspectives for applications in horticulture. *Environ. Exp. Bot.* **2016**, *121*, 22–38. [CrossRef]
21. Sergejeva, D.; Alsina, I.; Duma, M.; Dubova, L.; Augspole, I.; Erdberga, I.; Berzina, K. Evaluation of different lighting sources on the growth and chemical composition of lettuce. *Agron. Res.* **2018**, *16*, 892–899. [CrossRef]
22. Johkan, M.; Shoji, K.; Goto, F.; Hahida, S.; Yoshihara, T. Effect of green light wavelength and intensity on photomorphogenesis and photosynthesis in Lactuca sativa. *Environ. Exp. Bot.* **2012**, *75*, 128–133. [CrossRef]
23. Jishi, T.; Kimura, K.; Matsuda, R.; Fujiwara, K. Effects of temporally shifted irradiation of blue and red LED light on cos lettuce growth and morphology. *Sci. Hortic.* **2016**, *198*, 227–232. [CrossRef]
24. Martínez-García, J.F.; Gallemí, M.; Molina-Contreras, M.J.; Llorente, B.; Bevilaqua, M.R.R.; Quail, P.H. The Shade Avoidance Syndrome in Arabidopsis: The Antagonistic Role of Phytochrome A and B Differentiates Vegetation Proximity and Canopy Shade. *PLoS ONE* **2014**, *9*, e109275. [CrossRef]
25. Chen, M.; Chory, J.; Fankhauser, C. Light Signal Transduction in Higher Plants. *Annu. Rev. Genet.* **2004**, *38*, 87–117. [CrossRef] [PubMed]
26. Galvão, V.C.; Fankhauser, C. Sensing the light environment in plants: Photoreceptors and early signaling steps. *Curr. Opin. Neurobiol.* **2015**, *34*, 46–53. [CrossRef]
27. Ahmad, M.; Cashmore, A.R. HY4 gene of A. thaliana encodes a protein with characteristics of a blue-light photoreceptor. *Nat. Cell Biol.* **1993**, *366*, 162–166. [CrossRef] [PubMed]
28. Paik, I.; Huq, E. Plant photoreceptors: Multi-functional sensory proteins and their signaling networks. *Semin. Cell Dev. Biol.* **2019**, *92*, 114–121. [CrossRef] [PubMed]
29. Tilbrook, K.; Arongaus, A.B.; Binkert, M.; Heijde, M.; Yin, R.; Ulm, R. The UVR8 UV-B Photoreceptor: Perception, Signaling and Response. *Arab. Book* **2013**, *11*, e0164. [CrossRef]
30. Yang, Y.; Yang, X.; Jang, Z.; Chen, Z.; Ruo, X.; Jin, W.; Wu, Y.; Shi, X.; Xu, M. UV RESISTANCE LOCUS 8 From Chrysanthemum morifolium Ramat (CmUVR8) Plays Important Roles in UV-B Signal Transduction and UV-B-Induced Accumulation of Flavonoids. *Front. Plant Sci.* **2018**, *9*, 955. [CrossRef]
31. Kelly, N.; Choe, D.; Meng, Q.; Runkle, E.S. Promotion of lettuce growth under an increasing daily light integral depends on the combination of the photosynthetic photon flux density and photoperiod. *Sci. Hortic.* **2020**, *272*, 109565. [CrossRef]
32. Wojciechowska, R.; Długosz-Grochowska, O.; Kołton, A.; Żupnik, M. Effects of LED supplemental lighting on yield and some quality parameters of lamb's lettuce grown in two winter cycles. *Sci. Hortic.* **2015**, *187*, 80–86. [CrossRef]
33. Holmes, M.G.; Smith, H. The function of phytochrome in the natural environment—ii. The influence of vegetation canopies on the spectral energy distribution of natural daylight. *Photochem. Photobiol.* **1977**, *25*, 539–545. [CrossRef]
34. Saha, S.; Monroe, A.; Day, M.R. Growth, yield, plant quality and nutrition of basil (Ocimum basilicum L.) under soilless agricultural systems. *Ann. Agric. Sci.* **2016**, *61*, 181–186. [CrossRef]
35. Naderianfar, M.; Azizi, M.; Koohestani, S. Determination of water-yield basil function under deficit irrigation conditions and use of nano fertilizer. *Prog. Agric. Eng. Sci.* **2017**, *13*, 51–68. [CrossRef]
36. Wang, X.; Wang, S.; George, T.S.; Deng, Z.; Zhang, W.; Fan, X.; Lv, M. Effects of schedules of subsurface drip irrigation with air injection on water consumption, yield components and water use efficiency of tomato in a greenhouse in the North China Plain. *Sci. Hortic.* **2020**, *269*, 109396. [CrossRef]
37. Craven, D.; Gulamhussein, S.; Berlyn, G. Physiological and anatomical responses of Acacia koa (Gray) seedlings to varying light and drought conditions. *Environ. Exp. Bot.* **2010**, *69*, 205–213. [CrossRef]
38. Piovene, C.; Orsini, F.; Bosi, S.; Sanoubar, R.; Bregola, V.; Dinelli, G.; Gianquinto, G. Optimal red:blue ratio in led lighting for nutraceutical indoor horticulture. *Sci. Hortic.* **2015**, *193*, 202–208. [CrossRef]

39. Zhang, M.; Park, Y.; Runkle, E.S. Regulation of extension growth and flowering of seedlings by blue radiation and the red to far-red ratio of sole-source lighting. *Sci. Hortic.* **2020**, *272*, 109478. [CrossRef]
40. Zhang, M.; Whitman, C.M.; Runkle, E.S. Manipulating growth, color, and taste attributes of fresh cut lettuce by greenhouse supplemental lighting. *Sci. Hortic.* **2019**, *252*, 274–282. [CrossRef]
41. Lee, M.-J.; Son, K.-H.; Oh, M.-M. Increase in biomass and bioactive compounds in lettuce under various ratios of red to far-red LED light supplemented with blue LED light. *Hortic. Environ. Biotechnol.* **2016**, *57*, 139–147. [CrossRef]
42. Murchie, E.H.; Pinto, M.; Horton, P. Agriculture and the new challenges for photosynthesis research. *New Phytol.* **2008**, *181*, 532–552. [CrossRef]
43. Emerson, R.; Chalmers, R.; Cederstrand, C. Some Factors Influencing The Long-Wave Limit of Photosynthesis. *Proc. Natl. Acad. Sci. USA* **1957**, *43*, 133–143. [CrossRef]
44. Li, Q.; Kubota, C. Effects of supplemental light quality on growth and phytochemicals of baby leaf lettuce. *Environ. Exp. Bot.* **2009**, *67*, 59–64. [CrossRef]
45. Kim, H.-J.; Yang, T.; Choi, S.; Wang, Y.-J.; Lin, M.-Y.; Liceaga, A.M. Supplemental intracanopy far-red radiation to red LED light improves fruit quality attributes of greenhouse tomatoes. *Sci. Hortic.* **2020**, *261*, 108985. [CrossRef]
46. Chang, X.; Alderson, P.G.; Wright, C.J. Solar irradiance level alters the growth of basil (Ocimum basilicum L.) and its content of volatile oils. *Environ. Exp. Bot.* **2008**, *63*, 216–223. [CrossRef]
47. Salisbury, F.B.; Ross, C.W. *Plant Physiology*; Wadsworth Pub. Com., Inc.: Belmont, CA, USA, 1992.
48. Tarakanov, I.; Yakovleva, O.; Konovalova, I.; Paliutina, G.; Anisimov, A. Light-emitting diodes: On the way to combinatorial lighting technologies for basic research and crop production. *Acta Hortic.* **2012**, 171–178. [CrossRef]
49. Naznin, M.T.; Lefsrud, M.; Gravel, V.; Azad, O.K. Blue Light added with Red LEDs Enhance Growth Characteristics, Pigments Content, and Antioxidant Capacity in Lettuce, Spinach, Kale, Basil, and Sweet Pepper in a Controlled Environment. *Plants* **2019**, *8*, 93. [CrossRef]
50. Hosseini, A.; Mehrjerdi, M.Z.; Aliniaeifard, S.; Seif, M. Photosynthetic and growth responses of green and purple basil plants under different spectral compositions. *Physiol. Mol. Biol. Plants* **2019**, *25*, 741–752. [CrossRef]
51. Legris, M.; Ince, Y.Ç.; Fankhauser, C. Molecular mechanisms underlying phytochrome-controlled morphogenesis in plants. *Nat. Commun.* **2019**, *10*, 5219. [CrossRef]
52. Montesano, F.F.; Van Iersel, M.W.; Boari, F.; Cantore, V.; D'Amato, G.; Parente, A. Sensor-based irrigation management of soilless basil using a new smart irrigation system: Effects of set-point on plant physiological responses and crop performance. *Agric. Water Manag.* **2018**, *203*, 20–29. [CrossRef]
53. Onwuka, B.M.B. Effects of Soil Temperature on Some Soil Properties and Plant Growth. *Adv. Plants Agric. Res.* **2018**, *8*, 1–5. [CrossRef]
54. Pennisi, G.; Orsini, F.; Blasioli, S.; Cellini, A.; Crepaldi, A.; Braschi, I.; Spinelli, F.; Nicola, S.; Fernandez, J.A.; Stanghellini, C.; et al. Resource use efficiency of indoor lettuce (Lactuca sativa L.) cultivation as affected by red:blue ratio provided by LED lighting. *Sci. Rep.* **2019**, *9*, 1–11. [CrossRef] [PubMed]

plants

MDPI

Article

Modelling of Soybean (*Glycine max* (L.) Merr.) Response to Blue Light Intensity in Controlled Environments

Tina Hitz, Simone Graeff-Hönninger and Sebastian Munz *

Institute of Crop Science, Cropping Systems and Modelling, University of Hohenheim, 70599 Stuttgart, Germany; tina.hitz@uni-hohenheim.de (T.H.); graeff@uni-hohenheim.de (S.G.-H.)

* Correspondence: s.munz@uni-hohenheim.de; Tel.: +49-711-22359

Received: 29 October 2020; Accepted: 9 December 2020; Published: 11 December 2020

Abstract: Low photosynthetic photon flux density (PPFD) under shade is associated with low blue photon flux density (BPFD), which independent from PPFD can induce shade responses, e.g., elongation growth. In this study, the response of soybean to six levels of BPFD under constant PPFD from LED lighting was investigated with regard to morphology, biomass and photosynthesis to increase the knowledge for optimizing the intensity of BPFD for a speed breeding system. The results showed that low BPFD increased plant height, leaf area and biomass and decreased leaf mass ratio. Photosynthetic rate and internode diameter were not influenced. A functional structural plant model of soybean was calibrated with the experimental data. A response function for internode length to the perceived BPFD by the internodes was derived from simulations and integrated into the model. With the aim to optimize lighting for a speed breeding system, simulations with alternative lighting scenarios indicated that decreasing BPFD during the growth period and using different chamber material with a higher reflectance could reduce energy consumption by 7% compared to the experimental setup, while inducing short soybean plants.

Keywords: photomorphogenesis; blue photon flux density; functional structural plant modelling; indoor farming; LED lighting

1. Introduction

In horticulture and indoor farming, LEDs have several advantages e.g., they save energy, emit less heat and have a long lifetime [1,2]. A spectrum can be designed depending on the response of the specific crop and the production aim. However, to fully exploit the spectral flexibility of LED lighting an increased knowledge of the spectral effects on plant morphology and growth is required [3]. Energy consumption can also be considered during spectral optimization as this can vary between spectra depending on the LED types [4]. A higher energy consumption of red than of blue LEDs has been reported [4,5], but theoretically the energy consumption of blue LEDs is higher than of red LEDs due to the higher energy level per photon of shorter than of longer wavelengths [6].

The advantages of LED lighting can be used in speed breeding, a breeding system developed particularly for growth chambers. The aim of a speed breeding system is to grow many generations per year to shorten the time for developing new cultivars. For instance, in a speed breeding system for several cereals, pea and chickpea six generations can be grown per year [7]. For a more efficient use of space, plants can be grown in a multi-layer system. For these systems, short plants are desirable to increase the number of layers of plants and hereby the possibility to include more genotypes at the same time. Therefore, a spectrum for speed breeding should not delay seed setting (many generations) and induce a shorter plant height to cultivate in more layers (many genotypes). These requirements deviate from other indoor plant productions aiming to increase resource use efficiency considering other properties, such as yield and nutritional value [5]. Recently, a speed breeding protocol for soybean was developed using LED lighting. Red and blue light was found not to influence flowering time and was recommended to induce short compact plants. However, only two ratios of red and blue light (1:1 and 2:1) were studied [8].

The spectral light environment is perceived by the plant photoreceptors, which in a natural environment induce morphological changes such as those that express shade adaptations [9]. Shaded plants experience a reduced red to far-red ratio perceived by phytochrome [10] and a reduced photosynthetic photon flux density (PPFD). The latter is associated with a reduced blue photon flux density (BPFD) perceived by cryptochrome [11]. Typical shade responses of soybean are elongated internodes and petioles, increased specific leaf area (SLA) and decreased biomass and internode diameter [12–14]. Under LED lighting, BPFD can be reduced by lowering the ratio of blue light in the spectrum without a simultaneous reduction of PPFD. By reducing BPFD, some morphological shade responses, e.g., increased height, can be triggered also under constant PPFD [15,16]. For soybean, earlier studies found an increased plant height with decreasing BPFD [17–19] showing that high BPFD can be applied to induce short soybean plants, but these studies used a broad spectrum and included only one treatment [17] with a blue light ratio over 30%. None of these studies derived a response function to BPFD for soybean height under LED lighting with narrow peaks and none focused on blue light ratios between 15–78%. Earlier studies in other species than soybean also explored relatively low BPFD ratios (<50%) with the aim of avoiding extreme elongation under sole red LED lighting [15] or explored an intermediate BPFD to maximize biomass [20]. The aim in the present study was to reduce plant height to its minimum under a high BPFD.

Beside the influences on plant growth through the perceptions of photoreceptors, the light spectra can also influence the photosynthetic rate. Whereby, carbon assimilation can differ depending on the spectrum even under a constant PPFD. Photosynthetic pigments of plants absorb light mainly within the range of wavelength from 400 to 700 nm. The photosynthetic most effective part is considered to be the light within the red range (600–700 nm) due to a better balance of excitation between photosystem I and II and due to a more effective transfer between the red light absorbing chlorophylls than from the blue light absorbing carotenoids to chlorophyll [21]. Despite this, several studies measuring photosynthesis on plants grown under different light spectra found similar rates of photosynthesis under spectra with different ratios of light within the blue range (400–500 nm) [22–24].

The optimization of light spectrum for a specific crop and production system is very time-consuming given the many aspects that have to be considered, e.g., light quality, intensity and day length. Also, the transfer of knowledge between studies and into practice can be impaired by variability in several factors, e.g., plant density, type of light source and dimensions of the climate chambers. In this context, functional structural plant (FSP) modeling can assist as a tool for optimization of crop production and understanding of plant responses to its environment. An FSP model simulates plant growth and development, while considering its architectural appearance, by responding to the experienced environment on the individual organ level [25,26]. Hereby, responses can be related to the actually perceived spectrum of individual organs. The perceived spectral light environment can differ from the environment above the canopy and between phytomers due to self-shading and light reflection, as other studies found focusing on PPFD [27–29] or the red to far red ratio [30–34].

Earlier FSP models using artificial light sources for indoor plant production addressed the light regime for greenhouse production [35–37], while only one study used an FSP model with LEDs being the only light source [38]. An FSP model within an LED chamber can be a tool to reduce the amount of necessary experiments for spectral optimization and assist in the understanding of the plant response to the indoor environment and in the transfer of knowledge between studies and into practice.

The aim of this study was to find an optimal BPFD inducing short soybean plants under a narrow peaked red and blue LED spectrum, also considering energy consumption. We hypothesize that an optimum BPFD for minimum plant height, not influencing flowering time, could be determined with a combination of experiments and FSP modelling. The objectives were to (i) examine the influence of different levels of BPFD under constant PPFD on soybean biomass, photosynthesis and morphology and (ii) calibrate an FSP model of soybean and integrate a response function to BPFD for internode length and (iii) to find by simulation the minimum BPFD to reduce plant height and energy consumption.

2. Results

2.1. Experimental Data—Plant Scale

Biomass and leaf area per plant showed similar differences among the six light treatments with a BPFD of 60, 110, 160, 210, 260 and 310 μmol m^{-2} s^{-1} (B60–B310). The treatments B110–B160 resulted in the highest values and B210–B310 in the lowest values, whereas plant height consistently decreased with increasing BPFD (Figure 1).

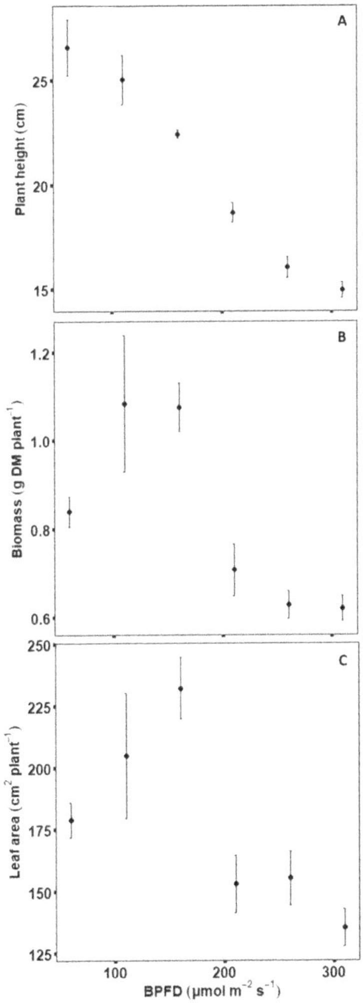

Figure 1. Final plant height (**A**), biomass (**B**) and leaf area (**C**) under different blue photosynthetic flux densities (BPFD). Error bars indicate standard error of the mean (n = 8).

Plant height responded rapidly after beginning of the experiment and the differences between treatments became more pronounced over time with differences between the highest and lowest values of 34%, 46%, and 34% on day 9 and 77% 75% and 72% on day 23 for plant height, biomass, and leaf area, respectively (Supplementary Materials, Figure S1).

2.2. Experimental Data—Phytomer Scale

2.2.1. Biomass

At the third phytomer, significant differences between treatments were found for biomass of internodes and leaf laminas. An increase in BPFD decreased the biomass of internodes and leaf laminas with the minimum of 0.022 and 0.078 g under B260 and the maximum of 0.046 and 0.136 g under B160. The same tendency was found for biomass of the second internode and the petiole, with the latter being less expressed with no significant differences between treatments. The leaf mass ratio (LMR) differed significantly between BPFD levels with a minimum value of 0.64 under B160 increasing to 0.69 under B60 and 0.71 under B310. The internode mass ratio of the stalk (IMRS) decreased from 0.72 under B260 to 0.63 under B60 (Table 1).

Table 1. Least square means of the final measurement of organ biomass, leaf mass ratio (LMR) and internode mass ratio of the stalk (IMRS) at phytomer scale under different blue photosynthetic flux densities (BPFD). Letters indicate significant differences between treatments ($p < 0.05$).

Phytomer Level	Organ/Ratio	Treatment					
		B60	B110	B160	B210	B260	B310
Second	Internode (g)	0.082 a	0.082 a	0.076 a,b	0.058 b,c	0.047 c	0.049 c
Third	Internode (g)	0.041 a	0.044 a	0.046 a	0.030 b	0.022 b	0.023 b
	Petiole (g)	0.016 a	0.018 a	0.018 a	0.015 a	0.013 a	0.013 a
	Leaf lamina (g)	0.127 a	0.136 a,b	0.116 a,c	0.100 a,c	0.078 c	0.081 b,c
	LMR	0.69 a,b	0.68 b,c	0.64 c	0.67 b,c	0.71 a	0.71 a
	IMRS	0.72 a	0.72 a	0.71 a	0.68 a,b	0.62 b	0.63 b

2.2.2. Leaf Morphology and Physiology

At the third phytomer, the SLA significantly increased from 303 under B110 to 346 $cm^2\ g^{-1}$ under B310. No significant differences were observed for carbon assimilation, but there was a slight reduction with increased BPFD from 28.81 $\mu mol\ CO_2\ m^{-2}\ s^{-1}$ under B110 to the minimum assimilation of 26.53 $\mu mol\ CO_2\ m^{-2}\ s^{-1}$ under B310. SPAD values did not differ significantly between treatments and showed no tendency (Table 2).

Table 2. Least square means of the final measurement of specific leaf area (SLA), carbon assimilation (*A*) and SPAD at phytomer scale under different blue photosynthetic flux densities (BPFD). Letters indicate significant differences between treatments ($p < 0.05$).

Phytomer Level	Measurement	Treatment					
		B60	B110	B160	B210	B260	B310
Third	SLA ($cm^2\ g^{-1}$)	324.35 a,b	303.04 b	327.68 a,b	327.02 a,b	341.41 a,b	346.38 a
	SPAD value	26.51 a	30.08 a	31.65 a	28.81 a	30.37 a	30.92 a
Youngest fully developed	A ($\mu mol\ CO_2\ m^{-2}\ s^{-1}$)	27.49 a	28.81 a	27.78 a	27.03 a	26.79 a	26.53 a

2.2.3. Elongation

The internode responded more to a decrease in BPFD at the third than the second phytomer with a length of 2.14 cm under B310 and 3.81 cm under B60, corresponding to a 78% increase (Figure 2; Table 3). Whereas, the second internode increased from 4.33 to 6.86 cm, corresponding to 59%. The response of the petiole was smaller with a length increase from 4.46 cm under B310 to 5.61 cm under B160, corresponding to a 26% increase. The tendency differed from that of the internodes with a maximum length under B160 and an insignificant decrease until B60. Increasing BPFD from B160 also decreased the length of the petiole, but with no significant differences from B210 to B310. Length of the leaf lamina did not respond significantly but had a tendency to decrease from B160 to B310, similarly to the petiole. The internode elongation was not accompanied by a reduced internode diameter, which showed no significant differences with a slight tendency of responding similar to the petiole.

Figure 2. Soybean grown under B310 (**A**) and B60 (**B**).

Table 3. Least square means of the final measurement for lengths of internodes, petioles and leaves and diameter of internodes at phytomer scale under different blue photosynthetic flux densities (BPFD). Lower case letters indicate significant differences between treatments ($p < 0.05$).

Phytomer Level	Organ	Treatment					
		B60	B110	B160	B210	B260	B310
Second	Internode (cm)	6.86 ^{a}	5.86 ^{b}	5.39 ^{b}	4.70 ^{c}	4.26 ^{c}	4.33 ^{c}
Third	Internode (cm)	3.81 ^{a}	3.41 ^{a,b}	3.28 ^{a,b}	2.75 ^{b,c}	2.28 ^{c}	2.14 ^{c}
	Petiole (cm)	5.11 ^{a,b}	5.08 ^{a,b}	5.61 ^{a}	4.80 ^{b,c}	4.72 ^{b,c}	4.46 ^{c}
	Leaf lamina (cm)	5.25 ^{a}	5.07 ^{a}	5.27 ^{a}	4.58 ^{a}	4.49 ^{a}	4.48 ^{a}
	Internode diameter (mm)	3.09 ^{a}	3.29 ^{a}	3.52 ^{a}	3.41 ^{a}	3.46 ^{a}	3.28 ^{a}

2.2.4. Growth Dynamics

Growth of the individual organs was fitted to the beta-function and parameters for the third phytomer showed significant differences for internode and petiole, but not for the leaf lamina. The absolute differences of the parameters for all three organs were relatively small and did not show any tendency to change with decreased BPFD (Table 4).

Table 4. Least square means of estimated parameters of the beta-function. Time of elongation (te) and time of maximum elongation (tm) for internode, petiole and leaf at the third phytomer under different blue photosynthetic flux densities (BPFD). Letters indicate significant differences between treatments ($p < 0.05$).

Organ of Third Phytomer	Parameter	Treatment					
		B60	B110	B160	B210	B260	B310
Internode	te	13.96 ^{b}	14.39 ^{a,b}	14.21 ^{b}	14.41 ^{a,b}	14.67 ^{a}	14.36 ^{a,b}
	tm	5.23 ^{a}	6.86 ^{a}	6.60 ^{a}	6.33 ^{a}	6.60 ^{a}	6.13 ^{a}
Petiole	te	15.76 ^{c}	16.03 ^{b,c}	15.85 ^{c}	16.47 ^{a,b}	16.39 ^{a,c}	16.68 ^{a}
	tm	9.79 ^{b}	10.25 ^{a,b}	10.18 ^{a,b}	10.25 ^{a,b}	10.74 ^{a}	10.70 ^{a}
Leaf lamina	te	13.94 ^{a}	14.12 ^{a}	14.04 ^{a}	14.11 ^{a}	14.19 ^{a}	13.89 ^{a}
	tm	5.19 ^{a}	5.04 ^{a}	4.57 ^{a}	4.55 ^{a}	5.50 ^{a}	4.95 ^{a}

2.2.5. Energy Consumption

The highest energy consumption was measured under a high BPFD (Table 5). The consumption increased from 94.4 W under B60 to 107.2 W under B310 corresponding to an increase of 14% under B310.

Table 5. Measured energy consumption (Watt) of the LED chambers under different blue photosynthetic flux densities (BPFD).

Treatment	Energy Consumption (W)
B60	94.4
B110	95.1
B160	96.4
B210	97.6
B260	101.7
B310	107.2

2.3. Modelling

Simulations based on the found parameters of the beta-function resulted in simulated length of internodes, petioles and leaf laminas following the measurements well (exemplified by the treatments B60 and B310 in Supplementary Materials, Figure S2).

2.3.1. Blue Light Response Function of Internodes

Based on the simulated $BPFD_{per}$, the relative elongation response of the second and third internode were closer to each other compared to using the emitted BPFD (Figure 3). Especially at high BPFD levels, the response of the two internodes was close, implying a common response function to $BPFD_{per}$.

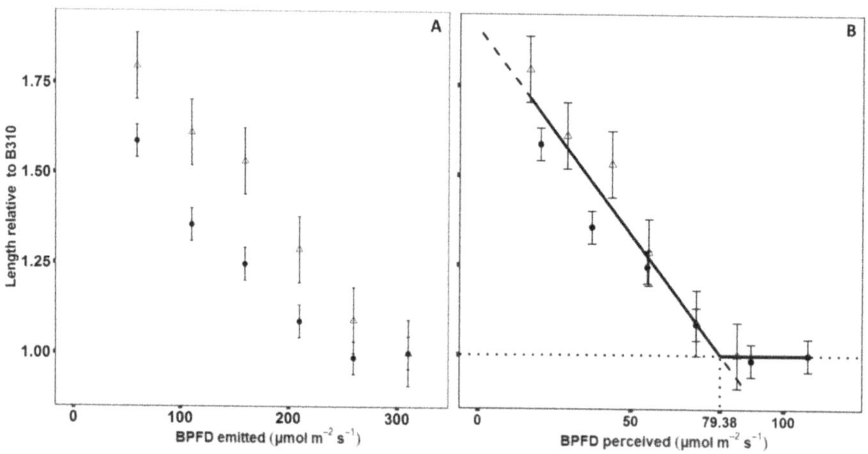

Figure 3. Least square mean of length of second (point) and third (triangle) internode relative to B310 in response to BPFD emitted by the LED modules (**A**) or simulated BPFD perceived by the internode (**B**). Dashed line showing the function fitted to relative internode lengths higher than one and dotted lines showing the interception of the function with 1. Black line showing the final response function to BPFD. Error bars indicate standard error of the LS-mean ($n = 8$).

Due to the relatively small differences between the parameters t_e and t_m of the beta-function and no clear tendencies in their response to BPFD, only differences in the final length of the internode (L_{max}) were considered in the response function. Internodes with a relative length to B310 below one were considered to have no elongation response to $BPFD_{per}$ and a common function for final internode length was fitted to the $BPFD_{per}$ of the internodes under treatments with a relative length to B310 higher than one (Figure 3B).

The common function for the relative length of the second and third internode was:

$$L_{rel} = -0.01 * BPFD_{per} + 1.91, \quad (1)$$

The interception of this function with one was at 79.38 μmol m^{-2} s^{-1} $BPFD_{per}$, which was hereby the minimum amount of BPFD that an internode should perceive to express no elongation response to $BPFD_{per}$.

This resulted in the BPFD response function:

$$Internode\ length = L_{min}\left(1 + \left(79.38 - BPFD_{per}\right) * 0.01\right),\ 79.38 - BPFD_{per} > 0, \quad (2)$$

where L_{min} is the final internode length with no elongation response to $BPFD_{per}$ and $BPFD_{per}$ is the BPFD perceived by the internode. The black line in Figure 3B shows the response function in the range from the minimum (16.78 μmol m^{-2} s^{-1}) to maximum (108.13 μmol m^{-2} s^{-1}) perceived light during the simulations.

2.3.2. Evaluation and Light Optimization

During the simulations based on the found response function, L_{min} and the growth parameters t_e and t_m were set according to the treatment B310 (baseline scenario). The simulated height until the third internode fitted well with the measurement at the last day, which was also used for parameterization of the model. This shows that the response function was well integrated in the model. Comparing to earlier measurement days which were not used for parameterization of the model, the simulations had a tendency of underestimating the height under low BPFD levels (B60–B110). Importantly for the alternative scenarios the simulated height until third internode fitted well under the higher BPFD levels (B160–B310) (Figure 4).

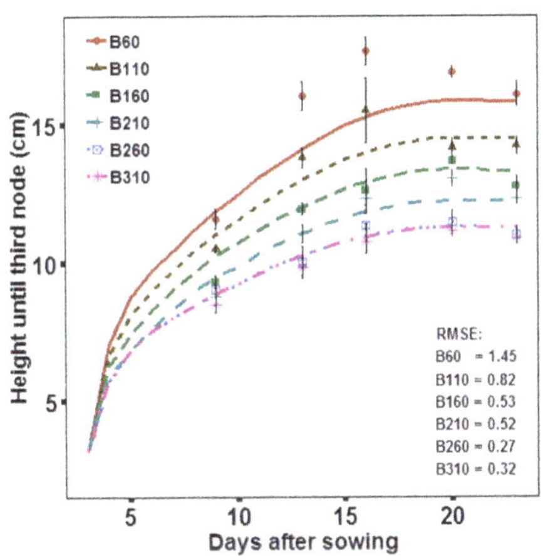

Figure 4. Simulated (line) and measured (points) plant height until the third node and root mean square error (RMSE) between simulations and measurements. Error bars indicate standard error of the mean (day 9–20: $n = 4$, day 23: $n = 8$).

The simulations from the first scenario with a reflective surface of pots, soil and bottom resulted in an increase of perceived BPFD. The total height until the third node decreased under all treatments (Figure 5A) compared with the experimental chamber design (Figure 4). In the experimental design, the minimum height was reached between B260 and B310, while in the alternative scenario it was reached between B210 and B260. The simulated length of the second and third internode (Figure 5B,C) showed that the shorter height was a result of an increase in the perceived BPFD of the second internode, where the minimum length was already reached between B160 and B210 (Figure 5B).

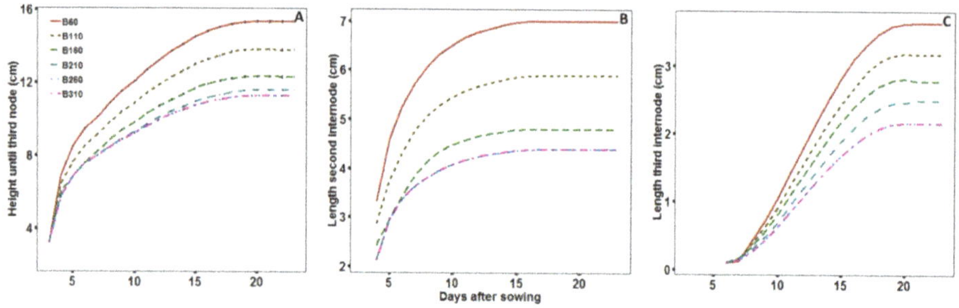

Figure 5. Simulated height until the third node (**A**) and length of second (**B**) and third (**C**) internode with a reflective surface of pots, soil and bottom.

The optimization indicated by the first scenario was applied in the second scenario by optimizing BPFD during the growth period. When the third internode started to develop the treatment changed from B210 to B260 and hereby increased BPFD. The results of the second scenario showed that increasing BPFD on day nine resulted in the minimum length of both internodes and the minimum height of 11.28 cm until the third node. The reduction in the average BPFD emitted by the LED modules resulted in a reduction of energy consumption from 107.2 W in the experimental scenario to 101.7 W in the first and 100.1 W (−7%) in the second scenario (Table 6).

Table 6. Average energy consumption during 23 days of growth for the three simulated scenarios to reach the minimum plant height.

Scenario	Light Spectra	Average Energy Consumption (W)
Experimental	B310	107.2
First scenario	B260	101.7
Second scenario	B210/B260	100.1

3. Discussion

3.1. Biomass and Photosynthesis

During the experiments, data on photosynthesis and biomass was collected. This data shows the response of carbon assimilation and translocation to BPFD, which is of minor importance in a speed breeding system, but of interest for improving yield in indoor farming.

No significant influence on carbon assimilation per leaf area was observed with increased BPFD, which is in agreement with earlier studies, although red light is considered to be the most effective for photosynthesis [21,39]. An increased maximum assimilation with increased BPFD in cucumber was associated with an increased leaf thickness [16,22]. Similar, the tendency of decreased assimilation in this study was associated with thinner leaves under high BPFD. In ice plant He et al. [23] found no change in saturated assimilation between BPFD ratios of 10 and 100%. Although an increasing BPFD ratio from 0 to 20% increased photosynthesis in lettuce, it dropped again at 30% [24]. In this study, the lowest BPFD ratio was 15% under the B60 treatment and an effect below this ratio cannot be excluded.

The decrease in biomass at high BPFD found in this study was most probably related to similar differences in leaf area, which decreased light interception and consequently carbon assimilation per plant. Another reason could be an increased root biomass, but earlier studies found no change in the biomass ratio of soybean under BPFD ratios of 10 and 25% [19]. In addition, an influence of

BPFD on the assimilation over time could reduce biomass. For instance, the photosynthetic rate of tomato decreased more in the afternoon under monochromatic red and blue light than under a broader spectra [40].

Increased LMR under high BPFD confirming earlier results [18,19] indicated a reduced carbon export from the leaves. In tomato, light spectra also influenced the ratio of carbon export from the leaves, but not in agreement with this study as export increases under monochromatic blue and orange light at intermediate PPFD [40]. This can be caused by different responses comparing monochromatic spectra with broader spectra exploring ratios between wavelengths. A decreased fraction of the carbon translocated from the leaves to the stem (internode and petiole) was located in the internodes (low IMRS) under increased BPFD. These results of biomass proportion between organs showed, that an elongation response to reduced BPFD increased the translocation of carbon from the leaves to the stem, but with a higher priority of internodes than petioles. Extensions of the FSP model could assist in the exploration of carbon assimilation and translocation between organs following a similar approach as Bongers et al. [33] combining response functions to light environment with increased carbon demand of specific organs.

3.2. Response to BPFD Under Shade

Low BPFD in Nature is associated with low PPFD and low red to far-red ratio under shade, which trigger morphological responses, e.g., by interactions of the photoreceptors cryptochrome and phytochrome, to increase light interception [41]. The performed experiments represented unnatural spectra that do not occur in nature and hereby show the response of soybean to BPFD without interactions with PPFD and red to far-red ratio.

The elongation response of internode and petiole to low BPFD was in accordance with a shade avoidance response of soybean to low PPFD [12,42] and show that low BPFD can trigger the response also under high PPFD and in the absence of far-red light. The stronger response of internodes than of petioles supports earlier indications of internode elongation being the main shade avoidance response to low PPFD (associated with low BPFD), whereas petiole elongation responded strongly to low red to far-red ratio [12]. The slight decrease in SLA under low BPFD in this study is not in accordance with earlier studies in soybean, which found no response to BPFD in SLA under high PPFD [18,19]. This could be an effect of the lower maximum BPFD ratios applied in earlier studies. Cucumber under low BPFD responded with an increased SLA, which indicates differences between species or an effect of the lower light intensity (100 μmol m^{-2} s^{-1}) applied in these studies [16,22]. Decreased SLA and unchanged internode diameter under low BPFD differ from the soybean response to low PPFD resulting in increased SLA [12,18,19] and decreased internode diameter [12,42]. This indicated that SLA and internode diameter are not regulated by the perception of low BPFD associated with low PPFD, but instead supports earlier studies indicating that SLA is regulated e.g., by sugar signaling [43–45].

3.3. BPFD Response Function

A linear function described well the response to BPFD and was applied for the simulations. Kahlen and Stützel [46] also applied a linear response to PPFD and red to far-red ratio for modeling the response of cucumber to light environment. Other studies found a non-linear response function to BPFD for stem length of soybean [17–19]. This can be due to lower BPFD levels in these studies (BPFD levels < 5%) based on which a non-linear function could be fitted [19]. A continuation of the function in the present study below a BPFD ratio of 15% could evolve non-linear, but in the context of speed breeding this low BPFD levels are not important as this would result in tall plants. For the speed breeding system, it was important to determine the point of a saturated response to BPFD to reach short plants and reduce the BPFD to reduce energy consumption. A saturated response to emitted BPFD was reached under treatments between 210 and 310 μmol m^{-2} s^{-1} in the experimental setup. Two earlier studies on soybean found a saturated response already under 30–50 μmol m^{-2} s^{-1} BPFD [17,18], whereas one study also found an effect from higher BPFD (130 μmol m^{-2} s^{-1}) [19], indicating interactions with other factors resulting in these discrepancies. One aspect could be the light spectrum, as earlier studies used broader spectra containing green and far-red light [17–19] and additionally included UV-A light in the BPFD [19]. Green light can influence cryptochrome antagonistic to blue light and especially under high PPFD [47]. The addition of green light to a red and blue spectrum increased plant height of soybean under a PPFD of 200 μmol m^{-2} s^{-1} but had no influence under 500 μmol m^{-2} s^{-1} [48]. Far-red light can lead to an increase in plant height by reducing the red to far-red ratio perceived by phytochrome as shown for soybean by adding far-red light to a broad light spectrum [12]. In addition, a broader spectrum within the blue range can influence the magnitude of the blue light effect on cryptochrome. For hypocotyl elongation of *Arabidopsis thaliana* (L.) Heynh., the action spectrum of cryptochrome to monochromatic light did not change within the range 390–530 nm, but an increased stability of CRY2 protein was observed under monochromatic light compared to a broader blue spectrum [49]. These differences in the reactions under narrow peaks compared to the reactions during the response to high PPFD, here imitated with high BPFD, indicated that a broader spectrum within the blue range could affect the BPFD level necessary to avoid an elongation response.

In the experiments, the elongation response of the third internode to low BPFD was slightly stronger than at the second internode, and a higher BPFD level was necessary to achieve the minimum length of the third internode. Simulations indicated that this was due to self-shading, which was larger at the third than the second internode. Based on the simulated $BPFD_{per}$, a common response function was found for the second and third internode. This emphasizes the importance of knowing the perceived light environment at organ-level, e.g., as in this study by means of simulations with an FSP model, as it enables a better evaluation of the influence from the light microclimate than relating the response directly to the light emitted from the light source [46].

The parameters t_e and t_m of the beta-function were in most cases not significantly different between treatments and no trend was present, which indicated that a common parameter could be used for all levels of BPFD only changing L_{max} according to the $BPFD_{per}$. This was confirmed by the accurate simulations of internode length at all BPFD levels based on t_e and t_m found under B310. Importantly, the small difference in height between B310 and B260 were well simulated, showing that the model was very useful to determine the necessary BPFD to reach the minimum height.

3.4. Optimization of Light Spectrum

The decreased biomass with increased BPFD is in accordance with earlier studies in soybean [19] and other species [5,23]. This results in a decreased efficiency of the applied PPFD in indoor farming producing biomass, but in a speed breeding system there are no apparent advantages of a high biomass. Further consideration for a spectral optimization would be whether the minimum necessary BPFD found here could be reduced through other light microclimatic factors. Light intensity would be an important factor to determine possible interactions between absolute and relative amount of BPFD on morphology and interactions between PPFD and BPFD on photosynthesis. Further studies could also investigate whether the necessary BPFD could be reduced with an increased effect on cryptochrome with a broader blue spectrum or the addition of other wavelengths.

The alternative scenarios showed that the amount of necessary BPFD of the emitted light could be reduced through increased reflection of the bottom and soil and by changing the amount of BPFD during the growth period. Within the used LED modules, the blue LEDs had a higher energy consumption than the red LEDs, as expected from theory [6]. Simulations with BPFD levels optimized for the alternative chamber design showed the potential to decrease energy consumption. Additionally, decreased BPFD can increase water use efficiency by decreasing stomatal conductance [5]. The simulations showed a high potential for light optimization in indoor crop production and speed breeding as the model can be adjusted to the dimensions, LED types and placements and reflective properties for a system-specific recommendation for the light spectrum. Further development of the model could include response functions to more wavelengths and light intensities and make the model sink-source driven [33].

4. Materials and Methods

4.1. Experimental Setup

Soybean plants were grown inside three LED chambers (Compled Solutions GmbH, Dresden, Germany) with the dimensions: 1.1 m high, 0.5 m wide and 0.7 m deep inside a larger climate chamber at the University of Hohenheim (Germany). The LED chambers had openings at the top and bottom enabling ventilation to keep a constant temperature around 27 °C. Seeds of the soybean (*Glycine max* (L.) Merr.) cultivar Merlin (Saatbau Linz eG, Leonding, Austria) were inoculated (Soya BeanInoculant, Legume Technology Ltd., Nottinghamshire, UK) and sown in a mixture of peat substrate (Substrat 5 + Perlite; Klasmann-Deilmann GmbH, Geeste, Germany). The initial three plants per pot were thinned to one plant according to homogenous development on day nine by the start of the plant measurements. Twelve pots (9.5 × 9.0 × 9.0 cm) were distributed evenly within each chamber (0.35 m^2) resulting in a plant density of around 34 plants m^{-2}. The twelve pots were placed in a common tray and irrigated regularly to avoid water limitations. The experiment consisted of four runs within three LED growth chambers to achieve two repetitions for each of the six light treatments.

4.2. Light Treatments

Within each chamber, four LED modules (Sunsim VIS_v3; Compled Solutions GmbH, Dresden, Germany) were placed which allowed to adjust light intensity for the different wavelength ranges. The applied light treatments were comprised only of red and blue light, with one peak in the blue range at 440 nm and two peaks in the red range at 620 and 640 nm (Figure 6). The day length was set to 10 h and all treatments had a PPFD of 400 µmol m^{-2} s^{-1} to test the influence of different levels of BPFD independent from changes in PPFD. The spectrum of the six light treatments had a BPFD of 60, 110, 160, 210, 260 and 310 µmol m^{-2} s^{-1} (B60–B310) (Figure 6), while the remaining PPFD was delivered by the red LEDs. The light spectrum did not include any far-red light to exclude differences in phytochrome-mediated responses between the light treatments. Setting of the spectral intensity of the treatments was performed at 80 cm distance from the LED modules according to the measurements from a FLAME-S-XR1-ES spectrometer (Ocean Optics Germany GmbH, Ostfildern, Germany). The spectrometer measured in the range from 200–1025 nm with a resolution of around 2 nm and was equipped with a collimating lens (74-UV-MP) and a right-angle reflector with cosine corrector (74-90-UV-CC3). The photon flux density was recorded for 400–700 nm (PPFD), 400–500 nm (BPFD) and 600–700 nm (red light).

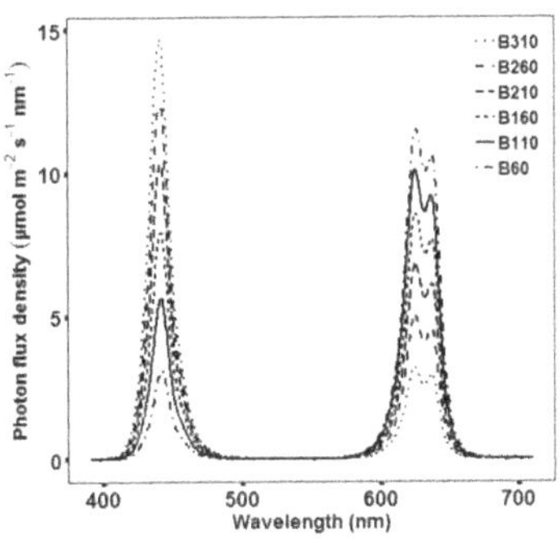

Figure 6. The measured spectrum of the six treatments with a BPFD of 60, 110, 160, 210, 260 and 310 µmol m^{-2} s^{-1}.

Energy consumption of the treatments was measured with a volt-ohm meter (Voltcraft, Energy Check 3000, Conrad Electronic SE, Wernberg-Köblitz, Germany). The measurements were reasonable as compared to the estimated energy consumption given by the software of the LED chambers.

4.3. Plant Measurements

Biomass and leaf area were measured on five dates during each run. For the first four measurements (on day 9, 13, 16 and 20), two plants were randomly selected, and the remaining four plants were used for the final measurement on day 23. On each of the five dates, total plant height (from soil to apical bud), and leaf area and biomass of each phytomer of the two/four plants were determined. Leaf area was estimated with ImageJ [50] from pictures of the leaves and dry mass was measured separately for internodes, petioles and leaf laminas after drying for at least 48 h at 60 °C until constant weight.

On day 23—when start of flowering was observed under all treatments—additional measurements of the photosynthetic rate and SPAD values were performed on the remaining four plants. The SPAD, which is representative for chlorophyll content, was measured using a SPAD meter (SPAD 502 Plus, Konica Minolta, Inc., Tokyo, Japan) and the photosynthetic rate was measured on the youngest fully developed leaf on each of the four plants per light treatment with a LCpro-SD portable system (ADC BioScientific Ltd., Hoddesdon, UK). Measurements were performed under ambient conditions within the chambers (clear glass cover to measure under the applied light treatments). Values were recorded when a steady photosynthetic rate was reached (after around 20 min).

LMR was calculated from leaf biomass/above ground biomass and IMRS was calculated from internode biomass/biomass of stalk (internode plus petiole).

Morphological measurements were performed at seven dates (day 9, 11, 13, 16, 18, 20, and 23) on the four plants within each chamber used for the final measurements. These measurements were used for calibration of the FSP model of soybean and for statistical analysis of the influence of BPFD on growth dynamics. They comprised length and diameter of internodes and petioles, length and width of leaflets, angle between internodes and petioles and angles of the leaf lamina. The latter comprising the lamina inclination measured from the base to the tip of the lamina, and the rotation angle around the midrib. To describe the unfolding of the leaf lamina, the angle between the midrib and each of the two halves of the leaf lamina was determined. Diameters were measured with a caliper, length and width with a ruler and angles with a protractor.

4.4. Statistical Design and Analysis

Measurements of growth dynamics of internodes, petioles and leaflets of each phytomer were used to fit the beta-function (Equation (3)) [51]:

$$
\begin{aligned}
L(t) &= L_{max}\left(1+\frac{t_e-t}{t_e-t_m}\right)\left(\frac{t}{t_e}\right)^{\frac{t_e}{t_e-t_m}} \quad 0 \le t_m < t_e \\
L(t > t_e) &= L_{max}
\end{aligned}
\tag{3}
$$

where $L(t)$ is the size at day t, L_{max} is the final size, t_e is the day when the final size is reached and t_m is the day on which the growth rate peaks. The parameters L_{max}, t_e and t_m were estimated with the nls-function in the R-package stats [52].

To determine the plants used for the destructive measurements during each experiment, randomizations were performed within two blocks (plant location). The first block comprised the first and last row and the second block the two center rows (Figure 7).

The six light treatments of the experiment were performed with two replicates. Given three LED chambers, three out of the six light treatments could be tested in the same run, i.e., each replicate comprised two runs, resulting in four runs in total (Supplementary Materials, S1). For the arrangement of treatments within the LED chambers over time and space, an α-design with two replicates and a block (time) size of two was used. The effect of BPFD on biomass, morphology, leaf physiology and parameters of the beta-function was tested. The second and mostly third phytomer (hypocotyl counted as first phytomer) were chosen for specific analysis, because they comprised the most comprehensive measurements from beginning to end of growth. According to the experimental design, the following mixed model was used to analyze the data in the SAS® software (SAS Institute, Inc., Cary, NC, USA):

$$y_{iklmn} = \mu + b_k + i_{kl} + p_{klm} + r_{klmn} + \tau_i + e_{iklmn} \tag{4}$$

where μ is the intercept, b_k is the fixed effect of the kth complete replicate, i_{kl} is the random effect of the lth incomplete block (time) within the kth replicate, p_{klm} is the random effect of the mth chamber within the lth run, r_{klmn} is the random effect of the nth block (plant location) within the mth chamber of the lth run and the kth replicate, τ_i is the main effect of the ith light treatment, and e_{iklmn} is the error effect of observation y_{iklmn} with homogeneous variance. Residuals were checked graphically for normal distribution and homogeneous variance. After finding significant effects via F-test, a multiple t-test to compare least square means was used to create a letter display [53]. Note that least square means are presented in the results section as data was not balanced, because only three out of six light treatments were tested within each run. Least square means are based on model (Equation (4)) to adjust for block effects.

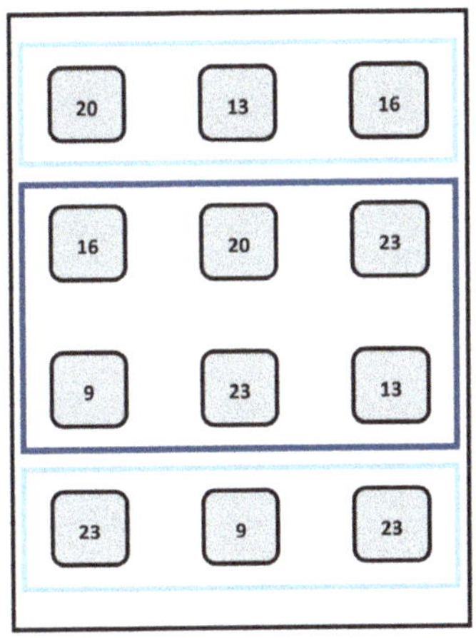

Figure 7. Illustration of the randomization for the plants used for the destructive measurements within the first (light blue square) and second (dark blue square) block (plant location). The numbers exemplarily show the day of the destructive measurements the plants were used for.

4.5. FSP Model

An existing 3D model of the LED growth chamber [54] in the modelling platform GroIMP [55] was used. The virtual LED chamber can be adjusted in its dimensions and the placement of the LED modules and proved to simulate the spectral light distribution with a high accuracy [54]. The single LED types are defined by their spectral and physical light distribution and total emitted power. Then, individual LEDs can be placed according to their position within the LED module. The simulations of the spectral light distribution were performed with the integrated spectral Monte-Carlo ray tracer GPUFlux [56] set to a spectral resolution of 5 nm within the 400–800 nm range. The optical properties (reflection, absorption and transmittance) of the sidewalls were zero transmission and an absorption of 0.02 within the 400–600 and 700–800 range and 0.04 within the 600–700 range.

The optical properties of the chamber were not changed from the setting in the original model [54] as the side wall material was the same. The optical properties of the soybean leaf and the substrate were set in a 5 nm resolution according to measurements from a typical soybean leaf ([57], Supplementary Materials, Figure S3) and peat [58]. The adjustments of the original model of the virtual LED chamber were location and intensity of the individual LEDs and the location of the LED modules. The intensity of the virtual LEDs were parameterized to emit the same intensity of red and blue light at 80 cm distance as in the experimental treatments (Supplementary Materials, Figure S4) using virtual sensors [54]. The virtual LEDs were set to emit 25 million rays with a maximum of 50 reflections, ensuring that all rays were absorbed by an object or reflected outside of the virtual chamber before reaching the maximum number of reflections.

Within the virtual LED chamber, an FSP model of soybean was constructed based on the generic model FSPM-P [27]. Internodes and petioles of the virtual plants were constructed as simple cylinder objects, while the shape of the leaf lamina was triangulated, based on a picture of a soybean leaflet and was composed of 42 triangles (Supplementary Materials, Figure S5).

Each leaflet was constructed from two half leaflets enabling unfolding of the leaflet from the midrib according to measurements (Supplementary Materials, Figure S5A,B). Simultaneously with the unfolding of the two leaflet halves, the leaflet moved from a vertical position with the leaflet tip pointing upwards towards a final inclination according to the measurements from leaflet base to tip and rotation around the midrib according to measurements from one side of the leaflet to the other (Supplementary Materials, Table S2; Figure S5C). The leaflet unfolded with 0.7°/hour.

The virtual plants grew according to the found growth parameters of the beta-function to simulate the plant structure according to the experimental observations. Additional inputs taken from the experiment were final organ length, petiole angles and length to diameter ratio of internodes for each treatment and ratio between length of side leaflets and center leaflet of the trifoliate leaf, leaflet length to leaflet area ratio and length to diameter ratio of petioles as an average of all treatments (Supplementary Materials, Table S2).

According to the experimental design, twelve virtual plants were simulated at the beginning, and then during the simulation two of them were randomly chosen within the two blocks to be taken out of the virtual scene on day 9, 13, 16, and 20, respectively.

4.6. Response Function

The response function was derived and integrated following four steps: (1) fit beta-function for all treatments, (2) run FSP model to obtain perceived BPFD of internodes for each treatment, (3) derive response function across all treatments for internode length in dependence of simulated perceived BPFD and (4) integrate the derived response function into the model for all treatments.

Based on the experimental observations, the parameters for the beta-function were derived for each light treatment. Based on these parameters, the FSP model of soybean dynamically simulated the plant architecture over time under each of the six light treatments.

At this stage, the growth of the internodes stopped at a final length according to the measurements. During the simulations, BPFD perceived by the internodes was recorded for each hour. The simulated perceived BPFD was used to fit a response function to the internode length observed under the different BPFD treatments:

$$Internode\ length = L_{min}\left(1 + \left(BPFD_{min} - BPFD_{per}\right)a\right) \quad BPFD_{min} - BPFD_{per} > 0 \tag{5}$$

where L_{min} is the minimum possible length of the internode, $BPFD_{min}$ is the lowest BPFD giving L_{min}, $BPFD_{per}$ is the average perceived BPFD of the internode during the first four days of growth and a is the slope of the response to $BPFD_{per}$. The first four days of growth were used due to a very rapid growth inhibiting effect of blue light [59] and according to the results of Kahlen and Stützel [46] who found four successive days starting one week before reaching maximum growth rate to be particularly sensitive to changes in PPFD.

The found BPFD response function was integrated in the FSP model to determine L_{max} of the beta-function and the final internode length. They were hereby simulated in dependence of the perceived BPFD during the simulations and at this stage no longer determined by experimental measurements. Internode two until nine elongated according to the integrated response function, with L_{min}, and t_m set for each internode according to the found parameter values for the treatment

B310 (baseline scenario with shortest internodes). Because the first internode (hypocotyl) grew before thinning and beginning of experimental measurements, its length was set to grow to the final average length of all treatments.

4.7. Model Evaluation and Alternative Scenarios

The light simulations were evaluated in an earlier study by comparing light measurements and simulations within a soybean canopy grown under two treatments identical to this study (B110 and B160) [38]. The integration of the parameters of the beta-function used to calibrate the dynamic model in this study was evaluated by comparing length over time for internodes, petioles and leaf laminas at all phytomer levels under all treatments. Model simulations after integration of the response function to blue light were evaluated by comparing the measured and simulated plant height until the third phytomer (hypocotyl counted as first phytomer). The comparison included the eight plants per chamber selected for the first four dates of destructive measurements, which were not used for the model parameterization.

Then, the model was applied for spectral optimization with the aim of minimizing BPFD emitted by the LED modules to reduce energy consumption, but still reach the minimum internode length. The first alternative scenario was run with a different virtual LED chamber design for evaluating the effect on the perceived BPFD. The chamber design was changed by setting the reflection of the bottom, pots and substrate to the same level as the sidewalls of the chamber. This was chosen for simulating a situation similar to e.g., hydroponics with plants placed in more reflective containers than the substrate and black pots in the experiment. This change in chamber design was expected to increase the perceived BPFD and hereby reduce BPFD emitted by the LEDs that is necessary to induce short plants. Simulations from the first scenario indicated that the spectrum could be optimized according to the developmental stage. The suggested optimization was applied in the second scenario by changing the emitted BPFD during the growth period.

5. Conclusions

The length of internodes and petioles increased under low BPFD, similar to the shade response under low PPFD, whereas the limited response of SLA and internode diameter indicated that the shade responses of these might not be regulated by cryptochrome. Further studies could investigate alternative regulation of these together with extended photosynthetic measurements over time to increase the understanding of carbon assimilation and translocation under different BPFD levels. Several aspects of the exact spectral effects on morphology and physiology should be further investigated, both for narrow peaks independent and the interactions with broader spectra.

Internode length dependent on perceived BPFD was well simulated in the FSP model and the simulations gave an increased insight into the response of the second and third internode based on the perceived BPFD. The model was a useful tool to determine the minimum necessary BPFD within an alternative chamber environment. Modelling with an FSP can be applied for further optimizations of indoor plant production implementing advances in knowledge of spectral effects on plant morphology and physiology.

Supplementary Materials: The following are available online at http://www.mdpi.com/2223-7747/9/12/1757/s1, Figure S1: Plant height (A), biomass (B) and leaf area (C) per plant and under different blue photosynthetic flux densities (BPFD). Error bars indicate standard error of the mean (day 9–20: $n = 4$, day 23: $n = 8$), Figure S2: Simulated (line) and measured (points) length of petioles, internodes and leaf laminas under the treatments B310 and B60, Figure S3: The absorption, reflection and transmission of radiation (%, relative to the incident radiation) from 400–700 nm by a soybean leaf used for the optical properties of the simulated soybean leaves. Data taken from Kasperbauer (1987), Figure S4: The simulated spectra (total PFFD of 400 μmol m^{-2} s^{-1}) of the six treatments with a simulated BPFD of 60, 110, 160, 210, 260 and 310 μmol m^{-2} s^{-1}, Figure S5: Visualizations of the simulated unfolding (A, B) and fully developed (C) trifoliate leaf, Table S1: The spread of the six treatments within three chambers over time, Table S2: Model inputs to determine ratios and angles of organs.

Author Contributions: Conceptualization, T.H. and S.M.; methodology, T.H. and S.M.; software, T.H.; validation, T.H.; formal analysis, T.H.; investigation, T.H.; data curation, T.H.; writing—original draft preparation, T.H.; writing—review and editing, T.H., S.M. and S.G.-H.; visualization, T.H.; supervision, S.M. and S.G.-H.; project administration, T.H., S.M. and S.G.-H.; funding acquisition, S.M. and S.G.-H. All authors have read and agreed to the published version of the manuscript.

Funding: This research was funded by the German Federal Ministry for Economic Affairs and Energy according to a decision of the German Federal Parliament within the Central Innovation Program for SMEs (ZF4279901CR6).

Conflicts of Interest: The authors declare no conflict of interest and the funders had no role in the design of the study; in the collection, analyses, or interpretation of data; in the writing of the manuscript, or in the decision to publish the results.

References

1. Bantis, F.; Smirnakou, S.; Ouzounis, T.; Koukounaras, A.; Ntagkas, N.; Radoglou, K. Current status and recent achievements in the field of horticulture with the use of light-emitting diodes (LEDs). *Sci. Hortic.* **2018**, *235*, 437–451. [CrossRef]
2. Singh, D.; Basu, C.; Meinhardt-Wollweber, M.; Roth, B. LEDs for energy efficient greenhouse lighting. *Renew. Sustain. Energy Rev.* **2015**, *49*, 139–147. [CrossRef]
3. Olle, M.; Viršile, A. The effects of light-emitting diode lighting on greenhouse plant growth and quality. *Agric. Food Sci.* **2013**, *22*, 223–234. [CrossRef]
4. Ahlman, L.; Bånkestad, D.; Wik, T. Using chlorophyll a fluorescence gains to optimize LED light spectrum for short term photosynthesis. *Comput. Electron. Agric.* **2017**, *142*, 224–234. [CrossRef]
5. Pennisi, G.; Blasioli, S.; Cellini, A.; Maia, L.; Crepaldi, A.; Braschi, I.; Spinelli, F.; Nicola, S.; Fernandez, J.A.; Stanghellini, C.; et al. Unraveling the role of red:blue LED lights on resource use efficiency and nutritional properties of indoor grown sweet basil. *Front. Plant Sci.* **2019**, *10*, 305. [CrossRef]
6. Schulze, P.S.C.; Barreira, L.A.; Pereira, H.G.C.; Perales, J.A.; Varela, J.C.S. Light emitting diodes (LEDs) applied to microalgal production. *Trends Biotechnol.* **2014**, *32*, 422–430. [CrossRef]
7. Watson, A.; Ghosh, S.; Williams, M.J.; Cuddy, W.S.; Simmonds, J.; Rey, M.-D.; Asyraf Md Hatta, M.; Hinchliffe, A.; Steed, A.; Reynolds, D.; et al. Speed breeding is a powerful tool to accelerate crop research and breeding. *Nat. Plants* **2018**, *4*, 23–29. [CrossRef]
8. Jähne, F.; Hahn, V.; Würschum, T.; Leiser, W.L. Speed breeding short-day crops by LED-controlled light schemes. *Theor. Appl. Genet.* **2020**, *133*, 2335–2342. [CrossRef]
9. Fraser, D.P.; Hayes, S.; Franklin, K.A. Photoreceptor crosstalk in shade avoidance. *Curr. Opin. Plant Biol.* **2016**, *33*, 1–7. [CrossRef]
10. Li, J.; Li, G.; Wang, H.; Deng, X.W. Phytochrome signaling mechanisms. *Arab. Book* **2011**, *9*, e0148. [CrossRef]
11. Yu, X.; Liu, H.; Klejnot, J.; Lin, C. The cryptochrome blue light receptors. *Arab. Book* **2010**, *8*, e0135. [CrossRef]
12. Hitz, T.; Hartung, J.; Graeff-Hönninger, S.; Munz, S. Morphological response of soybean (Glycine max (L.) Merr.) cultivars to light intensity and red to far-red ratio. *Agronomy* **2019**, *9*, 428. [CrossRef]
13. Yang, F.; Huang, S.; Gao, R.; Liu, W.; Yong, T.; Wang, X.; Wu, X.; Yang, W. Growth of soybean seedlings in relay strip intercropping systems in relation to light quantity and red:far-red ratio. *Field Crop. Res.* **2014**, *155*, 245–253. [CrossRef]
14. Gong, W.Z.; Jiang, C.D.; Wu, Y.S.; Chen, H.H.; Liu, W.Y.; Yang, W.Y. Tolerance vs. avoidance: Two strategies of soybean (Glycine max) seedlings in response to shade in intercropping. *Photosynthetica* **2015**, *53*, 259–268. [CrossRef]
15. Wollaeger, H.M.; Runkle, E.S. Growth and acclimation of impatiens, salvia, petunia, and tomato seedlings to blue and red light. *HortScience* **2015**, *50*, 522–529. [CrossRef]
16. Hernández, R.; Kubota, C. Physiological responses of cucumber seedlings under different blue and red photon flux ratios using LEDs. *Environ. Exp. Bot.* **2016**, *121*, 66–74. [CrossRef]
17. Wheeler, R.M.; Mackowiak, C.L.; Sager, J.C. Soybean stem growth under high-pressure sodium with supplemental blue lighting. *Agron. J.* **1991**, *83*, 903–906. [CrossRef]
18. Cope, K.R.; Bugbee, B. Spectral effects of three types of white light-emitting diodes on plant growth and development: Absolute versus relative amounts of blue light. *HortScience* **2013**, *48*, 504–509. [CrossRef]
19. Dougher, T.A.; Bugbee, B. Differences in the response of wheat, soybean and lettuce to reduced blue radiation. *Photochem. Photobiol.* **2001**, *73*, 199–207. [CrossRef]

20. Naznin, M.; Lefsrud, M.; Gravel, V.; Azad, M. Blue light added with red LEDs enhance growth characteristics, pigments content, and antioxidant capacity in lettuce, spinach, kale, basil, and sweet pepper in a controlled environment. *Plants* **2019**, *8*, 93. [CrossRef]
21. Hogewoning, S.W.; Wientjes, E.; Douwstra, P.; Trouwborst, G.; van Ieperen, W.; Croce, R.; Harbinson, J. Photosynthetic quantum yield dynamics: From photosystems to leaves. *Plant Cell* **2012**, *24*, 1921–1935. [CrossRef]
22. Hogewoning, S.W.; Trouwborst, G.; Maljaars, H.; Poorter, H.; van Ieperen, W.; Harbinson, J. Blue light dose-responses of leaf photosynthesis, morphology, and chemical composition of Cucumis sativus grown under different combinations of red and blue light. *J. Exp. Bot.* **2010**, *61*, 3107–3117. [CrossRef]
23. He, J.; Qin, L.; Chong, E.L.C.; Choong, T.-W.; Lee, S.K. Plant growth and photosynthetic characteristics of mesembryanthemum crystallinum grown aeroponically under different blue- and red-LEDs. *Front. Plant Sci.* **2017**, *8*, 361. [CrossRef]
24. Kang, W.H.; Park, J.S.; Park, K.S.; Son, J.E. Leaf photosynthetic rate, growth, and morphology of lettuce under different fractions of red, blue, and green light from light-emitting diodes (LEDs). *Hortic. Environ. Biotechnol.* **2016**, *57*, 573–579. [CrossRef]
25. Vos, J.; Evers, J.B.; Buck-Sorlin, G.H.; Andrieu, B.; Chelle, M.; de Visser, P.H.B. Functional-structural plant modelling: A new versatile tool in crop science. *J. Exp. Bot.* **2010**, *61*, 2101–2115. [CrossRef]
26. Evers, J.B. *Simulating Crop Growth and Development Using Functional-Structural Plant Modeling*; Springer: Dordrecht, The Netherlands, 2016; pp. 219–236.
27. Henke, M.; Kurth, W.; Buck-Sorlin, G.H. FSPM-P: Towards a general functional-structural plant model for robust and comprehensive model development. *Front. Comput. Sci.* **2016**, *10*, 1103–1117. [CrossRef]
28. Burgess, A.J.; Retkute, R.; Pound, M.P.; Mayes, S.; Murchie, E.H. Image-based 3D canopy reconstruction to determine potential productivity in complex multi-species crop systems. *Ann. Bot.* **2017**, *119*, 517–532. [CrossRef]
29. Evers, J.B.; Bastiaans, L. Quantifying the effect of crop spatial arrangement on weed suppression using functional-structural plant modelling. *J. Plant Res.* **2016**, *129*, 339–351. [CrossRef]
30. Chelle, M.; Evers, J.B.; Combes, D.; Varlet-Grancher, C.; Vos, J.; Andrieu, B. Simulation of the three-dimensional distribution of the red:far-red ratio within crop canopies. *N. Phytol.* **2007**, *176*, 223–234. [CrossRef]
31. Evers, J.B.; Vos, J.; Chelle, M.; Andrieu, B.; Fournier, C.; Struik, P.C. Simulating the effects of localized red:far-red ratio on tillering in spring wheat (Triticum aestivum) using a three-dimensional virtual plant model. *N. Phytol.* **2007**, *176*, 325–336. [CrossRef]
32. Buck-Sorlin, G.; Hemmerling, R.; Kniemeyer, O.; Burema, B.; Kurth, W. A rule-based model of barley morphogenesis, with special respect to shading and gibberellic acid signal transduction. *Ann. Bot.* **2008**, *101*, 1109–1123. [CrossRef] [PubMed]
33. Bongers, F.J.; Pierik, R.; Anten, N.P.R.; Evers, J.B. Subtle variation in shade avoidance responses may have profound consequences for plant competitiveness. *Ann. Bot.* **2018**, *121*, 863–873. [CrossRef] [PubMed]
34. Gautier, H.; Měch, R.; Prusinkiewicz, P.; Varlet-Grancher, C. 3D Architectural Modelling of Aerial Photomorphogenesis in White Clover (*Trifolium repens* L.) using L-systems. *Ann. Bot.* **2000**, *85*, 359–370. [CrossRef]
35. Buck-Sorlin, G.; de Visser, P.H.; Henke, M.; Sarlikioti, V.; van der Heijden, G.W.; Marcelis, L.F.; Vos, J. Towards a functional–structural plant model of cut-rose: Simulation of light environment, light absorption, photosynthesis and interference with the plant structure. *Ann. Bot.* **2011**, *108*, 1121–1134. [CrossRef] [PubMed]
36. Dieleman, J.A.; De Visser, P.H.B.; Meinen, E.; Grit, J.G.; Dueck, T.A. integrating morphological and physiological responses of tomato plants to light quality to the crop level by 3D modeling. *Front. Plant Sci.* **2019**, *10*, 839. [CrossRef]
37. Kalaitzoglou, P.; van Ieperen, W.; Harbinson, J.; van der Meer, M.; Martinakos, S.; Weerheim, K.; Nicole, C.C.S.; Marcelis, L.F.M. Effects of continuous or end-of-day far-red light on tomato plant growth, morphology, light absorption, and fruit production. *Front. Plant Sci.* **2019**, *10*, 322. [CrossRef]
38. Hitz, T.; Henke, M.; Graeff-Honninger, S.; Munz, S. Simulating light spectrum within a soybean canopy in an LED growth chamber. In Proceedings of the 6th International Symposium on Plant Growth Modeling, Simulation, Visualization and Applications (PMA), Hefei, China, 4–8 November 2018; pp. 120–125.

39. McCree, K.J. The action spectrum, absorptance and quantum yield of photosynthesis in crop plants. *Agric. Meteorol.* **1972**, *9*, 191–216. [CrossRef]
40. Lanoue, J.; Leonardos, E.D.; Grodzinski, B. Effects of light quality and intensity on diurnal patterns and rates of photo-assimilate translocation and transpiration in tomato leaves. *Front. Plant Sci.* **2018**, *9*, 756. [CrossRef]
41. Pierik, R.; de Wit, M. Shade avoidance: Phytochrome signaling and other aboveground neighbor detection cues. *J. Exp. Bot.* **2014**, *65*, 2815–2824. [CrossRef]
42. Feng, L.; Raza, M.A.; Li, Z.; Chen, Y.; Khalid, M.H.B.; Du, J.; Liu, W.; Wu, X.; Song, C.; Yu, L.; et al. The influence of light intensity and leaf movement on photosynthesis characteristics and carbon balance of soybean. *Front. Plant Sci.* **2019**, *9*, 1952. [CrossRef]
43. Park, Y.; Runkle, E.S. Far-red radiation and photosynthetic photon flux density independently regulate seedling growth but interactively regulate flowering. *Environ. Exp. Bot.* **2018**, *155*, 206–216. [CrossRef]
44. Kim, G.-T.; Yano, S.; Kozuka, T.; Tsukaya, H. Photomorphogenesis of leaves: Shade-avoidance and differentiation of sun and shade leaves. *Photochem. Photobiol. Sci.* **2005**, *4*, 770. [CrossRef] [PubMed]
45. Weston, E.; Thorogood, K.; Vinti, G.; López-Juez, E. Light quantity controls leaf-cell and chloroplast development in Arabidopsis thaliana wild type and blue-light-perception mutants. *Planta* **2000**, *211*, 807–815. [CrossRef] [PubMed]
46. Kahlen, K.; Stützel, H. Modelling photo-modulated internode elongation in growing glasshouse cucumber canopies. *New Phytol.* **2011**, *190*, 697–708. [CrossRef] [PubMed]
47. Sellaro, R.; Crepy, M.; Trupkin, S.A.; Karayekov, E.; Buchovsky, A.S.; Rossi, C.; Casal, J.J. Cryptochrome as a Sensor of the blue/green ratio of natural radiation in Arabidopsis. *Plant Physiol.* **2010**, *154*, 401–409. [CrossRef]
48. Snowden, M.C.; Cope, K.R.; Bugbee, B. Sensitivity of seven diverse species to blue and green light: interactions with photon flux. *PLoS ONE* **2016**, *11*, e0163121. [CrossRef]
49. Ahmad, M.; Grancher, N.; Heil, M.; Black, R.C.; Giovani, B.; Galland, P.; Lardemer, D. Action spectrum for cryptochrome-dependent hypocotyl growth inhibition in Arabidopsis. *Plant Physiol.* **2002**, *129*, 774–785. [CrossRef]
50. Schneider, C.A.; Rasband, W.S.; Eliceiri, K.W. NIH Image to ImageJ: 25 years of image analysis. *Nat. Methods.* **2012**, *9*, 671–675. [CrossRef]
51. Yin, X.; Goudriaan, J.; Lantinga, E.A.; Vos, J.; Spiertz, H.J. A flexible sigmoid function of determinate growth. *Ann. Bot.* **2003**, *91*, 361–371. [CrossRef]
52. R Core Team. *R: A Language and Environment for Statisticalcomputing*; R Foundation for Statistical Computing: Vienna, Austria, 2019; Available online: https://www.R-project.org/ (accessed on 10 December 2020).
53. Piepho, H.-P. An algorithm for a letter-based representation of all-pairwise comparisons. *J. Comput. Graph. Stat.* **2004**, *13*, 456–466. [CrossRef]
54. Hitz, T.; Henke, M.; Graeff-Hönninger, S.; Munz, S. Three-dimensional simulation of light spectrum and intensity within an LED growth chamber. *Comput. Electron. Agric.* **2019**, *156*, 540–548. [CrossRef]
55. Henke, M.; Buck-Sorlin, G.H. Using a full spectral raytracer for calculating light microclimate in functional-structural plant modelling. *Comput. Inform.* **2018**, *36*, 1492–1522. [CrossRef]
56. Van Antwerpen, D.G. Unbiased Physically Based Rendering on the GPU. Master's Thesis, Delft University of Technology, Delft, The Netherlands, 2011.
57. Kasperbauer, M.J. Far-red light reflection from green leaves and effects on phytochrome-mediated assimilate partitioning under field conditions. *Plant Physiol.* **1987**, *85*, 350–354. [CrossRef] [PubMed]
58. Jacquemoud, S.; Baret, F.; Hanocq, J.F. Modeling spectral and bidirectional soil reflectance. *Remote Sens. Environ.* **1992**, *41*, 123–132. [CrossRef]
59. Parks, B.M.; Folta, K.M.; Spalding, E.P. Photocontrol of stem growth. *Curr. Opin. Plant Biol.* **2001**, *4*, 436–440. [CrossRef]

Publisher's Note: MDPI stays neutral with regard to jurisdictional claims in published maps and institutional affiliations.

plants

MDPI

Article

How Supplementary or Night-Interrupting Low-Intensity Blue Light Affects the Flower Induction in Chrysanthemum, a Qualitative Short-Day Plant

Yoo Gyeong Park [1] and Byoung Ryong Jeong [1,2,3,*]

[1] Institute of Agriculture and Life Science, Gyeongsang National University, Jinju 52828, Korea; ygpark615@gmail.com

[2] Division of Applied Life Science (BK21 Plus Program), Graduate School, Gyeongsang National University, Jinju 52828, Korea

[3] Research Institute of Life Science, Gyeongsang National University, Jinju 52828, Korea

* Correspondence: brjeong@gmail.com; Tel.: +82-010-6751-5489

Received: 19 October 2020; Accepted: 27 November 2020; Published: 2 December 2020

Abstract: This research examined the effects of the supplementary or night-interrupting (NI) blue (B) light supplied at a low intensity on the flowering, gene expression, and morphogenesis of chrysanthemum, a qualitative short-day plant. White (W) light-emitting diodes (LEDs) were used to provide light with a photosynthetic photon flux density (PPFD) of 180 $\mu mol \cdot m^{-2} \cdot s^{-1}$ during the photoperiod to grow the plants in a plant factory. The control group was constructed with plants that were exposed to a 10-h short day (SD10) treatment without any blue light. The B light in this research was used for 4 h to either (1) extend the photoperiod for plants at the end of a 9-h short day (SD) treatment as the sole light source (SD9 + 4B), (2) provide night interruption (NI) to plants in the 13-h long-day (LD) treatment (LD13 + NI − 4B), (3) provide NI to plants in the 10-h SD treatment (SD10 + NI − 4B), or (4) supplement the W LEDs at the end of a 13-h LD treatment (LD13 + 4B). Blue LEDs were used to provide the supplementary/NI light at 10 $\mu mol \cdot m^{-2} \cdot s^{-1}$ PPFD. The LD13 + NI − 4B treatment resulted in the greatest plant height, followed by LD13 + 4B. Plants in all treatments flowered. It is noteworthy that despite the fact that chrysanthemum is a qualitative SD plant, chrysanthemum plants flowered when grown in the LD13 + 4B and LD13 + NI − 4B treatments. Plants grown in the LD13 + 4B had the greatest number of flowers. Plants grown in the LD13 + 4B treatment had the highest expression levels of the *cryptochrome 1*, *phytochrome A*, and *phytochrome B* genes. The results of this study indicate that a 4-h supplementation of B light during the photoperiod increases flower bud formation and promotes flowering, and presents a possibility as an alternative method to using blackout curtains in LD seasons to practically induce flowering. The B light application methods to induce flowering in SD plants requires further research.

Keywords: blue LED; flower bud formation; number of flowers; photoperiod

1. Introduction

Plants adapt to the signals, such as the light quality, they perceive from the environment and accordingly modify their biological cycles [1]. Different types of photoreceptors, such as cryptochromes and phytochromes, enable plants to perceive changes in the light quality [2,3]. Throughout their lifecycle, the growth and development of plants are influenced by the photoreceptors. Photoreceptors monitor the light environment and also help plants time key developmental transitions, such as flowering and seed germination [4]. Phytochrome is a photoreceptor that primarily absorbs red (R) and far-red (Fr) lights, while cryptochrome is a photoreceptor that primarily absorbs ultraviolet-A (UV-A) and blue (B) lights, both of which help regulate flowering [5]. Multiple cryptochrome (*CRY1* and

CRY2) and phytochrome (*PHYA*, *PHYB*, *PHYC*, *PHYD*, and *PHYE*) varieties can exist, depending on the species [6,7].

Light supplementation is often utilized for enhancing the quality of seedlings and rooted cuttings [8]. Photoperiod manipulation can reduce the production time and improve the overall crop quality to reduce production costs [9]. Light supplementation may take the form of supplementary light in a background of natural light, or additional light that extends the day length [8]. Night interruption (NI) interrupts a length of dark period with lighting, thus creating modified long-day (LD) conditions [10,11].

Studies have reported that B light negatively affects stem elongation and leads to a reduced leaf area [12–16]. Senger [17] found that blue light played a pivotal role in chloroplast development and formation, as well as the stomatal opening. It has been suggested that photoreceptors related to B light played a part in the flowering process [18,19]. Jeong et al. [20] reported that supplementary blue light at least in part promotes the elongation of stems and internodes without inhibiting the flower bud formation. In the short-day (SD) plant chrysanthemum, NI with B light did not effectively inhibit flowering, although B light is part of visible light [21,22]. Our previous study [11] split the traditional 4-h NI into two 2-h periods and shifted the NI light quality to examine how these changes affect the flowering and morphogenesis of chrysanthemum. They found out that B, Fr, R, and white (W) lights used in the first 2 h of the NI did not affect the morphogenesis nor flowering, while the same lights used in the last 2 h of the NI significantly impacted the morphogenesis and flowering. In addition, they discovered that flowering was induced in all NI treatments concluding with a blue light. Hence, we hypothesized that blue light at a low intensity supplemented to either LD or SD conditions may induce flowering in SD plants. Therefore, this study examined the effects of low-intensity (10 μmol m^{-2} s^{-1} PPFD) blue light used as supplementary or NI light on the flowering, gene expression, and morphogenesis in chrysanthemum 'Gaya Yellow' (a qualitative SD plant).

2. Materials and Methods

2.1. Growth Conditions and Plant Materials

Chrysanthemum (*Dendranthema grandiflorum* 'Gaya Yellow') spray-type cuttings were stuck in plug trays with 50 cells each filled with a commercial Tosilee Medium (Shinan Grow Company, Jinju, Korea). The cuttings were subsequently put on a glasshouse bench to root. The cuttings were relocated 12 days after they were stuck, when they have rooted, to a closed walk-in growth chamber that is 7700 cm by 2500 cm by 2695 cm in size. There, the plants were acclimatized to 20 ± 1 °C, 60% ± 10% RH, and 140 μmol·m^{-2}·s^{-1} PPFD supplied with F48T12-CW-VHO fluorescent lamps (Philips Co., Ltd., Eindhoven, The Netherlands). The closed walk-in growth chamber was constructed such that numerous uniformly distributed holes allowed conditioned air to blow horizontally into the growing spaces. CO_2 was supplemented from a compressed gas tank to maintain an atmospheric concentration of 350 ± 50 μmol· mol^{-1}. The plants, after 11 days of acclimatization (the 16-h LD) in the growth chamber, were approximately 7.0 cm in height and were subjected to the photoperiodic light treatments. After being planted, the chrysanthemums were fertigated once a day (from 9:00 a.m. to 10:0 a.m.) throughout the experiment with a greenhouse multipurpose nutrient solution [11]. A 3-replication randomized complete block design was employed with a total of 6 plants for each treatment, with 2 plants in each replication. Within a controlled environment, the photoperiodic light treatments were randomly located in between replications to minimize the effects of the light treatment positioning.

2.2. Photoperiodic Light Treatments

Plants were grown with light at an intensity of 180 μmol m^{-2} s^{-1} PPFD provided by white MEF50120 LEDs (More Electronics Co. Ltd., Changwon, Korea) (Figure 1A). The different photoperiods used in this experiment, as well as the lighted period during the NI (referred to as 'photoperiod' hereafter) were as follows: B light with a wavelength of 450 nm was used for 4 h to either (1) extend the photoperiod at the end of a 9-h SD as the sole light source (SD9 + 4B), (2) provide NI following the 13-h LD (LD13 + NI − 4B), (3) provide NI after the 10-h SD (SD10 + NI − 4B), or (4) supplement W LEDs at the end of a 13-h LD (LD13 + 4B) (Figures 1B and 2). The control was constructed by exposing the plants to a 10-h short-day treatment (SD10) without B light. B light at an intensity of 10 ± 3 μmol· m^{-2}·s^{-1} PPFD was provided by LEDs for the photoperiodic light treatments. A HD2102.1 digital photometer (Delta OHM, Padova, Italy) measured the average PPFD 20 cm above the bench top, for each light treatment. The lighting was adjusted such that the same PPFD levels were provided to the plants regardless of the light treatment. A USB 2000 Fiber Optic Spectrometer (Ocean Optics Inc., Dunedin, FL, USA; detects wavelengths between 200 to 1000 nm) scanned the spectral distribution in 1-nm wavelength intervals 25 cm above the bench top.

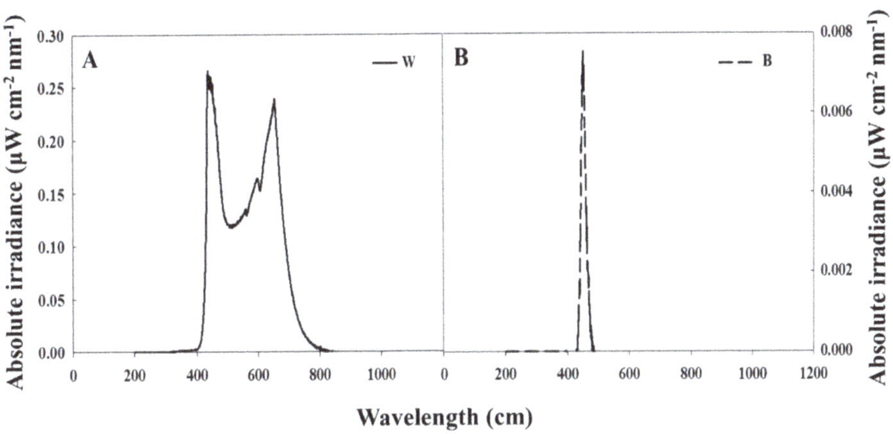

Figure 1. The spectral distribution of lights used in this experiment: daily W light provided by white LEDs (**A**) and B light from blue LEDs used as the supplementary and night-interrupting light (**B**).

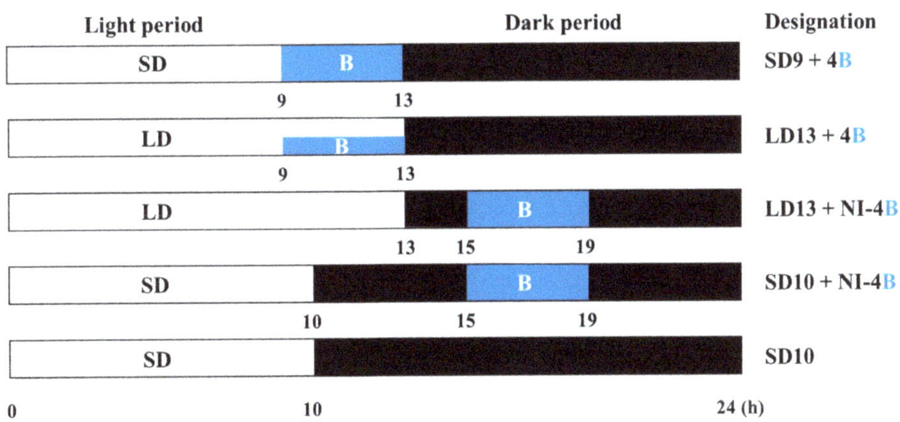

Figure 2. Supplementary and night-interrupting blue (B) light schemes employed in this study. B light was used for 4 h to either (1) extend the photoperiod at the end of a 9-h SD as the sole light source (SD9 + 4B), (2) provide NI following a 13-h LD (LD13 + NI − 4B), (3) provide NI after a 10-h SD (SD10 + NI − 4B), or (4) supplement the W light at the end of a 13-h LD (LD13 + 4B). Plants in the control were grown with a 10- hour SD treatment (SD10) without any B light.

2.3. Data Collection and Statistical Analysis

The dry mass, number of leaves per plant, number of nodes per plant, number of flowers per plant, plant height, leaf area, chlorophyll content, percent flowering, days of treatment needed to visible flower bud or days to visible buds (DVB), flower width, and photoreceptor gene expressions were measured after 41 days of the photoperiodic light treatments. All leaves with a length greater than 1 cm in were counted to determine the number of leaves per plant. Divided samples of the shoot and root were dried at 70 °C for 72 h in a Venticell-222 drying oven (MMM Medcenter Einrichtungen GmbH., Munich, Germany) before the dry mass measurements were taken with an EW 220-3NM electronic scale (Kern and Sohn GmbH., Balingen, Germany). Leaf area measurements were taken with a LI-3000 leaf area meter (LI-COR Inc., Lincoln, NE, USA). The chlorophyll concentration was estimated from 10-mg samples of fresh, young, and fully developed leaves. Chlorophyll was extracted with 80% acetone at 4 °C. A Biochrom Libra S22 spectrophotometer (Biochrom Co. Ltd., Holliston, MA, USA) measured the absorbance of the supernatant at 645 and 663 nm, after the extracted chlorophyll was centrifuged at 3000 rpm. Calculations were performed according to the method described by Dere et al. [23]. The statistical analysis was performed with the SAS 9.1 software (SAS Institute Inc., Cary, NC, USA). An analysis of variance (ANOVA) and Tukey's multiple range test were performed with the results of this study. SigmaPlot 12.0 (Systat Software Inc., San Jose, CA, USA) was used for graphing.

2.4. Total RNA Isolation, cDNA Synthesis, and Real-Time Polymerase Chain Reaction (PCR) of Selected Genes

After 20 days of the photoperiodic light treatments, plants started displaying visible flower buds and the most recently matured 10 leaves per plant were collected for total RNA extraction. The latest leaf to be matured was collected an hour after the daily photoperiodic treatments began, at 9:00 a.m. This collection time was chosen because the photosynthetic rates are high at this time of the day. Equal amounts of cDNA using primers of *cryptochrome 1* (*CRY1*), *phytochrome A* (*PHYA*), and *phytochrome B* (*PHYB*), whose sequences are shown in Table 1, were used to perform the independent PCRs. As actin is frequently used to normalize molecular expression studies, it was used as an internal control. The $2^{-\Delta\Delta Ct}$ method [24] was used to determine the relative expression levels of each gene. At each sampling date, the individual gene expression levels in the plants grown with the light treatments were divided by the mean gene expression levels for plants in the control (SD10). The total RNA isolation and real-time quantitative PCR analysis of the selected genes were performed according to the method described in Park et al. [11].

Table 1. The primers used to quantify the gene expression levels.

Gene	Accession no.	Forward Primer	Reverse Primer
CRY1	NM_116961	5′-CGTAAGGGATCACCGAGTAAAG-3′	5′-CTTTTAGGTGGGAGTTGTGGAG-3′
PHYA	EU915082	5′-GACAGTGTCAGGCTTCAACAAG-3′	5′-ACCACCAGTGTGTGTTATCCTG-3′
PHYB	NM_127435	5′-GTGCTAGGGAGATTACGCTTTC-3′	5′-CCAGCTTCTGAGACTGAACAGA-3′
Actin	AB205087	5′-CGTTTGGATCTTGCTGGTCG-3′	5′-CAGGACATCTGAAACGCTCA-3′

3. Results

3.1. Morphogenesis

It was observed that the supplementary and night-interrupting blue light increased the plant heights in this study (Figure 3A). Plants grown in LD13 + NI − 4B had the greatest height (Figure 3A), where it was 22% greater than that of plants grown in SD10. Additionally, it was observed that even plants in SD9 + 4B had a greater mean height than those in SD10.

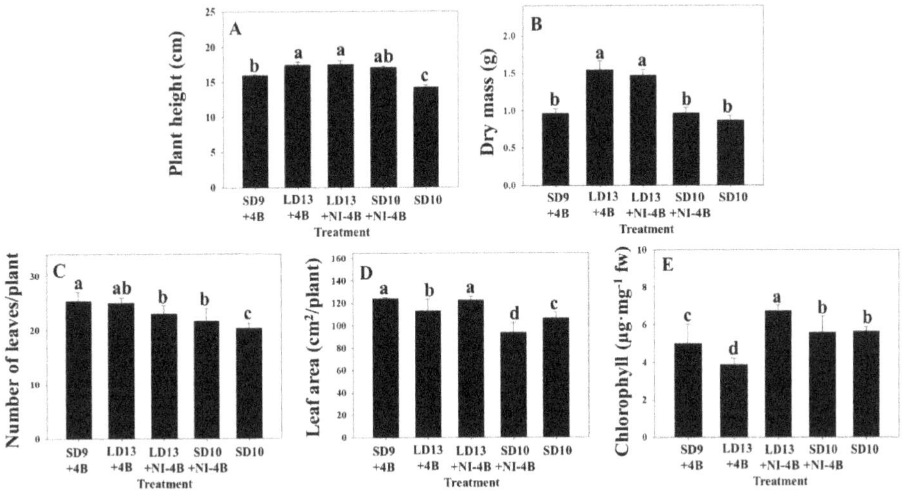

Figure 3. The effects of the supplementary and night-interrupting B light on the plant height (**A**), dry mass (**B**), number of leaves per plant (**C**), leaf area per plant (**D**), and chlorophyll levels (**E**) in *D. grandiflorum* 'Gaya Yellow'. The control was constructed by exposing plants to a 10-h SD treatment (SD10) without any B light. Data are the mean ± S.E of the 3 biological replicates. Means accompanied by different letters significantly differ ($p < 0.05$) according to Tukey's studentized range test at a 5% significance level.

The dry mass of plants grown under all photoperiodic treatments was greater compared to that of the plants in the SD10 control (Figure 3B). Increasing the photoperiod, as with LD13 + 4B and LD13 + NI − 4B, significantly increased the dry mass of the plants in this study. The other treatments, SD9 + 4B and SD10 + NI − 4B, were not as effective as the LD treatments in increasing the dry mass (Figure 3B).

Plants in SD9 + 4B had the greatest number of leaves per plant while those in SD10 had the lowest number of leaves per plant (Figure 3C). The average leaf area was the greatest for plants in LD13 + NI − 4B and the smallest for plants in SD10 + NI − 4B (Figure 3D). The leaf area per plant was 12% for plants in SD10 + NI − 4B when compared to that for plants in SD10 (Figure 3D). Furthermore, all B light treatments except for SD10 + NI − 4B increased the leaf area compared to the control (Figure 3D). The chlorophyll levels were the lowest for plants in LD13 + 4B and the highest for plants in LD13 + NI − 4B (Figure 3E). Plants in LD13 + 4B had 32% lower chlorophyll contents than plants in SD10 did (Figure 3E).

3.2. Flowering and Gene Expression

The flowering percentage of plants was 100% in all treatments (Table 2 and Figure 4). The fastest flowering induction was observed for plants in the control (SD10). It is noteworthy that plants in LD13 + 4B and LD13 + NI − 4B flowered, despite the fact that chrysanthemum is a qualitative SD plant (Table 2 and Figure 4).

Table 2. The effects of the supplementary and night-interrupting 10 μmol·m^{-2}·s^{-1} PPFD B light on the flowering characteristics of chrysanthemum (*D. grandiflorum* 'Gaya Yellow'), after 41 days of exposure to the photoperiodic light treatments.

Treatment [z]	Flowering (%)	DVB [y] (Day)	No. of Flowers/Plant	Flower width (cm)
SD9 + 4B	100	17.7 c [x]	11.0 c	2.6 b
LD13 + 4B	100	22.5 b	21.3 a	0.7 c
LD13 + NI − 4B	100	28.7 a	15.3 b	0.5 d
SD10 + NI − 4B	100	18.0 c	15.0 b	2.8 a
SD10	100	17.5 c	11.0 c	2.9 a
F-test		***	***	***

[z] See Figure 2 for details on the photoperiodic treatments with B light. [y] Days of treatment to visible flower bud or days to visible buds. [x] Mean separation within columns by Duncan's multiple range test at a 5% level. ***: Significant at $p \leq 0.001$.

Figure 4. The effects of the supplementary and night-interrupting 10 μmol·m^{-2}·s^{-1} PPFD B light on the flowering of chrysanthemum (*D. grandiflorum* 'Gaya Yellow'), after 41 days of exposure to the photoperiodic light treatments: side view (**A**) and top view (**B**) (see Figure 2 for details on the photoperiodic treatments with B light).

Plants in SD10 had the smallest DVB whereas plants in LD13 + NI − 4B had the greatest DVB (Table 2). The DVB was observed to increase as the photoperiod increased (Table 2). The DVB of plants in LD13 + 4B was smaller than those of plants in LD13 + NI − 4B. Interestingly, plants in LD13 + 4B had 93% more flowers per plant compared to plants in the SD10 control. Plants in the SD10 control and SD9 + 4B had the lowest number of flowers (Table 2). Plants in the SD10 control had the greatest flower width (Table 2).

The photoreceptor gene expression (*PHYA*, *PHYB*, and *CRY1*) in response to the B light was also analyzed (Figure 5). Plants in LD13 + 4B had the greatest expression levels of *PHYA* and *PHYB*, followed by plants in SD10 + NI − 4B (Figure 5). *PHYA* had the lowest expression levels in plants in LD13 + NI − 4B (Figure 5). *PHYB* was the least expressed in plants in LD13 + NI − 4B (Figure 5). Plants in LD13 + NI − 4B had significantly higher *CRY1* expression levels compared to plants in the other treatments (Figure 5). *CRY1* was the least expressed in plants in LD13 + 4B (Figure 5).

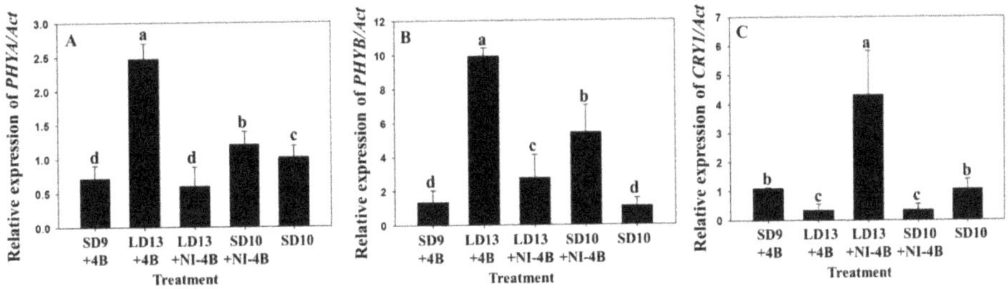

Figure 5. The effects of the supplementary and night-interrupting 10 μmol·m^{-2}·s^{-1} PPFD B light on the relative gene expression levels of) *D. grandiflorum* 'Gaya Yellow' determined by real-time PCR of *PHYA* (**A**), *PHYB* (**B**), and *CRY1* (**C**). (Details of the NI light qualities are presented in Figure 2). At each sampling date, the individual gene expression levels for the plants in the photoperiodic light treatments were divided by the mean gene expression level for plants in the SD10 control. The data are presented as the mean ± S.E of the 3 biological replicates. Means accompanied by different letters indicate significant differences ($p < 0.05$), according to Tukey's studentized range test at a 5% significance level.

4. Discussion

4.1. Plant Height

Different studies have observed that blue light acts to limit the elongation of the petiole, stem, and hypocotyl in various horticultural species, such as chrysanthemum, lettuce, pepper, and soybean [25–32]. Normally, increasing blue light decreases the stem length to a maximum threshold level [32]. Schuerger et al. [30] observed that blue light for 12 h a day plays a role in changing the stem anatomy, inhibiting the growth, and determining the morphogenetic characteristics of pepper plants. Furthermore, Khattak and Pearson [33] found that B light during the photoperiod in low-light environments resulted in reduced plant heights. Cryptochromes are also known to influence the stem elongation, and various of plants exhibit suppressed shoot elongation in response to B light in a 12-h day [26]. However, these photomorphogenic responses are different for different species. Previous studies used B light during the photoperiod to control the morphogenesis, while the current study used B light as a supplement or for NI to control morphogenesis and flowering.

All the photoperiodic light treatments considered in this study resulted in greater plant heights than that observed in the SD10 control (Figure 3A). This indicates that blue light may be used in the production of cut chrysanthemum flowers, as longer stems are considered to be of better quality. Kong et al. [34] stated that the increased elongation growth of plants in response to B light is linked to lower phytochrome activity, and is a shade-avoidance response, where different species have different sensitivities. These results agree with those of Jeong et al. [20], where it was found that an extended photoperiod with blue light promoted stem elongation of chrysanthemum. Longer photoperiods are known to be associated with the presence of higher gibberellin levels, which enhance stem elongation in chrysanthemums [20,35]. In many species, including salvia and marigold, B light was more effective than R light in increasing the shoot elongation [36]. Muleo and Morini [37] reported that internode extension on the stem leader in apple was inhibited by B LED, which determined the lowest values among all the light qualities tested. The differing responses of different plants to B light indicates that a species' responses to a specific light quality cannot necessarily be predicted on the basis of responses of other species.

4.2. Dry Mass and Leaf Growth

In this study, plants in LD13 + 4B and LD13 + NI − 4B had greater shoot and root dry masses compared to plants in the SD10 control (Figure 3B). These results indicate that a prolonged photoperiod contributed to the dry matter production. Moreover, B light supplementation increases the photosynthetic carbon assimilation and may also allow greenhouse crops to accumulate a greater biomass [38].

Plants in all the photoperiodic light treatments had a higher number of leaves per plant compared to plants in the control, and plants in SD9 + 4B had the greatest number of leaves (Figure 3C). Plants in SD10 + NI-4B had a smaller mean leaf area than plants in the SD10 control did (Figure 3D), resulting from shorter leaf lengths and widths (data not shown). All other treatments with B light led to a greater leaf area than that of plants in the SD10 control. Wang et al. [39] reported similar results, where light treatments with weak 50 μmol·m^{-2}·s^{-1} PPFD B light lead to increases in the leaf area. Honecke et al. [26] proposed that B light is required during the photoperiod for the normal growth of lettuce seedlings grown under R LEDs; if the B light level was low, long, narrow leaves developed. Iacona and Muleo [40] reported that total leaf area per plant in cherry rootstock 'Colt' was significantly greater in B LED-exposed plants than other treatments. However, these photomorphogenic responses are specific to the particular species. Dougher and Bugbee [41] reported that increasing the B light proportion resulted in decreasing leaf area in soybean, while in lettuce, increasing the B light proportion resulted in increasing the leaf area. Eskins [42] observed that the *Arabidopsis thaliana* leaf area was negatively correlated with the B light proportion, as a high-intensity B light irradiance corresponded to the development of small leaves.

Gang et al. [43] observed that the chlorophyll levels increased as plants transitioned from the vegetative to the reproductive growth, and decreased during maturation. Correspondingly, the lower chlorophyll content of plants in LD13 + 4B compared to that of plants in the other treatments in this study (Figure 3E) may be due to the continued maturation after the plants transition from the vegetative to reproductive growth.

4.3. Expression of Genes Related to the Morphogenesis and Flowering

The expression levels of *PHYA* and *PHYB* were the highest for plants in LD13 + 4B, and the expression level of *CRY1* was the greatest for plants in LD13 + NI − 4B. Plants in these two treatments were also the tallest. It has been reported that cryptochromes and phytochromes affect the height of chrysanthemums [33]. In *Arabidopsis*, high *PHYB* levels can increase the expression of *AtGAox2*, which controls the synthesis of gibberellins (GAs) [35]. Furthermore, it has been verified that both phytochromes and cryptochromes play a part in the regulation of the plant hormone GA levels [35,44]. Thus, it is speculated that the high expression levels of *PHYA*, *PHYB*, and *CRY1* found in plants grown in LD13 + 4B and LD13 + NI − 4B may promote the synthesis of GAs and eventually result in greater plant heights.

It is well known that photoreceptors related to B light were involved in the flowering process [18,20]. The CRY1 and CRY2 both mediate the flowering promotion by B light [45]. *PHYA* mediates the flowering promotion by Fr light, and *PHYB* mediates the flowering inhibition by R light in *Arabidopsis* [46–48]. Although *PHYA* and *PHYB* are R light receptors, it has also been shown that they also function under B light in *Arabidopsis* [49], and it has been proven that either *PHYA* or *PHYB*, as well as cryptochromes, were required for responses to B light [24,46,50]. In this study, the number of flowers per plant was shown to increase with the B light treatments. This may be attributed to the high *CRY1* expression levels. Similarly, Park et al. [11] reported that a greater number of flowers per plant was observed with light shifting from B during the NI, which may be attributed to a high light energy induction as well as shade avoidance responses, a behavior where plants evade darkness by lengthening the internodes. In rice, NI with B light delayed the flowering time, but this delay was not reproduced in the *PHYB-1* mutant [51], which means *PHYB* is a negative regulator for the flowering time. It was also observed that while chrysanthemum is a qualitative SD plant, those in the LD13 + 4B and LD13 + NI − 4B treatments still flowered. This indicates that high *PHYA* and *CRY1* expression levels may induce flowering. However, further research is necessary to verify this speculation.

In summary, B light resulted in a greater height and promoted the flowering in chrysanthemum. The results of this study illustrate that a 4-h B light supplementation during the photoperiod promoted flowering and increased the number of flower buds formed. Hence, B light supplementation may be an optimal technique to induce flowering, and can be practically applied to commercial cultivation of SD plants. This study suggests that B light supplementation is an alternative practical technique to induce flowering in SD plants to using blackout curtains during LD seasons. Further research is still needed to optimize B light supplementation techniques for flowering induction of SD plants.

Author Contributions: B.R.J. and Y.G.P. conceived and designed the experiments; Y.G.P. performed the experiments; B.R.J. and Y.G.P. analyzed the data; B.R.J. and Y.G.P. contributed reagents/materials/analysis tools; B.R.J. and Y.G.P. wrote the paper. All authors have read and agreed to the published version of the manuscript.

Funding: The APC was funded by Gyeongsagn National University.

Acknowledgments: The authors express their gratitude to Sowbiya Muneer, Prabhakaran Soundararajan, and Young Don Chin for their assistance.

Conflicts of Interest: The authors declare no conflict of interest.

References

1. Demotes-Mainard, S.; Péron, T.; Corot, A.; Bertheloot, J.; Gourrierec, J.G.-L.; Pelleschi-Travier, S.; Crespel, L.; Morel, P.; Huché-Thélier, L.; Boumaza, R.; et al. Plant responses to red and far-red lights, applications in horticulture. *Environ. Exp. Bot.* **2016**, *121*, 4–21. [CrossRef]
2. Galvão, V.C.; Fankhauser, C. Sensing the light environment in plants: Photoreceptors and early signaling steps. *Curr. Opin. Neurobiol.* **2015**, *34*, 46–53. [CrossRef]
3. Huché-Thélier, L.; Crespel, L.; Gourrierec, J.G.-L.; Morel, P.; Sakr, S.; LeDuc, N. Light signaling and plant responses to blue and UV radiations—Perspectives for applications in horticulture. *Environ. Exp. Bot.* **2016**, *121*, 22–38. [CrossRef]
4. Kami, C.; Lorrain, S.; Hornitschek, P.; Fankhauser, C. Light-regulated plant growth and development. *Curr. Top. Dev. Biol.* **2010**, *91*, 29–66. [CrossRef]
5. Cashmore, A.R.; Jarillo, J.A.; Wu, Y.-J.; Liu, D. Cryptochromes: Blue light receptors for plants and animals. *Science* **1999**, *284*, 760–765. [CrossRef]
6. Clack, T.; Mathews, S.; Sharrock, R.A. The phytochrome apoprotein family inArabidopsis is encoded by five genes: The sequences and expression of PHYD and PHYE. *Plant Mol. Biol.* **1994**, *25*, 413–427. [CrossRef]
7. Sharrock, R.A.; Quail, P.H. Novel phytochrome sequences in *Arabidopsis thaliana*: Structure, evolution, and differential expression of a plant regulatory photoreceptor family. *Genes Dev.* **1989**, *3*, 1745–1757. [CrossRef]
8. Wang, K.; Weng, Q.; Huang, L.; Wang, K.; Deng, J.; Jiang, R.; Ye, Z.; Gan, M. A new source of multi-spectral high spatial resolution night-time light imagery—JL1-3B. *Remote Sens. Environ.* **2018**, *215*, 300–312. [CrossRef]
9. Runkle, E.S.; Heins, R. Manipulating the light environment to control flowering and morphogenesis of herbaceous plants. *Acta Hortic.* **2006**, *711*, 51–60. [CrossRef]
10. Yamada, A.; Tanigawa, T.; Suyama, T.; Matsuno, T.; Kunitake, T. Night break treatment using different light sources promotes or delays growth and flowering of *Eustoma grandiflorum* (Raf.) Shinn. *J. Jpn. Soc. Hortic. Sci.* **2008**, *77*, 69–74. [CrossRef]
11. Park, Y.G.; Muneer, S.; Jeong, B.R. Morphogenesis, flowering, and gene expression of *Dendranthema grandiflorum* in response to shift in light quality of night interruption. *Int. J. Mol. Sci.* **2015**, *16*, 16497–16513. [CrossRef]
12. Appelgren, M. Effects of light quality on stem elongation of *Pelargonium* in vitro. *Sci. Hortic.* **1991**, *45*, 345–351. [CrossRef]
13. Dougher, T.A.; Bugbee, B. Long-term blue light effects on the histology of lettuce and soybean leaves and stems. *J. Am. Soc. Hortic. Sci.* **2004**, *129*, 467–472. [CrossRef]
14. Folta, K.M.; Lieg, E.J.; Durham, T.; Spalding, E.P. Primary inhibition of hypocotyl growth and phototropism depend differently on phototropin-mediated increases in cytoplasmic calcium induced by blue light. *Plant Physiol.* **2003**, *133*, 1464–1470. [CrossRef] [PubMed]
15. Kubota, C.; Rajapakse, N.C.; Young, R.E. Carbohydrate status and transplant quality of micropropagated broccoli plantlets stored under different light environments. *Postharvest Biol. Technol.* **1997**, *12*, 165–173. [CrossRef]

16. Rajapakse, N.C.; McMahon, M.J.; Kelly, J.W. End of day far-red light reverses height reduction of chrysanthemum induced by CuSO4 spectral filters. *Sci. Hortic.* **1993**, *53*, 249–259. [CrossRef]
17. Senger, H. The effect of blue light on plants and microorganisms. *Photochem. Photobiol.* **1982**, *35*, 911–920. [CrossRef]
18. Fankhauser, C.; Ulm, R. Light-regulated interactions with SPA proteins underlie cryptochrome-mediated gene expression. *Genes Dev.* **2011**, *25*, 1004–1009. [CrossRef]
19. Hirose, F.; Shinomura, T.; Tanabata, T.; Shimada, H.; Takano, M. Involvement of rice cryptochromes in de-etiolation responses and flowering. *Plant Cell Physiol.* **2006**, *47*, 915–925. [CrossRef]
20. Jeong, S.W.; Hogewoning, S.W.; Van Ieperen, W. Responses of supplemental blue light on flowering and stem extension growth of cut chrysanthemum. *Sci. Hortic.* **2014**, *165*, 69–74. [CrossRef]
21. Higuchi, Y.; Sumitomo, K.; Oda, A.; Shimizu, H.; Hisamatsu, T. Day light quality affects the night-break response in the short-day plant chrysanthemum, suggesting differential phytochrome-mediated regulation of flowering. *J. Plant Physiol.* **2012**, *169*, 1789–1796. [CrossRef] [PubMed]
22. Park, Y.G.; Jeong, B.R. Night interruption light quality changes morphogenesis, flowering, and gene expression in *Dendranthema grandiflorum*. *Hortic. Environ. Biotechnol.* **2019**, *60*, 167–173. [CrossRef]
23. Dere, S.; Gunes, T.; Sivaci, R. Spectrophotometric determination of chlorophyll—a, b and total carotenoid contents of some algae species using different solvents. *Turk. J. Bot.* **1998**, *22*, 13–17.
24. Livak, K.J.; Schmittgen, T.D. Analysis of relative gene expression data using real-time quantitative PCR and the $2^{-\Delta\Delta Ct}$ method. *Methods* **2001**, *25*, 402–408. [CrossRef] [PubMed]
25. Brown, C.S.; Schuerger, A.C.; Sager, J.C. Growth and photomorphogenesis of pepper plants under red light-emitting diodes with supplemental blue or far-red lighting. *J. Am. Soc. Hortic. Sci.* **1995**, *120*, 808–813. [CrossRef]
26. Hoenecke, M.; Bula, R.; Tibbitts, T. Importance of 'blue' photon levels for lettuce seedlings grown under red-light-emitting diodes. *HortScience* **1992**, *27*, 427–430. [CrossRef]
27. Holmes, M.G. Action spectra for changes in the "high irradiance reaction" in hypocotyls of Sinapis alba L. *Planta* **1981**, *153*, 267–272. [CrossRef]
28. Kim, S.-J.; Hahn, E.-J.; Heo, J.-W.; Ali, M.B. Effects of LEDs on net photosynthetic rate, growth and leaf stomata of chrysanthemum plantlets in vitro. *Sci. Hortic.* **2004**, *101*, 143–151. [CrossRef]
29. Oyaert, E.; Volckaert, E.; DeBergh, P. Growth of chrysanthemum under coloured plastic films with different light qualities and quantities. *Sci. Hortic.* **1999**, *79*, 195–205. [CrossRef]
30. Schuerger, A.C.; Brown, C.S.; Stryjewski, E.C. Anatomical features of pepper plants (*Capsicum annuum* L.) grown under red light-emitting diodes supplemented with blue or far-red light. *Ann. Bot.* **1997**, *79*, 273–282. [CrossRef]
31. Shimizu, H.; Ma, Z.; Tazawa, S.; Douzono, M.; Runkle, E.S.; Heins, R. Blue light inhibits stem elongation of chrysanthemum. *Acta Hortic.* **2006**, *711*, 363–368. [CrossRef]
32. Wheeler, R.M.; Mackowiak, C.L.; Sager, J.C. Soybean stem growth under high-pressure sodium with supplemental blue lighting. *Agron. J.* **1991**, *83*, 903–906. [CrossRef]
33. Khattak, A.M.; Pearson, S. Spectral filters and temperature effects on the growth and development of chrysanthemums under low light integral. *Plant Growth Regul.* **2006**, *49*, 61–68. [CrossRef]
34. Kong, Y.; Stasiak, M.; Dixon, M.A.; Zheng, Y. Blue light associated with low phytochrome activity can promote elongation growth as shade-avoidance response: A comparison with red light in four bedding plant species. *Environ. Exp. Bot.* **2018**, *155*, 345–359. [CrossRef]
35. Xu, Y.L.; Gage, D.A.; Zeevaart, J. Gibberellins and stem growth in *Arabidopsis thaliana*. Effects of photoperiod on expression of the GA4 and GA5 loci. *Plant Physiol.* **1997**, *114*, 1471–1476. [CrossRef]
36. Heo, J.; Lee, C.; Chakrabarty, D.; Paek, K.-Y. Growth responses of marigold and salvia bedding plants as affected by monochromic or mixture radiation provided by a light-emitting diode (LED). *Plant Growth Regul.* **2002**, *38*, 225–230. [CrossRef]
37. Muleo, R.; Morini, S. Light quality regulates shoot cluster growth and development of MM106 apple genotype in in vitro culture. *Sci. Hortic.* **2006**, *108*, 364–370. [CrossRef]
38. Hao, X.; Papadopoulos, A.P. Effects of supplemental lighting and cover materials on growth, photosynthesis, biomass partitioning, early yield and quality of greenhouse cucumber. *Sci. Hortic.* **1999**, *80*, 1–18. [CrossRef]

39. Wang, X.Y.; Xu, X.M.; Cui, J. The importance of blue light for leaf area expansion, development of photosynthetic apparatus, and chloroplast ultrastructure of Cucumis sativus grown under weak light. *Photosynthetica* **2015**, *53*, 213–222. [CrossRef]
40. Iacona, C.; Muleo, R. Light quality affects in vitro adventitious rooting and ex vitro performance of cherry rootstock Colt. *Sci. Hortic.* **2010**, *125*, 630–636. [CrossRef]
41. Dougher, T.A.; Bugbee, B.G. Difference in response of wheat, soybean and lettuce to reduced blue radiation. *J. Photochem. Photobiol.* **2011**, *73*, 199–207. [CrossRef]
42. Eskins, K. Light-quality effects on *Arabidopsis* development. Red, blue and far-red regulation and morphology. *Physiol. Plant.* **1992**, *86*, 439–444. [CrossRef]
43. Gang, Z.; Yu, T.; Banghua, Y.; Xiaolei, L. A study on the relationship between the chlorophyll content and the yield of plant of tartary buckwheat. In Proceedings of the 5th International Symposium on Buckwheat, Taiyuan, China, 20–26 August 1992; pp. 122–126.
44. Zhao, X.-Y.; Yu, X.-H.; Liu, X.; Lin, C. Light regulation of gibberellins metabolism in seedling development. *J. Integr. Plant Biol.* **2007**, *49*, 21–27. [CrossRef]
45. Mockler, T.; Yang, H.; Yu, X.; Parikh, D.; Cheng, Y.-C.; Dolan, S.; Lin, C. Regulation of photoperiodic flowering by *Arabidopsis* photoreceptors. *Proc. Natl. Acad. Sci. USA* **2003**, *100*, 2140–2145. [CrossRef] [PubMed]
46. Johnson, E.; Bradley, M.; Harberd, N.P.; Whitelam, G.C. Photoresponses of light-grown phyA mutants of *Arabidopsis* (Phytochrome A is required for the perception of daylength extensions). *Plant Physiol.* **1994**, *105*, 141–149. [CrossRef]
47. Mockler, T.C.; Guo, H.; Yang, H.; Lin, D.C. Antagonistic actions of *Arabidopsis* cryptochromes and phytochrome B in the regulation of floral induction. *Development* **1999**, *126*, 2073–2082.
48. Franklin, K.A.; Praekelt, U.; Stoddart, W.M.; Billingham, O.E.; Halliday, K.J.; Whitelam, G.C. Phytochromes B, D, and E act redundantly to control multiple physiological responses in *Arabidopsis*. *Plant Physiol.* **2003**, *131*, 1340–1346. [CrossRef]
49. Shinomura, T.; Nagatani, A.; Hanzawa, H.; Kubota, M.; Watanabe, M.; Furuya, M. Action spectra for phytochrome A- and B-specific photoinduction of seed germination in *Arabidopsis thaliana*. *Proc. Natl. Acad. Sci. USA* **1996**, *93*, 8129–8133. [CrossRef]
50. Usami, T.; Mochizuki, N.; Kondo, M.; Nishimura, M.; Nagatani, A. Cryptochromes and phytochromes synergistically regulate *Arabidopsis* root greening under blue light. *Plant Cell Physiol.* **2004**, *45*, 1798–1808. [CrossRef]
51. Ishikawa, R.; Shinomura, T.; Takano, M.; Shimamoto, K. Phytochrome dependent quantitative control of Hd3a transcription is the basis of the night break effect in rice flowering. *Genes Genet. Syst.* **2009**, *84*, 179–184. [CrossRef]

plants

MDPI

Article

Changes in Beneficial *C*-glycosylflavones and Policosanol Content in Wheat and Barley Sprouts Subjected to Differential LED Light Conditions

Muthusamy Muthusamy [1,†], Jong Hee Kim [1,2,†], Suk Hee Kim [1], Joo Yeol Kim [1], Jeong Wook Heo [3], HanGyeol Lee [4], Kwang-Sik Lee [4], Woo Duck Seo [4], Soyoung Park [1], Jin A Kim [1] and Soo In Lee [1,*]

1 Department of Agricultural Biotechnology, National Institute of Agricultural Sciences (NAS), RDA, Jeonju 54874, Korea; biotech.muthu@gmail.com (M.M.); gllmon@naver.com (J.H.K.); hbhb0@naver.com (S.H.K.); rlawnduf@korea.kr (J.Y.K.); psy0203@korea.kr (S.P.); jakim72@korea.kr (J.A.K.)
2 Division of Horticultural Biotechnology, Hankyung National University, Anseong 17579, Korea
3 Department of Agricultural Engineering, National Institute of Agricultural Sciences (NAS), RDA, Jeonju 54874, Korea; wooncho@korea.kr
4 Division of Crop Foundation, National Institute of Crop Science (NICS), RDA, Wanju 55365, Korea; gajae93@gmail.com (H.L.); kslee840118@gmail.com (K.-S.L.); swd2002@korea.kr (W.D.S.)
* Correspondence: silee@korea.kr; Tel.: +82-63-238-4618; Fax: +82-63-238-4604
† These authors contributed equally to this work.

Received: 24 September 2020; Accepted: 4 November 2020; Published: 6 November 2020

Abstract: The spectral quality and intensity of light, photoperiodism, and other environmental factors have profound impacts on the metabolic composition of light-dependent higher plants. Hence, we investigate the effects of fluorescent light (96 μmol m^{-2}s^{-1}) and white (100 μmol m^{-2}s^{-1}), blue (100 μmol m^{-2}s^{-1}), and red (93 μmol m^{-2}s^{-1}) light-emitting diode (LED) light irradiation on the *C*-glycosylflavone and policosanol contents in young seedlings of wheat and barley. Ultra-high-performance liquid chromatography (UHPLC) analyses of *C*-glycosylflavone contents in barley reveal that the saponarin content is significantly enhanced under blue LED light irradiation. Under similar conditions, isoorientin and isoschaftoside contents are improved in wheat seedlings. The contents of these *C*-glycosylflavones differed along with the light quality and growth period. The highest accumulation was observed in sprouts after three days under blue LED light irradiation. GC/MS analyses of policosanol contents showed that 1-hexacosanol (C26:o–OH) in barley and 1-octacosanol (C28:o–OH) in wheat seedlings were reduced under LED light irradiation, compared to seedlings under fluorescent light conditions. Nonetheless, the policosanol contents gradually improved with the extension of growth times and treatments, irrespective of the light quality. Additionally, a positive correlation was observed between the expression pattern of biosynthesis-related genes and the respective metabolite content in barley. This study demonstrates that blue LED light irradiation is useful in maximizing the *C*-glycosylflavone content in barley and wheat sprouts.

Keywords: saponarin; isoorientin; hexacosanol; octacosanol; fatty acyl-coenzyme A reductase (FAR)

1. Introduction

Wheat (*Triticum aestivum* L.) is one of the staple food grains for approximately 40% of the global population [1]. It ranks third in terms of global production and its nutritional importance in the human diet has long been investigated [2]. Similarly, barley (*Hordeum vulgare* L.) is the fourth most important cereal crop for both humans and animals worldwide, having the highest dietary fiber content [3,4]. Barley is rich in biologically active molecules/metabolites, which are essential for plants. These metabolites have the potential to exhibit health benefits in the human diet. Barely or its extracts have shown powerful antioxidant effects as dietary supplements for humans. These antioxidant effects are mainly attributable to the presence of saponarin, lutonarin, and hexacosanol molecules [3]. Barley grass also possesses numerous other phytonutrients, including gamma-aminobutyric acid (GABA), flavonoids, proteins, minerals, pigments, vitamins (A, B1, C, and E), dietary fiber, polysaccharides, alkaloids, and polyphenols [4]. Recent reports discussing the broad therapeutic roles of functional ingredients or derived components of barley suggest that it may be the best fit in the modern human diet as a functional food [5–7]. Barley saponarin has several health benefits, including anti-inflammatory response [4,8], prevention of bacterial infections [9], regulation of glucose homeostasis, insulin sensitivity [10], reducing low-density lipoprotein (LDL) cholesterol [4], and anti-carcinogenic responses [11]. Similarly, isoorientin from wheat acts as an anti-cancer compound [12] and also possesses anti-inflammatory, antibacterial, antiviral, antiplatelet [13], and antioxidant activities [14]. Wheat (or its derived products) also possesses several beneficial bioactive molecules, including pelargonidin and cyanidin derivatives [15], essential amino acids, fatty acids, flavonoids (e.g., rutin, quercetin, and catechin), vitamin C [16], and policosanols [17]. The *C*-glycosylflavone and policosanol content in barley vary with growth duration [18]. Reports have claimed that sprouts produce higher concentrations of health-promoting molecules than grains [19].

Sprouting implies a series of active biochemical, metabolic, and physiological processes, resulting in the release of active nutrients (e.g., free amino acids and lipid catabolism) for growing plant tissues [18,19]. These metabolites often possess potential health benefits for humans [18,20]. Thus, sprouting is considered one of the easiest natural strategies to enhance nutritional profiles with healthy attributes [20]. Owing to their nutritive values, sprouting seeds has recently received growing interest. Meanwhile, researchers have attempted to identify the presence of novel functional ingredients of sprouts under varying growth and environmental conditions [18,19]. Plants increase the production of a variety of metabolites, in order to mitigate the effects of adverse environmental factors, such as drought [21], salinity [22], high-intensity light or artificial lighting [23], temperature [24], and elevated CO_2 levels [25]. Therefore, effective management and/or the controlled application of physical energy forms (e.g., light, temperature, and water) may serve as a viable option to enhance the accumulation of health-promoting compounds in sprouts, which has been shown to be successful in previous attempts [25–27]. In terms of physical energy forms, light irradiation has been employed in different sprouting seeds, in order to increase metabolites with health-promoting benefits [23,28]. The availability of artificial lighting resources (e.g., light-emitting diodes (LEDs)) renders the possibility of studying the effects of specific light on the concentrations of biologically important metabolites in plants. Herein, the potential effects of a fluorescent lamp (FL) and different spectra of LED light irradiations (white, blue, and red) on beneficial metabolite content were investigated in barley and wheat sprouts. Additionally, we attempted to identify and profile the expression patterns of metabolite biosynthesis-related genes of barley sprouts. This study facilitates understanding of the differential responses relating to *C*-glycosylflavones and policosanols in wheat and barley sprouts.

2. Results

2.1. Changes of *C*-glycosylflavone Content in Barley and Wheat Seedlings Exposed to Differential LED Light Irradiation

The changes in saponarin (barley), isoorientin, and isoschaftoside (wheat) content in sprouts treated with different light qualities (FL and white, blue, or red LED irradiation) were measured using ultra-high-performance liquid chromatography (UHPLC) in plant materials harvested after 3–9 days of treatment. The UHPLC results of barley and wheat sprouts revealed that blue LED light irradiation increased the *C*-glycosylflavone content more than other light conditions. The LED irradiation differentially influenced the saponarin content in the barley sprouts. In comparison with FL, LED light irradiation significantly altered the content: Blue light irradiation prominently improved the saponarin content (51.7%–57.7%) across all growth stages and irradiation periods (Figure 1A). The highest concentration of saponarin was observed in sprouts after 3 days of blue LED light irradiation. Interestingly, sprouts treated with red LED light irradiation demonstrated a significant reduction in saponarin content, compared to their respective controls, after 3, 5, 7, and 9 days of irradiation. Conversely, white LED irradiation for three consecutive days did not alter the content; however, on day 5, it statistically significantly increased the contents. The extension of white LED treatment for 7 or 9 days resulted in a reduction of saponarin content. Moreover, regardless of the lighting resource or quality, a consistent reduction in saponarin content was observed in the growing sprouts. Among the growth times and light qualities tested in this study, the highest content of saponarin was observed in blue LED irradiated sprouts on the 3rd day, while red LED radiation remarkably reduced the content in all treatments and sprout growth periods.

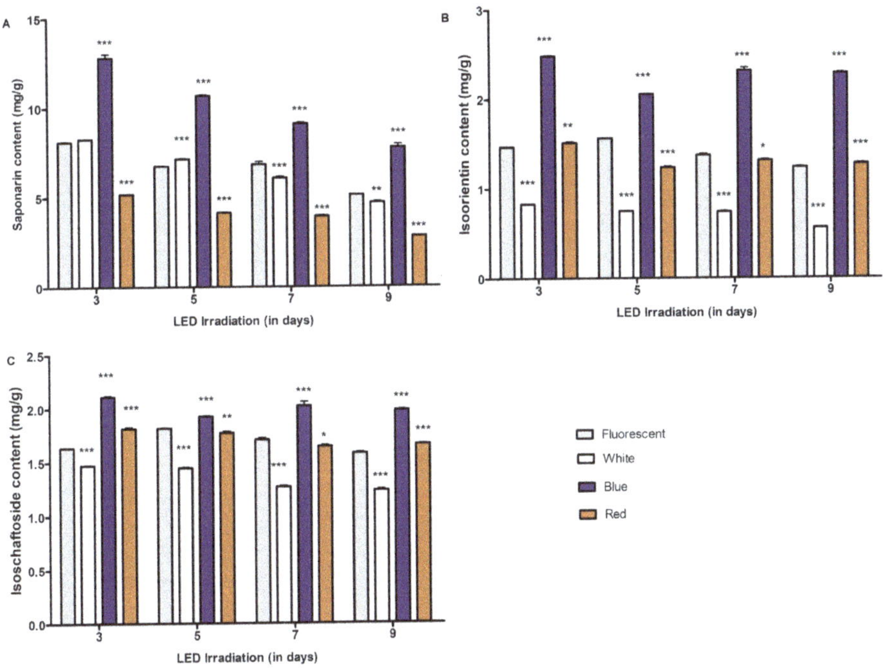

Figure 1. *C*-glycosylflavone content in young barley and wheat seedlings subjected to differential light qualities. (**A**) represents the saponarin content of barley sprouts (mg/g dry weight (DW)) under different light and growth periods, while (**B**,**C**) represent the isoorientin and isoschaftoside contents (mg/g DW), respectively, of wheat sprouts. * ($p < 0.05$), ** ($p < 0.001$), and *** ($p < 0.0001$) indicate the statistical significance.

Isoschaftoside and isoorientin are the major flavone-C-glycosides (C-glycosylflavones) frequently reported in wheat and its derived products [29]. In this study, we found that these metabolites were significantly altered in wheat sprouts exposed to LED light irradiation over 3–9 days (Figure 1B,C). LED light irradiation significantly altered the metabolite concentration in barley and wheat sprouts, compared to their content in sprouts treated with traditional fluorescent lamp light conditions. Blue LED light irradiation markedly improved the concentration of isoschaftoside as well as isoorientin, compared to control (FL) or other (white and red) LED treatments. The highest mean concentrations of 2.1 and 2.47 mg (per g dry weight (DW)) of isoschaftoside and isoorientin, respectively, were observed in seedlings subjected to 3 days of blue LED light irradiation. Under similar conditions, the seedlings treated with FL accumulated 1.63 mg and 1.46 mg (per g DW) of isoschaftoside and isoorientin, respectively (Figure 1B,C). On day 3, a slight improvement in isoschaftoside (11.04%) and isoorientin (2.78%) content was also observed in red LED light irradiated sprouts. However, 5 and 7 days of red LED light irradiation significantly reduced the isoorientin and isoschaftoside contents in sprouts. Compared to FL, a maximum of 21.15% reduction in isoorientin content was observed after five days of red LED light irradiation. Under similar conditions, a 2.25% reduction was noted for isoschaftoside content (Figure 1C). Conversely, 9 days of red light irradiation increased the isoorientin (3.15%) and isoschaftoside (5.06%) contents in sprouts. Compared to FL, white LED light irradiation led to reductions in the isoschaftoside (9.81–22.1%) and isoorientin (43.84–55.3%) levels across all growth times. In terms of sprout growth periods, the highest accumulation of metabolites was observed after 3 days of light treatment (Figure 1 and Figure S1).

2.2. Effect of LED Light Irradiation on Major Long-Chain Fatty-Alcohol (Policosanol) Biosynthesis in Barley and Wheat Seedlings

Crude extracts comprised of policosanols from barley and wheat sprouts grown under FL, white, blue, and red LED light irradiations for 3–9 days were profiled and quantified by gas chromatography coupled with mass spectroscopy (GC/MS). GC/MS results of policosanol content in barley and wheat sprouts indicated that growth time was an influential factor in determining the hexacosanol and octacosanol content in these sprouts. Hexacosanol, a major constituent of the policosanol profile of barley seedlings, was altered during LED light irradiation in accordance with sprout growth periods. In general, the hexacosanol content (μg/g DW) in LED light treated samples was significantly lower than that in seedlings treated for 3 or 9 days with FL (Figure 2A). The comparative analysis of hexacosanol content between FL and LED irradiation at the third day of treatment showed that hexacosanol content was significantly reduced for white and blue LED light irradiations. Under these conditions, a maximum reduction of 43.3% was noted for blue LED light irradiation, while white light treatment showed a reduction of 36.9% in hexacosanol content. Conversely, on days 5 and 7, white and blue LED light irradiation did not alter the hexacosanol concentration significantly, in comparison with that of control sprouts. Altogether, the results indicated that the changes in hexacosanol content were not consistent with light qualities, indicating the light quality is not the only factor influencing changes in hexacosanol biosynthesis. Interestingly, hexacosanol content in sprouts treated with either white or blue LED irradiation for 5, 7, and 9 days was higher than the hexacosanol content in sprouts treated only for three days under similar conditions. Likewise, the octacosanol content (μg/g DW) in wheat seedlings was analyzed using GC/MS (Figure S2). The results showed that the octacosanol content was altered irregularly by different LED light qualities (Figure 2B). A comparative analysis of FL and LED treatments indicated that blue LED light irradiation for 3 and 9 days enhanced the octacosanol content in wheat sprouts. In comparison with FL, white light irradiation also increased the content in sprouts exclusively on day 3. Conversely, red LED irradiation resulted in octacosanol content reduction (on day 5). Nonetheless, the magnitude of change in content was not consistent for any of the LED treatments.

2.3. Effect of Growth Periods on C-glycosylflavones and Policosanol Content in Young Barley and Wheat Seedlings

To understand the effect of sprout growth periods (3–9 days) on the metabolite content (saponarin, isoorientin, isoschaftoside, hexacosanol, and octacosanol) under fluorescent and LED light irradiation, we conducted a further investigation. Of all the growth times, the saponarin content of barley and isoorientin/isoschaftoside contents of wheat seedlings were found to be the highest on the 3rd day of the treatment (Figure S3A–C). The contents of these *C*-glycosylflavones showed a declining trend with an increase in growth times (i.e., after 5, 7, and 9 days) in barley and wheat seedlings (Figure S3A–C). However, the magnitude of reduction in metabolite content varied with different light qualities and metabolites in these sprouts. The saponarin content of barley was reduced by 37.96%, 43.58%, 39.43%, and 45.93% for FL, white, blue, and red LED light irradiation, respectively, after the further growth period of 9 days. A statistically significant negative correlation (−0.93 to −0.99) between saponarin content and sprout growth period was observed during FL and LED light (white, blue, and red) irradiation (Table 1). A similar attempt to establish relationships between isoorientin content and growth periods in wheat sprouts under FL and LED treatment showed that isoorientin accumulation was negatively correlated with growth periods under specific light (FL, white, and red) treatments. Furthermore, isoschaftoside accumulation is negatively regulated under white and red LED light treatments across growth periods. Nonetheless, their relationships in blue LED irradiated sprouts were inconclusive.

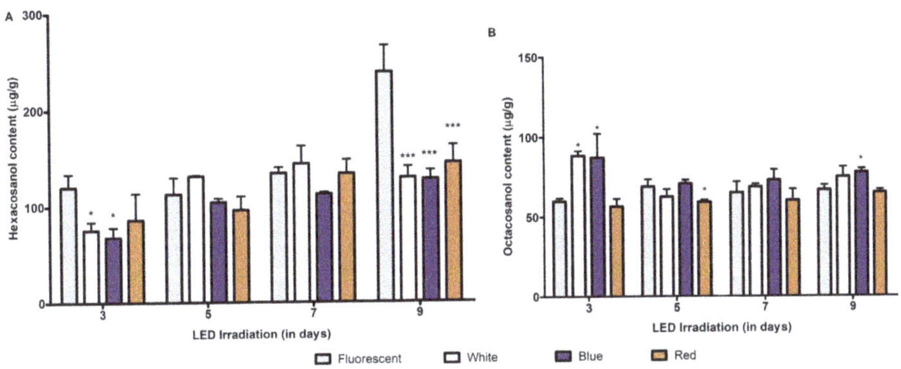

Figure 2. Policosanol content (µg/g DW) in barley and wheat seedlings during different growth periods and light conditions. (**A**) denotes hexacosanol (major policosanol) content in barley sprouts, whereas (**B**) denotes octacosanol content in wheat sprouts. * ($p < 0.05$), and *** ($p < 0.0001$) indicate the statistical significance.

Table 1. Analysis of correlation statistics (Pearson correlation) among metabolite content, growth periods, and their respective biosynthesis-related gene expression patterns across different light qualities.

	Metabolite Content vs. Gene Expression							
	HvOGT1	*HvFNSII*	*HvCHS1*	*HvFAR2*	*HvFAR3*	*HvFAR4*	*HvFAR5*	*HvFAR6*
Saponarin	0.68 (p = 0.01) **	−0.53 (p = 0.07)	0.13 (p = 0.68)	–	–	–	–	–
Hexacosanol	–	–	–	−0.59 (p = 0.04) *	0.67 (p= 0.01) **	−0.62 (p = 0.03) *	−0.76 (p = 0.004) **	−0.69 (p = 0.01) **

	Growth periods (–9 days) vs. metabolite content			
	Fluorescent	White	Blue	Red
Saponarin	−0.93 (p = 0.00001) ****	−0.99 (p = 0.00001) ****	−0.99 (p = 0.00001) ****	−0.97 (p = 0.00001) ****
Isoorientin	−0.82 (p = 0.001) ***	−0.92 (p = 0.00002) ****	−0.23 (p = 0.47)	−0.66 (p = 0.01) **
Isoschaftoside	−0.32 (p = 0.31)	−0.94 (p = 0.00001) ****	−0.47 (p = 0.12)	−0.90 (p = 0.00006) ****
Hexacosanol	0.79 (p = 0.001) ***	0.69 (p = 0.01) **	0.91 (p = 0.00004) ****	0.81 (p = 0.001) ***
Octacosanol	−0.07 (p = 0.82)	−0.43 (p = 0.16)	−0.17 (p = 0.57)	−0.04 (p = 0.90)

* $p < 0.05$, ** $p < 0.001$, *** $p < 0.0001$, **** $p < 0.00001$ indicates statistical significance (replicates were included for measuring Pearson correlation).

In contrast to C-glycosylflavone content, the hexacosanol content in barley sprouts gradually increased with the extension of the growth period. In fact, the highest accumulation of hexacosanol was observed in barley sprouts after 9 days of FL and LED light (white, blue, and red) irradiation treatments (Figure S3D). The correlation coefficient (Pearson) analysis showed that positive correlations (0.69 to 0.91) existed between hexacosanol content and growth period, which was statistically significant in both FL and LED irradiated seedlings (Table 1). Unlike hexacosanol, the positive relationship between octacosanol content and wheat growth period was not clear (Figure S3E). As indicated in Figure S3E, the growth period did not impose any significant effect on octacosanol content under white, red, and blue LED or FL light irradiation. Nonetheless, a comparison of growth periods under white LED irradiation showed significant differences in octacosanol concentration in wheat sprouts.

2.4. Expression Profiling of Potential Genes Involved in the Biosynthesis of C-glycosylflavones/Flavonoids and Policosanols in Barley Seedlings under Different Light Conditions and Growth Times

The homology-based gene identification approach was used to predict the potential genes associated with the biosynthesis of flavones/flavonoids and long-chain fatty alcohols in the barley genome. The mRNA sequences of potential candidate genes were designated according to their highest sequence similarity to known genes of other crops (Table S1). As shown in Figure S3, sprout growth periods were one of the factors that greatly influenced the saponarin and hexacosanol contents; thus, we selected the growth periods associated with high accumulation of these metabolites for investigation of expression analyses of flavonoid and fatty alcohol biosynthesis-related genes. The expression levels of *HvCHS1*, *HvFNSII*, and *HvOGT1* for flavonoid biosynthesis and *HvFAR2*, *HvFAR3*, *HvFAR4*, *HvFAR5*, and *HvFAR6* for policosanol biosynthesis were significantly altered under FL, white, blue, and red light irradiation (Figure 3A–H). The expression pattern of *HvOGT1* under differential light conditions was positively correlated (0.68) with saponarin content in barley (Table 1). In terms of light responses, the expression of *FNSII* was significantly upregulated in red LED irradiated seedlings, while its expression was downregulated by white LED light, in comparison with FL or blue LED light. The light-responsive expression of *CHS1* was slightly changed under LED irradiation, but the changes were negligible (Figure 3C). In terms of policosanol biosynthesis-related gene expression, the expression pattern of *HvFAR3* had a positive correspondence (0.67) with hexacosanol accumulation (Table 1). The *HvFAR3* expression was reduced under white and blue LED light irradiation, while it was unaltered in red LED light irradiated seedlings. Interestingly, the expression patterns of *HvFAR2*, *HvFAR4*, *HvFAR5*, and *HvFAR6* were negatively correlated (−0.59 to −0.76) with hexacosanol content in barley sprouts. Their expressions were upregulated by one or more LED light irradiations. *HvFAR2* and *HvFAR5* showed their highest expression under red LED light irradiation (Figure 3D,F–H). Likewise, the highest expression of *HvFAR6* was observed under blue LED irradiation. It is clear that *HvFAR3* is possibly involved in hexacosanol biosynthesis in barley sprouts.

2.5. Influence of LED-Light Irradiation on Seedling Growth

In most of the LED treatments in wheat seedlings, there were no statistically significant differences in leaf length, compared to control (FL) seedlings (Figure 4). However, the root growth of wheat sprouts treated with LED light irradiation for 3 days was significantly reduced, compared to that of the controls. Interestingly, further LED treatment on 5 and 7 days did not produce a significant impact on root growth, except under blue LED irradiation on day 7. Nonetheless, continuing white and red LED irradiation up to 9 days reduced root growth, which was statistically significant. In terms of light quality, the highest reduction in root growth was observed under white LED irradiation. Like wheat sprouts, the root growth in barley sprouts was altered by specific light qualities, in accordance with the growth period. Red LED irradiation consistently reduced barley root growth across all growth periods. White or blue LED irradiation also resulted in root growth reduction in one or more growth periods. Although initial treatment with red LED irradiation (on day 3) reduced leaf growth in barley sprouts, the extending the irradiation for 7 or 9 days had a positive impact on barley leaf growth. LED treatment in wheat sprouts showed that the leaf growth was mostly not altered.

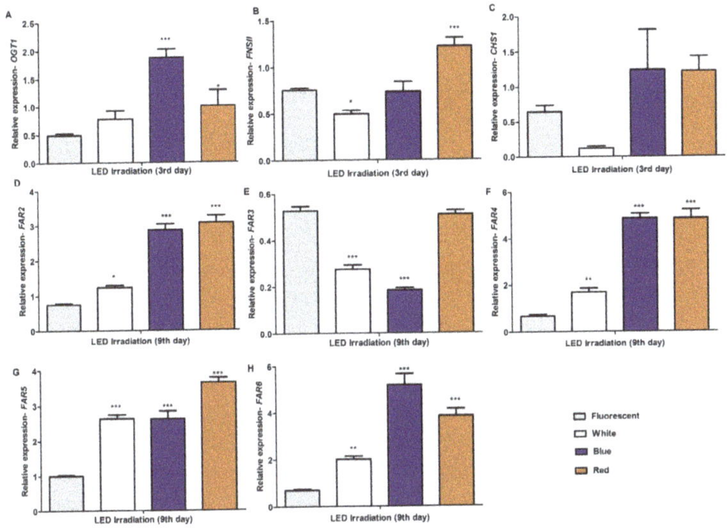

Figure 3. Relative quantification of expression changes in flavonoid and policosanol biosynthesis-related genes in barley sprouts. (**A–C**) represent the relative expression levels of *UDP-Glc: Isovitexin 7-O-glucosyltransferase 1 (OGT1)*, *flavone synthase II (FNSII)*, and *chalcone synthase 1 (CHS1)*, respectively. Likewise, (**D–H**) represent the expression patterns of different classes of *fatty acyl-coenzyme A reductase (FAR)* genes. The results represent the qRT-PCR-based relative quantification of genes in barley sprouts exposed to fluorescent and LED (white, blue, and red) light irradiations. The gene expression was normalized using the internal control *HvActin*. * ($p < 0.05$), ** ($p < 0.001$), and *** ($p < 0.0001$) indicate the statistical significance.

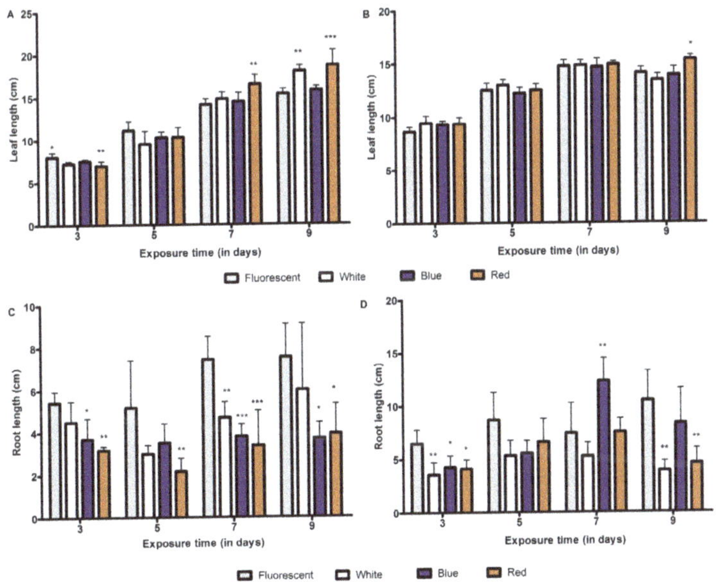

Figure 4. Leaf and root growth parameters of fluorescent and light-emitting diode (LED) light (white, blue, and red) irradiated barley and wheat seedlings at different growth periods. (**A**,**C**) represent barley growth parameters, while (**B**,**D**) represent the growth parameters of wheat sprouts. * ($p < 0.05$), ** ($p < 0.001$), and *** ($p < 0.0001$) indicate the statistical significance.

3. Discussion

Phytochromes and cryptochromes are the specialized photoreceptors of plants that sense the spectral quality and quantity, transducing the light signal to regulate genes responsible for secondary metabolite production [30]. Therefore, it is possible to determine the metabolic composition or to enhance the nutritional functionality of the target crops through selective application of light resources and photoperiodism. The application of LEDs for special metabolite production is considered promising, where it has been shown that the metabolic profiles also depend on several other factors, including crop genetics [18,26,31]. The increasing application of LED irradiation sources for the development of designer foods/functional foods may revolutionize the food industry. In this study, we attempted to enhance the *C*-glycosylated flavones/flavonoids and policosanol contents in barley and wheat sprouts using varying light qualities.

C-glycosylated flavones constitute the major portion of flavonoids found in barley seedlings [32]. Saponarin (isovitexin-7-O-glucoside) is a major *C*-glycosylated flavone, which is naturally present in young barley seedlings [11]. Among cellular organelles, saponarin is efficiently stored in vacuoles [33] and high accumulation is typically observed in primary leaves [4]. Similarly, isoorientin and isoschaftoside are the major *C*-glycosylated flavones often reported in wheat seedlings [29]. These *C*-glycosylflavones have potential roles as beneficial flavonoids in the human diet [4,8,34]. In this study, we found that the *C*-glycosylflavone (saponarin, isoorientin, and isoschaftoside) content was high in the early growth stages of seedlings, where the maximum accumulation was induced by blue LED light irradiation. We found an inverse relationship between *C*-glycosylated flavone content and growth times, indicating that the *C*-glycosylflavone content remains high in young sprouts. In a previous study investigating barley sprouts (13–56 days post-sprouting), it was stated that the saponarin content continued to decline with increasing growth periods [9]. Now, it is clear that the saponarin content in barley sprouts starts declining just three days post-sprouting. In terms of light quality, blue LED light had a positive impact on saponarin content. Blue LED light showed the highest accumulation (57.7% and 68.68% than that in the control, respectively) in barley and wheat sprouts. On the other hand, the impact of red LED light irradiation on saponarin content in barley, and isoorientin and isoschaftoside in wheat sprouts, seemed to differ. In barley, the saponarin content was reduced, while the levels of isoorientin and isoschaftoside were increased in wheat sprouts, suggesting that the effect of red LED light is specific to metabolites and/or crops.

Policosanol (PC) is another beneficial metabolite in the human diet, which is frequently found in cuticular waxes in primary leaves of young cereal sprouts. It represents a mixture of long-chain fatty alcohols (20–36 carbon) mostly comprised of docosanol (C22), tetracosanol (C24), hexacosanol (C26), octacosanol (C28), and triacontanol (C30) [35,36]. Octacosanol and policosanol (long-chain saturated fatty alcohols) are useful in preventing high-fat diet-induced obesity [37]. Owing to its importance in lowering blood cholesterol and protection from platelet aggregation, it has been commercialized in the health industry for a long time [35]. Among the PCs, hexacosanol (C26) and octacosanol (C28) are often observed in barley and wheat sprouts [36,38]. LED light irradiation showed a differential influence on the policosanol content in barley and wheat sprouts, suggesting that the LED response is likely specific to either metabolites or crops. Compared to FL conditions, LED light irradiation reduced the hexacosanol content in barley, while a similar condition in wheat sprouts showed an irregular pattern of octacosanol accumulation; suggesting that factors other than light quality also influence its content in wheat sprouts. Interestingly, in most cases, the policosanol content in barley and wheat sprouts gradually increased with the extension of growth time. Statistical analyses confirmed that the saponarin content under LED treatment was negatively correlated with barley growth periods. A similar negative relationship between content and growth period was also evident in wheat sprouts, albeit restricted to white and/or red LED light treatments. Unlike *C*-glycosylflavone, the hexacosanol content in sprouts appeared to have a positive correlation with sprout growth periods, suggesting the importance of growth level in the determination of policosanol content in barley. Our study corroborates previous findings which have reported a positive correspondence between hexacosanol and growth periods in barley [36]. However, a similar correlation pattern was not observed between octacosanol content and growth periods in FL, white, and blue LED irradiated wheat sprouts, possibly suggesting that this relationship might be specific to crop genetics. Altogether, our results show that light qualities and growth periods are two crucial factors in determining the C-glycosylflavone and policosanol contents in barley and wheat sprouts. In addition, light conditions are important parameters in determining the photomorphogenesis of plants [39]. Herein, we found that red LED light irradiation mostly reduced the root growth of sprouts, while white and blue LED light mediated root growth inconsistently across growth periods, indicating the role of other factors regulating growth parameters. The LED light responses of leaves of young seedlings also indicated that LED irradiation does not induce a regular growth pattern for leaves in barley and wheat sprouts. At this stage, it is difficult to conclude the growth impact of LED light irradiation, as this study utilized low intensity LED spectra. Further studies concerning the selection of optimum light intensity and LED spectra for enhancing sprout growth are, therefore, essential.

An interesting observation is that comparison of the metabolite accumulation trends revealed the C-glycosylflavones and policosanol contents to have an inverse relationship in barley sprouts. It is unknown whether this is due to a balancing act of metabolic pathways or just an influencing act of sprout growth periods. It warrants further in-depth studies, in order to understand the underlying mechanisms of metabolite biosynthesis. The molecular basis for the specialized metabolite accumulation to a particular physiological condition is mainly due to changes in the expression pattern of one or more specialized biosynthetic genes. Lee et al. [11] suggested that the expression pattern of UDP-Glc: Isovitexin 7-O-glucosyltransferase (OGT) is likely responsible for saponarin biosynthesis in barley. *OGT* is responsible for the conversion of isovitexin to saponarin in barley [33]. There are two classes of OGT (*OGT1* and *OGT2*) found in the barley genome, where a change in the expression level of *OGT1* has been associated with saponarin concentration [11]. In the present study, we found that a positive correlation exists between the *HvOGT1* expression pattern and saponarin accumulation, suggesting the possibility of *OGT1* involvement in saponarin biosynthesis. Both metabolite accumulation and *HvOGT1* expression were found to be highest after 3 days of treatment. It is clear that blue LED light irradiation accumulated saponarin by upregulating *HvOGT1* expression in barley sprouts. The flavonoid biosynthesis pathway gene, *HvCHS1*, is not linked with saponarin accumulation, indicating the possibility of a specific pathway controlling saponarin biosynthesis in barley. Studies on other crops have shown that fatty acyl-coenzyme A reductases (FAR) are involved in long-chain primary alcohol biosynthesis, part of the cuticular waxes found in leaf surfaces [40]. Studies have also shown that the number of *FAR* genes and their functions may vary, according to species-specific genetics. In other crops, it is evident that *FAR2* and *FAR3* are responsible for the biosynthesis of C_{26} and C_{28} primary alcohols [41]. Another report has claimed that at least five *FARs* are responsible for primary alcohol (C_{16} to C_{28}) biosynthesis in *Aegilops tauschii* [42]. Until recently, there has been no information about *HvFARs* and their potential role in hexacosanol biosynthesis. Identification of the *FARs* responsible for policosanol biosynthesis is inevitable for tailoring metabolic pathways towards enhanced production. Hence, we used the homology-based gene identification method to predict the gene sequences of *HvFAR2*, *HvFAR3*, *HvFAR4*, *HvFAR5*, and *HvFAR6* from available barley genome information. We also measured their expression changes using quantitative PCR during differential light treatments and growth periods, in order to infer their relationship to hexacosanol biosynthesis. Of all the *HvFARs* analyzed in this study, the expression changes of *HvFAR3* were positively associated with hexacosanol accumulation, suggesting their involvement in hexacosanol biosynthesis. Other *HvFARs* did not have a positive correlation with hexacosanol content, suggesting that they may not be associated with its biosynthesis. In this study, we identified the potential candidate genes involved in saponarin and hexacosanol biosynthesis. These genes can be effectively used to enhance the metabolite concentration by means of genetic manipulation. We also provided an expression atlas of *HvFARs* during LED light irradiation treatment in sprouts, which may be useful in future studies associated with other policosanol biosynthesis routes. A system-wide identification and characterization would add more information about the genetic factors controlling metabolite biosynthesis in barley sprouts.

To conclude, we showed that light qualities and growth times are crucial factors determining the contents of *C*-glycosylflavones and policosanols in barley and wheat sprouts. Blue LED light may be useful for increasing the *C*-glycosylflavone contents in cereal sprouts. Regardless of the light quality, management of growth time of sprouts is essential for policosanol content. This study will help to maximize the beneficial flavonoids and policosanol contents in cereal sprouts through LED applications in the future.

4. Materials and Methods

4.1. Plant Materials, Growth Condition, and LED Treatment

The barley (*Hordeum vulgare* L. variety "Keunalbori No.1") and wheat (*Triticum asetivum* L. variety "Baekchalmil") seeds used by this study were sourced from the National Institute of Crop Science (Korea). The seeds at full maturity were separated by soaking thrice in tap water. Then, visually healthy seeds were once again soaked separately in distilled water for one day. For germination, seeds were sown on 16 plastic sprout cultivating trays (90 cm by 30 cm; ~180 g per tray) and maintained in greenhouse conditions (25 °C, 16 h photoperiod) for one day. Four trays with sprouting seeds (per treatment) were moved to individual growth chambers (Mokmin Co., Ltd, Suwon, South Korea) equipped with FL or different spectra LED light irradiations (470 nm for blue, 380 nm for white, and 660 nm for red) over 9 days (Figure 5). Sprouts were watered at regular intervals. The photosynthetic photon flux density (PPFD) of FL (96 µmol $m^{-2}s^{-1}$), blue (100 µmol $m^{-2}s^{-1}$), white (100 µmol $m^{-2}s^{-1}$), and red (93 µmol $m^{-2}s^{-1}$) lights was measured at plant level by using a quantum meter (Apogee Instruments, Logan, USA). Fresh leaves and coleoptiles of each treatment were harvested simultaneously at 3, 5, 7, and 9 days after irradiation (DAI). All samples were frozen in liquid nitrogen and stored at −80 °C for RNA extraction and metabolomic analyses. All sprouting seeds throughout the treatments were maintained in a growth chamber with 16 h photoperiods, 60–80% humidity, and 22–25 °C.

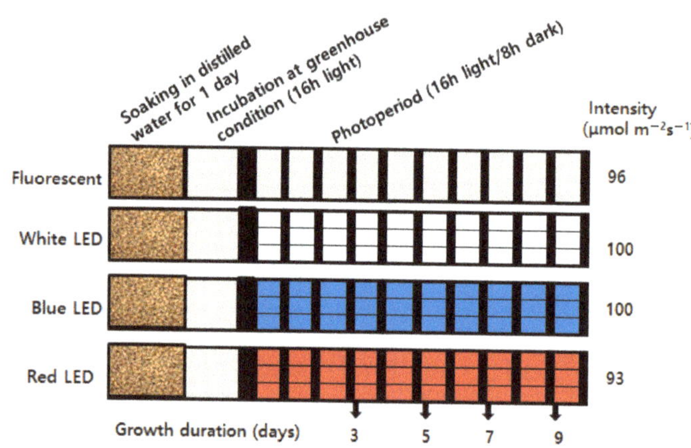

Figure 5. Schematic representation of experimental design used for LED treatment on barley and wheat seedlings. The diagram illustrates the germination method, growth period, and light quality and intensity.

4.2. Extraction of C-glycosylated Flavones and Measurement

The extraction method of Lee et al. [11] was used for the preparation of crude extracts from dried barley and wheat seedlings. An ultra-high-performance liquid chromatograph equipped with UV detectors (Dionex Ultimate 3000; Thermo Scientific, Waltham, MA, USA) and reversed-phase

HPLC column (ACQUITY UPLC BEH C18, 2.1 mm by 100 mm) was utilized for separation and quantitative analyses [8]. One gram of freeze-dried and chopped seedlings from each treatment was extracted by treating with either 50% ethanol (barley) or 100% methanol (wheat) on a shaker for 24 h at room temperature. The ethanolic or methanolic extracts were then filtered through 0.2 μm syringe filters. Following evaporation under vacuum, the extracts were dissolved in 10% dimethyl sulfoxide (DMSO) containing 50% ethanol and 1.3 μL was injected into the column for separation and detection at 325 nm. The mobile phase was comprised of trifluoroacetic acid (TFA) (0.1%) in water (A) and acetonitrile (B) with a flow rate of 0.5 mL/min, which was applied for separation of the analyte. Saponarin, isoorientin, and isoschaftoside molecules were identified by comparing the retention times to those of standards (obtained from Extrasynthese, Lyon, France and NICS, Jeonju, Korea). A standard calibration curve was prepared by plotting the peak area (y) of the chromatogram and the respective concentrations (31.25, 62.5, 125, 250, 500, and 1000 μg/mL) (x). The equations of the calibration curves for saponarin, isoorientin, and isoschaftoside were $y = 0.0876x + 0.3514$ ($r^2 = 0.999$), $y = 0.0917x - 0.5536$ ($r^2 = 0.998$), and $y = 0.0.0547x + 0.22935$ ($r^2 = 0.999$), respectively. The metabolite content (mg/g DW) from three technical replicates of biologically independent samples were used as input for the statistical analyses.

4.3. Extraction and Quantification of Policosanols from Barley and Wheat Seedlings

The preparation of barley and wheat crude extracts, policosanol standards, GC/MS parameters, and the quantification method of GC/MS as described elsewhere [17,36], was used in this study. Briefly, 1 g of freeze-dried and chopped samples collected from all treatments was extracted separately into 10 mL of hexane on a shaker for 24 h at room temperature. The supernatant of the mixture was collected by centrifugation at 3000 g for 3 min, filtered through a syringe filter with a pore size of 0.45 μm (Whatman Inc., Maidstone, UK), and kept under vacuum conditions until the hexane was completely removed. To the final extract, 250 μL of N-Methyl-N-trimethylsilyfluoroacetamide (MSTFA) and 0.5 mL of chloroform was added and stirred for 15 min at 60 °C. Chloroform was added to make up one ml of sample and, then, one μL was injected into the gas chromatograph using an auto sampler with a split ratio of 1:5. The GC was equipped with an HP-5MS UI (diphenyl 5%-dimethylsiloxane 95% co-polymer) capillary column (30 m by 0.25 μm by 0.25 μm film thickness) and a 5977A series mass spectroscopy (Agilent Technologies, Palo Alto, CA). The policosanol content was quantified according to the methods of Ra et al. [17]. The policosanol standards—Eicosanol (C_{20}), heneicosanol (C_{21}), docosanol (C_{22}), tricosanol (C_{23}), tetracosanol (C_{24}), hexacosanol (C_{26}), heptacosanol (C_{27}), octacosanol (C_{28}), and triacontanol (C_{30})—hexane, and chloroform solutions were purchased from Sigma (Sigma-Aldrich, St. Louis, MO).

4.4. Identification of Saponarin and Policosanol Biosynthetic-Related Genes

In general, flavone/flavonoid and long-chain fatty alcohol (or primary alcohol) biosynthesis pathways comprised of one or more enzymes and their information in barley are limited. Hence, the gene names and sequence information of other crops were searched for sequence homologies to unearth orthologous transcripts (full and partial) in *Hordeum vulgare* and *Triticum asetivum* species. The gene sequences were retrieved from NCBI Gene [43], PlantsDB [44], and GrainGenes [45] databases. For annotation, potential orthologous gene sequences were searched for the presence of conserved domains using the NCBI Conserved Domains [46] database. Transcripts encoding for flavone/flavonoid-related—UDP-Glc: *OGT1*, flavone synthase II (*FNSII*), and chalcone synthase 1 (*CHS1*)—and fatty alcohol-related—fatty acyl-CoA reductase 2 (*FAR2*), *FAR3*, *FAR4*, *FAR5*, and *FAR6*—biosynthesis were identified in barley. The designated gene sequences are listed in Table S1, which were utilized in primer synthesis for qRT-PCR-based gene expression profiling.

4.5. Analysis of Expression Pattern of Biosynthetic Related Genes in Seedlings Subjected to LED Light Irradiation

Approximately 100 mg of powdered leaf samples of five or more seedlings were utilized for total RNA extraction using an RNeasy Plant Mini kit (Qiagen, Hilden, Germany). First-strand cDNA was synthesized from DNA decontaminated total RNA (5 μg) of each treatment using amfiRivert cDNA Synthesis Platinum Master mix (GenDepot, Katy, TX, USA). The diluted cDNA (1:10) was used as a template, along with gene-specific primers (Table S1) and AccuPower 2× GreenStar Master Mix (Bioneer, Daejeon, Korea), for relative quantification of saponarin, policosanol, and general flavonoid biosynthesis-related genes in quantitative RT-PCR with CFX96™ Real-Time PCR Detection System (Bio-Rad, Hercules, CA, USA). *HvActin* was used as an internal control. The relative quantification of gene expression was calculated using the $2^{-\Delta\Delta CT}$ method.

4.6. Phenotyping of LED-Treated Cereal Sprouts

The effect of LED light irradiation on the growth of wheat and barley sprouts was evaluated from five or more seedlings of each treatment, collected after 3–9 days of treatment. The length of all primary leaves and roots were measured. The mean values of growth parameters were compared between treatments and the mean differences were evaluated by statistical significance analysis.

4.7. Statistical Analyses

All treatments and experiments in this study had at least three independent biological and technical replicates. The mean values are presented in graphs drawn using the GraphPad Prism 5 software, while standard deviations are represented as error bars. One-way analysis of variance (one-way ANOVA) and Tukey's HSD test were carried out to assess the significant variations existing between the treatments and analyses performed in this study. Statistics by ANOVA test are shown; * $p < 0.05$, ** $p < 0.001$, and *** $p < 0.0001$. Pearson correlation coefficients and their significance were measured using the Social Science Statistics [47].

Supplementary Materials: The following are available online at http://www.mdpi.com/2223-7747/9/11/1502/s1, Figure S1: Representative chromatogram of saponarin and hexacosanol of barley seedlings exposed to differential LED light treatments. Figure S2: Chromatogram of isoorientin, isoschaftoside, and octacosanol of wheat seedlings exposed to differential LED light treatments. Figure S3: Effect of growth periods on the C-glycosylflavone and policosanol contents in fluorescent and LED light irradiated barley and wheat sprouts. Table S1: List of genes, respective gene and protein IDs, primer sequences, and their annealing temperatures used in the quantitative RT-PCR assay.

Author Contributions: Conceptualization, S.I.L.; investigation, M.M., J.H.K., S.H.K., J.Y.K., J.W.H., H.L., K.-S.L., W.D.S., S.P., J.A.K.; writing—original draft preparation, M.M.; writing—review and editing, M.M. and S.I.L. All authors have read and agreed to the published version of the manuscript.

Funding: This research was funded by the Rural Development Administration (Korea) through the Rural Program for Agricultural Science and Technology Development, grant number PJ01495701 and Cooperative Research Program for Agricultural Science and Technology, grant number PJ01421201.

Conflicts of Interest: The authors declare that they have no conflict of interest.

References

1. Giraldo, P.; Benavente, E.; Manzano-Agugliaro, F.; Gimenez, E. Worldwide research trends on wheat and barley: A bibliometric comparative analysis. *Agronomy* **2019**, *9*, 352. [CrossRef]
2. Shewry, P.R.; Hey, S.J. The contribution of wheat to human diet and health. *Food Energy Secur.* **2015**, *4*, 178–202. [CrossRef] [PubMed]
3. Byun, A.R.; Chun, H.; Lee, J.; Lee, S.W.; Lee, H.S.; Shim, K.W. Effects of a dietary supplement with barley sprout extract on blood cholesterol metabolism. *Evid. Based Complement. Altern. Med.* **2015**, *473056*, 1–7. [CrossRef] [PubMed]
4. Zeng, Y.; Pu, X.; Yang, J.; Du, J.; Yang, X.; Li, X.; Li, L.; Zhou, Y.; Yang, T. Preventive and therapeutic role of functional ingredients of barley grass for chronic diseases in human beings. *Oxid. Med. Cell. Longev.* **2018**, *2018*, 3232080. [CrossRef] [PubMed]

5. Lee, J.H.; Jia, Y.; Thach, T.T.; Han, Y.; Kim, B.; Wu, C.; Kim, Y.; Seo, W.D.; Lee, S.J. Hexacosanol reduces plasma and hepatic cholesterol by activation of AMP-activated protein kinase and suppression of sterol regulatory element-binding protein-2 in HepG2 and C57BL/6J mice. *Nutr. Res.* **2017**, *43*, 89–99. [CrossRef]
6. Lee, Y.H.; Kim, S.H.; Lee, S.; Kim, K.M.; Jung, J.C.; Son, T.G.; Ki, S.H.; Seo, W.D.; Kwak, J.H.; Hong, J.T.; et al. Antioxidant effect of barley sprout extract via enhancement of Nuclear Factor-Erythroid 2 Related Factor 2 activity and glutathione synthesis. *Nutrients* **2017**, *9*, 1252. [CrossRef] [PubMed]
7. Kim, Y.J.; Hwang, S.H.; Jia, Y.; Seo, W.D.; Lee, S.J. Barley sprout extracts reduce hepatic lipid accumulation in ethanol-fed mice by activating hepatic AMP-activated protein kinase. *Food Res. Int.* **2017**, *101*, 209–217. [CrossRef]
8. Seo, K.H.; Park, M.J.; Ra, J.E.; Han, S.I.; Nam, M.H.; Kim, J.H.; Lee, J.H.; Seo, W.D. Saponarin from barley sprouts inhibits NF-κB and MAPK on LPS-induced RAW 264.7 cells. *Food Funct.* **2014**, *5*, 3005–3013. [CrossRef]
9. Park, M.J.; Seo, W.D.; Kang, Y.H. The antioxidant properties of four korean barley cultivars at different harvest times and profiling of major metabolites. *J. Agric. Sci.* **2015**, *7*, 94. [CrossRef]
10. Seo, W.; Hae, J.; Jia, Y.; Wu, C.; Lee, S. Saponarin activates AMPK in a calcium-dependent manner and suppresses gluconeogenesis and increases glucose uptake via phosphorylation of CRTC2 and HDAC5. *Bioorg. Med. Chem. Lett.* **2015**, *25*, 5237–5242. [CrossRef]
11. Lee, H.; Ra, S.W.J.; Woo, K.L.; Seo, D.; Hwan, J. Saponarin content and biosynthesis-related gene expression in young barley (*Hordeum vulgare* L.) seedlings. *J. Plant Biotechnol.* **2019**, *2818*, 247–254. [CrossRef]
12. Ye, T.; Su, J.; Huang, C.; Yu, D.; Dai, S.; Huang, X.; Chen, B.; Zhou, M. Isoorientin induces apoptosis, decreases invasiveness, and downregulates VEGF secretion by activating AMPK signaling in pancreatic cancer cells. *OncoTargets Targets* **2016**, *9*, 7481–7492. [CrossRef]
13. Xiao, J.; Capanoglu, E.; Jassbi, A.R.; Miron, A. Advance on the flavonoid C-glycosides and health benefits. *Crit. Rev. Food Sci. Nutr.* **2016**, *56*, S29–S45. [CrossRef]
14. Nam, T.G.; Kim, D.-O.; Eom, S.H. Effects of light sources on major flavonoids and antioxidant activity in common buckwheat sprouts. *Food Sci. Biotechnol.* **2018**, *27*, 169–176. [CrossRef] [PubMed]
15. Sytar, O.; Bosko, P.; Živčák, M.; Brestic, M.; Smetanska, I. Bioactive phytochemicals and antioxidant properties of the grains and sprouts of colored wheat genotypes. *Molecules* **2018**, *23*, 2282. [CrossRef]
16. Benincasa, P.; Falcinelli, B.; Lutts, S.; Stagnari, F.; Galieni, A. Sprouted grains: A comprehensive review. *Nutrients* **2019**, *11*, 421. [CrossRef] [PubMed]
17. Ra, J.; Woo, S.; Lee, K.; Ja, M.; Young, H.; Mi, H.; Chung, I.; Hyun, D.; Hwan, J.; Duck, W. Policosanol profiles and adenosine 5′-monophosphate-activated protein kinase (AMPK) activation potential of Korean wheat seedling extracts according to cultivar and growth time. *Food Chem.* **2020**, *317*, 126388. [CrossRef] [PubMed]
18. Lemmens, E.; Moroni, A.V.; Pagand, J.; Heirbaut, P.; Ritala, A.; Karlen, Y.; Kim-Anne, L.; Van den Broeck, H.C.; Brouns, F.J.P.H.; De Brier, N.; et al. Impact of cereal seed sprouting on its nutritional and technological properties: A Critical Review. *Compr. Rev. Food Sci. Food Saf.* **2019**, *18*, 305–328. [CrossRef]
19. Rico, D.; Peñas, E.; del Carmen García, M.; Martínez-Villaluenga, C.; Rai, D.K.; Birsan, R.I.; Frias, J.; Martín-Diana, A.B. Sprouted barley flour as a nutritious and functional ingredient. *Foods* **2020**, *9*, 296. [CrossRef] [PubMed]
20. Aborus, N.E.; Canadanovi, J.; Saponjac, V.T. Powdered barley sprouts: Composition, functionality and polyphenol digestibility. *Int. J. Food Sci. Technol.* **2017**, *52*, 231–238. [CrossRef]
21. Wu, X.; Cai, K.; Zhang, G.; Zeng, F. Metabolite profiling of barley grains subjected to water stress: To explain the genotypic difference in drought-induced impacts on malting quality. *Front. Plant. Sci.* **2017**, *8*, 1547. [CrossRef]
22. Wang, Y.; Zeng, X.; Xu, Q.; Mei, X.; Yuan, H.; Jiabu, D.; Sang, Z.; Nyima, T. Metabolite profiling in two contrasting Tibetan hulless barley cultivars revealed the core salt-responsive metabolome and key salt-tolerance biomarkers. *AoB Plants* **2019**, *11*, plz021. [CrossRef]
23. Kang, C.H.; Yoon, E.K.; Muthusamy, M.; Kim, J.A.; Jeong, M.J.; Lee, S.I. Blue LED light irradiation enhances L-ascorbic acid content while reducing reactive oxygen species accumulation in Chinese cabbage seedlings. *Sci. Hortic.* **2020**, *261*, 108924. [CrossRef]
24. Cheng, C.; Liu, Y.; Fang, W.; Tao, J.; Yang, Z.; Yin, Y. iTRAQ-based proteomic and physiological analyses of mustard sprouts in response to heat stress. *RSC Adv.* **2020**, *10*, 6052–6062. [CrossRef]

25. Muthusamy, M.; Hwang, J.E.; Kim, S.H.; Kim, J.A.; Jeong, M.J.; Park, H.C.; Lee, S.I. Elevated carbon dioxide significantly improves ascorbic acid content, antioxidative properties and restricted biomass production in cruciferous vegetable seedlings. *Plant. Biotechnol. Rep.* **2019**, *13*, 293–304. [CrossRef]
26. Tuan, P.A.; Thwe, A.A.; Kim, Y.B.; Kim, J.K.; Kim, S.J.; Lee, S.; Chung, S.O.; Park, S.U. Effects of white, blue, and red light-emitting diodes on carotenoid biosynthetic gene expression levels and carotenoid accumulation in sprouts of tartary buckwheat *(Fagopyrum tataricum* gaertn.). *J. Agric. Food Chem.* **2013**, *61*, 12356–12361. [CrossRef]
27. Meng, T.; Nakamura, E.; Irino, N.; Joshi, K.R.; Devkota, H.P.; Yahara, S.; Kondo, R. Effects of irradiation with light of different photon densities on the growth of young green barley plants. *Agric. Sci.* **2015**, *06*, 208–216. [CrossRef]
28. Hasan, M.M.; Bashir, T.; Bae, H. Use of utrasonication tchnology for the icreased poduction of pant scondary mtabolites. *Molecules* **2017**, *22*, 1046. [CrossRef]
29. Kowalska, I.; Jedrejek, D.; Jonczyk, K.; Stochmal, A. UPLC-PDA-ESI-MS analysis and TLC-DPPH activity of wheat varieties. *Acta Chromatogr.* **2019**, *31*, 151–156. [CrossRef]
30. Kopsell, D.A.; Sams, C.E. Increases in shoot tissue pigments, glucosinolates, and mineral elements in sprouting broccoli after exposure to short-duration blue light from light emitting diodes. *J. Am. Soc. Hortic. Sci.* **2013**, *138*, 31–37. [CrossRef]
31. Ghimire, B.K.; Yu, C.Y.; Chung, I.M. Assessment of the phenolic profile, antimicrobial activity and oxidative stability of transgenic Perilla frutescens L.overexpressing tocopherol methyltransferase (γ-tmt) gene. *Plant. Physiol. Biochem.* **2017**, *118*, 77–87. [CrossRef]
32. Brauch, D.; Porzel, A.; Schumann, E.; Pillen, K.; Mock, H.P. Changes in isovitexin-O-glycosylation during the development of young barley plants. *Phytochemistry* **2018**, *148*, 11–20. [CrossRef]
33. Marinova, K.; Kleinschmidt, K.; Weissenböck, G.; Klein, M. Flavonoid biosynthesis in barley primary leaves requires the presence of the vacuole and controls the activity of vacuolar flavonoid transport. *Plant. Physiol.* **2007**, *144*, 432–444. [CrossRef]
34. Kamiyama, M.; Shibamoto, T. Flavonoids with potent antioxidant activity found in young green barley leaves. *J. Agric. Food Chem.* **2012**, *60*, 6260–6267. [CrossRef]
35. Irmak, S.; Dunford, N.T.; Milligan, J. Policosanol contents of beeswax, sugar cane and wheat extracts. *Food Chem.* **2006**, *95*, 312–318. [CrossRef]
36. Seo, W.D.; Yuk, H.J.; Curtis-long, M.J.; Jang, K.C.; Lee, J.H.; Han, S.; Kang, H.W.; Nam, M.H.; Lee, S.; Lee, J.H.; et al. Effect of the growth stage and cultivar on policosanol profiles of barley sprouts and their adenosine 5′-monophosphate-activated protein kinase activation. *J. Agric. Food Chem.* **2013**, *61*, 1117–1123. [CrossRef]
37. Sharma, R.; Matsuzaka, T.; Kaushik, M.K.; Sugasawa, T.; Ohno, H.; Wang, Y.; Motomura, K.; Shimura, T.; Okajima, Y.; Mizunoe, Y.; et al. Octacosanol and policosanol prevent high-fat diet-induced obesity and metabolic disorders by activating brown adipose tissue and improving liver metabolism. *Sci. Rep.* **2019**, *9*, 5169. [CrossRef]
38. Richardson, A.; Franke, R.; Kerstiens, G.; Jarvis, M.; Schreiber, L.; Fricke, W. Cuticular wax deposition in growing barley (*Hordeum vulgare*) leaves commences in relation to the point of emergence of epidermal cells from the sheaths of older leaves. *Planta* **2005**, *222*, 472–483. [CrossRef]
39. Gil, K.E.; Park, C.M. Thermal adaptation and plasticity of the plant circadian clock. *New Phytol.* **2019**, *221*, 1215–1229. [CrossRef]
40. Wang, Y.; Wang, M.; Sun, Y.; Wang, Y.; Li, T.; Chai, G.; Jiang, W.; Shan, L.; Li, C.; Xiao, E.; et al. FAR5, a fatty acyl-coenzyme A reductase, is involved in primary alcohol biosynthesis of the leaf blade cuticular wax in wheat (*Triticum aestivum* L.). *J. Exp. Bot.* **2015**, *66*, 1165–1178. [CrossRef]
41. Wang, Y.; Sun, Y.; You, Q.; Luo, W.; Wang, C.; Zhao, S.; Chai, G.; Li, T.; Shi, X.; Li, C.; et al. Three Fatty Acyl-Coenzyme A Reductases, BdFAR1, BdFAR2 and BdFAR3, are involved in cuticular wax primary alcohol biosynthesis in *Brachypodium distachyon*. *Plant. Cell Physiol.* **2018**, *59*, 527–543. [CrossRef]
42. Wang, M.; Wu, H.; Xu, J.; Li, C.; Wang, Y.; Wang, Z. Five fatty acyl-coenzyme a reductases are involved in the biosynthesis of primary alcohols in *Aegilops tauschii* leaves. *Front. Plant Sci.* **2017**, *8*, 1012. [CrossRef]
43. NCBI Gene. Available online: https://www.ncbi.nlm.nih.gov/gene (accessed on 7 August 2020).
44. PlantsDB. Available online: http://pgsb.helmholtz-muenchen.de/plant/search.jsp (accessed on 7 August 2020).
45. GrainGenes. Available online: https://wheat.pw.usda.gov/blast/ (accessed on 7 August 2020).

46. NCBI Conserved Domains. Available online: https://www.ncbi.nlm.nih.gov/Structure/cdd/wrpsb.cgi (accessed on 8 August 2020).
47. Social Science Statistics. Available online: https://www.socscistatistics.com/tests/pearson/default.aspx. (accessed on 26 October 2020).

Publisher's Note: MDPI stays neutral with regard to jurisdictional claims in published maps and institutional affiliations.

plants

Article

LED Lights Affecting Morphogenesis and Isosteroidal Alkaloid Contents in *Fritillaria cirrhosa* D. Don—An Important Chinese Medicinal Herb

Chia-Chen Chen [1], Maw-Rong Lee [2], Chi-Rei Wu [1], Hsin-Ju Ke [2], Hui-Min Xie [3], Hsin-Sheng Tsay [4], Dinesh Chandra Agrawal [4,*] and Hung-Chi Chang [5,*]

1 Department of Chinese Pharmaceutical Sciences and Chinese Medicine Resources, China Medical University, Taichung 40402, Taiwan; ferny1010@yahoo.com.tw (C.-C.C.); crw@mail.cmu.edu.tw (C.-R.W.)
2 Department of Chemistry, National Chung-Hsing University, Taichung 40227, Taiwan; mrlee@dragon.nchu.edu.tw (M.-R.L.); qaznancy@gmail.com (H.-J.K.)
3 Nin Jiom Pharmaceutical Co. Ltd., Taipei 108024, Taiwan; patricia@ninjiom.com.tw
4 Department of Applied Chemistry, Chaoyang University of Technology, Taichung 41349, Taiwan; hstsay@cyut.edu.tw
5 Department of Golden-Ager Industry Management, Chaoyang University of Technology, Taichung 41349, Taiwan
* Correspondence: dcagrawal@gm.cyut.edu.tw (D.C.A.); changhungchi@cyut.edu.tw (H.-C.C.); Tel.: +886-4-23323000 (ext. 4238) (D.C.A.); +886-4-23323000 (ext. 5345) (H.-C.C.)

Received: 12 August 2020; Accepted: 7 October 2020; Published: 13 October 2020

Abstract: Investigations were carried out to study the effects of light-emitting diode (LED) lights on growth and development of isosteroidal alkaloids in embryogenic calli of *Fritillaria cirrhosa* D. Don, an important traditional Chinese medicine herb. Calli were cultured in glass bottles, each containing 100 mL of Murashige and Skoog's basal medium supplemented with 2% sucrose and 0.4% gellan gum powder, a gelling agent. These bottles were incubated in a specially designed plant growth chamber equipped with eight different LED lights consisting of single or combinations of four different light spectra emitting blue (450 nm), green (525 nm), red (660 nm), and far-red (730 nm) light. After three months of incubation, morphological changes in embryogenic calli were recorded, and LC-MS/MS analysis of cultures was carried out for peimisine, sipeimine, peiminine, and peimine. The highest number of somatic embryos and the maximum fresh weight was recorded in calli incubated under red (9R), infrared (9IR), and a combination of red+blue+infrared (3R3B3IR), respectively, in decreasing order. The highest contents of peimisine, peiminine, and peimine were recorded under red (9R) and infrared (9IR) lights, respectively. Eight LED lights had significant effects on the morphogenesis of embryogenic calli of *F. cirrhosa* D. Don and contents of isosteroidal alkaloids.

Keywords: *Fritillaria cirrhosa* D. Don; alkaloid content; callus; in vitro culture; LED lights; light intensity

1. Introduction

Fritillaria, a bulbiferous and perennial monocot plant genus, belongs to the family Liliaceae. The genus consists of about 130 species distributed in the temperate regions of Central Asia and the Mediterranean region [1]. Though some *Fritillaria* species are grown as ornamental plants, several *Fritillaria* species possess valuable medicinal properties. *Fritillaria* bulbs composed of fleshy, farinaceous scales constitute essential plant parts and have been used to relieve cough for centuries [2].

In different *Fritillaria* species, a majority of bioactive compounds (86%) identified so far (~130) consist of isosteroidal alkaloid skeletons [3]. Alkaloids in *Fritillaria* bulbs are the main bioactive compounds responsible for relief from coughs [3,4]. However, the quantities and types of alkaloids vary depending on species [1,3–5].

In a recent study, it was found that peimine, an alkaloid from *Fritillaria*, blocked the Nav1.7 ion channel and inhibited the Kv1.3 ion channel in HEK 293 cell lines, indicating that the compound has a role in relieving pain and possesses anti-inflammatory properties [6]. More recently, Liu and co-workers investigated the potential effect and mechanism of six isosteroidal alkaloids on oxidative stress. The findings showed that *F. cirrhosa* D. Don bulbs might play a protective role in cellular oxidative stress by activating the Nrf2-mediated antioxidant pathway [7].

China is the center of diversity of the *Fritillaria* genus. *F. cirrhosa* D. Don (FC) is an important traditional Chinese medicine commonly known as "Chuanbeimu" (川貝母), and is one of the most exploited species. Due to scarcity, the price of wild *F. cirrhosa* D. Don bulbs escalated almost nine times from $60 to $560 USD between 2002–2017 [8]. Due to the excessive collection of FC bulbs from natural habitats, the species is now under protection [9]. Therefore, alternative propagation methods of FC bulbs and the production of critical isosteroidal alkaloids by tissue culture techniques must be optimized.

The culture of plant tissues and organs is an important bio-technique to produce secondary plant metabolites under controlled environmental conditions in a laboratory. Under culture conditions, callus (an undifferentiated mass of cells) can easily be induced, practically from any living plant part. Induced callus can be cultured and multiplied in vitro on a defined nutrient medium with or without plant growth regulators. There are several advantages of using callus cultures as a source of valuable secondary metabolites, including (i) ease of induction and multiplication; (ii) production of bioactive compounds throughout the year independent of season; (iii) whole plants do not need to be cultivated, especially rare and endangered species; (iv) amenability of scaling by bioreactors for mass production, etc. More recently, we have reported on the micropropagation of bulblets and the production of isosteroidal alkaloids in tissue culture-derived materials of *F. cirrhosa* D. Don [10]. Several recent studies have demonstrated that the light quality not only affects morphogenetic responses in plants but has significant effects on their physiological processes, including metabolic pathways and the production of secondary metabolites [11,12].

A recent review listed several studies on the effects of light quality on the production of secondary metabolites in different plant species [11]. In the present study, four isosteroidal alkaloids (peimisine, sipeimine, peiminine, and peimine) were analyzed considering their therapeutic effects. Peiminine (Pm) is one of the major isosteroidal alkaloids in *Fritillaria* that is reported to have extensive pharmacological activities, including anti-inflammatory [6], anti-cancer [13], and antioxidant [14] capabilities. In another report, sipeimine and peiminine from bulbs of *F. wabuensis* inhibited pro-inflammatory mediators in lipopolysaccharide (LPS) stimulated RAW 264.7 macrophage cells [15]. More recently, it was demonstrated that peimine relieved inflammatory effects in IL-1β-induced chondrocytes, indicating that peimine might be a potential therapeutic agent for osteoarthritis [16]. Antitussive, expectorant, and anti-inflammatory effects of several alkaloids, including sipeimine, chuanbeinone, peiminine, and peimine isolated from Bulbus *Fritillaria cirrhosa* were demonstrated in mice by using a phytochemical method [17]. A most recent study confirmed the anti-cancer effects of sipeimine obtained from bulbs of *F. cirrhosa* against non-small cell lung cancer (NSCLC) both in vivo and in vitro [18]. This anti-cancer property of sipeimine is largely due to anti-inflammation action affected by NF-κB inhibition, making it a potential drug candidate for treating cancer at early stages [18]. Recently, Yin and co-workers have reported several therapeutic properties of peimine, including anti-cancer, anti-inflammatory, antitussive, expectorant, and sedative [19]. A more recent study has also demonstrated cough relief by peimine by affecting the systemic network of proteins and pathways [20].

The objective of the present study is to investigate the effects of different LED lights on growth and development in embryogenic calli and the contents of four isosteroidal alkaloids (peimisine, sipeimine, peiminine, and peimine) in in vitro cultures of *F. cirrhosa* D. Don. Findings in the study may be of help to produce certain alkaloids under laboratory conditions irrespective of the season and thus avoid having to collect *F. cirrhosa* D. Don bulbs from the wild.

2. Material and Methods

2.1. Callus Multiplication

Callus obtained in our previous experiments, as reported earlier [10], was further multiplied in the liquid medium. Callus was cultured in 125 mL Erlenmeyer flasks, each containing 20 mL basal salts and vitamins of Murashige and Skoog [21] medium (MSBM) supplemented with 2,4-D (0.5 mg/L) and 2% sucrose (Sigma). The culture flasks were placed on an orbital shaker (Model SK-302A, Sun Kaun Instruments Co., Taichung, Taiwan) set at 100 rpm and incubated at 25 ± 1 °C in the dark.

2.2. Influence of Different Light Spectra on Morphogenesis in Embryogenic Calli and Contents of Isosteroidal Alkaloids

To investigate the effects of LED lights on the morphogenesis of embryogenic calli and the contents of isosteroidal alkaloids, embryogenic calli from liquid cultures were taken out and kept for 1 min on sterilized filter paper in a laminar flow before inoculation to glass bottles. Callus (3.0 g) was cultured in glass bottles (650 mL capacity), each containing 100 mL of MSBM medium with 2% sucrose and 0.4% gellan gum powder (GPP), a gelling agent (PhytoTechnology Laboratories®, USA). The pH of the medium was adjusted to 5.7 ± 0.1 before the addition of GPP and autoclaving at 1.05 kg/cm for 15 min. To facilitate LED light exposure to cultures, each bottle was closed with a piece of transparent, autoclavable plastic sheet. These bottles were incubated in a specially designed plant growth chamber equipped with eight different LED lights (Nano Bio Light Technology Co., Ltd., Taiwan). The chamber had two tiers, and each tier had four partitions (Figure 1a). Culture bottles were kept in these eight sections and exposed to different light spectra by eight specially designed LED lids (CW-5000K, WW-2700K, 8R1B, 7R1G1B, 3R3B3IR, 6R, 6B, and 6IR) (Figure 1b). LED lids CW-5000K and WW-2700K represented cool (C) and warm (W) white (W) light, while 5000K and 2700K represented color temperature, respectively. As reported in a previous study from our laboratory [22], six of these LED lids emitted single or combinations of four different light spectra with wavelengths such as blue (450 nm), green (525 nm), red (660 nm), and far-red (730 nm). The symbols in each LED lid code and the spectral distribution (quantum ratio) in eight LED lids were as follows: blue (B), green (G), red (R), infrared (IR), CW-5000K (28:43:29:0), WW-2700K (8:46:46:0), 8R1B (16:0:84:0), 7R1G1B (17:9:74:0), 3R3B3IR (57:0:43:37), 9R (0:0:100:0), 9B (100:0:0:0), and 9IR (0:0:0:100). In this work, the number (9, 7, 3, 1) in each LED lid code represents the number of LED chips in a particular lid. The light intensity of each LED lid was as follows: CW-5000K* (57 μmol m^{-2} s^{-1}); WW-2700K (56 μmol m^{-2} s^{-1}); 7R1G1B (56 μmol m^{-2} s^{-1}); 8R1B (57 μmol m^{-2} s^{-1}); 9B (57 μmol m^{-2} s^{-1}); 9R (56); 9IR (10 μmol m^{-2} s^{-1}); 3R3B3IR (56 μmol m^{-2} s^{-1}).

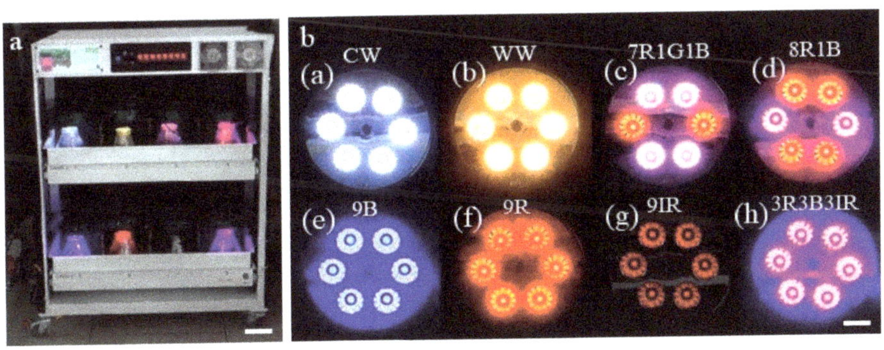

Figure 1. (**a**) Plant growth chamber with LED lights. Bar = 8.3 cm; (**b**) Eight different LED lights: (**a**) CW-5000 K, (**b**) WW-2700 K, (**c**) 7R1G1B, (**d**) 8R1B, (**e**) 9B, (**f**) 9R, (**g**) 9IR, (**h**) 3R3B3IR. Bar = 2 cm. Red (R): 660 nm; green (G): 525 nm; blue (B): 450 nm; infrared (IR): 730 nm.

The LED growth chamber was set on a 16 h light and 8 h dark cycle, kept in a culture room set at 25 ± 2 °C, and fully covered by a thick, dark cloth to cut off outside light. Morphological features of embryogenic callus under each light condition were recorded after three months of incubation. In addition, cultures under different LED lights were analyzed by LC-MS/MS for contents of peimisine, sipeimine, peiminine, and peimine.

2.3. Development of Bulblets from Somatic Embryos

For the development of bulblets, single somatic embryos (SEs), clusters with five somatic embryos each, and a single embryo with cotyledonary leaf obtained from cultures under CW5000K or WW2700K lights were further transferred to fresh MSBM medium in glass bottles as described in Section 2.2. Each bottle contained seven explants. These bottles were incubated in a culture room with a temperature set at 25 ± 2 °C, and under light (white fluorescent tubes, illumination intensity of 34 μmol m^{-2} s^{-1}) with a 16 h light and 8 h dark cycle. Observations of bulblet development in all three types of SEs were recorded after three months of incubation.

Microphotographs of cultures were taken by a digital camera (Nikon D90, Tokyo, Japan). For SEM, samples of bulblets were frozen in liquid nitrogen and observed using a scanning electron microscope (JEOL JSM-6330F, Tokyo, Japan).

2.4. Chemicals and Other Materials

Peimisine, sipeimine, peiminine, and peimine standards were procured from SunHank Technology Co., Ltd., Taiwan. Ammonium formate, sodium acetate, and octadecyltrimethoxysilane were purchased from Sigma-Aldrich (St. Louis, MO, USA). Ammonium hydroxide solution, ferric chloride hexahydrate, ethylene glycol, and acetonitrile were obtained from Fluka (Steinheim, Germany), Alfa Aesar (Heysham, UK), Acros Organics (Morris County, NJ, USA), and J. T. Baker (Phillipsburg, NJ, USA), respectively. Methanol and ethanol were purchased from Merck (Darmstadt, Germany). Ultrapure water (18.2 MΩ) was freshly obtained from a Millipore Simplicity system (MilliporeSigma, Bedford, MA, USA). The stock solutions (1 mg/mL) of each analyte were prepared in methanol separately and stored in the dark at −30 °C. The working solutions were prepared by diluting the stock solutions before use.

2.5. Preparation of Fe_3O_4@C_{18} Nanoparticle Composite

The adsorbent (magnetic nanoparticles, Fe_3O_4@C18) used for magnetic solid-phase extraction was synthesized according to the literature [23,24] with minor modification. Briefly, ferric chloride hexahydrate (2.7 g) and sodium acetate (7.2 g) were dissolved entirely in ethylene glycol (100 mL) before being poured into a Teflon-lined stainless steel autoclave heated at 200 °C for 8 h. The resultant Fe_3O_4 nanoparticles were washed with ethanol and dried. Afterward, the Fe_3O_4 nanoparticles (10 mg) were dispersed in a mixture of 970 μL ethanol, 10 μL water, and 20 μL octadecyltrimethoxysilane through sonication for 30 s, followed by shaking for 8 h at 45 °C. The derivatized Fe_3O_4@C18 was washed with ethanol, water, and acetonitrile three times, respectively. The final product was resuspended in 1 mL of acetonitrile (10 mg/mL).

2.6. Extraction Procedure

Ultra-pure water (2 mL) was added to each 0.1 g powdered sample of in vitro culture, 3-month-old in vitro derived bulblets, and 3-year-old wild type commercial bulbs before vortexing for 1 min and letting stand for 5 min. After centrifugation at 4000 rpm for 5 min, 0.8 mL of supernatant was collected. Then, 8 μL internal standard (50 μg/mL), 50 μL ammonia solution, and 140 μL Fe_3O_4@C18 were added into the supernatant with a vortex treatment for 30 s then left to stand still for 5 min. With the help of an external magnet, the analyte-adsorbed Fe_3O_4@C18 was rapidly separated from the supernatant and vortexed with 100 μL acetonitrile for 30 s to elute the analytes. After that, the external magnet was used to settle the magnetic adsorbent while the elution solution was filtered through a 0.22 μm PTFE (polytetrafluoroethylene) membrane for LC-MS/MS analysis.

2.7. LC-MS/MS Conditions

The analysis was performed with a Surveyor LC-MS/MS system (Thermo Scientific, Waltham, CA, USA). The chromatographic separation was achieved by using a Thermo Scientific Accucore C18 column (2.1 × 100 mm, 2.6 μm) at a constant column temperature of 30 °C with a flow rate of 0.2 mL/min. The mobile phase A was 10 mM ammonium formate aqueous solution containing 0.1% ammonia solution, and mobile phase B was methanol. The gradient elution program was started at 80% A for 1 min, dropped to 35% A in 1 min, decreased to 20% A in 2 min, held for 5 min, decreased to 10% A within 2 min, held for 3 min, resumed at 80% A within 1 min, and kept constant for 5 min. Figure 2 shows the typically extracted ion chromatograms of the mixed standard solution of peimisine, sipeimine, peiminine, and peimine at the concentration of 0.5 μg/mL.

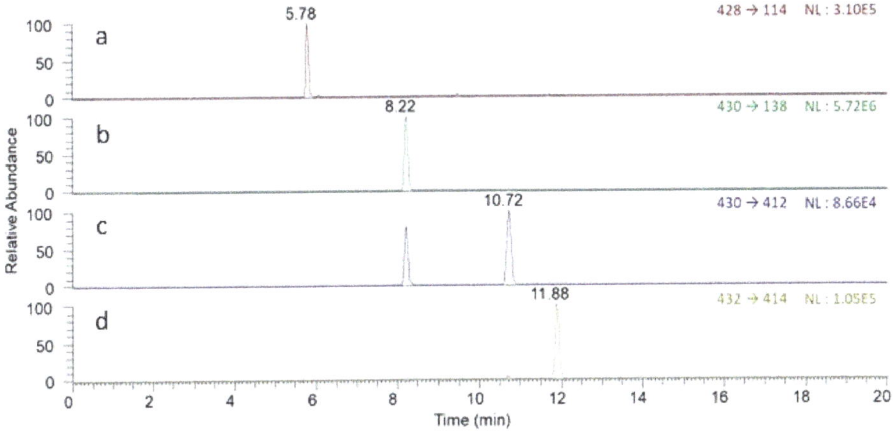

Figure 2. Extracted ion chromatograms of isosteroidal alkaloid standards (0.5 μg/mL): (**a**) Peimisine, (**b**) Sipeimine, (**c**) Peiminine, and (**d**) Peimine.

The mass spectrometer was equipped with an electrospray ionization (ESI) interface operating in positive ion mode. The selected reaction monitoring (SRM) mode was used to acquire the mass spectrometric data. The full width at half maximum (FWHM) of Q1 and Q3 was 0.7 for both. The optimal ESI source parameters were set as follows: the capillary temperature was 300 °C; spray voltage was 4600 V; the pressure of sheath gas, aux gas, and ion sweep gas was maintained at 45, 10, and 10 arb units, respectively. The ion transitions and optimal collision energy of selected reaction monitoring chosen for quantitative analysis were as follows: m/z 428→m/z 114 (44 eV) for peimisine; m/z 430→m/z 138 (48 eV) for sipeimine; m/z 430→m/z 412 (38 eV) for peiminine; m/z 432→m/z 414 (40 eV) for peimine. The retention time, protonated molecule ions (represented as [M+H]+), and the analytical parameters of the developed method for analysis of four alkaloids are listed in Table 1, and chemical structures are shown in Figure 3.

Table 1. LC-MS/MS based method development and validation for four standards.

Marker Compounds	tR [a] (min)	Mass/Charge (m/z)	Linearity and Range			Sensitivity [c]	
			Regression Equation [b]	Correlation Coefficient (r^2)	Linear Range (μg/g)	LOD (ng/g)	LOQ (ng/g)
Peimisine	5.78	428.316	y = 0.0396x + 0.0630	0.9923	0.1–40	0.01	0.04
Sipeimine	8.22	430.332	y = 0.5426x + 0.1531	0.9968	0.1–40	0.02	0.06
Peiminine	10.72	430.332	y = 0.0047x + 0.0010	0.9967	0.1–40	0.97	3.23
Peimine	11.88	432.347	y = 0.0067x + 0.0016	0.9977	0.1–40	0.56	1.88

[a] tR: Retention time. [b] The regression equations are presented as y = mx+c, and y and x are defined as peak area and concentration of the compound, respectively. [c] LOD: Limit of detection, S/N = 3; LOQ: limit of quantification, S/N = 10.

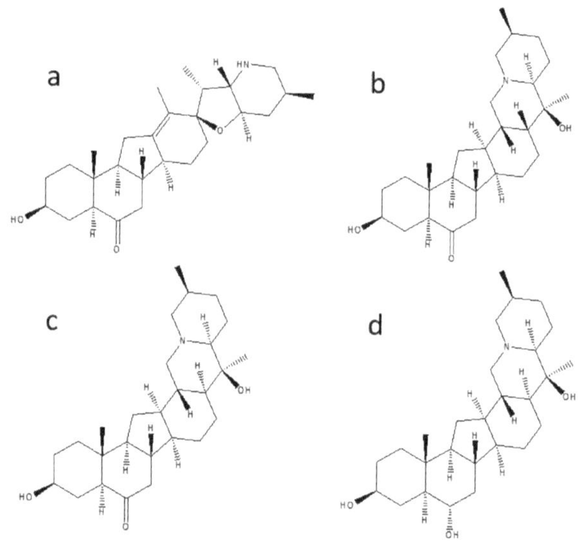

Figure 3. Chemical structures of four isosteroidal alkaloids: (**a**) Peimisine, (**b**) Sipeimine, (**c**) Peimine, (**d**) Peiminine.

2.8. Statistical Analysis

Software SAS 9.1 was used for statistical analysis. Data were subjected to the least significant difference (LSD) test at a 5% probability level ($p < 0.05$) wherever possible. In Tables 2 and 3, the number of replicates is 3 (three bottles under each LED light treatment). In Table 4, the number of replicates is 21 (each bottle had 7 explants, and each treatment had 3 bottles). The experiments were repeated three times, including LC-MS/MS analysis.

Table 2. Influence of different LED lights on the growth and development of embryogenic callus cultures *.

LED Light Treatment	Av No. of Somatic Embryos/Bottle	Av No. Of Somatic Embryos with Cotyledonary Leaves/Bottle	Av Total Fresh Weight of Cultures/Bottle (g)	Morphological Features of Cultures (Somatic Embryos = SEs)
CW-5000K *	63.7 ± 7.6 ^{c} **	12.0 ± 6.0 ^{abc}	12.82 ± 1.09 ^{c}	SEs at all stages from globular to mature, SEs with long cotyledonary leaves. Color of cultures is light green.
WW-2700K	122.0 ± 65.0 ^{bc}	22.3 ± 13.9 ^{a}	13.64 ± 2.82 ^{c}	SE stages are similar to CW-5000K. Color of cultures is light green.
7R1G1B	108.0 ± 38.2 ^{bc}	8.5 ± 3.5 ^{bc}	12.99 ± 2.02 ^{c}	A majority of SEs in early stages, including globular shapes. Cotyledonary leaves absent. Color of cultures is light green to light brown.
8R1B	157.3 ± 9.3 ^{ab}	16.7 ± 8.3 ^{ab}	15.23 ± 0.97 ^{abc}	Stages similar to CW-5000K. Color of cultures is light green.
9B	169.0 ± 66.1 ^{ab}	3.7 ± 2.3 ^{c}	14.61 ± 1.02 ^{bc}	SE stages similar to CW-5000K, but cotyledonary leaves shorter in length. Color of cultures is light green to light yellow.
9R	223.7 ± 57.5 ^{a}	5.3 ± 4.9 ^{c}	17.92 ± 0.77 ^{a}	A majority of SEs in the early stages, including globular shapes. Cotyledonary leaves absent. Color of cultures is white.
9IR	231.3 ± 62.3 ^{a}	4.7 ± 0.6 ^{c}	16.67 ± 1.85 ^{ab}	SE stages and color of cultures similar to 9R. Only a few cotyledonary leaves are seen. Color of cultures is white.
3R3B3IR	230.7 ± 23.4 ^{a}	4.3 ± 2.5 ^{c}	17.56 ± 2.35 ^{a}	SE at all stages from globular to mature but cotyledonary leaves shorter in length. Color of cultures is dark green.

* Murashige and Skoog's basal medium supplemented with 2% sucrose and 0.4% GPP. pH = 5.7 ± 0.1. Observations recorded after three months of incubation. ** Mean ± standard error. Means within each column followed by the same letter(s) are not significantly different at 5% level by Fisher's protected LSD test.

Table 3. LC-MS/MS analysis of four isosteroidal alkaloids in in vitro cultures exposed to eight different LED lights, in vitro derived bulblets (3 months old), and commercial *Fritillaria cirrhosa* D. Don bulbs (wild type, three years old).

LED Light Treatment	Plant Material	Isosteroidal Alkaloids (μg/g/dw)				Total of Four Alkaloids (μg/g/dw)
		Peimisine	Sipeimine	Peimine	Peiminine	
CW-5000K	In vitro cultures	ND *	ND	ND	0.12 ± 0.20 [b] **	0.12 ± 0.20 [c] **
WW-2700K	In vitro cultures	ND	ND	ND	0.28 ± 0.27 [b]	0.28 ± 0.27 [c]
7R1G1B	In vitro cultures	ND	ND	ND	0.19 ± 0.32 [b]	0.19 ± 0.32 [c]
8R1B	In vitro cultures	ND	ND	ND	ND	0.00 ± 0.00 [c]
9B	In vitro cultures	ND	ND	ND	0.60 ± 0.43 [b]	0.65 ± 0.45 [c]
9R	In vitro cultures	3.65 ± 1.68 [b] **	ND	0.38 ± 0.11 [a]	2.40 ± 0.30 [a]	6.42 ± 2.06 [b]
9IR	In vitro cultures	3.22 ± 3.28 [b]	ND	0.05 ± 0.09 [b]	2.21 ± 0.87 [a]	5.48 ± 3.21 [b]
3R3B3IR	In vitro cultures	ND	ND	ND	0.26 ± 0.24 [b]	0.26 ± 0.24 [c]
Fluorescent tube	In vitro bulblets (3 months old)	0.91 ± 0.97 [b]	ND	ND	2.98 ± 1.09 [a]	3.90 ± 1.51 [bc]
Natural habitat	Commercial bulbs (wild type, 3 years old)	68.4 ± 7.8 [a]	0.6 ± 0.4 [a]	ND	ND	69.0 ± 7.4 [a]

* ND: Not detected. ** Mean ± standard error. Means within each column followed by the same letter(s) are not significantly different at 5% level by Fisher's protected LSD test.

Table 4. Development of bulblets in single embryo, a cluster of five embryos, and single embryo with cotyledonary leaf (3–6 cm long) in *Fritillaria cirrhosa* D. Don.

Type of Somatic Embryo (SE) *	Percentage of Response (%)	Av No. of Bulblets/SE
Single embryo	90.0 ± 10.0 [a] **	4.7 ± 1.3 [a] **
Cluster of five embryos	86.7 ± 12.0 [a]	3.3 ± 1.5 [ab]
Embryo with the cotyledonary leaf	43.6 ± 29.0 [b]	1.1 ± 0.7 [b]

* Culture medium MSBM supplemented with 2% sucrose, 0.4% GPP. Observations recorded after three months of incubation. ** Mean ± standard error. Means within each column followed by the same letter(s) are not significantly different at 5% level by Fisher's protected LSD test.

3. Results and Discussion

3.1. Callus Proliferation

Callus of *F. cirrhosa* cultured in Murashige and Skoog's liquid medium with 2,4-D, under agitated conditions in dark incubation for six weeks, proliferated readily. Callus not only grew in volume but also became embryogenic since early stages of embryos were observed (Figure 4). There are several reports from our laboratory where various secondary metabolites have been obtained from different culture systems [25], including callus cultures of several medicinally important plant species, e.g., *Salvia miltiorrhiza* Bunge [26] and *Saussurea involucrata* Kar. et Kir. [27]. These and numerous other reports demonstrate that callus cultures have tremendous potential for the sustainable and large-scale production of secondary metabolites used in pharmaceuticals. Biotechnological applications of plant callus cultures have been recently reviewed [28], and, according to the author, the full potential of callus plant culture technology has not yet been exploited. Callus cultures and suspension cell cultures offer a wide range of applications in agriculture and horticulture, including for Chinese medicinal plants. Genetically transformed callus cultures cannot only be used for the synthesis of bioactive compounds but also for the development of plants with traits [28].

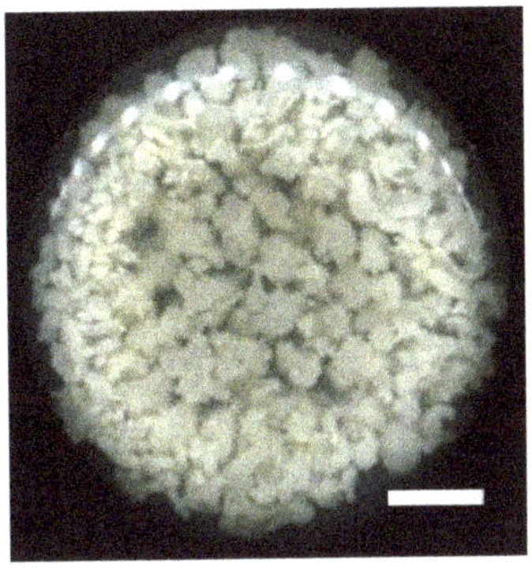

Figure 4. Embryogenic calli (EC) of *F. cirrhosa* D. Don growing in Murashige and Skoog's liquid basal medium supplemented with 2,4-D (0.5 mg/mL) and 2% sucrose. The image is a photograph taken by a scanner of the bottom of the culture flask after six weeks of incubation (bar = 1.2 cm).

3.2. Influence of LED Lights on Morphogenesis of Embryogenic Calli (EC) of F. cirrhosa

The eight LED lights had significant effects on the growth and development of embryogenic calli (EC) of *F. cirrhosa* (Table 2). Although the development of somatic embryos (SEs) in embryogenic calli were recorded under all the light treatments, the maximum number of SEs was recorded under red (9R, 223.7), infrared (9IR, 231.3), and a combination of red+blue+infrared light (3R3B3IR, 230.7), respectively. Among the red, infrared, and a combination of red+blue+infrared LED lights, there was a low significant difference concerning the number of SEs and the number of SEs with cotyledonary leaves (Table 2). LED lights also influenced the number of embryos with the development of cotyledonary leaves. Calli exposed to white light (WW-2700K) developed a maximum number of SEs with cotyledonary leaves (22.3), followed by 8R1B (16.7) and CW-5000K (12.0), respectively. SEs under LED lights 9B, 9R, 9IR, and 3R3B3IR developed a reduced number of cotyledonary leaves (in the range of 3.7 to 5.3). There was a low significant difference in the total fresh weight of EC among different LED light treatments. The maximum total fresh weights (FW) of EC biomass, i.e., 17.92 g and 16.67 g, were recorded with treatments of red (9R) and a combination of red+blue+infrared light (3R3B3IR), respectively. Eight LED lights also influenced the growth and development, and morphological features of somatic embryos. Embryogenic calli exposed to different LED lights developed somatic embryos in different developmental stages from early globular to mature SEs with cotyledonary leaves (Figure 5a–i). However, the average number of SEs varied depending upon the LED light. LED light WW-2700K developed the maximum number of SEs with cotyledonary leaves (22.3), followed by 8R1B (16.7) (Table 2). Different LED lights also affected the overall color of the embryogenic calli mass (Figure 5a–h). Under red (9R) and infrared (9IR) treatments, cultures turned white (Figure 5f,g), while under red+blue+infrared (3R3B3IR), these were dark green (Figure 5h).

Light is one of the essential components required by plants for photosynthesis. However, its quantity and duration (photoperiod) drastically affect plant growth and development [29]. Fluorescent tubes are the most common lighting source in a culture room in a typical tissue culture set up. However, in the recent past, due to several advantages, more advanced light-emitting diodes (LEDs) have been used as a source of light. LEDs are relatively cool, emit light of specific wavelengths (spectra), are much smaller in size, and are more durable compared to conventional ones [30]. Due to their efficiency in growth and development, LED lights have been used for micropropagation of many horticultural and agricultural crops [31–33]. Depending on requirements, different types of growth chambers equipped with LED lights can be designed. Since the supply of specific light spectra can be controlled in plant tissue culture systems via LED lights, the effects of individual or combinations of light spectra on plant growth and development can be investigated appropriately as in the present study. Several studies on the influence of LED lighting on plant growth, physiology, and secondary metabolism have been reviewed [11,34]. Recently, Pedmale and co-workers reported that the quality of light affects plant growth and development by the regulation of different mechanisms, including the selective activation of light receptors, such as phytochromes by red and far-red light, cryptochromes and phototropins by blue light, and UV-B receptors by ultraviolet light [35]. In a previous study in our laboratory, LED lights affected the development of somatic embryos and callus proliferation (fresh weight) in *Peucedanum japonicum* Thunb [22]. Several studies have demonstrated that the quality, duration, and intensity of red, infrared, blue, and ultraviolet light can have a profound influence on plants by activating or deactivating physiological reactions and controlling their growth and development [36–38]. These studies confirm with many other reports that LED lights are more efficient in plant growth compared to fluorescent lamps. Similar to our results, the beneficial effects of some LED light sources on the induction of embryogenesis in *Oncidium* have been reported [39]. Due to these beneficial effects, LED light systems are being increasingly used to boost the horticulture industry in Taiwan and several other countries [40–42].

Figure 5. Influence of different LED lights on the growth and development of embryogenic calli of *F. cirrhosa* D. Don: (**a**) CW-5000K; (**b**) WW-2700K; (**c**) 7R1G1B; (**d**) 8R1B; (**e**) 9B; (**f**) 9R; (**g**) 9IR; (**h**) 3R3B3IR; (**i**) Different stages of somatic embryos. For microphotographs, cultures are taken out of the bottles and transferred to sterilized Petri dishes. g: globular embryos; m: mature embryos; c: cotyledonary leaf; ec: embryogenic callus. (a–f, h, bar = 0.57 cm; g, bar = 1.4 cm). Culture medium is Murashige and Skoog's basal medium supplemented with 2% sucrose and 0.4% GPP. Culture vessels are 650 mL glass bottles, each containing 100 mL medium. Observations recorded after three months of incubation in a specially designed LED light growth chamber.

3.3. *Influence of LED Lights on the Contents of Isosteroidal Alkaloids in In Vitro Cultures of F. cirrhosa*

Responses of LED lights drastically varied among the four alkaloids. Out of four alkaloids tested (by LC-MS/MS) in in vitro cultures of *F. cirrhosa,* the most noticeable effects of eight LED lights were recorded in the case of peiminine (Table 3). The maximum peiminine in cultures (2.40 μg/g/dw) was detected under red (9R) light, followed by infrared (9IR) (2.21 μg/g/dw). Peiminine content in cultures exposed to LED lights CW-5000K, WW-2700K, 7R1G1B, and 3R3B3IR was in the range of (0.12–0.28 μg/g/dw). In in vitro-derived bulblets (3 months old), peiminine content (2.98 μg/g/dw) was detected in the sample grown under fluorescent light. However, this alkaloid could not be detected in cultures exposed to LED lights with 8R1B or in commercial bulbs (3 years old, wild type). Alkaloid sipeimine could not be detected in any in vitro culture or under any light treatment. The only material in which sipeimine in low quantities (0.6 μg/g/dw) could be detected was commercial bulbs (Table 3). Contents of another alkaloid peimisine in cultures were recorded only under two LED light treatments, 9R (3.65 μg/g/dw) followed by 9IR (3.22 μg/g/dw). Between two types of bulbs, the maximum peimisine was noted in commercial bulbs (68.4 μg/g/dw), followed by in vitro-derived bulblets (0.91 μg/g/dw). Similar to peimisine, the presence of peimine was found only under two LED light treatments, 9R (0.38 μg/g/dw) and 9IR (0.05 μg/g/dw), though the quantities were much lower compared to peimisine. Like peiminine, peimine was not detected in commercial bulbs (wild types).

Light quality affects the photochemical control of gene expression in various metabolic pathways, affecting the synthesis of nucleic acids, amino acids, organic acids, and sugars, etc., which are essential not only for cell growth and development but also for cell maintenance [43,44]. The response of plant cells to stress and their reorientation to developmental programs results in the expression of protein kinases, transcription factors, and structural genes that contribute to the adaptation [45]. Similar to the present study, the beneficial effects of LED lights on the contents of bioactive compounds in *Peucedanum japonicum* Thunb have been reported in our laboratory [22].

3.4. *Development of Bulblets in Somatic Embryos of F. cirrhosa*

Bulblet formation was observed in all three types of somatic embryos taken from cultures under LED lights CW5000K or WW2700K, as described in Section 2.3 (Table 4; Figure 6a–c; Figure 7a–b). However, the response percentage and number of bulblets varied depending upon the developmental stage of the somatic embryos. The highest response of bulblet formation (90%) was recorded with single embryos with an average of 4.7 bulblets/SE. In the case of a cluster of five embryos, the percentage of response and the average number of bulblets were 86.7 and 3.3 bulblets/SE, respectively. The lowest response (43.6%) and the least number of bulblets (1.1/SE) were recorded in SEs with cotyledonary leaf. However, in this case, some secondary somatic embryos at the base of cotyledonary leaf were observed, though cotyledonary leaf did not grow further, and withered and dried. Single embryos grew further in size and developed multiple bulblets without much callus growth. The cluster of five embryos grew in size and developed further embryogenic callus and secondary somatic embryos.

Since bulbs constitute the most critical parts of *Fritillaria* plants and are the primary source of isosteroidal alkaloids in *Fritillaria* species used in traditional Chinese medicine (TCM), the production of bulbs in *F. cirrhosa* by tissue culture technology is highly desirable. The objective of the present study was to investigate the effects of LED lights on the growth and development of embryogenic callus and the analysis of alkaloid contents in cultures; however, the development of bulblets in somatic embryos is an important observation in the study because in natural conditions, one bulb typically develops into a single seedling and it takes about 5–6 years to grow into an appropriate size [46]. It has also been reported that isosteroidal alkaloids in *F. cirrhosa* bulbs are greatly influenced by environmental conditions, plant age, and harvest times [47]. Recently, Chang and co-workers in our laboratory reported an efficient micropropagation method of bulblet production in *F. cirrhosa* and also the presence of some isosteroidal alkaloids in tissue culture-derived bulblets and callus [10].

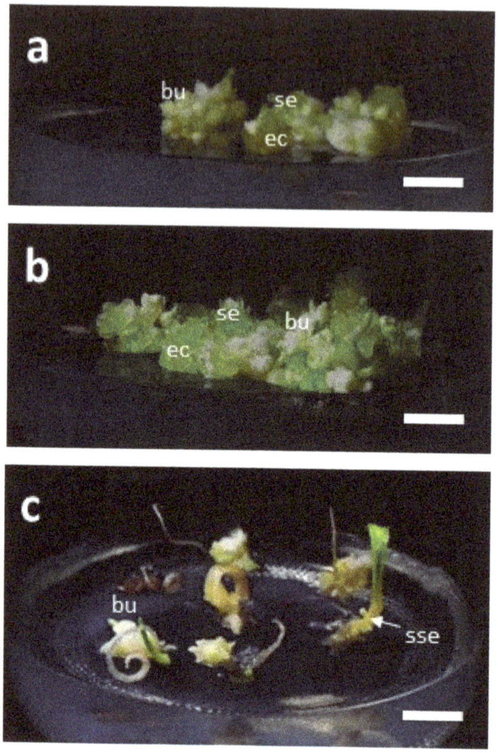

Figure 6. Development of bulblets in somatic embryos of *F. cirrhosa* D. Don (photographs taken after three months of incubation): (**a**) Bulblets from a single embryo, (**b**) bulblets in a cluster of five embryos, (**c**) bulblets in embryo with cotyledon leaf. bu: bulblet; se: somatic embryo; sse: secondary somatic embryo; ec: embryogenic callus. Bar = 0.8 cm.

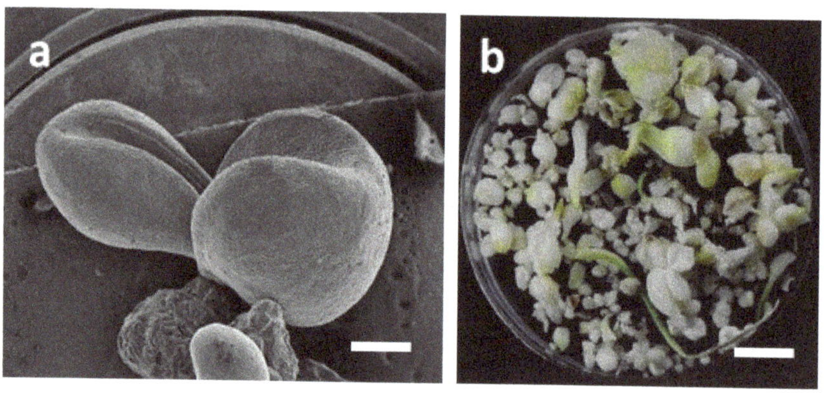

Figure 7. Development of bulblets from somatic embryos of *F. cirrhosa* D. Don: (**a**) Scanning electron microscopy (SEM) of bulblets (bar = 0.77 mm), (**b**) different stages of growth of culture-derived bulblets of *F. cirrhosa* (bar = 1.0 cm).

4. Conclusions

Eight LED lights influenced the growth and development, and morphological features of somatic embryos of *F. cirrhosa* D. Don. Embryogenic calli exposed to different LED lights developed somatic embryos in different developmental stages from early globular to mature SEs with cotyledonary leaves. The average number of somatic embryos that developed varied depending upon the wavelength of light emitted by the LEDs. The maximum number of SEs was recorded under red light spectra, followed by infrared and a combination of red/blue/infrared light spectra, respectively. Concerning alkaloids, the most significant effects were recorded with red and infrared light spectra in which peimisine, peimine, and peiminine were recorded. Sipcimine was not detected in any culture. Results obtained in the study indicate that red and infrared light spectra may be useful to obtain peimisine, peimine, and peiminine from callus cultures of *F. cirrhosa* D. Don. However, further research is needed to boost the quantities of these compounds in cultures and also for optimization to scale up production in bioreactors for commercial feasibility. Further research is also required for the optimization of large qualities of bulblet production in cultures. The production of important and precious alkaloids under a laboratory set up may reduce our dependence on natural materials from the wild and may help in the conservation of vulnerable plant species, such as *F. cirrhosa* D. Don.

Author Contributions: Conceptualization, H.-C.C.; methodology, C.-C.C. and H.-J.K.; software, C.-C.C. and H.-J.K.; validation, H.-S.T., C.-R.W. and M.-R.L.; formal analysis, H.-C.C.; investigation, C.-C.C.; resources, H.-C.C. and H.-M.X.; data curation, H.-C.C.; writing—Original draft preparation, D.C.A.; writing—Review and editing, D.C.A. and H.-C.C.; visualization, C.-C.C.; supervision, H.-C.C.; project administration, H.-C.C.; funding acquisition, H.-C.C. All authors have read and agreed to publish this version of the manuscript.

Funding: This research was funded by the Ministry of Science and Technology (MOST), Taiwan. Grant numbers 106-2313-B-324-001-MY2, 108-2313-B-324-001, and 109-2628-B-324 -001.

Acknowledgments: Financial support by the Ministry of Science and Technology, grant number 106-2313-B-324-001-MY2, 108-2313-B-324-001, and 109-2628-B-324 -001 is gratefully acknowledged.

Conflicts of Interest: The authors declare that they have no competing interests among them.

References

1. Lin, G.; Li, P.; Li, S.L.; Chan, S.W. Chromatographic analysis of *Fritillaria* isosteroidal alkaloids, the active ingredients of Beimu, the antitussive traditional Chinese medicinal herb. *J. Chromatogr. A* **2001**, *935*, 321–338. [CrossRef]
2. The State Pharmacopoeia Commission of P.R. China. *Pharmacopoeia of the People's Republic of China*, 10th ed.; China Medical Science Press: Beijing, China, 2015; p. 37.
3. Li, H.J.; Jiang, Y.; Li, P. Chemistry, bioactivity and geographical diversity of steroidal alkaloids from the Liliaceae family. *Nat. Prod. Rep.* **2006**, *23*, 735–752. [CrossRef]
4. Li, K.; Wu, W.; Zheng, Y.; Dai, Y.; Xiang, L.; Liao, K. Genetic diversity of Fritillaria from Sichuan province based on ISSR. *China J. Chin. Mater. Med.* **2009**, *34*, 2149–2154.
5. Ding, K.; Lin, G.; Ho, Y.P.; Cheng, T.Y.; Li, P. Prederivatization and high-performance liquid chromatographic analysis of alkaloids of bulbs of *Fritillaria*. *J. Pharm. Sci.* **1996**, *85*, 1174–1179. [CrossRef] [PubMed]
6. Xu, J.; Zhao, W.; Pan, L.; Zhang, A.; Chen, Q.; Xu, K.; Lu, H.; Chen, Y. Peimine, a main active ingredient of *Fritillaria*, exhibits anti-inflammatory and pain suppression properties at the cellular level. *Fitoterapia* **2016**, *111*, 1–6. [CrossRef] [PubMed]
7. Liu, S.; Yang, T.; Ming, T.W.; Gaun, T.K.W.; Zhou, T.; Wang, S.; Ye, B. Isosteroid alkaloids from *Fritillaria cirrhosa* bulbs as inhibitors of cigarette smoke-induced oxidative stress. *Fitoterapia* **2020**, *140*, 104434. [CrossRef]
8. Cunningham, A.B.; Brinckmann, J.A.; Pei, S.J.; Luo, P.; Schippmann, U.; Long, X.; Bi, Y.F. High altitude species, high profits: Can the trade in wild harvested *Fritillaria cirrhosa* (Liliaceae) be sustained? *J. Ethnopharmacol.* **2018**, *223*, 142–151. [CrossRef]
9. Zhang, D.; Gao, L.; Yang, Y. Genetic diversity and structure of a traditional Chinese medicinal plant species, *Fritillaria cirrhosa* (Liliaceae) in southwest China and implications for its conservation. *Biochem. Syst. Ecol.* **2010**, *38*, 236–242. [CrossRef]

10. Chang, H.C.; Xie, H.M.; Lee, M.R.; Lin, C.Y.; Yip, M.K.; Agrawal, D.C.; Tsay, H.S. In vitro propagation of bulblets and LC–MS/MS analysis of isosteroidal alkaloids in tissue culture derived materials of Chinese medicinal herb *Fritillaria cirrhosa* D. Don. *Bot. Stud.* **2020**, *61*, 9. [CrossRef]
11. Batista, D.S.; Felipe, S.H.S.; Silva, T.D.; de Castro, K.M.; Mamedes-Rodrigues, T.C.; Miranda, N.A.; Ríos-Ríos, A.M.; Faria, D.V.; Fortini, E.A.; Chagas, K.; et al. Light quality in plant tissue culture: Does it matter? *In Vitro Cell. Dev. Biol.-Plant* **2018**, *54*, 195–215. [CrossRef]
12. Kozai, T. Why LED lighting for urban agriculture? In *LED Lighting for Urban Agriculture*, 1st ed.; Kozai, T., Fujiwara, K., Runkle, E., Eds.; Springer: Singapore, 2016; pp. 3–18.
13. Xue, Y.; Gu, H.L. Determination of peimine and peiminine in *Fritillaria thunbergii* by HPLC-ELSD. *Acta Pharm. Sin.* **2005**, *40*, 550–552.
14. Ruan, X.; Yang, L.; Cui, W.X.; Zhang, M.X.; Li, Z.H.; Liu, B.; Wan, Q. Optimization of supercritical fluid extraction of total Alkaloids, Peimisine, Peimine and Peiminine from the Bulb of *Fritillaria thunbergii* Miq, and evaluation of antioxidant activities of the extracts. *Materials* **2016**, *9*, 524. [CrossRef] [PubMed]
15. Wu, K.; Mo, C.; Xiao, H.; Jiang, Y.; Ye, B.; Wang, S. Imperialine and verticinone from bulbs of Fritillaria wabuensis inhibit pro-inflammatory mediators in LPS stimulated RAW 264.7 macrophages. *Planta Med.* **2015**, *81*, 821–829. [CrossRef] [PubMed]
16. Luo, Z.; Zheng, B.; Jiang, B.; Xue, X.; Xue, E.; Zhou, Y. Peiminine inhibits the IL-1β induced inflammatory response in mouse articular chondrocytes and ameliorates murine osteoarthritis. *Food Funct.* **2019**, *10*, 2198–2208. [CrossRef]
17. Wang, D.; Zhu, J.; Wang, S.; Wang, X.; Ou, Y.; Wei, D.; Li, X. Antitussive, expectorant and anti-inflammatory alkaloids from Bulbus Fritillariae Cirrhosae. *Fitoterapia* **2011**, *82*, 1290–1294. [CrossRef]
18. Lin, Q.; Qu, M.; Patra, H.K.; He, S.; Wang, L.; Hua, X.; Xiao, L.; Fu, Y.; Gong, T.; He, Q.; et al. Mechanistic and therapeutic study of novel anti-tumor function of natural compound imperialine for treating non-small cell lung cancer. *J. Ethnopharmacol.* **2020**, *247*, 112283. [CrossRef]
19. Yin, Z.; Zhang, J.; Guo, Q.; Chen, L.; Zhang, W.; Kang, W. Pharmacological effects of verticine: Current status. *Evid. Based Complement. Altern. Med.* **2019**, *2019*, 2394605. [CrossRef]
20. Zhang, L.; Cui, M.; Chen, S. Identification of the molecular mechanisms of peimine in the treatment of cough using computational target fishing. *Molecules* **2020**, *25*, 1105. [CrossRef]
21. Murashige, T.; Skoog, F. A revised medium for rapid growth and bioassays with tobacco culture. *Physiol. Plant* **1962**, *15*, 473–497. [CrossRef]
22. Chen, C.C.; Agrawal, D.C.; Lee, M.R.; Lee, R.J.; Kuo, C.L.; Wu, C.R.; Tsay, H.S.; Chang, H.C. Influence of LED light spectra on in vitro somatic embryogenesis and LC–MS analysis of chlorogenic acid and rutin in *Peucedanum japonicum* Thunb.: A medicinal herb. *Bot. Stud.* **2016**, *57*, 9. [CrossRef]
23. Hsiao, H.H.; Hsieh, H.Y.; Chou, C.C.; Lin, S.Y.; Wang, A.H.J.; Khoo, K.H. Concerted experimental approach for sequential mapping of peptides and phosphopeptides using C18-functionalized magnetic nanoparticles. *J. Proteome Res.* **2007**, *6*, 1313–1324. [CrossRef] [PubMed]
24. Yu, X.; Liu, H.; Diao, J.; Sun, Y.; Wang, Y. Magnetic molecularly imprinted polymer nanoparticles for separating aromatic amines from azo dyes–Synthesis, characterization and application. *Sep. Purif. Technol.* **2018**, *204*, 213–219. [CrossRef]
25. Agrawal, D.C.; Chang, H.C.; Chen, C.C.; Kuo, C.L.; Tsay, H.S. Biotechnology of medicinal plants and fungi in Taiwan—Production of bioactive secondary metabolites in vitro culture systems. In *Medicinal Plants and Fungi: Recent Advances in Research and Development*, 1st ed.; Agrawal, D.C., Chang, H.C., Chen, C.C., Kuo, C.L., Tsay, H.S., Eds.; Springer Nature: Singapore, 2017; pp. 459–483.
26. Wu, C.T.; Vanisree, M.; Satish, M.N.; Chen, C.L.; Yang, T.F.; Tsay, H.S. Isolation and quantitative analysis of cryptotanshinone, an active quinoid diterpene formed in the callus of *Salvia miltiorrhiza* Bunge. *Biol. Pharm. Bull.* **2003**, *26*, 845–848. [CrossRef] [PubMed]
27. Kuo, C.L.; Agrawal, D.C.; Chang, H.C.; Chiu, Y.T.; Huang, C.P.; Chen, Y.L.; Huang, S.H.; Tsay, H.S. In vitro culture and production of syringin and rutin in *Saussurea involucrata* (Kar. et Kir.)—An endangered medicinal plant. *Bot. Stud.* **2015**, *56*, 12. [CrossRef]
28. Efferth, T. Biotechnology applications of plant callus cultures. *Engineering* **2019**, *5*, 50–59. [CrossRef]
29. Higuchi, Y.; Hisamatsu, T. Light acts as a signal for regulation of growth and development. In *LED Lighting for Urban Agriculture*, 1st ed.; Kozai, T., Fujiwara, K., Runkle, E., Eds.; Springer: Singapore, 2016; pp. 57–73.

30. Kim, S.J.; Hahn, E.J.; Heo, J.W.; Paek, K.Y. Effects of LEDs on net photosynthetic rate, growth and leaf stomata of chrysanthemum plantlets in vitro. *Sci. Hortic.* **2004**, *101*, 143–151. [CrossRef]
31. Li, H.M.; Xu, Z.G.; Tang, C.M. Effect of light-emitting diodes on growth and morphogenesis of upland cotton (*Gossypium hirsutum* L.) plantlets in vitro. *Plant Cell Tissue Organ Cult.* **2010**, *103*, 155–163. [CrossRef]
32. Nhut, D.T.; Takamura, T.; Watanabe, H.; Okamoto, K.; Tanaka, M. Responses of strawberry plantlets cultured in vitro under super bright red and blue light-emitting diodes (LEDs). *Plant Cell Tissue Organ Cult.* **2003**, *73*, 43–52. [CrossRef]
33. Saebo, A.; Krekling, T.; Appelgren, M. Light quality affects photosynthesis and leaf anatomy of birch plantlets in vitro. *Plant Cell Tissue Organ Cult.* **1995**, *41*, 177–185. [CrossRef]
34. Olle, M.; Viršile, A. The effects of light-emitting diode lighting on greenhouse plant growth and quality. *Agric. Food Sci.* **2013**, *22*, 223–234. [CrossRef]
35. Pedmale, U.V.; Huang, S.C.; Zander, M.; Cole, B.J.; Hetzel, J.; Ljung, K.; Reis, P.A.B.; Sridevi, P.; Nito, K.; Nery, J.R.; et al. Cryptochromes interact directly with PIFs to control plant growth in limiting blue light. *Cell* **2016**, *164*, 233–245. [CrossRef] [PubMed]
36. Briggs, W.R.; Olney, M.A. Photoreceptors in plant photomorphogenesis to date, five phytochromes, two cryptochromes, one phototropin, and one superchrome. *Plant Physiol.* **2001**, *125*, 85–88. [CrossRef] [PubMed]
37. Briggs, W.R.; Beck, C.F.; Cashmore, A.R.; Christie, J.M.; Hughes, J.; Jarillo, J.A.; Kagawa, T.; Kanegae, H.; Liscum, E.; Nagatani, A.; et al. The phototropin family of photoreceptors. *Plant Cell* **2001**, *13*, 993–997. [CrossRef] [PubMed]
38. Clouse, S.D. Integration of light and brassinosteroid signals in etiolated seedling growth. *Trends Plant Sci.* **2001**, *6*, 443–445. [CrossRef]
39. Chung, J.P.; Huang, C.Y.; Dai, T.E. Spectral effects on embryogenesis and plantlet growth of Oncidium 'Gower Ramsey'. *Sci. Hortic.* **2010**, *124*, 511–516. [CrossRef]
40. Agarwal, A.; Gupta, S.D. Impact of light-emitting diodes (LEDs) and its potential on plant growth and development in controlled environment plant production system. *Curr. Biotechnol.* **2016**, *5*, 28–43. [CrossRef]
41. Bello-Bello, J.J.; Pérez-Sato, J.A.; Cruz-Cruz, C.A.; Martínez-Estrada, E. Light-emitting diodes: Progress in plant micropropagation. In *Chlorophyll*; Jacob-Lopes, E., Zepka, L.Q., Queiroz, M.I., Eds.; InTech: Rijeka, Croatia, 2017; pp. 93–103.
42. Fang, W.; Chen, C.C.; Lee, Y.Y.; Chang, M.Y. Development of LED lids for tissue culture lighting. *Acta Hortic.* **2011**, *907*, 397–402. [CrossRef]
43. Cepauskas, D.; Miliute, I.; Staniene, G.; Gelvonauskiene, D.; Stanys, V.; Jesaitis, A.J.; Baniulis, D. Characterization of apple NADPH oxidase genes and their expression associated with oxidative stress in shoot culture in vitro. *Plant Cell Tissue Organ Cult.* **2016**, *124*, 621–633. [CrossRef]
44. Li, C.X.; Xu, Z.G.; Dong, R.Q.; Chang, S.X.; Wang, L.Z.; Khalil-Ur-Rehman, M.; Tao, J.M. An RNA-Seq analysis of grape plantlets grown in vitro reveals different responses to blue, green, red led light, and white fluorescent light. *Front. Plant Sci.* **2017**, *8*, 78. [CrossRef]
45. Neelakandan, A.K.; Wang, K. Recent progress in the understanding of tissue culture-induced genome level changes in plants and potential applications. *Plant Cell Rep.* **2012**, *31*, 597–620. [CrossRef]
46. Ruan, X.; Cui, W.X.; Yang, L.; Li, Z.H.; Liu, B.; Wang, Q. Extraction of total alkaloids, peimine and peiminine from the flower of *Fritillaria thunbergii* Miq using supercritical carbon dioxide. *J. CO_2 Util.* **2017**, *18*, 283–293. [CrossRef]
47. Konchar, K.; Li, X.L.; Yang, Y.P.; Emshwiller, E. Phytochemical variation in *Fritillaria cirrhosa* D. Don (Chuan Bei Mu) in relation to plant reproductive stage and timing of harvest. *Econ. Bot.* **2011**, *65*, 283. [CrossRef]